AF807274

Lecture Notes in Civil Engineering

Volume 153

Series Editors

Marco di Prisco, Politecnico di Milano, Milano, Italy

Sheng-Hong Chen, School of Water Resources and Hydropower Engineering, Wuhan University, Wuhan, China

Ioannis Vayas, Institute of Steel Structures, National Technical University of Athens, Athens, Greece

Sanjay Kumar Shukla, School of Engineering, Edith Cowan University, Joondalup, WA, Australia

Anuj Sharma, Iowa State University, Ames, IA, USA

Nagesh Kumar, Department of Civil Engineering, Indian Institute of Science Bangalore, Bengaluru, Karnataka, India

Chien Ming Wang, School of Civil Engineering, The University of Queensland, Brisbane, QLD, Australia

Lecture Notes in Civil Engineering (LNCE) publishes the latest developments in Civil Engineering - quickly, informally and in top quality. Though original research reported in proceedings and post-proceedings represents the core of LNCE, edited volumes of exceptionally high quality and interest may also be considered for publication. Volumes published in LNCE embrace all aspects and subfields of, as well as new challenges in, Civil Engineering. Topics in the series include:

- Construction and Structural Mechanics
- Building Materials
- Concrete, Steel and Timber Structures
- Geotechnical Engineering
- Earthquake Engineering
- Coastal Engineering
- Ocean and Offshore Engineering; Ships and Floating Structures
- Hydraulics, Hydrology and Water Resources Engineering
- Environmental Engineering and Sustainability
- Structural Health and Monitoring
- Surveying and Geographical Information Systems
- Indoor Environments
- Transportation and Traffic
- Risk Analysis
- Safety and Security

To submit a proposal or request further information, please contact the appropriate Springer Editor:

- Pierpaolo Riva at pierpaolo.riva@springer.com (Europe and Americas);
- Swati Meherishi at swati.meherishi@springer.com (Asia - except China, and Australia, New Zealand);
- Wayne Hu at wayne.hu@springer.com (China).

All books in the series now indexed by Scopus and EI Compendex database!

More information about this series at http://www.springer.com/series/15087

José C. Matos · Paulo B. Lourenço ·
Daniel V. Oliveira · Jorge Branco · Dirk Proske ·
Rui A. Silva · Hélder S. Sousa
Editors

18th International Probabilistic Workshop

IPW 2020

Springer

Editors
José C. Matos
ISISE
Department of Civil Engineering
University of Minho
Guimarães, Portugal

Paulo B. Lourenço
ISISE
Department of Civil Engineering
University of Minho
Guimarães, Portugal

Daniel V. Oliveira
ISISE
Department of Civil Engineering
University of Minho
Guimarães, Portugal

Jorge Branco
ISISE
Department of Civil Engineering
University of Minho
Guimarães, Portugal

Dirk Proske
Department of Architecture
Wood and Civil Engineering
Bern University of Applied Sciences
Burgdorf, Switzerland

Rui A. Silva
ISISE
Department of Civil Engineering
University of Minho
Guimarães, Portugal

Hélder S. Sousa
ISISE
Department of Civil Engineering
University of Minho
Guimarães, Portugal

ISSN 2366-2557 ISSN 2366-2565 (electronic)
Lecture Notes in Civil Engineering
ISBN 978-3-030-73615-6 ISBN 978-3-030-73616-3 (eBook)
https://doi.org/10.1007/978-3-030-73616-3

© The Editor(s) (if applicable) and The Author(s), under exclusive license to Springer Nature Switzerland AG 2021
This work is subject to copyright. All rights are solely and exclusively licensed by the Publisher, whether the whole or part of the material is concerned, specifically the rights of translation, reprinting, reuse of illustrations, recitation, broadcasting, reproduction on microfilms or in any other physical way, and transmission or information storage and retrieval, electronic adaptation, computer software, or by similar or dissimilar methodology now known or hereafter developed.
The use of general descriptive names, registered names, trademarks, service marks, etc. in this publication does not imply, even in the absence of a specific statement, that such names are exempt from the relevant protective laws and regulations and therefore free for general use.
The publisher, the authors and the editors are safe to assume that the advice and information in this book are believed to be true and accurate at the date of publication. Neither the publisher nor the authors or the editors give a warranty, expressed or implied, with respect to the material contained herein or for any errors or omissions that may have been made. The publisher remains neutral with regard to jurisdictional claims in published maps and institutional affiliations.

This Springer imprint is published by the registered company Springer Nature Switzerland AG
The registered company address is: Gewerbestrasse 11, 6330 Cham, Switzerland

Committees

Organizing Committee

José C. Matos, University of Minho, Portugal, Chair
Paulo B. Lourenço, University of Minho, Portugal, Chair
Dirk Proske, Axpo Power AG, Switzerland
Jorge M. Branco, University of Minho, Portugal
Hélder S. Sousa, University of Minho, Portugal

Scientific Committee

Daniel V. Oliveira, University of Minho, Portugal, Chair
Angelo Palos Teixeira, University of Lisbon, Portugal, Vice-Chair
Laura Caldeira, National Laboratory for Civil Engineering, Portugal, Vice-Chair
Luís Andrade Ferreira, University of Porto, Portugal, Vice-Chair
Rui A. Silva, University of Minho, Portugal, Secretary

Members

Abel Henriques, Portugal
Alfred Strauss, Austria
André Orcesi, France
Andre T. Beck, Brazil
António Cândido, Portugal
Bernt J. Leira, Norway
Bram van den Eijnden, The Netherlands
Bruno Sudret, Switzerland
Carmen Andrade, Spain

Dan M. Frangopol, USA
Daniel Straub, Germany
Daniil Yurchenko, UK
Dimitri Val, UK
Dirk Proske, Switzerland
Edoardo Patelli, UK
Eduardo Cavaco, Portugal
Emilio Bastidas-Arteaga, France
Frank Coolen, UK
Hélder S. Sousa, Portugal
Ioannis Kougioumtzoglou, USA
Javier Ortega, Portugal
Jianye Ching, Taiwan
Joan R. Casas, Spain
Joaquim Tinoco, Portugal
Jochen Kohler, Norway
Jorge M. Branco, Portugal
José C. Matos, Portugal
Junho Song, South Korea
Konrad Bergmeister, Austria
Leonardo G. Rodrigues, Portugal
Luís C. Neves, UK
Luís C. Silva, Portugal
Mário Coelho, Portugal
Mark Stewart, Australia
Matthias Voigt, Germany
Mauricio Sánchez-Silva, Colombia
Maximilian Huber, Austria
Micaela Demichela, Italy
Michael Beer, Germany
Michael H. Faber, Denmark
Michel Ghosn, USA
Miroslav Sýkora, Czech Republic
Mitsuyoshi Akiyama, Japan
Mohamed Eid, France
Panagiotis Spyridis, Germany
Paula V. Ferreira, Portugal
Paulo B. Lourenço, Portugal
Peter Mark, Germany
Pieter van Gelder, The Netherlands
Radomir Pukl, Czech Republic
Raphael Steenbergen, The Netherlands
Raquel Menezes, Portugal
Rita Bento, Portugal
Robby Caspeele, Belgium

Roman Wan-Wendner, Belgium
Tiago M. Ferreira, Portugal
Timo Schweckendiek, Netherlands
Tom Lahmer, Germany
Vasily Demyanov, UK
Vikram Pakrashi, Ireland
Xin Ruan, China

Sponsors

Workshop Organization

University of Minho
School of Engineering

isise Institute for Sustainability and Innovation in Structural Engineering

Supporting Associations

EUROSTRUCT

RISCOS
ASSOCIAÇÃO PORTUGUESA DE RISCOS, PREVENÇÃO E SEGURANÇA

MUNICÍPIO DE GUIMARÃES

Preface

The need to handle uncertainty and to make informed decisions renders evident the importance of the probabilistic and reliability topics. This can be seen in the most recent advances on the topic of the existing infrastructure maintenance and management, especially those related to safety and security under extreme events. Additionally, it is well-known that climate change issues are becoming even more relevant, with an impact on society, mostly affecting the likelihood and consequences of some natural hazards. Indeed, there is a need to develop deeper studies on data science, as well as on its application to system analysis, combining probabilistic and reliability tools to face the huge uncertainty.

The International Probabilistic Workshop (IPW) series started in 2003 as the Dresden Probabilistic Symposium at the Technical University of Dresden, repeated in 2004. In 2005, the 3rd edition held in Vienna was renamed as International Probabilistic Workshop. The previous IPWs took place in Berlin (2006), Ghent (2007), Darmstadt (2008), Delft (2009), Szcecin (2010), Braunschweig (2011), Stuttgart (2012), Brno (2013), Weimar (2014), Liverpool (2015), Ghent (2016), Dresden (2017), Vienna (2018) and Edinburgh (2019).

The IPW2020 (18th edition) was planned to take place in September 2020 at the University of Minho, Guimarães, Portugal. Unfortunately, the worldwide COVID-19 pandemic forced the postponement of the event to May 2021 and the adoption of an online format. Nevertheless, the scientific value and quantity of contributions (65 papers from 27 countries covering different probabilistic calculation methods) ensure the high quality of this Workshop, keeping the same scientific level as the previous ones.

The editors would like to thank all authors, keynote speakers, organizers of special sessions and participants for their valuable contributions, members of the Scientific

Committee for their meticulous work and the Workshop Secretariat for the dedicated teamwork, particularly during this exceptional pandemic period.

Guimarães, Portugal	José C. Matos
Guimarães, Portugal	Paulo B. Lourenço
Guimarães, Portugal	Daniel V. Oliveira
Guimarães, Portugal	Jorge Branco
Burgdorf, Switzerland	Dirk Proske
Guimarães, Portugal	Rui A. Silva
Guimarães, Portugal	Hélder S. Sousa

Contents

Keynote Papers

Decision Analysis Applied to Natural Hazards

Herbert H. Einstein and Rita L. Sousa

Abstract Formal methods to handle decision-making under uncertainty that have been created for business management lend themselves to applications in many other areas, in which uncertainties play a major role. Hence, the authors and their co-workers have applied decision analysis to landslides since the 1980′s but many other approaches to landslide assessment and management have in principle done so. The keynote lecture itself will illustrate the application of decision analysis with many examples. For this reason, we concentrate in this paper on the principles of decision-making under uncertainty and the concept of using these principles in hazard and risk analysis of natural threats. We also like to note that what we present here is a summary of our past work. The paper starts with an introduction to the decision-making process and its application to natural threats. Risk management of natural threats is then demonstrated in detail with decision trees and Bayesian networks. This leads to sensitivity analyses to determine which risk management action is most effective.

Keywords Natural threats · Landslides · Decision making · Bayesian networks

1 Introduction

Uncertain events can be formally handled by decision-making under uncertainty that was developed for business management [1]. Given the uncertainty of many natural events, it is, therefore, quite logical to apply methods of decision-making under uncertainty to natural threats such as landslides, floods and wildfires, for instance. The authors of this paper have developed and applied these decision-making processes to landslides (e.g. [2, 3]). This involved the use of classic decision tree procedures that were extended to include warning systems. Very importantly, an alternative approach

H. H. Einstein (✉)
Massachusetts Institute of Technology, Cambridge, MA, USA
e-mail: einstein@mit.edu

R. L. Sousa
Stevens Institute of Technology, Hoboken, NJ, USA

© The Author(s), under exclusive license to Springer Nature Switzerland AG 2021
J. C. Matos et al. (eds.), *18th International Probabilistic Workshop*, Lecture Notes in Civil Engineering 153, https://doi.org/10.1007/978-3-030-73616-3_1

using Bayesian networks was then developed [4]. This paper, therefore, will first introduce the reader to the principles of decision-making under uncertainty (Sect. 2) and then comment on the formalization of the threat assessment process and how to incorporate it in the decision-making process (Sect. 3). This will be followed by showing examples of decision trees (Sect. 4), the use of Bayesian networks (Sect. 5) and end with conclusions (Sect. 6).

2 Decision-Making Under Uncertainty

Figure 1 is a schematic of decision-making under uncertainty based on the original development at the Harvard Business School [1]. As can be seen, the process can lead directly to the result of accepting the risk or to an updating cycle. The updating cycle on the left side relates to obtaining and using additional information or to managing the risk. The information model on the right side can be used to decide if it is worthwhile to collect additional information or not. Sousa et al. [5, 6] have

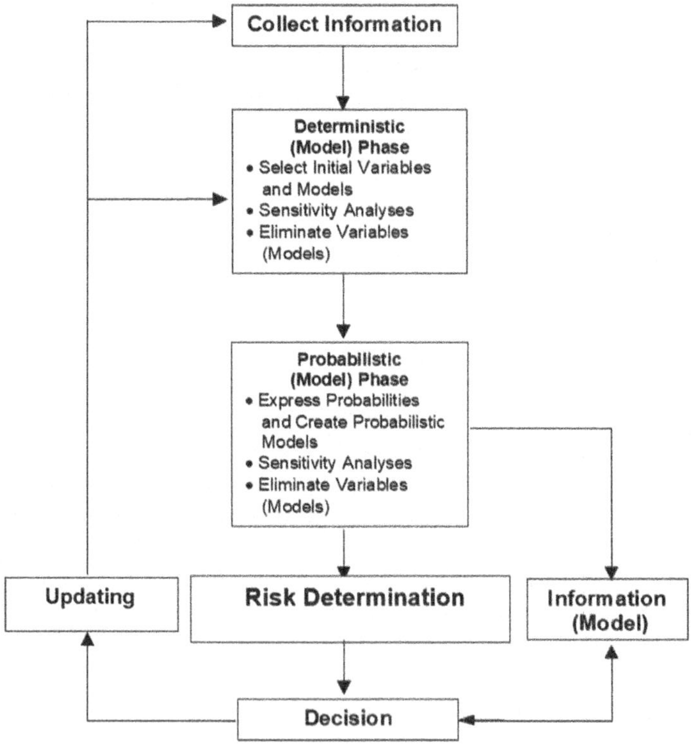

Fig. 1 Decision analysis cycle | Decision: accept risk or "Update" | Update: collect more information and/or manage risk

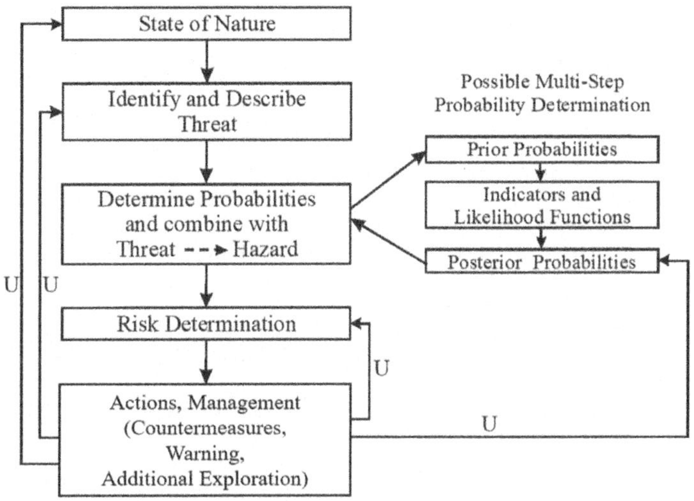

Fig. 2 Decision analysis cycle applied to natural threats | U = Updating

applied and explained the use of such information models in the context of natural hazards and tunneling.

The decision process of Fig. 1 can be expanded and adapted to dealing with natural threats as shown in Fig. 2. The expansion contains details on the decision in form of different actions in the context of risk management.

3 Formalization of the Threat Assessment Process

The terms threat, hazard and risk have already been used in Fig. 2, and they need to be formally defined. This is first done through the verbal expressions of Table 1 that lists the definitions as formulated by the Technical Committee No. 32 of the

Table 1 Definitions (Based on glossary of TC 32 of the ISSMGE)

Term	Definition
Threat (Danger)	Natural phenomenon that could lead to damage. Described by geometry, mechanical and other characteristics. Can be an existing one, or a potential one, such as a rockfall. No forecasting
Hazard	Probability that a particular threat (danger) occurs within a given period of time
Risk	Measure of the probability and severity of an adverse effect to life, health, property, or the environment Risk = Hazard × Potential Worth of Loss

Table 2 Other important concepts

Concept	Definition
Consequence	Result of a hazard being realized
Damage	Another way of expressing detrimental consequences
Vulnerability	– Often expressed on a scale of 0 (no loss) to 1 (total loss) – Expresses the fact that even if a threat materializes, it is not necessarily 100% certain that the consequences materialize – Can be formulated as a conditional probability

ISSMGE. In addition, several other concepts (terms) need to be used, and they are listed in Table 2.

The expressions in Table 1 and Table 2 can be used in the formal decision-making process discussed in Sect. 4.

4 Decision-Making Process

The intent is to make a decision in the context of risk management (recall Fig. 2). Before doing so, it is important to point out that very often it is better to work with hazard than with risk. The latter requires that one expresses the consequences with a value. Although this value can be qualitative or quantitative it can be often problematic e.g. if one deals with lives. Hazard to lives can be dealt with the so-called FN charts [7, 8] as shown in Fig. 3 for Hong Kong. The frequency (F) of events is the hazard and it is subjectively related to the number of fatalities (N).

If one goes all the way to risk (see also Table 1):

$$\begin{aligned} \text{Risk} &= \text{Probability of Threat} \times \text{Worth of Loss} \\ &= \text{Hazard} \times \text{Worth of Loss} \\ &= \text{Hazard} \times \text{Consequences} \end{aligned} \quad (1)$$

This can be expressed as:

$$R = P[T] \times u(X_i) \quad (2)$$

where

R Risk
P[T] Probability of Threat = Hazard
$u(\underline{X_i})$ Utility of the consequence, where $(\underline{X_i})$ is a vector of attributes if one uses a multiattribute approach [9, 10]

As indicated in Table 2 the fact that the consequences are uncertain is reflected by vulnerability, which can be expressed by the conditional probability $P[X_i|T]$ and

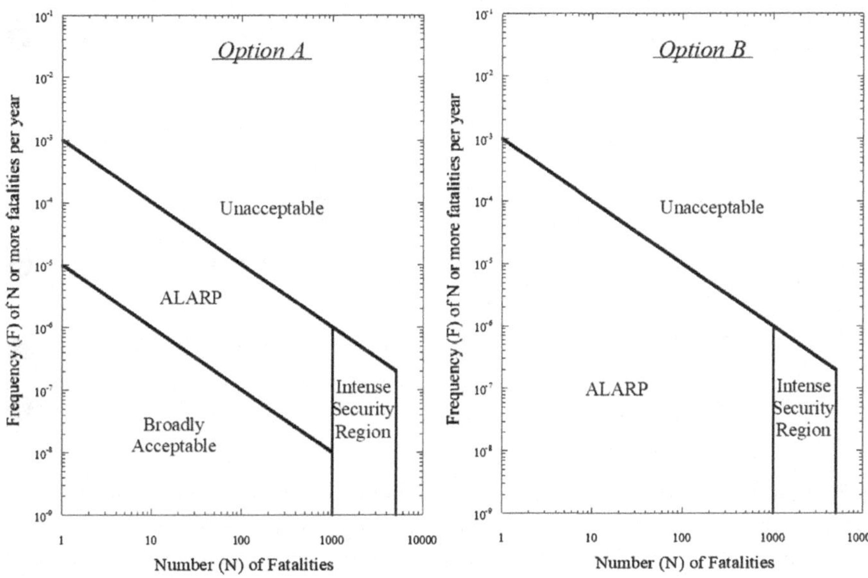

Fig. 3 Consideration of life losses with F-N charts. Example from Hong Kong [7] | ALARP = As Low As Reasonably Practical

thus risk is:

$$R = P[T] \times P[X_i|T] \times u(X_i) \tag{3}$$

One can manage risk in the following manner:

- No action
- Active countermeasures reduce P[T] i.e. the hazard
- Passive countermeasure reduce $P[X_i|T]$ i.e. the vulnerability
- Warning systems also reduce $P[X_i|T]$ i.e. the vulnerability.

Clearly combinations of all the above are possible.

5 Decision Trees

The management actions and their "cost" will produce what we term as "modified risk". If the modified risk is smaller than the original one, it is worthwhile to take the management action. All this will now be shown in detail with decision trees related to the typical management actions. Figure 4 shows the overall decision tree that includes all actions.

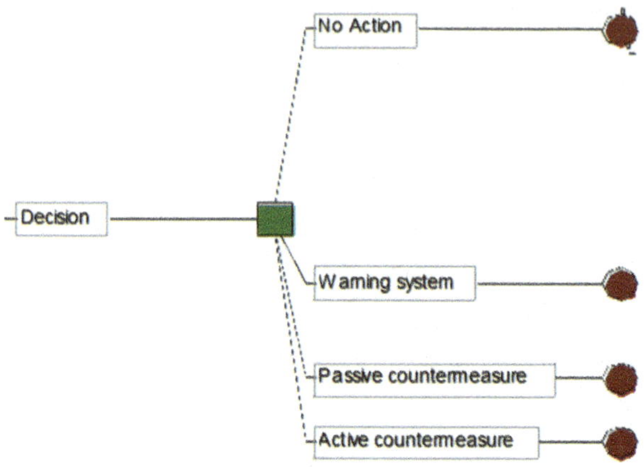

Fig. 4 Decision tree tool showing possible actions

The first possibility is "no-action" for which the decision tree is shown in Fig. 5. With this tree we also introduce some basic concepts and assumptions: The hazard model represents the probability P[T] that the threat occurs. The specific numbers (20.7, 79.3%) can be obtained e.g. with a probabilistic slope stability analysis. The vulnerability model provides the probability $P[X_i|T]$ that a consequence materializes if the threat occurs. The numbers used here are subjective estimates. Finally, one needs to associate costs with consequences, which is done in the consequence model. It is important to realize that vulnerability and consequence depend on each other. This is expressed here by having smaller vulnerability (40%) for the higher consequence costs (−20,000). These costs are here in terms of utilities. The total risk of no action is then obtained by multiplying and summing [(0.5x − 10,000) + (0.4x − 20,000)] × 0.207 = −2691.

This "no action risk" is the "original risk" R that will be compared to modified risks R′ reflecting active or passive management actions. These management actions have a cost that needs to be included when determining the modified risk, as will be seen in the following.

With *active countermeasures* one reduces the probabilities of the threat from P[Threat] (20.7%) to P′ [Threat] (5.2%). This reflects, for instance, the effect of stabilizing a slope.

The stabilizing measures do have a cost that need to be considered. The modified risk will then be:

$$R' = u(C_{ac}) + P'[\text{Threat}] \times P[X_i|\text{Threat}] \times u(X_i) \tag{4}$$

where C_{ac} = cost of countermeasures.

Figure 6 presents the decision tree for active countermeasures. Different from the tree for no action it now includes the cost of countermeasures "-2000" and the lower probability of the threat. The multiplying and summing is as before leading to a slightly lower modified risk R′ = - 2672.75 compared to the original (no action) risk R = −2691.

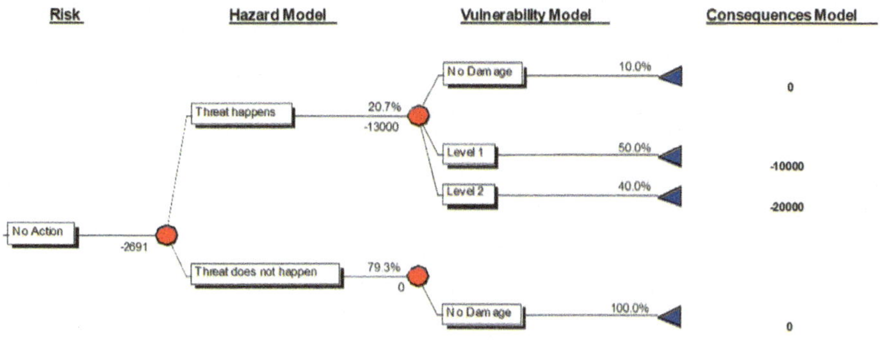

Fig. 5 Decision tree—no action

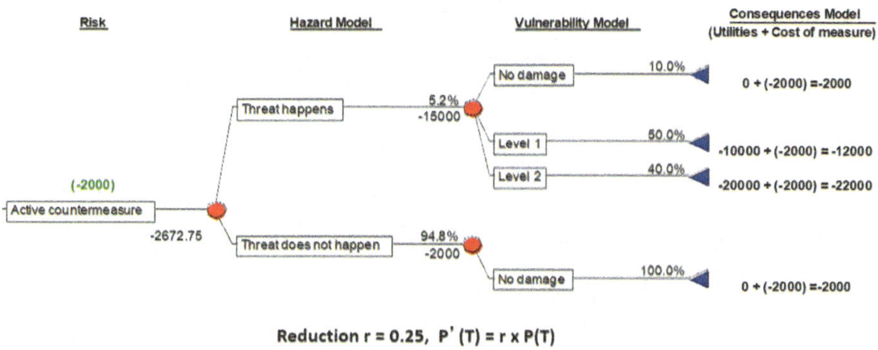

Fig. 6 Decision tree—active countermeasures.

Passive countermeasures reduce the vulnerability e.g. a protective shed against rockfall consequences. In the modified risk R', the hazard P[T] will be the same as for no action but the vulnerability will change to $P'[X_i|\text{Threat}]$ and thus R' will be:

$$R' = u(C_{pc}) + P[\text{Threat}] \times P'[X_i|\text{Threat}\} \times u(X_i) \tag{5}$$

where C_{pc} = cost of passive countermeasures. In the corresponding decision tree (Fig. 7) the vulnerabilities reflect the fact that the countermeasures reduce the probability of damage occurring and correspondingly increase the probability of no damage. With the numbers shown in Fig. 7 one obtains a modified risk of $R' = 2864.6$ that is higher than what resulted from active countermeasures.

Warning systems are also a kind of passive countermeasures. Many such systems exist, notably the tsunami warning systems in Japan and the Caribbean as well as avalanche warning systems in Switzerland [11] and Norway. Figure 8 shows how such systems fit into the overall decision-making process. The important component of warning systems is the trigger and this also complicates the decision-making process. Specifically, the reliability of the warning system that can be expressed

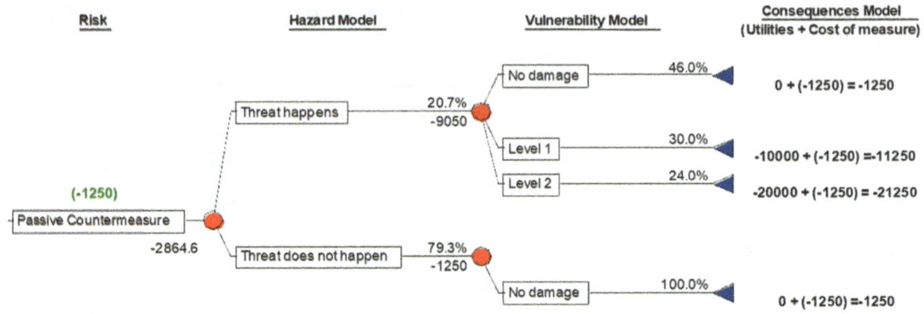

Fig. 7 Decision tree—passive countermeasures.

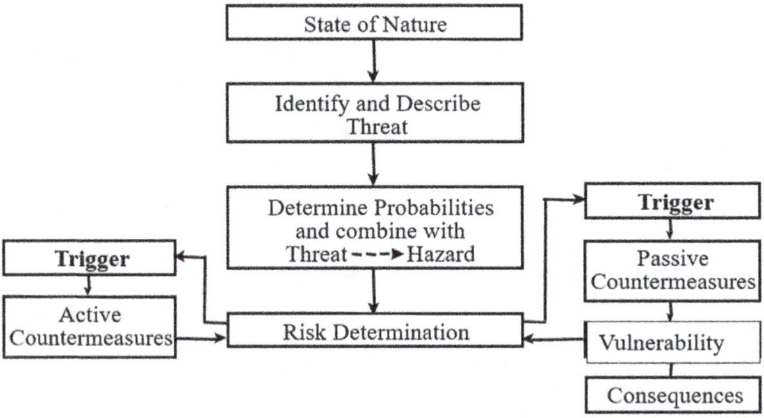

Fig. 8 Decision cycle for natural threats with warning system. | The "trigger" initiates countermeasures

in form of a reliability matrix (Fig. 9) needs to be included. In all decisions with countermeasures (active, passive, warning systems) it is also possible to include the effectiveness of countermeasures.

The decision trees show that there are sets of branches for each decision model. In the complete tree and going from right to left these models are "consequences", "vulnerability", "hazard", and "reliability". The number of trees increases if other models such as "effectiveness of countermeasures and multiple dependent hazards (e.g. earthquake or rainfall causing landslides) are included. In the extreme case one may thus end up with tens of branches. While informative since one can follow the decision process, it becomes visually difficult to fully capture the process.

Fig. 9 Reliability matrix: shows probability that alarm is triggered if threat occurs

Reliability matrix

	Reality	
	Threat	No Threat
Alarm	0.9	0.1
No Alarm	0.1	0.9

6 Bayesian Networks

This can be remedied by using Bayesian networks [4], a probabilistic graphical model, that represents a set of random variables and their conditional dependencies via a directed acyclic graph. Figure 10 represents a generic **BN**. In this BN one has 5 random variables: X1, X2, X3, X4 and X5, represented by the nodes of the graph, and several edges that represent the conditional dependencies between variables. For example X2 has two parent nodes X1 and X4, so X2 conditionally depends on X1 and X4. On the other hand, for example, the random value X3 is conditionally independent of X4. Attached to each node of the BN are prior probability distributions (in the case of random variables without parent nodes) and conditional probability distributions for all the other nodes. Bayesian networks represent joint probability distributions in a compact and factorized way, by taking advantage of conditional independence, considering that not all variables depend on each other (i.e. do not have edges connecting all variables). In Fig. 11 the results of using Bayesian networks for the previously described cases using decision trees are given in table form and the results are summarized in Fig. 12.

In the discussion so far we assessed probabilities to demonstrate what can be done in decision-making under uncertainty. What is particularly interesting is the possibility to conduct sensitivity analyses to determine how the results i.e. the risk expressed in utilities will change if the underlying probabilities change. An example is shown in Fig. 13 in which P[T] the hazard is varied. For low P[T] no action results while warning systems are recommended for higher P[T].

$$P\,(X_1, X_2, \ldots, X_n) = \prod_{i=1}^{n} P(X_i|\pi_i)$$

X_i: Random variables in the BN

π_i: Parents nodes of X_i

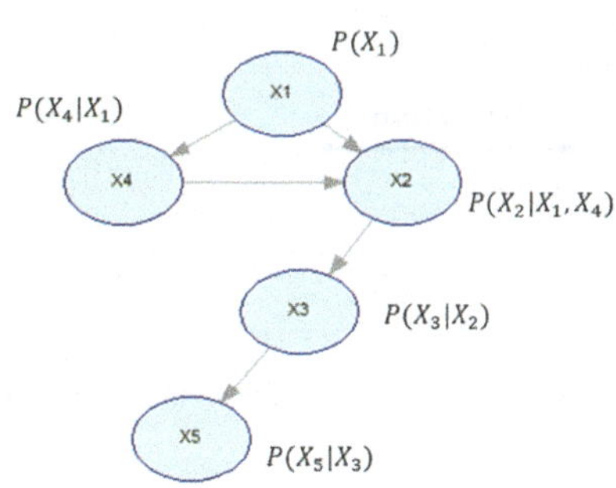

Fig. 10 Bayesian networks are a concise representation of joint probability

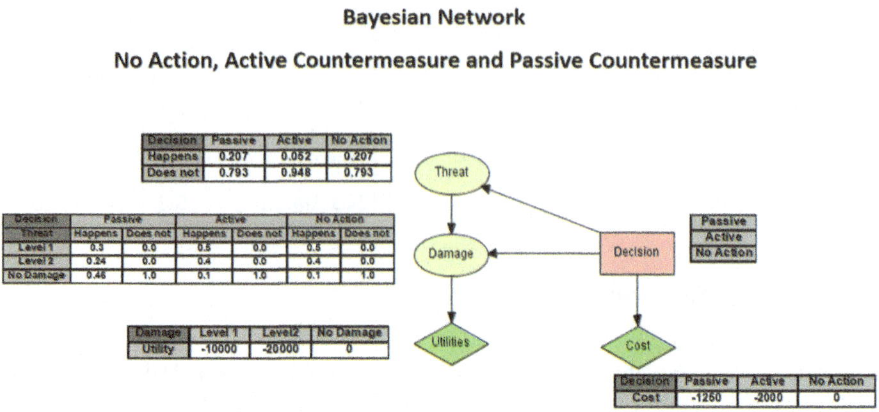

Fig. 11 Bayesian network applied to management of risk caused by natural threats

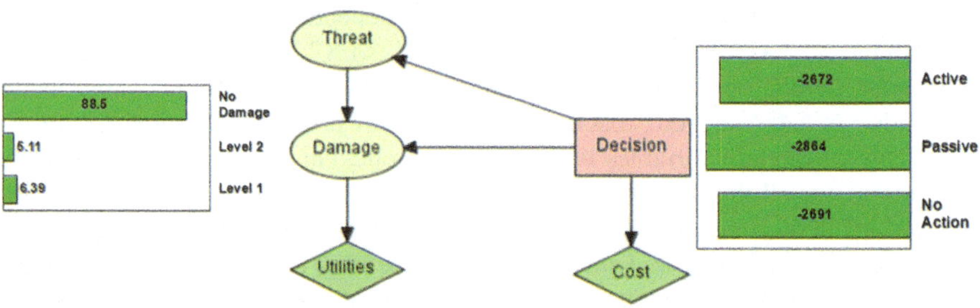

Fig. 12 Bayesian network applied to management of risk caused by natural threats-results

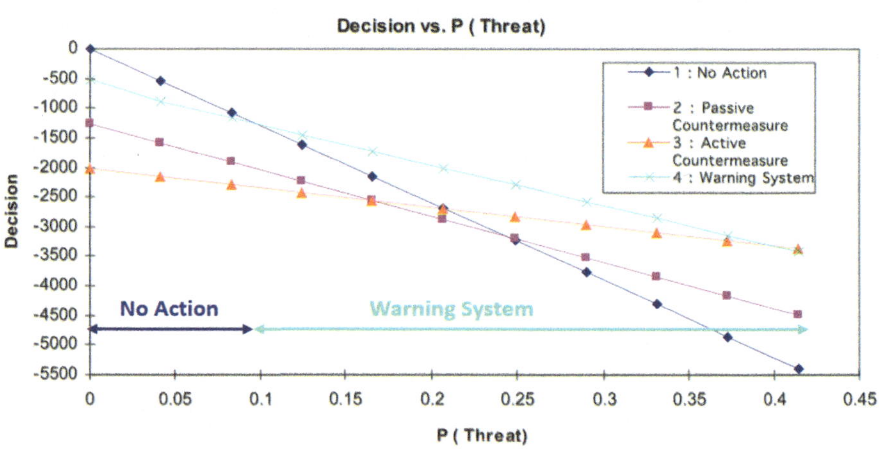

Fig. 13 Sensitivity analysis—different actions depending on probability of threat

7 Conclusions

Natural threats are characterized by uncertainty regarding temporal occurrence, spatial extent and many other aspects. Using probabilistic methods to describe the uncertainties is therefore common. It is then also logical to use methods of decision-making under uncertainty to assess and manage the threats and their consequences. Over the years the authors of this paper have developed decision-making approaches mostly regarding landslides. The keynote presentation and this paper summarize these approaches, which use decision trees and Bayesian Networks. This paper in essence provides a succinct guideline on how to use the decision-making approaches. The keynote presentation will then build on this with applications to practical cases mostly involving landslides but also other natural threats.

References

1. Pratt, J., Raiffa, H., & Schlaifer, R. (1965, 2008). *Introduction to statistical decision theory.*
2. Einstein, H. H., & Sousa, R. L. (2012). Risk in slopes. In L. Ribeiro e Sousa, E. Vargas Jr., M.M. Fernandes, & R. Azevedo (Eds.), *Innovative Numerical Modelling in Geomechanics, Chapter 11* (pp. 201–210). CRC Press. ISBN 9780415616614.
3. Einstein, H. H., Sousa, R., Karam, K., Manzella, I., & Kveldsvik, V. (2010). Rock slopes from mechanics to decision making. Keynote paper. In *Proceedings of the ISRM International Symposium—EUROCK 2010*, 15–18 June, Lausanne, Switzerland. ISRM-EUROCK-2010-001.
4. Sousa, R. L. (2010). *Risk analysis for tunneling projects* (Ph.D. dissertation). Massachusetts Institute of Technology. https://hdl.handle.net/1721.1/58282.
5. Sousa, R. L., Karam, K., & Einstein, H. H. (2014). Exploration analysis for landslide risk management. *Georisk, 8*(3), 155–170. https://doi.org/10.1080/17499518.2014.958174soli.
6. Sousa, R. L., Karam, K., Costa, A. L., Einstein, H. H. (2016). Exploration and decision-making in geotechnical engineering—A case study. In *10th Anniversary Special Issue of Georisk* (Vol. 11, No. 1, pp. 129–145).
7. Ho, K., Leroi, E., & Roberts, B. (2000). Quantitative. In *Proceedings of International Conference on Geotechnical and Geological Engineering GEOENG 200. Melbourne Risk Assessment—Applications, Myths and Future direction.*
8. Health and Safety Executive (HSE). (1984). *Control of Industrial Major Hazards.*
9. Keeney, R. L., & Raiffa, H. (1976). *Decision analysis with multiple conflicting objectives.* Wiley.
10. Baecher, G. B. (1981). *Risk screening for civil facilities* (20 p). Department of Civil Engineering, Massachusetts Institute of Technology CER-81-9.
11. Bründl, M., Etter, H. J, Steiniger, M., Klingler, Ch., Rhyner, J., & Ammann, W. J. (2004). IFKIS (Interkantonales Frühwarn und Kriseninformationssystem)-a basis for managing avalanche risk in settlements and on roads in Switzerland. *Natural Hazards and Earth Sciences, 4.*

Probabilistic Seismic Risk Assessment of School Buildings

Ricardo Monteiro

Abstract The inadequate behavior of existing school buildings observed during past earthquakes in Italy have underlined the need to accurately understand their seismic performance. In order to do so, different metrics can be adopted to characterize their seismic response, either more focused on structural aspects or economic variables. This paper assesses the seismic risk level for three case study school buildings, representing the main typologies found within the Italian school building stock, and comments on the eventual need for retrofitting. A probabilistic-based earthquake engineering (PBEE) performance assessment is carried out using detailed numerical models, analyzed under ground motion records of increasing intensity, to quantify risk-based decision variables, such as expected annual loss and mean annual frequency of collapse. As an alternative to the detailed PBEE framework, a simplified seismic risk classification framework, recently applied in Italy, was also implemented. Different uncertainty parameters are included in the risk estimation frameworks, with a view also to future large-scale implementation of cost-benefit analyses. Lastly, one of the school buildings is further analyzed to understand the impact of the structural modelling uncertainty in the risk estimates and the consequent need for its proper consideration. The results show how the simplified risk classification framework is, as expected, conservative with respect to the detailed component-based approach, as well as the need for retrofitting of some of the building structural systems.

Keywords Risk assessment · Seismic retrofit · Cost-benefit analysis · Loss estimation · Modelling uncertainty

1 Introduction

Extensive damage and structural collapse observed in Italian school buildings during past seismic events have pointed out the need for seismic risk mitigation programs.

R. Monteiro (✉)
University School of Advanced Studies IUSS, Pavia, Italy
e-mail: ricardo.monteiro@iusspavia.it

© The Author(s), under exclusive license to Springer Nature Switzerland AG 2021
J. C. Matos et al. (eds.), *18th International Probabilistic Workshop*, Lecture Notes in Civil Engineering 153, https://doi.org/10.1007/978-3-030-73616-3_2

These should identify the most vulnerable building typologies and reduce the earthquake-related economic losses and casualties through adequate seismic retrofit strategies. The collapse of a school in San Giuliano di Puglia during the 2002 Molise earthquake in Italy, which caused 30 fatalities, is a key example of the seismic vulnerability of the Italian existing school building stock [1]. Recent studies have also pointed out the importance of non-structural elements in achieving adequate seismic performance levels for an entire building system [2–4]. De Angelis and Pecce [5] reported the death of a student caused by the collapse of a classroom ceiling on November 22nd, 2008 at the Darwin High School in Rivoli, Italy and proposed a simplified methodology to assess the safety of non-structural elements installed in school buildings. Based on these considerations, the need for a seismic risk identification scheme for Italian school buildings comprising both structural and non-structural elements appears evident. Grant et al. [6] developed a risk-management framework to prioritize rehabilitation interventions for Italian school buildings; once the more vulnerable structures are identified. Furthermore, the seismic risk classification guidelines recently introduced in Italy [7] provide a simplified method that classifies existing buildings before and after strengthening interventions. The use of these guidelines may result in tax deductions as an incentive to improve the seismic safety of the existing Italian school building stock, leading to increased awareness of seismic safety and the importance of adequate seismic retrofit among citizens.

To contribute to this important issue, the European Centre for Training and Research in Earthquake Engineering (EUCENTRE) conducted "Progetto Scuole", a research project aimed at investigating the seismic vulnerability of Italian school buildings. A comprehensive database was developed for approximately 49,000 school buildings in Italy by Borzi et al. [8]. Data related to structural behavior, as well as other features concerning school organizations, was collected. From the database, it was observed that approximately 80% of school buildings in Italy are made of unreinforced masonry (URM) and reinforced concrete frames with masonry infill (RC), whereas the remaining 20% are characterized by other typologies, such as precast structures (PC), steel constructions or mixed assemblies [9]. The knowledge of the main features of the existing school building stock allowed the identification of representative case study school buildings in order to perform detailed loss estimation studies, to be used in future identification of adequate retrofit strategies.

The well-known performance-based earthquake engineering (PBEE) methodology, proposed by Cornell and Krawinkler [3], and subsequently developed by the Pacific Earthquake Engineering Research Center (PEER) in California as the PEER-PBEE methodology, is applied in a systematic fashion in this study to perform the seismic loss assessment [4] of three case study school buildings, representative of different structural typologies, namely RC frames with masonry infill, URM buildings and PC structures. As reported by Taghavi and Miranda [10], the initial monetary investment in non-structural elements for office/schools, hotels, and hospitals buildings can reach up to 60–90% of the total building value.

In this study, the complete seismic loss assessment of the aforementioned three case-study school buildings, belonging to the most common typologies of the Italian existing school building stock, is presented. A detailed inventory of structural and

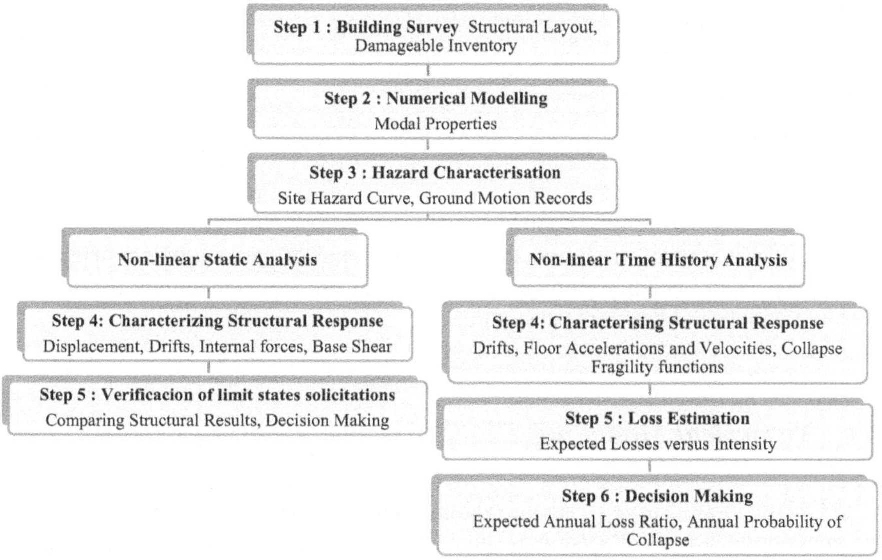

Fig. 1 Steps of the performance assessment, based on the PEER-PBEE methodology [3], applied to the case study school buildings [9]

non-structural elements was developed during in-situ surveys to enable the implementation of the component-based seismic loss assessment foreseen by the FEMA P58 procedure, using more advanced metrics. Figure 1 summarizes the steps that were followed.

The results of this detailed exercise are then compared with those obtained through nonlinear static analyses [12] to evaluate the structural performance at different limit states defined by the Italian National Code (NTC 2018) and to apply the seismic classification guidelines recently introduced in Italy [7].

2 Case-Study School Buildings

The data collected in the EUCENTRE database of school buildings [8], as well as the results of previous studies [13, 14], were used to identify the main features of existing school buildings in Italy and the main sources of structural vulnerability. The construction period was found to be a preliminary indicator of the seismic vulnerability. Table 1 lists the construction typology, number of stories and constructions periods of the three case-study school buildings extracted from the database and analyzed in this study. An in-situ building survey was carried out for each building to gather information and create an inventory of damageable structural and non-structural elements. From a structural point of view, all the information required to identify the main deficiencies of the buildings was collected, including also the

Table 1 General information for case study school buildings, adapted from [9]

Typology	Label	No. of storeys	Construction period
Reinforced concrete frame with masonry infill	RC	3	1960s
Unreinforced masonry	URM	2	1900s
Precast RC frame	PC	2	1980s

possible degradation of structural elements and, for the PC case study school building, the details on the beam-column connections and on the connections between the cladding panels and the frame structure.

2.1 Numerical Modelling

Based on the in-situ surveys, advanced non-linear numerical models were developed to simulate the structural seismic response of the three case study school buildings. The main features of the numerical models are reported in the next sections.

2.1.1 Reinforced Concrete Case Study School Building

The OpenSees software [15] was used to develop a numerical model of the RC school building. To account for all possible deficiencies related to RC structures designed before the 1970s in Italy, the modelling recommendations by O'Reilly and Sullivan [16] were followed. Beam and column members were modelled as force-based beam-column elements with a modified Radau plastic hinge integration scheme, as suggested by Scott and Fenves [17] that provides a lumped plasticity component. Frame elements included a post-peak strength and stiffness degradation, while the non-linear behavior of beam-column joints was simulated using zero-length elements, as illustrated in Fig. 2. The slab was assumed to be rigid based on its structural configuration. The stair cases were modelled using elastic frame elements to consider the potential shear failure of the surrounding columns. The numerical model also included the effect of exterior masonry infill walls. As shown in Fig. 2, masonry infill walls were represented through an equivalent diagonal strut.

2.1.2 Precast Concrete Case Study School Building

The numerical model of the PC case study school building was also developed in OpenSees [15]. The structural system comprises precast columns that support precast beams in the longitudinal direction. The absence of precast beams in the transverse direction was confirmed from the in-situ survey. The precast columns were modelled with a lumped plasticity approach following Haselton et al. [18] recommendations.

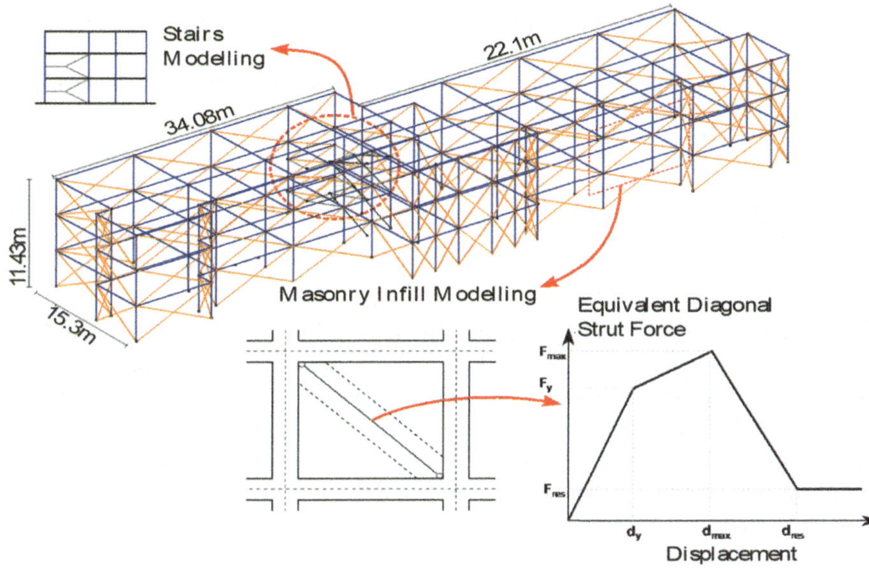

Fig. 2 Main features of the numerical model of the RC case study school building, adapted from [9]

Due to the lack of continuity in the beam-column joints, no moment transfer was assumed between adjacent beams. For this reason, the beams were modelled as elastic elements. The slab's bending stiffness was explicitly modelled to account for the lack of structural connections and the absence of beams in the transverse direction. A detailed numerical model was developed to simulate the beam-column connections through rigid elements to represent the depth of the beam and zero-length elements at its top and bottom to simulate a gap effect and contact seat, respectively. The cladding panels were incorporated into the numerical model following the recommendations by Belleri et al. [19].

2.1.3 Masonry Case Study School Building

The URM case study school building was modelled in TreMuri [20], a specialized software suitable for the seismic analysis of 3D masonry buildings. The model uses an equivalent frame approach to simulate the behavior of the building, which takes into account the two main components in a masonry wall: the piers and the spandrels. The piers act as the main vertically resisting elements, while the spandrels couple the response of two adjacent piers. The non-linear macro-element implemented in TreMuri allows two main failure modes to be simulated: (1) flexural failure, expressed as rocking and crushing mechanisms; and (2) diagonal cracking and shear sliding to account for shear failure. The failure of the panels is defined in terms of a drift limit. If the maximum story drift in a pier is achieved, the element becomes a strut,

Table 2 Translational elastic mode periods of the numerical models and adopted conditioning periods

School building	Longitudinal mode period ($T_{1,X}$) (s)	Transverse mode period ($T_{1,Y}$) (s)	Arithmetic mean period ($T_{average}$) (s)	Conditioning period (T*) (s)
RC	0.36	0.61	0.49	0.50
PC	1.10	1.11	1.11	1.00
URM	0.22	0.49	0.36	0.20

meaning that shear and bending capacity are reduced to zero, while the axial load is still supported. Following NTC 2018 [21], it was considered reasonable to set the drift limit for shear and bending failure at 0.4% and 0.8%, respectively.

Second-order geometrical P-Δ effects were taken into account for all the case study buildings. The inherent damping in the URM school building was defined through a 5% tangent stiffness proportional Rayleigh damping model, whereas constant 5% critical damping to all modes of vibration was assumed for the RC and PC school buildings. Cracked section stiffness was assumed in the numerical analyses. Table 2 reports the results of the eigenvalue analyses carried out on the three case study school buildings. The elastic fundamental periods reported in Table 2 were used to characterize, by different conditioning periods, the hazard curves necessary for the analyses, as described in the next section.

2.2 *Seismic Hazard Characterization*

Probabilistic seismic hazard analysis (PSHA) was conducted for the city of Cassino, Italy—the actual location of one of the selected case study school buildings. Adopting the hazard model proposed by Meletti et al. [22], this location is characterized by a peak ground acceleration (PGA) of 0.21g for a return period of 475 years. The sets of ground motion records were chosen to match a conditional spectrum generated with the REASSESS software tool [23]. A total of 22 pairs of ground motion records in two horizontal components were taken from the PEER NGA-West 2 database [24]. Furthermore, the spectral acceleration, Sa(T*) at a conditioning period, T*, was chosen as the intensity measure (IM) for the hazard curve. The arithmetic mean of the two computed orthogonal fundamental periods was used to define T*, as suggested in FEMA P58 [4] and reported in Table 2 for each case study school building. The hazard curves for each conditioning period T* are shown in Fig. 3.

Moreover, for the structural performance assessment, the uniform hazard spectra (UHS) were also calculated at different return periods for the selected site. The following four return periods were considered according to the prescriptions provided by NTC 2018 [21]: 45, 75, 712 and 1463 years, corresponding respectively, to SLO: operational limit state, SLD: damage limitation limit state, SLV: life safety limit state

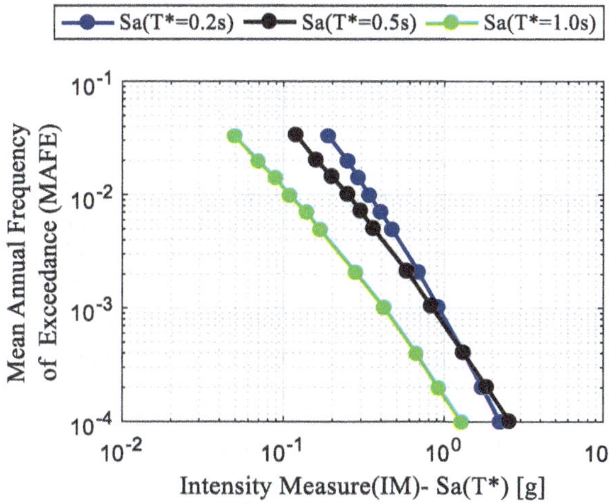

Fig. 3 Hazard curves at conditioning periods T* for Cassino, Italy

and SLC: collapse prevention limit state, for a building class III with a nominal life of 75 years, which would correspond to a school building.

2.3 Structural Response

The structural performance of the case study school buildings was assessed through nonlinear static and dynamic analyses. Figure 4 shows the static pushover curves of these buildings, expressed in terms of base shear coefficient (ratio between the base shear capacity and seismic weight of the building) and the building roof drift (roof displacement/total building height). The performance points corresponding to the four limit states, also plotted in Fig. 4, were identified according to the following criteria: SLO as two thirds of the suggested drift for limiting the damage to non-structural elements specified in NTC 2018 [21]; SLD as the minimum between the drift at incipient yielding in a structural element and the recommended drift limit [16]; SLV as the drift at maximum lateral capacity; and SLC as the drift after reaching a drop of 20% of SLV lateral capacity.

The SLD drift limits are described as 0.5% for a building with rigid partitions (adopted for the RC and PC school buildings), and 0.2% for unreinforced-masonry structures (adopted for the URM school building). Due to the type of failure mechanism observed in the URM case study school building, the SLV and SLC limit states for this typology were assumed to be achieved when the building reached a drift of 0.4% and 0.6%, respectively. These values, suggested by Morandi et al. [25], are conservative for describing the incremental damage undergone by URM structures when shear is the dominant failure mode. For the bare RC building, the SLV and SLC limit states were determined based on the capacity of the bare RC frame structure, following the recommendations established by NTC 2018 [21] and Eurocode [26].

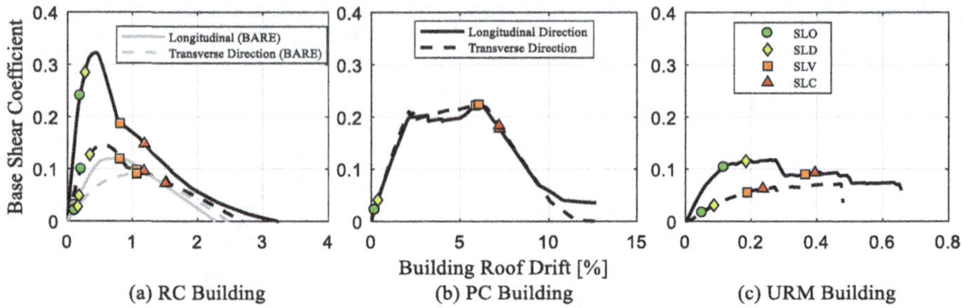

Fig. 4 Static pushover curves for the case study school buildings, where the base shear is normalized by the total building weight and the point at which each of the limit states are exceeded is marked

The lateral capacity of this assembly in both directions is significantly different if the masonry infills are considered or not (i.e. BARE in Fig. 4a), while for the PC case study school building the lateral capacities in the two directions are similar. The PC case study school building shows the highest displacement capacity, while the URM case study school building is characterized by the lowest lateral strength and deformation capacities.

In order to evaluate the structural performance of the case study school buildings, the N2 method [27] was implemented, considering the structural requirements of NTC 2018 [21]. The serviceability limit states (SLO and SLD) were assessed according to the drift limits proposed by NTC 2018; this verification is illustrated in Fig. 5a. On the other hand, the requirements for the ultimate limit states (SLV and SLC) imply maintaining the vertical stability and the development of a ductile mechanism, which avoids soft story or weak-story failure and promotes the strength hierarchy criteria presented by Tasligedik et al. [28].

Figure 5 illustrates the distributions of maximum story drifts along the building height for each return period considered. The RC case study school building is the only structure meeting the drift criteria at the serviceability limit states, while the PC and URM case study school buildings largely exceed the drift limits, being more detrimental in the case of the PC structure. In terms of the ultimate limit states, the RC and PC case study school buildings exhibit drift concentrations, which can be related to the lack of lateral story strength and/or stiffness. No results are shown in Fig. 5 for the URM case study school building at these ultimate limit states since its pushover curve does not intersect with the life safety demand intensity, meaning that it is expected that it would have collapsed already. These results highlight the very high seismic vulnerability of this building.

The local response of the structural elements was also investigated with a view to identify, in subsequent studies, the best retrofit strategies for the case study school buildings. The performance assessment illustrated in Fig. 5 evaluates the main code requirements [21], demonstrating that in many cases the buildings are not meeting the requisites for diverse limits states. Consequently, retrofit interventions will need to be considered to satisfy the code provisions.

Nonlinear response history analyses (NRHA) were then conducted following the multiple-stripe analysis (MSA) methodology, using the ground motion records described in Sect. 2.2. Figure 6 illustrates the response of the different case study school buildings in terms of median value of the maximum peak story drifts (PSD) and peak floor accelerations (PFA), along the building height, in both principal directions. The drift demands for the RC and PC case study school buildings are larger than that of the URM case study school building. This highlights the flexibility of these buildings when compared to the URM structure, which is stiffer. The RC case study school building is both stiffer and stronger than the PC case study school building due to the presence of the masonry infills. This can also be verified through the translational mode periods listed in Table 2 and the drift profiles plotted in Fig. 5.

The results of the NRHA were used to construct the collapse fragility function for each building, considering uncertainty due to record-to-record variability through the use of ground motion sets and amplifying it to account for modelling uncertainty. The number of collapses at each intensity level was expressed as a fraction of the total number of records and then used to compute the probability of collapse. These collapse probability data points were then fitted with a lognormal distribution through the maximum likelihood method outlined by Baker [29]. This method is described by

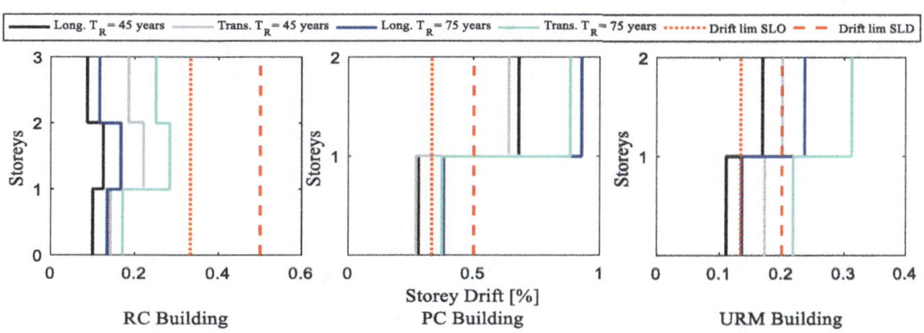

(a) Maximum storeys drifts for the serviceability limit states (SLO, SLD)

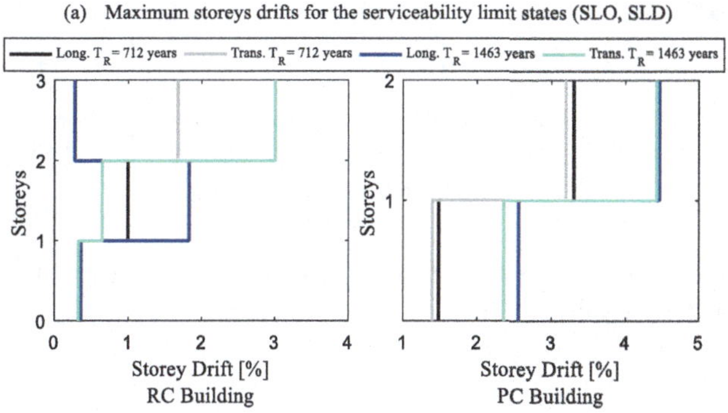

(b) Maximum storeys drifts for the ultimate limit states (SLV, SLC), no plot is shown for the URM building since the drift demands exceed the maximum drift capacity.

Fig. 5 Maximum storey drifts in case study school buildings at different limits state intensities specified in NTC 2018 [21] a median collapse intensity, θ, and a logarithmic standard deviation, β_R associated with the record-to-record variability as presented in Fig. 7. The numerical values for θ and β_R associated with these collapse fragility functions are listed in Table 3 along with the collapse margin ratios (CMRs), defined as the ratio between the median collapse intensity and the intensity at the collapse prevention limit state (spectral intensity at $T_R = 1463$ years listed in Table 3). These results along with other sources of uncertainties (β_{MU} and β_T) will be discussed in the next section. For the RC school building, collapse was assumed to have occurred when the story drift exceeded 5% at any level of the building in either direction. The same drift limit was assumed to define the probability of collapse for the PC case study school building, considering that reaching 5% drift involves excessive columns' plastic hinge rotations and unseating of the beams from the column corbels. For the URM case study school building, collapse was evaluated in terms of failure of the pier elements due to shear or flexure mechanisms. The maximum drift limit assumed for reaching collapse was taken as 0.5%, as defined in the NTC 2018 [21].

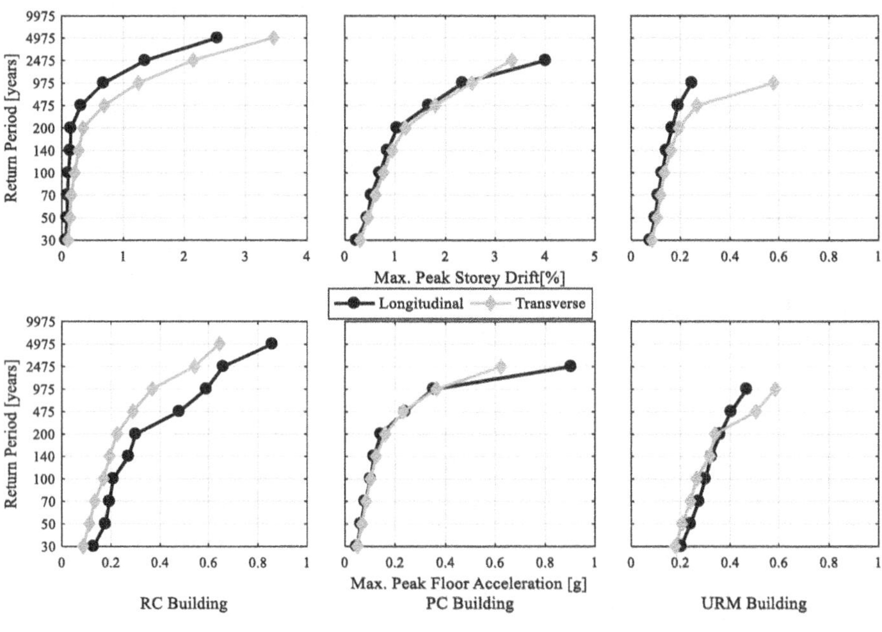

Fig. 6 Median peak story drifts and peak floor accelerations over the height of the case study school buildings in both principal directions

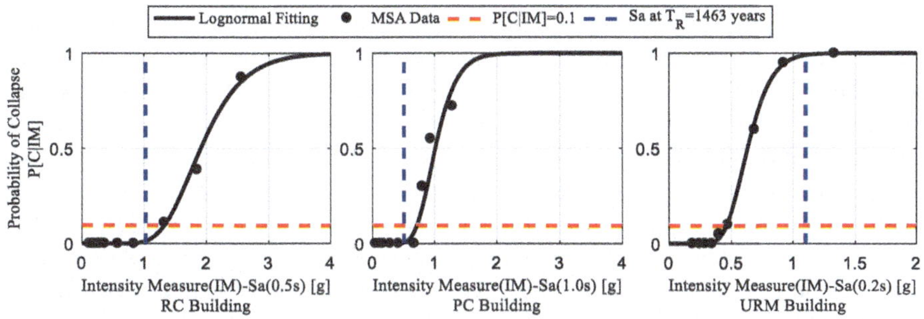

Fig. 7 Collapse fragility functions for the case study school buildings

3 Detailed Component-Based Loss Estimation and Collapse Performance

A loss estimation assessment was carried out for each case-study school building. The total expected losses at different intensity levels $E[L_T|IM]$ were determined. The epistemic uncertainty, β_{MU}, was introduced in the assessment to consider the modelling uncertainties, whereas the aleatory uncertainty, β_R, was associated with the record-to-record variability. These values are presented in Table 3. The epistemic uncertainty was adopted from FEMA P58 guidelines [4] according to each building typology. The total dispersion β_T was obtained as the square root sum of the squares (SRSS) of β_{MU} and β_R. The loss calculation was performed in the software PACT [30] for 11 return periods resulting in 200 realizations. Figure 8a illustrates the vulnerability curve describing the loss ratio associated with each return period. The expected losses reach the replacement cost at a different return period for each school building. For the URM school building, the expected losses equal the replacement cost (i.e. expected loss ratio = 1) near the 475-year return period. For the RC and PC case study school buildings, the expected loss ratio reaches a value equal to unity near the 2475-year return period.

The collapse performance of each case study school building was assessed according to FEMA P695 by verifying that the probability of collapse under

Table 3 Median collapse intensities, θ, dispersion due to record-to-record variability, β_R, dispersion due to model uncertainty, β_{MU}, total dispersion, β_T, median collapse intensity at collapse prevention limit stated, S_a at T_R = 1463 years and collapse margin ration, CMR, for each case study school building

School building	Median IM, θ (g)	Dispersion, β_R	Dispersion, β_{MU}	Dispersion, β_T	Sa at T_R = 1463 years	Collapse margin ratio (CMR)
RC	1.91	0.28	0.15	0.32	1.02	1.87
PC	1.01	0.27	0.35	0.44	0.50	2.01
URM	0.63	0.24	0.20	0.31	1.10	0.57

maximum considerable earthquake (MCE) ground motions (P[C|MCE] is less than 10%. The return period associated with the MCE intensity level was defined according to NTC 2018 and is equal to 1463 years. The CMR values, presented in Table 3, are very similar for the RC and PC school buildings (approximately 2.0) meaning that these buildings present a considerable safety margin against collapse. On the other hand, the CMR for the URM school building is lower than unity, highlighting once again the building's vulnerability to collapse (i.e. P[C|MCE] > 50%). Another approach to assess the collapse performance is to integrate the collapse fragility curve over the entire hazard curve to obtain the mean annual frequency of collapse (MAFC), which is presented in Fig. 8b for each school building. Dolšek et al. [31] reviewed typical acceptable MAFC limits obtained from various studies available in the literature and noted that this limit is in the range of 10^{-5} to 10^{-4}, which are represented by the red dotted horizontal lines in Fig. 8b. It is clear that the performance of the URM case study school buildings is unacceptable. Furthermore, even if not as prominently, the RC and PC case study school buildings also fail to meet the acceptable MAFC limits as well. Based on these considerations, the seismic performance of the case-study school buildings should be improved through appropriate structural retrofit interventions to meet code requirements and evaluate their influence on the collapse performance.

The obtained EALs, computed using Eq. (1), are listed in Table 4 for each school building along with the assumed replacement cost of the building (RepC). The URM case study school building is the most vulnerable in terms of EAL, followed by

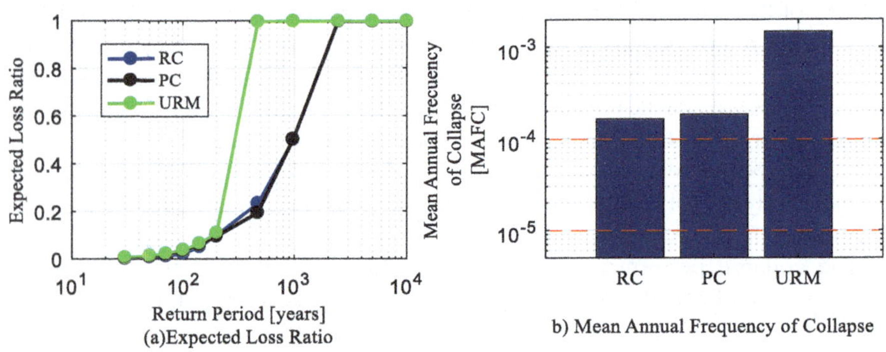

Fig. 8 Loss estimation of case study school buildings, **a** expected annual loss vs return period and **b** mean annual frequency of collapse (range of acceptable MAFC limits indicated by horizontal dotted red lines)

Table 4 Expected annual loss ratios, EAL, and total replacement cost, RepC, for each case study building.

School building	EAL (%)	RepC (€)
RC	0.27	3,929,937
PC	0.27	4,212,616
URM	0.43	2,075,892

the RC and PC case study school buildings. The EALs reported in this study are in line with other results available in the literature. For example, Cardone and Perrone [49] examined the performance of an existing RC frame building with masonry infill located in L'Aquila and reported EAL values between 0.75% and 1.07%. Considering that L'Aquila is characterized by a much higher seismicity with respect to the Cassino site considered herein, the findings of Cardone and Perrone [32] are in line with the EAL value reported in this study. A similar range was also highlighted in Perrone et al. [33] for infilled RC frames located Italy. Similarly, Sousa and Monteiro [34] have analyzed pre-1970 RC frames with masonry infills in different Italian locations characterized by low to high seismicity, and obtained EAL ranging from 0.2 to 0.5%; hence, again, in agreement with the results presented here. Ottonelli et al. [35] examined two case study URM buildings located in L'Aquila and reported EAL values between 0.55 and 0.68%, which again align reasonably well with the findings herein, considering the relative differences in seismicity. Lastly, Cornali et al. [36] examined existing PC frame buildings in Italy and reported EAL ratio between 0.51 and 0.71%, which again are in line with the results obtained in this study.

$$\mathrm{EAL} = \int \mathrm{E}[\mathrm{L_T}|\mathrm{IM}]\left|\frac{\mathrm{d}\lambda}{\mathrm{dIM}}\right|\mathrm{dIM} \tag{1}$$

4 Simplified Assessment Through Risk Classification Framework

A document of particular relevance to the case study school buildings considered herein is the recent Italian seismic risk classification guidelines detailed in Decreto Ministeriale (D.M.) 58/2017 [7]. These guidelines aim to incorporate some of the more recent advancements in the field of seismic risk assessment into a procedure that is both straightforward to implement and integrates well with existing codes in Italy. The guidelines focus on two specific aspects regarding buildings; life safety and expected annual loss, and provide a classification system with which practitioners can assess the current status of buildings and demonstrate improvements in seismic performance via different retrofitting measures. The procedure is illustrated in Fig. 9 and shows how only a pushover analysis is required to identify the different limit states described in Italian national code [21] for each building typology. These four limit states are outlined qualitatively in NTC 2008 as: 'Operational' (SLO); 'Damage Control' (SLD); 'Life Safety'; and 'Collapse Prevention' (SLC).

By identifying these four limit states for a case study building and converting it to an equivalent single degree-of-freedom (SDOF) system, as shown in Fig. 9b, c, the intensity required to develop each of these is identified in Fig. 9d. This intensity is defined in terms of the peak ground acceleration (PGA) of the code-defined acceleration spectrum. With this intensity, the mean annual frequency of exceedance

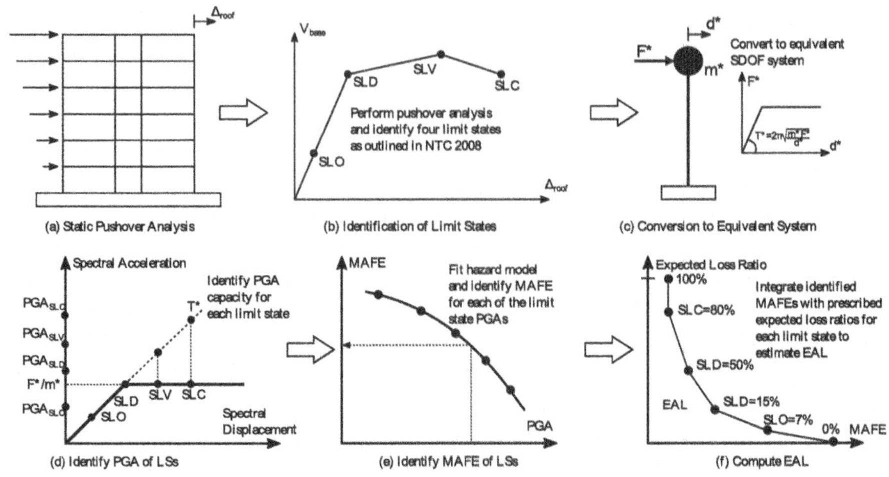

Fig. 9 Illustration of various steps within the Italian seismic risk classification scheme described in [7]

(MAFE) is determined from a hazard model fitted to the available seismic hazard information since this information is typically available in discrete values of return period, as shown in Fig. 9e. Once the MAFE for each of the limit states are established, these are then integrated with prescribed values of expected loss ratio for each limit state outlined in D.M. 58/2017 to compute the EAL as the area under the loss curve illustrated in Fig. 9f. This approach is quite simple as it requires the analyst to conduct just a pushover analysis and eliminates the need for many of the steps involved in the PEER-PBEE loss estimation methodology discussed in Sect. 3. The end result of the guidelines is that an EAL is computed and classified within a letter. In addition to the EAL-based score that classifies the seismic performance in terms of economic loss, another score is attributed based on the structural safety of the building. This is determined based on the ratio of the PGA required to develop the SLV limit state (PGASLV in Fig. 9d) to the PGA demand that structures are designed for at the same limit state. For example, a regular structure with a design life of 50 years would have a design return period of 475 years (10% probability of exceedance in 50 years), whereas school buildings would be designed for a return period of 712 years (10% probability of exceedance in 75 years).

Using the demand to capacity ratio computed as function of the PGA at the SLV limit state, termed IS-V, another letter-based rank is attributed to the building and the overall ranking is determined as the more critical of the EAL-based and life safety-based ranks, which are listed in Table 5.

It was also deemed reasonable to assume that the selected school buildings may exist in any location throughout Italy hence an additional site, characterized by medium-low seismic hazard intensity level (medium-low and medium-high), corresponding to the city of Ancona was selected and the PSHA performed. The first site (Ancona) is characterized by a PGA on stiff soil equal to 0.16g for a 10% probability

Table 5 Seismic performance classification ranking system as a function of EAL and IS-V prescribed in [7]

EAL classification range	Life safety index classification range	Classification ranking
EAL ≤ 0.5%	1.00 ≤ IS-V	A +
0.5% ≤ EAL < 1.0%	0.80 ≤ IS-V < 1.00	A
1.0% ≤ EAL < 1.5%	0.60 ≤ IS-V < 0.80	B
1.5% ≤ EAL < 2.5%	0.45 ≤ IS-V < 0.60	C
2.5% ≤ EAL < 3.5%	0.30 ≤ IS-V < 0.45	D
3.5% ≤ EAL < 4.5%	0.15 ≤ IS-V < 0.30	E
4.5% ≤ EAL < 7.0%	IS-V < 0.15	F
7.0% ≤ EAL		G

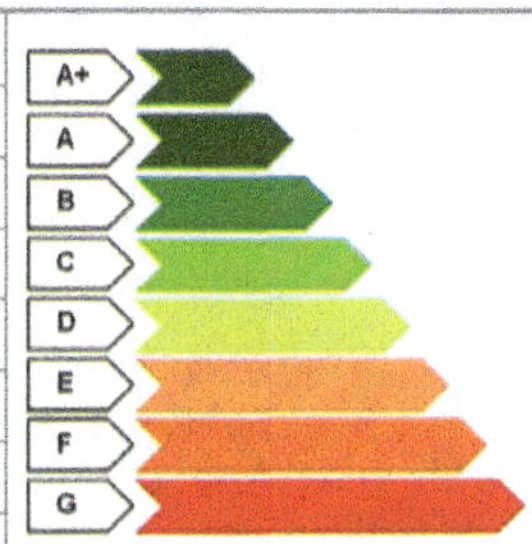

of exceedance in 50 years (or 475 years return period), while the corresponding PGA value for the second site (Cassino) is 0.21g.

Following the simplified procedure outlined in D.M. 58/2017 and utilizing the relevant analysis methods from NTC 2008, this simplified method was also applied to the three case study school buildings at both site locations to establish their seismic performance within these classification guidelines. Their performances were also compared with those computed following the rigorous approach outlined in Sect. 3. Each of the building models detailed in Sect. 2 was analyzed using static pushover analyses, their limit states identified and their equivalent SDOF systems determined. This was performed for the two principal directions of each school building and the PGA to exceed each limit state in either direction of the school building was determined, with the lesser of the two PGAs being adopted for final evaluation. The MAFE for each limit state were then determined, the EAL computed and the final values are reported in Table 6. In addition, the life safety index (IS-V) was also computed as the ratio of the PGA_{SLV} determined in Fig. 9d and the PGA corresponding to a design return period of 712 years for school buildings. The scoring for both of these criteria was determined and the resulting overall seismic classification of the buildings is listed in Table 6.

By comparing the values presented in Table 6 first, it is clear that the life safety index is the governing criteria that determines the overall seismic classification in all but one case, with the URM buildings being observed to be particularly vulnerable when examining the IS-V indices. This is somewhat consistent with the previous findings presented in Sect. 3, where the collapse performance was seen to be critical for the URM buildings. The guidelines note that for a new structure designed using NTC 2008, the IS-V score ought to be greater than or equal to unity.

Comparing the EAL values reported in Table 6 with those computed using the rigorous approach, some discrepancy can be seen in the results plotted in Fig. 10. The overall magnitude of the EAL values computed using the simplified method is much higher than those computed in Sect. 3. While the overall magnitude of the EAL computed using the two methods differs, the overall trend and relative differences

Table 6 EAL and IS-V values computed using D.M. 58/2017 guidelines.

Site location	High seismicity			Medium seismicity		
Building typology	RC	URM	PC	RC	URM	PC
EAL	0.84%	1.01%	0.60%	0.60%	0.68%	0.48%
EAL classification	A	B	A	A	A	A+
IS-V	0.60	0.50	0.90	0.79	0.67	1.20
IS-V classification	C	C	A	B	B	A+
Overall classification	**C**	**C**	**A**	**B**	**B**	**A+**

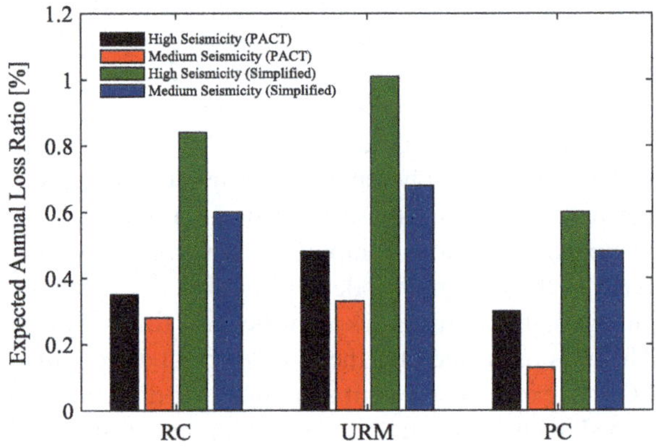

Fig. 10 Comparison of the EAL ratios from detailed analysis and those estimated from simplified analysis

between the different typologies and site locations remain the same. This suggests that, while the results obtained using the simplified method may not align closely with the values obtained from detailed analyses, the relative quantities remain similar and may still be used as a method with which to demonstrate an improvement in seismic performance. Nevertheless, even if conservative with respect to detailed analysis, the simplified seismic classification guidelines introduce more advanced means of quantifying seismic performance and offer a metric with which the overall seismic resilience of communities can be increased in addition to providing a motivation for stakeholders to upgrade the seismic performance of their buildings.

5 Impact of Modelling Uncertainty in Risk Estimations

The masonry-infilled school building is now considered to further scrutinize its seismic risk, given the variability typically observed in some of the required modelling parameters. The seismic assessment of existing masonry-infilled reinforced concrete (RC) buildings is a highly relevant issue in Italy and other Mediterranean countries, particularly in regions where a large part of the built environment was not designed according to modern seismic codes. In the past, masonry infills were generally considered as non-structural elements and were not accounted for in the design process. However, the results of experimental tests and past earthquake evidence demonstrated the influence of masonry infills on the global and local behavior of RC buildings. Furthermore, the acknowledgment of different construction practices points out the significant variability affecting the masonry infills in terms of material properties, thickness, presence of openings and manufacturing techniques. While the previous sections have considered the masonry-infilled school building with constant infill properties, this section accounts for the uncertainty related to the variability in the mechanical properties and the modelling of masonry infills and its impact on the corresponding risk estimates. Statistical data and results of experimental tests were used to identify the main masonry infill typologies adopted in the Italian context. Five different infill types are defined according to, amongst other parameters, their strength and the detailed and simplified risk analysis of the infilled RC building are repeated for each of them to understand the potential differences.

5.1 Masonry Infill Classification

As mentioned earlier, despite the high-level of uncertainty surrounding the masonry infill properties, constant mechanical and geometrical properties of the masonry infills are typically assumed in risk assessment studies both at single-building and regional scale. To the authors best knowledge, the results of in-situ tests on masonry infills are not available. The results of experimental studies [12], also reported in recent databases dealing with the definition of the mechanical properties of masonry infills [4], are taken into account to define the variability in the masonry infills' characterization, which is related to many parameters, such as the maximum shear strength of the panel, the stiffness of the panel, the relative stiffness between the panel and the surrounding frame, the vertical/horizontal compressive strength of the masonry, vertical/horizontal modulus of elasticity and thickness of the masonry. In this study, it was decided to proceed with a macro-level distinction of the infills in terms of shear strength. According to the results available in the literature [15], this parameter is one of the most important affecting the lateral capacity of RC frames and could also lead to local shear failure (e.g. shear failure of columns that were built without particular construction details). Five masonry infill typologies, from weak to strong, classified according to the shear capacity, were thus selected as representative of the existing masonry infill typologies used in RC residential buildings built in 1970–1980 in Italy. The results of three quasi-static cyclic tests [12] were used to evaluate the properties of the infill panels. The first three typologies were selected according to the classification proposed by Hak et al. [16]. These three typologies are referred to as infill type 1, type 2 and type 3. Infill type 1 is a weak masonry infill that was investigated in [12], in which the clay blocks and mortar were selected according to the building practice of the European earthquake-prone countries. This is a single leaf infill, constructed of horizontally hollowed brick with a 1.0cm thick plaster on each external side and thickness of 8.0cm. Infill type 2 is made up of double leaves, each constructed of horizontally hollowed bricks and 12.0cm of thickness, covered with a 1.0cm thick external plaster and divided by a cavity of 5.0cm; infill type 3 is a single leaf constructed by vertically hollowed bricks and 30.0cm thick. The other two masonry infills, termed type 4 and 5, were selected from the experimental tests provided in [13], respectively. Infill type 4 [13] is an ordinary single-leaf masonry infill of 35cm thickness, made up of vertically hollowed lightweight tongue and groove clay block units, while infill type 5 [14] is a traditional single-leaf masonry infill of 15 cm thickness, consisting of vertically hollowed clay block units. Considering the high variability in the selected infill typologies, it is believed that, even if some typologies that can be found in existing buildings are not represented, due for example to the variation in the thickness of the panels, the combination of five different masonry infill typologies is representative (Table 7).

5.2 *Detailed and Simplified Risk Analysis Using Different Masonry Infill Types*

The masonry-infilled RC school building was analyzed according to the methodologies described and applied in Sects. 3 and 4, assuming the five different masonry infill types outlined in Table 6. The results of the detailed component-based PEER approach are illustrated in Fig. 11.

It can be seen how important variations in the estimated expected annual losses occur when considering separately the drift-sensitive and acceleration-sensitive components, with variations of nearly 50%. This is somehow an expected outcome, given the growing stiffness given by the increase in the masonry infill strength (from Type1 to Type5). On the other hand, when combining the two types of elements, for this particular building and its component inventory, the differences somehow counterbalance and the total expected annual loss presents a more modest variation, in the order of 5–10%. Although this observation may seem reassuring with respect to the low impact of not carrying out an accurate characterization and modelling of the masonry infill properties, this can be the case for this particular building, in which the proportion of drift-sensitive and acceleration-sensitive elements is balanced. In other buildings (e.g. residential, industrial, hospitals, etc.) the observations might differ significantly and further research is necessary. Moreover, it is important to note that the fragility functions of the masonry infill panels themselves were kept constant throughout the five typologies, given that no specific studies were available to assign distinct fragility models. The differences in the estimated losses come,

Table 7 Representative masonry infills

References	Type	Macro classification	t_w (mm)	E_{wv} (MPa)	E_{wh} (MPa)	G_w (MPa)	f_{wv} (MPa)	f_{wlat} (MPa)	f_{wu} (MPa)
Calvi and Bolognini [4]	1	Weak	80	1873	991	1089	2.02	1.18	0.44
Hak et al. [55]	2	Weak-Medium	240	1873	991	1873	1.5	1.11	0.25
Hak et al. [55]	3	Medium-Strong	300	3240	1050	1296	3.51	1.5	0.3
Morandi et al. [5]	4	Medium-Strong	350	5299	494	2120	4.64	1.08	0.359
Cavaleri and Di Trapani [6]	5	Strong	150	6401	5038	2547	8.66	4.18	1.07

t_w: Thickness; E_{wv}: Elastic modulus vertical direction; E_{wh}: Elastic modulus horizontal direction; G_w: Shear modulus; f_{wv}: Vertical strength; f_{wlat}: Lateral strength; f_{wu}: Shear sliding strength

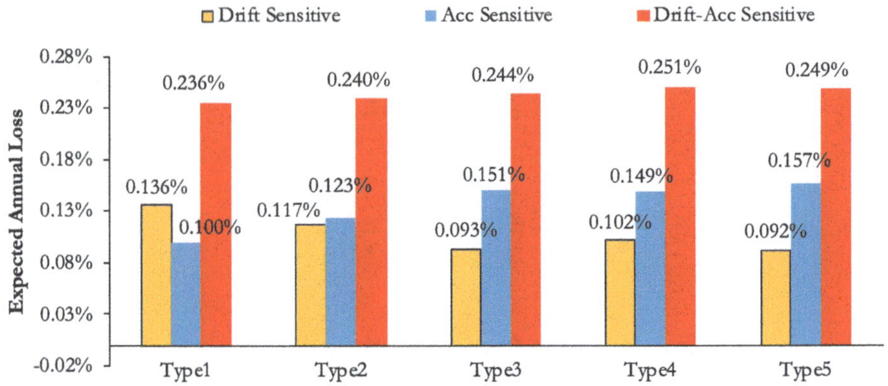

Fig. 11 Drift-sensitive, acceleration-sensitive and total expected annual losses, as a function of the masonry infill typology

therefore, solely from the reduced structural demand, caused by an increase in the strength of the infill panels.

Furthermore, Fig. 12 illustrates the obtained simplified variability metrics (mean and standard deviation) in the loss ratio (percentage of the replacement cost of the building) given by the variability in the masonry infill typology. The results are presented for the drift-sensitive components only (the ones with highest variation) and the total losses (drift-sensitive and acceleration-sensitive put together). The results highlight the larger dispersion around the mean loss ratios for higher return periods. Given that higher return periods have much lower probability of occurrence, the impact on the expected annual losses is less pronounced, as discussed in the previous paragraph.

For what concerns the simplified SISMABONUS approach, the estimated seismic risk classes when foreseeing the different types of masonry infill are outlined in Table 8. It can be seen that, contrarily to what was observed for the detailed component-based approach, according to the simplified SISMABONUS framework currently

Fig. 12 Dispersion (mean and standard deviation) of drift-sensitive and Total Expected Loss Ratio, for increasing return period

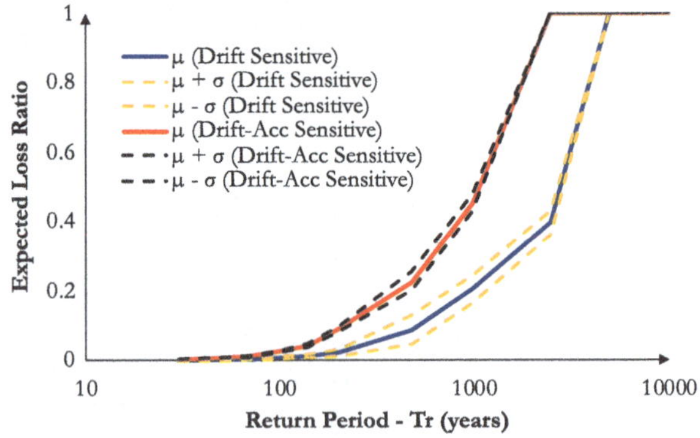

Table 8 Risk classification according to the simplified approach, as a function of infill type

Type	Macro classification	Risk class
1	Weak	C
2	Weak-Medium	C
3	Medium-Strong	C
4	Medium-Strong	C
5	Strong	B

implemented in Italy, the uncertainty surrounding the modelling of the masonry infills leads to a different risk class only when the strong masonry infill type is used. Such difference corresponds to an increase in the performance of one risk class. This outcome is very likely due to the simplified, conservative nature of the procedure, which relies only on the capacity curve of the building, which, after the collapse of the infills, becomes very similar for all the infill typologies, unless significant damage has been induced to the RC frames, for the case of the stronger infills. It is also worth noting how the risk class is nevertheless low (essentially C), which clearly indicates a need for retrofitting of the building.

6 Conclusions

The seismic risk of existing Italian school buildings of three different typologies has been analyzed and presented in this paper. A detailed seismic loss assessment framework, following the developments by the Pacific Earthquake Engineering Research Center on performance-based earthquake engineering, and the recent simplified Italian guidelines developed for seismic risk classification of existing buildings were both applied, as detailed and simplified approaches respectively, to the three school buildings in Italy to identify the most vulnerable structural typologies. The school buildings were selected from an Italian school database developed within the "Progetto Scuole" as well as during previous research projects made in collaboration with the Italian Department of Civil Protection. Each building had a different construction typology: 1) Unreinforced Masonry (URM), Reinforced Concrete (RC) and Precast (PC). Detailed in-situ surveys were performed to gather the information required for the numerical modelling as well as for the loss estimation assessment. Two sites characterized by different seismic hazard levels were initially selected to compare the results of the detailed analyses with those of the simplified methodology. The RC school building was then further analyzed under the consideration of five different masonry infill typologies, to understand the impact of the uncertainty modelling of this component type on the estimated losses.

The results highlighted the seismic vulnerability of existing school buildings in Italy both in terms of collapse capacity and expected annual loss (EAL). The URM was identified as the most vulnerable structural typology and the damage to non-structural elements tends to dominate the EAL for all analyzed school typologies, in particular at more frequent return periods of the seismic intensity. Comparing the findings of the detailed loss assessment with those of the recent seismic risk classification guidelines, introduced in Italy in 2017, a similar trend in terms of identification of more vulnerable typologies was observed. However, due to the simplified nature of the guidelines, the overall magnitude of the EAL was significantly overestimated with respect to the EAL evaluated using the detailed approach. Nevertheless, this simplified methodology is recognized as a useful tool to the future investigation of the seismic risk of school buildings at regional scale. A lower, but always considerable, seismic risk has been observed for RC school buildings, while the lowest seismic risk has been observed for PC school buildings. Finally, the additional analysis of the RC building showed that the uncertainty on the infill characterization has a relevant impact on both drift-sensitive and acceleration-sensitive components, although, when combined, the overall expected annual losses present lower variability with respect to the infill strength. Further research is required, through the extension to additional buildings, with different configurations and component inventories, to check these specific conclusions.

The results of this case-study application are considered particularly useful to governmental decision makers who would need to decide and justify the distribution of limited financial resources that aim to reduce the overall seismic risk of the Italian school building stock and the analysis framework can be easily reproduced and adapted to any context.

Acknowledgments The work presented in this paper has been developed within the framework of the project "Dipartimenti di Eccellenza", funded by the Italian Ministry of University and Research at the University School for Advanced Studied IUSS Pavia. The author expresses his gratitude to the other members of the research team, Andre Filiatrault, Daniele Perrone, Gerard O'Reilly, Wilson Carofilis and Gianrocco Mucedero, who greatly contributed to this work.

References

1. Borzi, B., Ceresa, P., Faravelli, M., Fiorini, E., & Onida, M. (2013). Seismic risk assessment of Italian school buildings. Computational methods in earthquake engineering. *Computational Methods in Applied Sciences, 30,* 317–344.
2. Calvi, M., Moratti, M., & Filiatrault, A. (2016). Studio della risposta di elementi non strutturali di edifici scolastici soggetti ad eventi sismici/role and importance of non-structural elements in the seismic vulnerability of school buildings (in Italian). *Progettazione sismica, 6*(3), 9–29.
3. Cornell, C. A., & Krawinkler, H. (2000). Progress and challenges in seismic performance assessment. *PEER Cent News, 3,* 1–2.
4. FEMA P58-1. (2012). *Seismic performance assessment of buildings: Volume 1—Methodology (P-58-1)* (Vol. 1). Washington, DC.
5. De Angelis, A., & Pecce, M. (2015). Seismic nonstructural vulnerability assessment in school buildings. *Natural Hazards, 79,* 1333–1358.
6. Grant, D., Bommer, J., Pinho, R., Calvi, M., Goretti, A., & Meroni, F. (2007). A prioritization scheme for seismic intervention in school buildings in Italy. *Earthquake Spectra, 23*(2), 291–314.

7. Ministry Decree. Decreto Legge n. 58 del 27/12/2017 Allegato A: linee guida per la classificazione del rischio sismico delle costruzioni. Italian Ministry of Inftastructures and Transport (in Italian).
8. Borzi, B., Ceresa, P., Faravelli, M., Fiorini, E., & Onida, M. (2011). Definition of a prioritization procedure for structural retrofitting of Italian school buildings. *COMPDYN 2011–3rd ECOMAS*, Corfu, Greece.
9. Perrone, D., O'Reilly, G. J., Monteiro, R., & Filiatrault, A. (2019). Assessing seismic risk in typical Italian school buildings: From in-situ survey to loss estimation. *International Journal of Disaster Risk Reduction, 44,* 101448.
10. Taghavi, S., & Miranda, E. (2013). *Response assessment of nonstructural building elements*. PEER report 2003/05, Berkeley, California.
11. O'Reilly, G. J., & Sullivan, T. J. (2018). Probabilistic seismic assessment and retrofit considerations for Italian RC frame buildings. *Bulletin of Earthquake Engineering, 16*(3), 1447–1485.
12. Cosenza, E., Del Vecchio, C., Di Ludovico, M., Dolce, M., Moroni, C., Prota, A., & Renzi, E. (2018). The Italian guidelines for seismic risk classification of constructions: Technical principles and validation. *Bulletin of Earthquake Engineering, 16*(12), 5905–5935.
13. Fiore, A., Mezzina, M., Porco, F., Raffaele, D., & Uva, G. (2017). Seismic safety assessment of school building in Puglia (Italy): Overview and cases studies. In *15th World Conference on Earthquake Engineering*, Lisbon, Portugal.
14. Buratti, N., Minghini, F., Ongaretto, E., Savoia, M., & Tullini, N. (2017). Empirical seismic fragility for the precast RC industrial buildings damaged by the 2012 Emilia (Italy) earthquakes. *Earthquake Engineering & Structural Dynamics, 46*(14), 2317–2335.
15. McKenna, F., Scott, M. H., & Fenves, G. L. (2010). Nonlinear finite-element analysis software architecture using object composition. *Journal of Computing in Civil Engineering, 24,* 95–107.
16. O'Reilly, G. J., & Sullivan, T. J. (2017). Modelling techniques for the seismic assessment of existing Italian RC frame structures. *Journal of Earthquake Engineering, 23*(8), 1262–1296.
17. Scott, M. H., & Fenves, G. L. (2006). Plastic hinge integration methods for force-based beam-column elements. *Journal of Structural Engineering, 132,* 244–252.
18. Haselton, C., Liel, A., Taylor, S., & Deierlein, G. (2008). Beam-column element model calibrated for predicting flexural response leading to global collapse of RC frame buildings. PEER repost 2007/03.
19. Belleri, A., Torquati, M., Marini, A., & Riva, P. (2016). Horizontal cladding panels: In-plane seismic performance in precast concrete buildings. *Bulletin of Earthquake Engineering, 14,* 1103–1129.
20. Lagomarsino, S., Penna, A., Galasco, A., & Cattari, S. (2013). TREMURI program: An equivalent frame model for the nonlinear seismic analysis of masonry buildings. *Engineering Structures, 56,* 1787–1799.
21. NTC. (2018). Norme Tecniche Per Le Costruzioni. Rome, Italy.
22. Meletti, C., Galadini, F., Valensise, G., Stucchi, M., Basili, R., Barba, S., et al. (2008). A seismic source zone model for the seismic hazard assessment of the Italian territory. *Tectonophysics, 450,* 85–108.
23. Iervolino, I., Chioccarelli, E., Cito, P. (2015). REASSESS V1.0: A computationally efficient software for probabilistic seismic hazard analysis. In: *COMPDYN 2015–5th ECCOMAS Thematic Conference on. Computational Methods in Structural Dynamics and Earthquake Engineering*, Crete Island, Greece.
24. Ancheta, T.D., Darragh, R.B., Stewart, J.P., Seyhan, E., Silva, W.J., & Chiou, B.S.J., et al. (2013). *PEER NGA-West2 database*. PEER Report 2013/03.
25. Morandi, P., Albanesi, L., Graziotti, F., Li Piani, T., Penna, A., & Magenes, G. (2018). Development of a dataset on the in-plane experimental response of URM piers with bricks and blocks. *Construction and Building Materials, 190,* 593–611.
26. EN 1998-3:2005. (2005). Eurocode 8: Design of structures for earthquake resistance—Part 3: Assessment and retrofit of buildings. Brussels, Belgium.

27. Fajfar, P. (2000). A nonlinear analysis method for performance based seismic design. *Earthquake Spectra, 16*(3), 573–592.
28. Tasligedik, S., Akguzel, U., Kam, W., & Pampanin, S. (2016). Strength hierarchy at reinforced concrete beam-column joints and global capacity. *Journal of Earthquake Engineering,* 1–34.
29. Baker, J. W. (2015). Efficient analytical fragility function fitting using dynamic structural analysis. *Earthquake Spectra, 31,* 579–599.
30. FEMA P58-3. (2012). Seismic *Performance Assessment of Buildings Volume 3—Performance Assessment Calculation Tool (PACT) Version 2.9.65 (FEMA P-58-3. 1)* (Vol. 3). Washington, DC, 2012.
31. Dolšek, M., Lazar Sinković, N., & Žižmond, J. (2017). IM-based and EDP-based decision models for the verification of the seismic collapse safety of buildings. *Earthquake Engineering & Structural Dynamics, 46,* 2665–2682.
32. Cardone, D., & Perrone, G. (2017). Damage and loss assessment of Pre-70 RC frame buildings with FEMA P-58. *Journal of Earthquake Engineering, 21,* 23–61.
33. Perrone, G., Cardone, D., O'Reilly, G. J., & Sullivan, T. J. (2019). Developing a direct approach for estimating expected annual losses of Italian buildings. *Journal of Earthquake Engineering,* 1–32.
34. Sousa, L., & Monteiro, R. (2018). Seismic retrofit options for non-structural building partition walls: Impact on loss estimation and cost-benefit analysis. *Engineering Structures, 161,* 8–27.
35. Ottonelli, D., Cattari, S., & Lagomarsino, S. (2016). Assessment and retrofit of masonry structures. In T. J. Sullivan, G. M. Calvi, & R. Monteiro (Eds.), *Towards simplified displacement-based loss assessment approaches* (pp. 5–62). Italy: Pavia.
36. Cornali, F., Belleri, A., & Riva, P. (2016). Assessment and retrofit of pre-cast concrete buildings. In T. J. Sullivan, G. M. Calvi, & R. Monteiro (Eds.), *Towards simplified displacement-based loss assessment approaches* (pp. 181–221). Italy: Pavia.

Towards Climate Change Adaptation of Existing and New Deteriorating Infrastructure

Emilio Bastidas-Arteaga

Abstract Infrastructure assets are essential components to the economical development of modern societies. They are designed to ensure target levels of serviceability and safety on the basis on past experiences and current knowledge on design, construction and maintenance practices. However, changes in climate could modify the lifetime performance of infrastructure by increasing or decreasing failure risks. Therefore, a rational and scientific approach is necessary to deal with the adaptation of existing and new deteriorating infrastructure in a comprehensive way. This keynote paper provides an overview of recent works on this area including: (1) assessment of climate change effects, (2) adaptation to new environmental conditions for future climate change scenarios and (3) decision-making under a changing climate. Several examples for different kind of deteriorating infrastructure assets are also presented and discussed in this paper.

Keywords Climate change · Adaptation · Infrastructure · Reliability

1 Introduction

Infrastructure assets are a key component for the development of modern societies. These assets are designed to provide specific services and are subjected to environmental or operational actions that could affect their serviceability and safety [1]. Among the environmental ones, deterioration processes (corrosion, fatigue, etc.) or extreme events (hurricanes, floods, winds, etc.) are examples of actions that decrease the infrastructure performance gradually or suddenly and could in some cases lead to structural failure.

Design and operation of infrastructure assets is mainly based on the past experiences and knowledge at the design time. Probabilistic approaches are generally directly or indirectly included in the design stage to account for uncertainties related, among others, with environmental actions. Nevertheless, studies on climate change

E. Bastidas-Arteaga (✉)
Laboratory of Engineering Sciences for Environment, UMR CNRS 7356, La Rochelle University, La Rochelle, France
e-mail: ebastida@univ-lr.fr

© The Author(s), under exclusive license to Springer Nature Switzerland AG 2021
J. C. Matos et al. (eds.), *18th International Probabilistic Workshop*, Lecture Notes in Civil Engineering 153, https://doi.org/10.1007/978-3-030-73616-3_3

announce variations (increase, decrease, frequency, intensity, etc. [2]) of the patterns of current weather that are not included in the design stage but that could affect the infrastructure performance. Therefore, a rational and scientific approach is necessary to deal with the adaptation of existing and new deterioration infrastructure in a comprehensive way [3].

Adaptation of infrastructure to new environmental conditions is not an easy decision-making task because hard policy choices are needed in the present to deal with consequences in the mid- (20–40 years) or long-term (more than 70 years). The choices should take into account current needs and uncertain and complex future risks under a changing climate. Within this context, this paper provides an overview of recent research works dealing with the topics of: (1) assessment of climate change effects (Sect. 2), (2) adaptation to climate change (Sect. 3), and (3) decision-making under a changing climate. Applications to various study cases including power distribution systems, bridges subjected to scour, and corroded reinforced concrete structures are also included in the paper.

2 Effects of Climate Change on Infrastructure

Climate change could affect serviceability and safety of infrastructure assets by three ways: (1) changes (increase/decrease) in the intensity and frequency of extreme events; (2) changes in the kinetics of deterioration rates; and (3) a combination of effects on extreme events and progressive deterioration.

The current patterns of extreme events that affect the structural integrity could be affected by climate change. For example, Iman [4] evaluated the effects of climate change on bridges subjected to scour. The development of scour holes, that could induce the bridge failure, is widely influenced by the river flow magnitude. This latter is affected, in turn, by modifications on precipitation patterns that could be affected by climate change. Figure 1 presents some projected changes in seasonal streamflow in various European rivers. It is observed in this figure that future climate conditions could increase or decrease the normalised discharge depending on the location. For some rivers (e.g. Loire, Danube, etc.) there would be future peak values of normalised discharges that will increase bridge scour risks. For other places, the change is not significant or there is a reduction of the normalised discharge that could reduce scour damage. These findings highlight that a comprehensive management of structures subjected to new environmental conditions affected by climate change should account for these local effects.

Climate change is also affecting the long-term performance of deteriorating infrastructure [5–11]. For example, Merschman et al. [5] studied the effect of variations of temperature and precipitations on the durability of timber poles for power distribution systems subjected to decay. In Fig. 2 are given the mean timber decay for timber poles placed in Miami and New York City and subjected to various climate change scenarios (RCP4.5 and RCP8.5). It is observed in this figure that without considering climate change, the local weather of each location provides a different timber

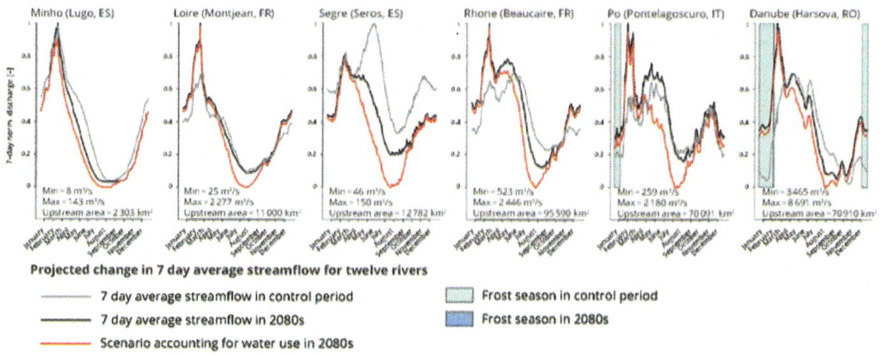

Fig. 1 Projected change in seasonal streamflow for rivers in Europe. *Source* European Environment Agency (EEA)

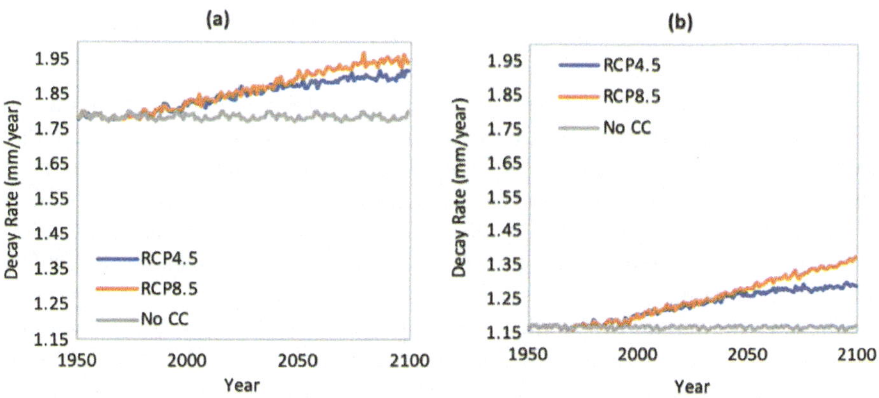

Fig. 2 Mean decay rate for **a** Miami and **b** New York City. Adapted from [5]

decay evolutions where the hotter and rainy weather in Miami increases the decay rate. It is also observed that higher decay rates are expected for the most pessimist scenario (RCP8.5) that announces higher temperature and precipitation by the end of this century. Thus, the lifetime assessment of ageing infrastructures should consider several (optimistic or pessimistic) climate change scenarios to give a wide overview of their potential effects that facilitate the formulation of robust design/maintenance solutions.

Merschman et al. [5] also considered the combination of progressive deterioration (timber decay) and extreme events (cyclones) on the assessment of climate change of timber poles. Cyclone's intensity and frequency could be affected by climate change; however, nowadays it is still very challenging to assess the extent of these changes for a particular location. Therefore, the probabilistic approach proposed by Merschman et al. [5] is useful to account for the uncertainties on these future predictions and allows to estimate the probability of failure when considering failure induced by

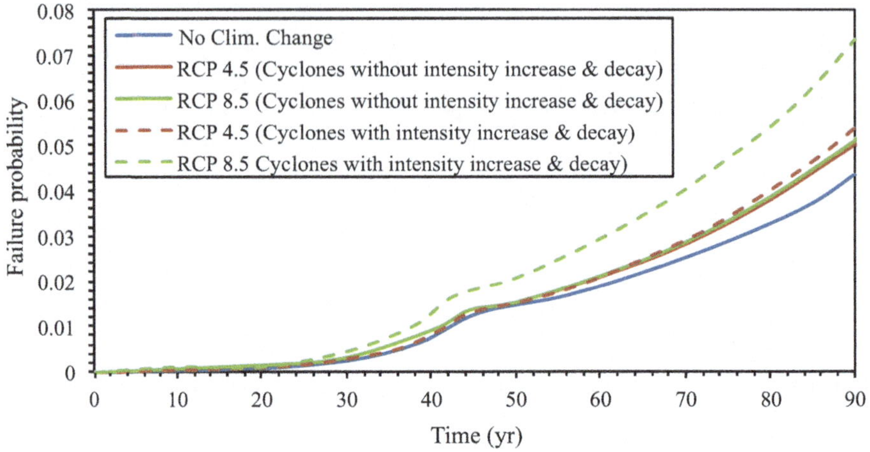

Fig. 3 Failure probability for timber poles located in New York City. Adapted from [5]

progressive decay and cyclones. In Fig. 3 are presented these failure probabilities for timber poles located on New York City by considering two climate change scenarios (RCP4.5 and RCP8.5) and two failure conditions: (1) decay and cyclones with intensity increase by climate change, and (2) decay and cyclones without intensity increase by climate change). In all cases, it is observed that climate change increases failure probability with the maximum values for the most pessimistic scenario RCP8.5. It is also noted that the changes of cyclones intensity due to climate change increase significantly the failure probability. Therefore, the consideration of the combined action of extreme events and progressive deterioration was crucial in this example to provide a more realistic assessment of the effects of climate change on the probability of failure of a timber pole.

3 Climate Adaptation

Adapting existing and new infrastructure assets to future uncertain environmental conditions is a challenging problem of decision-making under uncertainty. Towards this aim, Bastidas-Arteaga and Stewart [3] proposed a framework to estimate the cost-effectiveness of adaptation strategies that combines: (1) models able to estimate the effects of climate change, (2) stochastic approaches to account for the sources of uncertainty in the problem, (3) cost-benefit analysis to assess the cost-effectiveness of adaptation strategy under a given climate change scenario, (4) climate change predictions for a particular zone, and (5) the characteristics of the studied infrastructure that depend on the construction year, standards at the construction time, etc.

One of the key points in this framework is the evaluation of costs-effectiveness of adaptation strategies. Cost-benefit analysis in a probabilistic context could be applied to determine the benefit-to-cost ratio and probability of cost-effectiveness for adaptation strategies applied to both new and existing structures [12, 13]. The 'benefit' of an adaptation measure is the reduction in damages associated with the adaptation strategy, and the 'cost' is the cost of the adaptation strategy. The benefit-to-cost ratio $\mathrm{BCR}(T_t)$ over the service life period T_t is:

$$\mathrm{BCR}(T_t) = \frac{\mathrm{E}_{\mathrm{d\text{-}BAU}}(T_t)\Delta\mathrm{R}(T_t)}{\mathrm{E}_{\mathrm{a}}(T_t)}, \; \Delta\mathrm{R}(T_t) = \frac{\mathrm{E}_{\mathrm{d\text{-}BAU}}(T_t) - \mathrm{E}_{\mathrm{d\text{-}adapt}}(T_t)}{\mathrm{E}_{\mathrm{d\text{-}BAU}}(T_t)} \tag{1}$$

where $\mathrm{E}_{\mathrm{a}}(T_t)$ is the adaptation cost, $\mathrm{E}_{\mathrm{d\text{-}BAU}}(T_t)$ and $\mathrm{E}_{\mathrm{d\text{-}adapt}}(T_t)$ are the cumulative expected damage cost (economic risk) for no adaptation measures (business as usual BAU, or existing practice) and considering adaptation measures, respectively; $\Delta\mathrm{R}(T_t)$ represents the proportional reduction in expected costs due to an adaptation measure. $\mathrm{E}_{\mathrm{a}}(T_t)$, $\mathrm{E}_{\mathrm{d\text{-}BAU}}(T_t)$, $\mathrm{E}_{\mathrm{d\text{-}adapt}}(T_t)$ and $\Delta\mathrm{R}(T_t)$ may be computed from comprehensive models that take into account the effects of climate change [12, 13].

An adaptation strategy is cost-effective if the benefit-to-cost ratio is larger than one – i.e., $\mathrm{BCR}(T_t) > 1$. In addition, if probabilistic tools are used to propagate uncertainties in the cost-benefit analysis, it is possible to estimate the mean value of $\mathrm{BCR}(T_t) > 1$, as well as the probability that an adaptation measure be cost-effective, $\Pr(\mathrm{BCR}(T_t)) > 1$. These indicators are very useful to estimate the risk of adaptation investments under several climate change scenarios.

3.1 Adaptation of Existing Reinforced Concrete Infrastructure Subjected to Chloride-Induced Corrosion

Bastidas-Arteaga and Stewart [13] evaluated the cost-effectiveness of adaptation strategies for existing reinforced concrete structures located in Saint-Nazaire (France) under a splash and tidal exposure and designed according to different design standards. Table 1 presents the mean BCR for slabs built in different years under RCP4.5 and RCP8.5 future concentration scenarios. The adaptation strategies consisted on increasing the design concrete cover (c_a) by $\Delta c_a = 5$mm or $\Delta c_a = 10$mm for repairs carried out after the adaptation time $t_a = 2020$. The adaptation time is the year

Table 1 Mean $\mathrm{BCR}(T_t)$ for slabs built in different years and $t_a = 2020$

Construction year	Design concrete cover (c_a) (mm)	RCP4.5		RCP8.5	
		$\Delta c_a = 5$mm	$\Delta c_a = 10$mm	$\Delta c_a = 5$mm	$\Delta c_a = 10$mm
1970	40	0.8	0.7	0.8	0.7
1990	50	3.8	3.4	3.9	3.6
2010	55	4.6	4.3	4.7	4.5

after which repairs account for the extra concrete cover Δc_a. The service life period considered is $T_t = 100$ years.

The values of BCR(T_t) provided in Table 1 show that this indicator is less than one (1.0) for older structures and greater than one for recent structures, i.e. built in 1990 and 2010. A BCR(T_t) less than 1.0 implies that the adaptation measure is not cost-effective for old structures, built in 1970. Recent standards recommend larger design concrete covers and are therefore more cost-effective during the service life period. The increase of BCR(T_t) for recent structures is due, on the one hand, to the larger concrete cover recommended by the standards (Table 1) and/or considered by the adaptation measures. This means that a larger concrete cover is cost-effective for this splash and tidal exposure in Saint-Nazaire. On the other hand, larger BCR(T_t) values are also related to the increase of climate change effects on chloride ingress rates that justify the implementation of adaptation measures. Table 1 also shows that higher values of the mean BCR(T_t) are expected for the RCP8.5 scenario that imply more severe changes with respect to the actual climate (Fig. 4). The differences between the BCR(T_t) for both scenarios are slightly larger for recent structures because they will be exposed to larger climate variations that are more pronounced after 2050 for the RCP8.5 scenario (e.g., Fig. 4). These climate variations will accelerate chloride ingress, so the cost-effectiveness of adaptation strategies will also increase. In all cases, increasing cover by 10 mm is less cost-effective than a 5 mm increase in cover. Even if the risk reduction, ΔR(T_t), should be higher for $\Delta c_a = 10$ mm, the costs associated to this adaptation strategy are larger and thus reduce the mean BCR(T_t).

The effect of the time of adaptation on the mean BCR(T_t) and the probability that BCR(T_t) exceeds unity (Pr(BCR(T_t) > 1)) for slabs, concrete cover increase $\Delta c_a = 5$ mm and the RCP4.5 scenario is shown in Table 2. Note that the closer the adaptation year is to the end of the service life period, the lower the mean BCR and Pr(BCR(T_t) > 1) are. Of interest is that Pr(BCR > 1) only reaches a value of 59% when the mean BCR exceeds 4. This illustrates the high variability of damage risks caused by uncertainties of climate change projections, and variability of design parameters and deterioration processes.

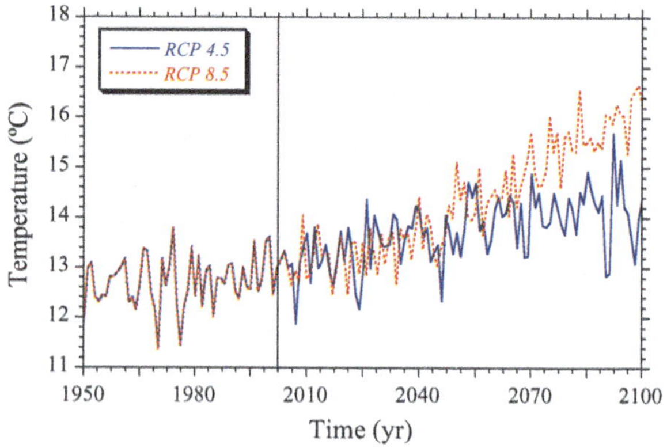

Fig. 4 Yearly temperature projections for Saint-Nazaire

Table 2 Mean BCR and Pr(BCR > 1) (within brackets) for slabs for various t_a, RCP 4.5 scenario and $\Delta c_a = 5$ mm

Construction year	Adaptation year (t_a)			
	2020	2040	2060	2080
1970	0.8 (6.1%)	0.05 (0.2%)	0 (0.0%)	–
1990	3.6 (43.5%)	1.4 (10.2%)	0.2 (0.7%)	0 (0.0%)
2010	4.6 (59.0%)	3.9 (44.7%)	1.7 (13.1%)	0.3 (0.9%)

These results could be used by structures' owners and other stakeholders to evaluate the benefits and risks of implementing adaptation strategies at various years. For example, it is observed that the mean BCR(T_t) and Pr(BCR(T_t) > 1) are small for older structures and therefore owners and stakeholders could prioritise investments in adaptation measures for more recent ones. These results could also be used to evaluate the impact of the adaptation year. For example, for structures built in 1990, if the owner or stakeholder decides to postpone the adaptation actions until 2040, the mean BCR(T_t) is about 1.4, which is still beneficial. However, the Pr(BCR(T_t) > 1) for this adaptation time is less than 11% indicating that the risks of having no benefits are high.

3.2 Adaptation of New Reinforced Concrete Infrastructure Subjected to Chloride-Induced Corrosion

This section focuses on the adaptation of new reinforced concrete structures placed in a chloride-contaminated environment under various exposures and climate change scenarios [12]. We particularly focus on structural reinforced concrete components subjected to atmospheric exposure to salt-spray (XS1 exposure according to the European Norm EN 206 [14]). The climatic conditions are defined by an oceanic environment placed at a middle latitude (i.e., Europe) where the yearly mean temperature and relative humidity vary between the intervals [5 °C; 25 °C] and [60%; 80%], respectively. The EN 206 [14] durability design requirements for a structural lifetime of 100 years and a rebar diameter of 16 mm are (1) 55 mm cover, and (2) 30 MPa concrete compressive strength. The adaption strategy will be to increase the concrete cover by $\Delta c_a = 5$ mm after the repair time, t_{rep}.

Using the probabilistic model of chloride-induced deterioration given in [15] are obtained the results given in Fig. 5 that present the time-dependent probability of severe cracking for various climate change scenarios. Figure 5 clearly shows that the rate of damage risk is highly dependent on climate change effects and environmental exposure. If there is no climate change, the probability of severe cracking increases with time and remains constant for all times of repair. However, if climate change reduces the environmental relative humidity, i.e. $\Delta RH = -10\%$ in 100 years, the chloride ingress mechanism slows down, and consequently, the probability of severe

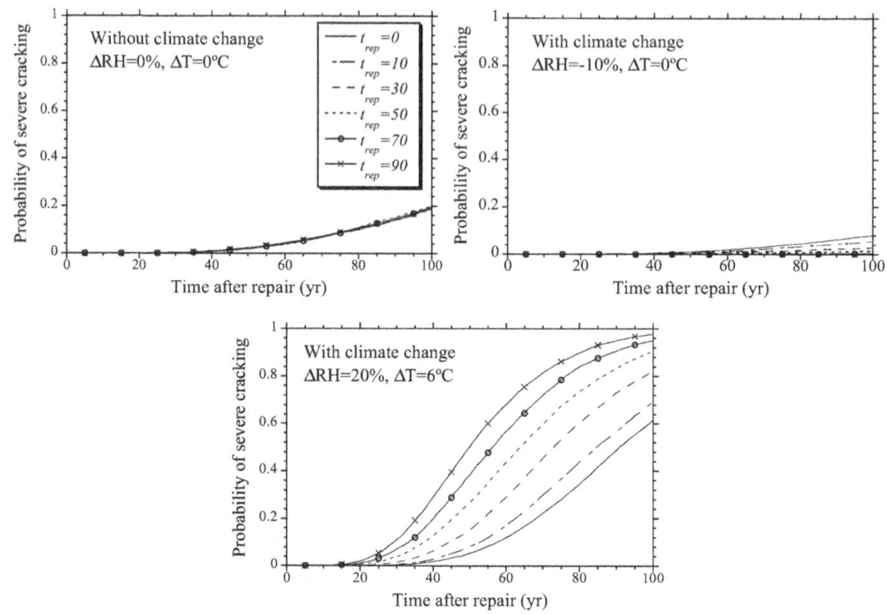

Fig. 5 Probability of several cracking after repair for various climate change scenarios

cracking decreases. For instance, for a structure with no repairs ($t_{rep} = 0$), the probability of damage decreases from 20% to 8% after 100 years of service. In this case, climate change has a 'positive effect' on reinforced concrete durability reducing by 60% the corrosion damage risk. An opposite behaviour is observed when climate change increases the temperature and relative humidity. For the same conditions, the probability of corrosion damage increases from 20 to 98%.

Table 3 summarises the mean BCR and Pr(BCR > 1) for various climate change scenarios and 300 mm reinforced concrete slabs. The results show that the mean of the BCR is highly dependent on both the exposure and the type of structural component. In some cases the mean BCR is lower than 1 indicating that the adaptation strategy is not cost-effective under given climate change scenarios. Similar behaviour is observed for the Pr(BCR > 1).

Higher temperature and relative humidity accelerate the deterioration processes by increasing the cost-effectiveness of the implementation of an adaptation measure. In Table 3, the mean BCR is only higher than one when climate change could induce increases of relative humidity equal or higher than 10% in 100 years. However, even under these scenarios, the Pr(BCR > 1) indicates that the risks associated to a bad investment are higher. It is noted that an increase of 5 mm cover provides higher estimates of BCR and Pr(BCR > 1). However, the likelihood that BCR > 1 is less than 60% even for a pessimistic (worst-case) climate change scenario of $\Delta RH = 20\%$ and $\Delta T = 6\ °C$.

Table 3 Mean BCR and Pr (BCR > 1) (within brackets) for slabs and several climate change scenarios

ΔRH (%)	ΔT = 0 °C (%)		ΔT = 2 °C (%)		ΔT = 4 °C (%)		ΔT = 6 °C (%)	
−10	0.33	(8)	0.53	(8)	0.57	(9)	0.57	(9)
0	0.68	(18)	0.91	(20)	0.94	(22)	0.95	(23)
10	1.17	(34)	1.27	(37)	1.35	(38)	1.32	(40)
20	1.65	(50)	1.70	(53)	1.72	(55)	1.76	(59)

4 Decision-Making Under A Changing Climate

Political, economic and social imperatives could affect the policy-making decision process. The World Resources Report [16] highlights five key elements for effective decision-making: Engagement; Decision relevant information; Institutional design; Tools for planning and policymaking; and Resources [3] that are presented and discussed in the next sections.

4.1 Engagement

Early and on-going public and/or private engagement is essential to effective adaptation. The first step to create this engagement will be to identify the different actors involved on the problem. In the case of adaptation of infrastructure and buildings, these actors (or stakeholders) are: governments, academia, owners, users, insurers, communities, designers and constructors. Engaging these actors will be crucial to prioritise needs, provide and recover information, determine acceptable risk levels, and choosing among the possible adaptation solutions. Discussions with these actors will also give them the possibility to express their views to increase the acceptability and applicability of adaptation measures.

Some challenges for improving public/private engagement in civil engineering are related with the type of climate hazard and affected infrastructure or buildings. Concerning the type of hazard, there is a high level of governmental engagement to implement initiatives to reduce vulnerability after an extreme events like floods [17, 18]. However, the lack of exchanges with the different actors involved in the problem such as the population (i.e., community), led to the implementation of expensive, unpopular and unsuccessful adaptation measures after storm Xynthia in France [19, 20]. For progressive climate-related hazards (acceleration of corrosion due to climate change), it is very difficult to create strong engagement with major decision-makers (government, owners) because climate change effects are expected in the long-term and decision-makers mainly focus on short- or mid-term problems.

Communication and education are also useful tools to increase the awareness of these actors concerning climate change risks. Future evolutions of civil engineering and public administration education programs should also consider that infrastructure and buildings would be subjected to a changing climate that will modify current practices of design, construction and management.

4.2 Decision-Relevant Information

Relevant information is required for different stages of decision-making. This information includes temperature, relative humidity, rainfall, wind speed, sea water level, etc. that is estimated from several global circulation models for different climate change scenarios. For the case studies presented in this paper, the information was provided from specialised websites or scientific partners and requires a given level of expertise to define if the databases are well suited for the applications. For example, downscaled data is necessary to obtain representative results in some cases. Under these conditions, it would be difficult for decision-makers to use it directly for assessing climate change effects and cost-effectiveness of adaptation strategies.

In addition to the above-mentioned climate hazards and specific technical information for infrastructure and buildings, other types of data such as demography, economy, social and environmental information are also needed. This information is crucial for vulnerability assessment as well as for understating if the adaptation measures could be locally implemented and accepted.

4.3 Institutional Design

The implementation of cost-effective adaptation strategies should be administratively supported at different institutional levels (ministries, public or private institutions, regions, states, municipalities). This will require evolutions on current practices to deal in a broad way with the problem. One of the major challenges is to maintain its consistency, completeness and integrity along with the national policies in terms of visions and goals. Towards this aim, the role of centralised agencies is crucial to coordinate these efforts in a comprehensive way.

It is expected that the work of centralised agencies will produce standardised procedures for decision-making under a changing climate sometime in the future. For the moment, some discussions for normalisation have started at national, European and International levels. In Europe CEN (European Committee for Standardisation) and CENELEC (European Committee for Electrotechnical Standardisation) launched the Adaptation to Climate Change Coordination Group (ACC-CG) in 2014 to coordinate standardisation activities in the field. An effort focused in construction, transport and energy sectors has bee initiated and a first guide for addressing climate change adaptation in standards has been produced in 2016 [21].

4.4 Tools for Planning and Policymaking

Advanced models able to simulate the behaviour of infrastructure and buildings under realistic environmental exposure have been employed in the examples mentioned in this paper to assess climate change effects and cost-effectiveness of adaptation strategies. In the current form, these models are useful to provide the information required for adaptation planning and policymaking in the studied applications. However, their use requires a certain level of expertise that is not currently easily accessible for non-academic users. In the future, it is expected that engineers, software and standards will provide the capability to directly integrate climate change issues into infrastructure planning. This will require implementing training programs as well as software development. In the meantime, collaborations between decision-makers, industry and academia will allow useful outputs to be produced to benefit decision-making.

4.5 Resources

In the aftermath of extreme events that directly affect communities such as floods, heatwaves and extreme winds, decision-makers urgently provide the necessary financial resources to restore the situation as before as these events. Or preferably, to reinstate infrastructure to a higher (less vulnerable) standard—i.e., to "build back better" [22]. Additional policies may also be adopted to reduce the vulnerability to future extreme events in the mid-term. Nevertheless, increasing the resilience of infrastructure and buildings to face climate change hazards would require targeted and sustained long-term funding. Public and/or private engagement motivated by financial incentives would help to promote the culture of long-term planning.

Prioritising spending is a complex task for decision-makers that should optimise their budget to deal with unexpected, short-, mid- and long-term expenditures. To help them in this task, a risk-based decision support containing the elements mentioned in Sect. 3 is useful for the evaluation of adaptation costs. This risk-based decision support is paramount for effective decision-making under a changing climate. Therefore, promoting and funding technical training on this subject among decision-makers will allow them to integrate climate risks into existing decision-making processes.

5 Conclusions

This paper summarised recent contributions to the field of climate adaptation of existing and new deteriorating infrastructure. Climate change could affect the serviceability and safety of infrastructure assets by modifying: (1) the intensity/frequency of extreme events, (2) the kinetics of the deterioration processes, and (3) a combination of both. Therefore, a widespread evaluation of the potential effects of climate change is crucial to carry out a realistic lifetime assessment under several climate change scenarios. Once this assessment is completed, adaptation becomes a problem of decision-making under uncertainty. Towards this aim, cost-benefit analysis is a very useful tool to evaluate the cost-effectiveness of adaptation strategies. Comprehensive models able to account for the effects of climate change in a probabilistic context are essential to support this cost-benefit analysis. The outcomes of this analysis provide valuable information to estimate the potential benefits and risks of climate adaptation investments. Besides these economic and technical considerations, decision-making requires: engagement of all the stakeholders related to the problem; relevant information for assessing climate change effects and for implementing adaptation measurements; institutional design that facilitate the implementation of adaptation measures; tools for planning and policymaking accessible for non specialists; and substantial/sustained financial resources.

Acknowledgments The work reported in this paper summarises multiple collaborations and discussions with M. Stewart, M. Sánchez-Silva, A. Chateauneuf, Y. Li, A. Salman, B. Imam, and J. Matos that are deeply acknowledged by the author.

This paper was carried out in the framework of the Strengthening the Territory's Resilience to Risks of Natural, Climate and Human Origin (SIRMA) project, which is co-financed by the European Regional Development Fund (ERDF) through INTERREG Atlantic Area Programme with application code: EAPA_826/2018. The sole responsibility for the content of this publication lies with the author. It does not necessarily reflect the opinion of the European Union. Neither the INTERREG Europe programme authorities are responsible for any use that may be made of the information contained therein.

References

1. Clifton, J. R. (1993). Predicting the service life of concrete. *MJ, 90*, 611–617. https://doi.org/10.14359/9756.
2. IPCC. (2013). Climate Change 2013: The physical science basis. *Contribution of Working Group I to the Fifth Assessment Report of the Intergovernmental Panel on Climate Change*. Cambridge, United Kingdom and New York, NY, USA: Cambridge University Press.
3. Bastidas-Arteaga, E., Stewart, M. G. (2019). *Climate adaptation engineering: Risks and economics for infrastructure decision-making*. Butterworth-Heinemann.
4. Imam B. (2019). Chapter six—Climate change impact for bridges subjected to scour and corrosion. In E. Bastidas-Arteaga & M. G. Stewar (Eds.), *Climate adaptation engineering* (pp. 165–206). Butterworth-Heinemann. https://doi.org/10.1016/B978-0-12-816782-3.00006-1.
5. Merschman, E., Salman, A. M., Bastidas-Arteaga, E., & Li, Y. (2020). Assessment of the effectiveness of wood pole repair using FRP considering the impact of climate change on decay and hurricane risk. *Advances in Climate Change Research* (Accepted).
6. Bastidas-Arteaga, E., Chateauneuf, A., Sánchez-Silva, M., Bressolette, P., & Schoefs, F. (2010). Influence of weather and global warming in chloride ingress into concrete: A stochastic approach. *Structural Safety, 32*, 238–249. https://doi.org/10.1016/j.strusafe.2010.03.002.
7. de Larrard, T., Bastidas-Arteaga, E., Duprat, F., & Schoefs, F. (2014). Effects of climate variations and global warming on the durability of RC structures subjected to carbonation. *Civil Engineering and Environmental Systems, 31*, 153–164. https://doi.org/10.1080/10286608.2014.913033.

8. Bastidas-Arteaga, E. (2018). Reliability of reinforced concrete structures subjected to corrosion-fatigue and climate change. *International Journal of Concrete Structures and Materials*, 12. https://doi.org/10.1186/s40069-018-0235-x.
9. Ryan, P. C., Stewart, M. G., Spencer, N., & Li, Y. (2014). Reliability assessment of power pole infrastructure incorporating deterioration and network maintenance. *Reliability Engineering & System Safety, 132,* 261–273. https://doi.org/10.1016/j.ress.2014.07.019.
10. Talukdar, S., & Banthia, N. (2016). Carbonation in concrete infrastructure in the context of global climate change: Model refinement and representative concentration pathway scenario evaluation. *Journal of Materials in Civil Engineering, 28,* 04015178. https://doi.org/10.1061/(ASCE)MT.1943-5533.0001438.
11. Salman, A. M. (2020). Optimization of condition-based maintenance of wood utility pole network subjected to hurricane hazard and climate change. *Frontiers in Built Environment, 6,* 18.
12. Bastidas-Arteaga, E., & Stewart, M. G. (2015). Damage risks and economic assessment of climate adaptation strategies for design of new concrete structures subject to chloride-induced corrosion. *Structural Safety, 52,* 40–53. https://doi.org/10.1016/j.strusafe.2014.10.005.
13. Bastidas-Arteaga, E., & Stewart, M. G. (2016). Economic assessment of climate adaptation strategies for existing reinforced concrete structures subjected to chloride-induced corrosion. *Structure and Infrastructure Engineering, 12,* 432–449. https://doi.org/10.1080/15732479.2015.1020499.
14. EN-206. (2000). *Concrete—Part 1: Specification, performance, production and conformity.* Comité Européen de Normalisation.
15. Bastidas-Arteaga, E., Chateauneuf, A., Sánchez-Silva, M., Bressolette, P., & Schoefs, F. (2011). A comprehensive probabilistic model of chloride ingress in unsaturated concrete. *Engineering Structures, 33,* 720–730. https://doi.org/10.1016/j.engstruct.2010.11.008.
16. World Resources Report. (2011). Decision making under a changing climate: Adaptation challenges and choices. United Nations.
17. Creach, A., Bastidas-Arteaga, E., Pardo, S., & Mercier, D. (2020). Vulnerability and costs of adaptation strategies for housing subjected to flood risks: Application to La Guérinière France. *Marine Policy, 117,* 103438. https://doi.org/10.1016/j.marpol.2019.02.010.
18. Creach, A., Bastidas-Arteaga, E., Pardo, S., & Mercier D. (2019). Chapter eight—Adaptation of residential buildings to coastal floods: strategies, costs and efficiency. In E. Bastidas-Arteaga, M. G. Stewar (Eds.), *Climate adaptation engineering* (pp. 245–270), Butterworth-Heinemann. https://doi.org/10.1016/B978-0-12-816782-3.00008-5.
19. Mercier, D., & Chadenas, C. (2012). La tempête Xynthia et la cartographie des « zones noires » sur le littoral français : analyse critique à partir de l'exemple de La Faute-sur-Mer (Vendée). *Norois, 222,* 45–60.
20. Pitié, C., & Puech, P. (2010). Expertise complémentaire des zones de solidarité délimitées en Vendée suite à la tempête Xynthia survenue dans la nuit du 27 au 28 février 2010. CGeDD/MEEDDM.
21. CEN-CENELEC. (2016). *Guide for addressing climate change adaptation in standards.* Committee for Standardization, European Committee for Electrotechnical Standardization.
22. NCCARF. (2017). *Climate proofing Australia's infrastructure.* Australia: Policy Brief 7, National Climate Change Adaptation Research Facility, Griffith University.

Papers

A DC Optimal Power Flow Approach to Quantify Operational Resilience in Power Grids

Zarif Ahmet Zaman and Edoardo Patelli

Abstract The primary objective of resilience engineering is to analyse and mitigate the risk of a system once a vulnerability has been triggered by an attack. Resilience is a multidimensional concept in the field of engineering and incorporates restoration in the form of a performance and time. Nodal restoration is a key factor in the analysis of resilience in systems, and the properties of the nodes can be analysed to assess the states on the system. The model proposed for the power grid to demonstrate the failure of the network has been used to simulate probability of contingencies on the system and applies a Sequential Monte Carlo simulation to simulate the energy supplied. Additionally, a weather model incorporating the effects of both severe winds and lightning storms has been applied to act as a trigger to the contingency. Once failure of one component has occurred, it cannot be repaired until the network's performance reaches zero. Given failure of all components, the network will immediately start its restoration phase, and using the same algorithm for optimal power flow calculations, a DC power flow approach is implemented to assess the energy supplied to the whole network in a transient model until the network's loads meet the demand criteria completely.

Keywords Power-grid · Risk · Resilience · DC-OPF · Uncertainty

1 Introduction

The power grid is an essential tool for modern society and its function is crucial for the wellbeing of people. A failure of the system can lead to major consequences, in a socio-economic scale. The assessment of reliability in power grid systems and the parameters incorporating reliability in the power grid such as availability, consequence modelling and energy not supplied has been an important field of research

Z. A. Zaman
Institute for Risk and Uncertainty, University of Liverpool, Liverpool, UK

E. Patelli (✉)
Department of Civil and Environmental Engineering, University of Strathclyde, Glasgow, UK
e-mail: edoardo.patelli@strath.ac.uk

© The Author(s), under exclusive license to Springer Nature Switzerland AG 2021

J. C. Matos et al. (eds.), *18th International Probabilistic Workshop*, Lecture Notes in Civil Engineering 153, https://doi.org/10.1007/978-3-030-73616-3_4

for the IEEE community. Past events such as the 2003 British National Power Grid Corporation outage which was responsible for the load loss of 724 MW, or approximately 20% of London's power load have costed the UK a significant economic burden [1].

The reliability of any system can be defined as the probability of success of the system at a given period of time, and the knowledge of reliability plays a role for system engineering to enable system maintenance planning to optimize risk mitigation [2]. System reliability can be thought of as a multidimensional analysis and incorporates many parameters. The user can analyse a single metric or multiple metrics simultaneously. The constraint with complex systems is developing the most computationally inexpensive technique and producing the most accurate results, with the aim to maximise the efficiency of the simulation.

Rochetta et al. developed a load flow approach to calculate failure probabilities from contingencies incorporating a wind model [3]. This was further developed with an artificial neural network surrogate model to act as a meta model for the analysis in order to minimise computational time when applied to AC optimal power flow calculations [4]. This model was developed on the basis of a severe weather model which was developed by Cadrini et al. [5] which combines stochastic extreme weather model and realistic power grid fault dynamics in order to model a restoration model quantified by sequential Monte Carlo. The constraint placed when applying this model is the high computational cost for the resilience function, especially when assessing networks with large scales of nodes.

There are various definitions of resilience available, both in a scientific context and a general context. The United Nations International Strategy for Disaster Reduction defines resilience as "The capacity of a system, community or society potentially exposed to hazards to adapt, by resisting or changing to reach and maintain an acceptable level of functioning and structure" [6]. However in a more specific context from that of extreme weather events, the definition of resilience can be thought of as "the network ability to withstand high impact low probability events, rapidly recovering and improving operations and structures to mitigate the impact of similar events in the future" [7]. Efforts placed on quantification of resilience analysis have been limited and have only been tested in the last 20 years. Additionally, such efforts placed into resilience analysis applied to the power grid have been performed, which includes various techniques such as transient performance modelling for the case study of typhoon Bolaven in South Korea [8]. However the authors mentioned that the limitations in their study included only computing resilience in the form of restorative and absorptive capacity without considering anticipated and adaptive capacities and also did not include a cost benefit analysis for the quantification of resilience in an economic sense. Panteli et al. developed a method to quantify resilience in the power function n with extreme weather events by developing the three phase resilience trapezoid [9]. This is an extension to the traditional resilience triangle developed in prior literature [10] which involves three stages to the disintegration, stagnation and recovery of the structure. The author divides resilience into two types, infrastructural and operational, stating that infrastructural resilience is in a more vulnerable condition given its recovery times and damage done to the system. Kim et. al developed a novel function to analyse the South Korean power grid network using cascading failure analysis using three different node centrality metrics; degree, clustering coefficient and betweenness [11]. A high clustering coefficient of a network indicates a more resilient network as it contains a higher redundancy potential as alternative paths in the network's nodes are present. Resilience has also been portrayed in the field of structural engineering [12] by associating a structural resilience index to for both a pre-event and post-event state. The arbitrary structural resilience index is conformed from certain parameters deduced by the nature of the structure as stated in the article.

1.1 Proposed Approach

This paper aims to apply a DC optimal power flow approach to quantify resilience in a simple power grid system after a network failure has occurred. The novel theme of this paper is the application of resilience as an extension to the weather induced model introduced in [3]. The chosen application for modelling will be MATLAB 2020b and the application will be case 9 as obtained from MATPower.

2 Resilience Model

The index of resilience chosen for the power grid system is the Expectation of Energy Not Supplied which is deemed to be the most appropriate performance and has historically been used as an indicator of reliability performance and can further be extended for resilience analysis. The equation listed below states the resilience index as a dividend of the load received and the expected load:

$$ENS = \sum_{t=1}^{T_{sim}} \sum_{i \in N} L_{cut,i,t} \cdot t \tag{1}$$

where T_{sim} is the simulation time and $L_{cut,I,t}$ is the load curtailed at each individual node during time t.

2.1 Optimization

In the case of optimization, the two models are the DC Optimal Power Flow approach and the AC Optimal Power Flow approach. In the real life power grid system, the electricity is generated in power plants using methods such as fossil fuels, converted fuels or geothermal steam and transfer this energy through the transmission network at high voltage using either DC or AC flow [13]. This high voltage steps down into a medium voltage range. The primary difference between the DC and AC optimal power flow models is the convexity. DC power is constantly in a steady state, and therefore is both a linear and convex optimization problem. However, AC optimal power flow calculations are non-linear and non-convex leading to much higher computational expense. It should also be noted that in high-fidelity models, DC optimal power flow are limited in terms of details for these grids as noted by [14]. This is due to DC optimal power flow models being an estimation of AC optimal power flow models and only accounts for active power, without reactive power in the model [15]. The equations for optimal power flow approach can be denoted below as obtained from [16]. The standard optimization vector is defined as:

$$\min_{x} f(x) \tag{2}$$

Subject to

$$g(x) = 0 \tag{3}$$

$$h(x) \leq 0 \tag{4}$$

$$x_{min} \leq x \leq x_{max} \tag{5}$$

The optimization vector for DC optimal power flow neglects reactive power and voltage magnitude and is defined as;

$$x = \begin{bmatrix} \theta \\ P_g \end{bmatrix} \tag{6}$$

Equation 2 is reduced to;

$$\min_{\theta, P_g} \sum_{i=1}^{n_g} f_P^i(p_g^i) \tag{7}$$

2.2 Load Contingencies

The representation of failure for this model will be in the form of contingencies. In this context, a contingency is defined as an event occurring that is not considered predictable at a given time. Contingencies when applied to the power grid network imply the network's architecture is the disruption of the load transfer from one bus to the next. This is commonly caused by a failure by extremely hot weather, system failure such as outages in loads and human errors [17].

2.3 Severe Weather Model

In extension to the contingencies faced in the model, a weather model has been proposed in the simulation algorithm to mimic the real-life application of an event. These events include lightning strikes, extremely high winds and natural disasters. The occurrence of normal weather conditions can be modelled as a homogeneous Poisson process. All equations for this weather model have been taken from [18].

$$P\left(N_f(t) = k\right) = \frac{[\lambda_n \cdot t]^k}{k!} e^{-\lambda_n \cdot t} \quad k = 0, 1, ..., N \tag{8}$$

Where $P(N_f(t){=}\ k)$ represents the probability that k failures happen within the network given the time (0, t) and $Nf(t)$ is the number of failures per kilometre of grid line. However, in a more realistic perspective, the weather model is more likely to affected by uncertainty. Therefore the occurrence of severe weather events is more suited to be modelled by a Non-homogeneous Poisson process:

$$P\left(N_f(t) = k\right) = \frac{[\lambda_n \cdot t]^k}{k!} e^{-\lambda_n \cdot t} \quad k = 0, 1, ..., N \tag{9}$$

In this case, $Ve(t)$ represents the time dependent probability of the event occurring and can be obtained applying the following equation:

$$V_e(t) = \int_t^0 v_e\left(t'\right) dt' \tag{10}$$

$v_e(t')$ is the rate at which the disturbance occurs. Given a severe weather occurrence in a storm consisting of severe winds and lightning strikes, the time of the event is obtained from data from previous events and will be carried out using probability distribution functions obtained from the variables listed in Table 1.

In the case of high winds, the windstorm intensity is obtained via the following equations:

$$W_\omega(t) = W_{crt} + \Delta_\omega(t) \tag{11}$$

where $W_w(t)$ is the wind speed intensity at time t for the and W_{crt} is a datum wind speed known as the critical wind speed set at 10 m/s. $\Delta_w(t)$ is the difference between the critical wind speed and the actual wind speed during the event. In terms of the lightning severe weather model, the intensity of a the weather event is set at the

Table 1 Variable attributes

	Distribution	Scale (a)	Shape (b)
D_ω	Weibull	9.86	1.17
D_{lg}	Weibull	0.96	0.85
$\Delta_\omega(t)$	Weibull	1.23	1.05
		Mean (μ_{Ng})	SD (σ_{Ng})
$N_g(t)$	Log-normal	−5.34	1.07

lightning strike ground density $N_g(t)$which takes the units of ground flashes per unit time and area [occh km^2] modelled with log-normal variability.

The table below shows the shape and scale factors for the respective variables:

Both high winds and lightning strikes are a cause of contingency and therefore it is crucial to define an equation which considers both contingencies to calculate the total failure rate:

$$\lambda(t) = \lambda_n + \lambda_\omega(W_\omega(t)) + \lambda_{lg}\left(N_g(t)\right) \tag{12}$$

λ_w is the total line failure contribution due to high wind measured per km and $\lambda_{(lg)}$is the lightning storms contribution. When considering individual lines, the contribution to line failure due to high winds can be denoted in the equation;

$$\lambda_\omega(W_\omega(t)) = \lambda_n\left(\frac{W_\omega(t)^2}{W_{crt}^2} - 1\right)\alpha_\omega \tag{13}$$

α_w is the regression parameter for failure data obtained from the literature. The line failure rate as a result of lightning can be denoted as:

$$\lambda_{lg}\left(N_g(t)\right) = \lambda_n \beta_{lg} N_g(t) \tag{14}$$

β_{lg} is the regression coefficient obtained from prior data [18].

2.4 Repair Speed

The model for recovery has been obtained from [5] and takes into consideration the efficiency of the repair crew as they are also affected by the adverse weather conditions. The assumptions in this model are:

i. Repair is initiated instantly after failure
ii. After a line is repaired, it is considered fully functional
iii. The transitional time between failure and repair is negligible.

$$v_{repair} = \begin{cases} \frac{v_{norm}}{1+\eta\cdot(W_{\omega}(t)-W_{crt})}, if\ W_{\omega}(t) \geq W_{crt}, N_g = 0 \\ \frac{v_{norm}}{1+\Psi\cdot N_g}, if\ W_{\omega}(t)\langle W_{crt}, N_g\rangle 0 \\ \frac{v_{norm}}{[1+\eta\cdot(W_{\omega}(t)-W_{crt})]+[1+\Psi\cdot N_g]}, if\ W_{\omega}(t) \geq W_{crt}, N_g > 0 \end{cases}$$

In this model and η are positive parameters and v_{norm} is set at 20[%/h]. The values for ψ and η are set to 40 and 0.4, respectively.

2.5 *Probabilistic Load Uncertainty*

It is important to quantify uncertainty in the model used which uses data for variability in average daily load demand. The aleatory uncertainty of the model can be considered by implementing a gaussian with parameters fitted on historical data;

$$f(L_i(t)) = \frac{1}{\sqrt{2\pi\sigma_{Li}(t)}} e^{-\frac{L_i(t)-\mu_{Li}(t)}{2\sigma L_i(t)^2}} \tag{15}$$

The parameters implemented are $L_i(t)$ is the transient node demand at node i during a specific hour of the day denoted as t. $\mu L_i(t)$ is the mean load value and $\sigma L_i(t)$ is the standard deviation of node i at time t. The gaussian will be applied at each output value with the standard deviation obtained from the equation subject to the variance from the parameters listed. An assumption of this model is that seasonal effects do not play a role in the parameter values.

3 Methodology

The proposed approach applied is a DC optimal power flow approach to quantify the energy not supplied during the severe weather contingency, which has been applied to quantify the resilience function of energy supplied after disaster through the same algorithm. The implementation of the methodology is applied using MATLAB 2020b, and the inputs are the parameters listed in Table 1. The network is presented with failure from a single continency simulated from the risk model combining high winds and lightning strikes from Eqs. 10–14 and is implemented on the power flow equations to calculate the loss of load for disaster, and following this, the energy supplied after the disaster has occurred. The simulation is repeated until the network's performance has been fully restored. The pseudocode below displays the steps of the proposed approach below;

Algorithm 1 DC Optimal Power Flow model

procedure *ENS* (*Risk Assessment based on DC-OPF*)
$Input = \{\lambda_n, \beta_{lg}, \alpha_w, W_{crt}, b_{Dw}, a_{Dlg}, b_{Dlg}, a_{\Delta w}, b_{\Delta w}, \nu_{norm}, \mu_{Ng}, \sigma_{Ng}\}$
t = 0, e = 1, f = 0
Sample failure events for $[0, T_{sim}]$
t = TTE(e)
if event **i** is a failure **then**
f = f + 1
Sample X_f and L_f for t
Update repair speed and time to repair
else Sample Te and (N_g, Δ_w)
Compute failure rates
if $t + TTF(f) > t + T_d$ **then**
Compute load
Compute ENS(f) using $L_{cut}(f)$ and $TTR(f)$
else set $t = t + T_d$ and $f = f + 1$
Sample X_f and L_f for t
Update repair speed and time to repair
OUTPUT - Energy Not Supplied

4 Case Study

This paper implements case 9 as an example from MATPower's default folders [19]. It is composed of a 9-node, 9-line network which is assumed to be equidistant in all lines. This file was chosen due to the ring style topology which represents a simplified version of a small landlocked country in nature (Fig. 1).

The power grid's network lines represent the various branches with a 10-mile length.The failure rates of each branch have been obtained from the original MATPower file and have been implemented in the table. The failure rate is given as a relative probability of a line contingency for each individual line (Table 2).

4.1 Results

The recovery time initiates after 1×10^{-4} s in the simulation and continues to restore the energy supplied to the nodes are fully recovered after 4×10^{-4} s, in which the system has fully recovered and therefore all the energy required for the nodes in the whole network is being supplied. The uncertainty applied from Eq. 15 shows the

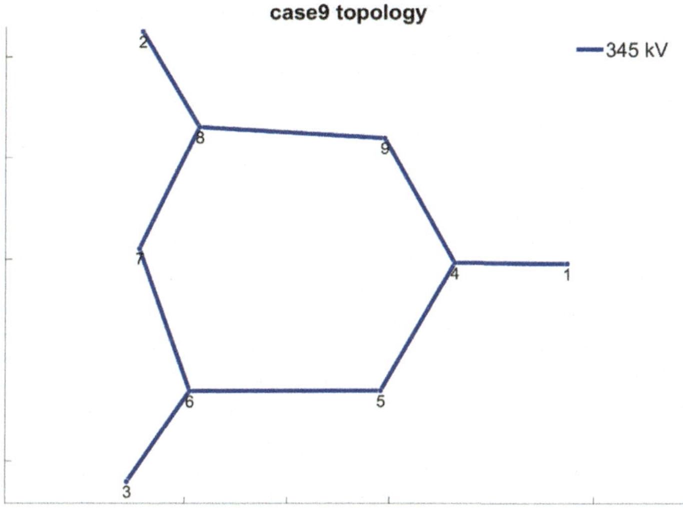

Fig. 1 Topology of case 9 network under matter to remain

Table 2 Branch failure rates [19]

Branch	Relative failure rate
1–4	0.1455
4–5	0.8693
5–6	0.5797
3–6	0.5499
6–7	0.1450
7–8	0.8530
8–2	0.6221
8–9	0.3510
9–4	0.5132

possible ranges of the energy supplied to the recovery function which also converges in the latter stages of the simulation. The model restores energy to each individual node simultaneously and therefore the restoration of all nodes improves rapidly during initial recovery, however, requires a start-up time in which no nodes are recovering. This initial period lasts less than 1 s of the simulation time and then increases rapidly. The drive for a lower range of uncertainty can be trialled by using more Monte Carlo simulations which are likely to sample more simulations on the same target output, energy not supplied leading to lower variance in results in output energies (Fig. 2).

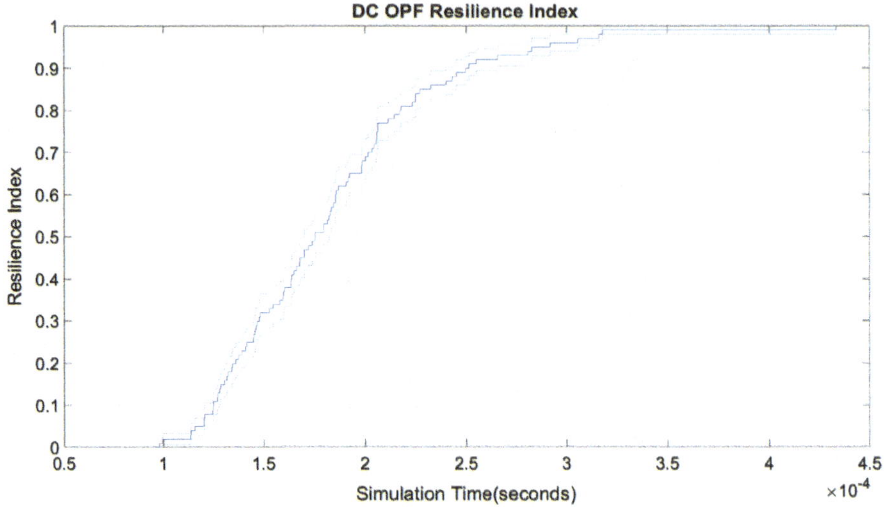

Fig. 2 Results for DC-OPF simulation under matter to remain

5 Conclusion

This paper demonstrates the application of a restoration function applied to a simple power network when DC optimal power flow is applied to the 9-node example provided in MATPower by applying a Monte Carlo approach. The work presented has innovated the weather model applied to contingencies in the general power grid to the application of resilience for the energy supplied after disaster. Further work that could be done on this topic includes developing a cost model for resilience quantification for the respective nodes in the network, and further expanding this application into three phase resilience models for realistic and complex networks using real time event timelines, rather than timelines based on simulation only.

References

1. Shuai, M., Chengzhi, W., Shiwen, Y., Hao, G., Jufang, Y., & Hui, H. (2018). Review on economic loss assessment of power outages. *Procedia Computer Science, 130*, 1158–1163. The 9th International Conference on Ambient Systems, Networks and Technologies (ANT 2018)/The 8th International Conference on Sustainable Energy Information Technology (SEIT-2018).
2. Koker, N. D., Viljoen, C., Lenner, R., & Jacobsz, S. (2020). Updating structural reliability efficiently using load measurement. *Structural Safety, 84*.
3. Rocchetta, R., Zio, E., & Patelli, E. (2018). A power-flow emulator approach for re-silience assessment of repairable power grids subject to weather-induced failures and data deficiency. *Applied Energy, 210*, 339–350. https://doi.org/10.1016/j.apenergy.2017.10.126.6.

4. Rocchetta, R., Bellani, L., Compare, M., Zio, E., & Patelli, E. (2019). A reinforcement learning framework for optimal operation and maintenance of power grids. *Applied Energy, 241*, 291–301. https://doi.org/10.1016/j.apenergy.2019.03.027.
5. Cadini, G. L., Agliardi, E., & Zio. (2017). A modeling and simula-tion framework for there liability/availabilityassessmentofapower transmission grid subject to cascading failures under extreme weather conditions. *Applied Energy, 185*, 267–279. https://doi.org/10.1016/j.apenergy.2016.10.086
6. I.P. on Climate Change, Climate Change. (2013). The Physical Science Basis: Working Group I Contribution to the Fifth Assessment Report of the Intergovernmental Panel on Climate Change, Cambridge University Press.
7. Panteli, M., & Mancarella, P. (2015). Influence of extreme weather and climate change on the resilience of power systems: Impacts and possible mitigation strategies. *Electric Power Systems Research, 27*, 259–270. https://doi.org/10.1016/j.epsr.2015.06.012.
8. Jufri, F. H., Widiputra, V., & Jung, J. (2019). State-of-the-art review on power grid resilience to extreme weather events: Definitions, frameworks, quantitative assessment methodologies, and enhancement strategies. *Applied Energy, 239*, 1049–1065. https://doi.org/10.1016/j.apenergy.2019.02.017
9. Panteli, M., Mancarella, P., Trakas, D. N., Kyriakides, E., & Hatziar-gyriou, N. D. (2017). Metrics and quantification of operational and infrastructure resilience in power systems. *IEEE Transactions on Power Systems, 32*(6), 4732–4742. https://doi.org/10.1109/TPWRS.2017.2664141.
10. Bruneau, M., Chang, S. E., Eguchi, R. T., Lee, G. C., O'Rourke, T. D., Reinhorn, A. M., et al. (2003). A framework to quantitatively assess and enhance the seismic resilience of communities. *Earthquake Spectra, 19*(4), 733–752. https://doi.org/10.1193/1.1623497.
11. Kim, D. H., Eisenberg, D. A., Chun, Y. H., & Park, J. (2017). Network topology and resilience analysis of South Korean power grid. *Physica A: Statistical Mechanics and its Applications, 465*, 13–24. https://doi.org/10.1016/j.physa.2016.08.002
12. Chiaia, B., Barchiesi, E., Biagi, V. D., & Placidi, L. (2019). A novel structural resilience index: Definition and applications to frame structures. *Mechanics Research Communications, 99*, 52–57. https://doi.org/10.1016/j.mechrescom.2019.03.007.
13. Amini, M. H., Bahrami, S., Kamyab, F., Mishra, S., Jaddivada, R., Boroo-jeni, K., Weng, P., & Xu, Y. (2018). Chapter 6—Decomposition methods for distributed optimal power flow: Panorama and case studies of the dc model. In: A. F. Zobaa, S. H. A. Aleem, & A. Y. Abdelaziz (Eds.), *Classical and recent aspects of power system optimization* (pp. 137–155). Academic Press. https://doi.org/10.1016/B978-0-12-812441-3.00006-9
14. Li, J., Tan, G., Cheng, B., Liu, D., & Pan, W. (2017). Transport mechanism of chitosan-n-acetylcysteine, chitosan oligosaccharides or carboxymethyl chitosan decorated coumarin-6 loaded nanostructured lipid carriers across the rabbitocular. *European Journal of Pharmaceutics and Biopharmaceutics, 120*, 89–97. https://doi.org/10.1016/j.ejpb.2017.08.013.
15. Hertem, D. V.: Usefulness of dc power flow for active power flow analysis with flow controlling devices. In: *IET Conference Proceedings* (pp. 58–62(4)).
16. Li, J., Shi, C., Chen, C., & Duenas-Osorio, L. (2018). A cascading failure model based on ac optimal power flow: Case study. *Physica A: Statistical Mechanics and its Applications, 508*, 313–323. https://doi.org/10.1016/j.physa.2018.05.081.
17. Golari, M., Fan, N., & Wang, J. (2014). Two-stage stochastic optimal is-landing operations under severe multiple contingencies in powergrids. *Electric Power Systems Research, 114*, 68–77. https://doi.org/10.1016/j.epsr.2014.04.007.
18. Alvehag, K., & Soder, L. (2011). A reliability model for distribution systems incorporating seasonal variations in severe weather. *IEEE Transactions on Power Delivery, 26*(2), 910–919. https://doi.org/10.1109/TPWRD.2010.2090363.
19. Zimmerman, R. D., Murillo-Sánchez, C. E., & Thomas, R. J. (2011). Matpower: Steady-state operations, planning, and analysis tools for power systems re-search and education. *IEEE Transactions on Power Systems, 26*(1), 12–19.

A Novel Analytical Method Set for Damage Control and Care-Process Management by the Cathedral of Milan

Francesco Canali, Lorenzo Cantini, Anthoula Konsta, and Stefano Della Torre

Abstract Since the introduction of expert systems for the preservation of Cultural Heritage, several research projects developed codified procedures for the condition assessment of the materials composing historical buildings. The identification of the decay typologies observed on the materials was considered a first step for supporting future interventions on the historical surfaces of the buildings. This kind of formal tool, providing data collected over the course of time by periodical survey campaigns, showed also other potentialities, like damage prediction and risk assessment. In recent years, the Veneranda Fabbrica del Duomo (VFD), the board managing the Cathedral of Milan, developed new approaches for preventing the risks connected to its rich apparatus of stone decorations. Based on the wide experience matured in a long-lasting calibration of good practices set for preserving the architectural features of the temple, the VFD adopted recently a new analytical procedure, set on behalf of the association of the Italian cathedrals (AFI), for detecting risk conditions and evaluating the evolution of decays and their potential consequences. The proposed method was studied by the authors, within the convention between Veneranda Fabbrica and Politecnico, in order to verify its reliability through several simulations of different scenarios. Moreover, this study pointed up the difficulties concerning an objective evaluation of the parameters on which the analytical procedure is based and therefore the need of defining criteria for an effective and reliable data gathering and processing to support decision-making. The expected results should provide alarm in case of dangerous scenarios and recommendations concerning the planning of preservation actions: the updating of the inspection interval, the necessity of further diagnostic investigations and the urgency of repair interventions.

Keywords Condition assessment · Risk assessment · Cathedral of Milan · Alert level · Analytical evaluation

F. Canali
Veneranda Fabbrica del Duomo, Milan, Italy

L. Cantini (✉) · A. Konsta · S. Della Torre
Department Architecture, Built Environment and Construction Engineering (DABC), Politecnico Di Milano, Milan, Italy
e-mail: lorenzo.cantini@polimi.it

© The Author(s), under exclusive license to Springer Nature Switzerland AG 2021
J. C. Matos et al. (eds.), *18th International Probabilistic Workshop*, Lecture Notes in Civil Engineering 153, https://doi.org/10.1007/978-3-030-73616-3_5

1 Introduction

The collaboration between Politecnico di Milano and the Veneranda Fabbrica del Duomo (VFD) is facing different topics concerning the practices for managing and preserving the Milan Cathedral. The research activities are focusing on the realization of a detailed 3D geometrical survey of the complex, static, dynamic and environmental monitoring, studies of the decay processes occurring on the stone materials, prevention strategies against risks.

The risks connected to monumental buildings were documented in several episodes. The vulnerability of the historical structures caused the collapse of the Civic Tower of Pavia, in 1989 and the partial failure of the masonry pillars of the Noto Cathedral, occurred in 1996. These two cases showed the effects of the long-term behaviour on load-bearing masonry elements [1]. Problems limited to the failure of some stone pieces, in 1999 a small marble stone (shorter than 20 cm) felt down from the vault of the New Sacristy, from a height of 28 m, into the complex of the Medici Chapels in Florence. In that case, no visitors were present in the monument, but in 2017, in Santa Croce Basilica, again in Florence, a corbel broke down and a piece of the stone felt down killing a Spanish tourist. This last event drove the association of the Italian cathedrals (AFI) to work on the development of an analytical tool for damage detection and risk interpretation set for the specific characteristics of ancient churches.

The analytical tool set by AFI is then described in its different evaluation steps. The procedure consists in a semi-qualitative analysis obtained by a first observation of the conditions of the investigated element and a second computational estimation of different parameters which will define its operative condition. The method is set for controlling the evolution of decays on the stone elements in order to define the correct timing between the periodical inspections on the building, according to a scale of 5 risk levels provided by the AFI analytical tool. The present work presents the results obtained by the application of a novel strategy for the condition assessment of the different elements composing the rich decorative apparatus of the Milan Cathedral (Fig. 1). The sculptures and the moldings composing the surfaces of the building are subjected to stone deterioration connected to chemical, physical and mechanical causes [2].

The procedure is set on the identification of the state of conservation of the selected elements. A first evaluation of the AFI procedure was obtained by some simulations aiming at observing the response of the system to the progressive increasing of the decay gravity for some categories of the evaluation method. The obtained alert levels and the corresponding agenda for the inspections and the maintenance interventions are the ultimate safety indicators proposed by AFI.

Fig. 1 Some features of the plastic decorations of the Milan Cathedral

2 Historic Buildings and the Analysis of Damage and Decay

2.1 Evaluation and Interpretation of the Observed Damages

Thanks to several efforts matured in architectural preservation field, guided procedures for the analysis of the material decay were introduced in national and international recommendations. The establishment of a shared methodology for decay and damage detection drove to the elaboration of a common lexicon, introducing the definitions of the decays for natural [3] and artificial stones [4].

The Damage Atlas published in 1998 and the Masonry Damage Diagnostic System (MDDS) were elaborated within a research project on brick masonry degradation supported by The European Commission [5]. The aim of the project was to improve our knowledge of the environmental effects on brick masonry deterioration and assist practitioners in the diagnosis process through the evaluation and interpretation of the observed damages [6].

As about the expert system MDDS, its structure was based on the logic procedures followed by an expert, divided in 5 steps: (1) Identification of the visible damage; (2) Analysis of the environmental circumstances; (3) Hypothesis on the damaging processes; (4) Scientific control of hypothesis; (5) Diagnosis of damage causes [7].

Further research in recent years regards the development of a Structural Damage Atlas that will finally be inserted in the MDDS [7]. Gathering together and ordering the current knowledge on the mechanical behaviour of brick- and stonework masonry under different actions, caused by sudden events such as earthquakes, floods etc. or by long term phenomena such as soil settlement, heavy loads and lack of maintenance, formed a basis for the definition of typical structural damage patterns.

2.2 Description And Prediction Of Damages In Different Scenarios

During the last years, the studies on the effects of the environmental factors on the cultural heritage have drawn attention to the issues related to climate change and the new pressures that will be caused by variations in temperature and precipitation, changes in soil conditions, groundwater and sea level, and extreme climatic events [8].

The Noah's Ark project has developed quantitative models for the global climate change impacts on the deterioration of different heritage materials on European geographical scale [8]. The results of the research project, gathered in different types of maps (climate maps, heritage climate maps, damage maps, risk maps), correspond to the methodological approach adopted. [9].

Within the Climate for Culture European project [10] research on climate change impacts has further extended to the impacts on indoor environments in historic buildings and their collections. In order to pass from outdoor to indoor climate change, climate modelling was combined with building simulation tools. Within the project, buildings were classified according to volume, window area, structure and moisture buffering performance and indoor climate and risk maps were produced for all the building types, with or without active climate control.

The Cultural Heritage Protection against Flood (CHEF) project [11] carried out a comprehensive study of protective measures, before, during, and after a flood, by investigating a large number of case studies and taking into consideration the wide diversity of situations that depends on flood characteristics, materials, structures and sites. The analysis of different cases contributed to the identification of the various damage processes and the classification of typical damages to cultural heritage on various scales, such as movable objects, buildings, heritage sites, cities and entire landscapes.

The research conducted within New Integrated Knowledge Based Approaches to the Protection of Cultural Heritage from Earthquake-Induced Risk (NIKER) project developed an integrated methodology for the improvement of the seismic behaviour of historic buildings. For this purpose, a structured catalogue linking earthquake induced failure mechanisms, construction typologies, structural elements and materials, interventions and assessment techniques was created [12].

2.3 Continuous Monitoring: Early Detection of Risks and Decay Evolution Modelling

The Smart Monitoring of Historic Structures SMooHs project [13] focused on the improvement and the effective use of monitoring systems for early detection of risks and the taking of prompt actions, as well as for the understanding of the long term effects of deterioration processes and the planning of adequate measures.

The main issues that were addressed by the project concern: the development of monitoring systems using wireless networks of miniature and robust sensors for minimally invasive installation at historic buildings; the monitoring of the most significant parameters (temperature, humidity, air velocity, strain and crack opening, acoustic emissions, vibration, inclination, chemical attack due to gases or salts, ambient and UV light) for the understanding of deterioration processes; data processing based on the built-in material deterioration models able to inform practitioners about changes and increasing risks and to support actions; development of modular and open source software that can be continuously updated [14].

Another research programme focused on the study of material degradation over time in relation to environmental parameters by using non-invasive techniques and passive monitoring combined with mathematical models is the Italian Research Unit for Integrated and Predictive Systems [15].

Among the various analytical approaches, an operative protocol for decay analysis and vulnerability identification was proposed by [16, 17], aiming at evaluating the risks referred to cultural heritage. The proposed analysis is divided in three steps:

1. Decay investigation, organized in the following inspection levels: visual inspection, decay typology, decay nature, decay intrinsic factors, and decay extrinsic factors.
2. Identification of the Risk Indicators (RI) associated to each form of decay. Each risk is defined according to a scale from 1 to 5 in order to indicate if the decay is acceptable or not. The RI are divided in 7 main classes: microstructure parameters variation, soluble salts concentration, determination of the soluble anions fraction, surface decay mapping, ultrasonic velocity propagation of the material, environment characterization, and evaluation of the surface typology.
3. Evaluation of the final risk index of the building, according to the above mentioned quantitative and qualitative parameters.

The risk index could be used as a part of a documentation system, organized as a control procedure addressed to support decision makers and the strategical interventions requested. This last example presents an analytical approach with some similarities to the analytical tool for damage control proposed by AFI and recently adopted by the Veneranda Fabbrica del Duomo (VFD). The problem connected to risk assessment and damage prevention became priority, in Italy, in 2017, when a Spanish tourist died after being hit by a piece of stone coming from the pulvinus of a column in Santa Croce, in Florence. Since that episode, the association collecting the managing boards of the Italian Cathedrals (AFI) introduced a shared analytical procedure for improving the risk assessment.

3 The AFI Novel Analytical Tool for Damage Control

The experts involved in the AFI association, set a shared procedure for the conservation of these monuments, based on a rigorous constant control. The Veneranda

Fabbrica del Duomo, for example, introduced since the 1960s, after important restoration works on the Cathedral of Milan, a periodical inspection of its parts. The observations are carried out by using different supports, like aerial platforms, moving from the top to the base of specific sectors of the building, following a precise order to cover the entire surfaces of the monument, along external and internal sides. This condition allowed introducing the experimentation of the new procedure for improving the damage control of local elements and the risk prediction.

3.1 The Semi-quantitative Evaluation Procedure Set by AFI

The evaluation procedure is carried out in two steps. A first evaluation of the problems is set by filling information in a form divided in 5 voices. This is the inspection form, designed for simple users, like the workers of the Veneranda Fabbrica, prepared staff with specific competences on the building characteristics. The indications collected in the inspection forms are then processed in an electronic sheet by the expert members of the technical office of the management board, like surveyors, architects and engineers. The aim of the procedure is to identify a final safety level for each analysed element, in order to drive future decisions on the conservation design of the building. The procedure is set to be repeated along the time, aiming at detecting the eventual evolution of the decay observed on a specific architectural element.

3.2 The Parametrization of the Evaluation Procedure

The categories proposed by the first part of the analytical method are addressed to define a numerical value for the operative state of the investigated element.

The first effort requested to the operator is to subdivide the building into main units and sub-classes. After defining the elements composing the considered unit, the inspection form requires the following evaluations:

- General state of conservation, defined by (a) condition state, (b) presence of new cracks, (c) presence of previous cracks, (d) falling effects of the element or its components.
- Damage level of the new cracks (DL), expressed by the 5 levels of severity.
- Maintenance timing, defined as the periodical interventions planned on the element.
- Inspection timing, defined as the expected period between the routine controls on the element.

Table 1 presents the detailed parametrization of the AFI procedure.

The recorded parameters are used in evaluating the Alert Level (AL) for each specific analyzed element. The alert level is defined as the operative state (S_e) of the element and is determined combining the parameters describing the damage

Table 1 The evaluation parameters used by the AFI procedure for risk identification

Category	Parameter	Coefficient description
Element inspection timing	periodicity of the inspection (T_a)	Expressed in years, indicates the recurrent period of control
Condition assessment	General state of conservation (a)	Optimum (1.0); good (0.8); moderate (0.6); poor (0.4); bad (0.2);
	New cracks (b)	No (1); Yes (0.8)
	Previous cracks (c)	Absent (1.0); slight (0.8); moderate (0.6); serious (0.4); deep (0.2)
	Elements falling effects (d)	Negligible (1.0); slow (0.8); moderate (0.6); serious (0.4); deep (0.2);
Damage level of the new cracks	Damage level (DL)	Deep (0.2); serious (0.4); moderate (0.6); slight (0.8); absent (1.0)
Maintenance and care	Maintenance nominal life (V_m)	Expressed in years, V_m represents the duration from the last maintenance intervention and the planned further one
	Time form the last maintenance intervention (T_u)	Expressed in years
	Investigations level (L_c)	Exhaustive (1.0); moderate (0.9); minimum (0.8);
	Inspection period (DT)	Respect to the inspection date (T), indicates the period from the previous control carried out on the element (T_0): $DT = T - T_0$

level observed on that element and the respect of the planned time of inspection and maintenance.

The inspection timing factor (F_t) indicates the respect of the planned periods for the inspections of the element under evaluation. It is obtained as reported in 1.

$$F_t = \left[e^{\frac{(DT - T_a)}{DT}} \right]^{\left(\frac{1}{V_m}\right)} \tag{1}$$

where DT is the difference from the date of the inspection (T) and the date of the previous one (T_0); Ta is the planned inspection timing and V_m is the supposed nominal life of the maintenance intervention.

The maintenance timing factor (f_m) refers to the failure in respecting the timing of the maintenance interventions. It is expressed as reported in 2.

Table 2 Determination of the alert level

Alert level	determination	Inspection timing
1	$75\% < S_e$	T_a
2	$45\% < S_e < 75\%$	T_a
3	$20\% < S_e < 45\%$	T_a
4	$10\% < S_e < 20\%$	$0.75 \cdot T_a$
5	$S_e < 10\%$	$0.5 \cdot T_a$

$$F_m = \left[e^{\frac{(T_u - V_m)}{T_u}}\right]^{\left(\frac{1}{V_m}\right)} \tag{2}$$

where T_u is the period spent from the last maintenance intervention and V_m is its supposed nominal life. This parameter is derived by the empirical experience matured by the experts involved in the conservation process and is also based on the rich documentary heritage of the archives belonging to the managing boards of the cathedrals.

The aim of the proposed procedure is to evaluate the operative condition (S_e) of the analyzed element as indicated in 3. The result provided by S_e allows the remodeling of the time interval for the inspections, according to the gravity condition of the element.

$$S_e = \frac{[a + (b \cdot c) + d]}{3} \cdot L_c \cdot \frac{\min(DL)}{F_t \cdot F_m} \tag{3}$$

According to the value obtained by the operative condition S_e, the alert level is defined as reported in Table 2. With the alert level 1, 2 and 3, the inspection timing T_a is not modified. In case of alert levels 4 and 5, the period planned for the inspection is reduced. The logic of the described procedure is to guarantee the constant check for those elements subjected to a recognize alteration process that can drive to the appearance of real dynamics of decay.

4 Experimentation and Discussion

The investigation procedure set by AFI is oriented to a precise definition of the periodical inspections on the elements composing the building. The idea is to use the planned audits on the building elements for the remodulation of the planned control activities in those cases presenting unexpected anomalies. This method should guarantee the constant control on the worsening processes occurring to the architectural elements.

As described before, the gravity of the damage is linked to the reliability of the planned activities: the inspection timing (F_t) and the maintenance timing (F_m). These two factors are determined as coefficients having the structure of the function

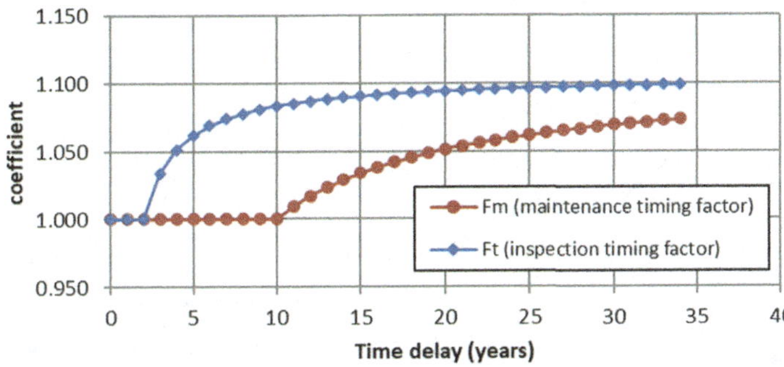

Fig. 2 Trend of the inspection timing factor and the maintenance timing factor according to the delay of the planned activities

F_e reported in 4. The coefficient b quantifies the respect of the targets for which the system is designed. This parameter represents the mean time to failure, intended as the recurring time required by a specific damage condition to reappear again after its repairing.

$$F_e = e^{\left(\frac{x}{a}\right)^b} \tag{4}$$

where a is the delay in the inspection or in the maintenance and b is the path of the curve obtained by $1/V_m$.

Considering a time for the inspections $T_a = 2$ years and a time for the maintenance nominal life $V_m = 10$ years, both parameters have an exponential growth when the delay is higher than 1 year, as described in Fig. 2. The analytical tool promotes the risk assessment by increasing the F_t index when the respect of the inspection activities is lacking. In this case, the impact on the evaluation of the risk level is really relevant.

A first simulation was carried out considering a nominal life $V_m = 30$ years for all the investigated elements, according to the main interventions reported in the archive documents of the VFD. For each simulation, one of the parameters of the AFI evaluation tool was modified from positive to negative condition, whereas the other parameters remained constant. The results showed the key role of the information connected to the crack pattern. Even if the cracks are considered to have a low level of gravity, the obtained level of alerts drive to a reduction of the timing for the inspections and the maintenance interventions.

The AFI inspection form is not set for an exhaustive evaluation of other decay pathologies. The influence of the effects produced by other decay categories, like erosion, black crusts or pulverization, are part of the parameter a, general conservation condition, connected to parameter d, effects of the element collapse. Among the various parameters, the estimation damage level attributed to the new cracks of a component (DL) should also plays a key role for the activation of the alarm imposing a rapid control given by the analytical tool. These three voices can affect the estimation of the operative condition (S_e) of the materials. Considering a nominal life

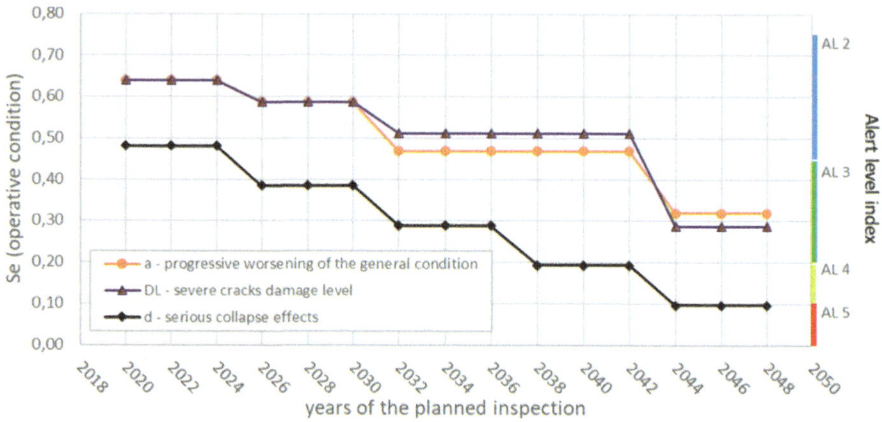

Fig. 3 Influence of 3 parameters in the evaluation of the operative condition of an element along a nominal life period of 30 years

of the maintenance intervention of 30 years and simulating a delay in the planned maintenance on a specific element, Fig. 3 compares the impact of parameters *a*, *d* and *DL* on the evaluation of its operative condition and the connected alert level.

The elaboration shows that for a contained increasing of the general decay condition of the element (*a*) in the first 22 years, the alert level remains 2, indicating that no maintenance interventions have to be rescheduled respect to the ones already planned. The same condition characterizes also the severe cracks persistence (*DL*). Only after more than 23 years these two parameters reach the alert level 3, a risk index imposing shorter periods between the inspections and urgent maintenance interventions. The parameter *d*, related to the effects given by an eventual collapse of the component, has higher consequences in the rearrangement of the planned maintenance activities: even a moderate risk, if connected to low consequences due to a possible impact by a falling object, is displayed in the alert level 3 and in about 12 years it moves to alert level 4.

The proposed procedure assumes a reliable result when the damage is evident, characterized by cracks and the riskiness of a detachment with the further fall of the element is identified with a high impact. The problem remains for those components that could present cracks, but a low risk in terms of collapse effects. Respect to the more accurate evaluation of the decays proposed for example by the vulnerability analysis described in [16, 17], the presence of cracks on a stone element remains a phenomena requiring only common periodical visual inspections, for the AFI method. The analytical tool does not provide any linkage between cracks and the origin of the tensions appeared into the stone element taken into consideration. The mechanical origin of that tensional state could derive by the combination of different materials, reacting with not uniform deformations, for example, to the exposition to environmental factors (temperature, moisture, etc.). In Milan Cathedral, this kind of dynamic occurs commonly on the stone coating containing metal elements: if water can get in contact with the metal bars reinforcing the structures, the dilatation determined by the oxidation process can produce serious stresses on the stone materials, like severe cracks and detachments.

The lower impact of other evidences given by the decay traces on the masonry elements, like the subjective estimation of the decay gravity, remain an issue requiring further improvements. The black crusts afflicting several stones of the cathedral, for example, present different effects that a non-expert observer could not correctly identify: from the deposits of the calcite crystals dissolved by acid rains getting in contact with the smog in the air, to the detachments caused by the increasing of the porosity into the stone substrate driving the material to a higher sensibility respect to frost cycles. A more detailed evaluation of the decay could be proposed for the determination of the a parameter, concerning the general state of conservation of the element, in order to reach a more reliable connection between the manifestations of the decays and their causes.

5 Conclusions

The novel analytical analysis procedure for the evaluation of the risks connected to architectural heritage offers an interesting opportunity for comparing a new method matured among the very high competences developed by the experts of the AFI association with the rich scenario of the projects concerning damage prediction and risk assessment.

The AFI procedure is structured in a series of steps giving a significant relevance on cracks severity and maintenance timing according to the periodical inspections on the architectural elements composing the building. The incidence of a wrong subjective interpretation of other decays is also consistent and could drive to a underestimation of the alert level, with important consequences on the setting of both inspections and maintenance interventions. Among the various parameters of the analytical procedure, according to other evaluation methods, like the one promoted in [17], future improvements for the AFI method could be focused on a more reliable identification of the general state of conservation of the investigated element. The guided inspection procedure could be implemented by a more specific decay identification, useful for applying new correction factors in the described analytical formulas that could be tested in future simulations.

Acknowledgements Authors wish to thank Riccardo Mattavelli, student from the School of Architecture Urban Planning Construction Engineering of Politecnico di Milano, for the support given to the research.

References

1. Garavaglia, E., Anzani, A., & Binda, L. (2008). Fragility curve probabilistic model applied to durability and long term mechanical damages of masonry. In *Materials and structures/Matériaux et Constructions* (Vol. 41, Issue 4, pp. 733–749), RILEM Pub. s.a.r.l., Springer.
2. Della Torre, S., & Cantini, L. (2018). Damage control, preservation procedures and durability studies: an investigation approach through the Milan Duomo Cathedral Archives. In G. Milani, G., Taliercio, A., & Garrity, S. (Eds), *Proceedings of the 10th International Masonry Conference* (pp. 2114–2123).
3. ICOMOS-ISCS: illustrated glossary on stone deterioration patterns. (2008). *ICOMOS. International Scientific Committee for Stone*. Available at https://iscs.icomos.org/glossary.html
4. Franke, L., & Schumann, I. (1998). *Damage Atlas. Classification and analyses of damage patterns found in brick masonry* (Vol. 2). Project EV5V-CT92–0108. Research Report N°8.
5. Van Hees, R., Binda, L., Van Balen, K., & Naldini, S. (1996). Expert system for evaluation of deterioration of ancient brick masonry structures. *The Science of the Total Environment, 189*(190), 247–254.
6. Van Hees, R. P. J., van der Klugt, L. J. A. R., Naldini, S., Van Balen, K., Mateus, J., Binda, L., Baronio, G., Franke, L., & Schumann, I. (1996). The masonry damage diagnostic system. In *8th International Congress on 'Deterioration and Conservation of Stone* (Vol. 3, pp. 1737–1739), Berlin.
7. Binda, L., Saisi, A., de Vent, I. A. E., van Hees, R. P. J., & Naldini, S. (2010). Structural damage in masonry: Description and interpretation of crack patterns as basis for finding the damage causes. *Restoration of Buildings and Monuments, Bauinstandsetzen und Bauenkmalpflege, 16*(2), 1–22.
8. Sabbioni, C., Cassar, M., Brimblecombe, P., & Lefevre R. A. (Eds.). (2008). *Vulnerability of cultural heritage to climate change*. Council of Europe Report AP/CAT(2008) 44.
9. Sabbioni, C., Brimblecombe, P., & Cassar, M. (Eds.). (2012). *The atlas of climate change impact on European cultural heritage: Scientific analysis and management strategies*. Anthem Press
10. climateforculture.eu, official website of Climate for Culture project [online]. Available at https://www.climateforculture.eu/. Accessed February 2020.
11. Drdacky, M., Binda, L., Hennen, I. C., Kopp, C., Lanza, L. G., & Helmerich, R. (2011). *CHEF—Cultural Heritage Protection against Flooding*. Prague.
12. Cantini, L., da Porto, F., Giacometti, G., Lorenzoni, F., Saisi, A., & Valluzzi, M. R. (2012). Creation of a structured catalogue. In J. Jasieńko (Ed.), *Proceedings of the International Conference on Structural Analysis of Historical Constructions, SAHC 2012* (Vol. I, pp. 2861–2869).
13. Rajčić, V., Pascale, G., Simon, S., Krüger, M., Troi, A., Colla, C., & Lukomski, M. (2012). Smart monitoring of historic structures FP7 SmooHS Project. In R. Žarnic, V. Rajcic, B. Vodopivec (Eds.), *Heritage protection, from documentation to interventions. Proceedings of the EU-CHIC international conference on cultural heritage preservation* (pp. 213–216)
14. Lukomski, M. (Ed.). (no date). Monitoring of historic structures. European Commission research project: SMOOHS Smart Monitoring of Historic Structures, Guidelines.
15. Calabrese, M., Odisio, N., Appolonia, L., Bernagozzi, A., Christille, J. M., Glarey, A., et al. (2018). Integrated and predictive systems for preventive conservation. *Studies in Conservation, 63*(sup1), 43–50.
16. Kioussi, A., Karoglou, M., Labropoulos, K., Bakolas, A., & Moropoulou, A. (2013). Integrated documentation protocols enabling decision making in cultural heritage protection. *Journal of Cultural Heritage, 14S*(2013), 141–146.

17. Kioussi, A., Karoglou, M., Bakolas, A., Labropoulos, K., & Moropoulou, A. (2013). Documentation protocols to generate risk indicators regarding degradation processes for cultural heritage risk evaluation. In *International Archives of the Photogrammetry, Remote Sensing and Spatial Information Sciences, vol. XL-5/W2, 2013, XXIV International CIPA Symposium* (pp. 379–384).

A Quick Criterion for Calculating Waiting Phenomena at Intersections

Raffaele Mauro, Marco Guerrieri, and Andrea Pompigna

Abstract Calculations of queues length and waiting times at intersections are essential to evaluate the quality of circulation at road junctions (Level of Service, LOS). These calculations are carried out with the theory of waiting phenomena (probabilistic and/or deterministic queue theory) and different models are adopted, depending on whether the operating conditions of the traffic are stationary or not. In technical practice for some time, both for sub-saturation and over-saturation situations for the intersection arms, the formulations of the time-dependent queues obtained with the so-called criterion of coordinates have been used. Depending on the degree of saturation of an input arm (traffic intensity), this criterion allows the transition from probabilistic solutions to deterministic ones. In the paper, after a brief review of time-dependent solutions, a quick criterion is provided for calculating the length of queues and waiting times in the event of peak traffic—as well as the duration of the effects of the latter—obtained under specific characteristics of the arrival processes at the intersection; a demonstration is given of how this criterion leads to solutions conforming to the deterministic type; estimates of the errors, which arise from the criterion developed in this paper to replace a time-dependent formulation, are provided in terms of confidence intervals with varying the degree of saturation.

Keywords Queuing theory · Non-stationary queues · Time-dependent queues · Renewal processes · Unsignalized intersections

1 Introduction

In Fig. 1 we consider the basic case of a road intersection in which only two traffic streams interact with each other. The major flow Q on the major street crosses the intersection, while the minor flow q on the minor street turns right.

Q has priority (priority flow) over q (non-priority flow). For this rule of priority of Q over q, the vehicles of flow q can form a queue. The interaction between q and Q can be modeled with the simplest of the queue models [1]. The waiting system of

R. Mauro · M. Guerrieri · A. Pompigna (✉)
DICAM University of Trento, Trento, Italy

© The Author(s), under exclusive license to Springer Nature Switzerland AG 2021

J. C. Matos et al. (eds.), *18th International Probabilistic Workshop*, Lecture Notes in Civil Engineering 153, https://doi.org/10.1007/978-3-030-73616-3_6

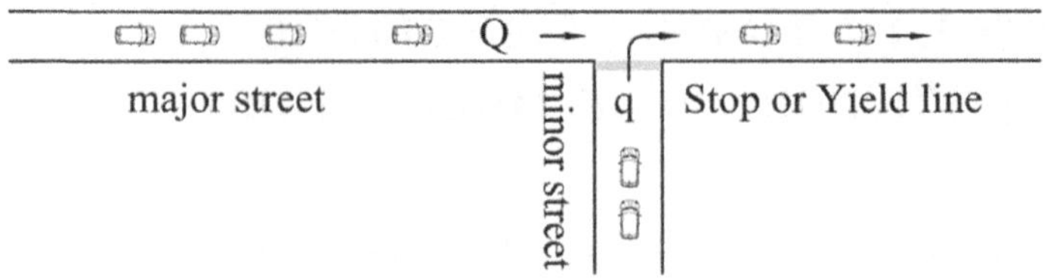

Fig. 1 Example of traffic flows with different priority at an intersection

Fig. 1 has a single service point (at Stop or Yield line) and a single serving channel (minor street). The discipline of the system service is First In–First Out (FIFO). The intersection of Fig. 1 can be studied with the mathematical queue theory [2, 3]. The theory of queues includes deterministic and probabilistic models to calculate waiting system states. Generally, the state variables $E\ [L_q]$ and $E\ [W_q]$ are used. $E\ [L_q]$ is the average length of the queue. $E\ [W_q]$ is the average waiting time in the queue. Other state variables are also used, which are deduced directly from $E\ [L_q]$ and $E\ [W_q]$. In a waiting system, inputs and outputs are sequences of events over time. The incoming flows are the arrivals in the system. The outgoing flows are the departures from the system (from the queue). Deterministic solutions are used when the count $x_t = A(t)$ of arrivals and the count $y_t = D(t)$ of departures are regular ($VAR[x_t] = VAR[y_t] = 0$) [4]. The cumulated functions $A(t)$ and $D(t)$ are step-functions of time t, with inter-arrival and departure times of constant amplitude over certain time intervals. So if we interpolate these step-functions with continuous functions of time, the rates $dA(t)/dt$ and $dD(t)/dt$ do not have random fluctuations. In deterministic solutions $E\ [L_q]$ and $E\ [W_q]$ are functions of time, like the other state variables. Thus, deterministic solutions are called time-dependent solutions. The probabilistic solutions of queue theory are used for random arrivals and departures [2, 3]. In this case, the queue length values L_q in the time form the random processes of queue. Every value L_q has its own probability of occurring. If the probability law of L_q does not change over time, evidently also $E\ [L_q]$ and $E\ [W_q]$ do not change over time. In this case, the waiting system is stationary. If the probability law of the queue length L_q varies instant by instant, waiting system is not in a steady state, but in a transitory state. Probabilistic solutions for non-stationary states are time-dependent solutions because they are obtained by probability law of L_q as time functions. In general, probabilistic solutions for stationary conditions are very simple mathematical expressions and $E\ [L_q]$ and $E\ [W_q]$ can be calculated easily. On the contrary, the solutions for non-stationary conditions are not simple mathematical expressions and the calculation of $E\ [L_q]$ and $E\ [W_q]$ is also not immediate. Thus, probabilistic time-dependent solutions are of little use in practical applications and heuristic solutions have been obtained for queues in transient states.

2 Operating Conditions of Intersections

In Highway engineering, the system shown in Fig. 1 is of significant interest. All the methods for calculating the performances for the unsignalized intersections are derived from the simple waiting system of Fig. 1. The operating conditions of the intersections depend on the traffic demand that must be served (Q and q of Fig. 1). It is intuitive that this intersection is in a steady state if Q and q are constant during an infinite time interval. If c is the capacity of the turn right maneuver by the vehicles of q, for the steady state it must be $q < c$. In all the other conditions of Q, q and c, the intersection is not in a steady-state condition. In general, c depends on Q, on the geometric layout of the intersection G and on human factor parameters θ of the drivers at intersections, i.e. $c = f\ (Q, G, \theta)$. Steady-state conditions are not realistic in actual traffic operating conditions because traffic demand always varies over time. Therefore, also Q and q always fluctuate during the different time periods of the day. Moreover, for $\rho > 0.8 - 0.6$, the queue length and the waiting times by stationary solutions tend to become infinite [5]. ρ is the traffic intensity, i.e. is the ratio between the average service time $1/c$ and the average interval between two arrivals $1/q$. This result is clearly unrealistic, because the time intervals with $\rho > 0.8 - 0.6$ are of finite length, and so the queue cannot grow indefinitely. In technical practice, in steady-state conditions but with $\rho > 0.8 - 0.6$ and in no steady-state conditions, heuristic solutions [4] are usefully used to calculate the state variables of the waiting system of Fig. 1. As mentioned, the non-stationary probabilistic solutions are too complicated to be easily implemented in the calculations. Heuristic solutions are also time dependent, since they relate to non-stationary conditions. These heuristic solutions are obtained with the method of the coordinates transformation [6] from the stationary solutions tending asymptotically to the deterministic solutions. The heuristic solutions $E\ [\bullet] = f\ (\rho)$ are continuous functions of ρ. ρ continuously assumes values in the interval $[0, +\infty]$, so all the possible operating conditions of the intersection in Fig. 1 can be analyzed in a unitary way from a low traffic intensity ($\rho \ll 1$) to a congested traffic situation ($\rho \geq 1$) (Fig. 2), as in the case of traffic peak (Fig. 3).

Heuristic solutions also allow the study of the effects of traffic peaks. However, if traffic peaks involve high levels of congestion, arrivals in the queue and departures from the queue are less random and more regular. Thus, with increasing congestion ($\rho \to +\infty$) the heuristic solutions tend to asymptotically coincide with the deterministic solutions. In operational terms the intersection of Fig. 1 is congested if $q > c$ ($\rho = q/c > 1$), or if $q \leq c$ ($\rho = q/c \leq 1$) but the queue evolves in the time from an already long extension.

To study the effects of traffic peaks at intersections, or in other traffic waiting systems, other solutions for peak traffic (also asymptotic formulations) can be used. These formulations have simpler mathematics than heuristic formulations. Furthermore, there are interesting mathematical relationships for asymptotic and deterministic formulations for congested traffic. These issues are covered in the following points of this paper.

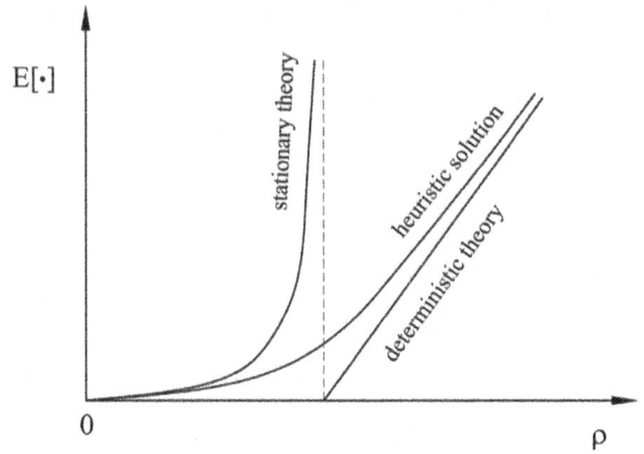

Fig. 2 Stationary, deterministic and heuristic solutions for continuous functions $E[\bullet] = f(\rho)$

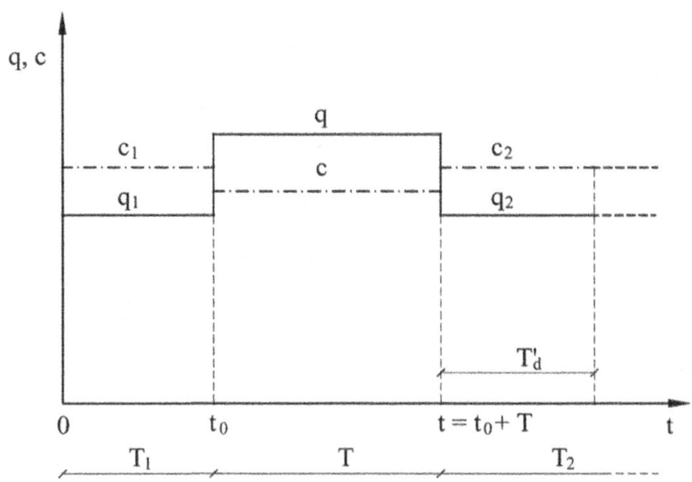

Fig. 3 Example case of a traffic peak

3 Approximate Solutions for Traffic Peaks

To analyze the intersection in Fig. 1 the state variables are:

- L number of vehicles in the system, i.e. waiting on the minor road. If the system is not empty, L is the number of vehicles in the queue L_q plus the vehicle in first position (i.e. Yield or Stop line):

$$L = L_q + 1 \tag{1}$$

- $\tau_y = s$ service time. s is the time spent by the vehicle in first position waiting to perform the maneuver;

- W_q time spent in queue. W_q is the time that passes between the instant in which a vehicle joins the queue and the instant in which it occupies the first position;
- W waiting time in the system. W is the total waiting time spent in the waiting line:

$$W = W_q + s \tag{2}$$

L_q, L, W_q, W and s are random variables. The mean values $E\ [L]$ and $E\ [W]$ are used in applications as state variables for the system. $E\ [L]$ and $E\ [W]$ are preferred to the mean values E $[L_q]$ and E $[W_q]$ for L_q and W_q because L and W contain more information than for L_q and W_q. Let be L at $t = 0$, L_0, $L_0 = 0$ (no vehicles initially observed in the system for a generic time interval [0, t]). To obtain the L equation (as function of time t), a balance equation can be written between the number x_t of vehicles arriving in the queue and the number y_t of vehicles leaving the queue at the end of a generic time interval [0, t]:

$$L_t = x_t - y_t \tag{3}$$

Now, let be the system characterized by a continuous and long queue with respect to the single vehicle. Also let be waiting times long if compared with the average service times. In this case, it may be appropriate and useful to replace these discontinuous trends with continuous functions that allow a "smoothing" over time. If x_t and y_t assumes sufficiently large values with respect to the unit, therefore, only small relative variations with respect to the average value can be expected. In these conditions, the negligible variations with respect to the average value allow to use a so-called first order approximation for x_t and y_t. Thus the continuous time and discrete valued random processes x_t and y_t are replaced with $A(t)$ and $D(t)$. $A(t)$ and $D(t)$ are continuous-time and continuous valued deterministic processes. Therefore, within this approximation and with $L_0 = 0$ we obtain:

$$L_t^* = A(t) - D(t) \tag{4}$$

where also L^*_t is a continuous and deterministic function of time. If $1/\tau_x(t) = q_t = dA(t)/dt$ and $1/\tau_y(t) = c_t = dD(t)/dt$, we have that $A_t = \int_0^t q(u)du$ and $Dt = \int_0^t c(u)du$. Assuming that $q_t = q$ and $c_t = c$ have constant values in the interval [0, t], in the more general case of non-zero initial queue L_0 we obtain:

$$L_t^* = L_0 + t(q-c) = t(\rho - 1)c \tag{5}$$

L^*_t represents the first order (or deterministic) fluid approximation for L_t [2, 3]. This approximation (i.e. $L_t \approx L^*_t$) only considers the accumulation that occurs due to the saturation of the system. In compliance with the Law of Large Numbers, its degree of approximation increases with increasing ρ over 1 and up to $+\infty$, as the basic hypotheses relating to the negligibility of the variations with respect to the average values are more sustainable. Still operating in the stochastic field, the problem of the peak of traffic can be tackled using the Renewal Theory [7]. Taking into account the balance Eq. (3) with a non-zero initial queue and under the same hypotheses of the fluid approximation, in terms of expected value we can write:

$$E[L_t] = E[L_0] + E[x_t] - E[y_t] \tag{6}$$

If the intersection is congested in [0, t] the queue is always present on the minor street. Under such conditions it is very likely that in the interval [0, t] the system is always busy. In this case, the times of departure of the vehicles that leave the system can be considered with the same probability distribution as the service times. The departure times are therefore independent from the arrival times so:

$$VAR[L_t] = VAR[L_0] + VAR[x_t] + VAR[y_t] \tag{7}$$

The renewal processes theory allows to express the values E [•] and VAR [•] for x_t and y_t, and finally for L_t. By a renewal processes, if vehicles arrive in the system and leave the system continuously (i denotes a generic vehicle), this processes can be interpreted as a sequence of replacements/substitutions with random times of replacements $\tau_{y,i}$ and random time of substitutions $\tau_{x,i}$, Let the random variables $\tau_{y,i}$ and $\tau_{x,i}$ be independent and identically distributed respectively with mean μ_y and μ_x and variance $\sigma^2{}_y$ and $\sigma^2{}_x$. It has been widely demonstrated that, whatever the distribution of $\tau_{y,i}$ and $\tau_{x,i}$, the realizations over time of the counting processes of departures y_t and arrivals x_t asymptotically follows (i.e. $t \to +\infty$) a normal distribution respectively with mean t/μ_y and t/μ_x and variance $t(\sigma^2{}_y/\mu^3{}_{y)}$ and $t(\sigma^2{}_x/\mu^3{}_x)$. With $C_y = \sigma_y/\mu_y$ ($C_x = \sigma_x/\mu_x$) we can write that for $t \to +\infty$:

$$y_t \sim N\left(t/\mu_y;\ C_y^2\, t/\mu_y\right) \tag{8}$$

$$x_t \sim N(t/\mu_x;\ C_x^2\, t/\mu_x) \tag{9}$$

Assuming that the parameters of the distributions do not vary over time, with $1/\mu_x = q$ and $\mu_y = E$ [s] we have:

$$y_t \sim N\left(t/s;\ C_y^2\, t/E[s]\right) \tag{10}$$

$$x_t \sim N\left(t\, q;\ C_x^2\, t\, q\right) \tag{11}$$

In view of this, Eqs. (6) and (7) approximate asymptotically with the following equations:

$$E[L_t] \sim E[L_0] + qt - t/E[s] = E[L_0] + t/E[s](\rho - 1) \tag{12}$$

$$VAR[L_t] \sim VAR[L_0] + C_x^2 t\, q + C_y^2 t/E[s] \tag{13}$$

The renewal processes theory proves that this asymptotic approximation for $t \to +\infty$ holds with good approximation if $t > \mu_y/C^2{}_y$ and $t > \mu_x/C^2{}_x$ [8]. It should be noted that the same result as in Eqs. (12) and (13) for the approximation with $\rho > 1$ and $t \to +\infty$ arises from a diffusive or second-order approach [2, 3], which proposes to approximate x_t and y_t (considered independent under the usual hypotheses of extended queue) as normally distributed, in accordance with the Central Limit Theorem [3]. An important result is proven for exponential distributions of arrival and departure inter-times (i.e. x_t and y_t realization of counting processes with Poissonian arrivals and departures). With reference to the system in Fig. 1, this is an M/M/1/∞/FIFO type queue system. In this case $C_y = C_x = 1$ and the approximations for $\rho > 1$ (and asymptotically $\to +\infty$) are exact for each value of t [8, 9]. In this case, in fact, y_t and x_t follow two Poisson distributions with mean $t\mu_y = t/E\,[s]$ and $t\mu_x = t\,q$. It turns out that:

$$E[L_t] = E[L_0] + qt - t/E[s] = E[L_0] + t(\rho - 1)/E[s] \tag{14}$$

$$VAR[L_t] = VAR[L_0] + t\,q + t/E[s] = VAR[L_0] + t(\rho + 1)/E[s] \tag{15}$$

4 Traffic Peaks Solution for Random Arrivals and Exponential Service Times

4.1 *Queue Growth in Saturation Conditions*

In the following, the simple traffic situation of Fig. 3 is considered for the intersection of Fig. 1. In Fig. 3 the minor flow $q_1 = q_2$ are constant and less than the capacities $c_1 = c_2$ before and after the peak period T. During the same peak period the flow q is greater than the flows before and after this period. Furthermore q during the peak period T exceeds capacity c. So in the interval T, i.e. between instants t_0 and $t_0 + T$, the intersection of Fig. 1 is subject to a traffic peak.

In this way the waiting system of Fig. 1 can be considered in a steady state at the instant t_0 starting from which the peak demand q occurs. The effects of the traffic peak on the intersection are affected beyond T, for a time interval T'_d starting from $t = t_0 + T$ and included in T_2. For the situation in Fig. 3, if L_0 is the steady state queue at the instant $t = 0$ before the start of the traffic peak, for Poissonian arrivals with rate q and service times distributed exponentially with mean $E\,[s]$ ($\rho_0 = q_1/c_1$), we have that:

$$E[L_t] = \rho_0/(1 - \rho_0) + t(\rho - 1)/E[s] \tag{16}$$

$$VAR[L_t] = \rho_0/(1 - \rho_0)^2 + t(\rho + 1)/E[s] \tag{17}$$

Equation (16) allows to obtain the expected value of the waiting time in the system W_t of a vehicle arriving at the instant t. This vehicle will see L_t vehicles waiting in queue, so neglecting that the vehicle currently in service at time t may be partially served, its waiting time S_{Lt} is equal to the sum of L_t exponential service times with mean $E[s]$. As is demonstrated for renewal processes [7], S_{Lt} is distributed according to an Erlang variable of parameters L_t and $1/E[s]$, with expected value $L_t E[s]$ and variance $L_t\,(E[s])^2$. For the hypotheses considered above, the expected value and the variance of the waiting time W_t of the vehicle arriving at the instant t conditioned with respect to the presence of total L_t vehicles in the system can be approximated as follows [9]:

$$E[W_t|\,L_t] \approx E[L_t]\,E[s] \tag{17}$$

$$VAR[W_t|\,L_t] \approx E[L_t](E[s])^2 \tag{18}$$

In this way the expected value for waiting time W_t, considering a vehicle arriving at the instant t (i.e., at the end of the peak period) is:

$$E[W_t] = E[s]\rho_0/(1-\rho_0) + t(\rho-1) \tag{19}$$

Furthermore, considering that [9]:

$$VAR[W_t] = E[L_t]\,VAR[W_t|\,L_t] + VAR[L_t]\,E[W_t|\,L_t] \tag{20}$$

using the previous equations we also obtain the approximation for the variance of W_t.

4.2 Saturation Queue Discharge

With reference to Fig. 3, we want to find the duration of the time interval T'_d starting from $t = t_0 + T$ and included in T_2. In this interval, the effects of peak traffic that arose during the T interval are exhausted. If $L_e = \rho_2/(1-\rho_2)$ is the steady state number of vehicle in the system with traffic demand q_2 and capacity c_2 ($\rho_2 = q_2/c_2$), using Eqs. (16) and (17) we can approximate $E[T'_d]$ and $VAR[T'_d]$. After some calculations, we have that [9]:

$$E\left[T'_d\right] = (E[L_T] - L_e)/(c_2(1-\rho_2)) \tag{21}$$

$$\begin{aligned} VAR\left[T'_d\right] &= 2\rho_2(E[L_T] - L_e)/(c_2^2(1-\rho_2)^3) \\ &\quad + (E[L_T] + VAR[L_T])/(c_2^2(1-\rho_2)^2) \end{aligned} \tag{22}$$

5 Comparisons Between Approximate and Heuristic Solutions

Firstly we propose some comparison tests for the queue growth between Approximate solutions (cf. 3.1) with Heuristic solutions (cf. 2). We refer to Kimber and Hollis [6] Heuristic formulation as proposed by [4, 10] with:

$$L_t = \frac{1}{2}\left[\sqrt{A^2 + B} - A\right] \tag{23}$$

where

$$A = (1 - \rho)ct + 1 - L_0 \tag{24}$$

$$B = 4(L_0 + \rho ct) \tag{25}$$

For the intersection of Fig. 1 we consider the traffic condition of Fig. 3, but in this case with $\rho_1 = 0$, therefore $L_0 = 0$. L_t is calculated with Eq. (16) and with Eqs. (23)–(25) for ρ ranging between 1.0 and 2.0, and $T = 10$ min. It is considered $1/E[s] = c = 0.278$ veh/s (1000 veh/h). Also, Fig. 4 shows the upper limit of Čebyšëv (*CSL*) interval at 85% for Eq. (16), considering the variance value obtained using Eq. (17).

The comparison shows the deviations between Heuristic and Approximate solutions which are more and more reduced with increasing the traffic congestion. The deviation values appear to be extremely low compared to CSL at 85%. For the analyzed case study, Fig. 5 shows the percentage deviation (plotted in logarithmic scale) between Heuristic and Approximate solution related to T in the range from 1 to 60 min. It is worth pointing out that in the time intervals usually used for the performance analysis of road intersections, the deviations between the two time-dependent solution types, already for $\rho = 1.2$ are less than 8 and 2.5% in case of $T = 15$ min and $T = 60$ min respectively.

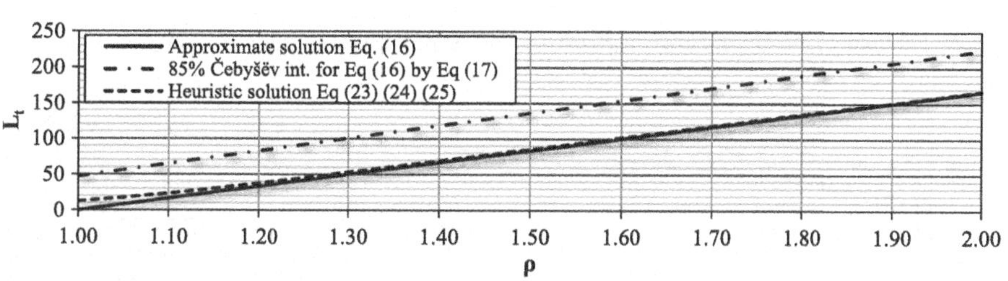

Fig. 4 Comparison between heuristic and approximate solutions

Figure 6 shows L_t values (plotted in logarithmic scale) obtained with the two time-dependent solution and the CSL at 85% in function of ρ, in the case of non-zero initial queue $L_0 = 9$ veh ($c_1 = 0.278$ veh/s $q_1 = 0.25$ veh/s, with $\rho_1 = 0.9$), under the condition $T = 15$ min and $1/E[s] = c = c_1$.

In accordance with L_t values of Fig. 6, the percentage deviation of Heuristic and Approximate solutions have been calculated in function of ρ ($\rho = 1.0$–2.0) and T ($T = 1$–60 min) as shown in Fig. 7a. Figure 7b reports similar assessments for the case $L_0 = 19$ veh ($c_1 = 0.556$ veh/s = 2000 veh/h and $q_1 = 0.528$ veh/s = 1900 veh/h, with $\rho_1 = 0.95$) and capacity $1/E[s] = c = c_1$. Thus, obviously, for fixed T and ρ values, the deviation between the two time-dependent solution types decreases as L_0 increases.

Finally, Fig. 8 shows the evolution of queue length in a traffic saturation state, for a time interval of 60 min, under the following conditions: initial queue $L_0 = 9$ veh, $c_1 = 0.278$ veh/s, $q_1 = 0.25$ veh/s ($\rho_1 = 0.9$), $1/E[s] = c = c_1$ and $q = 0.333$ veh/s = 1200 veh/h. Let be note here that for every solutions, the trend of L_k in Fig. 8 has been obtained considering subsequent regular intervals $k_i = 1, 2, \ldots 20$, each of 3 min. Once again, for the Approximate solution (Eq. 16) the 85% Čebyšëv interval values over the time (namely in function of k_i) are given. According to Eqs. (23), (24) and (25), in Fig. 8 for the Heuristic solution two trends are reported:

a) the first one with re-initialization of the queue at the beginning of each 3 min time interval respect to the initial equilibrium one ($L_0 = 9$ vehicles at the beginning of each interval);

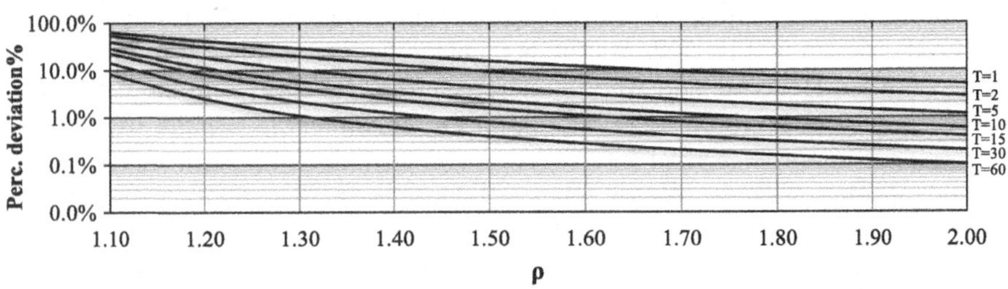

Fig. 5 Percentage deviation between Heuristic and Approximate solutions in function of ρ

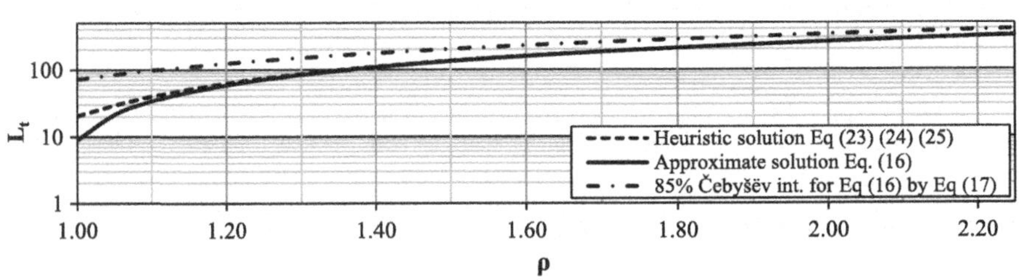

Fig. 6 L_t values in function of ρ (case of non-zero initial queue: $L_0 = 9$ veh)

b) the second one with the evolution of the queue at the beginning of each 3 min time interval considering the queue at the end of the previous interval [10].

As expected, the curves of Fig. 8 highlight a substantially convergence of the results obtained with the two different time-dependent solutions.

It is interesting to test the Eqs. (21) and (22) in order to evaluate the time interval T'_d obtained with the Approximate solution in comparison with the Heuristic one (queue discharge, cf. 3.2). In this regard, and in compliance with Fig. 3, the following traffic condition was analyzed: $q_1 = 300$ veh/h, $c_1 = 450$ veh/h ($\rho_1 = 0.667$); $q = 600$ veh/h, $c = 450$ veh/h ($\rho = 1.334$), $T = 10$ min and $q_2 = q_1 = 300$ veh/h, $c_2 = c_1 = 450$ veh/h ($\rho_1 = 0.667$). By means of Eqs. (23)–(25), a queue length $L_T = 29$ has been obtained. Using Eqs. (16) and (17), it results $E[L_T] = 27$ and $VAR[L_T] = 6$. These values demonstrate that the Approximate solution is very close to the Heuristic one.

Now we want to evaluate the time interval T'_d that, once the traffic peak period T is over, the system employs reducing the queue L_T to the stationary value for $\rho_1 = 0.667$, which is $L_e = 2$ veh. Using Eq. (21), it results $E[T'_d] = 600$ s and $(VAR[T'_d])^{1/2} = 277$ s.

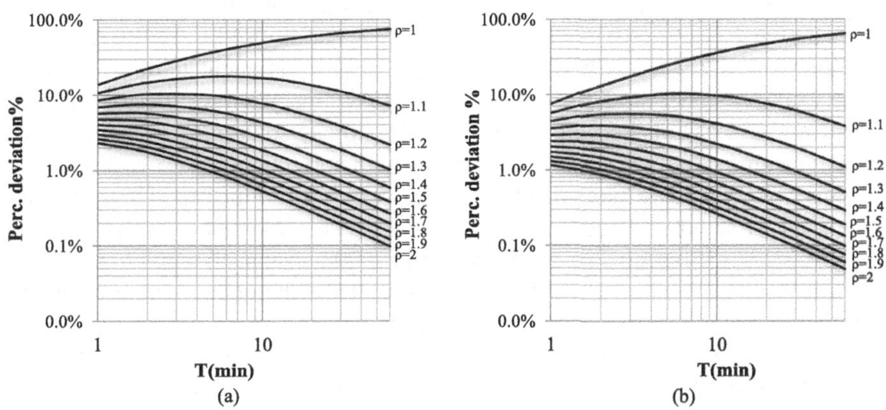

Fig. 7 Percentage deviation between the two time-dependent solutions (**a**: $L_0 = 9$ veh; **b**: $L_0 = 19$ veh)

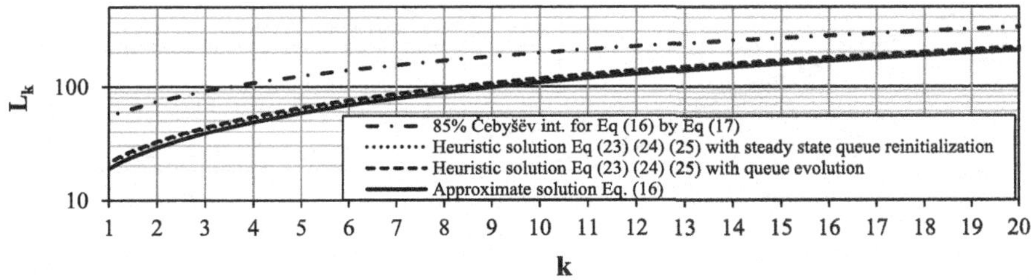

Fig. 8 Evolution of queue considering subsequent time intervals $k_i = 1, 2, \ldots 20$, each of 3 min

To estimate a realistic duration T'_d by Heuristic solutions, some corrections to increase accuracy of calculation of queue discharge are required [6]. In our case (e.g. $L_T > 2L_e$) we consider the following relationships [6]:

$$L_t = L_T + (\rho_2 - L_T/(L_T + 1))c_2 t \quad if \ \ 0 \le t \le t_C \tag{26}$$

$$L_t = 2L_e - \frac{1}{2}\left[\sqrt{A^2 + B} - A\right] \ \text{ if } t > t_C \tag{27}$$

with

$$A = (1 - \rho_2)(t - t_C)c_2 + 1 \tag{28}$$

$$B = 4\rho_2 c_2 (t - t_C) \tag{29}$$

$$t_C = (2L_e - L_T)/c_2(\rho_2 - L_T/(L_T + 1)) \tag{30}$$

For determining T'_d, it is essential to find the value of t for which Eq. (26) is satisfied with $L_t = L_e = 27$ veh for every $0 \le t \le t_C$, or else Eq. (27) with values according to Eqs. (28) and (29) if $t > t_C$. In both cases, t_C is calculated with Eq. (30). In the example under study, it results $t_C = 619$ s. Fig. 9 shows the values of L_t in function of t, calculated with Eq. (16) (Approximate solution) and using the set of relationships (26)–(30) (Heuristic equation with correction for queue discharge). Moreover, in Fig. 9 are plotted: the L_t values estimated with Eqs. (23)–(25); the value $E\ [T'_d] = 600$ s (from Eq. 21); the CSL 85%, obtained considering $(VAR\ [T'_d])^{1/2} = 277$ s (from Eq. 22). For $t = E[T'_d] = 600$ s, evaluated with Eq. (21)—for which it results E $[L_t] = L_e = 2$ veh with Eq. (16)—the queue length estimated with the corrected Heuristic solution is slightly below 5 veh, decreasing to 3 veh after just 2 min. Therefore, the value $E[T'_d]$ can be considered a good approximation of the time taken for the queue discharge due to the traffic peak. Based on these considerations, the proposed method (Approximate solution) turns out to be very accurate and even simpler than the Heuristic solution.

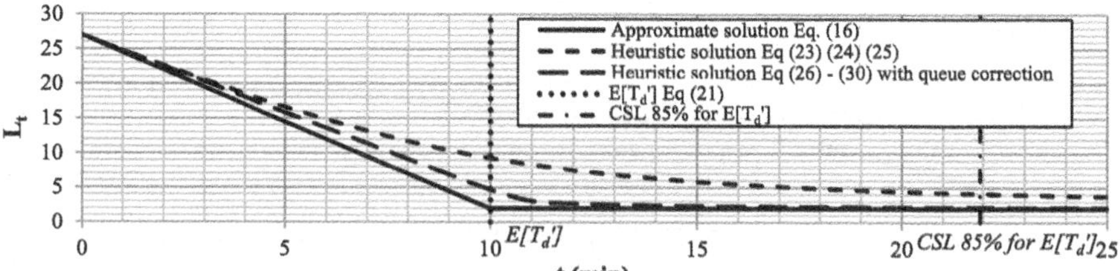

Fig. 9 Comparison among Heuristic solution with—without queue discharge correction and approx. solution

6 Conclusions

This article demonstrates that in congested traffic conditions, the approximate deterministic solutions for the calculation of the state variables of a road intersection can be deduced as an asymptotic result of the renewal theory. In congested traffic conditions, these approximate solutions have proven to be substantially coincident with the heuristic solutions by coordinates transformation method. The differences obtained by applying the two time-dependent solutions have been estimate as function of traffic intensity. These differences that we get, even if small, are to be considered in the context of the underlying uncertainty in road intersections calculation, primarily for traffic demand and driving behavior parameters. The results have shown that the proposed approximate solutions, marked by a reliable and rapid mathematical approach, could be widely and profitably used in traffic analysis. It should be noted that the solutions obtained were tested by taking comparison with theoretical queue models, on the basis of the most widespread and shared assumptions in the literature for the analysis of unsignalized intersections for arrival and service time probability distributions. The possibility of further testing the acceptability of the same assumptions through comparisons against real traffic scenarios in situations of congestion at unsignalized intersections represents an important topic that is interesting to deepen. This in the continuous search for calculation criteria that, as well as being quick and easy to formulate mathematically, are effectively able to represent real world occurrences.

References

1. Brilon, W. (2008). Delay at unsignalized intersections. *Transportation Research Record, 2071*, 98–108.
2. Kleinrock, L. (1975*). Queueing systems. Theory* (Vol. I). New York: Wiley.
3. Kleinrock, L. (1976). *Queueing systems: Computer applications* (Vol. 2). New York: Wiley.
4. Mauro, R. (2010). *Calculation of roundabouts: Capacity, waiting phenomena and reliability*. Berlin: Springer.
5. Louah, G. (1991). Priority intersection: modelling. In M. Papageorgiou (Ed.), *Concise encyclopedia of traffic & transportation systems*. Oxford: Pergamon Press.
6. Kimber, R. M., & Hollis, E. M. (1979). *Traffic queues and delays at road junctions*. Transport and Road Research Laboratory Report No. LR909. TRRL, Crowthorne, UK.
7. Cox, D. R. (1970). *Renewal theory* (p. 142). London: Methuen & Co. ISBN 0-412-20570-X.
8. Cox, D. R., & Smith, W. (1961). *Queues*. London: Chapman and Hall. ISBN-13: 978-0-412-10930-0.
9. Cox, D. R. (1955). The statistical analysis of congestion. *Journal of the Royal Statistical Society. Series A (General), 118*(3), 324–335.
10. Mauro, R., & Pompigna, A. (2020). State of the art and computational aspects of time-dependent waiting models for non-signalised intersections. *Journal of Traffic and Transportation Engineering*. https://doi.org/10.1016/j.jtte.2019.09.007.

A Reliability Based Crack Propagation Model for Reinforced Concrete Bridge Piers Subject to Vehicle Impact

Suman Roy and Andrew Sorensen

Abstract Bridges play a critical role in transportation infrastructure networks. As such, their ability to withstand hazardous loading scenarios is essential. The response of bridges and specifically of bridge piers to hazardous loading scenarios such as earthquake and blast have received significant attention. However, their behavior under vehicle impact has received less attention. This is significant due to the fact that the frequency of occurrence for vehicular impact is appreciably higher than that of earthquake and blast. Of larger significance however is the fact that bridge piers that have experienced impact loading may have reduced capacities to withstand secondary hazardous loadings. This multi-hazard loading scenario of bridge piers is ideally suited for reliability-based risk analysis methods for study; however, very few such studies exist. In this research, a reliability-based model is used to determine the crack propagation in circular reinforced concrete bridge piers at different strain rates. Crack propagation is an important characteristic for structural health and ability to withstand future loading. However, in the case of vehicular impact, only the deformation of the pier is typically taken into account for determine post impact serviceability. In the model, quasi-static to dynamic strain rates are considered for steel while only dynamic strain rates are considered for concrete. Using Monte Carlo simulation, crack propagation rates for both Grade 60 and Grade 80 reinforcing steel are developed. The Hasofer-Lind reliability method is then used to determine the subsequent reliability of the piers post-impact. Models representing the dynamic reliability indices validating limit state equations show persuasive and practical trends. Furthermore, the model can serve as a design tool in predicting serviceability as well as analyzing design scenarios in an economic and practical way.

Keywords Crack propagations · RC bridge pier · Vehicle impact Monte Carlo simulations · Reliability analysis

---

S. Roy (✉) · A. Sorensen
Department of Civil and Environmental Engineering, Utah State University, Logan, UT, USA

© The Author(s), under exclusive license to Springer Nature Switzerland AG 2021

J. C. Matos et al. (eds.), *18th International Probabilistic Workshop*, Lecture Notes in Civil Engineering 153, https://doi.org/10.1007/978-3-030-73616-3_7

1 Introduction

The increasingly widespread occurrence of vehicle-pier collision accidents, and blast due to terrorist attack present a significant threat to the safety and survivability of bridge structures subjected to these high strain rate loading events. Additionally, these loadings weaken the structure making them more vulnerable to subsequent loading scenarios such as seismic. This type of sequential multi-hazardous loading requires proper analysis techniques which can be used in normal practice for bridge pier design as well as for damage mitigation. Some accidental collisions result in severe damage to bridge structures, such as pier fracture and bridge collapse, while others cause localized concrete cracking at the specific impact location. To study the behaviors and failure modes of the impacted piers, it is necessary to develop utilitarian studies in order to accurately analyze the bridge-pier failure patterns and develop an insight for code calibration.

Reinforced concrete (RC) bridge piers are the first and most vulnerable structural element exposed to impact resulting in failure under static and dynamic load; and hence, crack propagation [1]. Current post impact rating models rely on the residual curvature and serviceability criteria to assign damage index levels [2]. Accurate prediction of crack propagation in a concrete structure is a crucial parameter for calculating reliability, improving durability, accurate structural health monitoring, and determining serviceability. Many investigations have been conducted to analyze structural performance under dynamic loading conditions and calibration followed by development of the respective codes has been undertaken by legitimate committees and standards bodies [3, 4].

To achieve rational, feasible, and reliable structural design approaches, prediction for the performance in terms of serviceability has become essential through its service life span and as such design rules stipulated in the codes should conform to corresponding performance criteria [5]. A number of analyses with numerical simulations have been carried out and have shown that crack propagation followed by the flexural deflection is frequently observed to be catastrophic for RC beam [6]. However, the pattern of the crack propagation in a concrete bridge pier caused by dynamic impact, appear to be due to localized action [6].

In the past couple of decades, the probabilistic assessment of highway bridges has developed rapidly in order to prioritize bridges for retrofit and rehabilitation based on their seismic risk [7]. Multi hazard sequential load predicting blast and impact has also been conducted [8]. Unfortunately, very few studies have focused on developing a holistic numerical approach to investigate frequently occurred damage as a function of crack propagation caused by vehicular impact, nor the validity of developed models for different concrete and steel grade combinations. This has led to a need to develop an innovative numerical limit state model envisaging dynamic crack propagation at different steel strain rates in order to assess the serviceability criteria. In this study, a numerical model predicting crack propagation for dynamic impact as a function of strain rates and yield strength of steel with the corresponding failure assessment has been explored by numerical simulation using 'Monte Carlo' method at different strain rates. In addition, reliability indices of the model are further investigated using the

Hasofer-Lind reliability index method [9]. Moreover, a proposed numerical model for static condition has been developed to evaluate target reliability indices to compare different combinations of concrete and steel grades in order to evaluate successful code calibration considering dynamic aspects for serviceability.

2 Methodology and Results

2.1 Determination of the Rate of Crack Propagations

RC bridge pier when experiences an impact from vehicular collision, the exposed part of the pier undergoes localized failure subject to its intensity. Failure can be occurred by experiencing cracks in concrete, sometimes penetrating beyond the sacrificial layer. Crack propagation due to impact is a highly mechanistic complex phenomenon which is dependent on the factors like concrete and steel grade, and the crack propagation patterns. This mechanism has been led by combination of strain rates in concrete and steel. In this study, a representative bridge pier is investigated for vehicular impact at different strain rates. Steel strain rates along with the modulus of elasticity controls the bridge pier performance in order to sustain the impact load.

Studies are required to incorporate the crack propagation due to vehicular dynamic impact on RC bridge pier as it is highly uncertain at high strain rates. Crack propagation in terms of depth and width has been investigated in this research for high strain rates in concrete and steel. In order to compute low impact duration, the rate of maximum crack-depth in normal concrete has been developed and is shown in Eq. 1 [10].

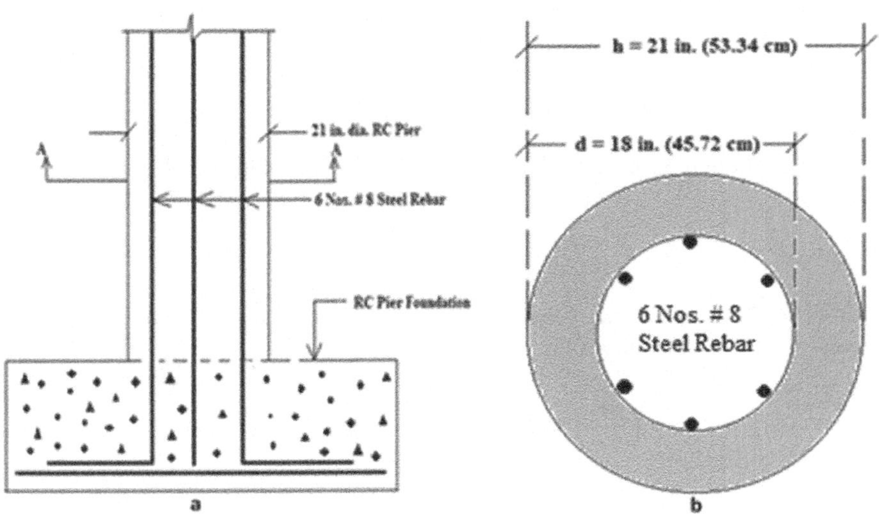

Fig. 1 (a) Representative RC bridge pier, (b) Section A-A

$$\sigma_y = \left(0.59 * \frac{M_{u1} h_{cr,lim}}{I_1^*}\right) * (h_{cr,lim}/t)^{1/2} * \alpha \tag{1}$$

Rearrangement of Eq. 1 yields Eq. 2.

$$\sigma_y = \left[\frac{(0.59 * M_{u1} * \alpha)}{I_1^* * t^{\frac{1}{2}}}\right] * (h_{cr,lim}/t)^{3/2} \tag{2}$$

In Eqs. 1 and 2, σ_y is the yield stress of tensile steel, M_{u1} is the flexural moment at extreme fiber of core cross-section, ρ is the steel ratio = 0.79 (in^2) × (6 nos.) / (π/4 × d^2) = 0.013 (=1.3%), I^*_1 = (π/64) * (D_g^4 – d_{core}^4) in^4 where, D_g and d_{core} are the gross (21 in. or 53.34 cm) and core (18 in. or 45.72 cm) diameters of the pier section respectively (as shown in Fig. 1). Replacing σ_y using high strain rate ($\acute{\varepsilon}$ = dε/dt) at steel due to impact consideration, and steel modulus of elasticity (E_s), and after rearranging Eq. 2, yields Eq. 3.

$$E_s * (\acute{\epsilon}) = E_s * \left(\frac{d\varepsilon}{dt}\right) = \left[(0.59 * M_{u1}) * \frac{\left(\frac{d\alpha}{dt}\right)}{(I_1^* * t^{\frac{1}{2}})}\right] * (h_{cr,lim}/t)^{3/2} \tag{3}$$

In Eqs. 1–3, 'α' is a factor expressed in Eq. 4 [10].

$$\alpha = \left(\frac{15.53}{\rho * \psi}\right) * \left(\frac{h_{cr,lim}}{h}\right) - 1.41/(\rho * \psi) \tag{4}$$

In Eq. 4, ψ (a factor) has been taken as 0.77 [10], $h_{cr,\,lim}$ is the limiting crack depth, and h is the depth of the cross-section (21 in. in Fig. 1). Hence, the product $(\rho * \psi)$ becomes 1.001. Area of steel rebar used for pier cross-section is 6 nos. # 8 steel rebar (4.74 in^2 or 30.58 cm^2).

Equation 4, after simplification and rearrangement, yields Eq. 5.

$$\alpha = 0.74 * h_{cr,lim} - 1.4086 \tag{5}$$

Taking differentiation of 'α' with respect to time (t), yields Eq. 6.

$$d\alpha/dt = 0.74 * (\frac{dh_{cr,lim}}{dt}) \tag{6}$$

Inserting dα/dt from Eq. 6 into Eq. 3, and after rearranging, yields Eq. 7, and expressed in terms of strain rates.

$$E_s * \left(\frac{d\varepsilon}{dt}\right) = \left(0.4366 * \frac{M_{u1}}{I_1^*}\right) * (h_{cr,lim}/t)^{5/2} \tag{7}$$

2.2 Computation of Cracking Moment

Cracking moment (M_{cr}) is determined using the standard equation (Eq. 8) and is appropriately applied when a RC pier experiences a collision at its exposed face [11].

$$M_{cr} = \frac{f_r * I_g}{y_t} \tag{8}$$

Using Eq. 8 and the dimensions shown in Fig. 1, f_r the modulus of rupture, is equal to $7.5*(f'_c)^{1/2}$ and becomes 410.791 psi (2.83 MPa) for $f'_c = 3000$ psi (20.68 MPa) The distance from the neutral axis to extreme tension fiber (y_t) is 21 in/2 = 10.5 in. (26.67 cm.) The gross moment of inertia, $I_g = \pi/64(21)^4$ in^4 which equates to 9547.801 in^4 (0.004 m^4) Considering these values, the cracking moment, M_{cr} becomes 373.54 kip-in. (42.204 kN-m).

Replacing M_{u1} by M_{cr}, and inserting the value of 29×10^{-6} psi (2×10^{-5} MPa) for E_s, the rearrangement of Eq. 7 yields Eq. 9.

$$\frac{dh_{cr,lim}}{dt} = 67.26 * (\frac{d\varepsilon}{dt})^{0.4} \tag{9}$$

In this study, quasi-static to dynamic strain rates are considered for the range from 10^{-4} to 10^{-8} [12].

2.3 Estimation of Rate of Crack-Widt

From ACI 318-14, static crack-width (w_c)can been evaluated using Eq. 10 [13].

$$w_c = 0.076 * \beta_h * f_s * (d_c * A)^{1/2} \tag{10}$$

In Eq. 10, f_s is the stress in steel, d_c is cover from extreme tension fiber to the outer face of the stirrup [d_c = Effective cover – stirrup diameter – 1/2(diameter of tension steel) = 3 in.—diameter of #4 bar—1/2 (diameter of #8 bar) = 2 in], and the cross sectional area A is equal to {6 in. × (2 × 4.71 in.)} is 56 in.^{2} (361.29 cm^2) However, rate of increase of crack-width (dw_c/dt) can be executed by taking first derivative of w_c with respect to time (t) at quasi-static strain rate and dynamic stress level of steel. This derivation is shown in Eq. 11.

$$\frac{dw_c}{dt} = 0.076 * \beta_h * \left(E_s * \frac{d\varepsilon}{dt}\right) * (d_c * A)^{1/2} \tag{11}$$

In Eq. 11, $d\varepsilon/dt$ is the dynamic strain rate in steel and β_h is given as (h – c)/(d – -c), where h = 21 in and d = 17 in (Fig. 1). However, in this study, β_h as a function of

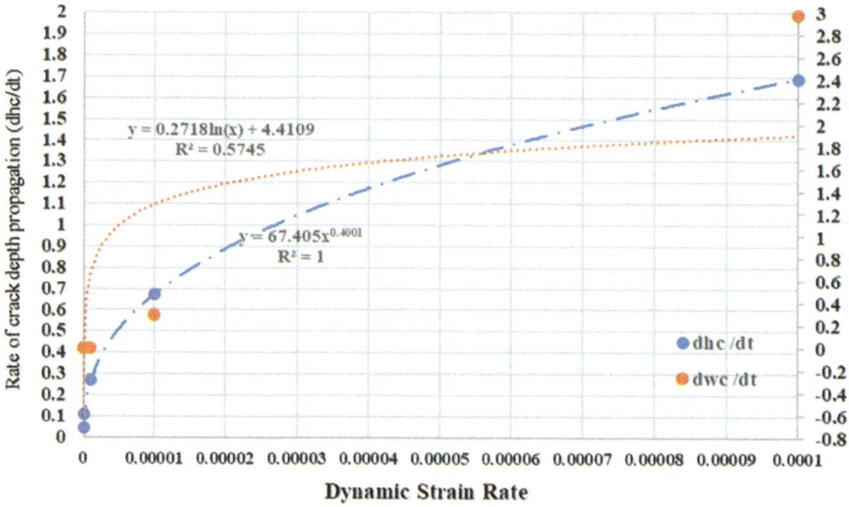

Fig. 2 Dynamic strain rate with rate of crack width and crack depth propagation

concrete is taken as non-deterministic, and strain rate (dε/dt) is considered for ranges from 10^{-4} to 10^{-8}. Furthermore, elastic modulus of steel (E_s) is also non-deterministic as it varies with the yield stress and strain rate.

Using the example column, In Eq. 11, c = a/β_1 and a = A_{st} * f_y/0.85(f'_c * h) where A_{st} = (3 × 0.79 in.^{2}) as 3 bars undergo tension and β_1 is 0.85. However, after computation, a and c become 2.65 in (67.31 mm) and 3.123 in (79.32 mm), respectively, and the nominal value of β_h ($\beta_{h,nominal}$) becomes 1.28.

Rearranging Eq. 11 yields Eq. 12.

$$\frac{dw_c}{dt} = 0.00103 * \beta_h * E_s * \left(\frac{d\varepsilon}{dt}\right) \quad (12)$$

Using Eq. 12, rates of crack propagation for depth and width can be computed, and their comparison are shown in Fig. 2. Figure 2 shows a potential comparison for rates of crack propagations in terms of crack-width and crack-depth plotted for different dynamic strain rates for grade 60 steel rebar.

2.4 Dynamic Increase Factor (DIF)

Dynamic Increase Factor (DIF) is a dimensionless number which describes how many times the deflections or stresses can be multiplied to the deflections or stresses caused by the static loads when dynamic load is applied. In other words, DIF is also specified as the ratio of dynamic strength over quasi-static strength in uniaxial compression or tension, is reported as a function of strain rate [14] However, DIFs considered in this study are compressive strength for concrete and tensile strength for steel. During high impact, concrete undergoes localized cracks which is a function of strain rate, whereas steel experiences deformation as a function of deterministic strain rates, and non-deterministic modulus of elasticity. So, DIF, for concrete and steel combining together, undergo complex mechanism and play significant role to assess crack propagation caused by dynamic impact.

2.4.1 Determination of DIF for Concrete in Compression and Steel in Tension

Dynamic increase factor for concrete (DIF_{con}) in compression has previously been described by [13] and is given in Eq. 13.

$$DIF_{con} = \left(\frac{\acute{\epsilon}_c}{\acute{\epsilon}_s}\right)^{1.206\alpha_c} \tag{13}$$

In Eq. 13, $\acute{\epsilon}_c$ and $\acute{\epsilon}_s$, are the quasi-static and static strain rates in concrete and steel, and are considered as 10^{-4} and 10^{-6} respectively [12]. However, α_c can be determined from Eq. 14 [15].

$$\alpha_c = [5 + 9 * (f_{cs}/f_{co})]^{-1} \tag{14}$$

In Eq. 14, f_{cs} is considered as 3 ksi, and f_{co} is taken as 1.45 ksi [15]. After computation from Eq. 14, α_c yields as 1.22, which has been taken as deterministic in this study as no randomness is involved and well predicted.

Dynamic increase factor for steel (DIF_s) in tension has been considered by [14], and expressed in Eq. 15.

$$DIF_s = (10^4 * \acute{\epsilon})^{\alpha_s} \tag{15}$$

In Eq. 15, $\acute{\epsilon}$ is the quasi-static to plastic strain rate (deterministic) for steel rebar varying from 10^{-4} to 10^{-8}. The magnification factor α_s, can be determined from the Eq. 16 [14].

$$\alpha_s = 0.074 - f_y/60 \tag{16}$$

In Eq. 16, f_y (yield stress of steel bar) is considered as 60 ksi, and hence deterministic. Steel strain-rate parameters for grade 60 (60 ksi) rebar within the range of 10^{-1}–10^{-4}, with the corresponding computed DIF_s, are shown in Fig. 3. On the other hand, streel strain-rate parameters (10^{-4}–10^{-8}) with the corresponding DIF_s for the higher yield strength 80 ksi rebar are shown in Fig. 4.

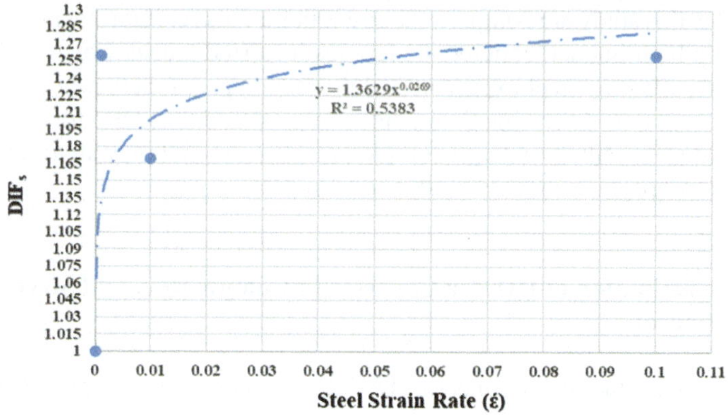

Fig. 3 DIF and $\acute{\varepsilon}_s$ (10^{-1}–10^{-4}) for 60 ksi yield strength

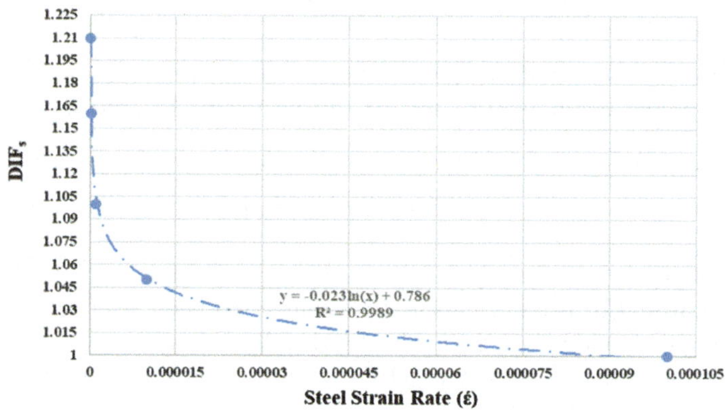

Fig. 4 DIF and $\acute{\varepsilon}_s$ (10^{-1}–10^{-4}) for 80 ksi yield strength

2.5 *Monte Carlo Model for Predicting Crack Width*

In this study, maximum permissible crack-width for serviceability limit state is taken as a value recommended by ACI 318-14 of 0.016 in (0.00063 mm) [16]. However, structural serviceability for the limit state function (G) comprises if the rate of crack-widths does not exceed permissible crack-widths. Equation 17 shows the limit state equation for serviceability for dynamic impact in terms of crack-widths evaluation.

$$G = 0.016 * DIF_{con} - (dw_c/dt) * DIF_s \tag{17}$$

Using Eq. 17 and substituting dw_c/dt from Eq. 12, yields Eq. 18.

$$G = 0.016 * DIF_{con} - 0.00103 * (d\varepsilon/dt) * \beta_h * E_s * DIF_s \tag{18}$$

Replacing strain rate ($d\varepsilon/dt$) as $\acute{\varepsilon}$ and after rearranging Eq. 18, yields Eq. 19.

$$G = 0.016 * DIF_{con} - 0.00103 * \left(\acute{\epsilon}\right) * \beta_h * E_s * DIF_s \tag{19}$$

Ten thousand Monte Carlo simulations are run for limit state function (G) and $\Phi^{-1}(P_i)$ are executed and plotted for different strain rates are shown in Figs. 5, 6, 7, 8 and 9, respectively. The probability of failure (P_i) and the inverse of the cumulative distribution function (CDF) for the P_i are plotted using ten thousand Monte Carlo simulations (statistical parameters mean and standard deviation converged at about this number of simulations) for the different strain rates (10^{-4}–10^{-8}). Random

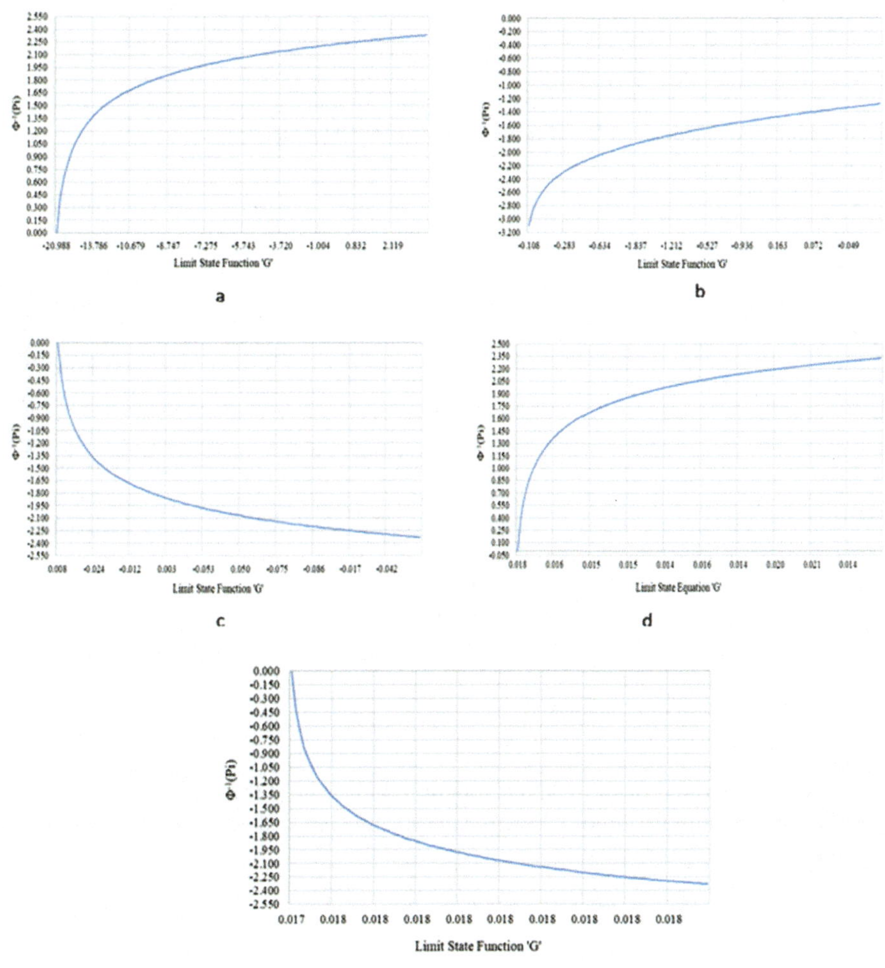

Fig. 5 Limit state function and $\Phi^{-1}(P_i)$ for strain rates at **a** 10^{-4}, **b** 10^{-5}, **c** 10^{-6}, **d** 10^{-7}, and **e** 10^{-8}

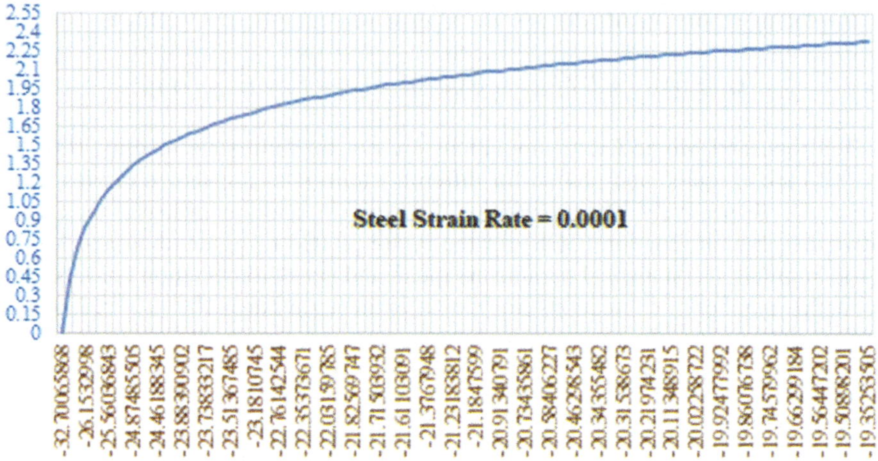

Fig. 6 Limit state function and $\Phi^{-1}(P_i)$ for 6 ksi concrete 60 ksi steel

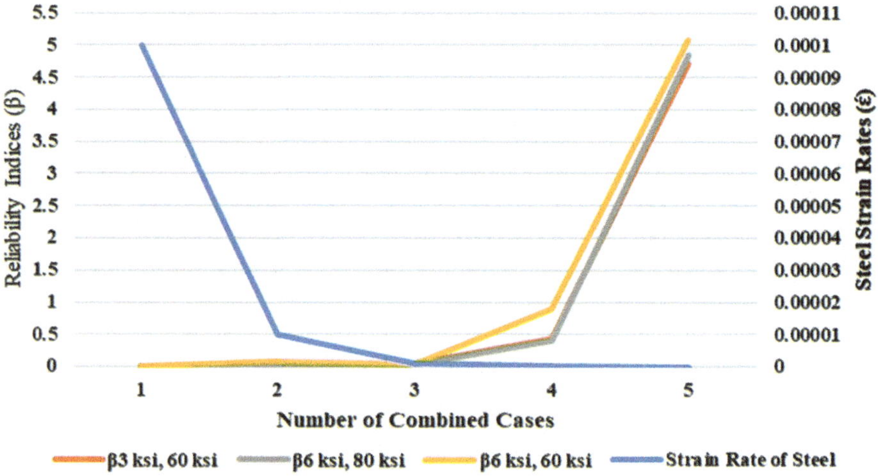

Fig. 7 Steel strain rates and combined β

numbers are generated from 'RAND' function. In Eq. 19, $\acute{\epsilon}$ is deterministic whereas β_h and E_s are random variables where mean (μ) and standard deviations (σ) are given in Table 1.

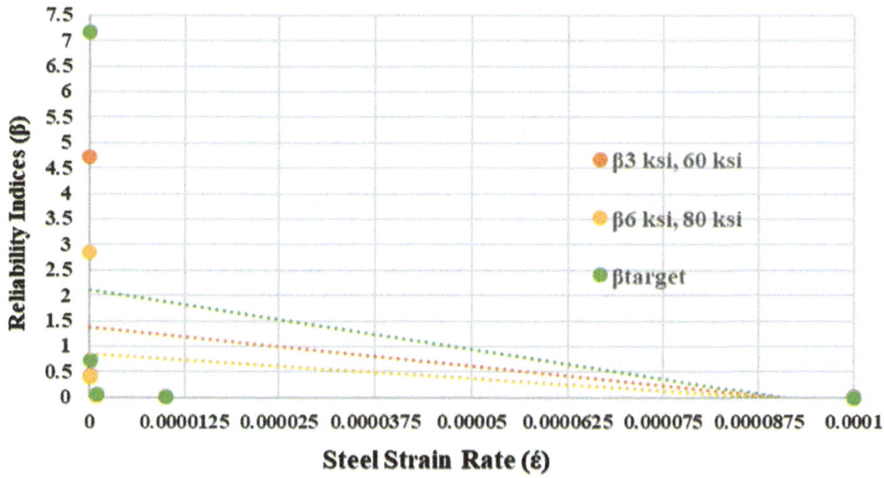

Fig. 8 Steel strain rates and β

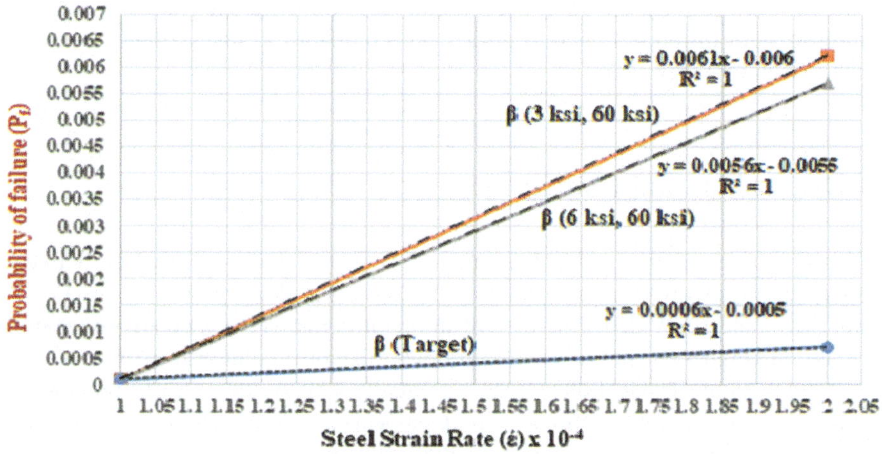

Fig. 9 Steel strain rate at 10^{-4} and β with probability of failure

2.6 *Predicting Reliability Indices for Dynamic Crack Propagation*

In this study, reliability prediction in terms of reliability indices (β) are evaluated using the Hasofer-Lind reliability index method. Test column reliability indices have been determined for 3 ksi (20.684 MPa) compressive concrete strength and 60 ksi (413.685 MPa) yield strength of steel. Additionally, reliability indices for 6 ksi (41.37 MPa) compressive concrete strength and 80 ksi (551.581 MPa) steel yield strength are also calculated. Reliability indices for both the cases are compared

Table 1 μ, V and σ for β_h and E_s

Variables	μ	V	σ
β_h (in)	1.6384	0.155	0.254
E_s (psi)	3E+07	1.1	3.4E+07

using limit state model shown in Eq. 15. Figure 10 shows the reliability indices for the both cases.

The mean (μ), variations (V), and standard deviations (σ) values for the non-deterministic parameters, β_h and E_s, are taken from a published source [8], and are shown in Table 1 [9].Target reliability indices (β_{target}) are determined using astatic condition model. Equation 20 evaluates target reliability indices (β_{target}) using the values from Table 1.

$$S = 0.016 * DIF_{con} - 0.00103 * (\dot{\epsilon}) * \beta_h * E_s * DIF_s \tag{20}$$

In Eq. 20, 'S' is the limit state equation for serviceability in terms of crack width for static condition. β_{target} has been computed using Hasofer-Lind reliability index model in limit state Eq. 20 [17]. Steel strain rates for different grades of steel are combined with different concrete grades are compared to assess their respective reliability indices (Fig. 11).

3 Discussions and Conclusions

In this research, an attempt has been undertaken to bridge a comparison between steel strain rates with the rate of crack-width and crack-depth propagations (Fig. 2) using Eqs. 9 and 12. With the increase of strain rates, rate of crack depth propagation increases at a higher rate initially and then slows to a more gradual rate. Conversely the rate of crack width propagation increases gradually with the increase of strain rates at a logarithmic increment. An inflexion point appears at a strain rate of approximately 0.000085. It can be further concluded that with the increase of strain rates, the propagation of cracks increases in a decremented way.

Monte Carlo simulation is undertaken for the limit state function (G) with the inverse of the cumulative distribution function (CDF) for the P_i, i.e., $\Phi^{-1}(P_i)$, for the strain rates at quasi-static to plastic strain rates, 10^{-7} to10^{-8} (Fig. 5d, e). Increased strain rates (10^{-7}–10^{-8}) show almost no possibility of crack propagation as crack is a function of time. On the other hand, strain rates of 10^{-4} to10^{-6} (Fig. 5a–d) provide a good understanding and an insight for predicting serviceability in terms of dynamic crack propagation.

Prediction of the reliability for the developed model (Eq. 15) has been computed using Hasofer-Lind reliability index method for different grades of concrete and steel strain rates combinations (Fig. 8). Target reliability indices (β_{target}) at different steel strain rates are determined in the static condition and compared with dynamic strain rates [18]. The dynamic reliability indices (β) for the combinations of 3 ksi (concrete) and 60 ksi (steel), and 6 ksi (concrete) and 80 ksi (steel) are computed as 1.5 and 1.1 respectively. Compared to the calculated β_{target} of 2.5 seems to be lower than the two combinations. However, calibration (β_{target} / β) for 3 ksi (concrete) and 60 ksi (steel), and 6 ksi (concrete) and 80 ksi (steel) are predicted to be 1.6 and 2.2, respectively from Fig. 8.

In Fig. 5, reliability indices for the 3 and 60 ksi combination at various strain rates seem more conservative than 6 and 60 ksi combination (Fig. 6), computed at a strain rate 10^{-4}, and much greater than that of target reliability index, showing a close proximity to the 6 ksi and 60 ksi. However, different combinations comprising models are suitable for the specific range of strain rates that steel undergoes. In addition, strain rates higher than 10^{-4} to 10^{-6}, for the combination of 6 and 60 ksi, behaves optimally well (Fig. 5a, c) whereas a combination of 6 and 80 ksi performs better at strain rates of 10^{-5} to 10^{-8} to predict the probability of failure (Fig. 5b, d).

This research is an attempt to provide an insight on recognizing the severity of damage incurred by dynamic impact. Assessment of the damaged column and its behavior in terms of severity are carried out to relate crack propagation rates at different strain rates of steel rebar, and DIF for both concrete and steel combinations, respectively. Numerical models are developed to recognize the serviceability criteria (crack propagation) for dynamic cracks using Monte Carlo simulation, which shows a well prediction of limit state equation corroborating crack propagation phenomenon.

In addition, reliability indices computed for different concrete and steel grades combinations, provide a close association with target reliability indices computed from the static condition. However, from the results, code calibrations of 1.6 and 2.2 seem to be highly recommended for dynamic condition based on the combinations of respective steel and concrete grades in order for the safe design. From the present research accomplished, a combination of at least 3 ksi grade concrete with 80 ksi steel can be well used for bridge pier to withstand the dynamic impact load. However, a combination of 6 ksi (concrete) and 60 ksi (steel) has been optimally proposed and high precision values are considered to understand reliability more scrupulously (Fig. 6), and are commended at steel strain rates of 10^{-4} to 10^{-6} to resist impact without failure.

The process outlined in this paper is limited to a specific case study of a representative pier. Further studies involving different materials and geometrical configurations are needed to enhance the use of the proposed framework as a design tool.

References

1. Shahbazpanahi, S., Hejazi, F., Paknahad, M., Rahimipour, A., & Nassimi, M. R. (2018). Modelling crack propagation in RC beam-column joints. *Teh Vjesn, 25,* 1183–1189.
2. Bao, X., & Li, B. (2010). Residual strength of blast damaged reinforced concrete columns. *International Journal of Impact Engineering, 37,* 295–308.
3. Parme, A. L. (1976). Recommendations for design of beam-column joints in monolithic reinforced concrete structures. *ACI Journal,* 375.

4. Code UB. (1997). In *International Conference of Building Officials*, Whittier CA.
5. Sheikh, M. N., & Légeron, F. (2010). Seismic performance-based design of bridges with quantitative local performance criteria. In *Proceedings, Annual Conference—Canadian Society for Civil Engineering* (Vol. 2, pp. 913–22).
6. An, X., Maekawa, K., & Okamura, H. (1997). Numerical simulation of size effect in shear strength of RC beams. *Doboku Gakkai Ronbunshu, 1997,* 297–316.
7. Choine, M. N., Kashani, M. M., Lowes, L. N., O'Connor, A., Crewe, A. J., Alexander, N. A., et al. (2016). Nonlinear dynamic analysis and seismic fragility assessment of a corrosion damaged integral bridge. *International Journal of Structural Integrity, 7,* 227.
8. Thomas, R. J., Steel, K., & Sorensen, A. D. (2018). Reliability analysis of circular reinforced concrete columns subject to sequential vehicular impact and blast loading. *Engineering Structures, 168,* 838–851.
9. Nowak, A. S., & Collins, K. R. (2012). *Reliability of structures*. CRC Press.
10. Jokūbaitis, V., & Juknevičius, L. (2013). Critical depth of normal cracks in reinforced concrete beams of rectangular cross-section. *Journal of Civil Engineering and Management, 19,* 583–590.
11. Mancuso, C., & Bartlett, F. M. (2017). ACI 318–14 criteria for computing instantaneous deflections. *ACI Structural Journal,* 114.
12. Cowper, G., & Symonds, P. (1957). Strain hardening and strain-rate effects in the impact loading of cantilever beam. *Division of Applied Mathematics at Brown University,* 1–46.
13. Allam, S. M., Shoukry, M. S., Rashad, G. E., & Hassan, A. S. (2012). Crack width evaluation for flexural RC members. *Alexandria Engineering Journal, 51,* 211–220.
14. Malvar, L. J., & Crawford, J. E. (1998). Dynamic increase factors for steel reinforcing bars. In *28th DDESB Semin*ar, Orlando, USA.
15. Malvar, L. J., & Crawford, J. E. *Dynamic increase factors for concrete*. Port Hhueneme, CA: Naval Facilities Engineering Service Center.
16. Abou-Zeid, M., Fowler, D. W., Nawy, E. G., Allen, J. H., Halvorsen, G. T., Poston, R. W., et al. (2001). Control of cracking in concrete structures. *Report, ACI Communications, 224,* 12–16.
17. Bastidas-Arteaga, E., & Soubra, A.-H. (2014). Reliability analysis methods. *ALERT Dr Sch 2014-Stochastic Anal Inverse Model,* 53–77.
18. Mestrovic, D., Cizmar, D., & Miculinic, L. (2008). Reliability of concrete columns under vehicle impact. In *10th International Conference on Shock & Impact Loads on Structures* (pp. 157–65).

Accounting for Joined Probabilities in Nation-Wide Flood Risk Profiles

Ferdinand Diermanse, Joost V. L. Beckers, Cathy Ansell, and Antoine Bavandi

Abstract A risk profile provides information about the probabilities of event impacts of varying magnitudes. In this study, a probabilistic framework is developed to derive a national-scale flood risk profile, which can be used for disaster risk management and financial planning. These applications typically require risk profiles over a wide range of return periods. For most countries, the historical record of flood impacts is limited to a few decades, insufficient to cover the longest return periods. To overcome this limitation, we developed a stochastic model that can generate arbitrarily long synthetic time series of flood events which have the same statistical characteristics as the historical time series. This includes the joint occurrence probabilities of flood events at different locations across the country. So, the probability of each pair of locations experiencing a flood event in the same event should be the same for the synthetic series as for the historic series. To this end, a novel approach based on 'simulated annealing' was implemented. Results show an almost exact reproduction of the statistical properties of the historical time series.

Keywords Joint probabilities · Risk profiles · Simulated annealing

1 Introduction

In order to increase the financial resilience of ASEAN + 3 members to climate and disaster risk, the Southeast Asia Disaster Risk Insurance Facility (SEADRIF) has been established by ASEAN + 3 in partnership with the World Bank. To support this endeavor, and to increase the financial resilience of Lao PDR, Cambodia and Myanmar against large-scale floods, the World Bank commissioned the development of tools to support a rapid response financing mechanism. Flood risk profiles for

F. Diermanse (✉) · J. V. L. Beckers
Deltares, Delft, The Netherlands
e-mail: ferdinand.diermanse@deltares.nl

C. Ansell · A. Bavandi
Disaster Risk Financing and Insurance Program, World Bank, Washington, DC, USA

© The Author(s), under exclusive license to Springer Nature Switzerland AG 2021

J. C. Matos et al. (eds.), *18th International Probabilistic Workshop*, Lecture Notes in Civil Engineering 153, https://doi.org/10.1007/978-3-030-73616-3_8

these three countries were required to inform the design of financial risk transfer instruments.

Disaster risk is often quantified in terms of "annual average affected population" or for financial purposes, "annual average losses". These are the long-term averages over many years of larger and smaller disasters as well as no-event years. However, for financial risk transfer, such as catastrophe insurance, a year loss table (YLT) or event loss table (ELT) is required due to the importance of the low frequency and high severity events. Crucially, these also provide information on the temporal and spatial variance of individual events which cannot be represented by an "average year". The historical record is often too short to act as an ELT and will not sufficiently represent the tail events, therefore a stochastic event set is required. We developed a methodology to derive a long synthetic time series of flood events to support SEADRIF countries.

2 Flood Modelling Concept

The flood modelling in this project is based on the concept of a single flood driver for a given type of flood and subarea. For example, a flood plain along a stretch of river is called a fluvial subarea. The local flood driver in this case is the river discharge. We assume the flood extent in the fluvial subarea is fully determined by the value of this flood driver. We defined four types of flooding and corresponding flood drivers: Fluvial flooding (river discharge), pluvial flooding (rainfall), tidal flooding (river water level) and coastal flooding (sea water level). The first step in the flood risk analysis is to identify and classify the subareas over the region of interest (typically a country). For each subarea, historical values for the flood driver are collected. Subsequently, the number of affected people for each flood map is calculated using the WorldPop population density grid [11].

The historical flood driver values include gauge readings from local hydrometeorological centers over the past few decades, simulated river discharges from a hydrologic model over a 35-year period (using 1979–2013 MSWEP meteorological input, see [1]), as well as storm surge levels from the Global Tide and Surge Reanalysis (GTSR) data set which also spans 35 years [9] augmented by observations and hydrodynamic simulation of historical cyclones.

3 Method for Generating Synthetic Time Series

The historical period of 35 years is sufficient for probabilistic assessment up to return periods of about 10 years, but not for the longer return periods (up to 1000 years) which are required for assessment of low frequency, high severity events. Therefore, we generate a long synthetic time series of flood events (characterized by flood driver values) that enables the analysis of higher return periods. Our methodology generates a synthetic time series which has the same statistical characteristics as the 35-year historical time series. This includes:

1. exceedance probabilities of annual maxima for each flood driver;
2. mutual correlations of annual maxima between all pairs of flood drivers;
3. probabilities of joint occurrences for all pairs of flood drivers, i.e. the probability that annual maxima of two flood drivers occur during the same event;
4. the influence of the joint occurrence on the correlation between the values of annual maxima (because if annual maxima occur during the same event, the mutual correlation is generally much higher).

A stochastic event sampling method was developed that reproduces these four statistical properties. This method consists of four components:

Component 1: deriving probability distributions of individual flood drivers;

Component 2: sampling of annual maxima of flood drivers;

Component 3: sampling of events;

Component 4: linking of annual maxima to event numbers.

These four components are detailed in the next four subsections.

3.1 *Component 1: Deriving Probability Distributions of Flood Drivers*

Extreme value distributions were derived for the various flood drivers, based on the available 35-year historical time series. For each flood driver, annual maximum values were selected, and an extreme value distribution function was fitted, applying fairly "standard" techniques such as described in Coles [2]. Figure 1 shows an example of a Gumbel fit on annual maximum discharges for Nam Khan River in Laos.

3.2 *Component 2: Sampling of Annual Maxima*

The second component of the sampling method concerns the sampling of annual maxima. In this step, the mutual correlation between the annual maxima of different flood drivers is taken into account. The correlation coefficient is derived for all flood driver pairs from the observed annual maximum values. This results in an $n * n$ covariance matrix, **C**, where n is the number of flood drivers ($n = 127$ for the study area).

To reproduce these correlations in the synthetic time series, a Gaussian Copula is applied in the sampling procedure (see e.g. [4, 8]). This method requires correlation matrix, $\boldsymbol{C}$, as input. As proven by Fang et al. [5], $\boldsymbol{C}$ should be taken equal to $\sin(\pi\tau/2)$, where τ is Kendall's rank correlation matrix. The procedure to generate correlated samples is as follows:

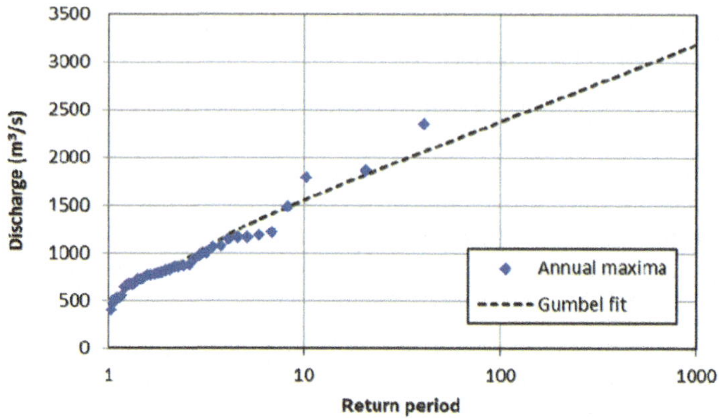

Fig. 1 Fitted Gumbel distribution function to simulated annual maximum discharges at Ban Mixay gauge station (Nam Khan River, Lao PDR)

1. Derive a matrix **P** for which: **PP′** = **C**, through Cholesky decomposition of correlation matrix **C** (see, e.g. [12]). Note: **P′** is the transpose of matrix **P**.
2. Sample values $u_1, \ldots ,u_n$ from the standard normal distribution function; store the results in a 1xn vector $\boldsymbol{u}$.
3. Compute: $\boldsymbol{u}^* = \boldsymbol{uP'}$.

The u*-values are subsequently transformed to "real-world" values of flood drivers, using the probability distribution functions of the individual flood drivers as derived in component 1:

$$\Phi(u_i*) = F_i(x_i) \tag{1}$$

Where Φ is the standard normal distribution function, u_i* is the sampled u*-value of the ith flood driver, x_i is the "real-world" realisation of the ith flood driver and F_i is the derived extreme value distribution function of x_i.

3.3 Component 3: Sampling of Events

The sampling method for annual maxima (AM) in the previous section creates a synthetic time series with correlated annual maxima for each flood driver. The correlation between AM refers to their value, not to their timing. Within a single year, the annual maxima of the flood drivers are not expected to all occur during the same event. Typically, there are several events per year and the AM are distributed over them. Since we are interested in *event* impacts, the relative timing of the maxima within a year also needs to be part of the sampling method. The generated AM are thus assigned to events and the number of annual maxima per event should be in accordance with the historical series. More specifically: the probability of each pair of flood drivers having an AM during the same event should be the same for the synthetic series as for the historical series. In the case study, there are 127 flood drivers, which means there are 127 * 126/2 = 8001 joined occurrence probabilities that need to be reproduced in the sampling procedure. For this challenging objective we developed a novel approach based on simulated annealing (Kirckpatrick et al., 1983).

3.3.1 Outline of the Method

The starting point of the analysis is a time series z_t, consisting of event numbers as derived from historical data. z_t is an $Y * n$ matrix, with Y being the number of years and n the number of flood drivers. If z_t (y,j) = 3, this means flood driver j had its AM in the 3rd "biggest" event of year y (note: in each year, events are ordered based on the number of flood drivers that had their AM occurring during the event). As an illustration, Fig. 2 shows the five biggest events in 1979 and 1980 and the flood drivers that had their AM during one of these events. The objective of the stochastic simulation method is to create a (lengthy) synthetic time series zs_t with similar characteristics as the historical time series z_t. Here, similarity refers to [1] the probability that AM of any two flood drivers, L_1 and L_2, occur during the same event and [2] the probability distribution of the number of flood drivers having their AM occur in the biggest event in the year.

To this end, a "cost" function, $G(zs_t, z_t)$, is defined that penalizes differences between the historical and synthetic series. G is formulated in such a way that it decreases if z_t and zs_t are in better agreement. Thus, function G needs to be minimized to obtain the best agreement between the historical and synthetic series. The choice of function G is critical to the performance of the procedure, both in terms of computation time and the quality of the end result. Both z_t and zs_t are matrices with the same number of columns, were each column represents a flood driver, but different numbers of rows, where each row represents a year. Matrix z_t has 35 rows, corresponding to the 35 years of observation, whereas zs_t has a user-defined number of rows (e.g. 10,000). Both z_t and zs_t contain event numbers, were 1 corresponds to the biggest event.

The event sampling procedure is based on the method of 'simulated annealing' (Kirckpatrick et al., 1983). Figure 3 shows the basic algorithm. The procedure starts with a randomly selected initial synthetic time series for which the cost function is evaluated. The elements of this synthetic time series are subsequently permutated in a (large) number of iteration steps, until a stop criterion is reached. In each iteration step, two elements of the time series are permutated to create a newly 'proposed' time series. For the simulated annealing procedure, we adopted the Matlab implementation of Joachim Vandekerckhove[1] and adapted it for our specific application. The algorithm is as follows:

[1] https://nl.mathworks.com/matlabcentral/fileexchange/10548-general-simulated-annealing-algorithm

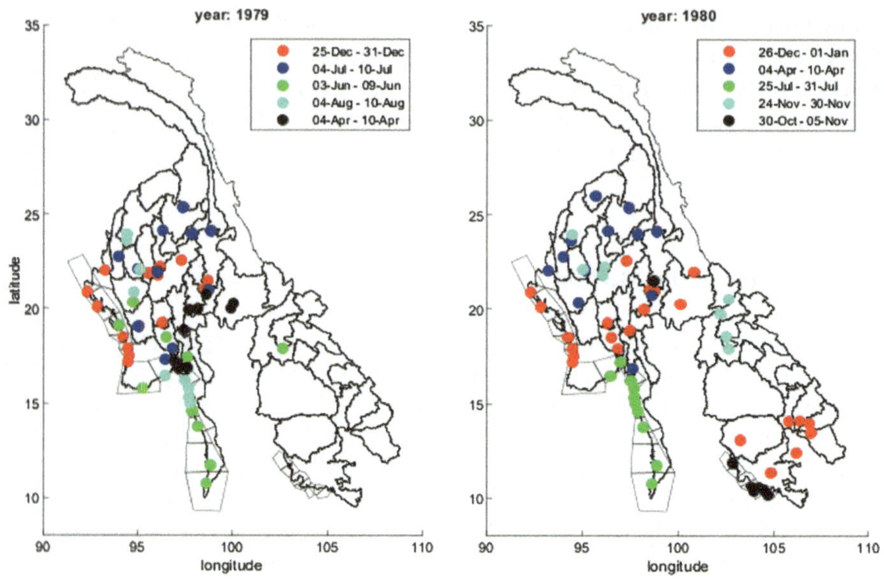

Fig. 2 Example of joint occurrences in the years 1979 and 1980. Clusters of flood drivers that experience the annual maximum within the same event have the same colour.

[1] start with an initial time series zs_t (0)

[2] evaluate cost function $G(zs_t, z_t)$

[3] start with an initial 'temperature' $T = T_0$

[4] Select an end temperature $T_e < T_0$ as stop criterion

[5] while $T > T_e$ do: (<u>*stop criterion 1*</u>)

[6] while *stop criterion 2* is not met do:

[7] Select a new time series zs_t* by randomly changing swapping 2 elements of of zs_t

[8] evaluate cost function $G(zs_t *, z_t)$

[9] accept zs_t* as the new solution, i.e.: $zs_t = zs_t$ *, with probability *p(T)* and reject it with probability 1-*p(T)*

[10] decrease *T*: $T = c*T$; with *c* a constant <1

This algorithm explores the $Y * n$-dimensional space of all possible outcomes, where *Y* is the number of years of the synthetic series and *n* is the number of flood drivers. The probability, $p(T)$, of accepting the proposal solution is decreasing to near-zero at the end of the procedure. Therefore, in the later phases of the procedure, the procedure has an increasing tendency to move into the direction of lower values of the cost function and to end up in a minimum. To prevent the procedure from ending up too early in a local minimum, a 'temperature' *T* is introduced, which allows for the solution to move to a higher value of the cost function. In the beginning of the procedure, the temperature is high, thereby increasing the probability of moving away

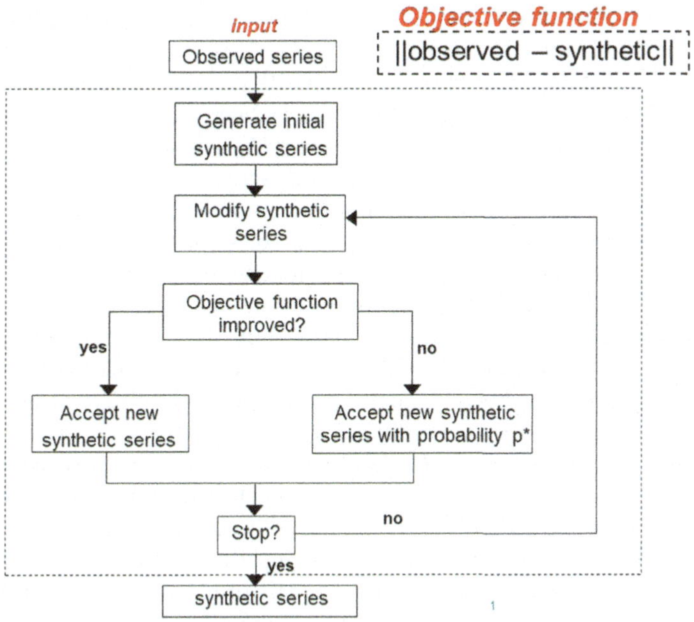

Fig. 3 Schematic view of the basic principle of simulated annealing

from a (local) minimum. The temperature is then slowly reduced, and the solution is forced to move towards a (local) minimum of the cost function. The procedure can be repeated several times with a different random seed to check if the same minimum is found each and every time. This is no guarantee for a global minimum, but it does provide more confidence.

3.3.2 Details of the Annealing Method

In this section, we describe some further details of the simulated annealing method step by step. The number below refer to the steps mentioned in Sect. 3.3.1.

[1] First, we have to define the number of years of the synthetic time series (Y) and the maximum number of events per year (N_e). The time series zs_t consists of event numbers 1 to $N_e + 1$. For example: if zs_t(y,j) = 5, this means flood driver j had its annual maximum in the 5th event of year y. Event number $N_e + 1$ represents the 'non-event', which means if zs_t(y,j) = $N_e + 1$, the annual maximum of flood driver j in year y is assumed to have taken place in isolation. In the initialization of zs_t, an integer number is randomly sampled from the range $[1, N_e + 1]$ for each combination of year y and flood driver j.

[2] The cost function $G(zs_t,zt)$ was chosen to be the sum of three functions G_1 - G_3. To compute these three functions, some pre-processing is required. First of all, the probability that annual maxima of any two flood drivers, L_1 and L_2, occur during the same event is estimated from the historic series. So, for example, if flood drivers L_1 and L_2, had their annual maxima occur in the same event in 10 out of 35 years, the estimated probability of joint occurrence is 10/35 ≈ 0.29. This is done for each pair of flood drivers, resulting in $M = n * (n - 1)/2$ percentages (n = the number of flood drivers), which are stored in an $M * 1$ vector H_o. The same computation is done for the current solution (zs_t) of the synthetic time series, and the results are stored in an $M * 1$ vector H_s. Subsequently, the absolute differences between H_o and H_s are computed and stored in an $M * 1$ vector D. Function G_1 is the maximum value of D. Function G_2 is the mean value of D. To compute function G_3, the maximum number of flood drivers with an annual maximum occurring in a single event is determined per year and stored in an Y * 1 vector. Subsequently, the mean over all the years is computed. In other words: the mean number of flood drivers that have their annual maximum occurring in the biggest event. This is done for the observed and synthetic series. Function G_3 quantifies the differences between the two.

[3 + 4 + 5] The starting value T_0 should be chosen in such a way that accepting a new time series with a higher ("worse") cost function than the current time series should be relatively large, whereas T_{end} should be chosen in such a way that accepting a new time series with a higher ("worse") cost function than the current time series should be close to 0. T_{end} should be several orders of magnitude smaller than T_0 to provide the method with a sufficient number of iterations to converge to a "good" result. The best choice of T_0 and T_{end} requires some insights in the cost function loss function $G(zs_t$,zt) and the speed with which it converges to the (local) minimum. In our case T_0 was set equal to 1 and T_{end} equal to 1E−8.

[6] Stop criterion 2 controls the number of iterations for a single temperature T. An obvious criterion is to set a maximum allowed number of iterations. Additional criteria can be to stop after a user-defined number of accepted new solutions and/or stop after a long successive series of rejected solutions. All three criteria have been implemented in the Matlab implementation of Joachim Vandekerckhove that was used as the basis of our method.

[7] A key step in the procedure is the selection of a new proposal time series. A straightforward method is to randomly select a specific flood driver in a specific year and to randomly generate a new event number for this flood driver. However, this approach led to very slow convergence of the procedure. To speed up the procedure, we implemented an alternative method in which we look for the combination of two flood drivers L_1 and L_2 that contribute most to the outcome of function $G_1(zs_t, z_t)$. In other words, the two flood drivers L_1 and L_2 for which the difference in computed joint occurrence probability between the observed series z_t and the synthetic series zs_t is the largest. We then change the event number of one of the flood drivers in such a way that the objective function is decreased. Note, however, that this approach is slightly in contrast with the concept of simulated annealing in which increases in the objective function in successive iterations should also be allowed to prevent it from converging too soon to a local minimum. Therefore, the final strategy was a mixture of both: with a probability p^* we apply the first method (random selection of flood driver) and with a probability 1 − p* we apply the second method (selection of flood driver that contributes most to the cost function). We found that a value of $p^* = 0.8$ in general gave most satisfactory results.

[8] The probability, p, of accepting a newly proposed solution, zs_t*, is set equal to:

$$p = \min\left(1, e^{-\frac{\Delta G}{kT}}\right); \ \Delta G = G\left(zs_t^*, z_t\right) - G(zs_t, z_t) \quad (2)$$

This means if $G(zs_t^*, z_t) \leq G(zs_t, z_t)$, the new solution is accepted with probability 1, whereas if $G(zs_t^*, z_t) > G(zs_t, z_t)$, the acceptance probability depends on the difference between $G(zs_t^*, z_t)$ and $G(zs_t, z_t)$, on the temperature T and constant k. To make this function generically applicable, cost function G is normalised by dividing it by G_0, i.e. by the value of the cost function in the first iteration. Constant k should preferably be inversely proportional to the total number of elements, N (N is equal to the number of years times the number of flood drivers in our case). We chose $k = 10/N$.

[9] The decrease in the value of T is taken care of as follows: $T = c * T$; with c a constant < 1. This means the temperature declines exponentially. We adopted a value of $c = 0.8$.

3.4 Component 4: Linking Annual Maxima to Event Numbers

The sampling procedures of component 2 and 3 are carried out independently from each other. That means the correlation between annual maxima of two flood drivers is not influenced by the fact whether these two maxima are observed during the same event. In reality, however, there is a relation between the two, as occurrence during the same event means there may be a common cause that is also likely to affect the magnitude of the annual maximum of both flood drivers. This is confirmed by an analysis of the data of the historical 35-year series. In the analysis we computed the correlation between rank numbers[2] of annual maxima for [A] all annual maxima occurring in the same year and [B] all annual maxima occurring in the same event. The table below shows the difference between the two is significant. Ignoring the relation between event numbers and (correlations between) annual maxima means the 'within-event-correlation' will be equal to the numbers shown under [A], whereas it should be equal to the numbers shown under [B].

Country	# Flood drivers	[A] Within year correlation	[B] Within event correlation
Cambodia	21	0.20	0.66
Lao PDR	22	0.24	0.35

(continued)

[2]Rank numbers are numbers from 1..35 indicating per flood driver the highest (1), second highest (2).. Lowest (35) annual maximum in the series of 35 years.

(continued)

Country	# Flood drivers	[A] Within year correlation	[B] Within event correlation
Myanmar	84	0.11	0.40
Combined	127	0.06	0.29

In order to take the relation between event numbers and annual maxima into account an additional simulated annealing procedure was implemented. In each iteration, a flood driver is randomly selected and the sampled annual maxima of this flood driver for two randomly selected years (as provided by component 2) are exchanged. This means the event numbers of these two annual maxima have been exchanged. This has an impact on the overall computed 'within event correlation'. A cost function is defined that quantifies the difference between the observed and synthetic 'within event correlation'. The iterations are carried out until the computed within event correlation is the same as the corresponding number of the observed data (see table above). It turned out that the method was capable of exactly reproducing these numbers.

4 Results

To test the applicability of the procedure, we verify if the relevant statistical properties of the historical times series are reproduced by the synthetic time series. Figure 4 shows the joint occurrence probabilities of annual maxima during an event for all flood driver pairs as derived from the historical series (vertical axis) and from a generated 10,000-year synthetic series (horizontal axis). The plot on the left shows results for Cambodia (21 flood drivers), the plot on the right shows results for the three countries combined (127 flood drivers). The figures show that the joint occurrence probabilities in the synthetic time series are in very good agreement with those in the historical time series.

The quantile plots in Fig. 5 compare the probability distributions of the number of flood drivers that had their annual maximum during the "biggest event" in the year, as derived from the historical series (horizontal axis) and from the 10,000-year synthetic series (vertical axis). The blue dots are all close to the line $y = x$. This shows the probability distribution of the number of locations in the largest event is very well captured in the synthetic series.

Figure 6 shows frequency curves of affected population that were derived from the synthetic series (red) and from the historical series (blue dots). The numbers were normalized by the 100-year return value as the actual numbers are not eligible for publication. The plot on the right is a zoomed version of the plot on the left. It shows that the derived frequency curves are well in accordance with the historical numbers, which is an essential validation of the method. The added value of the synthetic method is that it provides return values for much larger return periods, as can be seen from the left plot.

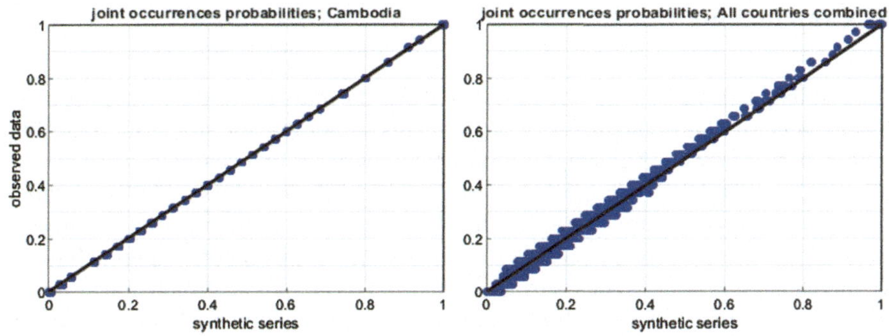

Fig. 4 Joint occurrence probabilities for all pairs of flood drivers in Cambodia (left) and all three countries combined (right). Synthetic time series (x-axis) compared to historical time series (y-axis)

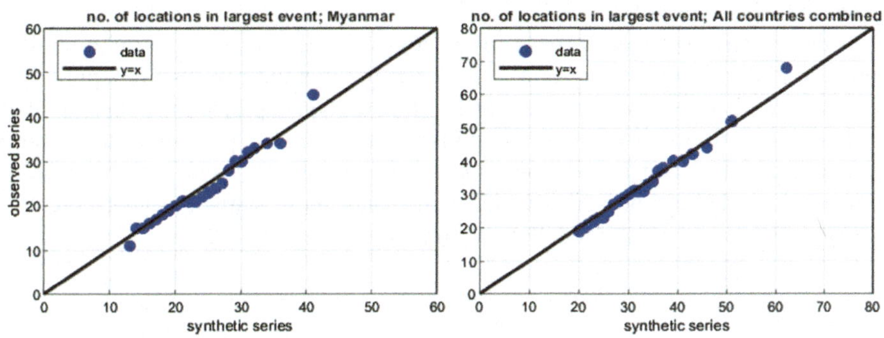

Fig. 5 Quantile plot for the number of flood drivers having their annual maximum during the largest event in Myanmar (left) and all three countries combined (right)

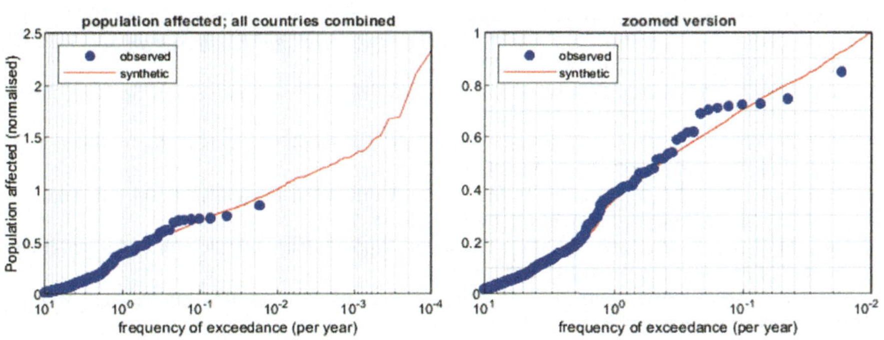

Fig. 6 Frequency curves (normalized) of population affected – empirical (blue) versus probabilistic (red). The plot on the right is a zoomed version of the plot on the left.

5 Conclusions

The results prove that the synthetic series of flood events have statistical properties that are very similar to the historical series. This shows that the stochastic sampling method performs well. The lengthy synthetic time series that can be generated with the stochastic model offers opportunities to provide an event loss table and detailed risk profile for various applications. The challenge of reproducing joint occurrence probabilities of ~8000 pairs of flood drivers was tackled by a novel approach based on simulated annealing (Kirckpatrick et al., 1983). One of the attractive features of this method is that multiple objective functions can be optimised simultaneously. This enabled the reproduction of several relevant statistical features of the historical time series in the synthetic time series. In this study, we have focused on population affected by flood events, but the methodology can easily be generalized to economic losses and other types of disasters.

In this paper, the objective was to generate a synthic time series with similar statistics as the historic time series. However, the method can also be applied to create synthetic time series that account for climate change projections. It is possible to choose/design virtually any set of statistics (for example perturbing the annual maxima frequency and correlations due to climate change) and to subsequently generate a synthetic time series which will match these statistics. That potential is very valuable to the risk modelling community.

References

1. Beck, H. E., van Dijk, A. I. J. M., Levizzani, V., Schellekens, J., Miralles, D. G., Martens, B., & de Roo, A. (2017). MSWEP: 3-hourly 0.25° global gridded precipitation (1979–2015) by merging gauge, satellite, and reanalysis data. *Hydrology and Earth System Sciences, 21*, 589–615. https://doi.org/10.5194/hess-21-589-2017
2. Coles, S. (2001). *An introduction to statistical modeling of extreme values*. Springer. ISBN 1-85233–459-2.
3. Deltares. (2018). Southeast Asia flood monitoring and risk assessment for regional DRF mechanism. Component 2 report. March 2018.
4. Diermanse, F. L. M., Geerse, C. P. M. (2012). Correlation models in flood risk analysis, reliability engineering and system safety (RESS). 64–72.
5. Fang, H., Fang, K., & Kotz, S. (2002) The meta-elliptical distributions with given marginals. *Journal of Multivariate Analysis, 82*(1), 1–16.
6. Holland, G. (1980). An analytical model of the wind and pressure profiles in hurricanes. *Monthly Weather Review, 108*, 1212–1218 (3, 11, 40)
7. Holland, G. (2008). A revised hurricane pressure–wind model. *Monthly Weather Review 136*, 3432–3445 (v, 17, 18)
8. Kaiser, H. F., & Dickman, K. (1962). Sample and population score matrices and sample correlation matrices from an arbitrary population correlation matrix. *Psychometrika, 27*(2), 179.
9. Muis, S., Verlaan, M., Winsemius, H. C., Aerts, J. C., & Ward, P. J. (2016). A global reanalysis of storm surges and extreme sea levels. *Nature Communications, 7*.
10. Schwarz, G. E. (1978). Estimating the dimension of a model. *Annals of statistics, 6*(2), 461–464, https://doi.org/10.1214/aos/1176344136.

11. Stevens, F. R., Gaughan, A. E., Linard, C., & Tatem, A. J. (2015). Disaggregating census data for population mapping using random forests with remotely-sensed and ancillary data. *PLoS ONE, 10,* e0107042.
12. Strang, G., (1982). *Linear algebra and its applications*, San Diego: Harcourt, Brace, Jovanovich, Publishers. ISBN 0-15-551005-3

An Adaptive Subset Simulation Algorithm for System Reliability Analysis with Discontinuous Limit States

Jianpeng Chan, Iason Papaioannou, and Daniel Straub

Abstract Efficient computational methods for system reliability assessment are of importance in many contexts, where crude Monte Carlo simulation is inefficient or infeasible. These methods include a variety of importance sampling techniques as well as subset simulation. Most of these methods function in an adaptive manner, whereby the sampling density gradually approaches the failure domain. The adaptation can work well when the limit state function describing system performance is continuous. However, many system reliability problems involve limit state functions that are non-continuous over the input sample space. Such situations occur in both connectivity- and flow-based problems, due to the binary or multi-state random variables entering the definition of the system performance or the discontinuous nature of the performance function. When solving this kind of problem, the standard subset simulation algorithm with fixed intermediate conditional probability and fixed number of samples per level can lead to significant errors, since the discontinuity of the output can result in an ambiguous definition of the sought percentile of the samples and, hence, of the intermediate domains. In this paper, we propose an adaptive subset simulation algorithm to determine the reliability of systems with discontinuous limit state functions. The proposed algorithm chooses the number of samples and the conditional probability adaptively. Numerical examples are provided to demonstrate the accuracy and efficiency of the proposed algorithm.

Keywords Subset simulation · System reliability analysis · Discontinuous limit state

1 Introduction

Infrastructure networks, such as power grids and water supply systems, deliver essential services to society. Failures of such networks can have severe consequences. Quantification of the probability of survival or, conversely, the probability of failure

J. Chan (✉) · I. Papaioannou · D. Straub
Engineering Risk Analysis Group, Technische Universität München, München, Germany
e-mail: jianpeng.chan@tum.de

© The Author(s), under exclusive license to Springer Nature Switzerland AG 2021

J. C. Matos et al. (eds.), *18th International Probabilistic Workshop*, Lecture Notes in Civil Engineering 153, https://doi.org/10.1007/978-3-030-73616-3_9

of such systems is essential in understanding and managing their reliability; this is the main purpose of network system reliability assessment.

The performance of the system can be assessed by the limit state function (LSF), also known as performance function, $g(\boldsymbol{X})$. $\boldsymbol{X}$ is an n-dimensional vector of random variables with joint cumulative distribution function (CDF) F_X and represents the uncertainty in the model. By convention, failure of the system occurs for all system states $\boldsymbol{x}$ for which $g(\boldsymbol{x}) \leq 0$. That is, $g(\boldsymbol{x})$ represents the 'distance' between the system state $\boldsymbol{x}$ and the failure surface, and hence can be regarded as the safety margin of state $\boldsymbol{x}$. The probability of failure of the system is defined as

$$\mathrm{P}_f \triangleq \mathrm{P}(g(\boldsymbol{X}) \leq 0) = \int\limits_{g(\boldsymbol{x}) \leq 0} \mathrm{d}F_X(\boldsymbol{x}). \tag{1}$$

Unlike in structural system reliability analysis, the vector of basic random variables $\boldsymbol{X}$ entering the definition of the LSF of network systems usually contains discrete random variables, which results in a discontinuous LSF. This is due to the fact that the performance of the network is often calculated through a function of a large number of binary or multi-state components. Moreover, real-word infrastructure networks are often designed to be highly reliable. This leads to high-dimensional reliability assessment problems with small failure probabilities [1].

Many methods have been proposed for evaluating system reliability, among which sampling based methods such as Monte Carlo simulation (MCS) and its variants feature prominently. For rare events simulation, crude MCS is inefficient and often infeasible when the LSF is expensive to compute. This is because the coefficient of variation of the crude Monte Carlo estimate is $\sqrt{(1 - \mathrm{P}_f)/N\mathrm{P}_f}$, with N denoting the sample size, and as a result, for small P_f the required sample size for an accurate estimate is very large. Therefore, advanced sampling techniques have been developed that decrease the required number of LSF evaluations for obtaining an accurate probability estimate. These techniques include a variety of importance sampling methods [2, 3] as well as subset simulation [4]. They mostly function in an adaptive manner, whereby the sampling gradually approaches the failure domain. The basic idea of importance sampling is to sample from a proposal distribution under which the rare event is more likely to happen and to correct the resulting bias in the estimate by multiplying each sample contribution in the estimator with the appropriate likelihood ratio [5]. In contrast, subset simulation expresses the probability of failure as a product of larger conditional probabilities of a set of intermediate nested events. This requires sampling conditional on the intermediate events, which is performed with Markov Chain Monte Carlo (MCMC) methods [6]. Subset simulation can be viewed as a generalization of the splitting method for static rare event simulation [7].

In the standard subset simulation algorithm [4], the intermediate failure events are chosen adaptively, so that the estimates of the conditional probabilities equal a predefined value p_0. This is achieved through generating a fixed number of samples in each conditional level, sorting the samples according to their LSF values and determining the p_0-percentile of the samples, which is set as the threshold defining the next failure event.

When solving network reliability problems, the discontinuous nature of the LSF can result in a large number of samples in a certain conditional level having the same LSF values. In such cases, the standard fixed effort subset simulation method will result in an ambiguous definition of the intermediate domains. In extreme conditions, all samples generated in a certain level might have the same LSF value, in which case the sample process can get stuck and might not reach the failure domain.

To address this issue, we introduce a novel variant of subset simulation, which chooses the number of samples per level and the respective conditional probability adaptively to ensure that an adequate number of samples fall in the subsequent intermediate domain. The performance of the method is illustrated by two numerical examples, a one-dimensional multi-state problem and a benchmark transmission power network system.

2 Brief Introduction of Standard Subset Simulation

2.1 *Brief Introduction of Subset Simulation*

The basic idea of subset simulation (or generalized splitting) is to express the rare failure event F as the intersection of a series of nested intermediate events $F_1 \supset F_2 \supset \cdots \supset F_m = F$. The failure probability is then written as

$$\mathrm{P}(F) = \prod_{i=1}^{m} \mathrm{P}(F_i | F_{i-1}) \tag{2}$$

where F_0 is the certain event. Ideally, the intermediate events are selected such that each conditional probability is appropriately large. In this way, the original problem of estimating a small probability is transformed to a sequence of m intermediate problems of evaluating larger conditional probabilities.

The estimation of each conditional probability $\mathrm{P}(F_i|F_{i-1})$ requires sampling from the distribution of the random variables conditional on F_{i-1}, denoted $Q(\boldsymbol{x}|F_{i-1})$, where $Q(\boldsymbol{x}|F_0) = F_X(\boldsymbol{x})$. $Q(\boldsymbol{x}|F_0)$ can be sampled by standard Monte Carlo sampling, but the distributions $Q(\cdot|F_i), i > 0$, are only known point-wise up to a normalizing constant and, hence, cannot be sampled directly. Therefore MCMC sampling is employed as an alternative. The sampling process in the jth sampling level is performed as follows: (1) Select the samples $\mathcal{P}^{(j-1)}$ from the $(j-1)$-th level that fall in F_j as the seeds $S^{(j)}$. ($\mathcal{P}^{(0)}$ are generated through Monte Carlo sampling) (2) From each seed, start a Markov chain that has the target distribution $Q(\cdot|F_j)$ as the stationary distribution, and record all the states as new samples $\mathcal{P}^{(j)}$. (3) Take the samples $\mathcal{P}^{(j)}$ located in F_{j+1} as new seeds $S^{(j+1)}$ and estimate $\mathrm{P}(F_{j+1}|F_j)$. The above three steps are repeated successively until F is approached. We note that the number of the samples per level or the sampling effort card($\mathcal{P}^{(j)}$) is usually fixed prior to the analysis.

Defining the intermediate events a priori is typically challenging. Hence, in standard subset simulation the intermediate failure events are chosen adaptively during the simulation such that each conditional probability equals a predefined constant p_0. This standard subset simulation approach is also termed (fixed effort) adaptive multilevel splitting [7]. In this variant, step (3) in the above sampling process is modified as follows: Order the samples $\mathcal{P}^{(j)}$ by their safety margins. The first p_0-percent of these sorted samples are then taken as seeds for the next sampling level and the safety margin of the p_0-percentile $g(\boldsymbol{x}_0^{(j)})$ is used to define the boundary of the intermediate domain, such that $F_{j+1} = \left\{\boldsymbol{x} : g(\boldsymbol{x}) \leq g\left(\boldsymbol{x}_0^{(j)}\right)\right\}$.

Various MCMC algorithms are proposed for constructing the Markov chains for subset simulation. These include the modified (or component-wise) Metropolis-Hasting [4] and the conditional sampling (CS) [6] algorithms, both developed to tackle high-dimensional problems. In this paper, we adopt the adaptive conditional sampling (aCS) as the MCMC algorithm [6]. This method is remarkably simple since it no longer involves the explicit choice of a proposal distribution [8]. Instead it adaptively tunes the correlation between candidate and current samples to achieve a near-optimal acceptance probability [6]. aCS is proposed for sampling in standard normal space, hence, it is necessary to transform the original sample space of $\boldsymbol{X}$ and define the reliability problem in standard normal space. This can be achieved by the Rosenblatt transformation [9]. We discuss this transformation in the next section, focusing on its implementation for discrete original sample spaces, which is particularly relevant for network reliability assessment. However, the proposed adaptive effort subset simulation algorithm can be implemented with any MCMC algorithm, including those that work in the original sample space.

2.2 Implementation in Standard Normal Space

Let $\boldsymbol{U}$ denote an n-dimensional random vector that has the independent standard normal distribution. One can define the reliability problem in the $\boldsymbol{U}$-space through an isoprobabilistic mapping $\boldsymbol{T} : \mathbb{R}^n \to \mathbb{R}^n$ such that

$$\mathrm{P}_f = \mathrm{P}(g(\boldsymbol{X}) \leq 0) = \mathrm{P}(G(\boldsymbol{U}) \leq 0) = \int_{G(\boldsymbol{u})\leq 0} \varphi_n(\boldsymbol{u})\mathrm{d}\boldsymbol{u} \quad (3)$$

where $G(\boldsymbol{U}) = g(\boldsymbol{T}(\boldsymbol{U}))$ and $\varphi_n(\boldsymbol{u})$ is the independent standard normal joint probability density function (PDF). The mapping $\boldsymbol{T}(\cdot)$ can be obtained by the Rosenblatt transformation, which is implemented as follows

$$\begin{aligned} x_1 &= F_{X_1}^{-1}(\Phi(u_1)) \\ x_2 &= F_{X_2|x_1}^{-1}(\Phi(u_2)) \\ &\vdots \\ x_m &= F_{X_m|x_1,\cdots,x_{m-1}}^{-1}(\Phi(u_m)) \end{aligned} \tag{4}$$

where Φrepresents the CDF of standard normal distribution and $F_{X_i|x_1,\cdots,x_{i-1}}(\cdot)$ denotes the conditional CDF of X_i given $X_1 = x_1, \cdots, X_{i-1} = x_{i-1}$. If any subset of $\boldsymbol{X}$ consists of discrete random variables, then it is possible that the functions $F_{X_i|x_1,\cdots,x_{i-1}}(\cdot)$ are not strictly invertible. Therefore, we use the following extended definition of the inverse of a CDF

$$F^{-1}(a) = \inf(x : F(x) \geq a) \tag{5}$$

We note that in such cases the Rosenblatt transformation is not one-to-one and, hence, the inverse mapping from $\boldsymbol{X}$ to $\boldsymbol{U}$ is not uniquely defined.

2.3 Statistics of the Subset Simulation Estimator

Assume that the intermediate events are defined prior to the simulation. In the Monte Carlo level, samples $\mathcal{P}^{(0)}$ are generated from $Q(\cdot|F_0)$ independently, and therefore the corresponding seeds $\mathcal{S}^{(1)}$ follow distribution $Q(\cdot|F_1)$. This will lead to so called perfect sampling when simulating the Markov chains in the next level. Since the chains have already reached the stationary state at the beginning, no burn-in period is needed, and all the samples $\mathcal{P}^{(1)}$ will follow $Q(\cdot|F_1)$. In this way, all samples $\mathcal{P}^{(j)}$ generated in any j-th level will follow the target distribution $Q(\cdot|F_j)$ and the corresponding estimator of the conditional probability $\widehat{\mathrm{P}}(F_{j+1}|F_j)$ will be unbiased. Moreover, [7] proves that the resulting failure probability estimator $\widehat{\mathrm{P}}_f(F)$ is also unbiased if both intermediate events and length of the Markov chain are predefined, i.e. if they are independent of the simulation process.

Since the intermediate events are usually selected adaptively and as a result, samples $\mathcal{P}^{(j)}$ will not completely follow the target distribution. Both conditional probability estimator and failure probability estimator will be slightly biased. Nevertheless, compared to the variance of the estimator, the squared bias is one order of magnitude smaller [4] and, hence, its contribution to the mean-square error (MSE) of the estimator is negligible, since the latter can be decomposed as

$$\mathrm{MSE}\left(\widehat{\mathrm{P}}_f\right) = \mathrm{Var}\left(\widehat{\mathrm{P}}_f\right) + \left(\mathrm{P_f} - \mathbb{E}\left(\widehat{\mathrm{P}}_f\right)\right)^2 \tag{6}$$

In other words, the error of the subset simulation is mainly due to the variance of the failure probability estimator rather than the bias. The most common and reliable way to calculate the variance $\mathrm{Var}(\widehat{\mathrm{P}}_f)$ is to run subset simulation several times and to use the sample variance as the unbiased estimation of the $\text{Var}(\widehat{\text{P}}_f)$. One can also evaluate the variance approximately through a single run of the subset simulation. More details can be found in [4, 6].

3 Adaptive Effort Subset Simulation Method

In each conditional level of the subset simulation method with fixed number of samples N and adaptive estimation of the intermediate events, the p_0-percentile of the safety margins of the samples $\mathcal{P}^{(j)}$, $g(\boldsymbol{x}_0^{(j)})$ is used to define the boundary of the intermediate domain. This works well when only a few samples are located on the boundary $g(\boldsymbol{x}) = g(\boldsymbol{x}_0^{(j)})$, i.e. a few samples have the same LSF value as the p_0-percentile. However, it may happen that many samples fall on this boundary, particularly in either of the following cases:

1. 1$\boldsymbol{X}$ includes discrete random variables.
2. The LSF is defined such that the probability measure of the set $\{\boldsymbol{x} : g(\boldsymbol{x}) = g(\boldsymbol{x}_0^{(j)})\}$ is positive.
   The parameters of the MCMC algorithm are inappropriately set, resulting in the candidates being rejected successively many times.

While case (3) can be avoided by an appropriate implementation of the algorithm, cases (1) and (2) are common in the context of network reliability assessment. This will result in an ambiguous definition of the intermediate domain F_{j+1} and can lead to an inaccurate estimate of the failure probability. In extreme situations, all samples generated in a certain level will have the same LSF value and the sample process can get stuck and never reach the failure domain.

To circumvent this problem and provide a clear (unambiguous) definition of the intermediate domains, we propose to discard all the samples on the boundary $g(\cdot) = g(\boldsymbol{x}_0^{(j)})$, and redefine the intermediate event F_{j+1} as $F_{j+1} = \left\{\boldsymbol{x} : g(\boldsymbol{x}) < g\left(\boldsymbol{x}_0^{(j)}\right)\right\}$. Then, we calculate the number of samples that fall in the domain F_{j+1} (number of the seeds). If this number is smaller than a predefined constant, we increase the sampling effort and append $\mathcal{P}^{(j)}$ with new samples. With a fixed F_{j+1} and increasing number of samples, the number of the seeds will keep increasing until the desired threshold is achieved. By doing this, for any state $\boldsymbol{x}$ in F_{j+1}, there exists $g(\boldsymbol{x}) < g(\boldsymbol{x}_0^{(j)}) < g(\boldsymbol{x}_0^{(j-1)})$, and thus $F_{j+1} \subset F_j$ is always true, which avoids a degeneracy of the sampling process. Even in the extreme case where all the samples in $\mathcal{P}^{(j)}$ have the same safety margin, the sampling process will keep moving forward towards the failure domain and will no longer get stuck in this level as in the standard subset simulation algorithm. Unlike standard subset simulation, the number of samples per level (sampling effort) is adapted throughout the simulation. We, hence, term the proposed approach *adaptive effort subset simulation method*.

In the following, we discuss the proposed adaptive effort subset simulation algorithm for implementation in standard normal space. The samples in each intermediate level are generated with the aCS algorithm. In addition to the initial number of samples per level N_0 and conditional probability p_0, the algorithm requires the choice of the parameter $tol \in (0, 1)$ that defines the minimum number of seeds through $tol \cdot N_0 \cdot p_0$. We have found that $tol \in (0.5, 0.8)$ is a good choice.

Adaptive effort subset simulation algorithm

$tol \in (0,1)$: Parameter which limits the minimum number of seeds inside each intermediate domain.

$p_0 \in (0,1)$: Parameter which represents the initial conditional probability when starting the iteration.

N_0: Parameter which represents initial number of samples when starting the iteration.

Set $= 0, b^{(0)} = \infty$.

while $b^{(l)} > 0$ **do**

Set $iter = 1$, $N^{(l,iter)} = N_0$ and $N_s^{(l,iter)} = 0$.

while $N_s^{(l,iter)} < tol \cdot N_0 \cdot p_0$ **do**

if $l = 0$

Sample $N^{(l,iter)}$ samples $\{\boldsymbol{u}_i^{(l,iter)}\}_{i=1}^{N^{(l,iter)}}$ from the standard normal distribution $\varphi_n(\boldsymbol{u})$.

else

Sample $N^{(l,iter)}$ samples $\{\boldsymbol{u}_i^{(l,iter)}\}_{i=1}^{N^{(l,iter)}}$ starting from the seeds $S^{(l)}$ with the aCS algorithm.

end if

Calculate the LSF values of the samples $\{G(\boldsymbol{u}_i^{(l,iter)})\}_{i=1}^{N^{(l,iter)}}$ and sort the samples, such that $G(\boldsymbol{u}_{i_j}^{(l,iter)}) \leqslant G(\boldsymbol{u}_{i_{j+1}}^{(l,iter)}), \forall j \in \{1,\ldots,N^{(l,iter)} - 1\}$. Denote the sorted samples as $\{\overline{\boldsymbol{u}}_i^{(l,iter)}\}_{i=1}^{N^{(l,iter)}}$ and let $b^{(l+1)} = G(\overline{u}_{p_0 N_0}^{(l,1)})$.

If $b^{(l+1)} \leqslant 0$

Set $b^{(l+1)} = 0$ and $N_s^{(l,iter)} \triangleq \sum_j \mathbb{I}\{G(\overline{u}_j^{(l,iter)}) \leqslant 0\}$.

else

$N_s^{(l,iter)} \triangleq \sum_j \mathbb{I}\{G(\overline{u}_j^{(l,iter)}) < b^{(l+1)}\}$.

end if

If $N_s^{(l,iter)} < tol \cdot N_0 \cdot p_0$

$N^{(l,iter+1)} = 2 \cdot N^{(l,iter)}$.

$iter = iter + 1$.

end if

end while

Define the intermediate failure event: $F^{(l+1)} \triangleq \{\boldsymbol{u}: G(\boldsymbol{u}) < b^{(l+1)}\}$.

Take the $S^{(l+1)} \triangleq \{\overline{u}_j^{(l,iter)}\}_{j=1}^{N_s^{(l,iter)}}$ as the seeds for the next level.

Calculate the conditional probability estimator: $\widehat{\mathbb{P}}(F^{(l+1)}|F^{(l)}) = \frac{N_s^{(l,iter)}}{N^{(l,iter)}}$.

$l = l + 1$.

end while

Estimate the failure probability:

$$\widehat{\mathbb{P}}(F) = \widehat{\mathbb{P}}(F^{(1)}) \prod_{j=2}^{l} \widehat{\mathbb{P}}(F^{(j)}|F^{(j-1)}).$$

4 Examples

4.1 Multistate Random Variable

Consider a discrete random variable X with 7 states $\{x_1, \ldots, x_7\}$. We consider two cases. In case 1, the CDF of X $F(\cdot)$ is set, such that $F(x_i)/F(x_{i+1}) \geq 0.1$, while in case 2, there is a big 'jump' between the third and the fourth state, i.e. $F(x_3)/F(x_4) \approx 1.67 \cdot 10^{-3}$. The CDF of X for the two considered cases is given in Table 1. The LSF is defined as $g(X) = X + 4$ such that the failure probability $\mathrm{P}(X < -4)$ equals 10^{-5} for the first case and $2 \cdot 10^{-5}$ for the second.

We implement standard subset simulation (SuS) and the proposed adaptive effort subset simulation (aE-SuS) respectively to evaluate the failure probability. For SuS, the sampling effort is fixed to 1000, and the conditional probability is 0.1. For aE-SuS, the parameters are set to be $tol = 0.5$, $N_0 = 1000$, $p_0 = 0.1$. Each method is run 1000 times to get the relative bias, coefficient of variation and average computation cost of the failure probability estimator. The results for case 1 and case 2 are shown in Tables 2 and 3 separately. In both cases, aE-SuS shows good accuracy, a negligible bias and a much smaller variance than the crude Monte Carlo results (shown in the red brackets). We note that the coefficient of variation of crude Monte Carlo is given for the same computational effort as the proposed aE-SuS method. In contrast, SuS gives the wrong estimate of the failure probability in the first case and falls into a dead loop in the second case.

Table 1 CDF of X

State	−6	−5	−3	−2	−1	0	1
CDF (case1)	1e−5	1e−4	1e−3	1e−2	1e−1	5e−1	1
CDF (case2)	1e−5	2e−5	5e−5	3e−2	1e−1	5e−1	1

Table 2 Statistical characteristics of the estimator (Case 1)

	Relative bias%	Coefficient of variation	Average computation effort
SuS	−97.8	3.747	7222
aE-SuS	−3.5	0.377(3.580)	7804

Table 3 Statistical characteristics of the estimator (Case 2)

	Relative bias%	Coefficient of variation	Average computation effort
SuS	/	/	/
aE-SuS	1.4	0.206(0.760)	86561

4.2 Power Network System

A transmission power network with the same topology as the IEEE39 bus benchmark system is considered here. It consists of 39 nodes and 43 weighted edges, whose weights represent the line reactance values. This example was previously investigated by Scherb et al. [10] to quantify the network reliability considering cascading effects and spatially distributed hazards, and by Rosero-Velasquez and Straub [11] to select representative failure scenarios. More details about the example can be found in these references.

The state of each node is considered as an independent Bernoulli random variable, with component failure probability set to 10^{-3}. The LSF is then defined as the function of the system state $\boldsymbol{x}$, which is a binary vector, as follows:

$$g(\boldsymbol{x}) = \frac{E(\boldsymbol{x})}{E(1)} - threshold \tag{7}$$

$$E(\boldsymbol{x}) = \frac{1}{|SN||TN|} \sum_{\substack{s \in SN, t \in TN \\ t \neq s}} eff_{st} \tag{8}$$

eff_{st} is the efficiency of the most efficient path from source node s to terminal node t and can be evaluated by adding up the reciprocals of the reactance values along that path. $E(\boldsymbol{x})$ is the efficiency of the whole system associated to the system state $\boldsymbol{x}$ (where the vector 1 is the intact system state) and is equal to the mean value of all the eff_{st} from each source node in set SN to each terminal nodes in set TN.

In order to model the cascading effects, Eq. (7) is modified to

$$g(\boldsymbol{x}) = \frac{E\left(\tilde{\boldsymbol{x}}\right)}{E(1)} - threshold \tag{9}$$

where $\tilde{\boldsymbol{x}}$ is the final system state after cascading effects. These are triggered by overloading in individual lines following initial failures, and are modeled following [10, 12].

The threshold is fixed to 0.3, which means the system fails when its efficiency is less than 30% of that of the intact system. We then apply aE-SuS algorithm to this problem and set the parameters $N = 2000$, $p_0 = 0.1$, $tol = 0.8$. Figure 1 shows the empirical CDF of $g(\boldsymbol{X})$ obtained by Monte Carlo Simulation and the aE-SuS algorithm respectively. The aE-SuS algorithm is run 25 times to obtain the mean value, 10 percentile and 90 percentile of the empirical CDF, while a single Monte Carlo Simulation run with 10^6 samples is carried out for validation.

The average computation cost of aE-SuS is 24437 calculations of the LSF $g(\cdot)$ and the relative bias of the failure probability is 9.17%, while the coefficient of variation is 0.57. Under the same computation cost, the coefficient of variation of

Fig. 1 Results obtained by aE-SuS for IEEE39 network.

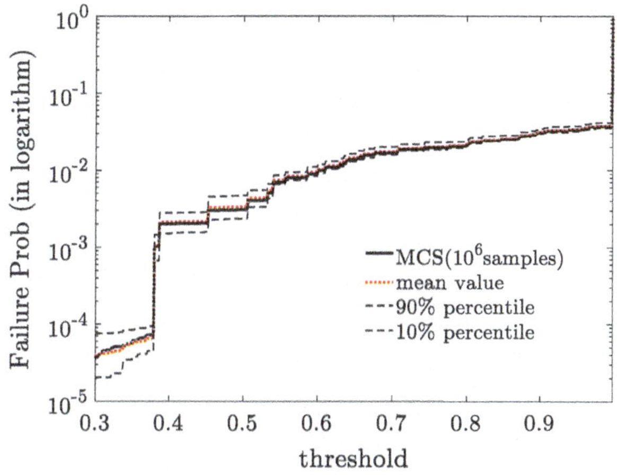

Monte Carlo Simulation is about 1.06 which is significantly larger than that of aE-SuS. The standard SuS algorithm is not applicable for this example due to the large jump in the CDF of the LSF below the threshold of 0.4.

5 Conclusions

We introduce an adaptive effort subset simulation algorithm that enables solving reliability problems with discontinuous limit states. Such problems often occur in network reliability assessment because of discrete random variables appearing in the input random vector or due to discontinuities in the function that defines the system performance. The proposed algorithm extends the applicability of standard subset simulation to problems where significant jumps in the distribution of the limit-state function occur. Numerical examples demonstrate the accuracy and efficiency of the proposed method and show that the method has increased efficiency compared to crude Monte Carlo in some problems where standard subset simulation fails to converge. For connectivity-based problems where the LSF can only take two possible values, the proposed algorithm will turn to crude Monte Carlo simulation and therefore becomes inefficient in rare event context.

References

1. Zuev, K. M., Wu, S., & Beck, J. L. (2015). General network reliability problem and its efficient solution by subset simulation. *Probabilistic Engineering Mechanics, 40,* 25–35.
2. Papaioannou, I., Geyer, S., & Straub, D. (2019). Improved cross entropy-based importance sampling with a flexible mixture model. *Reliability Engineering & System Safety, 191,* 106564.

3. Rubinstein, R. Y., & Kroese, D. P. (2016). *Simulation and the Monte Carlo Method*. Hoboken: Wiley.
4. Au, S. K., & Beck, J. L. (2001). Estimation of small failure probabilities in high dimensions by subset simulation. *Probabilistic Engineering Mechanics, 16*(4), 263–277.
5. Rubino, G., & Tuffin, B. (2009). *Rare Event Simulation Using Monte Carlo Method*. Chichester: Wiley.
6. Papaioannou, I., Betz, W., Zwirglmaier, K., & Straub, D. (2015). MCMC algorithms for subset simulation. *Probabilistic Engineering Mechanics, 41,* 89–103.
7. Botev, Z. I., & Kroese, D. P. (2012). Efficient Monte Carlo simulation via the generalized splitting method. *Statistics and Computing, 22*(1), 1–16.
8. Au, S. K. (2016). On MCMC algorithm for subset simulation. *Probabilistic Engineering Mechanics, 43,* 117–120.
9. Rosenblatt, M. (1952). Remarks on a multivariate transformation. *The Annals of Mathematical Statistics, 23*(3), 470–472.
10. Scherb, A., Garrè, L., & Straub, D. (2017). Reliability and component importance in networks subject to spatially distributed hazards followed by cascading failures. *ASCE-ASME Journal of Risk and Uncertainty in Engineering Systems, Part B: Mechanical Engineering, 3*(2), 021007.
11. Rosero-Velasquez, H., & Straub, D. (2019). Representative natural hazard scenarios for risk assessment of spatially distributed. In *Proceedings of the 29th European Safety and Reliability Conference* (pp. 1–7). Submitted.
12. Crucitti, P., Latora, V., & Marchiori, M. (2004). Model for cascading failures in complex networks. *Physical Review E, 69*(4), 045104.

An Efficient Solution for Reliability Analysis Considering Random Fields—Application to an Earth Dam

Xiangfeng Guo, Daniel Dias, and Qiujing Pan

Abstract Performing a reliability analysis using Monte Carlo Simulation (MCS) is usually time-consuming for cases with expensive-to-evaluate deterministic models or small failure probabilities. The computational burden of such analysis can be significantly alleviated by replacing the deterministic model with a meta-model. However, the meta-modeling techniques suffer from the curse of dimensionality issue. They are thus less efficient for geotechnical reliability analyses involving random fields (RF) since the considered problems are often high dimensional due to the RF discretization. This paper introduces a new procedure based on the Sparse Polynomial Chaos Expansions (SPCE) which can address the above-mentioned issues. It deals with high dimensional stochastic problems in two stages: the first stage consists in reducing the input dimension by the Sliced Inverse Regression (SIR), while the second stage constructs a SPCE with respect to the reduced dimension and then performs an MCS. Additionally, an adaptive experimental design technique is proposed for the construction of the SPCE model. The modified algorithm (termed as A-SPCE/SIR) is applied to an earth dam problem in which the cohesion and friction angle are modelled by lognormal RFs. The effects of the vertical autocorrelation distance and the input cross-correlation on the dam reliability are investigated. The efficiency and accuracy of the A-SPCE/SIR are highlighted by comparing with the direct MCS and a previous study.

Keywords Geotechnical reliability analysis · Random field · High dimension · Earth dam · Polynomial Chaos expansions

---

X. Guo (✉) · D. Dias
Univ. Grenoble Alpes, CNRS, Grenoble INP, 3SR, 38000 Grenoble, France
e-mail: Xiangfeng.guo@3sr-grenoble.fr

D. Dias
Geotechnical Expert, Antea Group, 92160 Antony, France

Q. Pan
School of Civil Engineering, Central South University, Changsha 410075, China

© The Author(s), under exclusive license to Springer Nature Switzerland AG 2021
J. C. Matos et al. (eds.), *18th International Probabilistic Workshop*, Lecture Notes in Civil Engineering 153, https://doi.org/10.1007/978-3-030-73616-3_10

1 Introduction

It is commonly recognized that soil properties exhibit spatial variation in space. The random field (RF) theory is usually used to model the soil spatial variability. Most geotechnical reliability problems are high dimensional since a high number of random variables (RVs) should be used for the discretization of RFs. The number of the required RVs, which represents the input dimension of a reliability analysis, could be dozens to thousands [1, 2]. The traditional First/Second Order Reliability Method cannot efficiently handle a large number of input variables and the provided failure probability (P_f) estimate could be questionable. Then, the sampling methods like Monte Carlo Simulation (MCS) and Subset Simulation (SS) are effective for high dimensional stochastic problems. However, MCS and SS are less efficient for cases with a low P_f or a time-consuming model. In order to alleviate the computational burden of a reliability analysis using sampling methods, the metamodeling technique (also known as surrogate modelling) was introduced. It consists in creating a surrogate of the original deterministic model and then performing an MCS or SS with the surrogate. Polynomial Chaos Expansion (PCE) is a popular tool for the metamodeling and has been used for the reliability analysis of different geotechnical works [1, 2]. However, PCE, like other metamodeling techniques, suffers from the 'curse of dimensionality'. The number of PCE terms exponentially grows with the number of input variables for a fixed order. This makes PCE to be less efficient for high dimensional problems. Additionally, the number of training samples necessary to build a satisfactory PCE remains an issue. The internal error estimate, such as the one of leave-one-out-error, gives a measure of the PCE model quality, but the provided P_f has no accuracy indicator.

This paper attempts to address the above-mentioned issues by proposing an efficient solution to the reliability analysis involving RFs. The proposed procedure combines the PCE with Sliced Inverse Regression (SIR)—a dimension reduction approach. This can avoid building a PCE considering a large number of input variables. Besides, an active learning process is coupled with PCE-SIR. It starts with an initial Design of Experiment (*DoE*) and gradually adds new samples. The added new samples are selected from a candidate pool and should be beneficial for the improvement of the constructed PCE in estimating P_f. The *DoE* enrichment process is stopped once satisfactory results are obtained. Therefore, the necessary size of the *DoE* can be determined automatically in the algorithm. The proposed procedure is applied to an earth dam problem in which two soil properties are modelled by lognormal RFs. It is also validated by comparing with a direct MCS and a previous study. Its accuracy and efficiency are highlighted by these comparison works.

2 Presentation of the Employed Methods

2.1 *Karhunen–Loève Expansions (KLE)*

A random field (RF) is able to describe the spatial correlation of a material property in different locations and represent its nonhomogeneous characteristic. In this work, the KLE is adopted since it is widely used for geotechnical reliability analyses and can lead to the minimal number of RVs involved in a RF discretization. In the KLE context, a stationary Gaussian RF $\boldsymbol{H}$ can be expressed as follows:

$$H(x_s) \approx \mu + \sigma \sum_{i=1}^{N_{KL}} \sqrt{\lambda_i \phi_i(x_s) \xi_i} \tag{1}$$

where $\boldsymbol{x}_s$ is the coordinate of an arbitrary point in the RF space, μ and σ represents respectively the mean and standard deviation of the RF, λ_i and φ_i are respectively the eigenvalues and eigenfunctions of the autocovariance function, ξ_i is a set of uncorrelated standard normal RVs and N_{KL} is used to truncate the KLE for practical applications. The autocovariance function is the autocorrelation function multiplied by the RF variance. In this work, the exponential autocorrelation function is used. The most important parameter in such functions is the autocorrelation distance: L_x and L_y (horizontal and vertical one) which is defined as the length leading to a decrease from 1 to 1/e for the autocorrelation function. Concerning the N_{KL}, it is determined by evaluating the error due to the truncation term. In this work, the variance-based error globally estimated in the RF domain is used [3].

2.2 *Polynomial Chaos Expansions (PCE)*

PCE is a powerful and efficient tool for the metamodeling which consists in building a surrogate of a complex computational model. It approximates a model response $\boldsymbol{Y}$ by finding a suitable basis of multivariate orthonormal polynomials with respect to the joint input probability density function (PDF) in the Hilbert space. The basic formula of PCE is given as below:

$$Y \approx \sum_{a \in N^M} k_\alpha \Psi_\alpha(\xi) \tag{2}$$

where $\boldsymbol{\xi}$ are independent RVs, $\boldsymbol{k}_\alpha$ are unknown coefficients to be computed with α being a multidimensional index and $\boldsymbol{\Psi}_\alpha$ are multivariate polynomials which are the tensor product of univariate orthonormal polynomials.

In this work, the Hermite polynomials in conjunction with standard normal RVs are used. The representation of Eq. (2) should be truncated to a finite number of terms for practical applications. To this end, the hyperbolic truncation scheme [1] is adopted in this study. Then, the unknown coefficients can be estimated by using the Least Angle Regression (LAR) method [4]. The accuracy of the truncated PCE can be assessed by computing the coefficient of determination R^2 and the Q^2 indicator [1]. R^2 is related to the empirical error using the model responses already existing in the design of experiment (*DoE*), while Q^2 is obtained by the leave-one-out cross-validation technique.

In order to further reduce the number of $\mathbf{\Psi}_\alpha$ after the truncate operation when the input dimension is high, the sparse PCE (SPCE) was proposed. It consists in building a suitable sparse basis instead of computing useless terms in the expansions that are eventually negligible. In this work, the LAR-based algorithm proposed in [4] is used to determine a sparse representation of PCE. It is noted that finding a sparse PCE aims at reducing the number of terms (unknown coefficients) in the metamodel to be constructed. This is different to the input dimension reduction technique (presented in next section) which focuses on reducing the number of input variables.

2.3 Sliced Inverse Regression (SIR)

This approach is based on the principle that a few linear combinations of original input variables could capture the essential information of model responses [1]. It aims to find the effective dimension reduction space by considering an inverse regression relation which regresses input variables against model responses. Figure 1 shows the basic concepts of SIR: starting from the original input RVs (left part), the eigenvectors (determined with a *DoE* within the SIR context) are able to reduce the input dimension from d to k (middle part), and then the model response can be represented by the new input RVs in a dimension-reduced space. In this paper, the algorithm presented in [1] is used to find these eigenvectors. An important parameter in the algorithm is

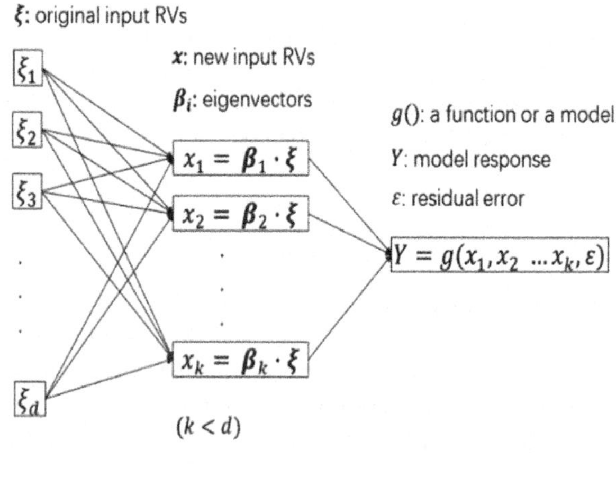

Fig. 1 Dimension reduction by SIR

the slice number *Nsir* for which the recommendations are: $10 \leq Nsir \leq 20$ for the cases with several hundred RVs and $20 \leq Nsir \leq 30$ when the number of input RVs is several thousand.

2.4 The Proposed Procedure

This section describes the proposed procedure which provides an efficient solution for the reliability analysis considering RFs. The procedure is based on an active learning metamodeling technique combining with a dimension reduction method. It starts by generating an initial *DoE* according to the input joint PDF and evaluating the model responses of each sample in the *DoE* using a deterministic computational model. Then, an SIR analysis is performed using the initial *DoE* and the model responses in order to reduce the input dimension. It is followed by constructing an SPCE model on the reduced input space. The next step is to add new samples into the current *DoE* and construct a new SPCE which is then coupled with an MCS for estimating the target reliability results. SIR is always conducted *a priori* to the SPCE construction so that the metamodeling is based on a reduced input space. Such a step of adding new samples and building new surrogate models is repeated until satisfactory results are obtained. Compared to the existing algorithms [1, 2], the main originality of the proposed procedure lies in using the adaptive *DoE* process for the SIR-aided SPCE training. It can reduce the number of deterministic calculations in a surrogate modelling for the high dimensional reliability analysis and lead to efficient P_f estimates. Additionally, two stopping conditions are proposed in this procedure in order to avoid an early stop with incorrect results and guarantee a convergence on the P_f estimation. More details about the procedure are given in Fig. 2 and Table 1. The proposed procedure is termed as A-SPCE/SIR.

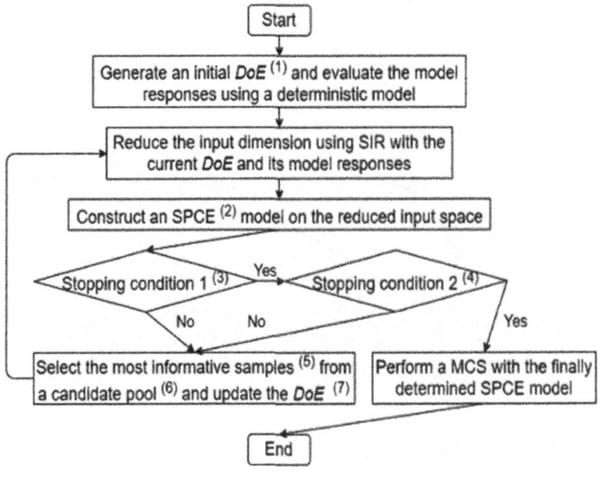

Fig. 2 Flowchart of the proposed procedure

Table 1 Detailed remarks as annotations to Fig. 2

(1)	The size N_{ini} of the initial *DoE* could be equal to *d* or higher (*d*: original input dimension) It is recommended to use the Latin Hypercube Sampling (LHS) for generating samples in the initial *DoE*
(2)	The algorithm presented in [1] is used to determine a sparse representation of PCE The optimal order of SPCE is determined by testing in a range (e.g. [2, 7]) [1]
(3)	Stopping condition 1 measures if the accuracy indicator Q^2 of the constructed SPCE model is higher than a target value ${Q^2}_t$
(4)	Stopping condition 2 evaluates the convergence of the P_f estimation by computing an error (Err_{con}) which is the maximum value of the relative errors calculated from all the possible pairs in a vector. The vector consists of the N_{s2} last P_f estimates in the adaptive *DoE* process. The condition will be satisfied if Err_{con} is lower than a given value Err_t
(5)	The samples to be added are selected by using the strategy of [5]
(6)	An MCS population is generated using the LHS as a candidate pool
(7)	The *DoE* is updated by adding the selected samples and their model responses

3 Application to an Earth Dam Problem

This section presents the application of the proposed procedure to an earth dam problem.

3.1 *Presentation of the Studied Dam and the Deterministic Model*

The studied dam is given in Fig. 3. It has a width of 10 m for the crest and a horizontal filter drain installed at the toe of the downstream slope. The soil is assumed to follow a linear elastic perfectly plastic behavior characterized by the Mohr Coulomb shear failure criterion. In this work, the dam stability issue will be analyzed by considering

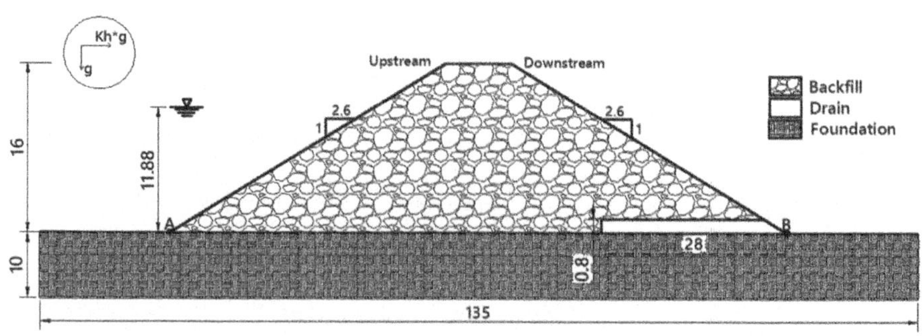

Fig. 3 Geometry of the studied dam (g: gravitational acceleration)

Table 2 Distribution parameters for c' and φ"

Soil property	Distribution	Mean	Coefficient of variation (CoV) (%)
Effective cohesion (kPa)	Lognormal	8.9	30
Friction angle (degree)	Lognormal	34.8	10

a constant water level of 11.88 m and a saturated flow. Additionally, a horizontal pseudo-static acceleration of 2.4 m/s^2 toward the downstream part is applied on the dam body. Concerning the input uncertainty modelling, two soil properties (effective cohesion c' and friction angle φ') of the compacted fill are modelled by lognormal RFs. The illustrative values for the distribution parameters of c' and φ' are given in Table 2. The uncertainties in the soil hydraulic parameters are not considered since the variation of the dam phreatic level in the downstream part is not significant due to the presence of the filter drain.

The deterministic model used in this work for estimating the dam factor of safety (FoS) is developed by using the idea of [2]. It combines three techniques: Morgenstern Price Method (MPM), Genetic Algorithm (GA) and a non-circular slip surface generation method. MPM is employed to compute the FoS of a given failure surface; GA aims at locating the most critical slip surface (i.e. minimum FoS) by performing an optimization work; the implementation of non-circular slip surfaces can lead to more rational failure mechanics for the cases of non-homogeneous soils. The principle of the model is to firstly generate a number of trial slip surfaces as an initial population, and then to determine the minimum FoS value by modelling a natural process along generations including reproduction, crossover, mutation and survivors' selection. The distribution of the pore water pressures inside the dam is given by a numerical model [6]. In this work, the developed deterministic model is termed as LEM-GA.

3.2 Generation of the Random Fields

This section presents the generation of the RFs used in the dam reliability analysis. Firstly, three calculation cases considering different autocorrelation distances or cross-correlation between c' and φ' are defined. Then, the truncation term N_{KL} in the KLE is determined for each case. The value of N_{KL} is also the input dimension for the following reliability analysis. Lastly, two RFs with different L_y values are presented. They are generated by using the KLE with the pre-defined parameters.

Table 3 presents the three cases to be analyzed. Different L_y and $\rho_{c\varphi}$ (cross-correlation coefficient between c' and φ' RFs) values are considered in order to investigate their effects on the dam reliability. The L_x is assumed to be 40 m for all the cases. Such a large value is adopted since earth dams are usually constructed by layers, so the soil properties are highly correlated in the horizontal direction

Table 3 Three reliability analysis cases considered in the work

Case	Soil variability	L_x (m)	L_y (m)	$\rho_{c\varphi}$	Input dimension
1	Table 2	40	40	0	60
2	Table 2	40	40	−0.5	60
3	Table 2	40	3	0	620

if the construction materials were well selected. Then, two L_y values are considered. Assuming a L_y of 40 m leads to an isotropic RF and represents a relatively homogeneous compacted fill, while considering a L_y of 3 m permits to account for a significantly varying soil along depth which corresponds to the case when the materials of different layers are from different borrow sites and/or they are not identically compacted. Besides, a negative correlation commonly exits between *c'* and *φ'*. Therefore, a $\rho_{c\varphi}$ of −0.5 is considered in Case 2. The input dimension in Table 3 will be explained later.

Figure 4 presents the determination of the truncation term N_{KL} in the KLE for two types of RFs: RF_a refers to an isotropic RFs with $L_x = L_y = 40$ m and RF_b represents an anisotropic RFs with $L_x = 40$ m and $L_y = 3$ m. The errors of the two RFs due to the truncation are plotted against the truncation term N_{KL} in Fig. 4. It can be observed that the error is decreased with increasing the N_{KL} and RF_b needs a larger N_{KL} to be lower than a given target error than RF_a. This means that more RVs are required to obtain a more accurate representation of KLE RF and the required RV number is larger for a smaller autocorrelation distance. In this work, the target error is 5%, then the finally determined N_{KL} is respectively equal to 30 and 310 for RF_a and RF_b. Therefore, the input dimension is 60 for Case 1 and 2, and 620 for Case 3 since two RFs (*c'* and *φ'*) should be generated for each simulation. When evaluating the error and generating the RFs, a Δd of 1 m [7] is added to each side of the necessary domain which should be 94 × 16 m (length AB × dam height). The aim is to avoid large errors at the boundaries of the RFs generated by series expansions methods (e.g. KLE).

As an illustration, one random realization of *c'* for both RF_a and RF_b is generated using the pre-defined parameters and is presented in Fig. 5. It can be observed that the upper graph shows uniform areas with similar *c'* values. This is due to the large autocorrelation distances considered in RF_a. On the contrary, the lower graph has less and smaller uniform zones and shows a significant variation in the vertical direction.

3.3 Reliability Analyses and Results

This section presents the dam reliability analysis for the three cases of Table 3 by using the proposed procedure (A-SPCE/SIR). The values for the user-defined parameters of the procedure are given as: $N_{ini}= d$; the slice number is 10 for Case 1 and 2,

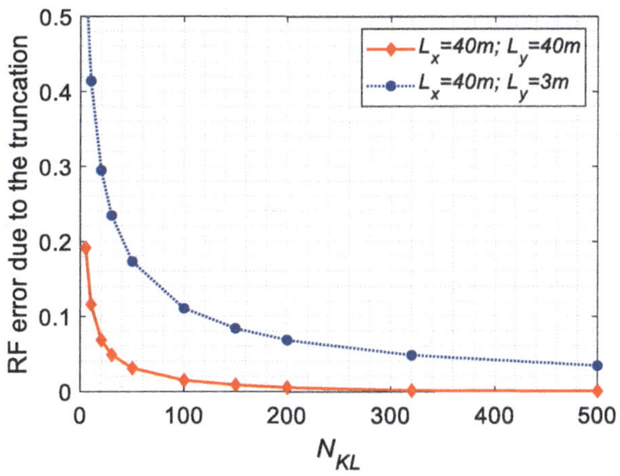

Fig. 4 Determination of the truncation term for the two types of RFs

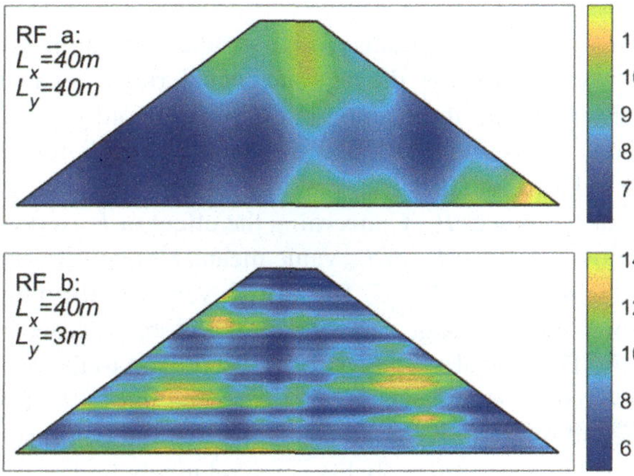

Fig. 5 An example for the two types of RFs (*c'*, in kPa)

and 20 for Case 3; the SPCE order is tested between 2 and 7; the target accuracy indicator $Q^2{}_t$ for the SPCE model is 0.95; the target error Err_t which measures the P_f convergence is 0.15; N_{s2} equals to 5; two samples are added at each iteration; the size for the MCS candidate pool is large enough so that the estimated P_f has a CoV lower than 5%. The dimension of the finally determined input space by SIR is smaller than 20 for all the three cases.

Figure 6 presents the main results provided by the A-SPCE/SIR for the three cases including P_f, FoS statistics and distribution. According to the PDF curves, the dam possible FoS is varied mainly between 0.9 and 1.5. A comparison among the three curves reveals that the dam FoS uncertainty/variation can be reduced by considering a negative correlation between *c'* and *φ'* or a smaller autocorrelation distance, since the PDFs of Case 2 and 3 are both taller and narrower than the one of Case 1. This leads to a decrease of one order of magnitude for the P_f from Case 1 to Case 2 or 3. It is found that the L_y or $\rho_{c\varphi}$ has negligible impacts on the mean value of FoS (μ_{FoS}),

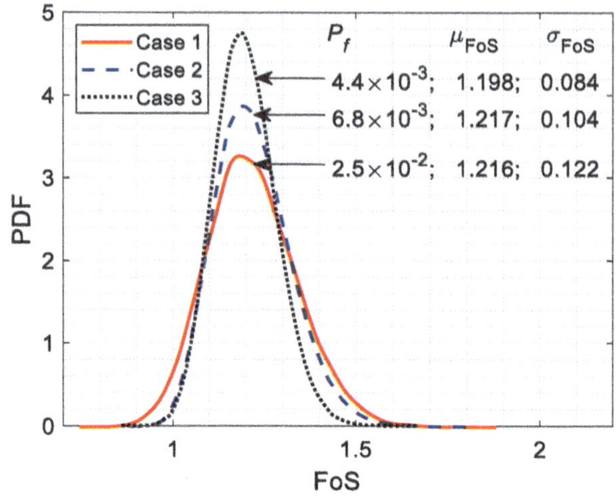

Fig. 6 Reliability analysis results of the considered three cases

but significantly influences the standard deviation (σ_{FoS}). For the effect of $\rho_{c\varphi}$, it can be explained by the fact that considering a negative cross-correlation can reduce the total input uncertainty and partially avoid generating a small value for both *c'* and φ' at a same location in one simulation. Therefore, the output variance can also be reduced (smaller σ_{FoS}) and the number of small FoS values is decreased which leads to a lower P_f. Concerning the effect of L_y, a possible interpretation is given as follows. A large L_x or L_y value means a great probability of forming large uniform areas as shown in Fig. 5. The global average of the field could be low, medium or high which means a large variation for the global average among different realizations of RFs. The global average is partially related to the estimated FoS so the latter could also have a large variation as evidenced in Fig. 6. Then, the P_f is higher since it is the tail probability of a distribution. On the contrary, for the case with a low L_x or L_y value, there are probably some relatively higher values generated in the neighbor of the area with low values and vice versa. As a result, the global average varies in a narrower range also the FoS, so the P_f is lower.

4 Accuracy and Efficiency of the Proposed Procedure

This section aims to assess the accuracy and efficiency of the proposed procedure A-SPCE/SIR. The effectiveness of the employed deterministic model LEM-GA is also evaluated.

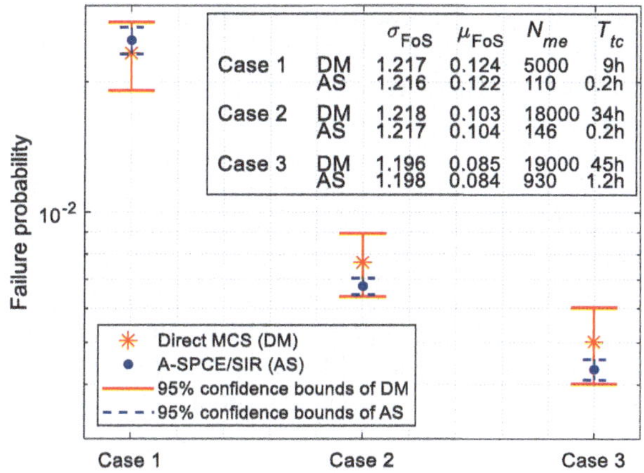

Fig. 7 Comparison between the A-SPCE/SIR and direct MCS for the three cases

4.1 Comparison with Direct MCS

A-SPCE/SIR is assessed by comparing the provided results with the ones of a direct MCS which means that the deterministic model is directly coupled with the MCS and no surrogate model is created. The efficiency of the proposed procedure can also be highlighted from such a comparison. The three cases defined in Table 3 are analyzed again but using the direct MCS in this section. The required number of the deterministic simulations, which is highly related to the total computation time of a direct MCS, is determined by searching a CoV of P_f (CoV_p) around 10%.

Figure 7 presents the comparison between the two methods in terms of P_f, FoS statistics, number of deterministic calculations (N_{me}) and total computational time (T_{tc}). Particularly, the 95% confidence bounds of the P_f estimate are also given. If the MCS size is large enough, the estimated P_f can be approximated by a normal distribution which makes the confidence bounds to be available. The T_{tc} is evaluated in a computer equipped with an CPU of Intel Xeon E5-2609 v4 1.7 GHz. A first review on the figure reveals that the two methods give similar P_f estimates for all the three cases, indicating a good accuracy of A-SPCE/SIR. Additionally, the P_f confidence bounds of the proposed procedure are covered by the ones of the direct MCS. In A-SPCE/SIR, it is acceptable to largely increase the MCS size in order to obtain a small CoV_p (around 3% in this work) since the SPCE-aided MCS is not time-consuming. However, much more computational efforts are required in the direct MCS if its size should be enlarged, so a CoV_p around 10% is adopted in this work. This finding and argument mean that a precise P_f with a small CoV_p could be easily obtained using A-SPCE/SIR. Concerning the FoS statistics, the results provided by the two methods show a good agreement with each other. Then, for the efficiency comparison, it is found that using A-SPCE/SIR can significantly reduce the total computational time compared to the direct MCS, for example from 45 to 1.2 h in Case 3. Such a time reduction is mainly due to the decrease in the number of calls to the deterministic model. Based on a limited number of model evaluations (e.g. 110 for Case 1), a surrogate can be constructed, then performing a surrogate-based MCS is very fast. Additionally, the employed active learning process further improves the efficiency in building a satisfactory SPCE model for the P_f estimate. In conclusion, the proposed procedure is able to provide accurate results in terms of P_f, μ_{FoS}, and σ_{FoS}, with a significantly reduced computational time. It is noted that the accuracy investigation is conducted with a P_f in the order of magnitude of 10^{-2} and 10^{-3}. For the cases with a very low P_f, even a surrogate-based MCS is time-consuming since a large number of realizations of RFs should be generated and evaluated. This is a common issue for metamodels when dealing with high dimensional problems. A possible solution is to replace the MCS in the A-SPCE/SIR with an advanced sampling method such as the Subset Simulation [2] for the P_f estimation, and this could be done in a future work to enhance the current algorithm.

4.2 Comparison with a Previous Study

The studied dam was firstly proposed and analyzed by [6] to investigate the effect of the soil spatial variability on earth dams. The SPCE/GSA method was used in [6] for the reliability analysis, which is also a metamodeling approach combined with a dimension reduction technique for high dimensional problems. Analysis N.1 of Table 5 in [6] is conducted again in this section with the proposed procedure using the same input parameters as in [6]. The aim is to show the efficiency of the A-SPCE/SIR by comparing with another advanced approach (SPCE/GSA). The deterministic model used in [6] is based on the finite difference method (FDM) combined with the strength reduction method (SRM). Therefore, the comparison also allows an assessment on the effectiveness of the introduced deterministic model (LEM-GA).

Table 4 presents the comparison results between the two studies. It is observed that the present study gives a P_f estimate very close to the one of [6]. This highlights again the result accuracy of the A-SPCE/SIR as also evidenced in Fig. 7. Additionally, such a good agreement between the two studies indicates that the LEM-GA is effective to estimate the dam FoS under the current calculation configuration (steady saturated flow + pseudo static loading). Using a simplified deterministic model (e.g. LEM-GA) is beneficial for a reliability analysis since it can reduce the total computational time. This strategy can thus be adopted in a preliminary design/assessment stage for efficiently obtaining first results. Then, a sophisticated model (FEM or FDM) is required in a next stage if complex conditions should be modelled (e.g. rapid drawdown and unsaturated flows) or multiple model responses (e.g. settlement and

Table 4 Comparison with a previous study

Source	Reliability method	Deterministic model	P_f	N_{me}
This paper	A-SPCE/SIR	LEM-GA	1.6×10^{-3}	511
[6]	SPCE/GSA	FDM-SRM	1.5×10^{-3}	3000

flow rate) are necessary. Table 4 also compares the number of model evaluations in the two studies. The results show that the proposed procedure leads to a significant decrease in the N_{me}. This is the main advantage of the A-SPCE/SIR. Contrary to the SPCE/GSA, no surrogate model is constructed with respect to the original input space (high dimension) and an adaptive *DoE* process is used in A-SPCE/SIR. These facts lead to the high efficiency of the proposed procedure as shown in Table 4.

5 Conclusions

This paper proposes an efficient solution to address high dimensional reliability problems. The proposed procedure A-SPCE/SIR is based on a combination of an active learning metamodeling technique (A-SPCE) and a dimension reduction approach (SIR). Using metamodels allows to alleviate the computational burden of a reliability analysis (e.g. MCS), and SIR aims to reduce the input dimension, so that the SPCE model needs to be trained on a limited number of RVs. The originality of the procedure lies in implementing an active learning process into the SIR-aided SPCE, which permits an automatic determination of the necessary *DoE* size and a continuous improvement of the P_f estimation. Particularly, two stopping conditions are proposed in this paper to enhance the process.

The A-SPCE/SIR is applied to an earth dam problem. *c'* and *φ'* of the compacted fill are modelled by means of lognormal RFs. Three cases are considered for the dam reliability analysis by using different L_y and $\beta_{c\varphi}$ values. It is found that both the L_y and $\beta_{c\varphi}$ has significant effect on the dam P_f. The dam FoS uncertainty can be reduced if a smaller L_y or a negative $\beta_{c\varphi}$ is used. A-SPCE/SIR is compared with the direct MCS within the three cases which include two input dimension scenarios: 60 and 620. The comparison shows that the proposed procedure can provide accurate results but with a significantly reduced computational time. For example, the A-SPCE/SIR can reduce the time from 45 h to 1.2 h for Case 3 compared to a direct MCS. Additionally, a comparison with a previous study is also carried out. The results highlight again the accuracy and efficiency of A-SPCE/SIR and confirms the effectiveness of the employed deterministic model.

This study also highlights some possible future works which are related to the following questions: (1) What will be the performance, especially the efficiency, of the A-SPCE/SIR when estimating very small P_f? (2) Can the A-SPCE/SIR effectively handle the cases with over thousand even ten thousand of RVs, which could happen if a large-scale 3D RF should be considered or small autocorrelation lengths are measured? These issues are expected to be addressed in future studies.

References

1. Pan, Q., & Dias, D. (2017). Sliced inverse regression-based sparse polynomial chaos expansions for reliability analysis in high dimensions. *Reliability Engineering and System Safety, 167,* 484–493.
2. Guo, X., Dias, D., Carvajal, C., Peyras, L., & Breul, P. (2019). A comparative study of different reliability methods for high dimensional stochastic problems related to earth dam stability analyses. *Engineering Structures, 188,* 591–602.
3. Pan, Q., & Dias, D. (2017). Probabilistic evaluation of tunnel face stability in spatially random soils using sparse polynomial chaos expansion with global sensitivity analysis. *Acta Geotechnica, 12,* 1415–1429.
4. Blatman, G., & Sudret, B. (2011). Adaptive sparse polynomial chaos expansion based on least angle regression. *Journal of computational Physics, 230*(6), 2345–2367.
5. Marelli, S., & Sudret, B. (2018). An active-learning algorithm that combines sparse polynomial chaos expansions and bootstrap for structural reliability analysis. *Structural Safety, 75,* 67–74.
6. Guo, X., Dias, D., & Pan, Q. (2019). Probabilistic stability analysis of an embankment dam considering soil spatial variability. *Computers and Geotechnics, 113,* 103093.
7. Guo, X., Du, D., & Dias, D. (2019). Reliability analysis of tunnel lining considering soil spatial variability. *Engineering Structures, 196,* 109332.

An Overview of Performance Predictive Models for Railway Track Assets in Europe

Maria José Morais, Hélder S. Sousa, and José C. Matos

Abstract A railway system degrades over time due to several factors such as aging, traffic conditions, usage, environmental conditions, natural and man-made hazards. Moreover, the lack or inadequate maintenance and restoration works may also contribute to the degradation process. In this aspect it is important to understand the performance of transportation infrastructures, the variables influencing its degradation, as well as the necessary actions to minimize the degradation process over time, improve the security of the users, minimize the environment impact as well as the associated costs. Thus, it is crucial to follow structured maintenance plans during the life cycle of the infrastructure supported by the forecasting of the degradation over time. This paper presents a brief description of the variables influencing the degradation of a rail-way system, and the way the performance of the railway track can be measured, within a probabilistic environment. The work developed in other transportation infrastructures, like roadway, is briefly presented for comparison purposes and benchmarking. It also presents an overview of the predictive models being used in railway systems, from the mechanistic to the data-driven models, where the statistical and artificial intelligence models are included.

Keywords Predictive models · Railway · Performance indicators · Probabilistic assessment

1 Introduction

Regular, planned and predetermined inspections and maintenance are essential to control the process of degradation of railway infrastructures and restore the damaged railway sections, thereby guaranteeing the reliability and availability of the railway track, as well as the passenger safety and comfort, not forgetting the cost reduction over the life cycle [1]. The management of railway infrastructures is supported by maintenance plans, that in turn are supported by quality indicators [2]. Just recently, with the aim of simplifying the communication between consultants, operators and

M. J. Morais (✉) · H. S. Sousa · J. C. Matos
ISISE, Department Civil Engineering, IB-S, University of Minho, Guimarães, Portugal

© The Author(s), under exclusive license to Springer Nature Switzerland AG 2021

J. C. Matos et al. (eds.), *18th International Probabilistic Workshop*, Lecture Notes in Civil Engineering 153, https://doi.org/10.1007/978-3-030-73616-3_11

owners, these indicators started to be called performance indicators [3, 4]. During the 12th Florence Rail Forum on the performance of the railway system, in 2016 [5] it was stressed the need to improve the performance of the European railways. For a more accurate maintenance, the railway track degradation must be predicted by using the appropriate models and methodologies [1] following a risk based approach with the identification of the possible risks and levels of risk. The causes of degradation of the track as well as the factors that could influence this degradation to happen should be known [6]. The aim of this paper is to present the most relevant factors affecting the degradation of the railway track as well as some possible damages, presenting also some relevant aspects to take into account. This will be made in parallel with the analysis of railway track through performance indicators based in railway standards, and the possible improvements that could be adopted from the work developed in other transportation infrastructures, as well as by providing an overview of predictive models being used in the railway system, namely with some operational aspects, possible variables to be used and the most relevant advantages and disadvantages. Special focus is given to the frameworks and management systems based on probabilistic methods.

2 Railway System and the Variables Influencing Its Degradation Over Time

The railway system is divided into railway infrastructure and rolling stock. The railway infrastructure involves the railway track (or simply track), the railway stations, signaling, the catenary system, the drainage system, among others. This paper focus on the first one, particularly in the railway track. The importance of the description of each component, within this paper, is mainly related to the use of Predictive models that are dependent on the component type as they present different deterioration models and patterns. For structural assessment purposes, the track is composed of the superstructure and the substructure [7]. The superstructure corresponds to the top of the track, consisting of rails, fastening system, rail pads and sleepers, while the substructure corresponds to the support and could consist of a ballast system (ballast, sub-ballast and subgrade) or a slab system [6, 8] and may be directly over the soil or crossing bridges, box culvert or tunnels.

Among the most relevant factors affecting the track degradation, the following ones can be highlighted. The track geometry has a heterogeneity along its path, thanks to the existence of straights, curves and crossovers. Besides that, there are discontinuities in the support conditions when the track crosses a bridge deck, a box culvert or a tunnel or when occurs a transition between a slab and a ballasted track [1, 6, 9]. The mentioned heterogeneity is characterized by a variation in the track stiffness [9], that can lead to differential settlements [6], non-uniform dynamic loading, corrugation, wear and fatigue failure of the rail, fatigue failure of the fastening system and cracking of the sleepers [9]. The track is influenced by the environment conditions

[8, 10] such as high temperatures, storm (or similar) with consequent heavy rainfall, and strong wind that can lead to flood, risk of destabilization of earthworks, drainage problems, faller trees or branches (in case they exist close by) and moisture degree of saturation more specifically in the case of the substructure [11].

As consequence of loading cycles from the train passage, a plastic differential settlement may occur, that over time can lead to deviations of the original track geometry, as well as breakage and shear deformation of the super and substructure [9]. In consequence an increasing of the acceleration of the train as well as an increasing of the dynamic forces caused by the train may occur [9]. The increasing of forces can speed up the degradation process of the track components, like for example vertical and lateral displacements. Differential lateral movements can compromise the lateral stability of the track and consequently track buckling [9]. In return, this track degradation process can cause the increase of the variation of the interaction between the track and the forces which in turn affects the performance of the track [7, 11].

According to literature, the track condition can be measured through five classes of variables (or parameters) [1, 12] (i) longitudinal level, (ii) alignment, (iii) gauge, (or vertical alignment), (iv) twist and (v) cant (or cross-level, or super elevation) (see Fig. 1). Even if the longitudinal level is considered as the critical factor, it is not realistic to consider only this parameter [1]. Therefore, results are more accurate if the analysis is made taking into consideration the combination between parameters [13].

Taking into account the heterogeneity of the track (in behavior and degradation) the track must be segmented into short length sections (or segment or maintenance units). The methodologies adopted are the division into constant length sections, like 100 or 200 m [1, 12] and the division based on similar structural, environmental, operational and maintenance history characteristics [1] being the last one more effective [1] but more complex.

It should be taken into account, for a realistic representation of the track, the influence of inspections, maintenance and renewal actions, like track accessibility and inspection frequency, rail lubrication, grinding and welding, ballast cleaning, tamping and stone blowing [8, 10]. For example, it is not realistic to assume that the tamping returns the track quality to its original condition [1]. Also, despites it contributes to the improvement of the track geometry condition, its use over the time can lead to the degradation of the ballast and consequently of the track geometry [10, 11].

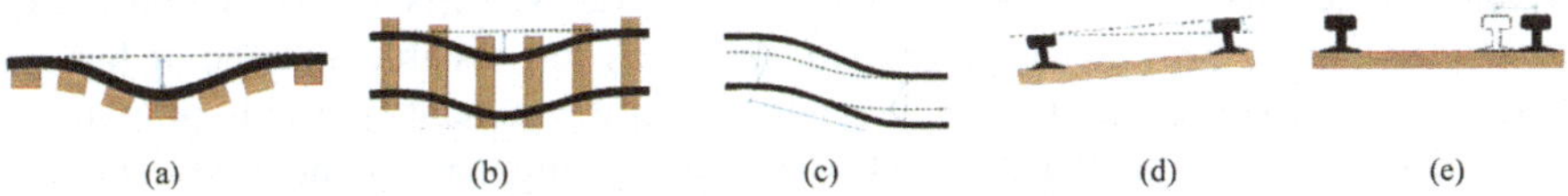

Fig. 1 Railway track quality parameters: **a** longitudinal level, **b** alignment, **c** twist, **d**, cant, **e** gauge (adapted from: [14])

There are other aspects that should be taken into account for legal and ethical aspects, such as reliability, economic loss, social impact, sustainability, not forgetting the implications for human (human injure and loss) [15]. Also, from the environmental and human healthy point of view, there are the pollution, noise and vibration produced during the use of the track.

3 Measurement of the Performance of a Railway Track

Performance indicators (PIs) capture the mechanical and technical properties of a infrastructure and its degradation process over its life cycle, considering also other aspects such as natural aging, quality of the material, serviceability, availability and robustness, sustainability, environmental efficiency (CO_2 foot-print), total life cycle costs and social indicators [3, 16]. PIs can be obtained through visual inspections, non-destructive tests or monitoring systems [2, 4, 16] and numerical and experimental modeling [17, 18]. They can be assessed through condition indices that can be obtained through ratio-based methods, weighted averaging approach, worst-conditioned component approach or qualitative methods [4]. This indices are then transformed into indexes, that are calculated based on the condition of the structural elements and the service provided by the asset, but can also be obtained through probabilistic models which calculates the failure probability [4]. Since PIs do not inform the limit values which indicates a fault or failure condition or what is intended to be obtained from an infrastructure (e.g. to be safe, to be available) performance goals need to be defined that in turn need to have performance criteria and thresholds [3, 4, 19].

Commonly, the railway infrastructure performance follows a RAMS analysis that covers the topics safety (S) and availability (A) that are based on reliability (R) and maintainability (M), more the operation and maintenance [20], taking also into account the Life cycle costs (LCC). These should follow the existing standards, that provide guidance in specifying and achieving these RAMS targets throughout the railway life cycle [21]. Among all the phases of the RAMS life cycle only those concerning the operation and maintenance and modification and retrofit are going to be covered in this paper.

The influencing parameters and RAMS formulas are given in the EN 50126-1 [21]. RAM mathematical formulas can be found in the EN 61703. Dedicated to the definitions and measurement for the technical, administrative and managerial areas of maintenance, it is worth to mention the EN 13306. The EN 15341, not dedicated exclusively to the railway system, gives the formulas and possible influencing factors, proposing a classification of PIs into groups. Figure 2 presents possible influencing parameters, as well as the RAMS and formulas according to the mentioned standards. Besides these, it is fundamental to consider also the Life cycle costs of the railway track and the influence of the track use for the environment, not directly addressed on the previous standards. According to the work presented by the Committee on Technical Cooperation in the Development of the Rail Transport System in 2016 concerning the RAMS and LCC [21], it should be followed the inspection, preventive maintenance, corrective maintenance, overhaul and cleaning costs for maintenance proposes, presenting some formulas that are shown in Fig. 3.

It is important to understand what is being done in other transport sectors, like roadway, with the aim of improving what is being done in the railway system. Looking at the work developed in the roadway sector, under COST-Action TU1406, the PIs for bridge assessment were clustered into five groups. (i) Reliability, (ii) Availability, (iii) Safety, (iv) Economy, and (v) Environment [2, 18]. This clustering was taken into account and can be adapted to the railway, particularly in the case of the PIs related with costs and environment.

4 Overview of the Predictive Models

Predictive models (or degradation models) are algorithms that analyse a set of data, identifying patterns and estimating the time until PIs reach limit values and by this way predicting the future condition of the track. This will contribute for an optimized management, allowing more efficient and integrated maintenance plans, regular inspections and monitoring.

From literature, the existing approaches to build a predictive model can be classified into (see Fig. 4), (i) Mechanistic (or Physical) and (ii) Data-driven Models existing also the (iii) Empirical Mechanistic Models [1, 24]. The Data-driven Models can be dived into Statistical and Artificial Intelligence Models [7, 11]. Each one of these approaches can be divided into various categories that in turn can be divided into various sub-categories [11].

These models need the definition, among other aspects, of the input and output variables. The first ones are the independent (or predictor) variables that are used to predict (or forecast) the second ones that are the dependent (or predicted) variables. The definition of these varies, not only according to data available and the results to be obtained, but also according to the type of model. A bigger number of independent variables can improve the accuracy and efficiency of prediction, the same way as a limited number of dependent variables [7]. Figure 5 presents, according to the literature [1, 7, 11, 25], the possible variables organized into groups.

4.1 Mechanistic and Empirical Mechanistic Approaches

In Mechanistic Models the properties of both track and train are based on laboratory experimental data, being also possible to take into consideration all the possible variables influencing the track degradation. This way the relation between the track and the vehicle can be properly clarified [1, 7]. However, these models do not deal well with the uncertainty of the behaviour of the track caused by the heterogeneity of the track, being also time consuming and complicated to have all the measurements

RAMS quantification		
Reliability	Description	Possibility of the track to perform a required function under given conditions and time interval.
	Parameters	λ(x); MUT; MTTF; MTBF; F(t); R(t)
	Formulas	$MTBF = \frac{TBF}{number\ of\ failures} = \frac{\sum(downtime\ start - uptime\ start)}{number\ of\ failures}$ $\lambda(x) = \sum \frac{1000000000\ hours}{MTBF}$ (FIT)
Availability	Description	Possibility of the track be in conditions to perform a required function under given conditions and time interval.
	Parameters	A; Ai; Ao; FA; SA
	Formulas	$A = \frac{MTBF}{(MTBF+MDT)}$ $MDT = \frac{\sum(Start\ up\ time - start\ of\ downtime)}{number\ of\ failures}$
Maintainability	Description	Possibility of a maintenance action be carried out at a track item under given conditions and time interval.
	Parameters	MDT; MTBM; MTBM(c); MTBM(p); MTTM; MTTM(c); MTTM(p); MTTR; MRT; FC; RC
	Formulas	$MTTR = \frac{\sum(end\ Corrective\ action - start\ Corrective\ action)}{number\ of\ Corrective\ action}$
Safety	Description	Freedom from unacceptable risk of harm.
	Parameters	h(t); pWSF; Active time to return to safe state
	Formulas	SIL = risk analysis where the risk associated with the hazards is calculated

Where:
λ(x) - Failure rate [1/time, 1/distance, 1/cycle]
TBF - Time Between Failures [time]
MTBF - Mean Time Between Failures (for repairable items) [time (distance, cycle)]
MTTF - Mean Time to Failure (for non-repairable items) [time (distance, cycle)]
MTTR - Mean Time to Repair [time] (for constant λ(x), constant repair rate and no preventative maintenance: MTTR=MDT
MDT - Mean Down Time [time (distance, cycle)]
MUT - Mean Up Time [time (distance, cycle)]
MTBM - Mean Time Between Maintenance [time (distance, cycle)]
MTBM(c), MTBM(p) - MTBM (corrective or preventive) [time]
MTTM - MTBM (corrective or preventive) [time]
MTTM(c), MTTM(p) - MTTM (corrective or preventive) [time]
MTTR - Mean Time to Maintain [time]
MRT - Mean Repair Time Fault [time]
FC - Fault Coverage Repair [dimensionless]
RC - Repair Coverage [dimensionless]
F(t) - Failure Probability [dimensionless]
R(t) - Reliability (Success Probability) [dimensionless]
Uptime - moment after an action on the previous failure
Downtime - moment of initiation of a failure
FIT - Failure in time
Active time to return to safe state [time]
A; Ai; Ao - Availability, inherent and operational [dimensionless]
FA - Fleet Availability [dimensionless]
SA - Schedule Adherence [dimensionless]
SIL - Safety Integrity Level
h(t) - Hazard rate [1/time, 1/distance, 1/cycle]
pWSF - Probability of wrong-side failure [dimension-less]

Fig. 2 RAMS quantification according to EN 50126-1 (adapted from: [9, 21]) and the terminologies according to the EN 13306 (adapted from: [22, 23])

of all the variables along all the track. This way, these models are considered suitable only for particular sections (and not for several sections) [1, 7, 11, 24]. Taking into account the mentioned disadvantages these models are rarely being used during the last years [7].

An alternative to these conventional Mechanistic Models are the Empirical Mechanistic Models, which are a combination of the mechanistic and the statistical models, being based on the behaviour of the system's components coupled with measurements, data records and observations. The advantage of these models compared to the conventional ones, is their ability to model the entire rail track.

Costs quantification		
Preventive maintenance costs	Formulas	$CY\_MP = \sum_{i=1}^{x} (No\_MP_i \cdot QT_i \cdot (CM\_MP_i + MH\_MP_i \cdot CMH))$
Corrective maintenance costs	Formulas	$CY\_MC = \sum_{i=1}^{x} (No\_MC_i \cdot QT_i \cdot (CM\_MC_i + MH\_MC_i \cdot CMH))$ $\quad No\_MC_i = IN\_FA_i \cdot OT$

Where:
QT = Total quantity of item
i = Maintenance action per item and failure mode
CMH = Costs per working-hour
IN_FAI = Failure rate
OT = Operating time or operating distance per life cycle (depending on the unit of the failure rate)
CY_MP = Preventive maintenance costs
CY_MP = Corrective maintenance costs
No_MP = Number of preventive maintenance actions per life cycle
No_MC = Number of corrective maintenance actions per life cycle
MH_MP = Working-hours per preventive action
MH_MC = Working-hours per corrective action
CM_MP = Average costs of material per preventive action
CM_MC = Average costs of material per corrective action

Fig. 3 Cost quantification according to committee on technical cooperation in the development of the rail transport system in 2016 concerning the RAMS and LCC (adapted from: [21–23])

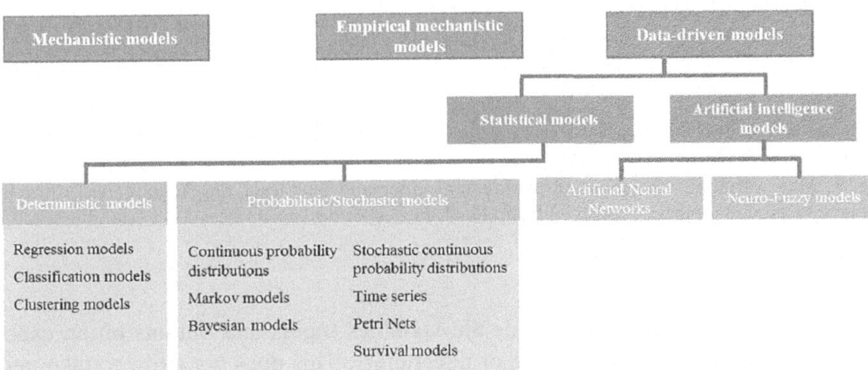

Fig. 4 Predictive model approaches (adapted from: [1, 7, 11, 25])

4.2 *Statistical Approach*

Unlike Mechanistic Models, Statistical (or empirical) Models do not consider the mechanical component of the track [11] and its interaction with the influencing factors [1], what can be seen as disadvantage. However, they are based in real data, what is considered an advantage, and in addition, they can work with a big quantity of data (both input and output), making them more accurate [1]. Soleimanmeigouni et al. [1] proposes the combination between Mechanistic and Statistical modeling. In these models the relationship between the factors influencing the track degradation, such as traffic, track components and maintenance variables, and its condition is obtained through the relation between the input variables (or descriptive factors) and output variables [1]. Statistical Models can be divided into Deterministic, Probabilistic and Stochastic Models [7, 11].

Influencing Variables	Predictive Models					
	MM	EMM	DM	PM	SM	AIM
Track Geometry	x	x	x	x	x	
Longitudinal level, alignment, gauge, cant and twist Breakage of the rail and settlement of the track						
Track Structure					x	
Type of rails, sleepers and fastening system Support and drainage system						
Track Quality Index		x		x		
Track Geometry Index + Track Structure Index						
Track Operations		x	x			
Train speed and track limit speed, traffic volume Axle weight and accumulated tonnage						
Environmental conditions				x		x
Temperature, snow and flood Soil type, falling rock, landslide						
Maintenance parameters		x	x	x	x	x
Time of inspection and renewal, Number of interventions, speed restrictions and track closures Maintenance actions as rail lubrification, grinding and welding, ballast cleaning, tamping and stone blowing						
Time		x			x	x

Notes:
MM - Mechanistic, EMM - Empirical mechanistic,
DM - Deterministic, PM - Probabilistic, SM - Stochastic and AIM - Artificial Intelligence models

Fig. 5 Main variables (or factors or parameters) influencing the track degradation, according to the type of predictive model (adapted from: [1, 7, 11, 25])

4.2.1 Deterministic Models

The Deterministic Models describes the variables inputs and outputs on an exact way, not involving any randomness or uncertainty. This does not allow to take into account possible errors and changes that can happen during measurements, as well as the variability in track performance (two similar track sections with the same type of use and maintenance may have different behaviours) [10]. This can lead to uncertainty in prediction. Besides that, these models do not apply in the same way the degradation rate on used tracks and maintained tracks, even if these different tracks are under the same loads [7]. Also, the interaction that could occur between degraded components of the infrastructure is not considered [26].

The Regression Models are a type of Deterministic Model that are generally used due to their simplicity in representing the underlying degradation path. However, they require a big quantity of measurement data to achieve an acceptable accuracy. Besides that, these models cannot be updated with new data and are independent of previous observations [1].

4.2.2 Probabilistic Models

Contrary to the Deterministic Models, Probabilistic ones involve randomness, i.e. the uncertainty is inherent to these models, and they consider the current condition state of the assets [26]. The heterogeneous degradation is represented by using a random effect for each section [1]. These models take use of distribution patterns to represent the probability of an event (e.g. failure) during an interval time [11]. The most used models are the Markov models and Bayesian Models.

- The **Markov Process** can be classified according to the nature of the time parameter and the state space. Taking into account the first one, a Markov process can be a discrete-time Markov process or a continuous-time Markov process. Relatively to the second one a Markov process can be discrete-state Markov process (also called Markov chain) or continuous-state Markov process [27]. The Markov chain consists of a set of transitions from one condition state for another, determined by a current state vector and a probability distribution, represented by a transition matrix. The current state vector, a N*1 matrix, corresponds to the possibility of starting at each one of the N possible states. The probability distribution is represented by a transition matrix N*N, where N is the number of possible conditions states (every possible state appears once in the rows and once on the columns), with each cell (i, j) corresponding to the probability of transitioning from state i to state j. This matrix satisfy the Markov property [28] (or memory-less property of the Markov process), i.e., the prediction of the future condition state is based only in the current state and not on the past state or the way the current state was achieved [11, 25, 26, 28], what does not represent the reality. The transition probabilities depend only of the current state, being independent of the age of the asset and its effects on the evolution of its degradation. This means that two assets, even if they have different ages, if they in the same current state, are going to have the same probability of changing to the next condition state [29]. Also, the transition probabilities need to be updated every time that new data is collected from a new inspection and or maintenance action what can be time consuming [26]. According [26], transition probabilities assume that the condition can keep the same or get worst, to avoid the difficulty in estimating the transition probability of assets where were performed conservation actions. These models have as limitation the fact that the transition between condition states must occur at a constant rate [7, 10]. Besides that, they do not consider in an efficiently way the effects caused by the interaction between different components with degradation [26] being limited to small track models [7].
- **Bayesian Models** (BN), like Hidden Markov Models (a particular case of Markov Models), are probabilistic graphical models, being composed of nodes and arcs (or edges or links). Here, the first one represents the random variables (or a set of variables) and the second one represents the conditional probabilistic relation between nodes. However, while in the Hidden Markov Models the arcs can go in both directions creating a cycle, in the BN the arcs can only go in one direction, not being possible to have a cycle, being this way named directed acyclic graph [29].

BN take use of the Bayes's Theorem, where the parameters, assumed as random variables, are quantified by a prior distribution (or previous distribution) combined with the likelihood to achieve the posterior distribution of the parameters. The prior distribution represents the "people beliefs" about the parameters before observing the data and the likelihood distribution represents the information given by that the observed data [11]. Hierarchical Bayesian Models (HBM) are a specific case of BN, formed by multiple sub-BN (or levels) integrated (or combined) in a hierarchical way to reach the posterior distribution by using the Bayes's Theorem, where the uncertainties in each sub-model are propagated from one level to the next. The implementation of the computational Method Markov Chain Monte Carlo (MCMC) allows to perform over and over the Bayes's Theorem [7, 11, 30], allowing the data to be updated over the time [11]. Markov Chain Monte Carlo is the application of the Monte Carlo (MC) Method by using the Markov Chain. The MC use the process of repeated random sampling to make numerical estimations of unknown parameters, being, this way, modelled the likelihoods of outcomes, which helps understanding the impact of risk and uncertainty in prediction and forecasting models. BN have the advantage of using and combining collected data (e.g. from inspections) with expert knowledge regarding variables on which no data exists. However, collecting and organizing expert knowledge (based in their belief) in a way that can be converted into probabilistic distributions can be difficult [31]. Another advantage of BN is the fact that, if some variables have a known state, it is possible to update the state of the remaining variables through an inference algorithm taking use of the Bayes' Theorem [32].

4.2.3 Stochastic Models

Stochastic Models, in opposite to deterministic ones and such as the probabilistic ones, involve randomness. The same set of input variables and initial conditions can result in different set of outputs, once it takes use of a probability distribution function [7, 11]. They aim in understanding the distribution of the degradation of the track over the time [7]. While the deterministic models are easier to use, the Stochastic ones are considered more realistic [11]. It is fundamental to consider the effect of the heterogeneity of the degradation along the track over the time due to the heterogeneity of the track in consequence of the variation in geometry, materials, traffic, environment and maintenance actions. In the following are briefly mentioned two of the different models according to [11] Time Series, Petri Nets and Survival Models.

- **Time Series** are a sequence of observations (or data collected) taken sequentially and at equally periods of time (e.g. monthly, annually), where the dependent variable is obtained in function of time, not existing an obvious independent variable. This allows to analyse the variable changes over time, being possible to analyse the past, monitor the present and predict the future [11]. According to the frequency of data different patterns can be observed and can be used to do the forecasting.

- **Petri Nets** are graphical-mathematical models composed of places and transitions, connected by arcs. Places, represented graphically by circles, simulates states, conditions or resources within the system to be modelled. Transitions, represented graphically by bars, simulates the events that occur in the system and may cause change in the condition of the system. Arcs connects places and transitions, and could be input and output arcs [11].
- **Survival models** (or failure time models) analyse the expected time until the occurrence of an event (e.g. damage, failure), having associated with the hazard function and the survival function, that analyse the probability of failure of the event (or asset) [11].

4.3 Artificial Intelligence Approach

Artificial Intelligence Models, take use of the knowledge of human brain behaviour [11], showing high accuracy compared to the mechanical and statistical approaches [7]. The models are trained with a bid quantity of data and them tested with another quantity of data, what has an impact in the accuracy of the model [7]. Once these models are recent, there is a lack of literature, what is considered as a disadvantage. Another disadvantage is that they lack the transparency that models like the mechanical or statistical ones have. These models can also present some difficulties in the calibration of the model parameters [7]. In the following are briefly mentioned two of the different models according to the literature, Artificial Neural Networks (ANNs) and Neuro-fuzzy Models (ANFIS) [7, 11].

- The **ANNs** consist in a neural network (group of independent neurons) that communicate with each other through weighted connections (synaptic weights). This network is trained, by attributing and changing the weight to the connections between the neurons until get closer to the desired outputs [14, 25].
- The **ANFIS** are a combination of ANNs with Fuzzy Inference System (FIS), integrating this way the neural networks and the Fuzzy Logic. The Fuzzy Logic uses the human decision making, working with all the possibilities between yes and no, in opposition to the conventional computer's logic, which simply uses the options of true and false that corresponds to the human's yes and no. The Fuzzy sets and the Fuzzy membership functions are the parameters of these models.

5 Conclusions

To predict the degradation of the railway track it is important to understand what influences its degradation, so it is possible to select the most appropriated performance indicators. This work can be improved by the experience gained in other infrastructures fields.

Besides that, it is important to understand which possible predictive models exists, their advantages and disadvantages, so this way, according to the available data and what is intended to predict, select the most appropriated model, so it can be possible to develop accurate maintenance plans.

It is fundamental to consider the effect of the heterogeneity of the degradation along the track over the time due to the heterogeneity of the track in consequence of the variation in geometry, materials, traffic, environment and maintenance actions. This is possible if Probabilistic and Stochastic Models are used. Using the Deterministic ones, it is not possible to consider this heterogeneity nor even measurement errors that could occur. Figure 5 shows the most important advantages and disadvantages, as well as the influencing parameters of the presented Predictive Models, as well as the most important Performance Indicators.

Acknowledgments This work was partly financed by FEDER funds through the Competitivity Factors Operational Programme—COMPETE and by national funds through FCT Foundation for

Predictive Models	Advantages	Disadvantages	Influencing Parameters
Mechanistic	- Use of few geometrical data - Based on the mechanical behaviour of the system's components	- Non-consideration of the uncertainty of the track behavior due to the heterogeneity of the track - Difficulty in quantifying the track and vehicle properties - Difficulty in understanding the interaction between the track components and properties	- Track Geometry
Statistical	- Capacity to work with a big quantity of data - Based in real data - Takes use of the distribution pattern to represent the probability of failure or disruption in a time interval	- Not based on the mechanical behaviour of the system's components	
Deterministic	- Easier to use	- Non-consideration of the randomness or uncertainty - Non-consideration of the possible interaction between degraded components	- Track Geometry - Track Operations - Maintenance parameters
Probabilistic / Stochastic	- Involve randomness - Non-consideration of the uncertainty of the track behavior due to the heterogeneity of the track - Consideration of the current state of the assets - More realistic	- Need for more statistical and computational ability	- Track Geometry - Environmental conditions - Track Geometry Index + Track Structure Index - Maintenance parameters
Artificial Intelligence	- Models trained and tested with a bid quantity of data	- Lack of information, since AI models are recent - Difficulty in calibrating the model parameters	- Environmental conditions - Maintenance parameters

Fig. 5 Advantages, disadvantages and parameters to take into consideration for predictive models

Science and Technology within the scope of the project POCI-01-0145-FEDER-007633. This work was supported by the European Commission-Shi. 2 Rail Program under the project "IN2TRACK2-826255-H2020-S2RJU-2018/H2020-S2RJU CFM-2018".

References

1. Soleimanmeigouni, I., Ahmadi, A., & Kumar, U. (2016). Track geometry degradation and maintenance modeling: A review. In *Proceedings of Institution of Mechanical Engineering Part F Journal of Rail Rapid Transit* (pp. 73–102). https://doi.org/10.1177/0954409716657849.
2. Campos e Matos, J., Casas, J. R., & Fernandes, S. (2016). Cost Action TU1406: Quality specifications for roadway bridges, standardization at a European level (BridgeSpec)—performance indicators, maintenance, monitoring safety, risk resilience bridge. Bridge Networks. In *Proceedings of 8th International Conference Bridge Maintenance, Safety Management* (pp. 935–942). IABMAS.
3. Strauss, A., Fernandes, S., Casas, J. R., Mold, L., & Matos, J. C. (2018). Quality specifications and performance indicators for road bridges in Europe, maintenance, safety, risk, management life-cycle performance bridge. In *Proceedings of 9th International Conference Bridge Maintenance, Safety Management* (pp. 1822–1831). IABMAS.
4. Stipanovic, J. R. C. I., Chatzi, E., Limongelli, M., Gavin, K., Allah Bukhsh, Z., Skaric Palic, S., Xenidis, Y., Imam, B., Anzlin, A., Zanini, M., Klanker, G., Hoj, N., Ademovic, N., & Matos, J. C. (2017). WG2 technical report: Performance goals for roadway bridges of COST ACTION TU 1406. ISBN: 978-3-900932-41-1.
5. Matthias Finger, D. K., & Bert, N. (2016). 12th European rail transport regulation summary. How to define, measure, and improve the performance of the European railway system? A summary of the presentations.
6. Ramos, A., Gomes Correia, A., Indraratna, B., Ngo, T., Calçada, R., & Costa, P. A. (2020). Mechanistic-empirical permanent deformation models: Laboratory testing, modelling and ranking. *Transportation Geotechnical, 23*. https://doi.org/10.1016/j.trgeo.2020.100326.
7. Akiyama, M., Frangopol, D. M., & Ishibashi, H. (2020). Toward life-cycle reliability-, risk- and resilience-based design and assessment of bridges and bridge networks under independent and interacting hazards: Emphasis on earthquake, tsunami and corrosion. *Structure and Infrastructure Engineering, 16*, 26–50. https://doi.org/10.1080/15732479.2019.1604770.
8. Elkhoury, N., Hitihamillage, L., Moridpour, S., & Robert, D. (2018). Degradation prediction of rail tracks: A review of the existing literature. *Open Transport Journal, 12*, 88–104. https://doi.org/10.2174/1874447801812010088.
9. Michas, G. (2012). Slab track systems for high-speed railways.
10. In2Rail—Shift2rail, I2R. (2015). Deliverable D3.4: Guideline for the evaluation and selection of innovative track solutions.
11. Audley, M., & Andrews, J. (2013). The effects of tamping on railway track geometry degradation. https://doi.org/https://doi.org/10.1177/0954409713480439.
12. Falamarzi, A., Moridpour, S., & Nazem, M. (2019). A review of rail track degradation prediction models. *Australian Journal of Civil Engineering, 17*, 152–166. https://doi.org/10.1080/14488353.2019.1667710.
13. Hingorani, R., Tanner, P., Prieto, M., & Lara, C. (2020). Consequence classes and associated models for predicting loss of life in collapse of building structures. *Structure Safety, 85*. https://doi.org/10.1016/j.strusafe.2019.101910.
14. Strauss, A., Vidovic, A., Zambon, I., Tanasic, N., & Matos, J. C. (2016). Performance indicators for roadway bridges, In *Maintenance, Monitoring Safety, Risk Resilience Bridge Bridge Networks—Proceedings of 8th International Conference Bridge Maintenance, Safety Management* (pp. 965–970). IABMAS.

15. Strauss, A., Mold, L., Bergmeister, K., Mandic, A., Matos, J. C., & Casas, J. R. (2019). Performance based design and assessment—Levels of indicators. In *Life-Cycle Analysis Assessment Civil Engineering Towards an Integration Vision—Proceedings 6th International Symposium Life-Cycle Civil Engineering* (pp. 1769–1778). IALCCE 2018.
16. Pakrashi, V., Wenzel, H., Matos, J., Casas, J., Strauss, A., Stipanovic, I., Haj-Din, R., Kedar, A., Guðmundsson, G., Limongelli, M.-P.-N., Xenidis, Y., & Palic, S. S. (2020). *WG5 Technical report: Drafting of guideline/Recommendations of Cost Action TU1406, 2019*. https://www.tu1406.eu/wp-content/uploads/2019/03/tu1406-wg5-report-final.pdf. Accessed January 17, 2020.
17. Strauss, A., Ivanković, A. M., Matos, J. C., & Casas, J. R. (2016). *WG1 technical report: Performance indicators for roadway bridges of cost action TU1406, 2016*. https://www.tu1406.eu/wp-content/uploads/2016/10/COST_TU1406_WG1_TECH_REPORT.pdf. Accessed January 17, 2020.
18. Bai, L., Liu, R., Sun, Q., Wang, F., & Xu, P. (2015). Markov-based model for the prediction of railway track irregularities. *Proceedings Institution Mechanical Engineering Part F Journal Rail Rapid Transit, 229*, 150–159. https://doi.org/10.1177/0954409713503460.
19. D'Angelo, G., Bressi, S., Giunta, M., Lo Presti, D., & Thom, N. (2018). Novel performance-based technique for predicting maintenance strategy of bitumen stabilised ballast. *Construction Building Materials, 161*, 1–8. https://doi.org/10.1016/j.conbuildmat.2017.11.115.
20. Matsumoto, M. (2008). Changing RAMS for railways: Proposals from Japan, JR EAST Technical Review, 5–8.
21. Reliability, Availability, Maintainability, Safety (RAMS) and Life Cycle Costs (LCC). (2017). Committee on technical cooperation in the development of the rail transport system/11th.
22. PRIME. (2018). *Platform of railway infrastructure managers in Europe, catalogue version 2.1 PRIME key performance indicators for performance benchmarking PRIME-Platform of Railway Infrastructure Managers in Europe*. https://webgate.ec.europa.eu/multisite/primeinfrastructure/sites/primeinfrastructure/files/12100105_prime_kpi_catalogue_2.1_final_20180530.pdf. Accessed February 5, 2020.
23. British Standard, BS EN 13306. (2018). *Maintenance—Maintenance terminology*.
24. UNIFE. (2016). IRIS international railway industry standard: GUIDELINE 4 : 2016 RAMS/LCC.
25. Monitoring, C. (2020). *Designing algorithms for condition monitoring and predictive maintenance* (pp. 1–4). https://www.mathworks.com/help/predmaint/gs/designing-algorithms-for-condition-monitoring-and-predictive-maintenance.html#mw_5d264a05-dce5-4ade-bb8e-82f3f34d2af0. Accessed February 10, 2020.
26. Soleimanmeigouni, I. (2020). *Predictive models for railway track geometry degradation*, Luleå University of Technology, Luleå, Sweden, 2019. www.LTU.se. Accessed April 1, 2020.
27. Shafahi, Y., Masoudi, P., & Hakhamaneshi, R. (2008). Track degradation prediction models, using Markov Chain, artificial neural and neuro-fuzzy network. In *8th World Congress Railway Research* (pp. 1–9), Seoul, Korea. https://www.railway-research.org/IMG/pdf/i.1.1.1.3.pdf.
28. Morcous, G., Rivard, H., & Hanna, A. M. (2002). Modeling bridge deterioration using case-based reasoning. *Journal of Infrastructure Systems, 8*, 86–95. https://doi.org/10.1061/(ASCE)1076-0342(2002)8:3(86).
29. Regado, T., Gonçalves, J. C. M. R. G., Tiago, B. G., Regado, & Matos, J. C. (2015). Desenvolvimento de um Modelo de Desempenho para Infraestruturas Ferroviárias aplicado à Linha Férrea. In *4° Congress Nac. Sobre Segurança e Conserv* (p. 135). Pontes - ASCP'2015. https://www.researchgate.net/publication/282654112_Desenvolvimento_de_um_Modelo_de_Desempenho_para_Infraestruturas_Ferroviarias_aplicado_a_Linha_Ferrea. Accessed February 21, 2020.
30. Zakeri, J. A., & Shahriari, S. (2012). Developing a deterioration probabilistic model for rail wear. *International Journal Traffic, 1*, 13–18. https://doi.org/10.5923/j.ijtte.20120102.02.
31. Santamaria, M., Fernandes, J., & Matos, J. C. (2019) Overview on performance predictive models—Application to bridge management systems. In *IABSE Symposium Guimaraes 2019 Towards a Resilient Built Environment Risk Asset Management—Report* (pp. 1222–1229).

32. Graves, T. L., & Hamada, M. S. (2009). A demonstration of modern Bayesian methods for assessing system reliability with multilevel data and for allocating resources. *International Journal Quality Statical Reliability, 2009,* 1–0. https://doi.org/10.1155/2009/754896.

Application of Fragility Analysis to Timber-Framed Structures for Seismic and Robustness Assessments

Leonardo G. Rodrigues, Jorge M. Branco, Luís A. C. Neves, and André R. Barbosa

Abstract In the past few years, the construction of multi-storey timber buildings has increased significantly in locations where high-intensity ground motions are likely to occur. On the other hand, the fast development of wood engineered products, as glued-laminated timber (GLT) and cross-laminated timber (CLT), has been challenging researchers to provide adequate guidelines for the design and assessment of structures built in seismic regions. Some guidelines and analysis methods considered in seismic design can improve robustness, commonly described as the ability of structures to sustain limited damage without disproportionate effects. This paper proposes a probabilistic methodology for seismic and robustness assessment of timber-framed structures. The seismic performance and the progressive collapse potential of a three-storey building are here exemplified through the proposed methodology, which accounts for uncertainties in mechanical properties of members and connections, as well as for external loads. The Latin Hypercube Sampling (LHS) was used in each assessment to generate a set of 1000 structural models. Each structural model corresponds to a realization of the random variables used to define the structural model. Incremental dynamic analyses were performed to develop seismic fragility curves for different damage levels. The fragility functions for robustness assessment were developed for distinct damage scenarios, exploiting the results of an alternate load path analysis (ALPA) that involved the performance of nonlinear static analyses (pushdown analyses). The methodology presented is suitable for risk-based assessments that consider the occurrence of different exposures, such as earthquakes, impacts, and explosions, while considering the direct and indirect consequences of failures. However, the methodology involves time-consuming analyses with distinct load scenarios, which can constitute a burdening task within a typical building design phase.

L. G. Rodrigues (✉) · J. M. Branco
ISISE, Department of Civil Engineering, University of Minho, Guimarães, Portugal

L. A. C. Neves
Resilience Engineering Research Group, University of Nottingham, Nottingham, UK

A. R. Barbosa
School of Civil and Construction Engineering, Oregon State Univ, Corvallis, OR 97331, USA

© The Author(s), under exclusive license to Springer Nature Switzerland AG 2021

J. C. Matos et al. (eds.), *18th International Probabilistic Workshop*, Lecture Notes in Civil Engineering 153, https://doi.org/10.1007/978-3-030-73616-3_12

Keywords Timber structures · Seismic assessment · Robustness assessment

1 Introduction

The performance of timber structures under intense earthquake ground shaking depends strongly on the type of failure modes, and in particular, their ductility. The failure of timber elements is usually brittle, whereas the failure of connections between timber elements can be ductile. Being constructed in seismic areas, the requirements promoted by the design codes are objectively stricter regarding connections and the sizes of timber members. Seismic design recommendations for timber structures focus on the formation of inelastic deformations of joints by increasing fastener slenderness, guaranteeing that failure occurs after yielding of fasteners. This measure aims to enhance the capacity of joints to withstand large inelastic deformations without rupture. Moreover, seismic resistant multi-storey buildings must have adequate bi-directional resistance and stiffness. Diaphragms are another key component since these allow the inertial forces to be transferred to the different vertical structural elements [1].

Robustness is another essential concept when studying multi-storey timber buildings, described as the ability of structures to avoid progressive collapse due to local damage resulting from an unpredictable event. Some of the aspects sought in seismic resistant systems in design codes include recommendations against progressive collapse due to unforeseen events such as explosions and impacts [2]. For multi-storey timber buildings, robustness is strongly dependent on the structural system and its ability to redistribute loads to undamaged parts, and thus, on its redundancy and ductility. After the loss of a load-bearing element, alternate load paths can be triggered through the capacity of connections to deform along with the development of tension stresses in beams and floors, the so-called catenary action. Moreover, in structures where long continuous beams are used, the catenary action can be substituted by an increment of bending forces, as the connections between these elements and the floor components withstand large deformations without experiencing brittle failure modes such as splitting, due to tension stresses perpendicular to the grain, and block shear. Another mechanism likely to occur is the membrane action of CLT floors, which is dependent on the inter-panel connections [3].

The main objective of this paper is to propose a probabilistic methodology for seismic and robustness assessment of timber structures leading to the development of fragility functions. A three-storey building is evaluated as an application example of the methodology presented.

2 Fragility Functions for Timber Structures

A fragility function F_R is defined as the conditional probability of the structural demand parameter (D) exceeds the structural capacity (C) for a given level of intensity measure (IM) of the hazard:

$$F_R = P[D > C|IM] \tag{1}$$

In the scope of the present work, the fragility curves addressed are obtained through analytical methods that include numerical analysis performance. Timber, as a natural material, presents a high variability in its mechanical properties, which contributes directly to uncertainties associated with the expected strength and deformation capacities of structures. In addition, it is also crucial to account for uncertainties related to external actions [3]. In terms of seismic assessment, the uncertainties in the demands are modelled by including ground motion record-to-record variability. In contrast, robustness assessments are associated with an initial damage imposed to the structure, which are assumed random by considering distinct scenarios. Additionally, in both types of analyses, the applied gravity loads are assumed as random variables. A considerable number of numerical simulations are typically needed to achieve robust fragility function results, requiring high computational costs. Such drawbacks can be partially overcome by using efficient sampling methods, such as Latin Hypercube Sampling, Importance Sampling, or others [4], and the use of high-throughput or high-performance computing [5].

2.1 *Fragility Functions for Seismic Assessment*

The methodology presented herein starts with the definition of the mechanical properties considered as random variables. Köhler et al. [6] propose probabilistic distributions and their statistical parameters for reference mechanical properties of timber (e.g., bending moment capacity, bending modulus of elasticity, and density). Additionally, it is crucial to consider the uncertainties related to connection response, where the performance of experimental tests can play a central role in the calibration of suitable numerical models [7]. A second phase involves the ground motion selection, which must consider the respective response spectra of the site, which can be retrieved from the standards and local documents. The different time-history records are usually scaled to elastic acceleration response spectra, whereas other sources of uncertainty can be considered, such as phasing and event duration [8]. The third phase includes the numerical modelling of the structural set, using simulation techniques, in a suitable nonlinear finite element analysis software. In the fourth step, the structural models are subjected to distinct accelerograms of increasing intensities (i.e., multi-record IDA). The fifth and final step corresponds to the definition of the seismic fragility functions, which is accomplished by fitting a lognormal distribution to the values of intensity measure (e.g., spectral acceleration) that caused the exceedance of a predefined demand threshold (e.g., drift, damage, or collapse).

2.2 Fragility Functions for Robustness Assessments

It is possible to understand that the concept of robustness is intrinsically connected with the terms "progressive collapse" and "disproportionate collapse." One considers that a progressive collapse refers to a series of components' failures that resulted from an unrelated initial component failure. It means that an initial failure triggered a chain reaction of failures throughout the structure. On the other hand, a disproportionate collapse occurs when the subsequent structural collapse is exaggerated ("disproportionate") compared to the initial failure. Therefore, a progressive collapse can be disproportionate, but the reverse is not necessarily true. To avoid ambiguities, Starossek and Haberland [9] proposed a definition of robustness using the probability of disproportionate collapse, which is given by:

$$P(DC) = P(DC|D) \cdot P(D|EX) \cdot P(EX) \quad (2)$$

where $P(EX)$ is the probability of occurrence of a particular exposure EX, $P(D|EX)$ is the conditional probability of initial damage D given the exposure EX, and $P(DC|D)$ is the conditional probability of disproportionate collapse DC given the initial damage D. In the scope of this work, the fragility functions focus exclusively on robustness analysis, which corresponds to the quantification of the conditional probability $P(DC|D)$. Failure and damage are terms used interchangeably herein when applied to structural components, and both mean that the performance criteria established were not fulfilled; e.g., a connection is damaged when it fails due to excessive deformation, or a column is damaged when the loading exceeds an ultimate shear capacity. When addressing the whole structure, failure shares the definition established for structural components, while the term damage refers to a deviation from the designed configuration that adversely affects functionality due to failure of a structural element.

As per seismic fragilities, the methodology starts with the determination of the mechanical properties considered as random variables. A second phase consists of the numerical modelling of a set of structures, using an adequate simulation technique, in a nonlinear finite element analysis software. The third phase includes the definition of distinct initial damage scenarios that can occur due to the exposures EX considered. The fourth step comprises an alternative load path analysis (ALPA), which aims to evaluate load redistribution through a pushdown analysis that starts with the notional removal of one or more elements to simulate the initial damage D [10]. The numerical analyses can be load controlled or displacement controlled and finish when the damage reaches an extent that is considered disproportionate to the original cause. The fifth and final step corresponds to the definition of the progressive collapse fragility function, which is accomplished by fitting a lognormal distribution to intensity measure values (e.g., dynamic load factor or vertical load applied) associated with disproportionate collapse.

3 Case Study: A Three-Storey Heavy-Timber Frame with Ring-Doweled Moment-Resisting Beam-Column Connections

The structure under analysis is a three-storey building constructed using two different lateral systems. The structure consists of three moment-resisting timber frames that ensure lateral resistance and stiffness on the x-z plane, as shown in Fig. 1a. Two braced timber frames provide lateral resistance and stiffness in the perpendicular direction, as presented in Fig. 1b. The structural system was designed in Callegari et al. [11], aiming at a ductility level compatible with High Ductility Class (DCH), as defined in EC8 [5]. For this study, the structure was re-evaluated using EC5 [12] and EC8 [5] for a site in Lisbon, Portugal.

3.1 Lateral Resisting Systems

The moment-resisting frames have continuous 13.2m long GL24h beams, as per EN14080 [11], with a 160 by 600 mm cross-section. The columns consist of two posts, each with a rectangular cross-section of 160 by 600 mm. The connections between columns and beams are executed with ring-dowelled joints with two layers of connectors. All dowels are from strength class 4.6 with a diameter of 12 mm. The connection used at the column bases consists of a hinged connection with a low rotational stiffness in the x-z plane. The lateral resisting system uses GL24h members interconnected through concealed steel plates and bolts in the perpendicular direction, with additional details provided in [14]. As presented in Fig. 1b, in the y-z plane, continuous beams with a 200 by 360 mm cross-section span over two 4m long bays. The columns have the same cross-sections as the beams, and their base shoes are built with slotted steel plates that are welded to a pinned support. The diagonals have a cross-section of 140 by 240 mm. The connection between the two different lateral systems is guaranteed through concealed steel plates and M12 bolts. Further information can be obtained in [14].

3.2 CLT Diaphragm

The CLT floor solution presented in Fig. 2a consists of 5-ply panels (nominally 140.0 mm thick) with the major strength along the building's Y-direction. The panels are manufactured with lumber from strength class C24. The CLT to beam connection is executed with ASSY plus VG 8 × 260 mm, as presented in Fig. 2b. The fasteners were designed to transfer inertial loads from the diaphragm to the seismic lateral resisting systems. The shear transfer between adjacent CLT panels is guaranteed through half-lap joints with 100 mm wide laps connected with ASSY 3.0 ECOFAST 8 × 120 mm fastened 100 mm on centre. The diaphragm moments are resisted by chord splices that consist of S235 structural steel plates (6.35 mm × 50.8 mm × 1500 mm) that meet the requirements of standard EN 10025:2004 [15]. These plates are fastened to the CLT panels with Simpson Strong-Tie 6.4 × 90 HEAVY-DUTY connectors spaced 60 mm on centre.

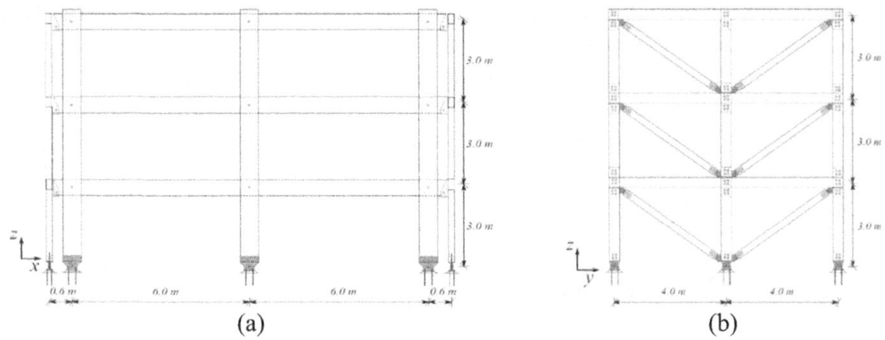

Fig. 1 Lateral resisting systems: **a** Moment-resisting frame. **b** Braced frame

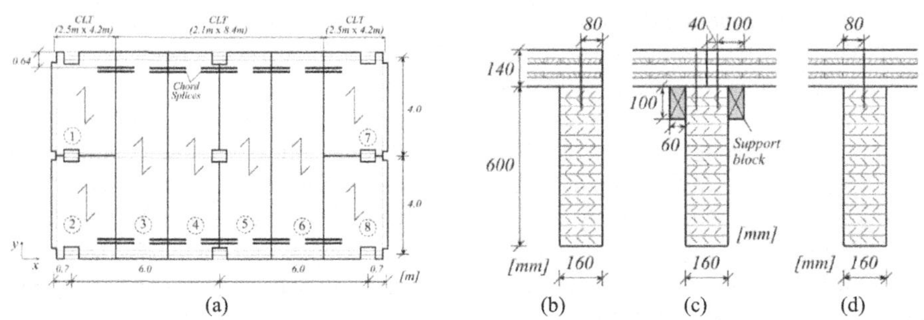

Fig. 2 CLT diaphragm: **a** Plan view. **b** CLT-to-beam connection at diaphragm edges. **c** Connection of an interrupted CLT panel to a central beam. **d** Connection of a continuous CLT panel to a central beam

3.3 Uncertainties in Material Properties

The Latin Hypercube Sampling (LHS) was used to generate 1000 structural models for each type of analysis performed in this work. Each model corresponds to a realization of the random variables assumed. The inherent uncertainties of timber, as a material, are considered by assigning different mechanical properties to each element. As presented in Table 1, seven random variables are evaluated for each timber element. The distribution parameters of the reference properties (bending strength, bending modulus of elasticity, and density) are computed based on characteristic values, defined in EN14080 [16] for the homogeneous GL24h strength class, and on the coefficients of variation and probabilistic distributions proposed in Köhler [6]. The expected values and the coefficients of variation of the remaining properties are computed according to the expressions proposed by [9] for Nordic softwood. Table 1 also presents the intra-element correlation coefficients considered, which were also taken from the work developed by Köhler [6]. Moreover, variability is also assumed for steel-to-timber connections with multiple bolt joints, overlap connections between CLT and glulam beams. The yielding force and the initial stiffness of each connection are modelled with lognormal distributions. Given the lack of experimental tests, a coefficient of variation equal to 15% is assumed for both properties. However, the uncertainties related to the mechanical models used shall be confirmed through testing in future studies. The remaining values used to establish the backbone curves are calculated by assuming ratios between each key point and the yielding point in terms of displacements and forces. These ratios were obtained through a fitting procedure based on experimental tests performed on similar connections. For the sake of brevity, further information on the calibration of Pinching4 can be obtained from [14]. The CLT floor members are modelled using an orthotropic material with deterministic mechanical properties [17].

4 Seismic Assessment

A set of 24 ground motion records was extracted from the PEER database (PEER 2012) and scaled to the spectra defined in EC8 [1]. Two types of seismic response spectra were considered when designing for a Lisbon site: a large magnitude and far-field earthquake (Type I) and a lower magnitude and near-field earthquake (Type II). The 5% linearly damped response spectra of the scaled ground motions are shown in Fig. 3. Additionally, the range of periods of interest (i.e., those within 0.3 and 3.0 times the median fundamental period of the structure) are also indicated in Fig. 3.

An incremental dynamic analysis (IDA) was performed on a set of 2D finite element models of the X-direction central moment-resisting frame (see Fig. 1a) in Opensees [18]. The beams and columns were modelled using linear elastic frame elements. The ring-dowelled connections were represented using zero-length elements and assigned Pinching4 force-deformation springs, including pinching,

Table 1 Random variables for timber material properties

Distribution parameters				Correlation matrix							
X	Dist	$E[X]$	$CoV[X]$		R_m	E_m	ρ_{den}	$R_{t,0}$	$R_{c,0}$	G_v	R_v
R_m	LN	31	0.15	R_m	1.0	0.8	0.6	0.8	0.8	0.4	0.4
E_m	LN	11,500	0.13	E_m		1.0	0.6	0.6	0.6	0.6	0.4
ρ_{den}	N	420	0.10	ρ_{den}			1.0	0.4	0.8	0.6	0.6
$R_{t,0}$	LN	18.6	0.18	$R_{t,0}$				1.0	0.5	0.4	0.6
$R_{c,0}$	LN	23.4	0.12	$R_{c,0}$					1.0	0.4	0.4
G_v	LN	718.8	0.13	G_v						1.0	0.6
R_v	LN	3.12	0.15	R_v							1.0

Description: R_m—bending strength // to the grain (N/mm^2); E_m—bending modulus of elasticity (N/mm^2); ρ_{den}—density (kg/m^3); $R_{t,0}$—tension strength // to the grain (N/mm^2); $R_{c,0}$—compression strength // to the grain (N/mm^2); G_v—shear modulus (N/mm^2); R_v—shear strength (N/mm^2)

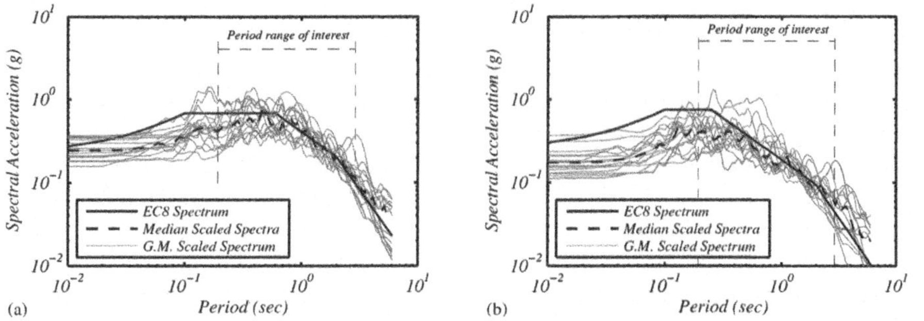

Fig. 3 Response spectra used to perform multi-record IDA: **a** Type I. **b** Type II

stiffness degradation, and strength degradation phenomena observed in the experimental results [19]. The Pinching4 model is defined by a response envelope, unloading and reloading rules, and three damage rules that control the evolution of deterioration mechanisms based on strength, stiffness, and energy dissipation. A total of 1000 samples were generated based on the assumed random variables and their distributions. Thus, a total of 24,000 curves were developed for the entire structure set. For simplicity, only the results obtained for a median structure are presented in this document. The IDA results obtained for the median structure (i.e., structure considering all random variables set equal to their median value) are shown in Fig. 4a. The fragility curves presented in Fig. 4b were obtained by fitting a lognormal distribution to the spectral accelerations that led to the exceedance of the Life Safety (LS) limit state threshold [13]. From the LS fragility functions, graphically represented in Fig. 4b, it can be observed that the limit state inter-storey drift ratios influence both the expected value and the coefficient of variation, indicating that their uncertainty should also be explicitly considered in future studies.

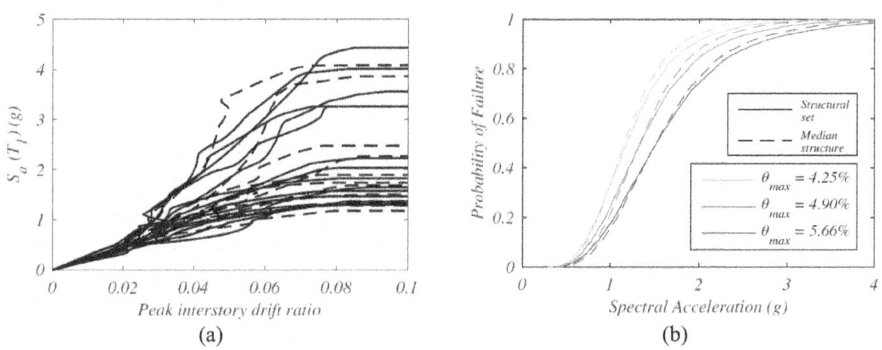

Fig. 4 Seismic assessment: **a** IDA curves obtained for the analysis of the median structure using the 24 ground motions selected. **b** Life safety fragility curves for distinct inter-storey drift levels θ_{max}.

5 Robustness Assessment

A 3D finite element model was used for the progressive collapse assessment. Linear elastic frame elements represent beams and columns, while elastic orthotropic ShellMITC4 four-node elements modelled the CLT panels. Geometric nonlinearities were incorporated in the form of P-Δ effects for columns and corotational transformation for beams. The connections were represented by zero-length elements that account for their nonlinear behaviour, incorporated through the Pinching4 material available in Opensees. The total number of pushdown analyses is given by the product of the number of models ($N_{sim} = 1000$) and the number of element removal scenarios considered ($N_{ele} = 4$); the four column removal scenarios considered are labelled scenarios D_1 to D_4 in Fig. 5. Thus, the progressive collapse assessment included a total of 4000 nonlinear static pushdown analyses. After each displacement increment step, a global failure assessment. It is worth noting that disproportionate collapse (DC) is assumed to occur when main beams, columns and connections, located at adjacent bays, reach their capacity. Although the structure was designed according to the Eurocodes, for the application of the ALPA method from the U.S. standard, US UFC 4-023-03 [20] was used. This choice was based on the fact that the latter code is more comprehensive and detailed than Eurocode 1-7 [21]. Hence, the load combination assumed for the ALPA method is given in the format:

$$G = \Omega[1.2DL + 0.5LL] \tag{3}$$

where DL is the dead load, LL is the live load, and Ω is a dynamic amplification factor. The dead loads are modelled with a lognormal probability distribution to avoid negative values; their mean value is assumed to be equal to the nominal value. The live load (LL) is modelled by a lognormal distribution with a mean value of 2.0 kN/m^2 and CoV equal to 0.30. From the performance of each pushdown analysis, an overload factor is calculated, which is given by the ratio between the load that a damaged structure is able to sustain (G_R) and its nominal gravity loads G_n ($\Omega = 1.0$).

The majority of the collapse cases were reached after the failure of one ring-dowelled connection. Several analyses returned the failure of one main beam due to excessive bending stresses. From the numerical results obtained, it is possible to state that the initial damage scenarios D_2 and D_3 led to collapse cases that did not exploit the ability of joints to deform. In these cases, the beams located just above the damaged elements sustained the loads and failed due to excessive bending stresses. The primary mechanism used to support the loads due to local failures was the beam action in the main GLT beams. Fragility functions, presented in Fig. 6, were defined for different damage state levels by fitting a lognormal distribution to the overload factor values that caused the exceeding of predefined demand threshold values.

The damage state levels were based on the rotational capacity of floor-to-beam connections and the extent of the damaged area. The first failure state (FF) is reached when any floor-to-beam connection exceeds its capacity. In turn, the collapse prevention (CP) limit state was exceeded when the damage was extended to an adjacent

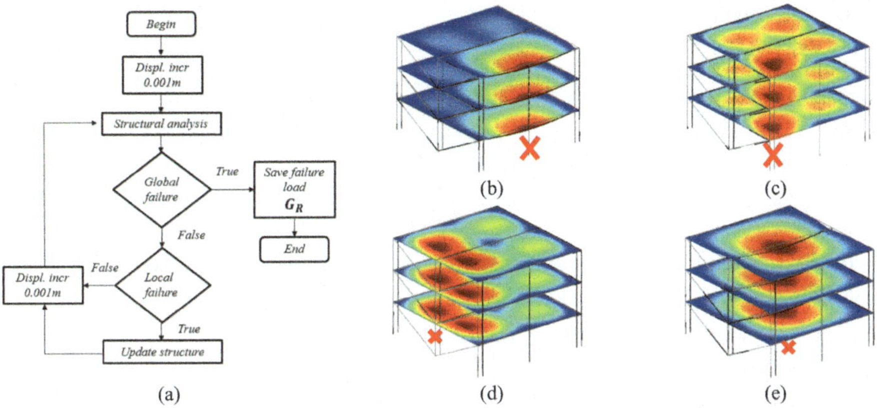

Fig. 5 Progressive collapse assessment: **a** pushdown procedure; **b** scenario D_1; **c** scenario D_2; **d** scenario D_3; **e** scenario D_4

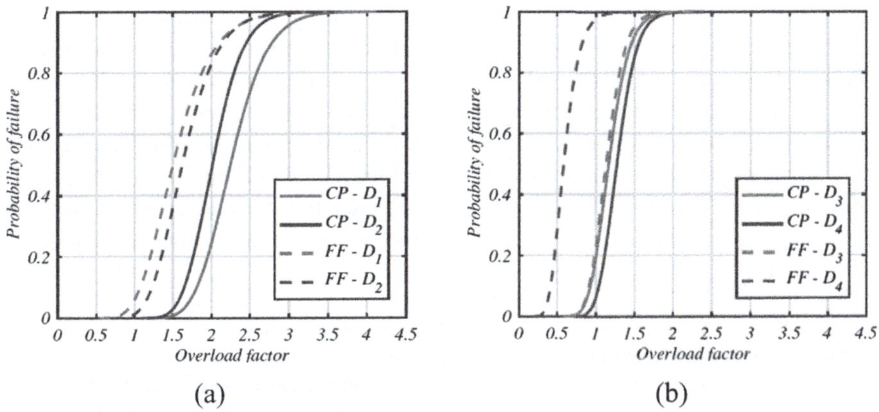

Fig. 6 Progressive collapse assessment: **a** probability density functions for maximum vertical displacements of the model structure for distinct elements loss scenarios; **b** pushdown curves for distinct scenarios

bay, which was verified when moment-resisting connections and glulam members reached their load bearing capacity. The vertical loads used to perform an ALPA method were multiplied by a dynamic amplification factor [22] equal to 2.0 for timber structures. To assess if the seismic design provisions used to design the structural system also guarantee structural robustness against progressive collapse, the probability of failure associated with an overload factor of 2.0 is evaluated for each damage scenario in Table 2.

From the results presented, it is possible to state that the application of seismic design provisions does not dismiss a progressive collapse assessment during. The main failure modes observed were related to bending stresses of GLT beams, followed

Table 2 Probability of failure (p_f) and respective reliability indexes (β) for an overload factor of 2.0

Damage scenario	Failure state	p_f	β
D_1	$P(CP\|D_1)$	0.135	1.103
D_2	$P(CP\|D_2)$	0.002	2.821
D_3	$P(CP\|D_3)$	0.980	−2.052
D_4	$P(CP\|D_4)$	0.439	0.153

by the loss of capacity of moment-resisting joints to transfer loads from GLT beams to columns.

6 Conclusions

The paper presented the implementation of a methodology suitable to perform structural assessments for seismic and robustness analyses. A prototype structure designed according to EC5 and EC8 for a site in Lisbon, Portugal, was presented as a case study. The methodology presented is suitable for risk-based assessments that consider different exposures, such as earthquakes, impacts, and explosions, while considering the direct and indirect consequences of failures. However, the methodology involves time-consuming analyses with distinct load scenarios, which can constitute a burdening task within a typical design phase of a building. Thus, the subsequent efforts can focus on developing software packages to alleviate the modelling and analysis time while still allowing for uncertainty propagation throughout the methodology.

Acknowledgments This work was partly financed by FEDER funds through the Competitiveness and Internationalization Operational Programme COMPETE, Portugal 2020, and by national funds through FCT—Foundation for Science and Technology within the Timquake project POCI-01-0145-FEDER-032031, and PhD grant PD/BD/113679/2015 included in the InfraRisk-PhD program.

References

1. CEN. (2013). EN 1998–1: Eurocode 8: Design of structures for earthquake resistance Part 1: General rules, seismic actions and rules for buildings. European Committee for Standardisation.
2. Branco, J. M., & Neves, L. A. (2011). Robustness of timber structures in seismic areas. *Engineering Structures, 33*(11), 3099–3105.
3. Huber, J. A., Ekevad, M., Girhammar, U. A., & Berg, S. (2019). Structural robustness and timber buildings—A review. *Wood Material Science & Engineering, 14*(2), 107–128.
4. Melchers, R. E., & Beck, A. T. (2018). *Structural reliability analysis and prediction.* Wiley.
5. Köhler, J., Sørensen, J. D., & Faber, M. H. (2007). Probabilistic modelling of timber structures. *Structural Safety, 29*(4), 255–267.

6. Casagrande, D., Bezzi, S., D'Arenzo, G., Schwendner, S., Polastri, A., Seim, W., & Piazza, M. (2020). A methodology to determine the seismic low-cycle fatigue strength of timber connections. *Construction and Building Materials, 231,* 117026.
7. Vamvatsikos, D., & Fragiadakis, M. (2010). Incremental dynamic analysis for estimating seismic performance sensitivity and uncertainty. *Earthquake Engineering & Structural Dynamics, 39*(2), 141–163.
8. Starossek, U., & Haberland, M. (2010). Disproportionate collapse: Terminology and procedures. *Journal of Performance of Constructed Facilities, 24*(6), 519–528.
9. Ellingwood, B. R., Smilowitz, R., Dusenberry, D. O., Duthinh, D., Lew, H. S., & Carino, N. J. (2007). Best practices for reducing the potential for progressive collapse in buildings.
10. Alam, M. S., & Barbosa, A. R. (2018). Probabilistic seismic demand assessment accounting for finite element model class uncertainty: Application to a code-designed URM infilled reinforced concrete frame building. *Earthquake Engineering & Structural Dynamics, 47*(15), 2901–2920.
11. Callegari, E. (2009). *Caratterizzazione del comportamento di telai sismoresistenti in legno lamellare, MS thesis (in Italian).* Trento, Italy: Universita degli Studi di Trento.
12. CEN. (2004a). EN 1995-1-1:2004—Eurocode 5: Design of timber structures—Part 1-1: General—Common rules and rules for buildings (Vol. 1).
13. Rodrigues, L. G., Branco, J. M., Neves, L. A., & Barbosa, A. R. (2018). Seismic assessment of a heavy-timber frame structure with ring-doweled moment-resisting connections. *Bulletin of Earthquake Engineering, 16*(3), 1341–1371.
14. Rodrigues, L. G. (2019). Robustness of multi-storey timber buildings in seismic regions, Ph.D. thesis, University of Minho, Guimarães, Portugal.
15. . (2004b), EN 10025-2:2004, European standard for hot-rolled structural steel. Part 2—Technical delivery conditions for non-alloy structural steels.
16. CEN. (2005a), EN 14080 Timber structures—Glued laminated timber Requirements. European Committee for Standardisation.
17. Blass, H. J., & Fellmoser, P. (2004). Design of solid wood panels with cross layers. In *Proceedings of the 8th World Conference on Timber Engineering 2014,* 14–17 June, Lahti, Finland.
18. McKenna, F. (2011). OpenSees: A framework for earthquake engineering simulation. *Computing in Science & Engineering, 13*(4), 58–66.
19. Polastri, A., Tomasi, R., Piazza, M., & Smith, I. (2013). Moment resisting dowelled joints in timber structures: Mechanical behaviour under cyclic tests. *Ingegneria Sismica, 30*(4), 72–81.
20. United States Department of Defense. (2016). UFC 4-023-0.3. Design of buildings to resist progressive collapse.
21. CEN. (2006). EN 1991-1-7 (2006) Actions on structures. Part 1-7: General actions—Accidental actions. European Committee for Standardisation.
22. ARUP. (2011). Review of international research on structural robustness and disproportionate collapse (Tech. Rep.), London.

Assessment of Design Concepts for Post-installed Punching Shear Retrofitting

Oladimeji B. Olalusi, Puneh Mowlavi, Nikolaos Mellios, and Panagiotis Spyridis

Abstract Punching shear is a brittle form of failure observed in reinforced concrete slab structures and occurs without any visible signs before failure. This phenomenon typically arises around the slab-column connections, due to transverse forces being highly concentrated in these areas and can cause that the column punches through the slab. This type of failure is very brittle. The unpredictability of its occurrence makes it a particularly critical and dangerous phenomenon. Several methods have been developed for retrofitting and strengthening existing flat slabs against punching shear failure using different reinforcement-types, like shear bolts, screw anchors or bonded anchors. These methods are called post-installed shear reinforcement for existing flat slab systems. This study aims to assess the safety and economic performance of the Eurocode 2 (EC2) design method for the design of post-installed reinforcement in an existing flat slab structure endangered by punching shear, using probabilistic analysis. The probabilistic analysis was conducted based on the Monte Carlo simulation technique implemented using a MATLAB code developed in the study. The reliability indices obtained for EC2 design procedure were found to be close to the EN 1990 target reliability level.

Keywords Punching shear · Flat slab · Eurocode · Reinforced concrete · Retrofitting · Probabilistic analysis

1 Introduction

Flat slabs are one of the most widely used concrete floor systems. The most critical aspects of a flat slab system are the column support joints. These areas are considered the starting point of a brittle and sudden failure caused by shear and flexural tension [1]. Generally, flat slabs are prone to punching failure, which occurs when a flat slab

O. B. Olalusi
University of KwaZulu-Natal, Durban, South Africa

P. Mowlavi · N. Mellios · P. Spyridis (✉)
Technical University of Dortmund, Dortmund, Germany
e-mail: panagiotis.spyridis@tu-dortmund.de

© The Author(s), under exclusive license to Springer Nature Switzerland AG 2021
J. C. Matos et al. (eds.), *18th International Probabilistic Workshop*, Lecture Notes in Civil Engineering 153, https://doi.org/10.1007/978-3-030-73616-3_13

system is overloaded, and the slab fails within a distance around the column. Novacek and Zich [2] described punching shear as a type of failure of reinforced concrete slabs due to shear forces. These forces are highly localised at column support points. Punching shear failure is very brittle and occurs without any visible signs before failure, which makes it a critical high phenomenon. Another significant issue is that the redistribution of the inner forces is minimal during the failure. This may lead to a progressive collapse of the structure [2], such as the collapse reported in Beutel [1] and Kunz et al. [3]. Many of such failures could be prevented if the concrete slabs were adequately retrofitted. Therefore, the design of flat slabs must be accorded proper and adequate attention to avoid failures.

Retrofitting existing structures is an important topic that has been given serious attention in the last decade. The older a building gets, the higher the need for inspections, health monitoring and maintenance. Nowadays built flat slabs are strengthened with shear reinforcement according to the existing code requirements to assure that their strength is sufficient to prevent failures. Existing older flat slabs supported by columns, however, must be strengthened against punching shear failure, in the case of insufficient existing strength. The reason for this can be attributed to higher code requirements as a result of increased knowledge gained in the past years on the topic. Other reasons are the change of use of the building and thus the increasing loads during the lifetime of the structure, but also construction and design errors [4].

Different methods have been proposed to improve and strengthen existing structures. One of these methods is called post-installed shear reinforcement for existing flat slab systems, where the used reinforcement-type is installed in the critical slab-column area of the slab. The methods use different reinforcement-types such as shear bolts, screw anchors or bonded anchors [5]. The estimation of the punching shear resistance depends on several variables (geometry, mechanical, material properties, etc.) with some degree of uncertainty. This study aims to investigate the safety performance of the European Eurocode 2 (EC2) design code [6, 7] for the design of post-installed reinforcement in flat slab structure endangered by punching shear, using probabilistic analysis. The probabilistic analysis is conducted based on the Monte Carlo simulation technique implemented using a MATLAB code developed in this study. This contribution intends to present a safety efficient method for the design of post-installed reinforcement in existing structures.

2 Punching Shear Resistance Formulations for Flats Slabs with Post-installed Retrofitting

The equations presented in this section are used to determine the punching shear capacity of a flat slab without shear reinforcement, and the contribution of the anchors used as post-installed reinforcement. These equations allow the characterization of the total resistance of the system.

2.1 Shear Resistance Formulations of the Eurocode [6, 7]

The characteristic resistance of a slab without punching shear reinforcement in $\left(\frac{MN}{m^2}\right)$ is expressed by Eq. (1).

$$v_{Rdc} = C_{Rd,c}k(100\rho_l f_{cm})^{\frac{1}{3}} \tag{1}$$

where f_{cm} = mean compressive concrete strength; $k = 1+\sqrt{\frac{200}{d}} \leq 2$ with d in (mm); $\rho_l = \sqrt{\rho_{ly}\rho_{lz}} \leq 0.02$ = longitudinal reinforcement ratio for the y- or z-axis; $C_{Rdc} = 0.18$ [6].

The equations of the anchors in tension according to the EC2 Part 4 [7] are expressed below.

The yield strength of an anchor (in kN) is expressed by Eq. (2).

$$N_{Rm,s} = A_i f_{yvm} \tag{2}$$

where A_i = cross-sectional area of the anchor in mm^2; f_{yvm} = mean yield strength of the anchor in $\frac{N}{mm^2}$.

The pull-out failure of an anchor in tension (in kN) is expressed by Eq. (3).

$$N_{Rm,p} = k_2 A_{brg} f_{cm} \tag{3a}$$

where $k_2 = 10.5$ (for anchors in uncracked concrete); d_h = head diameter in mm; d_a = shaft diameter in mm

$$A_{brg} = 0.25\left(d_h^2 - d_a^2\right) \tag{3b}$$

The bond strength of an anchor in tension (in kN) is expressed by Eq. (4).

$$N^0_{Rm,a} = \frac{h_{ef}\pi\Phi_1\tau_{bm}}{\alpha_1\alpha_2} \tag{4}$$

where h_{ef} is the effective length of the anchor in mm. Φ_1 is the concrete cover in mm. τ_{bm} is the strength of the capacity in $\frac{\text{N}}{\text{mm}^2}$. $\alpha_1 = 1.0$, $\alpha_2 = 1 - 0.15 * \frac{(c_d - \varphi_1)}{\varphi_1}$ [6]. $c_d = \min(\Phi_1, 0.5 * s)$. s is the distance of the anchors.

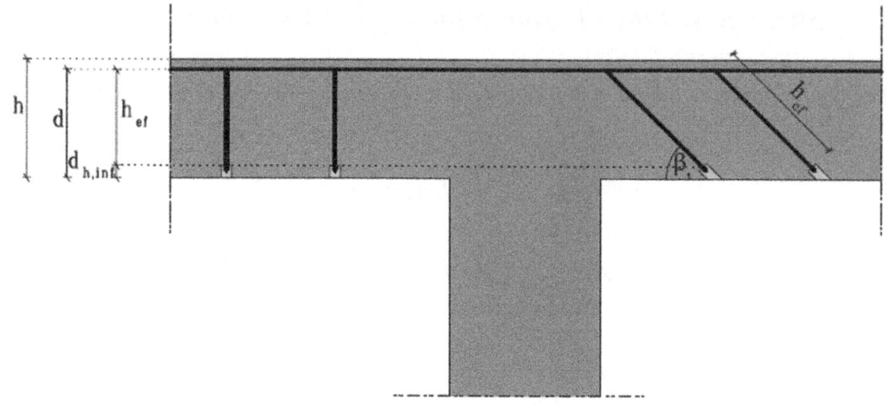

Fig. 1 Geometrical properties of a typical test case, also exemplarily indicating the anchor configuration

3 Probabilistic Analysis

3.1 *Parameters for the Punching Shear Test Cases*

Six scenarios of flat slabs with post-installed bonded anchors are chosen as test cases for the investigation. The post-installed bonded anchors in each test cases have a diameter of 20 mm and an anchoring plate with a diameter of 60 mm. The longitudinal reinforcement is chosen as B500 reinforcing steel bars. The material characteristics, the amount of longitudinal reinforcements, the height of the slab, the number of anchors and the angle of the anchors are varied in the chosen test cases, in order to assess their impact on the strength of the members (see Fig. 1). The material characteristics considered are the compressive strength of the concrete, the yield strength of both the transverse and longitudinal reinforcements and the bond strength of the adhesive. Table 1 details the parameters of the chosen test case. Figure 1 provides an overview of the geometrical properties of the test cases and explains the variables.

3.2 *Probability Models for the Basic Random Variables*

The variables of the test cases 1–6 presented in Table 1 is used in the probabilistic analysis. The variables are set as random variables that are normally distributed. The uncertainties in the variables (caused by time-dependent effects, inaccuracies or human errors) should be taken into account. Given this, the characteristic values of the variables (presented in Table 1) are converted to mean values. The values of the compressive strength of the concrete and the bond strength of the adhesive are

Table 1 Parameters for the test case (1–6)

	h (m)	d (m)	f_{ck} (MPa)	n (−)	$f_{yv,k}$ (MPa)	β_i (°)	$\tau_{b,k}$ (MPa)	A_s (cm^2)	$f_{yt,k}$ (MPa)
Case 1	0.35	0.32	C20/25	20	500	90	10	20	500
Case 2	0.35	0.32	C30/37	15	500	90	10	20	500
Case 3	0.30	0.27	C20/25	15	500	90	10	30	500
Case 4	0.30	0.27	C40/45	20	500	45	10	30	500
Case 5	0.25	0.22	C25/30	26	500	45	10	35	500
Case 6	0.25	0.22	C30/37	26	500	45	10	35	500

calculated according to EN 1990 [8]. The EN 1990 recommend values for calculating the mean values of normally distributed factors depending on the number of samples that should be used (Eqs. 5–7). For sample sizes larger than 30, the factor k_n shall be 1.64, according to EN 1990. This factor represents that the distance from the mean value of a normal distribution to the 5%-quantile of the distribution is 1.64 times the standard deviation, as shown in the equations below. This represents the overlap between the distribution of the load and the distribution of the resistance of a member (see Fig. 2).

$$X_m = X_k + k_n.\sigma \tag{5}$$

$$X_m = X_k + 1.64.\text{cov}.X_m \tag{6}$$

$$X_k = X_m - 1.64.\text{cov}.X_m \tag{7}$$

X_m is the mean value of the distribution. X_k represents the characteristic value, which equals the 5%-quantile of a normal distribution [8].

The coefficient of variation (cov) for the bond strength of an adhesive anchor τ_b is 0.10 [9]. The coefficient of variation for the yield strength of steel, according to the Federal International Federation for Structural Concrete [10] is 0.05. This could lead to higher variations in the resistance. The variation of the angle of the reinforcement depended on the chosen angle and was taken as 3° in this study. The cov used for the flexural reinforcement was taken as 0.05, and the concrete compressive strength was taken as 0.15, based on consideration of the values proposed by [11, 12]. The inaccuracies of the placing of the reinforcement are usually not higher than 10 mm, according to the Probabilistic Model Code [13]. The height of the specimens and the number of post-installed anchors are considered constant and deterministic. Table 2

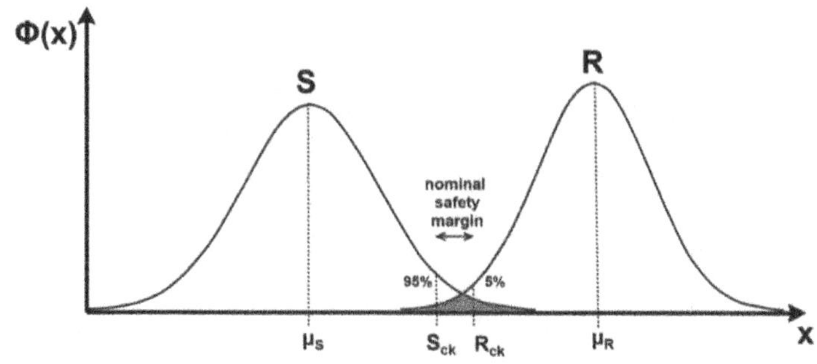

Fig. 2 Definition of safety margin, characteristic values, and basis of reliability concept

Table 2 Coefficients of variation and mean values of the variables used in the probabilistic calculation

Parameters (mean values)	h (m)	d (m)	f_c (MPa)	f_{yv} (MPa)	β_i (°)	τ_b (MPa)	A_s (cm^2)	f_{yt} (MPa)
Case 1	0.35	0.32	26.53	500	90	7.45	20	500
Case 2	0.35	0.32	39.79	500	90	7.45	20	500
Case 3	0.30	0.27	26.53	500	90	7.45	30	500
Case 4	0.30	0.27	53.05	500	45	7.45	30	500
Case 5	0.25	0.22	33.16	500	45	7.45	35	500
Case 6	0.25	0.22	39.79	500	45	7.45	35	500
cov	–	–	0.15	0.15	a	0.10	0.05	0.15

[a]The deviation in the installation is assumed to be 3° regardless of the mean value, i.e. cov equals 0.03 and 0.06 for installation at 90° and 45° respectively

shows the mean values and the coefficients of variation of the variables (calculated using Eq. 5) used in the probabilistic analysis.

3.3 Reliability Verification

In order to assess the safety and economic performance of members designed using the Eurocode design procedure, reliability verification, as presented in this section, was conducted [14]. The adequacy of design is confirmed if the limit states are not reached when the design values are introduced into the analysis models. The Eurocode demands that the design value of the resistance must be equal or higher than the design value of the load [8] (Eq. 8).

$$R_d \geq L_d \tag{8}$$

The design load calculated according to the Eurocode 2 requirements has a connection with the resistance. The design value for resistance may be obtained directly by dividing the characteristic value of a material or product resistance by 1.5 [15]. In order to obtain the design value of the load, the characteristic load is divided by 1.35, according to DIN EN 1990. Therefore, Eq. (8) is further expressed as (9).

$$R_d = \frac{R_k}{1.5} \geq L_d = 1.35.L_k \tag{9}$$

The characteristic value of the resistance equals to 5%-quantile of the distribution [8], which means that only 5% of the resistances are lower than the characteristic value. On the contrary, the characteristic value of the load equals the 95%-quantile of the distribution, since only 5% of the loads should be lower than this value. This way, one can ensure realistic values for both the resistance and the load. The 5%-quantile of the resistance and the 95%-quantile of the load are calculated according to Eqs. (10) and (11), respectively.

$$R_k = R_{5\%} = R_m - 1.64\sigma_R \tag{10}$$

$$L_k = L_{95\%} = L_m + 1.64\sigma_L \tag{11}$$

The standard deviation can be calculated by the multiplication of the coefficient of variation and the mean value of the distribution i.e. $\sigma_R = \text{cov}_R * R_m$ and $\sigma_L = \text{cov}_L * L_m$. The coefficient of variation for resistance cov_R was calculated in this study using the MATLAB code and was obtained as 0.10. The International Federation of Structural Concrete (*fib*) bulletin 80 [10] provides values for the coefficient of variation for basic variables in probabilistic models. The *fib* bulletin 80 gives a coefficient of variation of 0.10 for shear loads. With $\text{cov}_R = \text{cov}_L = 0.10$, therefore, Eqs. (7) and (8) can be expressed by Eqs. (12) and (13).

$$R_k = R_m - 0.164.R_m \tag{12}$$

$$L_k = L_m + 0.164.L_m \tag{13}$$

Replacing the characteristic values of the load and resistance in (9), the equation is further expressed by Eq. (14).

$$\frac{R_m - 0.164 * R_m}{1.5} \geq 1.35 * (L_m + 0.164 * L_m) \tag{14}$$

In order to obtain the mean value of the load, Eq. (14) is further solved to obtain Eq. (15). In this way, we get an equation, based on the mean resistance of a member, to calculate the mean value of the load that can be applied to each member.

$$L_m = 0.354 R_m \quad (15)$$

Considering the fact that Eqs. (12)–(15) are based on the Eurocode 0 and the Eurocode 2 design provisions, the resistance R_m required in Eq. (15) to estimate the mean value of the applied load is taken as the resistance of the Eurocode design formulation. This way, we have a load that is calculated for each test case individually, considering every variable. Hence, the probabilistic model for the load is obtained with a mean value of L_m, coefficient of variation of 0.10 and a normal distribution.

3.4 Determination of the Failure Probability

In order to confirm if the resistances values obtained according to the design guideline is sufficient and reasonable, the probability of failures is calculated. This is done by subtracting the total resistance value from the load value for each iteration, in accordance with the typical Limit State Equation $G = R - L$ [14]. When the value is positive, the resistance is higher than the load, representing Eq. (8). When the iterations yield negative values, a failure instance is marked because the load is higher than the total resistance of the slab. To estimate the probability of failure P_f out of several samples, the number of failures is divided by the number of samples.

The assessment is based on a target probability of failure $P_{f,T} = 1 \times 10^{-6}$ according to the EN 1990 for reliability class two (RC2) structure for a one-year reference period. This can be further explained as one failure in a million samples. The civil engineering industry also works with reliability index β [14], which is related to the probability of failure by the expression in (16).

$$\mathrm{P}_f = \Phi(-\beta) \quad (16)$$

where Φ is the cumulative distribution function of the standardised normal distribution.

The target probability of failure $P_{f,T} = 1 \times 10^{-6}$ for RC2 structure is equivalent to the target reliability index $\beta_T = 4.7$ [8]. The probability of failure and the connected reliability index is calculated for every column slab configuration of Table 1. The results are compared to the performance requirements recommended by the basis of design standards EN 1990 for RC2 structures.

4 Results and Discussions

4.1 Discussion of Estimated Deterministic and Probabilistic Resistances

The design values and the characteristic values obtained by the deterministic and the probabilistic calculations are presented in Fig. 3. The 5%-quantiles of the Eurocode 2 are 18 to 36% higher than the respective characteristic resistances. By comparing the design resistances with the 5%-quantiles, the design resistances are significantly lower. In addition to that, the ratio between the resistances are almost the same (with small variation) for the different test cases—the 5%-quantiles of the Eurocode are nearly two times the design resistance in most cases.

Test case 1 has the highest shear resistance compared to the other test cases. The high shear resistance obtained for test case 1 can be attributed to the combination of its large member size ($d = 0.32$) and the high number of anchors ($n = 20$) when compared to the other test cases. This is expected as the larger size of the member, and the high number of anchors implies a higher contribution from the concrete and anchor reinforcement, respectively, to total shear resistance.

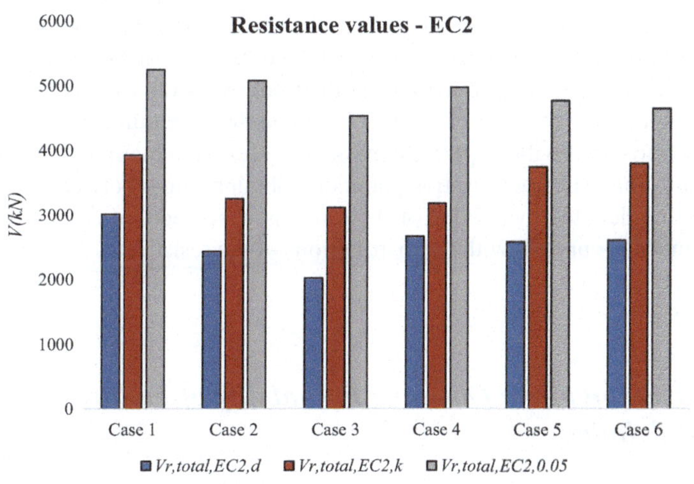

Fig. 3 Estimated total resistances in terms of design (Vr, total, EC2, d), characteristics (Vr, total, EC2, k) and 5%-quantiles values (5%-quantile values)

Table 3 Mean values μ and standard deviations σ of the total resistances $V_{r,total}$ according to EC2 codes (kN)

Case	K_n sample size	$V_{r,total,EC2}$ (kN)	
		μ	σ
1	10 Mio	6585.7	812.6
2	10 Mio	6299.9	744.8
3	10 Mio	5675.5	696.3
4	10 Mio	6420.9	882.4
5	10 Mio	6346.3	958.7
6	10 Mio	6196.6	942.4

4.2 Assessment of the Mean Values and Standard Deviations for the Resistance

The calculated mean values and standard deviations of the total resistances for the six scenarios are presented in Table 3. The total resistances include the contribution of the concrete V_c and the contribution of the shear reinforcement anchors V_{anchor}. Test cases 2 and 5 have the lowest and the highest coefficient of variations, respectively. The test cases differ majorly in terms of the flexural reinforcement ratio, the member size and number of anchors. Thus, it can be assumed that the variation in the various input parameters resulted in the variation in the resistance of the anchors.

The degree of fitting of the distributions of the obtained resistances was examined for unbounded normal and the lognormal distribution, taken as candidate distributions. A degree of the fitting is calculated based on the maximum likelihood estimation, in order to capture the appropriateness of the function mainly at the tails of the distributions. The fit characterisation parameter obtains values between 0 and 1, with an ideal fitting denoted by a value of 0.50. It was observed that neither the normal nor the lognormal curve fit with the distributions of the resistances for the 10 million samples.

4.3 Discussion of the Obtained Probability of Failure and β-Index

The derived probabilities of failure for the 10 million samples are presented in this section. The load applied on the samples is based on the approach of the Eurocode and was, therefore, kept the same for every sample. As shown in Table 4, the probabilities and the β-values differ for each test case. This trend of the result is attributed to the fact that different parameters influence and contribute to punching shear resistance, and thus have more considerable variation in the resistances. The failure probabilities of the first four cases of the Eurocode are smaller than 1×10^{-6} (the target probability of failure according to the EN 1990 RC2 structures). Reasonable consistency in the reliability of the Eurocode procedure is observed in Table 4. The highest reliability

Table 4 Probabilities of failure P_f and reliability index β for 10 Million samples

	EC2	
	P_f	β
Case 1	$5.0\ 10^{-7}$	4.89
Case 2	$1.0\ 10^{-7}$	5.19
Case 3	3.0×10^{-7}	4.99
Case 4	6.0×10^{-7}	4.85
Case 5	2.2×10^{-6}	4.59
Case 6	2.7×10^{-6}	4.54

is obtained for a test case with the largest member size and the lowest amount of reinforcements with $\beta = 5.19$ whereas the lowest reliability is obtained for a test case with the smallest size and highest amount of anchor reinforcement with $\beta = 4.54$.

Generally, EC2 has acceptable failure probabilities and thus, reliability indices above 4.7 target value (EN 1990 target reliability for reliability class RC2) in most cases. Considering the obtained results, the approach of the Eurocode has very reasonable and good results for post-installed anchors in flat slabs. Therefore, the Eurocode methods seem to be good design approximation of the design of retrofitting existing flat slabs and thus ensuring economy and enough safety against punching shear.

5 Conclusion

Flat slabs are slabs with low construction heights that are supported by columns or walls. These types of slabs are endangered by the risk of punching shear failure. This study aims to investigate the degree of the safety performance of EC2 method for the design of post-installed reinforcement in an existing flat slab structure endangered by punching shear, using probabilistic analysis.

- To assess the adequacy of the design concept, a probabilistic analysis was proposed. The investigation was conducted based on the Monte Carlo simulation technique implemented using a MATLAB code developed in this study. The main criterion considered to decide whether the resistances obtained from the design code is reasonable or not is the probability of failure or reliability index obtained for each test case. The obtained probability of failure is compared to the target probability of failure recommended by EN 1990 for RC2 structures.
- Assessment of the obtained results indicates that the reliability indices and probability of failure obtained for Eurocode method for post-installed anchors in flat slabs are in accordance with the target reliability requirement for Reliability Class 2 structures prescribed by the basis of design standards EN 1990, for most of the test cases considered in this study. Therefore, the Eurocode method seems to reflect reasonable design approximation of the design of retrofitting existing flat

slabs and thus ensuring enough safety against punching shear for the test cases investigated.

- This contribution established a safe method for the design of post-installed reinforcement in existing structures.

References

1. Beutel, R. R. (2002). *Durchstanzen schubbewehrter Flachdecken im Bereich von Innenstützen.* Northrhine-Westphalia. Aachen.
2. Novacek, J., & Zich, M. (2016, April 25). *Study of flat slabs strengthening against punching shear.* Brno, Czech Republic.
3. Kunz, J., Fernández Ruiz, M., & Muttoni, A. (2008). *Enhanced safety with post-installed punching shear reinforcement* (*No. CONF*). CRC Press.
4. Fernández Ruiz, M., Muttoni, A., & Kunz, J. (2010, July/August). Strengthening of flat slabs against punching shear using post-installed shear reinforcement. *ACI Structural Journal, 107*(4), 434–441.
5. Lapi, M., Ramos, A. P., & Orlando, M. (2019). Flat slab strengthening techniques against punching-shear. *Engineering Structures, 180,* 160–180.
6. DIN EN 1992-1-1. (2011). *Eurocode 2: design of concrete structures—Part 1–1: General rules and rules for buildings.* German version EN 1992-1-1:2001 + AC:2010. Beuth Verlag GmbH.
7. DIN EN 1992-1-4. (2019). *Eurocode 2: design of concrete structures—Part 4: Design of fastenings for use in concrete.* German version EN 1992-4:2018. Beuth Verlag GmbH.
8. DIN EN 1990. (2010). *Eurocode: Basis of structural design.* German version EN 1990:2002 + A1:2005 + A1:2005/AC:2010. Beuth Verlag GmbH.
9. Eligehausen, R., Appl, J. J., Bernhard, L., Meszaros, J., & Fuchs, W. (2004). Tragverhalten und Bemessung von Befestigungen mit Verbunddübeln unter Zugbeanspruchung. In K. Bergmeister, F. Flingerloos, & J.-D. Wörner (Eds.), *Beton-Kalender 2019* (pp. 561–571). Ernst & Sohn.
10. International Federation for Structural Concrete. (2016). A.2 Derivation of probabilistic models. In M. Prieto, P. Tanner, et al., *fib Bulletin 80—Partial safety factor method for existing concrete structures.* Document Competence Center Siegmar Kästl e.K.
11. Strauss, A., Zimmermann, T., Lehký, D., Novák, D., & Keršner, Z. (2014). Stochastic fracture-mechanical parameters for the performance-based design of concrete structures. *Structural Concrete, 15*(3), 380–394. https://doi.org/10.1002/suco.201300077.
12. Braml, T., Fischer, A., Keuser, M., & Schnell, J. (2009). Beurteilung der Zuverlässigkeit von Bestandstragwerken hinsichtlich einer Querkraftbeanspruchung. In K. Bergmeister, F. Fingerloos, & J.-D. Wörner (Eds.), *Beton-Kalender 2019* (pp. 799–812). Ernst & Sohn Verlag.
13. Joint Committee on Structural Safety. (2001). JCSS probabilistic model code part 3: Resistance models. In J. C. Safety (Ed.), *JCSS probabilistic model* (p. 3).
14. Olalusi, O. B., & Viljoen, C. (2020). Assessment of reliability of EN 1992-1-1 variable strut inclination method of shear design provisions for stirrup failure. *Structural Concrete, 21*(1), 303–315.
15. DIN EN 1992-1-1/NA. (2013). *National annex—Nationally determined parameters—Eurocode 2: Design of concrete structures—Part 1–1: General rules and rules for buildings.* Beuth Verlag GmbH.

At Issue: The Gaussian Autocorrelation Function

Marc A. Maes, Karl Breitung, and Markus R. Dann

Abstract This paper focuses on the use of Gaussian autocorrelation functions (ACF) in civil engineering applications involving random processes and random fields. It aims at debunking misgivings, verifying facts and figures, and formulating practical conclusions. A large majority of civil engineers active in random field modelling and reliability analysis is quite content to point out that the routine use of Gaussian autocorrelation functions is part of standard practice and perfectly harmless. A common approach in 2D random field problems, for instance, is to estimate an appropriate correlation length on some physical or empirical basis, and then plug it into a multivariate ACF that is both isotropic, and separable into a product of univariate ACFs: if both of these objectives are to be met, the Gaussian ACF naturally stands out as it is in fact the only real function to possess both of these properties. But as early as the nineteen-sixties, a substantive piece of electrical engineering literature pointed to "issues" and "red flags". The claim was that the Gaussian ACF produces unrealistic results, violates certain principles concerning both the modelling and the estimation of random properties, and runs into results that possibly defy common sense. Similarly, geostatisticians have been issuing warnings of hyper-predictability, super-smoothness, wildly underestimated estimation errors, and artificial results in applications such as spatial kriging using Gaussian ACFs, leading to the recommendation that the Gaussian model should never be used in practice. This paper revisits the use of the Gaussian ACF and presents a sober but principled look at the entire issue. Importantly, it also considers the pros and cons of replacement ACF models and adjusted ACF models. The paper includes examples and measurable outcomes with the aim of providing a fair assessment and justifiable recommendations.

Keywords Autocorrelation functions · Random fields · Random processes

---

M. A. Maes (✉) · M. R. Dann
University of Calgary, Calgary, Canada
e-mail: mamaes@ucalgary.ca

K. Breitung
Technical University of Munich, Munich, Germany

© The Author(s), under exclusive license to Springer Nature Switzerland AG 2021
J. C. Matos et al. (eds.), *18th International Probabilistic Workshop*, Lecture Notes in Civil Engineering 153, https://doi.org/10.1007/978-3-030-73616-3_14

1 Introduction

The selection of an analytical Gaussian autocorrelation (ACF) function of the type ce^{-at^2} where t is a spatial or temporal coordinate, $c > 0$, $a > 0$, is quite popular in civil engineering modelling for random processes and random fields. In fact, due to its ease of use and interpretation, the Gaussian ACF is often a prime candidate when data are sparse or inconclusive, yet one has a good "feeling" about suitable correlation properties.

However, since the nineteen-sixties a substantive piece of electrical engineering literature has pointed to "issues" and "red flags" (Kalaith [1]). The claim is that the Gaussian autocorrelation function produces unrealistic results, violates certain principles concerning both the modelling and the estimation of random properties, and runs into results that defy common sense.

Similarly, geostatisticians have warned of hyper-predictability and artificial results in applications such as spatial kriging using the Gaussian ACF, leading to the recommendation that the Gaussian model should never be used in practice (Matheron [2], Wackernagel [3]).

A large majority of civil engineers active in random field modelling and reliability is quite content to point out that the routine use of Gaussian ACFs is standard practice and perfectly harmless. In 2019 the Joint Committee of Structural Safety concurred with this sentiment saying it amounted to a "philosophical issue" and not per se a mathematical one.

At this point in time one may legitimately wonder what is going on. The present paper revisits the Gaussian ACF and presents a sober but principled look at the entire issue. Importantly, it also considers the pros and cons of "replacement" models suggested in literature. It touches on the statistical estimation of a suitable correlation model based on scarce empirical data. The paper includes examples and measurable outcomes with the aim of providing a fair assessment and justifiable recommendations.

2 The Extraordinary Sample Paths of the Gaussian ACF

For a stationary stochastic process $X(t)$ the ACF is defined as:

$$R(t) = E(X(t)X(0)) \tag{1}$$

The correlation length (or time) of the process, t_c, is defined as (Papoulis [4], p. 221):

$$t_c = \frac{1}{R(0)} \int_0^\infty R(\tau)d\tau \tag{2}$$

The Gaussian ACF with correlation length $t_c > 0$ and variance $\sigma^2 = R(0) > 0$, is as follows:

$$R(t) = \sigma^2 \exp\left(-\frac{\pi}{4}\left(\frac{t}{t_c}\right)^2\right) \tag{3}$$

Its name is somewhat unfortunate. It suggests that this model has some kind of importance in the same way as Gaussian random processes. Nothing could be farther from the truth.

First of all, it is important to note that R is infinitely differentiable, and correspondingly, all moments of the spectral density are finite, so that the corresponding process $X(t)$ has mean square derivatives of all orders (Lindgren [5]). In fact, an even stronger result holds: for any $t > 0$, as $n \to \infty$, the power series expansion of a sample path around 0, $\sum_{j=0}^{n} X^{(j)}(0)t^j/j!$ converges everywhere to $X(t)$ (Belayev [6], Papoulis [4]). That is, it is possible to predict $X(t)$ perfectly everywhere for all $t > 0$ based on observing $X(s)$ for all $s \in (-\epsilon, 0]$ for an arbitrary small $\epsilon > 0$.

Such behaviour must certainly be considered unrealistic for any kind of physical process. It means that the knowledge of the values of a sample path of $X(t)$ in any bounded interval is enough to find the values of it at all points anywhere in the universe.

The process $X(t)$ is therefore marked with "analytical" sample paths, caused essentially by the infinite differentiability of the Gaussian ACF (Breitung [7]). In fact, from a more general perspective, we need to point out that there exist 4 classes of stationary processes having analytic sample paths, as stated by Belyaev [6]: polynomials with random coefficients, trigonometric polynomials with random coefficients, processes with a bounded spectrum, and, finally, processes with a Gaussian ACF.

All of these processes possess smooth and analytical sample paths; for all of them, we can, from the values of the process and its derivatives at one point t in time, predict exactly the value of the process at every other time point $t + \tau$. In reality certainly all derivatives cannot be calculated exactly as numerical imprecision eventually takes over.

The same is also true in 2D or 3D applications. In fact, the Gaussian ACF enjoys even more popularity in higher dimensions due to the fact that it is very convenient to generate multivariate Gaussian ACFs by simply taking products of their univariate Gaussian version. This property is referred to as separability and, indeed, "separable" ACFs are easier to employ than isotropic ACFs, a fact which has further contributed to their popularity.

As it turns out, the only real functions on $\mathbb{R}^d$ that are both isotropic, and factor into positive definite functions of each coordinate are of the form $c\exp(-a|\mathbf{t}|^2)$ (Stein [8]) for all nonnegative a and c. However, we will show that, just as in the case of the one-dimensional setting, random fields modelled using multivariate ACFs are unrealistically smooth for physical phenomena.

3 Wildly Underestimated Estimation Errors

As a direct result of the super-smoothness of its analytical sample paths, statistical estimation involving spatial or temporal inter-/extrapolation, most notably kriging, suffers from the fact that the mean square errors (MSEs) of estimators are way too small. In other words, spatial variation is significantly underestimated (Wackernagel [3]).

The easiest way to illustrate this fact, is by contrasting the use of two ACFs:

- the ACF $R_0(t) = e^{-2|t|/t_c}\left(1 + 2\frac{|t|}{t_c}\right)$ which is in fact a Matérn 3/2 function i.e. a special case of the more general multiparameter Matérn class of functions (Stein [8] p. 48); because it has two derivatives at 0, it leads to processes that are exactly once mean square differentiable; and,
- the Gaussian ACF $R_1(t)$ as given by (3) with $\sigma^2 = 1$.

Both of these ACFs have unit variance and they share exactly the same correlation length t_c as given by (2). Both ACFs are plotted in Fig. 2. They are almost indistinguishable although R_0 has longer tails. Importantly, it is difficult from looking at these plots to see that R_1 is analytic whereas R_0 only has two derivatives at the origin. Incidentally, this also shows that plots of empirical ACFs are likely to be a poor way to distinguish between possible models for the ACF of a smooth process.

Now consider the t-axis laid out in Fig. 1. The central value $X(0)$ is estimated using ordinary kriging based on three sets of 10, 20, and 40 observations, respectively, spaced at intervals of 0.4 t_c, as shown in Fig. 1. To estimate $X(0)$ we use, in turn, the ACFs R_0 and R_1. Then we determine the MSEs of $X(0)$ as follows:

$$\text{MSE}_j = E_j\left(e_j(X(0)|\boldsymbol{x}_{\textbf{obs}})^2\right) \quad j = 0, 1 \tag{4}$$

where E_j denotes expectation using the ACF model R_j, e_j is the corresponding error on the estimator, and $\boldsymbol{x}_{\textbf{obs}}$ is each set of observations A, B, C as denoted in Fig. 1.

Table 1 shows both MSE_0 and MSE_1 for $X(0)$. The use of the Gaussian ACF R_1 suggests that there is hardly any error on the best linear estimate of $X(0)$. In fact, it is nearly 1000 times smaller than the kriging MSE_0 resulting from the use of R_0. This is implausibly small for any physical process, considering that the variance of the process var($X(0)$) $= 1$ and the correlation between $X(0)$ and the most nearby observation is, using (3), $R_1(0.4t_c) = \exp\left(-\frac{\pi}{4}(0.4)^2\right) = 0.882$.

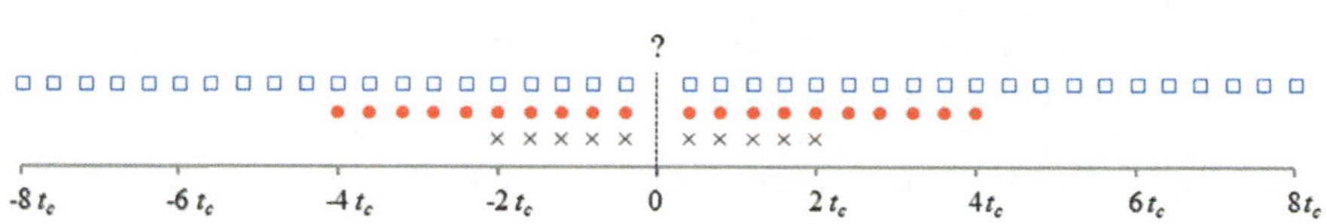

Fig. 1 Best linear unbiased estimation of a central value $X(0)$ using 3 sets of observations spaced $0.4t_c$ apart: Set A (×) 10 points in the interval $[-2t_c, +2t_c]$; Set B (red circle) 20 points in $[-4t_c, +4t_c]$; Set C (blue square) 40 points in $[-8t_c, +8t_c]$

Table 1 MSE of $X(0)$ based on the three sets shown in Fig. 1, using the 3/2-Matérn ACF R_0 and the Gaussian ACF R_1, both having the same correlation length t_c

	10-point set A	20-point set B	40-point set C
Using R_0	0.115	0.115	0.115
Using R_1	6.2×10^{-4}	1.9×10^{-4}	1.5×10^{-4}

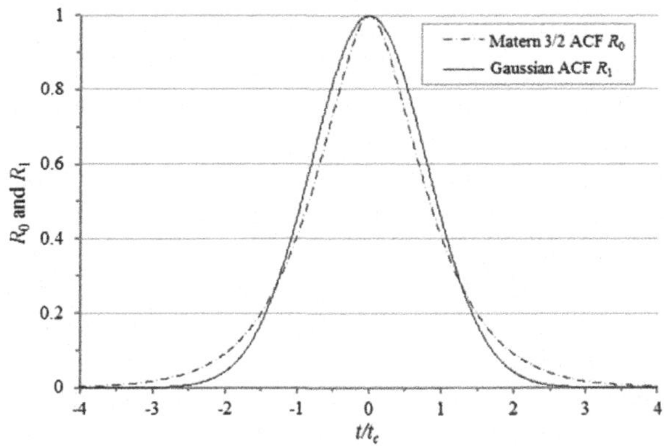

Fig. 2 Comparison of a Gaussian with a Matérn 3/2 autocovariance model, both having the same correlation length equal to t_c

In addition, while the Matérn 3/2 MSE_0 remains steady, as it should, when observations are added at the extreme ends of the data sets (inflating set A to B and then to C), the Gaussian MSE_1 improves by adding points to the outer fringes of the sets. For instance, by adding 20 observations (from set B to C) at a distance exceeding 4 correlation lengths from the centre, MSE_1 decreases by more than 20% from 1.9×10^{-4} to 1.5×10^{-4} even though the maximum correlation between $X(0)$ and any of the added observations is only equal to $R_1(4.4t_c) = 2.5 \times 10^{-7}$, which seems hardly credible.

Kriging weights and variances were studied in 1984 extensively in geostatistics by Rivoirard [9]. He showed that the use of Gaussian ACFs results in a whole range of "unnatural" effects in the context of exaggerated an/isotropy, axial structure, screening effects, negative kriging weights and incorrect local behaviour.

Also, for the above problem of estimating a central value $X(0)$ in a structure such as Fig. 1, Rivoirard [9] points out that linear estimation based on a set of n observed neighbouring values in a grid extending well beyond the correlation length, becomes a fully deterministic exercise in finite difference interpolation, i.e. setting all $\Delta^n X$ equal to zero in Newton's divided difference formula for the n-th degree interpolating polynomial. The infinite differentiability at the origin and the extraordinary light tail of the Gaussian $R_1(t)$ result in implausibly smooth sample paths and arbitrary small variances (Stein [8]).

4 Estimation Based on the Empirical ACF and Variograms

When unknown parameters of the covariance structure are estimated from the available data, the most commonly used method for predicting random fields and assessing the MSEs of these predictions is to estimate the second-order structure in some manner and then proceed as if this estimated second-order structure were the truth.

In geostatistics, best practice is to smoothen the empirical variogram—the ACFs more commonly used counterpart—and use it as the cornerstone of the predictive model. In civil engineering, the choice of the "best" ACF to fit a typically sparse empirical ACF is often problematic and therefore largely brushed over as inconsequential with respect to the remainder of the stochastic analysis. For instance, a common short cut is to eyeball the correlation length/time using some "reasonable" physical or empirical justification, and then plug this value in a simple parametric ACF model.

The construction of an empirical ACF is actually marked by strong correlations that exist in the empirical ACF at different distances. Therefore, a visually pleasing fit to an empirical ACF can appear quite regular and still be substantially in error. The fact that the empirical ACF has correlated values is well known (Stein [8]), but the consequences of potentially large correlations are not generally sufficiently appreciated. It is often a difficult psychological adjustment to look at an empirical ACF and its supposedly most pleasing best-fit ACF and recognize that the strong correlations present can easily result in a smooth empirical ACF that is very different from the "true" ACF.

To illustrate the above, we turn to a simulated $\boldsymbol{x}$ set of realizations of a zero-mean, unit-variance Gaussian random process $X(t)$ having the following Matérn 3/2 ACF:

$$R_0(t) = e^{-0.4|t|}(1 + 0.4|t|) \tag{5}$$

which has a correlation length $t_c = 5$. The 20 observations $\boldsymbol{x}$ (Stein [8]) in the interval $[-9.5, +9.5]$ in steps of $\Delta t = 1$ are shown in Fig. 3.

The empirical ACF $\hat{R}(t)$ is shown by (+) in Fig. 4 together with the original ACF (5) used to stimulate the 20 observations. Note that, already, rather large and apparently systematic differences can be spotted between R_0 and $\hat{R}$, due to the large correlations between any pair of neighbouring $\hat{R}(t)$ and $\hat{R}(t + \Delta t)$. In addition, we show the "restricted maximum likelihood" (RML, see Stein [8], p. 170) ACF-estimate $R_1(t)$. It can be seen that $R_1(t)$ may not offer a "pleasing" fit to the empirical date; accordingly, we add an additional eyeball estimate $R_2(t)$ which fits quite well at the shorter distances, see Fig. 4.

We now use the different ACF models to estimate by ordinary kriging the three "interpolated" $X(-4)$, $X(0)$ and $X(8)$ and the two "extrapolated" $X(10)$ and $X(10.5)$ as shown by the five vertical lines marked with "?" in Fig. 3, together with the MSEs of these five estimates: these are all included in Table 2.

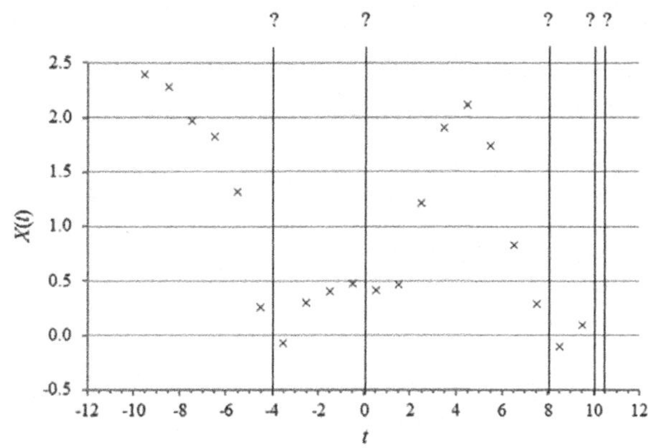

Fig. 3 Set (×) of 20 data for a Gaussian process with ACF given by (5)

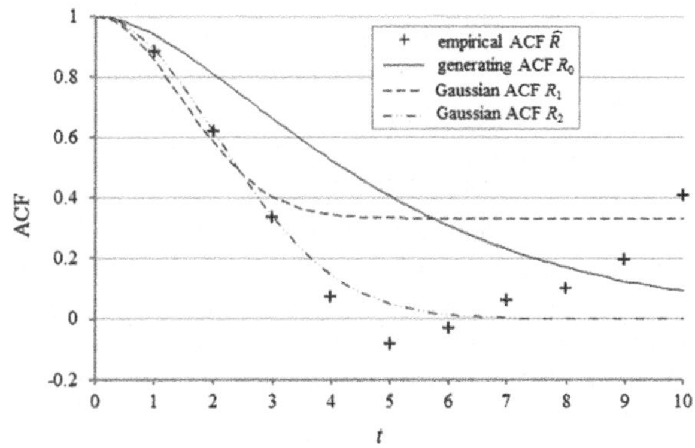

Fig. 4 Empirical ACF $\hat{R}(t)$ for the data shown in Fig. 3, together with the original generating ACF $R_0(t)$, the optimal RML estimated ACF $R_1(t)$, and the eyeball fit $R_2(t)$

Table 2 Estimation of interpolated $X(-4)$, $X(0)$, $X(8)$ and extrapolated $X(10)$, $X(10.5)$ using different ACFs

	Best linear estimate using			MSE of this estimate using		
	R_0	R_1	R_2	R_0	R_1	R_2
$X(-4)$	−0.04	−0.07	−0.07	2.74×10^{-3}	2.10×10^{-5}	5.48×10^{-9}
$X(0)$	0.48	0.47	0.47	2.74×10^{-3}	1.76×10^{-5}	2.23×10^{-9}
$X(8)$	0.10	0.14	0.18	2.78×10^{-3}	1.06×10^{-4}	3.75×10^{-7}
$X(10)$	0.23	0.51	1.24	1.82×10^{-2}	1.22×10^{-2}	4.14×10^{-4}
$X(10.5)$	0.41	0.94	3.23	7.66×10^{-2}	9.12×10^{-2}	5.86×10^{-3}

When it comes to interpolation, it appears that the choice of model hardly matters, except that the MSEs are once more implausibly small for the Gaussian R_1, and R_2, more so as the visual fit to the empirical ACF improves!

However, the situation turns around in the case of extrapolations: the estimated $X(10)$ and $X(10.5)$ become exceedingly "estranged" due to the "forced" Gaussian

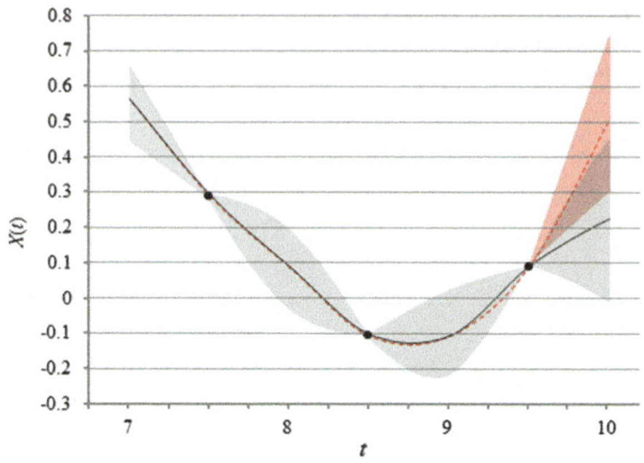

Fig. 5 Posterior $X|\boldsymbol{x}$ between $t = 7$ and $t = 10$: mean and 95% probability band; the black dots show the observed values; black/grey and red/orange denote the use of the Matérn 3/2 ACF R_0 and the RML Gaussian ACF R_1, respectively

continuity of all derivatives at the last observed data point at $t = 9.5$. And there, the maximum likelihood fit R_1 actually fares better than the eyeball fit R_2.

To exemplify this situation, we contrast in Fig. 5 the mean and the 95% confidence interval of the posterior $X(t)|\boldsymbol{x}$ in the interval $t = 7$ to $t = 10$: the grey line and shaded area refer to the Matérn 3/2 R_0 model and the pink dashed line and shaded area refer to the RML estimated R_1. It can be seen that as for as interpolation is concerned the pink shaded area is in fact totally lacking since the MSE for any interpolated value is much too small. But it explodes once the extrapolation starts. The posterior means themselves coincide pretty much for all ACF models to the left of the last data point (shown by black dots) but they diverge to its right, i.e. when $t > 9.5$.

5 Using the Gaussian ACF for Estimating 2D Random Fields

Similar conclusions also apply to higher dimensional random fields.

To illustrate this we turn to the iconic "elevation" data set from Wackernagel [3]: 52 altitude measurements (in ft) in a square region $\{0 \leq x_1 \leq 300\,\text{ft}, 0 \leq x_2 \leq 300\,\text{ft}\}$. These are located at the red dots $\boldsymbol{x} = (x_1, x_2)$ marked in Fig. 6.

The empirical ACF for these data is calculated and it is shown in Fig. 7.

We ran ordinary kriging using a variety of software packages (STK [10], Lataniotis et al. [11], SAS/STAT [12]) and options for statistical estimation/optimization:

(a) a Matérn 3/2 ACF of the type $R(t) = (1 + |t|/\theta)\exp(-|t|/\theta)$ where $t = |\boldsymbol{x} - \boldsymbol{x}'|$ and $\theta > 0$, using cross-validation estimation and genetic algorithm optimization;
(b) an exponential ACF of the type $R(t) = \exp(-|t|/\theta)$ using cross-validation estimation;
(c) a Gaussian ACF as in (3) using restricted maximum likelihood estimation;

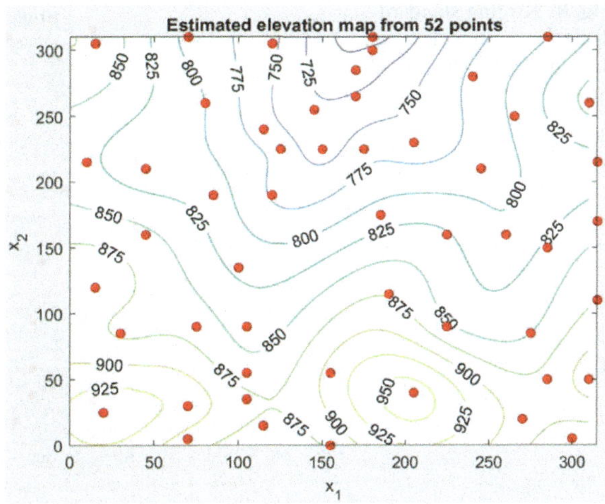

Fig. 6 Elevation data (red dots) in a 300 ft × 300 ft region (Wackernagel [3]). The coloured altitude lines show the ordinary kriging estimate using a Matérn 3/2 ACF

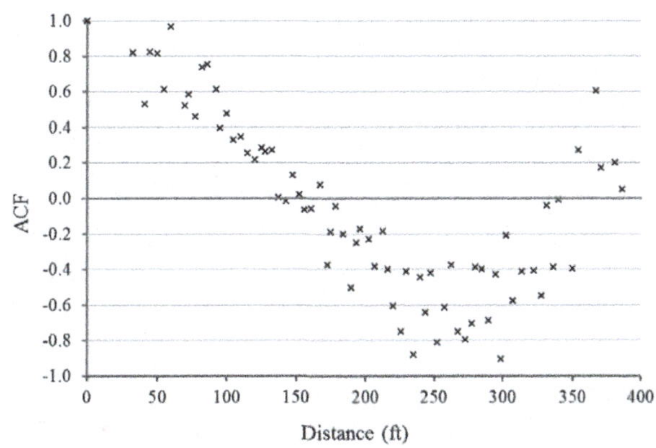

Fig. 7 Empirical ACF (bins of 5ft) for the elevation data in Fig. 6

(d) a Gaussian ACF including a small nugget at $t = 0$ equal to 1/1000 of the variance, using restricted maximum likelihood estimation and gradient-based optimization.

The estimated iso-elevation map also shown in Fig. 6 is the one corresponding with (a) above; however, it is very nearly identical to the maps that would be obtained using the other three ACF models except perhaps for some curl-up near the edges/corners of the square region in the case of (c) which confirms some of the difficulties associated with extrapolation discussed in Sect. 4.

Turning toward the MSE, or, rather, to its square root referred to in Figs. 8, 9, 10 and 11 as the kriging standard deviation σ_k, we obtain the four iso-line maps for ACF models (a), (b), (c), and (d).

The maps in Figs. 8 and 9 for the Matérn 3/2 and the exponential ACF look quite similar, although the "sinks" down to the value $\sigma_k = 0$ at each of the observation sites

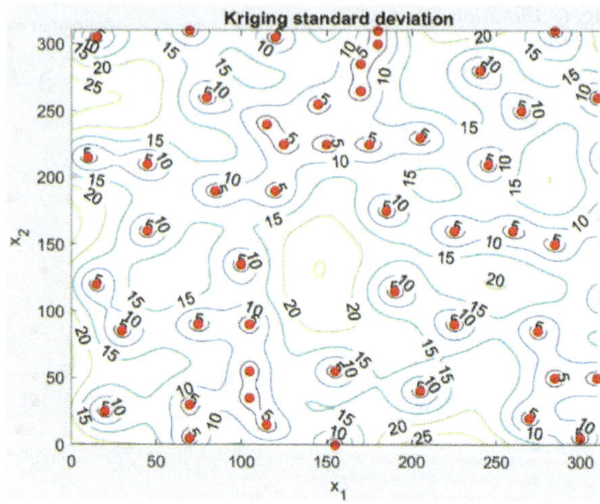

Fig. 8 Kriging standard deviation for the elevation data in Fig. 6: Matérn 3/2 ACF

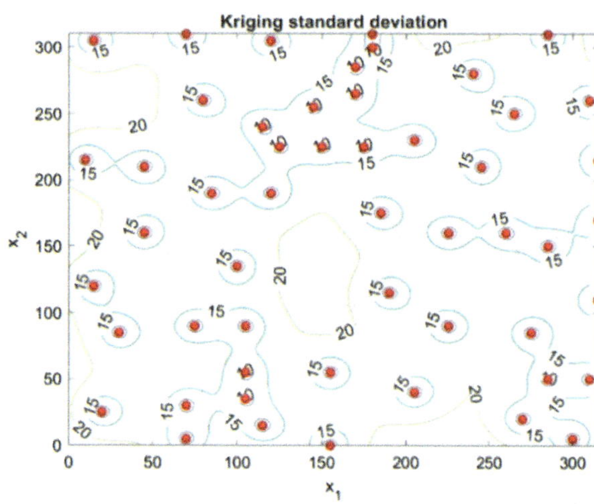

Fig. 9 Kriging standard deviation for the elevation data in Fig. 6: Exponential ACF

is somewhat narrower and steeper for (b) than for (a) due to the fewer derivatives at $t = 0$.

But, in the case of Gaussian ACF, the now familiar near-zero error map shown in Fig. 10 re-appears. It was this finding that led Matheron [2] and Wackernagel [3] to determine that, besides the fact that it is purely deterministic and predictable "just about anywhere in the universe using only a tiny small known area", it also has "unrealistic consequences for all kriging applications".

Finally, Fig. 11 shows the effect (d) of adding a small ad-hoc nugget to the Gaussian ACF, destroying its extreme inter-/extrapolative properties. The larger the nugget, the more marked this effect becomes. However, there appears to be no basis

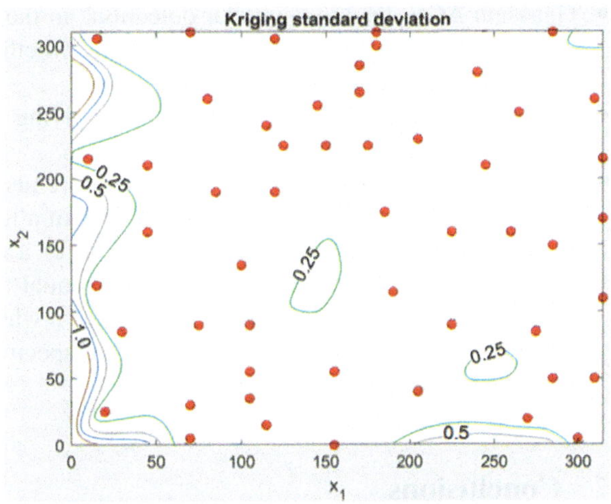

Fig. 10 Kriging standard deviation for the elevation data in Fig. 6: Gaussian ACF

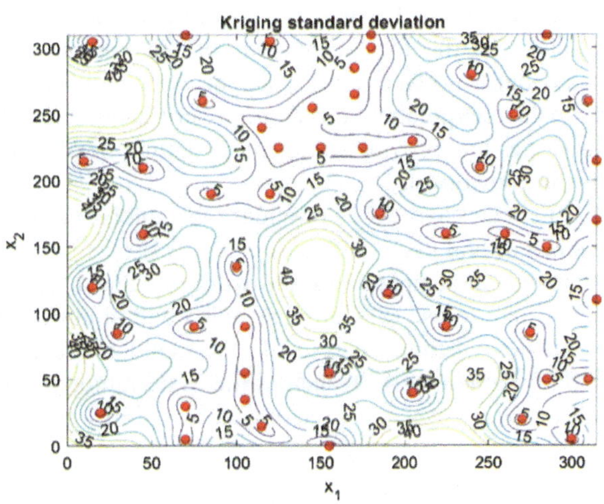

Fig. 11 Kriging standard deviation for the elevation data in Fig. 6: Gaussian ACF with a very small nugget

to estimate the "required" size of such a tiny nugget and there exists little justification for making it an element of best practice.

6 Additional Issues with the Use of a Gaussian ACF

An in-depth discussion by Breitung [7] also flags the following issues which are not explored further in the present paper:

- Gaussian ACFs lead to "singular detection" in the context of certain hypothesis tests for statistical estimation, e.g. deciding whether or not a stochastic process has zero mean (Kalaith [1]);
- the Wold decomposition (Lindgren [5]) involving Gaussian ACF consists solely of a singular component, see [7];
- a process with a Gaussian ACF, and therefore also a Gaussian spectral density (Lindgren [5]), can never arise as the output of a linear filter using a white noise input! This is because the system function for a random oscillator is always a complex rational function with numerator equal to 1 (Papoulis [4]). Therefore the spectral density of the output process for a white noise input, must also be a rational function and this conflicts with the spectral density function associated with the Gaussian ACF.

7 Conclusions

We have shown that the quality of spatial/temporal inter-/extrapolation depends strongly on the local behaviour of the random field/process. In practice, local behaviour is not known and must be estimated from the same data that will be used to do the interpolation. It is therefore critical to select ACF models for the covariance structures that provide sufficient modelling flexibility for the local behaviour of the qualities involved.

The Gaussian ACF provides absolutely no flexibility with regard to local behaviour and essentially assumes it is deterministic and known a priori. It results in analytical sample paths and wildly underestimated estimation errors. The alternatives are to use (a) Gaussian ACFs with a nugget, but only if empirical data are available to estimate and justify the size/dominance of such a nugget effect; or (b) a multi-parameter ACF such as the Matérn model, which includes a parameter that allows for a variable degree of differentiability of the random field.

References

1. Kalaith, T. (1966). Some results on singular detection. *Information and Control, 9,* 130–152.
2. Matheron, G. (1972). *Leçons sur les Fonctions Aléatoires d'Ordre 2.* Cours de l'Ecole des Mines, Fontainebleau, C-53, March.
3. Wackernagel, H. (2003). *Multivariate Geostatistics* (3rd ed.). Springer.
4. Papoulis, A. (1965). *Probability, random variables and stochastic processes* (1st ed.). McGraw-Hill Kogakusha Ltd.
5. Lindgren, G. (2013). *Stationary stochastic processes.* Chapman & Hall/CRC.
6. Belyaev, Y. K. (1959). Analytic random processes. *Theory of Probability and its Applications, 4,* 402–409.
7. Breitung, K. (2018). *Analytic autocorrelation functions: mathematical Lorelies.* https://www.researchgate.net/publication/327221256_Analytic_Autocorrelation_Functions_Mathematical_Lorelies

8. Stein, M. (1999). *Interpolation of spatial data.* Springer.
9. Rivoirard, J. (1984). *Le comportement des poids de Krigeage* [Doctoral thesis]. Sciences et Techniques Minières—Option Géostatique. École Nationale Supérieure des Mines de Paris, Paris, France.
10. STK. (2020). *The statistical toolkit project.* https://modal.lille.inria.fr/doc/html/
11. Lataniotis, C., Marelli, S., & Sudret, B. (2018). The Gaussian process modelling module in UQLab. *Journal of Soft Computing in Civil Engineering, 2*(3), 91–116.
12. SAS/STAT. (2020). © SAS Institute Inc., https://www.sas.com/en_ca/software/stat.html

Bridge Case Studies on the Assignment of Partial Safety Factors for the Assessment of Existing Structures

André Orcesi, Vazul Boros, Marija Kušter Marić, Ana Mandić Ivanković, Miroslav Sýkora, Robby Caspeele, Jochen Köhler, Alan O'Connor, Franziska Schmidt, Salvatore Di Bernardo, and Nisrine Makhoul

Abstract Aging bridges in combination with an ever-growing traffic volume are a matter of concern all over the world. Consequently, the reassessment of existing bridges is gaining importance rapidly. This paper presents two bridge case studies considered within the IABSE Task Group 1.3 "Calibration of Partial Safety Factors for the Assessment of Existing Bridges". The so-called design value method (DVM) and adjusted partial factor method (APFM), introduced in fib Bulletin 80 and both relying on a partial factor format, are considered in this paper. The objectives are (i) to illustrate how DVM and APFM can be used when specifying partial safety factors for assessment of existing bridges, and (ii) to discuss some of the assumptions that are implied by these methods. Two case studies are considered for illustration in

F. Schmidt
Université Gustave Eiffel, Champs-sur-Marne, France

V. Boros
Schömig-Plan Consulting Engineers, Kleinostheim, Germany

M. Kušter Marić · A. Mandić Ivanković
University of Zagreb, Zagreb, Croatia

M. Sýkora
Czech Technical University in Prague, Prague, Czech Republic

R. Caspeele
Ghent University, Ghent, Belgium

J. Köhler
Norwegian University of Science and Technology, Trondheim, Norway

A. O'Connor
Trinity College Dublin, Dublin, Ireland

S. Di Bernardo
Ciorba Group, Chicago, USA

N. Makhoul
Notre Dame University-Louaize, Zouk Mosbeh, Lebanon

A. Orcesi (✉)
Cerema, Champs-sur-Marne, France
e-mail: andre.orcesi@cerema.fr

© The Author(s), under exclusive license to Springer Nature Switzerland AG 2021
J. C. Matos et al. (eds.), *18th International Probabilistic Workshop*, Lecture Notes in Civil Engineering 153, https://doi.org/10.1007/978-3-030-73616-3_15

this paper: a single span reinforced concrete slab and a 3-span continuous reinforced concrete slab.

Keywords Existing structures · Bridges · Case study · Reinforced concrete · Partial safety factors

1 Introduction

The approach for the assessment of existing structures is in many aspects different from that for the design of new structures [1, 2]. The potentially available information about geometry, material properties, loading and environmental conditions is not the same for new and existing structures and this implies different levels of uncertainty. Reliability requirements might also be different for new and existing structures based on societal, cultural, economic, and sustainability considerations. A partial safety format applied for existing structures should cover the aforementioned differences in an appropriate way, to avoid non-effective decisions that in the end may have severe economic, environmental and socio-political consequences [2].

With the objective to promote the fields of structural reliability at an inter-association level, the IABSE TG1.3 is currently investigating how semi-probabilistic formats can be used for assessment of existing structures. In particular, the so-called Design Value Method (DVM), and the Adjusted Partial Factor Method (APFM) that were introduced within *fib* COM3 TG3.1 [3] are considered herein for specifying partial safety factors for reassessment.

This paper presents some first results of an ongoing work within IABSE TG1.3 and is organised as follows. First, DVM and APFM concepts are briefly revisited. Second, the two methods are applied on two different bridge case studies to illustrate how partial safety factors are specified. In the first case study, some measurements on the bridge are available, while the second case study only considers project documentation, without additional on-site measurements. Then the potential and issues of the two methods are discussed.

2 Methods for Updating Partial Factors in Bulletin 80

2.1 General Considerations

At present, existing structures are mostly verified using a partial factor format with identical safety factors as commonly applied in the design of new structures. Such assessments are often conservative and may lead to expensive structural repairs and interventions. However, especially if the existing structure is associated to material deterioration that cannot be observed directly, the uncertainties in the assessment situation are larger compared to the design situation and the application of the design partial safety factors might lead to non-conservative decision about the structure. The *fib* Bulletin 80 [3] provides a technical recommendation containing a format with adjusted partial factors for the assessment of existing concrete structures applicable to common situations re-assessment situations. The bulletin intends to be consistent with the background documents of the Eurocodes, ISO 2394:2015, CEB bulletins on the derivation of partial factors for concrete structures and JCSS publications [2, 4]. The following cases remain beyond the scope of the bulletin: structures severely affected by deterioration, non-linear FE models requiring specific safety formats, and compliance demonstrated by proof loading or monitoring. Such additional applications are however envisaged for the *fib* Model Code 2020 which is currently under development and takes basis in a similar philosophy.

When assessing existing structures, a partial factor format may be used, provided that additional information related to material parameters, loading conditions, localised structural defects, etc., is properly accounted for. Two methods relying on a partial safety factor format are provided in the bulletin: the design value method (DVM—general semi-probabilistic format for specifying partial factors, tied to ISO 2394 and EN 1990) and the adjusted partial factor method (APFM—partly a graphical procedure with pre-selected default values, based on DVM), both enabling the incorporation of specific semi-probabilistic aspects for existing structures.

The two methods utilise the generalised sensitivity factors for dominant and non-dominant resistance and load parameters, α_R and α_E respectively, as given in ISO 2394:2015 and EN 1990:2002 (see Table 1). Consideration of these generalised sensitivity factors makes it possible to adjust partial factors with respect to a modified reliability requirement β and the updated probabilistic representation of load and resistance variables. The simplifying assumption of the fixed sensitivity factors avoids the situation of having a wide range of different case-specific sensitivity factors which could make practical engineering applications cumbersome. The bulletin indicates β-values based on economic optimisation, and individual and group risk criteria on

Table 1 Basic variables and corresponding sensitivity factors as treated in detail in *fib* Bulletin 80

Basic variable	Distribution	Sensitivity factor, α
Concrete compressive strength, f_c	lognormal or normal (LN or N)	0.8
Yield strength of steel reinforcement, f_y	LN or N	0.8
Permanent action, G	N	−0.7 for unfavourable, 0.32 for favourable
Selected types of variable actions (snow, wind, imposed loads in buildings, and traffic loads on road bridges), Q	Gumbel (maxima)	−0.7
Resistance and load effect model uncertainties, θ_R and θ_E respectively	LN or N	0.32 for θ_R, −0.28 for θ_E

a lifetime basis; these levels being systematically lower compared to those provided in EN 1990:2002 for structural design.

DVM and APFM offer generally applicable equations and graphs to update partial factors for the basic variables and specified generalised sensitivity factors as indicated in Table 1. Specific equations and graphs are provided in [3]. Partial factors should be specified assuming appropriate distributions of the variables under consideration (based on prior information, or results of tests or the combination of both). In the bulletin the coefficient of variation, V, is considered as the major parameter for representing the new state of information about the variable. Furthermore, the partial factors can also be specified on the basis of modified reliability requirement and reference periods. Partial factors for model uncertainty are taken into account, relying on the assumption that resistance and load effect model uncertainties are the same for new and existing structures (i.e. assuming that the same models with same (epistemic) uncertainties are applied in design and assessment). These factors also include uncertainty in a decisive geometrical property such as the effective depth of a beam or the sectional area of a column.

2.2 *DVM Versus APFM*

The DVM proposes the fundamental basis for specifying partial factors. In contrast, the APFM provides adjustment factors to be applied on the partial factors for new structures in EN 1990:2002. The key inputs are the reliability requirement β, the coefficient of variation of the basic variable X, V_X and the reference period that corresponds to the reliability requirement β, t_{ref}.

Both methods essentially derive the partial factors for a material property and loads in a similar way while differing slightly in treating of model uncertainty factors. As the DVM can incorporate more structure-specific information (including possible biases in basic variables, i.e. deviations from common ratios between the mean and characteristic value of the basic variable), a specified reliability β may be achieved more accurately than for the APFM. The partial factors obtained by the DVM may be considerably different from those given in the Eurocodes if additional choices are made with respect to input parameters, i.e. additional measurements or investigations are undertaken.

The APFM is tailor-made for everyday use in practice and provides adjustment factors to be applied to the partial factors used in design. The adjustment factors take additional information into account when it is available. However, the adjustment factors are fixed, based on calibrated options for the input variables and hence also relying on these assumptions, while emphasising that collecting structure-specific information is strongly recommended.

The APFM may be seen to yield a more robust approach with respect to practical applications, i.e. with reduced influence of subjectivity, but also with limited regard to structure-specific conditions. The *fib* Bulletin 80 [3] provides the representative values of statistical characteristics of the basic variables listed in Table 1.

These simplified recommendations are deemed to cover most common cases and ensure the compatibility with Eurocodes. Due to the approximations accepted in the APFM, it often provides a more conservative format in comparison with the DVM. For the assessment of common structural elements, the APFM is particularly appropriate for normative purposes and calibration studies, considering the reduced amount of data required and the consistency with the Eurocode partial factors for the design of new structures. Nevertheless, recalibration is needed when the underlying Eurocode partial factors for structural design change. For important structures with large failure consequences and in the case of expensive upgrades, it is both justified and recommended to obtain additional case-specific information and to use a more advanced method for reliability verification. Use of the DVM is then recommended and if needed, higher level methods such reliability-based or risk-informed assessments might be conducted. Further to it, when reliability is significantly influenced by climatic actions, the use of the DVM is also recommended, provided that relevant measurements are available.

3 Case Study of Single Span Reinforced Concrete Slab

This case study corresponds to a highway bridge in Germany. The structure is a single span reinforced concrete slab of 75 cm thickness. The slab is connected to the abutments by concrete hinges, therefore it may be considered a simple supported beam with 10.6 m span (Fig. 1). The bridge was constructed approximately 50 years ago according to the German code in place at the time. Concrete grade corresponds to C20/25 according to the Eurocodes.

The bridge has been recalculated based on the reassessment guideline for existing bridges [5]. In addition to provisions regarding the consideration of outdated material grades, modified traffic load models and alternative design procedures, the guideline provides specifications for reduced partial safety factors. As a consequence among other alterations the traffic load model considered in the original structural design implied lower values than those currently required by the German national annex and

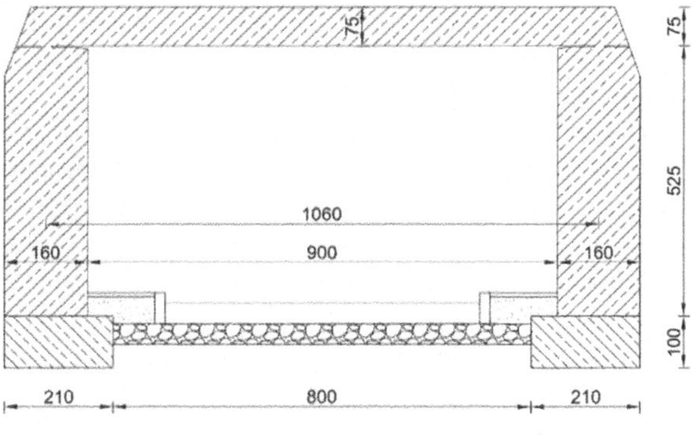

Fig. 1 Longitudinal section of the bridge selected as the first case study

Table 2 Partial factors for new structures and within German reassessment guideline [5]

Partial factors	New structure	German reassessment guideline
Concrete compressive strength	1.5	1.5
Reinforcing steel	1.15	1.05
Prestressing steel	1.15	1.1
Permanent loads	1.35	1.2
Traffic loads	1.35	1.5
Other live loads	1.5	1.5

some partial safety factors were also reduced. Table 2 details partial safety factors for new structures and those used for assessment of existing structures following [5]. It should be noted, that reduced partial safety factors for reinforcing and prestressing steel are only permitted for bending, if simultaneously the effective depth is modified unfavorably by 2.0 cm and 1.0 cm respectively. Furthermore, in the calculation of the design compressive strength the coefficient α_{cc} is 0.85 for Germany instead of the recommended value of 1.0 in the Eurocodes. The adjusted partial safety factor for permanent loads requires that the cross-section dimension and concrete density have been determined by measurements on the investigated structure. The apparent increase in the partial safety factor for traffic loads is misleading, since the guideline is based on the traffic loads of previous generations of the German standard where lower traffic loads were combined with a partial safety factor of 1.5. In contrast, the latest German national annex of EN 1991-2 assumes significantly higher traffic loads for road bridges, but also a partial safety factor of 1.35 (instead of 1.5), thus effectively emphasizing the serviceability limit states.

With these provisions the structure showed a just sufficient load bearing capacity both for bending and shear and could be approved for continued operation. If the currently valid traffic load model would be applied, the actions would exceed the resistances by approximately 15%.

It is quite common in Germany to obtain concrete core samples from bridges subjected to recalculation. These samples provide more specific data on the material properties of concrete but also on the density and therefore allow for an update on the available information both for actions and resistances. Consequently, the provisions of the *fib* Bulletin 80 [3] are demonstrated for the bridge based on the concrete compressive strength and the concrete density.

The target reliability index may be adjusted for existing structures. When relying on an economic consideration only, the *fib* Bulletin 80 [3] mentions the reduction of reliability indices related for the design working life accepted for new structures by about $\Delta\beta = 1.5$. Since the failure of the present bridge would have severe economic consequences, it may be categorised into consequences class CC3 and the minimum target reliability is $\beta_0 = 4.3 - 1.5 = 2.8$ (when considering as basic value the one provided in EN 1990:2002, see Table 3.2–1 in [3]). This value is considered in the following for illustration. It is based on economic optimization with a reference period equal to the remaining lifetime and is independent of the latter [3].

The six core samples taken from the structure have shown a mean value for in-situ compressive strength of 66.5 N/mm^2 and high coefficient of variation of 0.21. An updated characteristic value is $66.5 \times (1 - 2.18 \times 0.21) = 36.1$ N/mm^2, considering statistical uncertainties due to a small sample size of only six core samples by considering a value of $k_n = 2.18$ according to Table D1 in Annex D of EN 1990.

For DVM, the assumption of a lognormal distribution for concrete strength leads to the following partial safety factor:

$$\gamma_C = \gamma_{Rd} \cdot e^{\alpha_R \cdot \beta \cdot V_x - 1.645 \cdot V_x} = e^{0.4 \cdot 0.8 \cdot 2.8 \cdot 0.14} \cdot e^{0.8 \cdot 2.8 \cdot 0.21 - 1.645 \cdot 0.21} = 1.28 \quad (1)$$

where γ_{Rd} is the partial factor accounting for model uncertainty, α_R is the sensitivity factor for resistances, β is the reliability index and V_x is the coefficient of variation for the material property. In Eq. (1), one assumes a lognormal distribution for the model uncertainties which are expressed as $\gamma_{Rd} = \mu_{\theta_R}/\theta_{Rd} = e^{\alpha_{\theta_R} \cdot \beta \cdot V_{\theta_R}}$, assuming that $\mu_{\theta_R} = 1.0$ and that $V_{\theta_R} \approx 0.14$ for concrete, when geometrical uncertainties are significant [3].

Since there was no actual data on the density of the core samples for the selected case study, a data set with a mean of 2362 kg/m^3 and a coefficient of variation of 0.021 was considered from another similar structure for the purpose of demonstration. It shall be noted, that in a different study [6] a comparison of values for the concrete density for five different bridges recalculated in Germany showed a remarkable consistency in the coefficient of variation staying around 0.02. The partial safety factor for a normal distribution in case of a dominant variable based on the design value method is calculated as:

$$\begin{aligned} \gamma_G &= \frac{G_d}{G_k} \gamma_{Ed,G} = \frac{(1 - \alpha_E \cdot \beta \cdot V_G)}{(1 - k \cdot V_G)} e^{-\alpha_{\theta_E} \cdot \beta \cdot V_{\theta_E}} \\ &= \frac{1 - (-0.7) \cdot 2.8 \cdot 0.021}{1 - 0 \cdot 0.021} e^{0.4 \cdot 0.7 \cdot 2.8 \cdot 0.10} = 1.13 \end{aligned} \quad (2)$$

where $\gamma_{Ed,G}$ is the partial factor accounting for model uncertainty (assuming a lognormal distribution for model uncertainties and $V_{\theta_R} = 0.10$ for bending and shear [3]). It is highlighted that $k = 0$ in Eq. 2 as the *fib* Bulletin 80 [3] recommends this value for permanent actions (the mean value is chosen as the representative value). Note that commonly $V_G = 0.05$ is considered for self-weight of concrete structures.

In case of the adjusted partial factor method (APFM), the regular partial safety factors are modified by an adjustment factor ω_y. The adjustment factor is determined by the target reliability index and the ratio between the coefficient of variation for new structures and the actual value obtained on the structure. The adjustment factors for the current case study were calculated using the diagrams provided in the *fib* Bulletin 80 [3] as shown in Fig. 2 (and thus considering the limit for V_G''/V_G'). In the case of the self-weight of the bridge, in accordance with the recommendations of the bulletin, the diagrams were cut off at the lower limiting values. The resulting adjusted partial

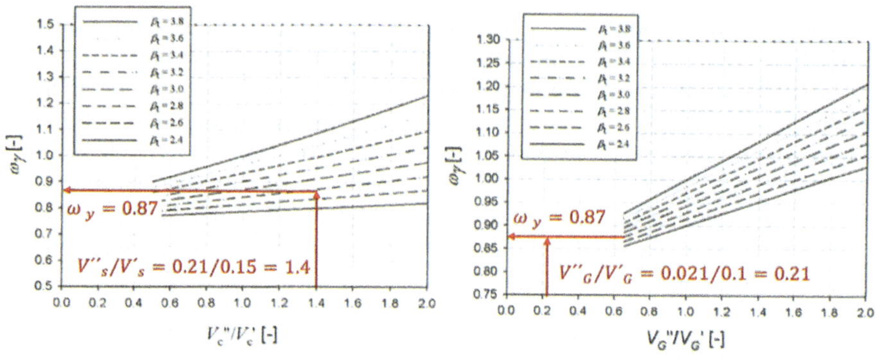

Fig. 2 Obtaining the adjustments factors for concrete and permanent actions for the APFM

safety factor for concrete is $\gamma_M = 0.87 \times 1.5 = 1.30$ and for self-weight $\gamma_G = 0.87 \times 1.35 = 1.18$.

4 Case Study of 3-Span Continuous Reinforced Concrete Slab

The second case study is 3-span continuous reinforced concrete slab located on the Croatian state road E65 near Posedarje town (Fig. 3), built in 1961. Detailed original design plans are available for all structural elements and amount of built-in reinforcement is checked, but no in situ measurements are taken in this study. Bridge spans are $9 + 15 + 9 = 33$ m. The slab is supported by concrete hinges on piers and the first (left) abutment, while a movable bearing is on the second abutment. Bridge

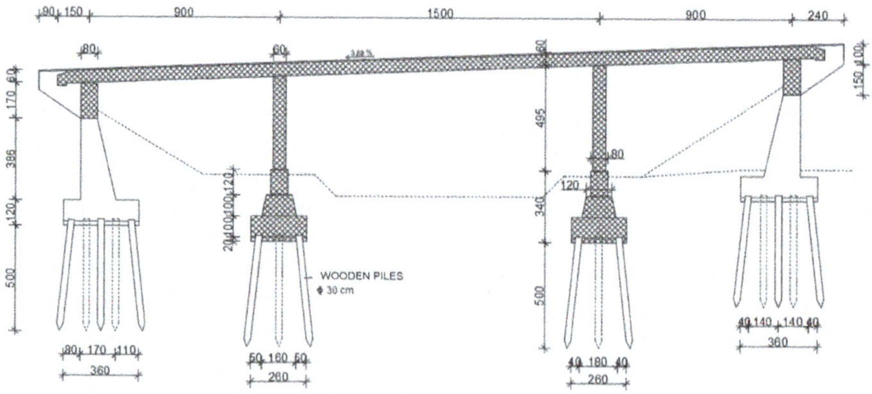

Fig. 3 Longitudinal section of the 3-span bridge selected as the second case study

slab is 0.60 m deep and 8.50 m wide. The bridge is designed according to the codes from 1940s. The designed concrete strength corresponds to C30/37 [7, 8].

In general, only visual inspections each six years are required for a bridge in good conditions without deficiencies (damage, cracks, corrosion, vibration). Additional destructive and non-destructive testing is needed in three cases: (i) for bridges of significant importance, large span or landmark bridges; (ii) deterioration or some deficiency is revealed by inspection, or (iii) after extreme events (e.g. earthquake, flood, etc.). This case study is focused on updating partial safety factors for the preliminary assessment based on no new material tests or traffic data.

The bridge is classified in Consequence Class CC3, the minimum target reliability is $\beta_0 = 4.3 - 1.5 = 2.8$ and two reference periods ($t_{\mathrm{ref},1} = 40$ years and $t_{\mathrm{ref},2} = 15$ years; both equal to remaining lifetimes in two alternatives under investigation) are assumed. The material factors for concrete (γ_{C}) and reinforcement (γ_{S}) are determined as follows:

$$\gamma_C = \gamma_{Rd} \cdot \gamma_c = e^{\alpha_{\theta Rc} \cdot \beta \cdot V_{\theta Rc}} \cdot \gamma_c = e^{0.4 \cdot 0.8 \cdot 2.8 \cdot 0.14} \cdot 1.09 = 1.13 \cdot 1.09 = 1.24 \quad (3)$$

$$\gamma_S = \gamma_{Rd} \cdot \gamma_s = e^{\alpha_{\theta Rs} \cdot \beta \cdot V_{\theta Rs}} \cdot \gamma_s = e^{0.4 \cdot 0.8 \cdot 2.8 \cdot 0.06} \cdot 1.03 = 1.06 \cdot 1.03 = 1.08 \quad (4)$$

where γ_{Rd} is the value for model uncertainties (concrete or steel) calculated by considering a lognormal distribution [3], and γ_{c} and γ_{s} are estimated according to Fig. 4 (left), considering $V_{\mathrm{c}} = 0.15$ and $V_{\mathrm{s}} = 0.05$ – basic values in [3].

Permanent action factor γ_G is calculated for unfavourable action and favourable load effects:

$$\gamma_G = \gamma_{Ed,G} \cdot \gamma_g = e^{-\alpha_{\theta E} \cdot \beta \cdot V_{\theta E}} \cdot \gamma_g = e^{0.4 \cdot 0.7 \cdot 2.8 \cdot 0.10} 1.19 = 1.08 \cdot 1.19 = 1.29 \quad \text{for an unfavourable action} \quad (5)$$

$$\gamma_G = \gamma_{Ed,G} \cdot \gamma_g = e^{-\alpha_{\theta E,fav} \cdot \beta \cdot V_{\theta E}} \cdot \gamma_g = e^{-0.4 \cdot 0.8 \cdot 2.8 \cdot 0.10} \cdot 0.92 = 0.91 \cdot 0.92 = 0.84 \quad \text{for a favourable action} \quad (6)$$

considering the recommended determination approach for $\gamma_{\mathrm{Ed},G}$ and $V_{\mathrm{g}} = 0.1$ (Fig. 4, middle and right).

Partial safety factors for variable action, γ_Q, is given as:

$$\gamma_Q = \gamma_{Ed,Q} \cdot \gamma_q = e^{-\alpha_{\theta E} \cdot \beta \cdot V_{\theta E}} \gamma_q \quad (7)$$

where model uncertainty factor is $\gamma_{\mathrm{Ed},Q} = 1.08$ for $\beta_0 = 2.8$ as $\gamma_{\mathrm{Ed},G}$ in Eq. 5 for the unfavourable permanent action effect. The factors γ_q are obtained from Fig. 5 considering $\beta_0 = 2.8$ and $t_{\mathrm{ref},1} = 40$ years or $t_{\mathrm{ref},2} = 15$ years considering the recommended values for V_T and V_{vb} (related to annual maxima of traffic load effect and basic wind velocity, respectively).

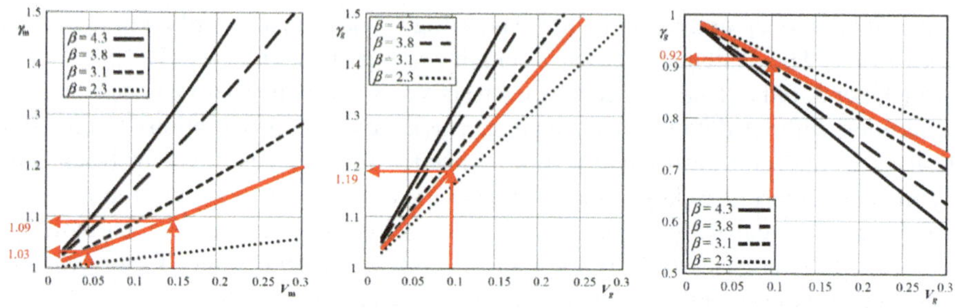

Fig. 4 Determination of γ_m ($V_C = 0.15$; $V_S = 0.05$; $\alpha_R = 0.8$ and $\beta = 2.8$) (left), the partial factor γ_g for unfavourable action ($V_g = 0.1$; $\alpha_E = -0.7$; $\beta = 2.8$) (middle) and for favourable action ($V_g = 0.1$; $\alpha_{E,\text{fav}} = 0.32$; $\beta = 2.8$)

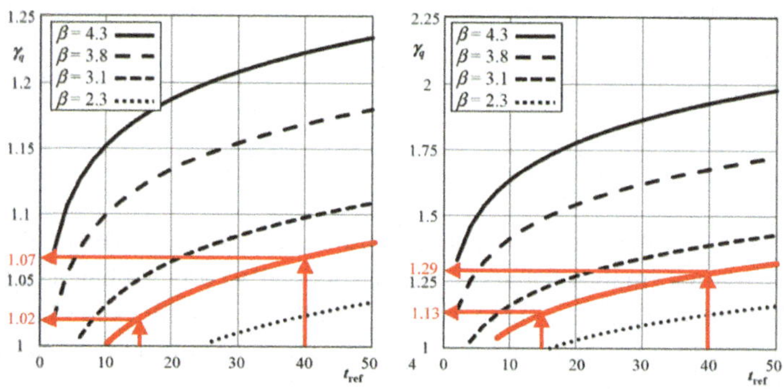

Fig. 5 Determination of the partial factors for: the traffic load, $\gamma_{q,t}$ ($V_T = 0.075$) (left); the wind action, $\gamma_{q,w}$ ($V_{vb} = 0.12$) (right) for $\beta = 2.8$, and $t_{\text{ref},1} = 40$ years or $t_{\text{ref},2} = 15$ years

According to APFM, the partial safety factors for existing structures can be generally determined as:

$$\gamma_X = \omega_\gamma \cdot \gamma_{X,new} \tag{8}$$

where ω_y is an adjustment factor and $\gamma_{X,\text{new}}$ is the partial factor for new structures given in the Eurocodes. The values of ratio between the coefficient of variation for new structures and the actual value obtained on the structure are assumed in the case study. Adjustment factors are determined graphically (Figs. 6 and 7).

The obtained factors according to the DVM and APFM methods are summarised in Table 3. Partial factors are reduced compared to the corresponding values for new structures. Factors for unfavourable permanent actions obtained by DVM and APFM are almost equal, while the APFM gives more conservative values of partial factors for traffic load and wind action in comparison to the DVM due to a simplified conservative treatment of model uncertainties.

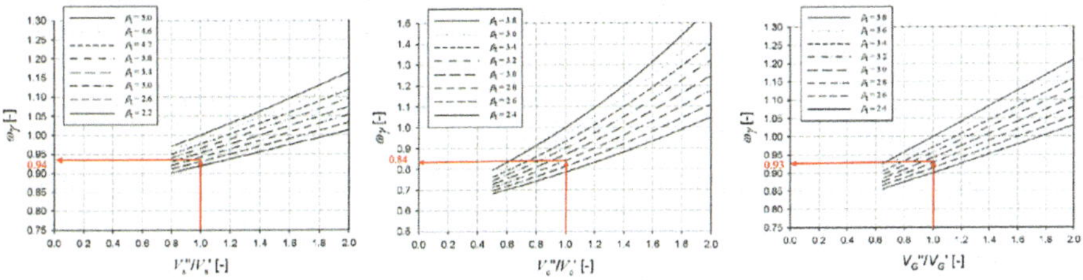

Fig. 6 Determination of the adjustment factors for steel $\omega_{y,s}$ (left), concrete $\omega_{y,c}$ (middle) and unfavourable permanent actions $\omega_{y,g}$ (right) according to the APFM ($\alpha_R = 0.8$; $\alpha_E = -0.7$; $\beta = 2.8$; $V_s''/V_s' = 1.0$; $V_c''/V_c' = 1.0$; $V_G''/V_G' = 1.0$; $V_G' = 0.1$)

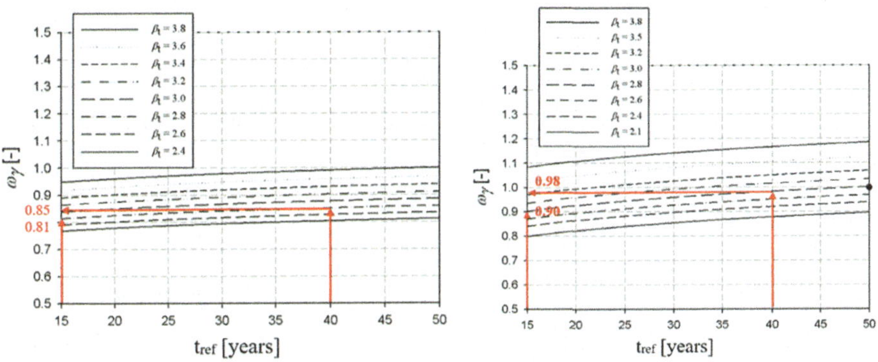

Fig. 7 Adjustment factors for variable actions: traffic loads $\omega_{y,t}$ (left) and wind actions $\omega_{y,sw}$ (right) according to the APFM – method A ($\alpha_E = -0.7$; $\beta = 2.8$; $t_{ref,1} = 40$ years or $t_{ref,2} = 15$ years; $V_{Qclim,I}''/V_{Qclim,I}' = 1.0$)

Analysis of the differences between γ_Q-factors of DVM and APFM (method A) reveals that the difference between model uncertainty factors $\gamma_{Ed,Q}$ is negligible ($\leq$0.5%) and the reasons for the difference needs to be found in the load factors γ_q.

DVM combines a time-variant component of the load (such as basic wind pressure) with time-invariant coefficients (e.g. shape, gust, roughness and other factors for wind pressure). The design value is then derived for given t_{ref} and β using recommended probabilistic models of the time-variant and time-invariant variables. The probabilistic models are simplified, but they are deemed to provide unbiased estimates, reflecting the existing knowledge. Consequently, use of the recommended models may not always lead to $\gamma_Q = 1.5$ (or 1.35 for road traffic load) for a "basic case" with $\beta = 3.8$ and $t_{ref} = 50$ years.

As an example let us consider a wind action effect for which the largest difference in γ_Q between DVM and APFM is indicated in Table 3. For the basic case ($\beta = 3.8$ and $t_{ref} = 50$ years), DVM leads to $\gamma_w = 1.75$ and $\gamma_W = 1.11 \times 1.75 = 1.94$. Changing β to 2.8 significantly decreases particularly γ_w. The partial factor becomes $\gamma_W = 1.08 \times 1.32 = 1.43$ for $\beta = 2.8$ and $t_{ref} = 50$ years, i.e. decrease by about

Table 3 Summarised results for modified partial factors according to the DVM and APFM methods ($\beta = 2.8$; $t_{\text{ref},1} = 40$ years and $t_{\text{ref},2} = 15$ years)

Partial factors	DVM t_{ref} (years) 40	DVM t_{ref} (years) 15	DVM Input assumptions	Adjustment/partial factors	APFM t_{ref} (years) 40	APFM t_{ref} (years) 15	APFM Input assumptions
γ_c	1.09		$V_{\text{m,C}} = 0.15$	$\omega_{\gamma,c}$	0.84		$V''_C / V'_C = 1$
γ_s	1.03		$V_{\text{m,S}} = 0.05$	$\omega_{\gamma,s}$	0.94		$V''_S / V'_S = 1$
γ_C	1.24		$V_{\theta R,\text{c}} = 0.14$	γ_C	1.26		
γ_S	1.08		$V_{\theta R,\text{s}} = 0.06$	γ_S	1.08		
γ_g	1.19		$\alpha_E = -0.7$	$\omega_{\gamma,G}$	0.93		$V''_S = V'_S = 0.1$
$\gamma_{g,fav}$	0.92		$\alpha_{E,\text{fav}} = 0.32$				
$\gamma_{q,w}$	1.29	1.13	$V_{vb} = 0.12$ (annual max.)	$\omega_{\gamma,W}$	0.98	0.90	$V'_{Q_{c\lim 1}} = V''_{Q_{c\lim 1}}$
$\gamma_{q,t}$	1.07	1.02	$V_T = 0.075$ (annual max.)	$\omega_{\gamma,t}$	0.85	0.81	$\beta' = 3.8$
γ_G	1.29		$\gamma_{Ed,G} = 1.07$	γ_G	1.26		
$\gamma_{G,fav}$	0.84		$\gamma_{Ed,G} = 1.00$				
$\gamma_{Q,w}$	1.39	1.22	$\gamma_{ED,Q} = 1.12$	$\gamma_{Q,W}$	1.47	1.35	
$\gamma_{Q,t}$	1.16	1.10		$\gamma_{Q,t}$	1.28	1.22	

25% in comparison to the basic case. Decreasing t_{ref} to 40 years or 15 years yields additional reductions by 3% or 15%, respectively.

APFM makes no explicit distinction between the time-invariant and time-variant components of a variable load effect. For imposed and traffic loads, the method is tuned to obtain $\gamma_Q = 1.5$ or 1.35, respectively, for the basic case. For snow and wind loads, it is recognised that $\gamma_Q = 1.5$ is mostly insufficient to reach $\beta = 3.8$. The recommendations in APFM have thus been established in a way that $\gamma_S = \gamma_W = 1.5$ is reached for $\beta = 2.8$, representing implicitly accepted risk in Eurocodes. The different assumptions in APFM lead to different sensitivity of γ_Q to changes in β and t_{ref}. For a wind load effect, APFM indicates $\gamma_W = 1.77$ for the basic case ($\omega_{y,sw} = 1.18$ when $\beta = 3.8$ and $t_{\text{ref}} = 50$ years). Decreasing β to 2.8 gives a smaller reduction (15%) than in the case of DVM. Decreasing t_{ref} to 40 years or 15 years yields again smaller reductions (by 2% or 10%, respectively) than for DVM.

5 Discussion and Conclusions

The performance assessment of existing structures can take place at various times during their service life and for various reasons: change in the use of the structure, evolution of environmental variables and their uncertainties as well as material properties, regulatory developments, doubts about serviceability or safety levels, damage observed in situ, etc. Considering the difficulties in reliability assessments of existing structures, the assessment process needs to be based on a rational approach whose degree of sophistication and related assumptions depend structure-specific conditions.

The two methods DVM and APFM provided in [3] are based on the same principles, introduced in 1980s as the design value format. The methods aim to cover most common cases and ensure the compatibility with Eurocodes. In this sense, they provide an interesting framework to specify partial factors for practitioners who use a semi-probabilistic format when assessing existing structures. The two case studies presented in this paper illustrate how partial factors would be specified in each case. The DVM and the APFM are applied and lead to slightly different results, particularly due to simplified conservative treatment of model uncertainties and variable load effects in the APFM. The differences between the methods become more significant when the target reliability and reference period substantially differ from the reference values, $\beta = 3.8$ and $t_{\text{ref}} = 50$ years. Further studies should investigate how the choice between these methods would change the result of the assessment process (after determining the load bearing capacity) and where the use of fixed partial factors with updated characteristic values (the procedure often adopted in standards) is sufficient.

A major simplification of the two methods is that statistical uncertainty associated with the new measurements on the existing structure is considered only in the estimate of characteristic values of basic variables. Particularly in the case of absent prior information, the effect of this simplification might be large, especially for small sample sizes that are typical for reassessment situations. Here, the application of the methods could lead to non-conservative estimates of structural reliability. Finally, as simplified methods, DVM and APFM are based on some assumptions on standardised sensitivity factors or types of probabilistic distributions. For a specific limit state, the real sensitivity factors might deviate to a certain extent (in particular for complex limit states with several variables). Further analyses should consider the effect of the reference period on the target reliability index—the use of the two methods should be explored for reliability verifications based on an annual reference period (as introduced e.g. in ISO 2394). A full probabilistic analysis should provide reference reliability levels to check the consistency of the simplified approaches.

Acknowledgements This study is an outcome of IABSE TG1.3 in collaboration with *fib* COM3 TG3.1. Some results of *fib* COM3 TG3.1 have been utilised. This study is partly based upon the work conducted in the framework of Croatian national project "Key Performance Indicators of Existing Bridges" supported by the University of Zagreb, and Czech national project 20-01781S supported by the Czech Science Foundation. The authors thank Dimitris Diamantidis from OTH Regensburg, Germany, for his valuable comments that helped to improve the quality of this paper.

References

1. JRC report of CEN/TC250/WG2 (2015) *New European technical rules for the assessment and retrofitting of existing structures*. ISBN 978-92-79-46023-4.
2. Diamantidis, D., et al. (2001). *Probabilistic assessment of existing structures*. RILEM Publications S.A.R.L.
3. Caspeele, R., Steenbergen, R., Sýkora, M., et al. (2017). *Partial factor methods for existing concrete structures* (*fib* Bulletin 80). Fédération internationale du béton (*fib*), *fib* COM3 TG3.1.
4. JCSS. (2001). *JCSS probabilistic model code (periodically updated online publication)*. Joint Committee on Structural Safety, www.jcss.byg.dtu.dk
5. Nachrechnungsrichtlinie. (2011). *Richtlinie zur Nachrechnung von Straßenbrücken im Bestand*. Bundesministerium für Verkehr, Bau und Stadtentwickelung, Bonn, Germany.
6. Boros, V. (2019). Reassessment of the partial safety factor for self weight of existing bridges. In *International Probabilistic Workshop 2019* (pp. 75–80). Edinburgh, United Kingdom, September 11–13, 2019.
7. Šram, S. (2002). *Bridge construction—Concrete bridges*. Golden marketing, Zagreb [in Croatian].
8. Grlić, S. (2019). Application of the partial factor methods on the Posedarje bridge [Master thesis]. Kušter Marić, M. University of Zagreb, Faculty of Civil Engineering, Zagreb [in Croatian].

Comparison of Measured and Simulated Traffic Loading based on BWIM Data from the Millau Viaduct

Marcel Nowak, Franziska Schmidt, and Oliver Fischer

Abstract Traffic load modeling on road bridges requires detailed information of both the actual traffic load impact and the resulting structural responses. Reviewing common approaches on the topic, two different principal strategies can be identified: direct analysis of sufficiently detailed measurement data—either records of the passing road traffic or the resulting structural response for the structure under investigation—, or evaluation of synthetic traffic generated by numerical traffic simulation based on statistical data representing the traffic characteristics at site. Ideally, both methods can be combined by calibrating a numerical simulation model by means of detailed measurement data (e.g. WIM or BWIM data), thus reducing the uncertainty in the numerical model while maintaining its high flexibility (e.g. evaluation of different structures or different traffic scenarios). Therefore, it is essential to identify those relevant aspects within a certain traffic record actually governing the part of bridge loading relevant for structural design and reassessment (i.e. extreme values of load effects), and accurately represent them in the numerical simulation model. With an extensive and detailed traffic data record at hand, covering a period of 180 days of BWIM measurements at the Millau Viaduct in Southern France, this study performs a comparison of traffic loading resulting from the recorded traffic stream directly and from traffic generated by numerical simulation. The comparison is performed for a set of different load effects and bridge lengths on different levels, regarding governing traffic characteristics and extreme values of resulting load effects. Based on the findings from this comparison, suggestions for appropriate modeling using numerical traffic simulations can be proposed.

Keywords Bridge loading · Traffic simulation · Bridge weigh-in-motion · Extreme load effects · Millau Viaduct

M. Nowak (✉) · O. Fischer
Chair of Concrete and Masonry Structures, Technical University of Munich, Munich, Germany
e-mail: marcel.nowak@tum.de

F. Schmidt
MAST, EMGCU, IFSTTAR, Université Paris-Est, Champs-sur-Marne, France

© The Author(s), under exclusive license to Springer Nature Switzerland AG 2021
J. C. Matos et al. (eds.), *18th International Probabilistic Workshop*, Lecture Notes in Civil Engineering 153, https://doi.org/10.1007/978-3-030-73616-3_16

1 Introduction

Road traffic is characterized by a high degree of diversity and complexity, with high volumes of truck traffic and different vehicle types on multiple traffic lanes. In addition, the bridge inventories of contemporary infrastructure networks show a large variety, with different types of structures and dimensions, enabling the presence of multiple vehicles on the bridge at the same time. Correspondingly, significant loading events for traffic load effects on a bridge can be manifold (e.g. multi-presence of trucks, exceptional permit trucks), and especially difficult to model and predict. Bridge loading is an apparently quite complex process, affected not just by the characteristics of road traffic at site, but also by the structure itself, and hence is always site-specific, or in other words with a distinct object reference [1].

Many of the common approaches for modeling traffic loading on bridges are based on numerical simulations, aiming for representing actual traffic and resulting load effects on a certain bridge structure in a realistic manner. For this purpose, it is essential to know those relevant aspects within typical road traffic actually governing the part of traffic loading relevant for structural design (i.e. extreme values of load effects), and accurately represent them in the simulation model. With an extensive and detailed traffic data record at hand, covering a period of 180 days of BWIM measurements at the Millau Viaduct in Southern France, this study performs a comparison of traffic loading resulting from the recorded traffic stream directly and from traffic generated by numerical simulation. The comparison is performed for a set of different load effects and bridge lengths on different levels, regarding governing traffic characteristics and extreme values of resulting load effects. Based on the findings from this comparison, suggestions for appropriate modeling using numerical traffic simulations can be proposed.

2 Traffic Load modeling

2.1 General Remarks

The principles of the bridge loading process can be broken down to a simple, exemplary model presented in Fig. 1. A loading event due to road traffic on a bridge structure is mainly characterized by the presence of (one or several) trucks on its deck, their positions along the roadway on the deck, and their respective axle weights.

The object reference inherent to the bridge loading process results in a large number of influencing factors. The characteristics of road traffic at site determine the actual load impact on the structure and are defined by many different parameters, describing the traffic volume and configuration, traffic flow and the single vehicles themselves. The structure itself affects the resulting response to a certain loading event by shape and dimensions of the influence functions corresponding to the load

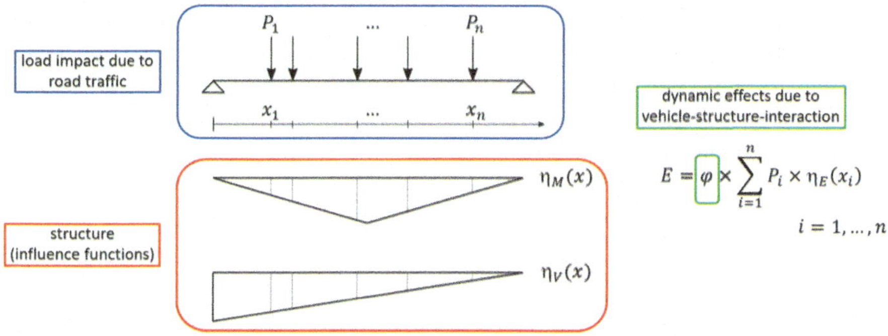

Fig. 1 Simple, exemplary model illustrating the principles of the bridge loading process

effects under consideration. They determine the possible number of vehicles simultaneously contributing to the structural response, and their respective share to it based on their position. Dynamic effects due to interaction between vehicles and structure can result in amplification of the traffic load effects, depending on the dynamic properties of the structure, the vehicles, and additional aspects, such as surface roughness of the roadway or velocity of the vehicles passing the bridge.

It would be desirable to define a theoretical statistical model for the bridge loading process, based on stochastic process theory. Due to the complexity inherent to the bridge loading process however, these attempts proved to be very challenging (e.g. [2]). Reviewing common approaches on the topic, two different principal strategies can be identified: direct analysis of sufficiently detailed monitoring data—either traffic monitoring recording the passing road traffic or bridge monitoring recording the resulting structural response for the structure under investigation—, or evaluation of synthetic traffic generated by numerical traffic simulation based on statistical data representing the traffic characteristics at site.

2.2 Analysis of Monitoring Data

If available, the analysis of sufficiently detailed monitoring data towards bridge loading is straightforward and comes with several advantages. Detailed data from traffic monitoring is nothing else but a record of the real, actual traffic on site. It can be analyzed directly towards bridge loading and resulting load effects, without further need for traffic modeling. Data from monitoring on the structure itself even provides direct information on the resulting traffic load effects, without even knowing the actual load impact from the passing road traffic causing them. The huge advantage of directly using monitoring data is its high accuracy (assuming proper acquisition and processing of the monitoring data). Tedious modeling work and necessary modeling assumption can be omitted. Monitoring data has the ability to represent the actual bridge loading process as close to reality as possible. This however is just true for bridge loading throughout the period of the monitoring. As these periods are usually rather small compared to the service life of a bridge, they hardly can be considered representative, not covering entirely the possible variety of bridge loading throughout the lifetime of a structure. Moreover, detailed monitoring data is often simply not available when analyzing certain structures for bridge loading.

2.3 Numerical Simulation

There are many different approaches for numerical traffic simulation within the context of bridge loading analysis, with varying degrees of detail and complexity. The method employed in this work is more general and somewhat simplified, but adopted from the procedures followed for background works on the development of load models in European standards. It generates streams of synthetic traffic applying random sampling based on statistical distribution models for relevant parameters describing traffic flow, traffic composition, and single vehicle characteristics [1]. The simulation predominantly neglects possible inter-vehicle correlations, and assumes statistical independence between single vehicles on same or adjacent traffic lanes. The generated traffic is evaluated using influence functions for load effects under consideration. The accuracy of this method heavily relies on the accuracy of the underlying statistical distribution models for single traffic parameters, and on reasonable modeling assumptions. Compared to the use of monitoring data, significantly larger inaccuracies in representing the actual bridge loading are to be expected. However, these drawbacks are usually outweighed by the high degree of flexibility this approach provides. Variable traffic scenarios and different locations within a structure can be analyzed regarding bridge loading over an arbitrary period. This makes numerical simulations especially suitable for analyzing bridge loading representative for the entire lifetime of a structure, usually resulting in more robust results.

3 Traffic Data from Millau Viaduct

The Millau Viaduct is a cable-stayed bridge crossing the valley of the River Tarn in Southern France. The structure consists of eight spans, with the six inner spans having a width of 342 m spans and the two end spans with a width of 204 m (see Fig. 2). Its steel orthotropic deck supports an important national highway, with a roadway of two traffic lanes for each direction. The bridge has been designed in accordance with European Standards (Eurocodes). It was brought into service in December 2004.

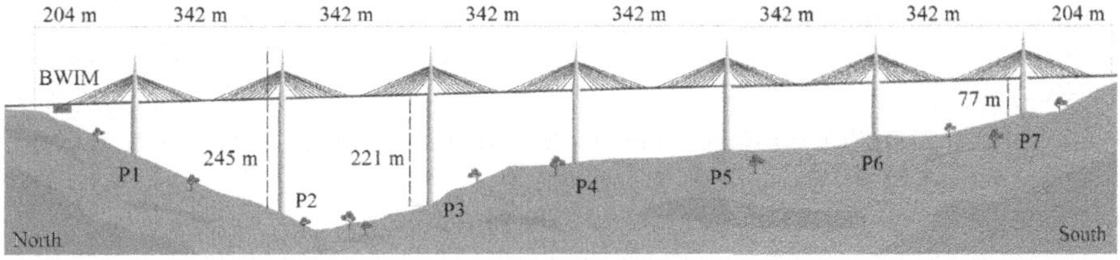

Fig. 2 View of Millau Viaduct, with location of BWIM system

3.1 *Traffic Monitoring*

In order to update the design model and detect the amount of heavy and overweighed vehicles, a monitoring campaign on the bridge was initiated. For this purpose, a bridge weigh-in-motion (BWIM) system was installed in the middle of the first span of the viaduct (see Fig. 2). It detects each passing vehicle on both traffic lanes for one of the driving directions, and provides information on following vehicle characteristics: total weight, axle weights and spacings (only up to 5 axles), vehicle length, number of axles, vehicle type, velocity, lane, and timestamp of passage (refer also to [3]). The data used for the scope of this paper covers the period from October 2016 until June 2017, with some breaks in between. In total, traffic records from 180 days are available. The data contains 162,555 truck vehicle records detected in this period.

3.2 *Data Preparation and Cleaning*

Before the BWIM traffic records can be analyzed towards bridge loading, some processing and cleaning of the data is required to remove nonrelevant and erroneous data samples. For this purpose, following rules are applied, mainly adopted from [4, 5], based on vehicles' axle weights and spacings, and on inter-vehicle gap distances (distance between last axle of preceding and first axle of following vehicle):

- Identification of "doubled" records (repeated recording of same vehicle or parts of it, such as last axle group in vehicle) for inter-vehicle gap distances <1.5 m (also negative values), deletion of the doubled record with smaller weights
- Identification of "separated" records (one vehicle split up into two records) for inter-vehicle gap distances within a range of 1.5–10.0 m, merging to one single record
- Removal of vehicles with a gross vehicle weight (GVW) <3.5 t
- Removal of vehicles with single axle weights >35 t or single axle weights >15 t constituting a ratio >85% of GVW
- Complement of axle weight and spacing for vehicles with a number of axles >5 (determination of the differences between total values (vehicle length and GVW) and sum of axle values (weights and spacing); even distribution of weight

difference to remaining axles; based on vehicle type even distribution of length difference to spacings between axle groups and fixed spacing of 1.4 m within axle groups)

- Identification of "incomplete" days as records that cover less than 22 h, deletion of entire record for these days
- Removal of traffic records from weekends and holidays
- Removal of traffic records with unusual traffic flow (e.g. entire traffic on the left lane maybe due to construction works or the like).

After this preparation and cleaning, a data set covering 111 working days and containing 99,389 truck vehicle records remains for further analysis.

3.3 Statistical Data Analysis of Traffic Record

Statistical data analysis is performed for the cleaned BWIM data set to derive suitable model descriptions, serving as input for subsequent numerical traffic simulations. Table 1 shows the results for parameter models for all relevant vehicle types (covering 99.4% of all recorded vehicles). The vehicle type represents the sequence of axle groups in a vehicle (e.g. vehicle type "113" consists of two single axles followed by an axle tridem). The total number of axles in a vehicle is computed by the cross sum of the type ID. The GVWs are modelled as bi- or tri-modal normal distributions, represented by model parameters mean (μ_i), standard deviation (σ_i) and modal contribution (ξ_i), obtained from model fit to the BWIM data. The distribution of GVW to the single axles and axle spacings are modelled with fixed values corresponding to the mean values obtained from the BWIM record.

The traffic volume (average daily traffic, ADT) and its distribution throughout the day are also modelled using fixed values represented by the mean values from the BWIM record (see Table 2). The vehicle velocity is set constant and uniform for all vehicles in the numerical traffic simulation, with an assumed value of 80 km/h.

For modeling the inter-vehicle distances, a simplified, traffic flow dependent lognormal distribution model is used (as proposed in [6]). For each block of 6 h and each separate traffic lane, the lognormal model is defined anew, and all vehicle distances in this block are sampled from this model. The expected value of the model is computed based on the traffic volume in the block, an average vehicle length (based on traffic composition and axle spacings), and an average vehicle velocity. The coefficient of variation is taken from a traffic flow dependent trend model. This trend model is obtained by analysis of the available BWIM traffic record. The entire record is separated into blocks of 6 h each, and for each block (and each separate traffic lane) the traffic volume is computed, and a lognormal distribution is fitted to the respective inter-vehicle distances. The trend model is determined by fitting a power law to the resulting data samples (see Fig. 3, which also shows the trend model "A61" used in [6] and derived for higher traffic volumes).

Table 1 Vehicle parameter models for relevant vehicle types of Millau BWIM record

Type ID	Ratio (%)	Axle weight distribution (%) / $\mu_i/\sigma_i/\xi_i$ for modes of GVW model (kN)	Axle spacing (m)
11	11.0	40.0/60.0	5.1
		38.3/3.3/0.34; 89.7/30.6/0.371; 54.1/28.1/0.29	
12	2.0	33.5/46.3/20.2	5.9/1.3
		88.6/36.9/0.17; 182.6/32.5/0.57; 231.3/54.4/0.26	
111	2.1	29.0/37.5/33.5	3.9/6.1
		46.4/8.5/0.55; 153.1/47.5/0.35; 227.8/68.0/0.10	
112	12.9	25.5/33.5/20.8/20.2	4.4/7.3/1.3
		172.4/44.2/0.29; 205.4/32.0/0.34; 288.4/39.4/0.37	
113	66.8	20.2/30.4/17.4/16.1/15.9	4.0/6.1/1.3/1.3
		204.3/37.8/0.27; 296.2/38.2/0.23; 396.5/33.0/0.50	
121	0.1	26.6/30.4/19.0/24.0	4.9/1.2/6.6
		147.7/52.3/0.38; 237.6/34.0/0.49; 328.6/67.3/0.13	
122	1.3	22.4/27.0/14.7/19.1/16.8	5.1/1.3/6.8/1.4
		241.2/49.5/0.66; 362.7/70.9/0.34	
123	0.3	16.3/17.4/17.7/17.6/15.1/15.9	3.7/1.2/6.3/1.3/1.3
		265.8/59.9/0.32; 392.7/50.6/0.37; 536.5/67.7/0.31	
211	0.2	8.7/12.2/32.7/46.4	1.2/3.3/7.4
		211.8/41.4/0.36; 304.3/31.5/0.64	
1111	1.8	24.2/34.9/21.8/19.1	4.7/6.3/3.7
		187.3/47.4/0.39; 213.9/29.6/0.47; 333.0/50.3/0.14	
1112	0.3	20.5/32.0/18.8/16.7/12.0	4.7/5.4/3.9/1.2
		205.5/33.3/0.41; 289.9/38.2/0.16; 390.1/36.7/0.43	
1121	0.2	18.9/30.0/16.1/17.3/17.7	4.2/5.8/1.2/2.9
		198.9/28.7/0.34; 284.7/33.0/0.27; 398.2/40.0/0.39	
1211	0.9	21.4/25.8/15.4/18.7/18.7	5.0/1.2/5.7/4.0
		228.5/23.9/0.42; 324.9/46.5/0.26; 332.1/87.0/0.32	
1212	0.1	17.3/23.1/15.2/17.4/14.8/12.2	4.7/1.2/4.9/4.2/1.2
		264.4/46.7/0.26; 401.8/29.6/0.66; 541.4/46.5/0.08	

Table 2 Mean values for traffic volume and distribution of Millau BWIM record

ADT (veh/d)		Ratio (%)	Distribution of ADT throughout the day (%)			
			0:00–6:00	6:00–12:00	12:00–18:00	18:00–0:00
895	Slow lane	99.2	10.3	29.5	38.1	22.1
	Fast lane	0.8	4.9	31.3	43.4	20.4

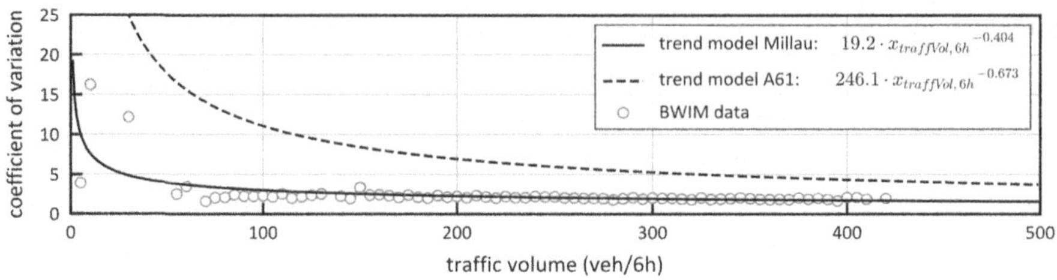

Fig. 3 Trend model for traffic flow dependent coefficient of variation of lognormal distribution models for inter-vehicle distances

4 Evaluation of Bridge Loading

4.1 Scope of Analysis

For evaluation of bridge loading, both recorded traffic from BWIM data (further referred to as "BWIM") and synthetic traffic from numerical traffic simulations are considered. For the latter, the simulation procedure as described in Sect. 2.3 is applied, using the parameter models derived from statistical data analysis of the BWIM data (see Sect. 3.3) and simulating traffic representative for the traffic monitored at Millau Viaduct (further referred to as "$\text{SIM}_{\text{Millau}}$"). For comparison purposes, an additional traffic scenario is considered for simulation, representing the case of having only partial information available for the traffic characteristics to be represented (which is often the case in engineering practice). It is assumed, that only traffic composition and traffic flow related data might be available from the site (e.g. based on automated counting stations), but further data on vehicle parameters (GVW, axle weight distribution, axle spacing) and inter-vehicle distance behavior might not be at hand. To compensate the missing parameter information, models derived from monitoring data from a different traffic site with supposedly similar traffic characteristics are assumed. In this case, parameter models for vehicles and inter-vehicle distance are adopted from federal highway A61 in Germany (refer to [6] and Fig. 3, this data was used for background works for calibration of the current load model LM1 from Eurocode 1 in Germany). For this traffic scenario (further referred to as "SIM_{A61}"), the traffic is modeled with vehicle types available from the A61 data (refer to Table 3 for considered traffic composition). In accordance with the available BWIM data of traffic on working days, both synthetic traffic streams are simulated for a period of 111 days.

Table 3 Traffic composition for scenario A61, vehicle types according to [6], with corresponding axle layout

Type ID	t8 (11)	t9 (12)	t33 (1111)	t41 (1211)	t97 (112)	t98 (113)
Ratio	11.0	2.0	2.0	1.5	16.4	67.1

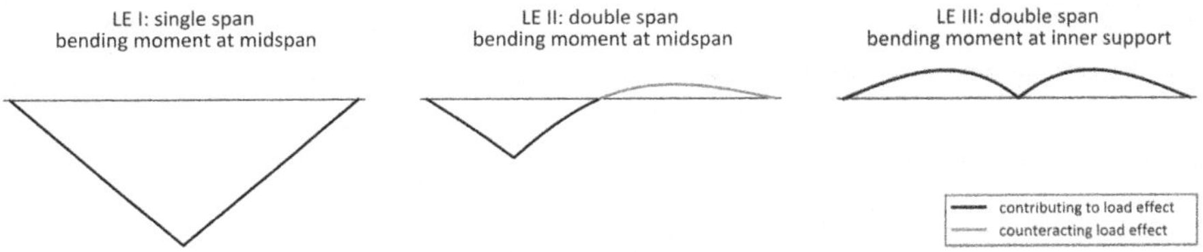

Fig. 4 Influence lines for load effects under consideration

All three traffic streams are evaluated for the following set of structures and load effects (refer to Fig. 4 for corresponding influence lines):

- single span, bending moment at midspan (LE I)
- double span, bending moment at midspan (LE II)
- double span, bending moment at inner support (LE III).

Evaluation is done for total bridge lengths of 20, 50 and 80 m. Possible effects of dynamic amplification are neglected for this analysis (traffic loads are considered static).

4.2 *Comparison of Governing Traffic Characteristics*

In a first step, simulated and recorded traffic are compared on the level of actual traffic load impact, independent of any structural response to it. As commented in Sect. 2.1, a loading event due to road traffic on a bridge structure is mainly characterized by the presence of (one or several) trucks on its deck, their positions along the roadway on the deck, and their respective weights. In this context, the traffic streams are evaluated and compared with regard to vehicle weights, inter-vehicle distances and possible multi-presence of vehicles on the bridge.

As the focus of this work is on extreme values of bridge loading, also the extreme values of recorded and simulated GVWs are compared. For this purpose, the upper 1000 weight values for each traffic stream are considered (refer to Fig. 5, left). The upper tails of SIM_{Millau} and SIM_{A61} are quite similar, however they are not able to

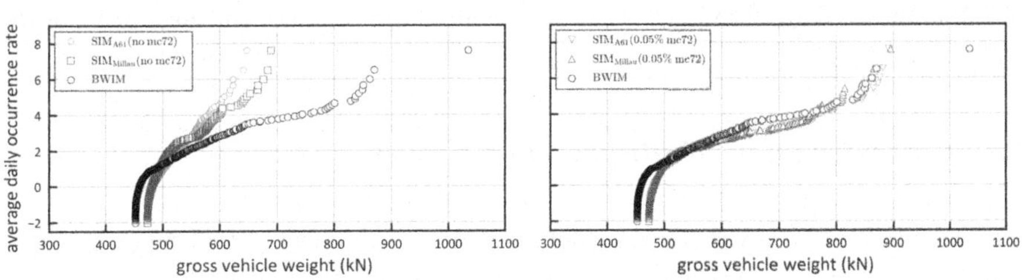

Fig. 5 Upper 1000 GVW values for recorded and simulated traffic, without (left) and including mobile crane mc72 (right)

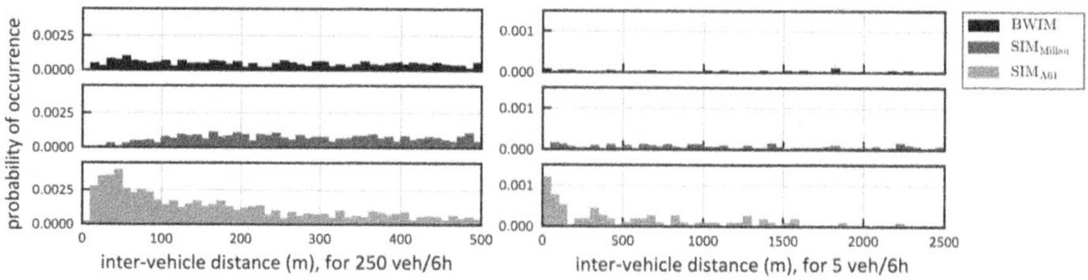

Fig. 6 Inter-vehicle distance distributions for recorded and simulated traffic for different traffic volumes

accurately represent the upper tail of GVWs from the actually recorded BWIM traffic. The reason for this might be the (compared to the total number of recorded vehicles) only small number of overloaded regular trucks or special permit trucks with extreme GVWs, that are difficult to cover with common weight parameter models derived from data analysis of extensive traffic data sets (normal distribution model fit is done to an entire data set, central area values dominate compared to extreme upper tail values). To correct this discrepancy, the simulation models are slightly modified. An additional vehicle type—a mobile crane with six axles and a mean GVW of 72 t ("mc72", refer to [6] for vehicle parameters)—is considered with a very low ratio of 0.05% (corresponding to a mean occurrence of once every second day) in the traffic. Due to the compact dimensions and high GVW values of this new vehicle type, the extreme weight values of the BWIM record can be covered quite well, with exception of one very heavy permit truck (see Fig. 5, right).

The inter-vehicle distance is modeled with a traffic flow dependent (traffic volume per 6 h block) lognormal distribution model for the simulated traffics. For the comparison with the recorded traffic, inter-vehicle distances for two different traffic volumes per 6 h—representative for slow lane (250 vehicles per 6 h) and fast lane (5 vehicles per 6 h)—are evaluated and compared with the simulation models (Fig. 6). For the traffic volume of 250 vehicles, the distance distribution of SIM_{Millau} follows the data from the BWIM record in general quite well, however it underestimates the occurrence of short distances up to 100 m (which might be critical for the bridge loading process). The SIM_{A61} model covers this short distance range better, however significantly overestimates it. For the traffic volume of 5 vehicles the data from the BWIM record is quite scarce. Both simulation models appear to cover the distances for the relevant ranges, with a tendency of overestimating the actual occurrence probability.

4.3 Comparison of Multi-presence Events

For evaluation of possible multi-presence of truck vehicles on the bridge, the definition of overlapping passage events (OPE) is introduced (in imitation of "truck convoys" in [2]). An OPE is defined as a sequence of consecutive vehicles crossing a bridge with at least one vehicle being present on the structure for each instance of

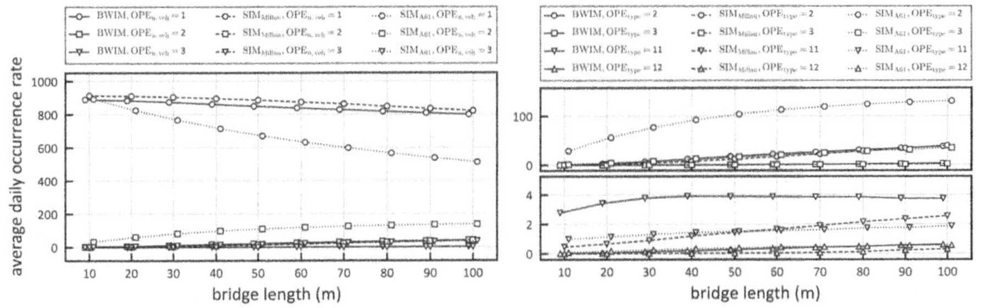

Fig. 7 OPE evaluation for recorded and simulated traffic, with classification according to joined number of vehicles (left) and number of vehicles per lane (right, with first digit for fast lane and second for slow lane)

time of the OPE. They are classified according to the number of vehicles in the corresponding event sequence, either joint or separate for each traffic lane. For a given traffic stream, they depend solely on the length of the influence line under consideration. As a matter of fact, the probability of multi-presence of truck vehicles on the bridge increases with increasing occurrence frequency of multi OPE and increasing number of vehicles in the event. Figure 7 shows the evaluation of the average daily occurrence of OPEs for simulated and recorded traffics. In general, SIM_{Millau} sligthtly underestimates, and SIM_{A61} (in most cases significantly) overestimates multi OPEs (with at least two vehicles) compared to the BWIM record. The only exception is the OPE type with one vehicle on each traffic lane (type 11), which is underestimated by both simulation models.

In a next step, the term of OPE for the traffic load impact is now connected to the term of bridge loading event (BLE) [7]. A BLE is always connected to an OPE, or in other words they are the same events but with different grounds for classification. BLEs refer to the number of vehicles contributing to (and hence present on the bridge at the moment of) the maximum value of a load effect during a specific OPE. A BLE class is always smaller than or equal to the corresponding OPE class. Hence, a BLE (or its classification) is always related to a specific load effect (influence line). BLEs represent a measure of actual multi-presence of truck vehicles on the bridge. Figure 8 shows the evaluation of average daily occurrence of multi BLEs (at least two vehicles), and the comparison to the occurrence rate of multi OPEs. It can be seen that the occurrence rates of BLEs are significantly lower compared to OPEs for the same class, for both recorded and simulated traffic and with differences between the single load effects. As before, SIM_{Millau} slightly underestimates the multi BLEs of the BWIM record, whereas SIM_{A61} significantly overestimates them.

In summary of this and the previous section (and presuming both simulation models are modified regarding the additional vehicle type mc72), simulation model SIM_{Millau} appears to show slight deficiencies in approximating the BWIM record, whereas simulation model SIM_{A61} seems to be a more conservative representation.

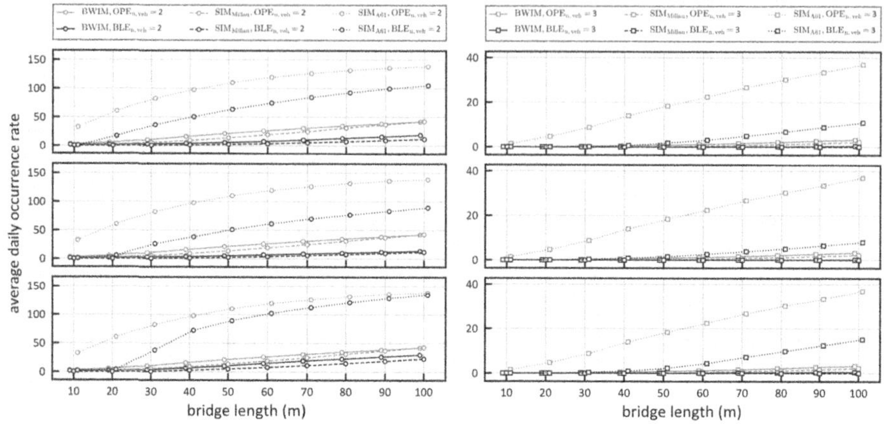

Fig. 8 Average occurrence of multi OPEs and multi BLEs for LE I (top), LE II (middle), and LE III (bottom)

4.4 Comparison of Extreme Load Effects

For traffic load modeling in bridge design and reassessment, extreme values of load effects due to road traffic are of interest. These are usually governed by a small number of extreme events, which are therefore essential to be accurately represented in simulation models. Figure 9 shows the block maxima data (with block size of one day) and contributing multi BLEs, for both recorded and simulated traffic. It can be observed that simulation model SIM_{A61} represents especially the upper tail of the block maxima data for most cases quite well (with exception of LE II for bridge length of 20 m), and appears a bit conservative for bridge length of 80 m. The contribution of multi BLEs (and especially BLEs with three or more vehicles) for lengths of 50 and 80 m is also higher than for BWIM. In contrast to that, simulation model SIM_{Millau} underestimates the contribution of multi BLEs in all cases. The model performs well for LE II (even though also underestimation for length of 20 m), also for the LE I with 20 m and 50 m. However, it shows quite bad performance for LE III for lengths 50 and 80 m.

To gain further insight into the performance of both simulation models for the different load effects and bridge lengths, BLEs causing the single block maxima are further analyzed with respect to multi-presence, vehicle weights, and load density on the bridge. For the multi-presence of truck vehicles, a contribution factor f_ξ related to each block maxima sample is computed (refer to Fig. 10). It is defined as the sum of the normalized contributions (with respect to largest contribution value) of all vehicles from the related BLE to the block maxima value. It is determined as follows:

$$f_\xi = \sum \xi_i / \xi_{\max} \tag{1}$$

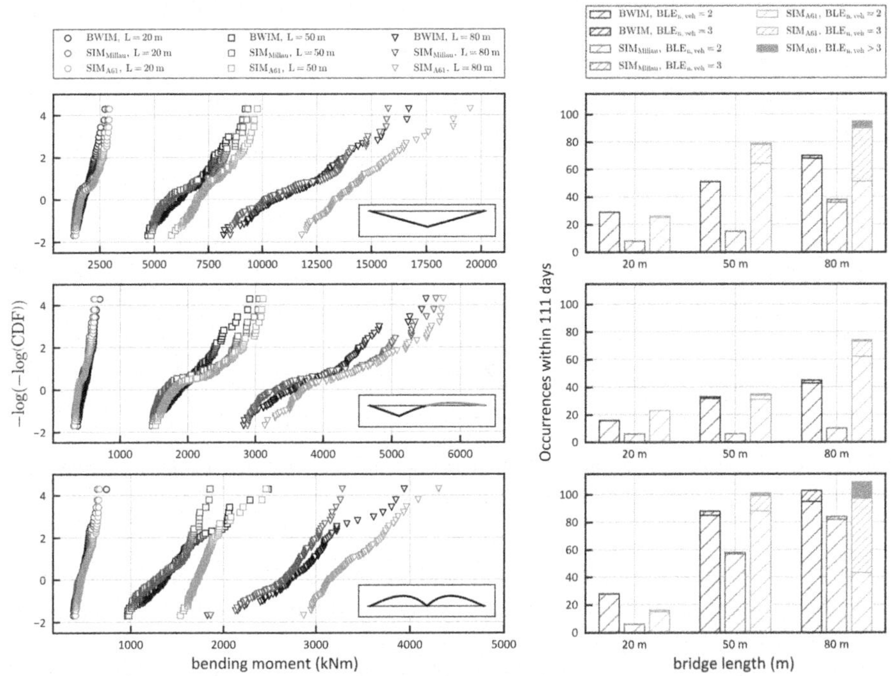

Fig. 9 Block maxima data (left) and contribution of multi BLEs (right) for recorded and simulated traffic for the investigated load effects and different bridge lengths

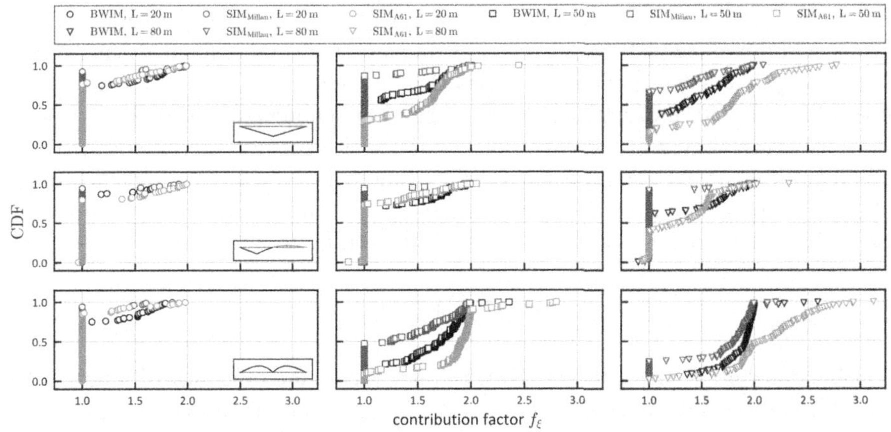

Fig. 10 Contribution factor for block maxima data for recorded and simulated traffic for the investigated load effects and different bridge lengths

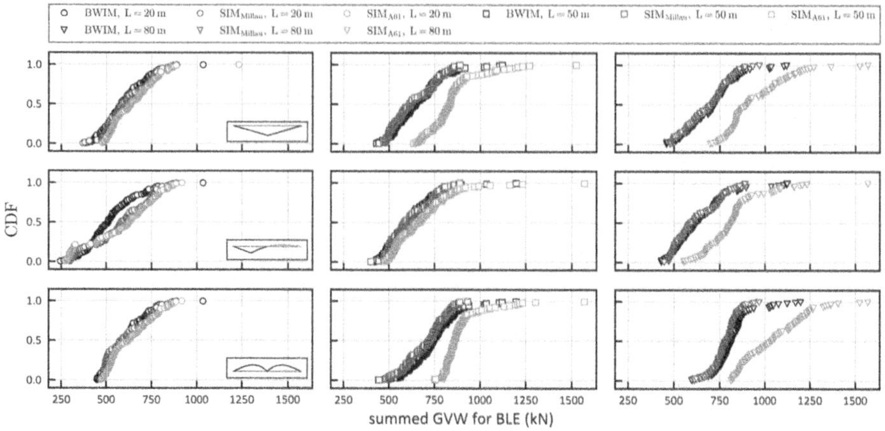

Fig. 11 Summed GVWs contributing to block maxima for recorded and simulated traffic for the investigated load effects and different bridge lengths

where $\xi_i = \text{LE}_i/\text{LE}$ is the contribution of load effect LE_i of the ith vehicle in the BLE to the block maxima value LE, and ξ_{max} is the maximum value of all contributions within the BLE.

The influence of vehicle weight is evaluated based on the summed GVWs of all vehicles from the BLE related to the block maxima value (refer to Fig. 11). The load density on the bridge related to a block maxima value is a bit more complex to describe and influenced by different aspects. On the one hand, the load density along the single vehicles (depending on their GVW, axle weight distribution and axle spacing) of the related BLE affects this parameter. On the other hand, the proximity of the single vehicles' positions in the BLE to their respective potentially most adverse location for the considered load effect is also influential. Both these aspects are jointly represented by the influence factors η_i of the single vehicles of the BLE, which multiplied with the respective vehicle weight G_i give the resulting load effects LE_i of the single vehicles. For further analysis, the sum f_η of the single influence factors for each block maxima value is computed (refer to Fig. 12):

$$f_\eta = \sum \text{LE}_i / G_i \tag{2}$$

The contribution factor in Fig. 10 illustrates nicely the influence of multi BLEs for different load effects and bridge lengths. It can be observed that block maxima for those load effects and bridge lengths mainly dominated by single BLEs in BWIM traffic are generally covered well by simulation model SIM$_{\text{Millau}}$ (see Fig. 9). This finding is supported by a good agreement between BWIM record and SIM$_{\text{Millau}}$ for the summed GVWs contributing to block maxima of these cases (see Fig. 11). For the remaining cases, multi-presence of trucks in block maxima related BLEs gets more dominant for the BWIM record, which cannot be represented accurately by SIM$_{\text{Millau}}$. This is also leading to larger discrepancies in the upper tail of the summed

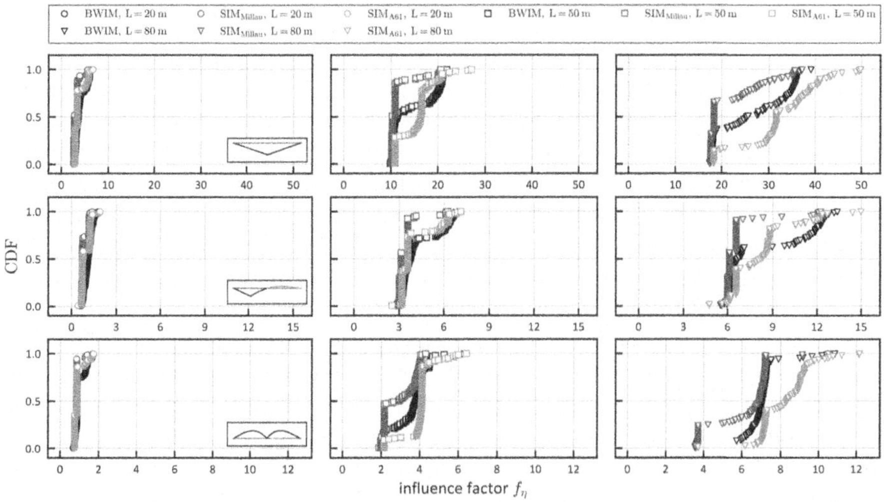

Fig. 12 Influence factor block maxima for recorded and simulated traffic for the investigated load effects and different bridge lengths

GVWs between BWIM record and $\text{SIM}_{\text{Millau}}$ for the respective cases. The fact that simulation model SIM_{A61} covers well the block maxima of the BWIM record in almost all the cases is supported by the contribution factor in Fig. 10, showing an even more dominant influence of multi-presence of truck vehicles for block maxima values compared to the BWIM record. This is also reflected by the summed GVWs in Fig. 11 being conservative for SIM_{A61} in most cases.

For the influence factor in Fig. 12, no immediate relation to the performance of the simulation models compared to the BWIM record can be observed. However, the underestimation of both simulation models for LE II for bridge length of 20 m could be interpreted using this factor. Contribution factors of this load effect in Fig. 10 seem quite similar for this load effect, the summed GVWs even would indicate the simulations models being conservative. However, the influence factor in Fig. 12 for this load effect is more unfavorable for the BWIM record compared to the simulation models. It appears that due to the short bridge length (and hence the even shorter lengths of positively signed section of influence line), the load density becomes the dominant parameter for this case.

5 Conclusion

Data from traffic monitoring (e.g. WIM or BWIM) accurately represents the traffic at the monitored site. However, it is of restricted use for traffic load modeling in many cases due to limited availability and monitoring periods. Common approaches for traffic load modeling rely on numerical simulations, aiming at representing actual traffic and resulting load effects in a realistic as possible manner. For this challenging task to fulfill, it is essential to know those governing aspects of traffic loading, and accurately represent them in the simulation model. This work presents a comprehensive comparison between bridge load modeling using traffic monitoring data and numerical traffic simulation. For this purpose, an extensive and detailed traffic data record obtained from BWIM measurements at the Millau Viaduct in South-ern France and covering a period of 180 days is analyzed. Simulated traffics and measured traffic record are evaluated towards bridge loading and compared with respect to governing traffic characteristics, multi-presence of truck vehicles on the bridge, and extreme values of load effects.

The results reveal two main aspects requiring careful attention for traffic load modeling: extreme vehicle weights due to overloading and permit trucks, and occurrence of multi-presence of trucks on the bridge. In this work, the former is accounted for by modification of the simulation models introducing an additional heavy and compact vehicle type (mobile crane with 6 axles) with a very low occurrence frequency. The latter constitutes a more challenging task. Multi-presence of trucks is directly related to the modeling of inter-vehicle distances. The simulation model based on the Millau data lacks of representing dominant events of multi-presence of truck vehicles on the bridge. This might be due to the somewhat simplified approach for modeling the inter-vehicle distances adopted in this work. The model is therefore just partially able to approximate extreme load effects from the BWIM record. Analysis with a more conservative inter-vehicle distance model leads to overestimation of short distances. This results in overestimation of multi-presence events and conservative results for extreme load effects in most of the considered cases.

The findings from this work are related to the characteristics of the traffic recorded at Millau Viaduct. Analysis should be expanded to further traffic sites with different characteristics to confirm the findings. Future works should also address improvements for modeling the inter-vehicle distance—especially in the short range—in the light of a more accurate representation of multi-presence truck events. An alternative could be an explicit simulation of multi-presence truck events considering additional information as input for traffic simulations [8].

References

1. Nowak, M., & Fischer, O. (2017). Objektspezifische Verkehrslastansätze für Straßenbrücken [Site-specific traffic load models for road bridges]. *Beton- und Stahlbetonbau, 112*(12), 804–814.
2. Croce, P., & Salvatore, W. (2001). Stochastic model for multilane traffic effects on bridges. *Journal of Bridge Engineering, 6*(2), 136–143.
3. Nesterova, M., Schmidt, F., & Soize, C. (2019). Probabilistic analysis of the effect of the combination of traffic and wind actions on a cable-stayed bridge. *Bridge Structures, 15*(3), 121–138.
4. Enright, B., & OBrien, E. J. (2011). *Cleaning weigh-in-motion data: Techniques and recommendations*. University College Dublin & Dublin Institute of Technology.
5. OBrien, E. J., & Enright, B. (2011). Modeling same-direction two-lane traffic for bridge loading. *Structural Safety, 33*(4–5), 296–304.

6. Freundt, U., & Böning, S. (2011). *Anpassung von DIN-Fachberichten „Brücken" an Eurocodes—Teil 1: DIN-FB 101: „Einwirkungen auf Brücken"*. Berichte der Bundesanstalt für Straßenwesen Heft B 77, Bundesanstalt für Straßenwesen.
7. Caprani, C. C. (2005). *Probabilistic analysis of highway bridge traffic loading* [Doctoral dissertation]. University College Dublin.
8. Nowak, M., Fischer, O., & Müller, A. (2020). Realitätsnahe Verkehrslastansätze für die Nachrechnung der Gänstorbrücke über die Donau [Realistic traffic load models for reassessment of Gänstor Bridge over the Danube]. *Beton-und Stahlbetonbau, 115*(2), 91–105.

Construction Risk Management in Portugal—Identification of the Tools/Techniques and Specific Risks in the Design and Construction Phases

António J. Marinho and João P. Couto

Abstract The environment of construction projects (CP) is especially important, as well as the context in which they started, developed, and completed. Its effects should be closely monitored, controlled whenever possible, and considered as a high-risk source. Despite the high impact of risks during the construction project life cycle, very few researches and papers do analyse them. So, the main purpose of this paper is to investigate the most relevant tools or techniques for risk management and identify the specific risks for designers and contractors that affect the phases of concept and execution of CP. So, a questionnaire was designed and administered to construction professionals (contractors and designers). The results show that the process of planning risk responses is the one evidencing less knowledge of the tools used by the construction professionals. Regarding designer specific risks in the project phase, the "problems of coordinating the various specialities" or "customer lacking the necessary experience or resources to support the project" have a high impact; in the construction phase, the "continuous change in the scope" of the project is the most impacting risk. For contractor specific risks, in the budgeting phase, "work's quality and value are insufficient for the cost" or "substandard budgeting documents" have a high impact; in the construction phase, the "low-skilled workforce" followed by "work's quality and value are insufficient for the cost" are the most impacting risks. Based on the results obtained, the findings of the research contribute to an understanding of the major tools and techniques used by designers and contractors for risk management and for the identification of the specific risks arising from life cycle phases that can affect a construction project in Portugal.

Keywords Risk management · Construction projects · Design and construction

---

A. J. Marinho (✉) · J. P. Couto
School of engineering, University of Minho, Campus of Azurém, 4804-533 Guimarães, Portugal

J. P. Couto
e-mail: jpc@civil.uminho.pt

© The Author(s), under exclusive license to Springer Nature Switzerland AG 2021

J. C. Matos et al. (eds.), *18th International Probabilistic Workshop*, Lecture Notes in Civil Engineering 153, https://doi.org/10.1007/978-3-030-73616-3_17

1 Introduction

The study of risks during the life cycle of construction projects (CP) is an issue of major interest in companies of the construction sector. Although project management (PM) in the sector has undergone a huge evolution in recent years, the lack of studies on risks during the life cycle has led to the failure of CPs. Those risks during the construction life cycle are defined as the project's total risks that could occur throughout all the phases of the project, including design, construction, operation and maintenance, repair and renewal, or even demolition. On the other hand, research has shown that there are several causes that determine the failure of a CP [1]. For example, according to Bekithemba et al. [2] and Szymański et al. [3], the main consequences of a CP failure are poor communication between stakeholders, poor leadership, failure in defining the project's scope, project managers with insufficient training and experience, and ineffective management of customer expectations.

A way to address this problem is to perform a good risk management (RM) for each construction project. Since risk definition depends on the perception of the sector and the project's goals, there are different definitions of risk for each case. So, risks can be defined as future issues that can be avoided or mitigated, rather than current problems that should be addressed immediately [4]. The environment of the final project is equally important, as well as the context in which it began, was developed and completed [5]. In PM, risk is an event or condition that occurs and may have a positive or negative effect. If the risk is successful, it will focus on one or more project goals, such as scope, cost, time and quality [6, 7].

The main goal of this work is to contribute to the scientific knowledge of RM that affects CP design and the construction phases. This paper includes two main contributions: (1) the identification of the techniques and tools used by designers and contractors to manage risks, and (2) the identification of the most significant risks for designers and contractors during the life cycle of construction projects in Portugal.

2 Literature Review

2.1 Research Criteria

Despite the high impact of risks during the life cycle phases of a CP, very few papers ever published analysed the main risks for designers and contractors during the construction project. Therefore, a comprehensive research was made by analysing recent papers, mostly from the period between 2013 and 2019, although some older papers were also cited due to important contributions to this review. We conducted a structured literature review and analysed the research already published in important journals, including Web of Science and Elsevier. The search for articles used the following keywords in different combinations: "risk management," "construction risk" and "risk life cycle construction".

The results yielded more than 311 articles from different publishers and then we categorized the relevant articles by declining studies that lacked a clear relationship with construction RM, thus limiting the search to 85 articles. The selection criteria filter was meant to identify and analyse empirical case studies that identify specific risks during the CP phases, and so we ended up with the 50 most relevant papers.

2.2 *Construction Risks*

Risk and problem are two words that are often confused with one another when it comes to their use. A risk is an uncertain event that has a probability associated with it, and a problem does not have this attribute; thus, issues are problems only when the project team must do something. While RM is a proactive activity, problem management is a reactive activity [8]. The focus on RM is the assessment of risks and implementation of a proper response. The goal is to achieve the largest sustainable value of all the activities of the organization, and RM increases the understanding of the potential benefit or weakness of the factors that can affect the organization, by increasing the probability of success and reducing the possibility of failure and the level of uncertainty associated with achieving the organization's goals [9]. Uncertainty is inherent to projects and refers to elements that change with time and therefore are difficult to predict and control. It can be described as the difference between the information one has and the information one needs to complete a task [10].

Throughout the design and construction phases, these processes interact with each other by exchanging information, with the former producing an output that becomes an input for the construction process. Besides, this interaction also produces several documents that help project planning [11, 12].

Regarding PM, Van Den Ende and Van Marrewijk [13] and Zeynalian et al. [14] conducted a study and verified that most of times the life cycle transition in projects remains little studied. They focused on the knowledge of the practice and the meaning of the procedures in the project's life cycle transition. They analysed four CPs and conducted 58 interviews and found out that the contribution to the dialogue of PM in temporary organizations lies in the conceptualization of procedures in the various phases of the life cycle. Another study, conducted by Qazi et al. [15], investigated current practices of RM in complex CPs by interviewing 13 multi-structured RM specialists and confirmed that managers depend on their intuition and past experience when dealing with risks.

2.3 *Tools or Techniques for Risk Management*

Project setbacks can be substantially reduced by adopting the correct risk methodology as an integral part of project planning. Recent history shows that planning and controlling a project's risks is fundamental to ensure high-quality project results [16].

Risks not recognized cannot be evaluated and addressed, and therefore an overall risk analysis becomes impossible. The task of RM is then to manage the essential risks as thoroughly as possible. The realism of risk estimates increases as the project evolves, but, despite this, the main decisions must be made at the beginning of the project's life cycle. Contingency steps must be implemented to respond to the risks, and not all the risks are completely recognizable, and additional risks may emerge as well [17, 18]. Risk identification determines what can happen and affect the project's goals and how these goals can be reached. Risk identification produces a register of the project's risks and their characteristics [19]. This risk register can be subsequently altered through a qualitative and quantitative risk analysis, risk response and risk monitoring processes.

Previously identified risks may disappear and other risks may occur [8]. There are several tools available to support the phases of the RM process. Raz and Michael [20] conducted a study designed to identify the most widely used tools, as well as the tools associated with successful PM in general and to successful RM in particular. Their study was based on a questionnaire administered to a sample of project managers from technology industries.

The results presented evidence that the most used tool in identifying risks is previously learned lessons. However, the Project Management Institute (PMI) published a practical guide for risk management in projects, evidencing the tools developed and widely used [5, 21]. Table 1 presents the most relevant tools/techniques used for risk management that were identified in the literature review.

Table 1 Tools/techniques used for risk management identified in the literature review

Tools/techniques		Authors
RM plan	Plan meetings	[5, 21]
	Plan analyses	[5, 20, 21]
Risk identification	Analysis of assumptions and constraints	[5, 21]
	Brainstorming	[5, 8, 21]
	Cause-effect diagrams	[5, 21]
	Checklists	[5, 8, 20, 21]
	Delphi technique	[5, 21]
	Document review	[5, 8, 21]
	FMEA/fault tree analysis	[5, 21]
	Previously learned lessons	[5, 20, 21]
	Risk breakdown structure (RBS)	[5, 21]
	Root cause analysis (RCA)	[5, 21]

(continued)

Table 1 (continued)

Tools/techniques		Authors
Qualitative and quantitative risk analysis	Estimation (probability/impact)	[5, 20, 21]
	Probability and impact matrix (PI Matrix)	[5, 20, 21]
	Decision tree analysis	[5, 20, 21]
	Expected monetary value (EMV)	[5, 21]
Risk response planning	Contingency plan	[5, 17, 18, 21, 22]
	Critical chain project management (CCPM)	[5, 21]
Risk monitoring	Risk analysis review (RAR)	[5, 20, 21]
	Risk reassessment	[5, 20–22]
	Meetings	[5, 20, 21]
	Risk audits	[5, 21]

2.4 *Specific Risks*

CPs are usually started in dynamic environments that may give rise to circumstances of high-risk uncertainty stemming from the accumulation of diverse interrelated parameters. A study carried out by Taylan et al. [23] allowed to evaluate analytical tools for CPs and their overall risks in situations marked by uncertainty. The study identifies the CPs' main risk criteria and evaluates these criteria via integrated hybrid methodologies. The study analysed 30 CPs about five main criteria: cost, time, quality, safety, and environment sustainability. The results showed that these new methodologies can assess the general risks of CPs, thus enabling to select the project that presents fewer risks via the contribution of the index of relative importance. The lists presented in Tables 2 and 3 show the specific risks identified in the literature review. Table 2 shows that the most frequent risks of the conceptual study and design phase referred in the literature regarding designers are "client lacking necessary experience or resources to support the project", "project's dimension/type is superior to the previous experience of the design team" and "continuous changes in project's scope". For constructors, there are only two risks identified in the literature review (see Table 2).

Table 3 shows that the most frequent risks for designers are "inaccurate/incorrect technical assumptions" and "lack of technical consultants specialized in critical aspects of the project" or "insufficient technical experience, design type and site resources". For constructors, the most relevant risks found are "low-skilled labour" and "unexpected problems regarding the soil of the work".

Table 2 Specific risks in the conceptual study and design phase

Type	Conceptual study and design phase	Authors
Designers	Client lacking the necessary experience or resources to support the project	[3, 22, 24, 25]
	Project's dimension/type is superior to the previous experience of the design team	[3, 22, 24, 26]
	Responsibility for the hiring method is not clear in the first phase	[24]
	Tender documents lacking the necessary quality	[24, 27]
	Problems of coordinating the various specialities	[22, 24]
	Continuous changes in project's scope	[3, 24, 27, 28]
	Inappropriate or inaccurate information/details of the tasks to be performed	[24, 27]
	Delays in obtaining customer agreement	[3, 24, 26]
	Project's scope exceeds available budget	[24]
	Uncertainty in total cost estimates due to uncertain amounts and unit prices	[24, 28]
	Incomplete calculation of project's cost and incorrect project scheduling	[24]
	Incompatibility with local standards and legislation	[22, 29]
	Inappropriate legislation	[22, 27]
	Lack of technical rigour	[25, 27]
Constructors	Substandard budgeting documents	[3, 22, 24, 25, 30]
	Insufficient or improperly executed architectural projects	[3, 14, 17, 24, 29–31]

Table 3 Specific risks during the construction phase

Type	Construction phase	Authors
Designers	Unforeseen soil conditions	[17, 22, 28, 29]
	Nonstandard contracts documentation	[24, 31]
	Inadequate and incomplete design	[3, 24, 27]
	Inaccurate/incorrect technical assumptions	[3, 24, 25, 27, 29]
	Insufficient technical experience, design type and site resources	[3, 24, 26, 27]
	Errors and omissions by consultants/contractors	[24]
	Lack of technical consultants specialized in critical aspects of the project	[24, 25, 27]
	Continuous changes in project's scope	[3, 24]
	Lack of interaction with local construction methods	[22, 25, 29]

(continued)

Table 3 (continued)

Type	Construction phase	Authors
	Incomplete calculation of project's cost and incorrect project scheduling	[24, 26]
	Customer lacking the necessary experience or resources to support the project	[24, 26]
	Project's scope is superior to the designer's previous experience	[22, 24, 25]
	Inappropriate legislation	[25]
Constructors	Nonstandard contracts documentation	[3, 24, 27, 28]
	Work's quality and value are insufficient for the cost	[3, 17, 24, 25, 28]
	Subcontractor or supplier unable to meet delivery times or costs	[3, 17, 22, 24, 27, 28]
	Insufficient documentation of the work plan	[14, 17, 22, 24, 25, 28]
	Lack of documentation coordination	[24, 28]
	Customer expectation management	[14, 25, 27, 28]
	Lack of budget	[3, 17, 22, 32]
	Communication risks among the main stakeholders	[25, 27, 28]
	Unexpected problems regarding the soil of the work	[3, 17, 22, 24, 25, 28, 30]
	Problems of managing the work	[3, 14, 24, 32, 33]
	Subcontractor or supplier unable to meet delivery times or costs	[3, 17, 22, 24, 25]
	Lack of monitoring by the designer	[14, 25, 27, 28]
	Accidents and injuries	[3, 22, 24, 29]
	Work with interruptions/productivity breaks	[24, 25, 33]
	Late payments	[24, 25, 28, 32]
	Low-skilled labour	[3, 14, 17, 24, 25, 27–29, 32]
	Destruction of worksite	[22, 24, 25]
	Availability of workforce, equipment, and material	[17, 28]
	Political, social, and economic risks (for instance, inflation)	[17, 25, 33, 34]
	Actual amounts of work	[3, 17, 25, 28]
	Poor or old equipment	[14, 27]
	Incorrect choice of equipment, materials, and construction techniques	[14, 24, 27]

3 Research Methodology

In this study, we adopted a comprehensive approach through the combination of a literature review with a questionnaire-based research. The questionnaire was divided into three sections. The **first Section** aimed to collect information and properly characterize the sample, namely the designer's and contractor's role, their experience in the field and whether they were knowledgeable in RM in CPs. In the **second Section**, respondents were asked to identify the tools they had previously used for RM processes (Sect. 2.3, Table 1). In the **third Section**, respondents were asked to assess the specific risks for designers or contractors during the design or construction phases (Sect. 2.3, Tables 2, and 3). The questionnaire ended with an open question asking the respondents to name other tools and other risks not contained in the list.

This section aims to hierarchize and assess the importance of risks in the design and construction phases. An online questionnaire was prepared for collecting qualitative and quantitative data based on the variables of the RM processes. Regarding the variable contractor and designer, we intended to identify the RM tools mostly used by them, and the specific risks identified in the CP by designers and contractors. Since it was not possible to study the entire Portuguese population of contractors and designers, we chose to analyse a random sample obtained through consulting both the IMPIC (Institute of Public Procurement, Real Estate and Construction) for general contractors with Class A building license, and the APPC (Portuguese Association of Designers and Consultants) for designers [35, 36]. We obtained a population of 190 contractors (to 94 answers and an confidence interval of 14.2%) and 132 designers (to 41 answers and confidence interval of 8.45%), totalizing 322 respondents and 135 answers (rate of 41.9%), which are acceptable rates in comparison to previous studies based on questionnaire surveys in the construction sector: for example, 7.4% in the study by Abdul-Rahman et al. [37], and 13.0% in the study by Jin et al. [38].

4 Data Analysis and Findings

4.1 Information About the Survey Participants

The questionnaire yielded 135 answers: 69.6% by designers and 30.4% by contractors. Regarding experience in the field, there was a higher concentration of respondents with 10 years plus experience (70.4% of the sample), 19.1% with experience between 5 and 10 years, 9.6% between 1 to 5 years, and 0.8% with less than 1-year experience. Regarding their knowledge of RM in CPs, 57.4% of respondents were not knowledgeable in the RM process and 42.6% reported having only knowledge of the risks related to Occupational Hygiene and Safety. In other words, all respondents lacked knowledge in the RM process.

4.2 Tools or Techniques for Risk Management

According to Table 4, the tools mostly used by **designers** in planning RM, were planning meetings (65%), followed by planning analyses (34%). However, 30% of the sample do not used any tools associated with this process. For the process of finding risks, the most used tools were checklists (42%) and previously learned lessons (37%). Similarly, 32% referred not using any tools during this process. Regarding the process of performing the qualitative and quantitative analysis of risks, 57% do not use any tool to manage the process, 30% use the estimate (probability/impact) and 25% use the probability and impact matrix. Regarding the process of planning risk responses, the contingency plan is 25% of the sample and 76% do not use any risk response tools. Meetings are the most widely used tool to check risks (48% of respondents), followed by risk revaluation (35%). 43% of the respondents do not use any tools for this process.

According to Table 4, the tools mostly used by **contractors** in planning RM were planning meetings (91%) and planning analyses (66%). The most used tools in finding risks were, previously learned lessons (83%), assumption and restriction analysis (54%), and brainstorming (37%). Regarding the process of performing qualitative and quantitative analysis of risks, 71% of the respondents use the estimation tool (probability/impact), and the probability and impact matrix (43%). Regarding the process of planning risk responses, the contingency plan stands for 80% of the sample and 17% do not use any risk response tools. The most widely used tool to monitor risks is meetings (83%), followed by risk reserve (29%), and risk revaluation (14%).

Table 4 Percentage of tools or techniques for RM identified by respondents

Tools/techniques		Designers (%)	Contractors (%)
RM plan	Plan meetings	65	91
	Plan analyses	34	66
	Benchmarking[a]	1	0
	Evaluation of the main activities[a]	4	0
	Overlap of the various specialties[a]	0	0
	Overall budget analysis[a]	0	0
	Does not use any	30	3
Risk identification	Analysis of assumptions and constraints	24	54
	Brainstorming	24	37
	Cause-effect diagrams	15	26
	Checklists	42	20
	Delphi technique	3	3

(continued)

Table 4 (continued)

Tools/techniques		Designers (%)	Contractors (%)
	Document review	29	26
	FMEA/fault tree analysis	9	6
	Previously learned lessons	37	83
	Risk breakdown structure (RBS)	6	0
	Root cause analysis	16	6
	Ishikawa diagram[a]	1	0
	Does not use any	32	3
Qualitative and quantitative risk analysis	Estimation (probability/impact)	30	71
	Probability and impact matrix (PI matrix)	25	43
	Decision tree analysis	18	11
	Expected monetary value (EMV)	6	3
	William T. fine method[a]	3	0
	Does not use any	57	11
Risk response planning	Contingency plan	25	80
	Critical chain PM (CCPM)	5	3
	Prevention plan[a]	1	0
	Risk map[a]	1	0
	Transfer of risk to subcontractors[a]	0	0
	Does not use any	76	17
Monitoring risks	Risk analysis review	5	29
	Risk reassessment	35	14
	Meetings	48	83
	Risk audits	23	17
	Does not use any	43	9

[a]Tools or techniques identified by respondents

4.3 *Specific Risks for Designers and Contractors*

Regarding the specific risks for designers, Table 5 shows that the most frequent risk during the project phase is the one associated with "problems in coordinating the various specialities" (59%). In second place comes the risk associated with "customer lacking experience or resources to support the project" (56%). Regarding the construction phase, the most frequent risk stems from "problems of coordinating the various specialties" (47%) and from "incorrect choice of equipment, materials, and construction tools" (39%).

Table 6 presents the results obtained regarding specific risks for contractors. The most frequent risk during the budgeting phase is the risk associated with "work's quality and value are insufficient for the costs" (69%); secondly, the risk associated with "substandard budgeting documents" (43%); and thirdly, "project's architecture and specialities are insufficient or poorly executed" (31%). Regarding the implementation phase (execution of the work), the most obvious risk is the one associated with "low-skilled workforce" (74%), followed by "work's quality and value are insufficient for the costs" (51%), and with "substandard budgeting documents" (46%).

Table 5 Specific risks for designers during design and construction phases

Specific risks for designers	Design (%)	Construction (%)
Customer lacking the necessary experience or resources to support the project	56	34
Project's dimension/type is superior to the designer's previous experience	27	23
Responsibility for the hiring method is not clear at the beginning	39	25
Tender documents lacking the necessary quality	41	24
Nonstandard contract documentation	24	20
Problems of coordinating the various specialities	59	47
Inadequate or inaccurate information/details of the tasks to be performed	27	32
Inadequate or incomplete design	24	18
Incomplete knowledge of site conditions	47	27
Inaccurate/wrong technical assumptions	32	28
Insufficient technical experience, design type and site resources	23	22
Incorrect choice of construction equipment, materials, and techniques	13	39
Incorrect estimates of geotechnical parameters, foundation, and structural modelling	32	25
Unavailability and/or incapacity of public services	29	19
Errors and omissions by consultants/contractors	28	27
Lack of technical consultants specialised in critical aspects of the project	34	25
Lack of owner involvement in architecture	33	8
Continuous changes to the project's scope	48	35
Delays in obtaining customer agreement	32	23

(continued)

Table 5 (continued)

Specific risks for designers	Design (%)	Construction (%)
Project's scope exceeds available budget	35	29
Uncertainty in total cost estimation due to uncertain quantities and unit prices during the first planning and design phase	27	15
Incorrect estimation of project's costs and scheduling	24	14
Incompatibility with local construction standards and legislation	30	13
Lack of interaction with municipal construction methods	15	19
Lack of technical rigor[a]	1	1
Absence of the director of work and lack of supervision[a]	1	1
Low-skilled contractors[a]	1	1
Inadequate national construction legislation[a]	1	1
Inadequate time frame to respond to errors and omissions[a]	1	1

[a]Specific risks found by respondents

Table 6 Specific risks for constructors during the budgeting and construction phases

Specific risks for constructors	Budgeting (%)	Construction (%)
Substandard budgeting documents	43	46
Nonstandard contract documentation	17	9
Work's quality and value are insufficient for the costs	69	51
Insufficient or improperly executed architectural project	31	37
Unexpected problems with the soil of the work	3	23
Problems in coordinating the work	3	26
Subcontractor or supplier(s) unable to meet delivery times or costs	6	31
Defective materials and subcontractors' poor constructive quality	0	37
Accidents and injuries	6	17
Time-out work/productivity breakdown	0	11
Late payments	0	20
Insufficient work plan documentation	11	11
Lack of documentation coordination	14	3
Low-skilled workforce	11	74
Destruction of the shipyard	3	9
Availability of workforce, equipment, and material	9	20
Inflation	0	9

(continued)

Table 6 (continued)

Specific risks for constructors	Budgeting (%)	Construction (%)
Actual amounts of work	3	17
Unforeseen soil conditions	0	6
Contingency margin	3	11

5 Conclusions

Managing CPs without managing the associated risks is a complex and, in most cases, impossible task. RM in CP is essential to the success of the project: planning, identifying, analysing, responding, implementing, and controlling risks. The need arose in the national context to evaluate the tools mostly used by construction professionals, namely designers and contractors, and evaluate the risks in the construction sector. The results obtained show that professionals in the Portuguese construction industry do not have knowledge of RM, despite their experience in the field. Construction professionals are using RM tools but are unaware of the associated processes. These risks are randomly managed and not addressed with a consistent structure.

In conclusion, the most used tool was to "plan meetings" regarding the process of planning RM, and construction professionals are less knowledgeable regarding the tools used for the process of planning risk responses. We also demonstrated that "previously learned lessons" and "checklists" are the most used tools for the process of identifying risks. For designers' specific risks in the planning phase, "customer's lacking experience or the resources needed to support the project" has a high impact. Regarding the construction phase, "continuous changes in the project's scope" also has a considerable impact. Regarding contractors' specific risks, "substandard budgeting documents" and "work's quality and value are insufficient for the costs" during the budgeting or construction phases have a high impact. Based on the results obtained, this paper also contributes to the theory and practice of PM with the systematization of the responses and types of risks identified in the various phases of construction. Briefly, the present work contributes to the identification of the tools used by designers and contractors and, also the identification of the most significant risks for designers and contractors in the Portuguese construction sector. The results obtained can contribute to a more efficient RM in CPs in Portugal.

As a proposal for future work, we suggest that construction companies implement the RM methodology present in the PMBOK guide®; the use of RM knowledge with parametric modelling tools of buildings, namely the BIM methodology; an analysis of the tools provided for the RM processes and verification of their applicability in the construction sector; and finally, an analysis of the impact of RM associated with knowledge management transfers through lessons previously learned in CPs.

References

1. Zidane, Y. J.-T., & Andersen, B. (2018). The top 10 universal delay factors in construction projects. *International Journal of Managing Projects in Business, 11*. https://doi.org/10.1108/IJMPB-05-2017-0052
2. Bekithemba, M., Godfrey, O. E., Cletus, M., & Adriaan, P. (2017). Profiling causative factors leading to construction project delays in the United Arab Emirates. *Engineering Construction and Architectural Management, 24,* 346–376. https://doi.org/10.1108/ECAM-05-2015-0072
3. Szymański, P., Szymanski, P., & Szymański, P. (2017). Risk management in construction projects. *Procedia Engineering, 208,* 174–182. https://doi.org/10.1016/j.proeng.2017.11.036
4. Chakrabarti, M. K. R. D. (2011). *ProjectManagement.com—Risk management: A real challenge for the construction sector* (2011).
5. PMI. (2009). *Practice standard project risk management* (p. 107). Project Management Institute.
6. dos Santos, R. B. P., Isaton, C., Jungles, A. E., & da Silva Júnior, O. F. P. (2015). Gerenciamento De Risco Na Construção Civil: Teoria X Prática, Simpósio Bras. *Gestão e Economia Da Construção,* 246–254.
7. Wang, N., Yao, S., Wu, G., & Chen, X. (2017). The role of project management in organisational sustainable growth of technology-based firms. *Technology in Society, 51,* 124–132. https://doi.org/10.1016/J.TECHSOC.2017.08.004
8. Caltrans. (2012). Project risk management handbook: A scalable approach. Caltrans Construction.
9. Airmic, A. & Irm, A. (2010). *Structured approach to enterprise risk management (ERM) and the requirements of ISO 31000.* https://doi.org/10.1016/j.solmat.2010.12.013
10. Osipova, E., & Eriksson, P. E. (2013). Balancing control and flexibility in joint risk management: Lessons learned from two construction projects. *International Journal of Project Management, 31,* 391–399. https://doi.org/10.1016/j.ijproman.2012.09.007
11. PMI. (2013). *Managing change in organizations—A practice guide.* Project Management Institute, Atlanta, USA. www.PMI.org
12. Silva, V. (2012). *Utilização de técnicas de Gestão de Projetos na análise de requisitos de projetos de software.* Universidade do Minho.
13. Van Den Ende, L., & Van Marrewijk, A. (2014). The ritualization of transitions in the project life cycle: A study of transition rituals in construction projects. *International Journal of Project Management, 32,* 1134–1145. https://doi.org/10.1016/j.ijproman.2014.02.007
14. Zeynalian, M., Trigunarsyah, B., & Ronagh, H. R. (2013). Modification of advanced programmatic risk analysis and management model for the whole project life cycle' s risks. *Journal of Construction Engineering and Management, 138,* 51–60. https://doi.org/10.1061/(ASCE)CO
15. Qazi, A., Quigley, J., Dickson, A., & Kirytopoulos, K. (2016). Project complexity and risk management (ProCRiM): Towards modelling project complexity driven risk paths in construction projects. *International Journal of Project Management, 34,* 1183–1198. https://doi.org/10.1016/j.ijproman.2016.05.008
16. Fontaine, M. (2016). Project risk management. *Enterprise Risk Management*, 47–58. https://doi.org/10.1016/B978-0-12-800633-7.00004-3
17. Mills, A. (2001). A systematic approach to risk management for construction. *Structural Survey, 19,* 245–252. https://doi.org/10.1108/02630800110412615
18. Schieg, M. (2012). Risk management in construction project management. *Journal of Business Economics and Management VI, I,* 77–83. https://doi.org/10.1080/16111699.2006.9636126
19. Ganbat, T., Chong, H.-Y., Liao, P.-C., Wu, Y.-D. (2018). A bibliometric review on risk management and building information modeling for international construction. *Advances in Civil Engineering.* https://doi.org/10.1155/2018/8351679
20. Raz, T., & Michael, E. (2001). Use and benefits of tools for project risk management. *International Journal of Project Management, 19,* 9–17. https://doi.org/10.1016/S0263-7863(99)00036-8

21. PMI. (2017). *Project management body of knowledge: A guide to the project management body of knowledge* (6^a Ed.). Project Management Institute. www.pmi.org
22. El-Karim, M. S. A. B. A., El Nawawy, O. A. M., & Abdel-Alim, A. M. (2015). Identification and assessment of risk factors affecting construction projects. HBRC Journal (2015). https://doi.org/10.1016/j.hbrcj.2015.05.001
23. Taylan, O., Bafail, A. O., Abdulaal, R. M. S., & Kabli, M. R. (2014). Construction projects selection and risk assessment by fuzzy AHP and fuzzy TOPSIS methodologies. *Applied Soft Computing Journal, 17,* 105–116. https://doi.org/10.1016/j.asoc.2014.01.003
24. Davies, D. (2002). Risk management. *Construction Best Practice, 18,* 414–420. https://doi.org/10.1016/S0267-3649(02)01108-1
25. Khodeir, L. M., & Mohamed, A. H. M. (2015) Identifying the latest risk probabilities affecting construction projects in Egypt according to political and economic variables. From January 2011 to January 2013. *HBRC Journal, 11,* 129–135. https://doi.org/10.1016/j.hbrcj.2014.03.007
26. Dziadosz, A., & Rejment, M. (2015). Risk analysis in construction project—Chosen methods. *Procedia Engineering, 122,* 258–265. https://doi.org/10.1016/j.proeng.2015.10.034
27. Arashpour, M., Abbasi, B., Arashpour, M., Reza Hosseini, M., & Yang, R. (2017). Integrated management of on-site, coordination and off-site uncertainty: Theorizing risk analysis within a hybrid project setting. *International Journal of Project Management, 35,* 647–655. https://doi.org/10.1016/j.ijproman.2017.02.016
28. Goh, C., Abdul-Rahman, H., & Abdul Samad, Z. (2013). Applying risk management workshop for a public construction project: Case study. *Journal of Construction Engineering and Management, 139,* 572–580. https://doi.org/10.1061/(ASCE)CO.1943-7862.0000599
29. Zhi, H. (1995). Risk management for overseas construction projects. *International Journal of Project Management, 13,* 231–237. https://doi.org/10.1016/0263-7863(95)00015-I
30. de Souza Pinto Barreto, F., & Andery, P. R. P. (2015). Contribuição à gestão de riscos no processo de projeto de incorporadoras de médio porte. *Ambiente Construído, 15,* 71–85. https://doi.org/10.1590/s1678-86212015000400040
31. Banaitiene Nerija, B. A. (2012). Risk management in construction projects. *Risk Management in Current Issues Challenges,* 429–448. https://doi.org/10.3846/20294913.2014.994582
32. Hwang, B.-G. G., Zhao, X., & Toh, L. P. (2014). Risk management in small construction projects in Singapore: Status, barriers and impact. *International Journal of Project Management, 32,* 116–124. https://doi.org/10.1016/j.ijproman.2013.01.007
33. Akintoye, A. S., & MacLeod, M. J. (1997). Risk analysis and management in construction. *International Journal of Project Management, 15,* 31–38. https://doi.org/10.1016/S0263-7863(96)00035-X
34. Queiróz, F. M. J. A, Casaque, C. R., Santos, C. M., & Sezar, M. G. R. (2003). *Gerenciamento de Riscos em Projetos de Construção Civil sob a Ótica dos Principais Stakeholders.* Fundação Instituto de Administração.
35. IMPIC. (2019). *Instituto dos Mercados Públicos, do Imobiliário e da Construção* (pp. 12570–12572). https://www.impic.pt. Accessed August 18, 2019.
36. Associação Portuguesa de Projetistas e Consultores, APPC. (2019).
37. Abdul-Rahman, H., Berawi, M. A., Berawi, A. R., Mohamed, O., Othman, M., & Yahya, I. A. (2006). Delay mitigation in the Malaysian construction industry. *Journal of Construction Engineering and Management, 132,* 125–133. https://doi.org/10.1061/(ASCE)0733-9364(2006)132:2(125)
38. Jin, R., Li, B., Zhou, T., Wanatowski, D., & Piroozfar, P. (2017). An empirical study of perceptions towards construction and demolition waste recycling and reuse in China. *Resources, Conservation and Recycling, 126,* 86–98. https://doi.org/10.1016/j.resconrec.2017.07.034

Cumulative Failure Probability of Deteriorating Structures: Can It Drop?

Ronald Schneider and Daniel Straub

Abstract The reliability of deteriorating structures at time t is quantified by the probability that failure occurs within the period leading up to time t. This probability is often referred to as cumulative failure probability and is equal to the cumulative distribution function of the time to failure. In structural reliability, an estimate of the cumulative failure probability is obtained based on probabilistic engineering models of the deterioration processes and structural performance. Information on the condition and the loading contained in inspection and monitoring data can be included in the probability estimate through Bayesian updating. Conditioning the probability of failure on the inspection or monitoring outcomes available at time t (e.g. detections or no detection of damages) can lead to a reduction in that probability. Such a drop in the cumulative failure probability might seem counterintuitive since the cumulative failure probability is a non-decreasing function of time. In this paper, we illustrate—with the help of a numerical example—that such a drop is possible because the cumulative probability before and after the updating is not based on the same information, hence not on the same probabilistic model.

Keywords Deterioration · Structural systems · Time-variant reliability · Bayesian updating · Inspection · Monitoring

1 Introduction

The capacity of deteriorating structural systems decreases with time. The loads on these structures also vary with time. Because of the uncertainties in (a) the evolution of the capacity of and (b) the time-variant loads on the structural systems, their time to failure (or lifetime) T_F is a random variable. T_F is probabilistically described by the

R. Schneider (✉)
Bundesanstalt für materialforschung und -prüfung (BAM), Berlin, Germany
e-mail: ronald.schneider@bam.de

D. Straub
Engineering Risk Analysis Group, Technische Universität München, Munich, Germany
e-mail: straub@tum.de

© The Author(s), under exclusive license to Springer Nature Switzerland AG 2021
J. C. Matos et al. (eds.), *18th International Probabilistic Workshop*, Lecture Notes in Civil Engineering 153, https://doi.org/10.1007/978-3-030-73616-3_18

cumulative distribution function (CDF) $F_{T_F}(t)$, which corresponds to the probability that failure occurs in the period leading up to time t, i.e. $F_{T_F}(t) = \Pr(T_F \leq t)$. In structural reliability, the event of failure of a structural system at any time up to t is commonly denoted by $F(t) = \{T_F \leq t\}$ and the probability $\Pr[F(t)] = \Pr(T_F \leq t) = F_{T_F}(t)$ is often called the cumulative failure probability or simply the failure probability at time t. All quantities that are commonly utilized to describe the time-dependent reliability of technical systems can be derived from $\Pr[F(t)]$ [1, 2]. This includes the probability density function (PDF) $f_{T_F}(t) = \mathrm{d}\Pr[F(t)]/\mathrm{d}t$ of the time to failure T_F, the reliability $Rel(t) = \Pr\left[\overline{F}(t)\right] = 1 - \Pr[F(t)]$ and the hazard function $h(t) = f_{T_F}(t)/Rel(t)$.

For technical systems for which enough records on past failures are available, the CDF $F_{T_F}(t)$ of the time to failure T_F or equivalently the failure probability $\Pr[F(t)]$ can be estimated purely based on data [1]. This approach—known as the actuarial approach—is not suitable for estimating the failure probability $\Pr[F(t)]$ of deteriorating structural systems, because failures of such systems are rare and hence no or only very limited data on failures are available. Furthermore, most structures are unique and, as a consequence, $\Pr[F(t)]$ varies strongly from one structure to another. Therefore, $\Pr[F(t)]$ is estimated based on probabilistic engineering models of the deterioration processes and structural performance [3, 4]. The estimate of the failure probability of deteriorating structural systems can be improved with information provided by in-service inspection and monitoring data using Bayesian methods [5–7]. All observations and measurements obtained up to time t_Z may be represented by a random vector $\mathbf{Z}(t_Z)$, and the updated failure probability is $\Pr[F(t)|\mathbf{Z}(t_Z) = \mathbf{z}(t_Z)]$.

In the literature, the updated probability of the failure event $F(t)$ of deteriorating structural systems is typically shown conditional on all data available up to time t, i.e. it is shown as $\Pr[F(t)|\mathbf{Z}(t) = \mathbf{z}(t)]$. This approach is known as filtering [8] and illustrated in Fig. 1b, which shows the filtered failure probability of the steel frame shown in Fig. 1a. The frame is subject to fatigue and inspected every 10 years. No fatigue cracks are detected during any of the inspections. The details of the example can be found in [9].

The results in Fig. 1 demonstrate that conditioning $\Pr[F(t)]$ on the data available up to time t can lead to a drop in that probability. Such a drop in the probability of the failure event $F(t)$ (or equivalently in the CDF $F_{T_F}(t)$ of the time to failure T_F) might seem counterintuitive since $\Pr[F(t)]$ is a non-decreasing function. In this paper, we illustrate with the help of a simple numerical example that such a drop can occur because the failure probability before and after the updating with new data is not based on the same information and thus not on the same probabilistic model.

2 Numerical Example: Steel Beam Subject to Corrosion

Consider the statically indeterminate steel I-beam shown in Fig. 2. The beam is characterized by its plastic moment capacity and subject to a time-variant (quasi-static) point load $S(t)$ and uniform corrosion. Corrosion starts at time $t = 0$ and the

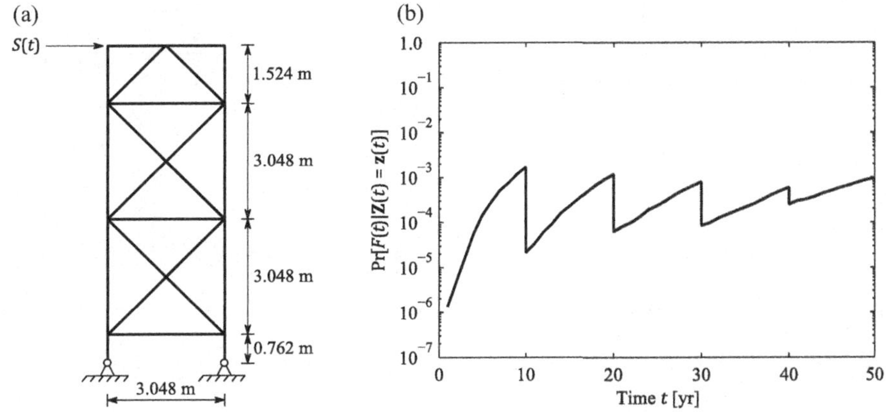

Fig. 1 **a** Steel frame subject to fatigue. **b** Filtered failure probability $\Pr[F(t)|\mathbf{Z}(t) = \mathbf{z}(t)]$ of the frame. The frame is inspected every 10 years. No fatigue cracks are detected. The details of the example can be found in [9]

corrosion depth D increases linearly with time t:

$$D(K, t) = K \cdot t \tag{1}$$

where K is the random corrosion rate, which is modeled as a time-invariant random variable.

The plastic section modulus W_{pl} at any position along the I-beam at time t can be expressed in function of the corrosion rate K as:

$$W_{pl}(K, t) = \underbrace{\left(h_0 - t_{f,0}\right) \cdot \left[t_{f,0} - 2D(K, t)\right] \cdot \left[w_0 - 2D(K, t)\right]}_{\text{flanges}} + \underbrace{\frac{1}{4}\left[h_0 - 2t_{f,0} + 2D(K, t)\right]^2 \cdot \left[t_{w,0} - 2D(K, t)\right]}_{web} \tag{2}$$

wherein h_0, w_0, $t_{f,0}$ and $t_{w,0}$ are the initial section height, flange width, flange thickness and web thickness (see Fig. 2).

The resulting plastic moment capacity R at any position along the I-beam at time t is:

$$R\left(F_y, K, t\right) = F_y \cdot W_{pl}(K, t) \tag{3}$$

where F_y is the random yield strength, which is also modeled as a time-invariant random variable.

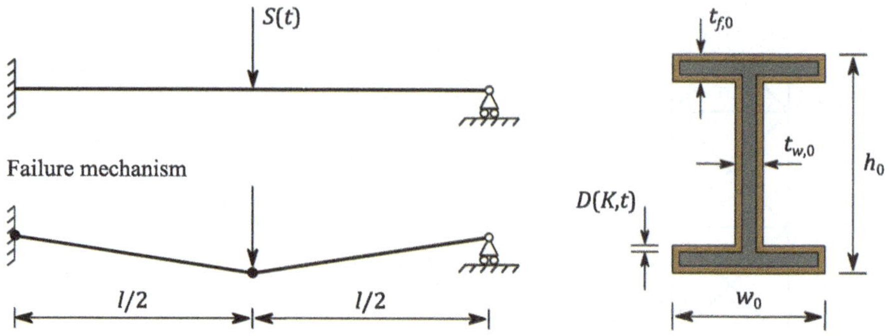

Fig. 2 Steel I-beam subject a time-variant load and uniform corrosion and its failure mechanism

2.1 Estimation of the Failure Probability

Failure of the beam is described at mechanism level [10]. Under the applied load, the failure mechanism shown Fig. 2 can form. The occurrence of this failure mechanism at time t is described by the following limit state function, which is derived using the principle of virtual work:

$$g(t) = 3 \cdot R\big(F_y, K, t\big) - 0.5 \cdot l \cdot S(t) \tag{4}$$

wherein l is the length of the beam.

The event of failure in the period leading up to time t can be defined as:

$$F(t) = \left\{\left[\min_{\tau \in [0,t]} g(\tau)\right] \le 0\right\} \tag{5}$$

The corresponding probability of failure of the beam up to time t is

$$\Pr[F(t)] = \Pr\left(\left[\min_{\tau \in [0,t]} g(\tau)\right] \le 0\right) \tag{6}$$

Equation (6) corresponds to a time-variant reliability problem.

In the current example, the deteriorating moment capacity $R\big(F_y, K, t\big)$ of the beam and the time-variant load $S(t)$ are statistically independent. The computation of the failure probability can thus be approached by discretizing time and transforming the time-variant reliability problem into a series of time-invariant reliability problems [2]. To this end, the service life T_{SL} of the beam is divided into $j = 1, \ldots, m$ intervals such that the jth interval corresponds to $t \in \big(t_{j-1}, t_j\big]$. The length of the interval is chosen to be one year. Furthermore, the load on the beam is represented by the maximum of $S(t)$ in each interval j

$$S_{max,j} = \max_{t \in (t_{j-1}, t_j]} S(t) \tag{7}$$

The event of failure in interval j (which neglects that failure may have occurred in any previous intervals) is denoted by F_j^*. In accordance with Eq. (4), the interval failure event F_j^* is described by the limit state function

$$g_j(F_y, K, S_{max,j}) = 3 \cdot R(F_y, K, t_j) - 0.5 \cdot l \cdot S_{max,j} \tag{8}$$

wherein $R(F_y, K, t_j)$ is the plastic moment capacity of the beam at the end of interval j.

The event of failure up to time t_j can now be defined as:

$$F(t_j) = F_1^* \cup F_2^* \cup \ldots \cup F_j^* \tag{9}$$

This event is described by the system limit state function

$$g_{1:j}(\mathbf{X}) = \min_{i \in \{1,\ldots,j\}} g_i(F_y, K, S_{max,i}) \tag{10}$$

wherein $\mathbf{X} = [F_y, K, \mathbf{S}_{max}^T]^T$ with $\mathbf{S}_{max} = [S_{max,1}, \ldots, S_{max,m}]^T$ is the vector of input random variables.

The probability of failure $\Pr[F(t_j)]$ up to time t_j is computed by integrating the joint probability density function (PDF) $f(\mathbf{x})$ of $\mathbf{X}$ over the failure domain

$$\Pr[F(t_j)] = \Pr(F_1^* \cup F_2^* \cup \ldots \cup F_j^*) = \int_{g_{1:j}(\mathbf{x}) \leq 0} f(\mathbf{x}) d\mathbf{x} \tag{11}$$

In the current example, the annual maxima $\mathbf{S}_{max}$ of the applied load are equi-correlated, identically distributed random variables with correlation coefficient $\rho_{S_{max}}$. The correlation model is equivalent to a hierarchical model, in which the annual maxima $\mathbf{S}_{max}$ are defined conditional on a common hyper-parameter representing a common influencing factor [7]. This type of correlation might arise from a common epistemic uncertainty such as, for example, statistical uncertainty in the probability distribution of $S_{max,j}$. The joint PDF $f(\mathbf{s}_{max})$ of $\mathbf{S}_{max}$ is modeled through the Gaussian copula (Nataf) model [11]. The yield strength F_y, the corrosion rate K and the annual maxima $\mathbf{S}_{max}$ are mutually statistically independent. The joint PDF $f(\mathbf{x})$ is thus obtained as the product of the marginal distributions of F_y, K and $\mathbf{S}_{max}$:

$$f(\mathbf{x}) = f(f_y) \cdot f(k) \cdot f(\mathbf{s}_{\max}) \tag{12}$$

In the current example, Monte Carlo simulation (MCS) is applied to evaluate the integral in Eq. (11), which is the simplest and most robust method for solving integrals of this type. The method is, however, inefficient in simulating probabilities of rare events. More efficient approaches for computing $\Pr\left[F\left(t_j\right)\right]$ are presented in [2]. MCS proceeds by separately generating samples $\left\{f_y^{(n)}\right\}_{n=1}^{N}$ from $f\left(f_y\right)$, samples $\left\{k^{(n)}\right\}_{n=1}^{N}$ from $f(\kappa)$ and samples $\left\{\mathbf{s}_{max}^{(n)}\right\}_{n=1}^{N}$ from $f\left(\mathbf{s}_{max}\right)$. The probability of failure $\Pr\left[F\left(t_j\right)\right]$ is then evaluated as:

$$\Pr\left[F\left(t_j\right)\right] \approx \frac{1}{N}\sum_{n=1}^{N}\mathbb{I}\left[\left[\min_{i\in\{1,\ldots,j\}} g_i\left(f_y^{(n)}, k^{(n)}, s_{max,i}^{(n)}\right)\right] \leq 0\right] \tag{13}$$

where $\mathbb{I}[\cdot]$ is the indicator function, which is equal to 1 if its argument is true and 0 otherwise. The Monte Carlo simulation is implemented such that the evaluations of the limit state function $\left\{g_i\left(f_y^{(n)}, k^{(n)}, s_{max,i}^{(n)}\right)\right\}_{n=1}^{N}$, $i = 1, \ldots, m$ are reused in the estimation $\Pr\left[F\left(t_j\right)\right]$ for different j.

Table 1 Probabilistic model of the parameters of the beam model

Parameter	Description	Distribution	Mean	Coefficient of variation
l	Beam length	Deterministic	10 m	
h_0	Initial section height	Deterministic	300 mm	
w_0	Initial flange width	Deterministic	150 mm	
$t_{f,0}$	Initial flange thickness	Deterministic	10.7 mm	
$t_{w,0}$	Initial web thickness	Deterministic	7.1 mm	
T_{SL}	Service life	Deterministic	30 yr	
F_y	Yield strength	Lognormal	240 N/mm^2	0.1
K	Corrosion rate	Lognormal	0.1 mm/yr	1.0
$S_{max,j}$	Maximum load	Lognormal	21.1 kN	0.4
$\rho_{S_{max}}$	Correlation coefficient among $S_{max,j}$	Deterministic	0.7	

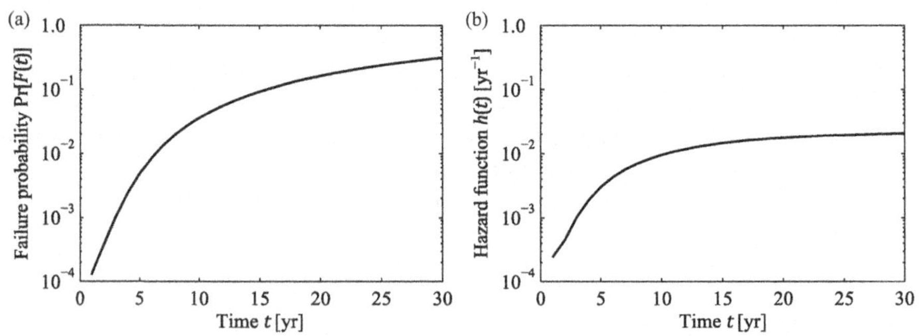

Fig. 3 **a** Estimate of the failure probability $\Pr[F(t)]$ and **b** corresponding hazard function $h(t)$ of the beam

The probabilistic model of the parameters of the beam model is summarized in Table 1. The mean of $S_{max,j}$ is chosen such that the interval failure probability of the undamaged beam is $\Pr\left(F_j^*\right) = 10^{-4}$.

The MCS is performed with $N = 10^7$ samples of the input random variables. The estimated probability of failure $\Pr[F(t)]$ of the beam is shown in Fig. 3a. For clarity, the uncertainty in the MCS estimate of the failure probability in terms of the confidence interval is not presented. Figure 3b shows the corresponding hazard function $h(t)$, which is computed from $\Pr[F(t)]$.

2.2 *Effect of Static Response Measurements on the Failure Probability Estimate*

Measurements of the static response of the beam contain indirect information on its condition because the stiffness of the beam is a function of the corrosion depth. This information can be applied to improve the estimate of the beam's failure probability. To illustrate the effect of static response measurements on the failure probability estimate, the beam is subject to a known point load s_{test} at time t_Z as shown in Fig. 4 and the end rotation Φ is measured.

The end rotation Φ at time t_Z can be predicted in function of the corrosion rate K based on linear elastic beam theory as follows:

$$\Phi(K, t_Z) = \frac{s_{test} \cdot a_{test}^2}{4 \cdot E \cdot I(K, t_Z)} \cdot \left(1 - \frac{a_{test}}{l}\right) \tag{14}$$

with

$$I(K, t_Z) = \underbrace{\frac{1}{6}[w_0 - 2D(K, t_Z)] \cdot \left[t_{f,0} - 2D(K, t_Z)\right]^3}_{\text{flanges}}$$

$$+ \underbrace{\frac{1}{2}\left(h_0 - t_{f,0}\right)^2 \cdot [w_0 - 2D(K, t_Z)] \cdot \left[t_{f,0} - 2D(K, t_Z)\right]}_{\text{flanges}}$$

$$+ \underbrace{\frac{1}{12}\left[t_{w,0} - 2D(K, t_Z)\right] \cdot \left[h_0 - 2t_{f,0} + 2D(K, t_Z)\right]^3}_{\text{web}} \tag{15}$$

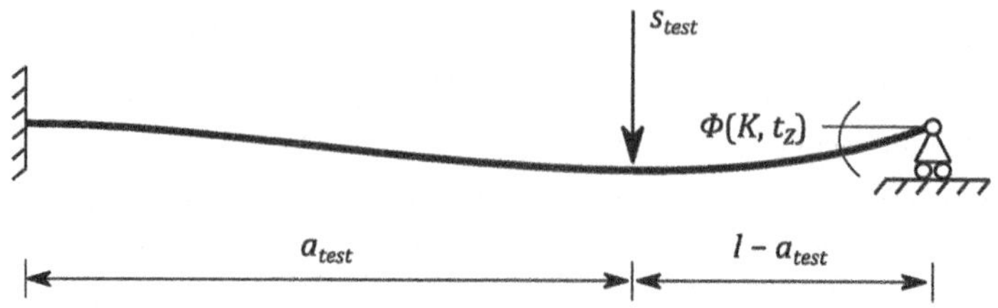

Fig. 4 Deformation of the beam during defined load test performed at time t_Z

wherein a_{test} is the position of the test load s_{test} (see Fig. 4), E is the Young's modulus, $I(K, t_Z)$ is the second moment of area at time t_Z and $D(K, t_Z)$ is the corrosion depth at time t_Z (see Eq. 1).

The measurement of the end rotation at time t_Z is probabilistically modeled by the random variable $Z(t_Z)$, which is linked with the model prediction of the end rotation $\Phi(K, t_Z)$ through an additive prediction error U:

$$Z(t_Z) = \Phi(K, t_Z) + U \tag{16}$$

The prediction error U jointly represents measurement uncertainty and modeling errors. It is probabilistically modeled by a Gaussian PDF $f_U(u)$ with zero mean and standard deviation σ_U.

In the current example, the measurement of the end rotation $Z(t_Z) = z(t_z)$ provides indirect information on the corrosion rate K. The relation between $Z(t_Z) = z(t_z)$ and K is probabilistically modeled through the likelihood function $L[k|z(t_Z)]$, which is proportional to the conditional PDF of $Z(t_Z) = z(t_z)$ given $K = k$. The likelihood function $L[k|z(t_Z)]$ is formulated based on the prediction error model defined in Eq. (7). From Eq. (7), it follows that $U = Z(t_Z) - \Phi(K, t_Z)$. Hence, the probability density of the difference between the measured and predicted end rotation $z(t_z) - \varphi(k, t_Z)$ is equal to the probability density of the prediction error U taking that value. The likelihood function $L[k|z(t_Z)]$ describing the measurement outcome $Z(t_Z) = z(t_z)$ is thus:

$$L[k|z(t_Z)] = f_U\left[z(t_z) - \varphi(k, t_Z)\right] = \frac{1}{\sigma_U\sqrt{2\pi}}\exp\left(-\frac{[z(t_Z) - \varphi(k, t_Z)]^2}{2\sigma_U^2}\right) \tag{17}$$

The information provided by the measurement outcome $Z(t_Z) = z(t_z)$ is included in the analysis through Bayesian updating of the PDF of the corrosion rate K:

$$f[k|z(t_Z)] \propto L[k|z(t_Z)] \cdot f(k) \tag{18}$$

The posterior joint PDF $f[\mathbf{x}|z(t_Z)]$ of the input random variables $\mathbf{X} = \left[F_y, K, \mathbf{S}_{max}^T\right]^T$ is

$$f[\mathbf{x}|z(t_Z)] = f\left(f_y\right) \cdot f[k|z(t_Z)] \cdot f(s_{\max}) \tag{19}$$

The probability of the failure event $F(t_j)$ conditional on the measurement outcome $Z(t_Z) = z(t_z)$ is obtained by replacing the prior PDF $f(\mathbf{x})$ of $\mathbf{X}$ with its posterior PDF $f[\mathbf{x}|z(t_Z)]$ in Eq. (11):

$$\Pr\left[F(t_j)|Z(t_Z) = z(t_Z)\right] = \Pr\left[F_1^* \cup \ldots \cup F_j^*|Z(t_Z) = z(t_Z)\right] = \int_{g_{1:j}(\mathbf{x}) \le 0} f[\mathbf{x}|z(t_Z)]\mathrm{d}\mathbf{x} \quad (20)$$

The integral in Eq. (20) is again evaluated using a MCS approach. To this end, standard MC sampling is applied to generate samples $\left\{f_y^{(n)}\right\}_{n=1}^N$ from $f(f_y)$ and samples $\left\{\mathbf{s}_{max}^{(n)}\right\}_{k=1}^N$ from $f(\mathbf{s}_{max})$. BUS (Bayesian updating with structural reliability methods) with subset simulation [12] is utilized to simulate conditional samples $\left\{k^{(n)}\right\}_{k=1}^N$ from $f[k|z(t_Z)]$. The updated failure probability $\Pr\left[F(t_j)|Z(t_Z) = z(t_Z)\right]$ is subsequently approximated as:

$$\Pr\left[F(t_j)|Z(t_Z) = z(t_Z)\right] \approx \frac{1}{N}\sum_{k=1}^{N} \mathbb{I}\left[\left[\min_{i \in \{1,\ldots,j\}} g_i\left(f_y^{(n)}, k^{(n)}, s_{max,i}^{(n)}\right)\right] \le 0\right] \quad (21)$$

In the current example, it is assumed that an end rotation $Z(t_Z) = 0.35$ mrad is measured when the beam is subject to the test load $s_{test} = 1$ kN at time $t_Z = 15$ yr. The additional parameters of the beam model are summarized in Table 2.

Figure 5 shows the prior and posterior CDF of the corrosion rate K. The latter is estimated based on 10^3 conditional samples of K generated from $f[k|z(t_Z)]$. The measurement leads to a reduction in the uncertainty in K. The posterior mean and standard deviation of K are estimated as 0.092 mm/yr and 0.033 mm/yr. In comparison, the prior mean and standard deviation of K are 0.1 mm/yr and 0.1 mm/yr (see Table 1).

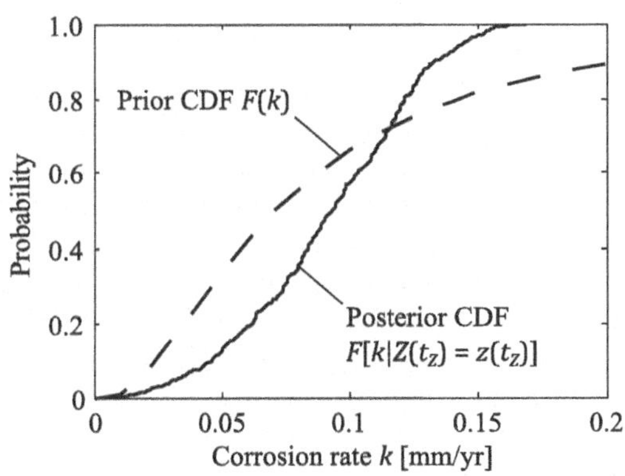

Fig. 5 Prior and (empirical) posterior CDF of the corrosion rate K

Table 2 Additional parameters of the beam model

Parameter	Description	Value
E	Young's modulus	$2.1 \cdot 10^5$ N/mm^2
s_{test}	Test load	1 kN
a_{test}	Position of the test load s_{test}	$2/3 \cdot l$ [m]
σ_U	Standard deviation of the prediction error U	0.05 mrad

The probabilities $\Pr\left[F\left(t_j\right)|Z(t_Z) = z(t_Z)\right]$, $j = 1, \ldots, m$ are estimated based on $N = 10^7$ samples of the input random variables. Figure 6 compares the prior and posteriors estimate of the failure probability, $\Pr[F(t)]$ and $\Pr[F(t)|Z(t_Z) = z(t_Z)]$, and additionally compares the estimates of the corresponding hazard functions $h(t)$ and $h[t|Z(t_Z) = z(t_Z)]$. The reduction in the uncertainty in the corrosion rate K leads to a reduction in the estimate of the failure probability and hazard function. Figure 7 presents the failure probability $\Pr[F(t)|Z(t) = z(t)]$ and the corresponding hazard function $h[t|Z(t) = z(t)]$ of the beam at time t conditional on the data available up to time t (filtering). When presenting the failure probability and failure rate in this way, a drop in both functions occurs at the time of the measurement $t_Z = 15$ yr.

3 Concluding Remarks

The numerical example in this paper illustrates the application of probabilistic analysis and structural reliability theory to the problem of assessing the reliability of a deteriorating structural system. In this approach, a deterministic engineering model combined with the prior probabilistic model $f(\mathbf{x})$ of the model parameters $\mathbf{X}$—the probabilistic engineering model—is applied to characterize the prior knowledge on the deteriorating structural system. Based on the probabilistic engineering model, the failure probability $\Pr[F(t)]$ at time t is estimated by means of time-variant structural reliability analysis.

The estimate of the failure probability $\Pr[F(t)]$ is an extrapolation from the domain of observation. In a probabilistic setting, in-service inspection and/or monitoring data obtained from the real structural system can be applied to improve the failure probability estimate using Bayesian analysis. The data obtained up to time t_Z can be probabilistically represented by a random vector $\mathbf{Z}(t_Z)$. In Bayesian analysis, the prior PDF $f(\mathbf{x})$ of $\mathbf{X}$ is updated with the obtained data $\mathbf{Z}(t_Z) = \mathbf{z}(t_Z)$ to the posterior PDF $f[\mathbf{x}|\mathbf{z}(t_Z)]$, which quantifies the updated knowledge on the deteriorating structural system. Subsequently, the failure probability estimate is also updated. This process is repeated each time new data becomes available.

As demonstrated in the numerical example, the updated estimate of the failure probability can be presented in two different ways: (a) The failure probability at time t can be—as illustrated in Fig. 6a—presented conditional on the fixed data set $\mathbf{Z}(t_Z) = \mathbf{z}(t_Z)$ available up to time t_Z, i.e. at each time t, the failure

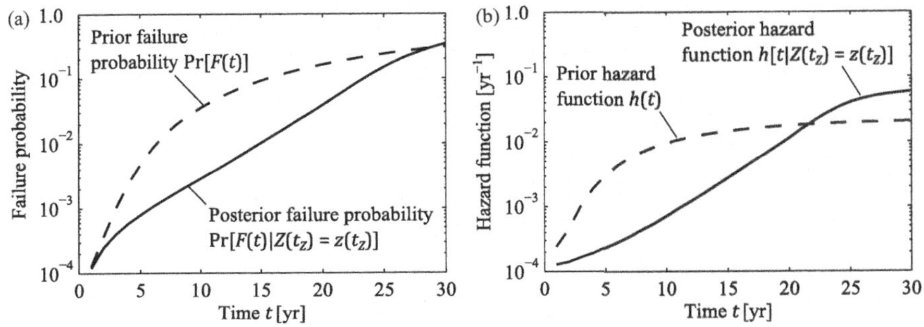

Fig. 6 Estimate of **a** failure probabilities $\Pr[F(t)]$ and $\Pr[F(t)|Z(t_Z) = z(t_Z)]$ and **b** corresponding hazard functions $h(t)$ and $h[t|Z(t_Z) = z(t_Z)]$ of the beam

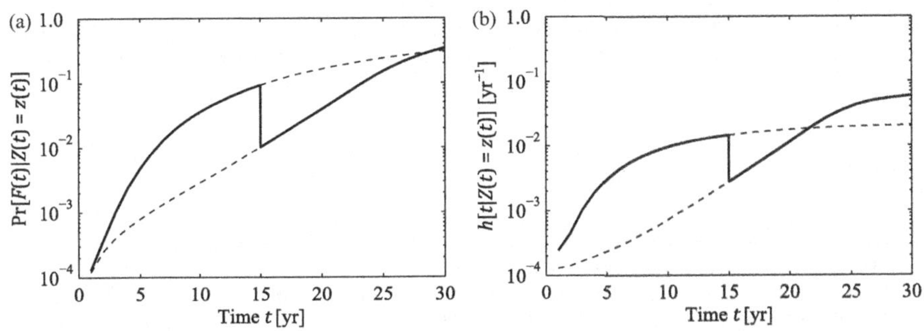

Fig. 7 **a** Filtered failure probability $\Pr[F(t)|Z(t) = z(t)]$ and **b** corresponding filtered hazard function $h[t|Z(t) = z(t)]$ of the beam

probability is estimated based on the conditional PDF $f[\mathbf{x}|\mathbf{z}(t_Z)]$ and shown as $\Pr[F(t)|\mathbf{Z}(t_Z) = \mathbf{z}(t_Z)]$. $\Pr[F(t)|\mathbf{Z}(t_Z) = \mathbf{z}(t_Z)]$ is a non-decreasing function of time t. (b) As illustrated in Fig. 7a, the failure probability at time t can be presented conditional on the data $\mathbf{Z}(t) = \mathbf{z}(t)$ available up to time t (filtering), i.e. at each time t, the failure probability is computed based on the conditional PDF $f[\mathbf{x}|\mathbf{z}(t)]$ and shown as $\Pr[F(t)|\mathbf{Z}(t) = \mathbf{z}(t)]$. In this case, if the data $\mathbf{Z}(t_1) = \mathbf{z}(t_1)$ available up to time t_1 is different from the data $\mathbf{Z}(t_2) = \mathbf{z}(t_2)$ available up to time t_2, the failure probabilities $\Pr[F(t_1)|\mathbf{Z}(t_1) = \mathbf{z}(t_1)]$ at time t_1 and $\Pr[F(t_2)|\mathbf{Z}(t_2) = \mathbf{z}(t_2)]$ at time t_2 are based on two different probabilistic models $f[\mathbf{x}|\mathbf{z}(t_1)]$ and $f[\mathbf{x}|\mathbf{z}(t_2)]$ of the model parameters $\mathbf{X}$. As a result, there can be a drop in the failure probability $\Pr[F(t)|\mathbf{Z}(t) = \mathbf{z}(t)]$.. The same behavior can also be observed for the hazard function, i.e. the failure rate conditioned on survival of the structure.

To further illustrate this point, Fig. 8 shows the filtered failure probability $\Pr[F(t)|\mathbf{Z}(t) = \mathbf{z}(t)]$ of the steel frame shown in Fig. 1a together with the corresponding failure probabilities $\Pr[F(t)|\mathbf{Z}(t_Z) = \mathbf{z}(t_Z)]$ conditional on the data available up to the inspection times $t_Z = 10, 20, 30, 40$ yr.

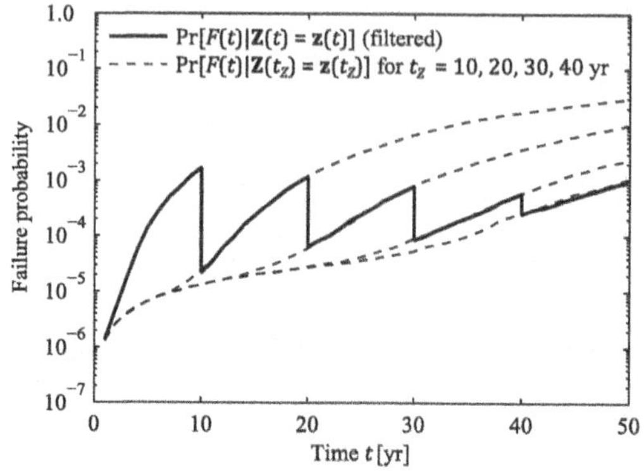

Fig. 8 Estimate of the failure probability of the steel frame shown in Fig. 1a conditional on inspection outcomes. The solid line is the filtered failure probability $\Pr[F(t)|\mathbf{Z}(t) = \mathbf{z}(t)]$. The dashed lines are the conditional failure probabilities $\Pr[F(t)|\mathbf{Z}(t_Z) = \mathbf{z}(t_Z)]$ for inspection times $t_Z = 10, 20, 30, 40$ yr. No fatigue cracks are detected. The details of the example can be found in [9]

References

1. Rausand, M., & Høyland, A. (2004). *System reliability theory: Models, statistical methods, and applications* (2nd edn.). New York: Wiley.
2. Straub, D., Schneider, R., Bismut, E., & Kim, H.-J. (2020). Reliability analysis of deteriorating structural systems. *Structural Safety, 82,* 101877.
3. Ditlevsen, O., & Madsen, H. O.(1996). *Structural reliability methods.* New York: Wiley.
4. Melchers, R. E. (1999). *Structural reliability analysis and prediction* (2nd ed.). New York: Wiley.
5. Madsen, H. O. (1987). Model updating in reliability theory. In *5th International Conference on Applications of Statistics and Probability in Civil Engineering (ICASP 5), Vancouver, Canada.*
6. Sindel, R., & Rackwitz, R. (1998). Problems and solution strategies in reliability updating. *Journal of Offshore Mechanics and Arctic Engineering, 120*(2), 109–114.
7. Straub, D. (2018). *Reliability assessment of deteriorating structures: Challenges and (some) solutions, 6th International Symposium on Life-Cycle Civil Engineering (IALCCE 2018).* Belgium: Ghent.
8. Straub, D. (2009). Stochastic modeling of deterioration processes through dynamic Bayesian networks. *Journal of Engineering Mechanics, 135*(10), 1089–1099.
9. Schneider, R. (2019). *Time-variant reliability of deteriorating structural systems conditional on inspection and monitoring data.* (Doctoral thesis) Technische Universität München, Germany.
10. Thoft-Christensen, P., & Murotsu, Y. (1986). *Application of structural systems reliability theory.* Berlin: Springer.
11. Liu, P.-L., & Der Kiureghian, A. (1986). Multivariate distribution models with prescribed marginals and covariances. *Probabilistic Engineering Mechanics, 1*(2), 105–112.
12. Straub, D., & Papaioannou, I. (2015). Bayesian updating with structural reliability methods. *Journal of Engineering Mechanics, 141*(3).

Development of Culvert Risk Condition Evaluation for Decision-Making Within Road Infrastructure Management

Fernando Sousa, Sara Dias, José C. Matos, and Aires Camões

Abstract Regarding road infrastructure management systems, culverts need to be assessed in order to avoid failures and road collapses. So, periodic inspections framework and condition rating implementation has an important role for life service estimation and reliability evaluation. In addition, the risk can be avoided through condition rating merged with culverts exposure and vulnerabilities. This will provide information to support decision-making and prioritize interventions. In this paper a new approach for decision-making process is presented taking into consideration the global risk index (α_G). The proposal includes a set of culverts descriptors, weight attribution and aggregation rules complying with external factors such as hazards, condition rates and consequences. Moreover, a case study with 25 different systems is conducted to qualitatively assess culverts global risk index and prioritize needed interventions.

Keywords Culverts management system · Risk assessment · Decision-making process · Hazards occurrence

1 Introduction

All around the world, road infrastructure systems are the most predominant networks by connecting villages, providing society with facilities and goods. So, road conservation agencies establish routines to manage the infrastructure serviceability, with special focus on pavement and bridges.

F. Sousa (✉) · S. Dias
Infrastructure Innovation and Management, Ascendi IGI, Matosinhos, Portugal

F. Sousa · J. C. Matos
Institute for Sustainability and Innovation in Structural Engineering (ISISE), University of Minho, Guimarães, Portugal

A. Camões
Centre for Territory, Environment and Construction (CTAC), University of Minho, Guimarães, Portugal

© The Author(s), under exclusive license to Springer Nature Switzerland AG 2021
J. C. Matos et al. (eds.), *18th International Probabilistic Workshop*, Lecture Notes in Civil Engineering 153, https://doi.org/10.1007/978-3-030-73616-3_19

However, in the past decade, agencies got more concerned about culverts because they let the water go through the roadbed preserving it from erosion and also, the failure of such construction may lead to the interruption of significant part of the road [1]. Deteriorated culverts and drainage structures requires the road conservation agencies to implement proper inventory and inspection programs [2].

In addition, culverts must support accidental conditions and other external actions, maintaining structural integrity and functionality. Maintaining these structures in a good state will avoid road collapses and traffic disruptions and therefore economic loss. Therefore, risk assessment must combine hazards or external actions, with the culvert's exposure, vulnerability, and the consequences to the road infrastructure. Providing trusty information to the decision-making process allowing to prioritize interventions and maintenance needs.

This paper aims to introduce a risk-based approach to support the decision-making process considering a qualitatively assessed Global Risk Index (α_G). Complying hazards, culvert's exposure and vulnerability, and its consequences. In this way, the proposed framework is presented including descriptors definition and aggregation rules, followed by concluding remarks and a brief analysis due a 25 culverts case of study.

2 Culvert Management System (CMS). Proposed Framework

The objective is to implement a Culverts Management System (CMS) in a roadway or railway, as a life cycle estimation based on the current condition rating and taking into account the design information and location [3]. Moreover, the risk assessment of the infrastructure during the remaining life service, with the possibility of hazards occurrence can influence transportation safety or induce the collapse [4].

In this way, a CMS framework is established and presented in Fig. 1 addressing three main steps along the management process and having two related operation cycles to provide reliable information.

The first cycle combines the two main steps Database Analysis and Qualitative Risk Assessment, having has objective the culverts prioritization and be a support to the decision-making process using the Global Risk Index (α_G) [5]. At the second cycle, the step Numerical Model Simulations is added to complement the qualitative analysis, performing the Ultimate Limit State (ULS) and Service Limit State (SLS) analysis, calculating the probability of failure through the Reliability Index (β_G) [6].

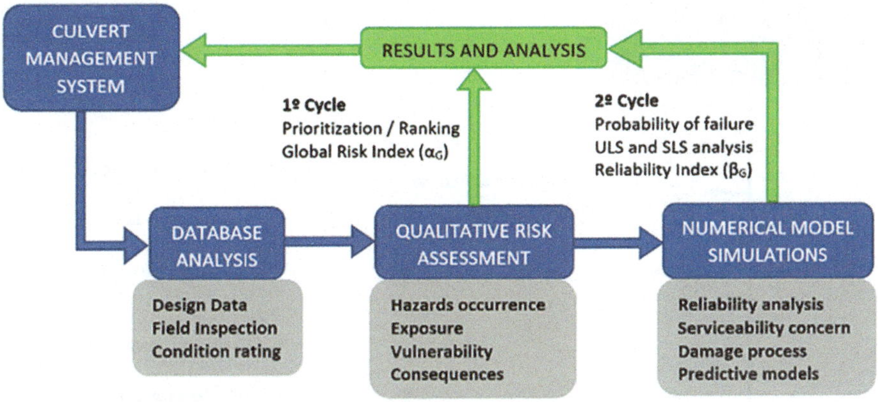

Fig. 1 Culvert management system framework

2.1 *Global Risk Index (α_G) Evaluation—Preliminary Risk Analysis*

To accomplish the main objective, a set of descriptors (α_i) organized by classes or partial risk indexes are defined. In this way, culvert's design data (design peak flow, drainage area, slope, etc.) and all information obtained during conception and construction (End treatments, materials, age, shape, length, span, etc.) are stored in the database. Likewise, field inspection by experts need to be considered in order to update all the available information about the culvert´s condition and provide an integral infrastructure characterization.

Considering all available information, the index flow chart is introduced in Fig. 2, where twelve descriptors (α_i) are grouped by External or Environmental actions (H), Structural condition and Reliability (E), and Human and Economic consequences (C).

Hence, the culverts global risk index (α_G) depends on three main risk elements as described by the equations

$$\alpha_G = \mathrm{H} \times \mathrm{E} \times \mathrm{C} \tag{1}$$

Where H represents hazards occurring, E is the culvert exposure given the current condition or existent damages, and C the consequences of damage to the system [7]. In this framework, as external actions (Partial Risk Index H) are considered the four main hazards that may affect a culvert. The exposure (Partial Risk Index E) is given by the culvert structural condition, level of damage and performance score, and material characteristics (e.g. ratio between culverts age and design life). Consequences (Partial Risk Index C), intend to reflect the direct losses due to culvert damages, and also the indirect ones that may result in traffic disruptions and change of the available transportation network (e.g. alternative road) [8].

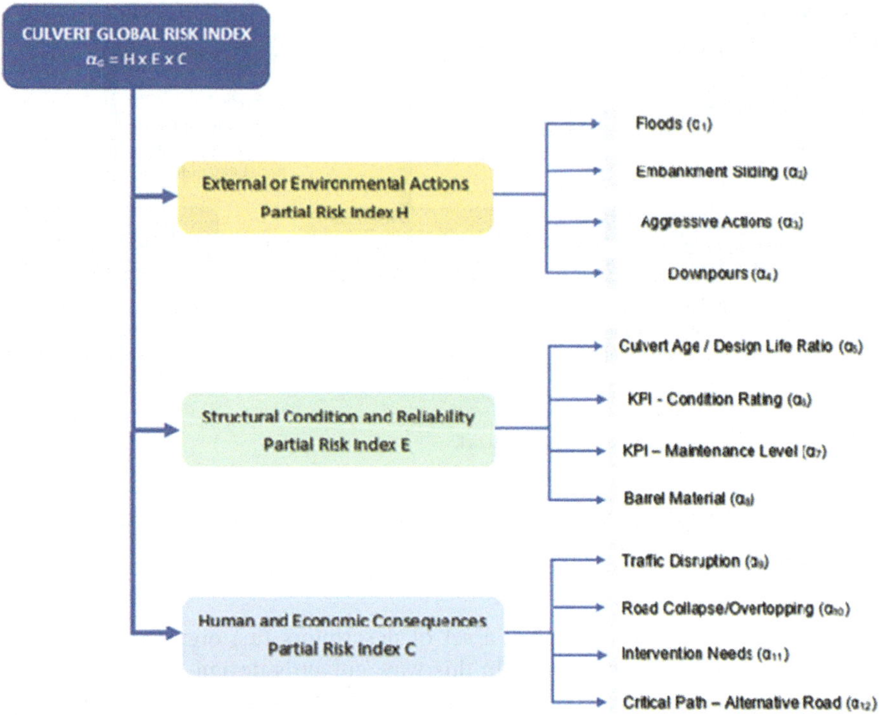

Fig. 2 Global risk index flow chart. Classes and descriptors definition

2.2 Descriptors and Partial Risk Index Definition

In order to obtain the global risk index, all the partial indexes associated to the descriptors are established in this section. For the three partial indexes Tables 1, 2 and 3 present the descriptors explanation with the considered support and assessment. Also, it is possible to understand that the partial risk index assess depends on design and field inspection data, but also historic data about hazards occurrence and culverts interventions. It is important to have a reliable information database and perform a critical analysis during the partial risk index assessment. Moreover, some aggregation rules are defined to calculate the partial risk indexes and the global risk index.

Almost all descriptors are assessed qualitatively and have five different levels: 1 is the lower and 5 the maximum risk impact. The partial index support differs taking into account all the available data, the expert judgement and the known impact to the culvert (e.g. Assess likelihood of a culvert flood event regarding the hydrological data, historic information, and field inspection report) [9].

Table 1 External or environmental actions (Partial Index H) descriptors definition

External or environmental actions (PI H)								
Descriptor	Floods α_1		Embankment sliding α_2		Aggressive actions α_3		Downpours α_4	
Description	Associated to the likelihood of occurrence floods larger than the design flood		Associated to the embankment characteristics and likelihood to slide due environmental actions		Associated with aggressive actions that could cause alterations of flow quality, sedimentation and contamination		Associated with culvert and watershed characteristics and likelihood to extreme flow erosive potential	
Support	Based on hydrological and design data		Based on landslides visual condition, type of material, slopes inclination		Based on visual inspection and expert judgement		Based on Inlet pressure, existence of structures for energy dissipation	
Assessment	Index	Conditions justifying	Index	Conditions justifying	Index	Condition justifying	Index	Condition justifying
	1	May occur in 100 years	1	Unlikely to occur	1	Clean water	1	May occur in 100 years
	2	May occur in 20 to 50 years	2	May occur due extreme events	3	Colored water	2	May occur in 20 to 50 years
	3	May occur in 10 to 20 years	3	May occur due intensive rainfall	5	Sewer water	3	May occur in 10 to 20 years
	4	May occur in 2 to 5 years	4	May occur during a rainfall			4	May occur in 2 to years
	5	This year or next	5	Can occur frequently			5	This year or next

Table 2 Structural condition/reliability (Partial index E) descriptors definition

Structural condition/reliability (PI E)								
Descriptor	Culverts age/design life ratio α_5		KPI—condition rating α_6		KPI—maintenance level α_7		Characteristics (Materials) α_8	
Description	Associated with culverts age and design life		Associated with visual detail inspection information input Condition Rating		Associated with visual routine inspection information input Maintenance Level		Associated with materials durability, vulnerability to chemical reactions and degradation	
Support	Based on the relation between culverts age and design life—R		Based on field inspection reports. Also, damages extension and location		Based on presence dense vegetation and flow's problematic sedimentation		Based on barrel material	
Assessment	Index	Conditions justifying	Index	Conditions justifying	Index	Conditions justifying	Index	Conditions justifying
	1	$R < 0.25$	1	No defects	1	No flow perturbation	1	PRFV, PP
	2	$0.25 < R < 0.50$	2	Minor defects, not urgent	2	Some flow perturbation	3	Reinforced Concrete
	3	$0.50 < R < 0.75$	3	Some damages to repair	3	Flow perturbation	5	Metallic
	4	$0.75 < R < 1.0$	4	Severe damage to repair				
	5	$R > 1.0$	5	Dangerous				

Table 3 Human/economic consequences (Partial index C) descriptors definition

Human/economic consequences (PI C)								
Descriptor	Traffic disruption α_9		Road collapse/overtopping α_{10}		Repair needs—investments α_{11}		Critical path—alternative road α_{12}	
Description	Associated to road service information		Associated to the overtopping risk based on embankment cover height		Associated to investments needs for culverts repair works or maintenance		Associated with user's alternative options in case of road collapse	
Support	Based on average daily traffic (ADT) and percentage of heavy traffic		In accordance with the embankment cover height (h)		Costs estimation (W) based on the type of intervention required, and past interventions		In accordance with the culvert's location and level of alternative roads existence	
Assessment	Index	Conditions justifying	Index	Conditions justifying	Index	Conditions justifying	Index	Conditions justifying
	1	ADT ≤ 7500	1	h > 20	1	W ≤ 10 000 €	1	Same level road
	2	7500 < ADT ≥ 15,000	2	13,5 < h ≥ 20	2	10 000 € < W ≥ 50 0000 €	2	Worst level road
	3	15,000 < ADT ≥ 25,000	3	6 < h ≥ 12	3	50 000 € < W ≥ 100 0000 €	3	No alternative road
	4	25,000 < ADT ≥ 40,000	4	2 < h ≥ 6	4	100 000 € < W ≥ 200 0000 €		
	5	ADT > 40,000	5	h ≤ 2	5	W > 200 000 €		

2.3 Aggregation Rules

There are two different types of aggregation in this framework, firstly between descriptors within the same partial risk index, secondly between the partials risks indexes.

For the partial risk indexes are calculated using Eqs. 2, 3, 4 as follows:

$$H = 0.2\alpha_1 + 0.15\alpha_2 + 0.3\alpha_3 + 0.35\alpha_4 \tag{2}$$

$$E = 0.15\alpha_5 + 0.5\alpha_6 + 0.1\alpha_7 + 0.25\alpha_8 \tag{3}$$

$$C = 0.15\alpha_9 + 0.35\alpha_{10} + 0.4\alpha_{11} + 0.1\alpha_{12} \tag{4}$$

where, H, E and C are the classes or partial risk indexes and α_i the descriptors. The relative weights are estimated considering the descriptors influence on the culvert functionality. For instance, at the partial risk index H the most important descriptors are α_3 (Aggressive actions) and α_4 (Downpours). Regarding the partial risk index E the main descriptor are the α_6 (KPI—Condition rating), based on the visual inspection damage report and culvert performance. Regarding partial risk index C, α_{10} (Road collapse/Overtopping) and α_{11} (Repair needs—Investments) are the most rated descriptors [10].

Finally, α_G is obtained by using the weighted product method applied to the basic risk equation, resulting as follows:

$$\alpha_G = 1.0\text{H} \times 1.50\text{E} \times 1.35\text{C} \tag{5}$$

Concerning Eq. 5, different weights are applied on partials risk index comparing to the basic risk equation. Aiming to give more emphasis to culvert data-related partial risk index. In other words, partial risk indexes associated to culvert structural condition (PI E) and consequences (PI C) are weighted, given the reliability of information available in the culvert's database to perform the assessment, when compared with external actions (PI H) [11].

3 Decision-Making Process. Culverts Prioritization

Nowadays, there is an effort to include risk management in asset management systems according to the ISO 55000 and ISO 31000 requirements. In this way, risk treatment actions should be concepted taking into account a previous risk identification, analysis and evaluation [12].

So, after the preliminary risk analysis performed by the aforementioned framework and α_G evaluation, it's possible to rank culverts and prioritize interventions.

Table 4 Culvert's decision-making process

α_G	Potential risk	Prioritization	Intervention requirements
$\alpha_G > 100$	Severe	Urgent	Immediately
$50 < \alpha_G < 100$	Significant	Critical	Less than 1 year
$25 < \alpha_G < 50$	Moderate	Alarming	Between 1 and 3 years
$10 < \alpha_G < 25$	Acceptable	Low	Monitoring
$\alpha_G < 10$	Minimal		

In the Table 4 a decision-making process is proposed regarding the α_G evaluation, establishing prioritization levels, and intervention requirements [13].

Likewise, five levels of culverts potential risk are introduced depending the α_G value, occurring the worst-case scenario when $\alpha_G > 100$. Regarding the culvert's prioritization, four levels are indicated, needing intervention schedule from $\alpha_G > 25$ corresponding to Alarming level. Culverts categorized in Critical level need to be repaired in less than a year, and the Urgent ones required immediate intervention. It is important refer that despite $\alpha_G < 25$ corresponding to Acceptable or Minimal levels of potential risk, culverts require continuous monitoring to refine the risk evaluation framework [14].

In the next chapter, a case of study is conducted with a sample of 25 culverts, presenting the preliminary risk analysis and the available visualization of the method results.

4 Case of Study—Application to 25 Culverts

The framework has been applied to 25 culverts from three different highways in center and north of Portugal and having condition rating (α_6) equal or higher than 2—Some damages to repair. In general terms, the sample is divided in two culvert types: Box Culvert and Pipe. The reinforced concrete box culvert represents forty-four percent of the sample, sixteen percent are corrugated steel pipes and forty percent are the traditional concrete pipes. In terms of design peak flow, sixty five percent have a design peak flow superior to 5.5 m^3/s.

Table 5 presents the results in terms of partial risk index and global risk index for each one of the culverts.

Applying this proposed framework to a set of 25 culverts it is possible to prioritize the interventions based on α_G and consequently establish an intervention plan. The Table 6 only represents the decision-making results for 10 of the culverts in order to characterize the sample.

Regarding the previous data and its intervention schedule, forty-eight percent of culverts need intervention within one to three years. Twenty-eight percent of

Table 5 Global risk index—partial index's

Label	Culvert type	Size/diameter	Partial index H	Partial index E	Partial index C	α_G
CLV63.1	Box culvert	1 □ 2 x 2	3.00	3.25	2.70	53.3
CLV66.1	Box culvert	1 □ 2 x 2	3.20	2.85	2.05	37.9
CLV103.1	Pipe	1 Ø 1000	1.70	2.50	2.30	19.8
CLV146.1	Pipe	1 Ø 2400	3.80	4.30	2.85	94.3
CLV119.2	Pipe	1 Ø 2930	1.50	3.10	1.70	16.0
CLV178.2	Box culvert	1 □ 2.5 x 2.5	2.15	2.75	1.85	22.1
CLV195.2	Pipe	1 Ø 1500	2.65	2.60	2.25	31.4
CLV205.2	Box culvert	1 □ 2 x 2	3.00	3.35	2.30	46.8
CLV254.2	Pipe	1 Ø 1000	3.35	3.10	3.40	71.5
CLV273.2	Pipe	1 Ø 1500	1.35	2.50	1.85	12.6
CLV63.3	Pipe	1 Ø 1000	2.30	3.50	2.80	45.6
CLV112.3	Box culvert	1 □ 2 x 2	2.75	3.25	3.25	58.8
CLV122.3	Box culvert	1 □ 2 x 1.8	1.70	3.25	3.25	36.4
CLV154.3	Box culvert	1 □ 2.1 x 2.2	3.85	3.75	3.35	97.9
CLV167.3	Box culvert	1 □ 2 x 2	2.30	2.75	2.15	27.5
CLV172.3	Box culvert	1 □ 2.5 x 2.5	2.35	2.60	2.85	35.3
CLV194.3	Box culvert	1 □ 2 x 2	1.85	3.25	2.50	30.4
CLV221.3	Pipe	2 Ø 2780	2.70	4.00	2.95	64.5
CLV223.3	Pipe	1 Ø 2860	3.70	4.40	3.35	110.4
CLV431.3	Pipe	1 Ø 1500	2.80	2.50	3.30	46.8
CLV505.3	Pipe	1 Ø 800	2.05	3.10	2.30	29.6
CLV509.3	Pipe	3 Ø 1500	1.95	2.50	2.35	23.2
CLV520.3	Box culvert	1 □ 4 x 4	3.40	3.25	3.25	72.7
CLV528.3	Pipe	1 Ø 800	1.35	3.50	3.35	32,1
CLV585.3	Pipe	1 Ø 1500	1.95	2.50	2.75	27.1

culverts need intervention in less than one year and twenty percent of the sample need to continue being monitored. At last, only four percent (Culvert CLV223.3) need immediate intervention. The next Fig. 3 shows the results of α_G in relation with the culvers prioritization levels. The y-axes represents the α_G result for each one of the culverts (x-axes).

Another interesting view about the results can be displayed, crossing α_G with any of the descriptors in an isolated way. Important conclusions can be made in terms of descriptors influence and establish relations between them. Besides, these results and charts analysis can be used by infrastructure managers as a decision-making support tool. In this way, the next Figs. 4 and 5 show the relation between global risk index α_G and KPI—Condition rating (α_6) and repair needs—Investments (α_{11}).

Table 6 Decision-making process based on α_G (10 culverts example)

Label	Decision-making process		
	Potential risk	Prioritization	Intervention requirements
CLV66.1	Moderate	Alarming	Between 1 and 3 years
CLV103.1	Acceptable	Low	Monitoring
CLV146.1	Significant	Critical	Less than 1 Year
CLV205.2	Moderate	Alarming	Between 1 and 3 years
CLV254.2	Significant	Critical	Less than 1 Year
CLV122.3	Moderate	Alarming	Between 1 and 3 years
CLV154.3	Significant	Critical	Less than 1 Year
CLV167.3	Moderate	Alarming	Between 1 to 3 years
CLV223.3	Severe	Urgent	Immediately
CLV520.3	Significant	Critical	Less than 1 Year
CLV585.3	Moderate	Alarming	Between 1 and 3 years

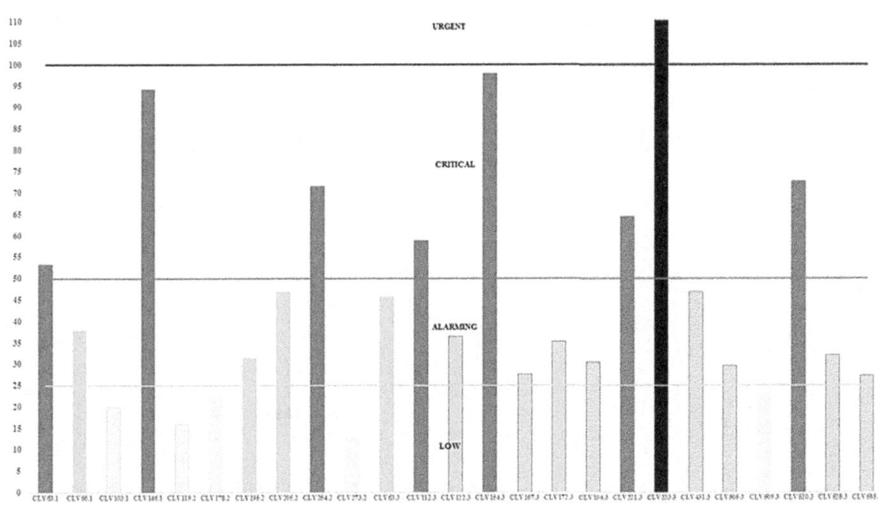

Fig. 3 Culverts Prioritization chart (α_G)

Fig. 4 Descriptors analysis charts, α_G versus α_6 and α_G versus α_{11}, respectively

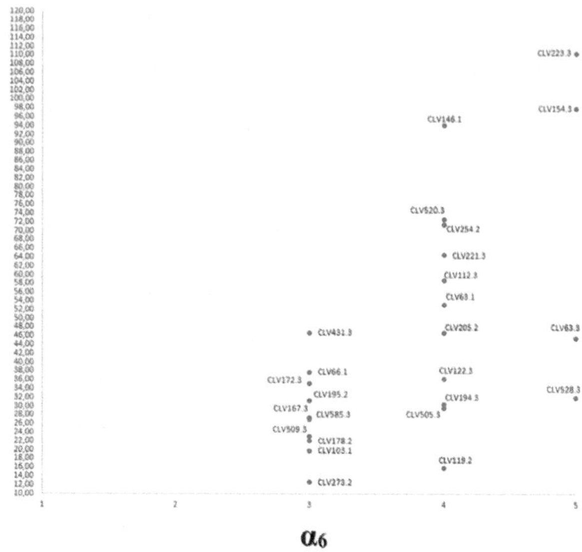

Fig. 5 Descriptors analysis charts, α_G versus α_6 and α_G versus α_{11}, respectively

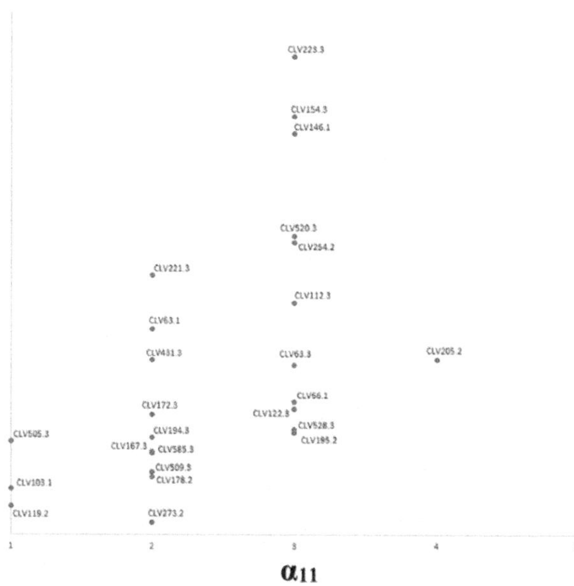

Observing data from the first scatter, it is possible to conclude that culverts with the highest values of KPI—Condition rating (α_6) tend to have highest global risk index. Important to refer that almost all culverts with high index α_6 result in $\alpha_G > 25$, and so in an Alarming prioritization level. Regarding the second scatter, it´s relevant to refer that, unlike the first scatter, culverts with the highest values of risk index are not necessarily culverts with the highest investment needs. For example, the

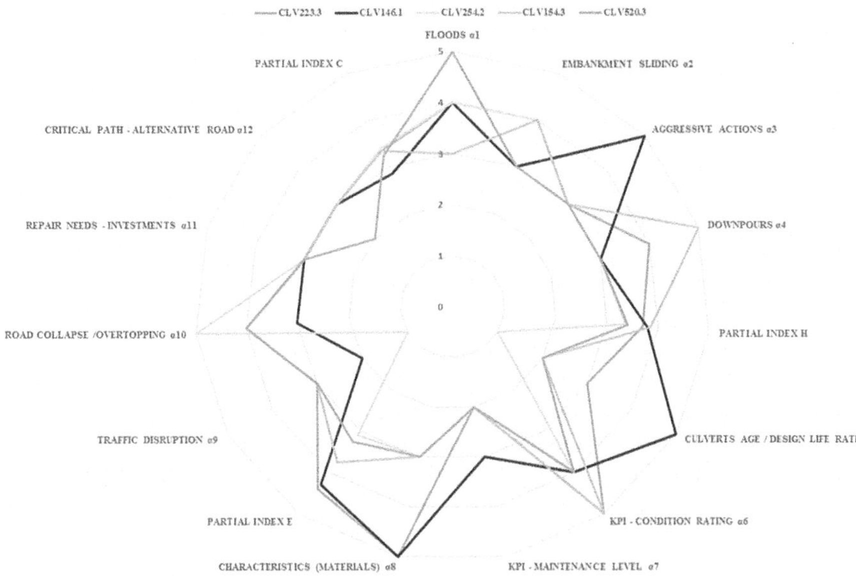

Fig. 6 Critical priority culverts analysis—descriptors

CLV223.3 has the highest global risk index, however, has a medium α_{11} index value. Comparing the first scatter with the second one, it is also possible to create links or connections to different descriptors. For instance, other descriptors like material characteristics α_8 or road collapse/overtopping α_{10} can be more preponderant to the analysis results.

Finally, Fig. 6 shows spider web chart of the descriptors applied to the more critical culverts.

Therefore, experts can evaluate which one of the descriptors must contribute and have paramount importance to the α_G value. Hence, it is attainable to relate the culverts descriptors and analyze which one has more relevance to the decision-making process. Furthermore, a quantitative analysis can be performed through ULS and SLS analysis in order to obtain the culvert's probability of failure.

5 Conclusions

This paper proposes a risk-based framework to support the decision-making process within a culvert management system. Three partial risk indexes are established with a set of twelve descriptors in order to obtain a single risk index α_G. By surveying all culverts related data available in databases and field inspection experience, descriptors weights and aggregations rules are defined. In this way, culverts condition rating and material characteristics, also road overtopping or collapse and investment needs are the more rated descriptors.

Likewise, depending on the culverts α_G value, different levels of potential risk and prioritization are defined to support infrastructure managers in the decision-making process. Intervention requirements can be automatically provided regarding the culverts' potential risk. This framework was applied to a case study with a 25 culverts sample, performing results analysis to the descriptors influence on the prioritization and intervention requirements. A set of charts are presented as a decision-making supporting tool, allowing different views of the culverts assessment in order of the partial risk indexes or descriptors.

To sum up, this framework enables culverts preliminary risk analysis considering three main risk elements: hazards, exposure and consequences. However, it is a qualitative base assessment and so dependant on the expert judgement and detail of field inspection. Also, the design data available plays an important role to the partial risk index evaluation. Furthermore, a quantitative risk analysis should be performed to complement culvert assessment, with special focus on the critical priority culverts ($\alpha_G > 50$).

References

1. Ryumin, M. G., & Shepitko, E. S. (2017). Forecast for terms of culvert inspection and repair. *Procedia Engineering, 189*(May), 598–604.
2. Najafi, M., & Bhattachar, D. V. (2011). Development of a culvert inventory and inspection framework for asset management of road structures. *J. King Saud University—Science, 23*(3), 243–254.
3. Meegoda, J. N., Juliano, T. M., Potts, L., Tang, C., & Marhaba, T. (2017). Implementation of a drainage information, analysis and management system. *Journal Traffic Transportation Engineering* (English Ed.)*, 4* (2), 165–177.
4. Misnevs, B., Melikyan, A., & Bazaras, D. (2015). Hazard assessment of weather factors for the occurrence of an emergency on the railway. *Procedia Computer Science, 77,* 40–47.
5. Sohler, F. A. S., & Caldeira, L. M. (2016). Safety of dams: A pathological approach of qualitative and quantitative risks. *Journal of Civil Engineering and Arquitecture, 10,* 1032–1051.
6. Casas, J. A., Hajdin, N. T. R., Kusar, M., Masovic, S., & Linneberg, P. (2018). TU1406—quality specifications for roadway bridges, standardization at a European level—WG3.
7. Masr, A., Ojrnsson, I., Ivanov, O. S. (2019). A qualitative prioritizations of the risks imposed on bridges due to climate changes. *IABSE Symposium "Towards a Resilient Built Environment Risk and Asset Management—Report*" (pp. 80–87).
8. Foucher, L., & Blés, T. (2019). Résilience des ouvrages de gestion des eaux au changement climatique - Méthodologie d'évaluation WATCH. *RGRA, 961,* 31–33.
9. Hunt, J. H., Zerges, S. M., et al. (2010). Culvert assessment and decision-making procedures manual for federal lands highway. FHWA-CFL/TD-10-005.
10. Statton, J., et al. (2010). Culvert risk assessment guideline. RTA, V3.02.
11. Lian, Y., & Yen, B. C. (2003). Comparison of risk calculation methods for a culvert. *Journal of Hydraulic Engineering Division of the American Society of Civil Engineers, 129*(2), 140–152.
12. Santiago, S. S. (2019). Brisa experience—Risk assessment and capital investment planning at Brisa. In *IABSE Symposium "Towards a Resilient Built Environment Risk and Asset Management—Report* (pp. 317–326).

13. Lajevardi, S., Matos, J., & Lourenço, P. B. (2019). Quality control index survey for railway bridge health monitoring. In: *IABSE Symposium Towards a Resilient Built Environment Risk and Asset Management—Report* (pp. 317–326).
14. Najafi, M., et al. (2008). An asset management approach for drainage infrastructures and culverts. CDFA 20.701.

Discussion of the Number of Risk Classes for Risk Based Maintenance

Dirk Proske and David Tschan

Abstract The importance and application of risk-based maintenance planning is growing in many areas, such as oil production or infrastructure management. For example, there exist already risk-based maintenance concepts for bridges and tunnels. However, these risk-based concepts require criteria for the decision of necessary actions. Such criteria must be risk classes in a risk-based maintenance concept. In this paper, the proposals for the number and limits of risk classes from different areas are compiled, discussed and recommendations for practice are given. However, experience has shown, that not only objective factors, but also subjective factors, such as the acceptance of risk classes by the inspectors, must be considered to achieve a successful application of risk-based maintenance.

Keywords Risk targets · Risk classes · Risk management

1 Introduction

For several decades, risk assessment has been gaining importance not only in engineering and the maintenance of infrastructure systems, but in practically all areas of social life. The success of risk assessment lies in its ability to incorporate influences and effects in decision-making which are not considered by other methods. However, and this must be pointed out again, risk assessment should not be the sole criterion for decision-making as shown by Arrow et al. [1].

To be a successful tool for decision making, not only must the methods for determining and assessing risks be fully developed and accepted, but also the limits of the risks proofs that lead to certain consequences must be clearly defined.

This paper deals with the definition of

1. the number of risk limits and risk classes,
2. the reference value of the limits,

D. Proske (✉) · D. Tschan
Department Architecture, Wood and Civil Engineering, Bern University of Applied Sciences, Burgdorf, Switzerland
e-mail: dirk.proske@bfh.ch

© The Author(s), under exclusive license to Springer Nature Switzerland AG 2021
J. C. Matos et al. (eds.), *18th International Probabilistic Workshop*, Lecture Notes in Civil Engineering 153, https://doi.org/10.1007/978-3-030-73616-3_20

3. the mathematical form of the limits, and
4. the dependence of the limits on the selected risk units and the object of protection.

The risk parameters are often defined in relation to the object of protection and safety, e.g. the risk of loss of life and limb, loss of animal life, environmental damage, economic loss, etc. Risk parameters for the loss of life and limb are mortalities, fatal accident rates, *F–N* diagrams, lost years of life, etc. In order to be able to compare the different objects of protection and safety, common risk parameters must be found. Usually monetary units are used for this purpose. Although the discussion in this paper is independent of the explicit definition of risk, this paper focuses on the mathematical-statistical definition of risk as product of event probability and frequency respectively and event damage. The paper is partly a meta-analysis which brings together recommendations from various references and it partly shows some alternatives which have not yet be fully investigated.

2 Investigation

2.1 Number of Limits and Risk Classes

The number of risk limits n is directly related to the number of risk classes ($n + 1$). A risk limit is a certain risk number, for example a mortality of 10^{-6} per year, which divides two risk classes and borders a risk class respectively. By risk classes we mean risk ranges that are associated with different actions. For example, risks above a certain value such as a certain mortality might not be acceptable, and all possible measures have to be taken. In other risk classes, no further measures need to be taken or further regulations from other areas apply. For example, the so-called de-minimis risk with 10^{-6} per year might be considered as upper limit of a risk class where no actions have to be taken. Well known is the consideration of two risk limits and three risk classes.

We have evaluated approx. 35 references from various fields of expertise regarding the proposed number of risk classes and present them in Fig. 1 as an absolute frequency distribution. In addition to engineering sciences, the data points also come from medicine, finance and disaster management. Table 1 provides further details on the individual data points and its origin. We have not included the full reference list due to page restrictions. Details can be provided by the authors. Most references are given in [2].

Figure 1 clearly shows that the vast majority of references and recommendations suggest three risk classes, although a relatively large number of references also recommend two risk classes. All other approaches are clearly behind these two approaches in terms of frequencies. The theoretical consideration of only one risk class for all risks would practically not correspond to a fixed limit value and thus no longer provide a basis for decision-making.

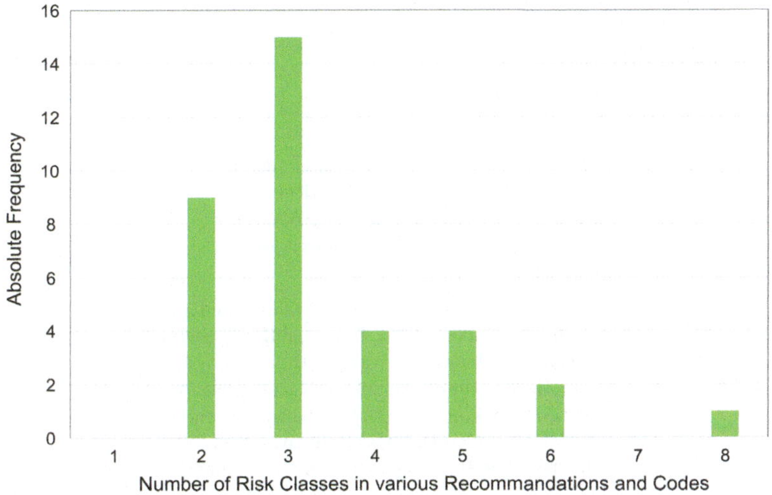

Fig. 1 Absolute frequency of risk classes in various references

The definition of two or three risk classes is linked to the ALARP (as low as reasonable possible, as low as practicable possible) region. The ALAPR region is a risk region and risk class respectively which does not explicitly state that further mitigation measures have to be carried out or that no further actions must be stated, it requires an assessment of mitigation measures.

In Faber et al. [3] it is stated that the ALARP principle is now best practice: "*the ALARP framework—the presently leading best practice format for life safety risk regulations.*" Other works [4–7] also show that the ALARP principle is state-of-the-art. These considerations and references indirectly confirm the application of two or three risk classes.

The application of a large number of risk classes, e.g. four to nine, requires the definition of further risk limit values. However, since the specification of limit values involves considerable discussion and development effort, the application of a large number of risk classes does not seem to bring any further advantage to decision-making.

Occasionally it is suggested not to define any limits and therefore introducing a limitless risk class. In this way, the ALARP principle would apply to all risks and would corresponds to an application of cost benefit analysis for all possible measures. However, legal problems and problems of practical feasibility arise here and prevent the application of this concept. One should keep in mind that a cost benefit analysis is not a substitution of a risk analysis. Even if the risks are represented in monetary units and not in loss of live and limb, they are given as absolute values whereas the cost benefit analysis only provides relative values.

Table 1 Risk classes in various references

Number of risk classes	References	Absolute frequency
0	None	
1	None	
2	Hongkong (1997), CEB-FIB Model-Code, VROM-Regel (Niederlande), CIRIA, Farmer Kurve (1967), Kinchin-Kurve (1982), Niederlande DG (1996), Western Australia, ASTRA	9
3	Eurocode Consequence Classes, Triage Medicine, Groningen-Curve (1978), Financial Risks, Swiss Störfallverordnung, Al-Wazeer (Brücken), Netherlands (1980), ACDS (1981), Oil platforms (1991), Hongkong (1993), Hongkong Chlorid (1997), HSE (2014), Swiss Railway, HSE (2001), Melchers (1990), Hongkong (Tai Lam Tunnel)	15
4	ASCE 7, Achs & Adam (Earthquakes Vienna), TÜV Süd (2014), Financial Risks	4
5	Disaster scale acc. to Gad-el-Hak, reliability classes GDR, financial risks, SBB conditions classes	4
6	Financial risks, Swiss railway intensity classes	2
7	None	0
8	Bradfords scale	1
9	None	0

2.2 *Reference Value of the Limits*

The risk thresholds and limits respectively can be absolute values, such as a maximum allowable mortality of 10^{-6} per year. In fact, most references known to the authors use such absolute values. However, there are also publications that specify relative values. As an example, Al-Wazeer is mentioned here [8]. There, the one hundred bridges with the highest risks are proposed as limit values. In some fields of specific risk parameters, however, relative values are widely used, such as the Lost Years of Life. Although there are target values for occupational health and safety, for example, most publications in this field use only relative references.

Some standards set a fixed limit value, such as the mortality rate of 10^{-5} per year, *and* relative values for the cost benefit analysis, such as Swiss code SIA 269/8 for existing structures under seismic loading or PLANAT, the Swiss recommendation for protection against natural hazards. Usually this approach can be applied to individual cases, not for are large number of objects as used in risk-based maintenance.

A special form of relative limits is the setting of fractile values based on the observed distribution of risk values. These fractile values can be regarded as relative or absolute values if they are not adjusted over many years. The limit is then

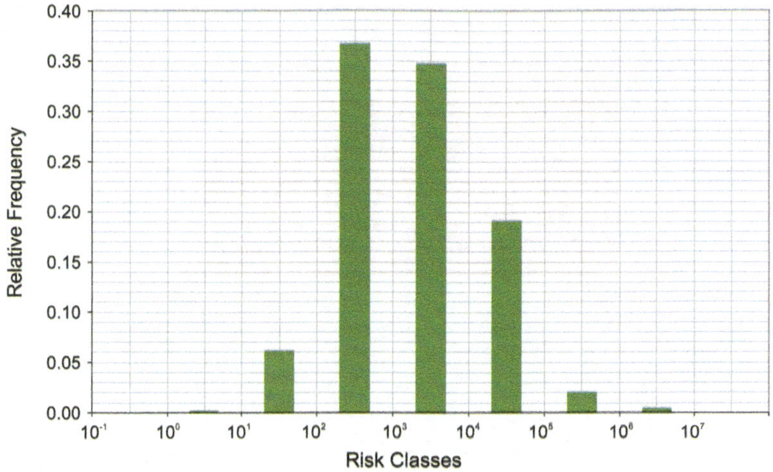

Fig. 2 Risk distribution of thousands of bridges of a large infrastructure operator in Switzerland

determined as follows:

$$R_{\text{Limit}} = m - k \times s \quad (1)$$

with

R_{Limit} as limit of the risk class
m as average or median value of all risk values
k as fractile value
s as standard deviation of all risk values.

Figure 2 shows the distribution of risk values for several thousand bridges of a large infrastructure operator in terms of monetary units. One can clearly see the distribution of the risk values. Risk limits can be defined here based on fractile values in the equation given above.

In fact, there are recommendations for the number of fatalities for different risks based on the above concept (Vrijling et al. [9]).

2.3 Mathematical form of the Limit

In the area of one-dimensional parameters, a number is sufficient as the mathematical form of the risk limit. Figure 3 shows the limits of different risk classes for different countries and different subject areas in terms of maximum annual mortality. Figure 4 visualizes the data from Fig. 3 as a frequency distribution of the logarithm of the mortality numbers. It becomes clear that most references suggest an annual mortality limit between 10^{-5} and 10^{-6}. However, significantly higher values can be observed

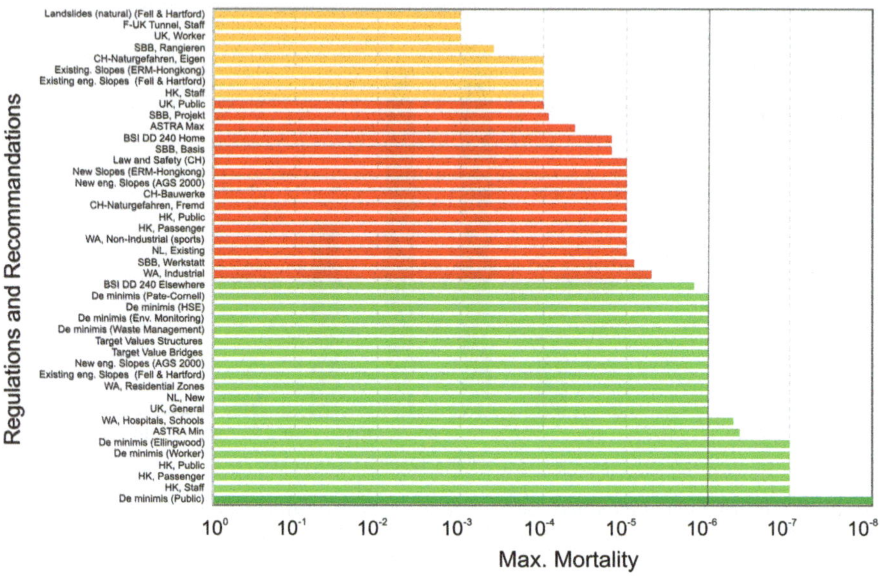

Fig. 3 Variety of risk limit values for mortalities [11–18]

for specific conditions, e.g. natural hazards, and in other countries. Also, higher risks limits are often not for the public but for specific professions.

In the area of two-dimensional risk parameters, the limit must correspond to a line. The line can be linear, multilinear or non-linear (quadratic function etc.). Figure 5 shows an *F-N* diagram with a variety of linear limit curves. The limit curves differ (a) in terms of the starting point, (b) in terms of the slope and (c) in terms of multilinearity. Also, some of the lines are based on risk concepts with only two risk classes whereas other lines belong to concepts with three risk classes. However, in general, the same diversity of risk limit values can be seen in Fig. 5 as in Fig. 3.

In addition, there are also suggestions of non-linear limit curves, see Fig. 6.

Three-dimensional risk parameters (e.g. frequency of occurrence, extent of damage, vulnerability) can also have limit surfaces. Examples are known.

2.4 Dependence on the Selected Risk Units and the Object to be Protected

The dependence of the risk limits, the risk classes and areas respectively on the selected object of protection can best be seen in the family of *F–N* curves [2, 10]. The units used there range from radiation doses and number of fatalities to monetary units (see Fig. 7). However, the transformation of a risk limit function from one unit to another might not be linear. Therefore, values and shapes of the risk limit function

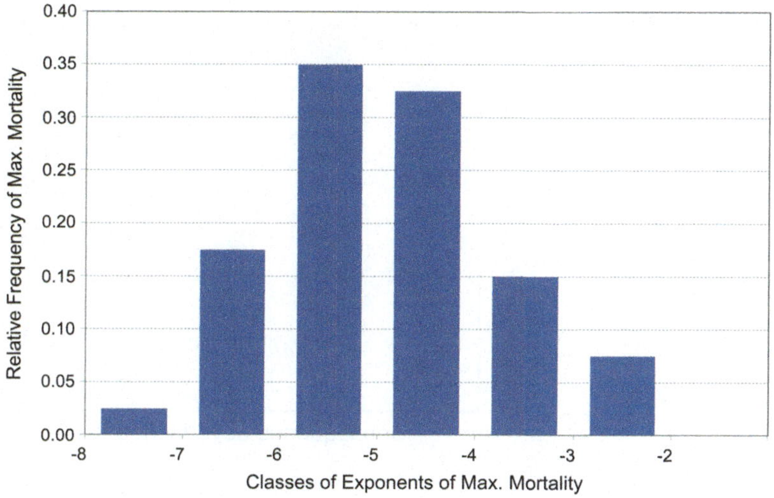

Fig. 4 Relative frequency distribution of the max. Mortality values according to Fig. 3

Fig. 5 Distribution of limit lines in *F*–*N* diagrams [2]

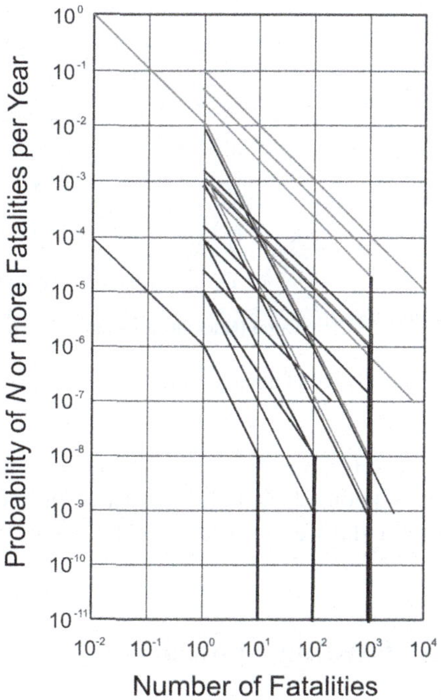

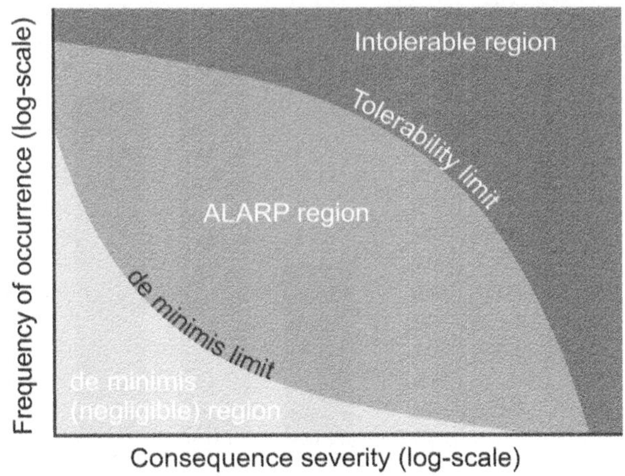

Fig. 6 Non-linear limit curves [18]

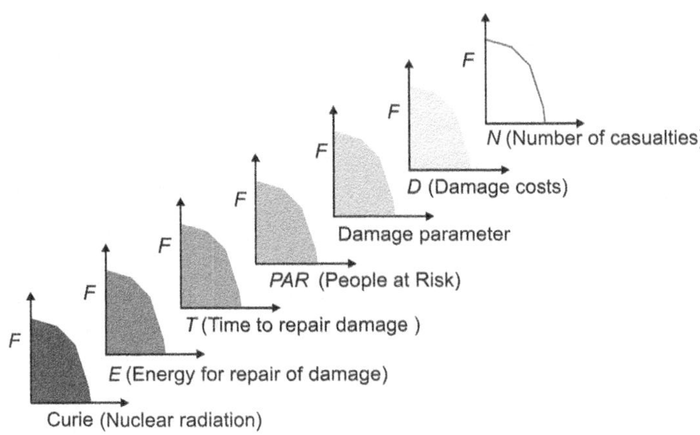

Fig. 7 Different units for limit curves in F–N diagrams [2]

may substantially change from one unit to another. A very ambiguous definition would consider a risk limit value in one unit and a second risk limit value in other unit and different risk aversion factors in the different units.

3 Existing Approaches

The determination of the amount, number, form, distribution value, and other parameters of the risk classes and their limits is not only a mathematical and engineering problem, but also involves issues of communication, jurisdiction and policy. From this point of view, the study presented here is one of a series of similar studies that examine the limits of risk classes based on former meta analyses, since the theoretical derivation of the limits of risk classes remains open.

We suggest the risk limits ca be determined via:

- Proof by falsification, i.e. risk limit values that have not proved to be effective in practice: example are the target core damage frequencies of first- and second-generation nuclear power plants. The disadvantage of this procedure is the great possibility of influence by politics, media and business, the fact that the decision is not taken into account in other areas of expertise and the dependence on single events.
- Proof by analogy: comparison of target values with other areas: As an example, see the Fig. 1 for the number of risk classes and the Figs. 3 and 5 for limit values. This is probably the most frequently used approach. However, it may incorporate large uncertainties, for example the counting of fatalities may differ between different areas of expertise.
- Proof by calculation: several methods and applications for the calculation of target values have been developed, such as the Life Quality Index (LQI) approach. The approach is not explained here in detail, see [5–7] for further information. In general, the application of life quality parameters has seen a strong and long-lasting growth in several areas such as social sciences, medicine and engineering. As advantage of this approach, parameters of interest from different areas can be considered in this approach and might provide a more realistic basis for decision makers. As disadvantage can be seen the large number of life quality parameters. In medicine more than 1000 life quality parameters have been introduced [2].
- Proof by observations of limit values in one field of expertise: Examples are road traffic fatality numbers since the 1970s or air traffic fatality numbers since the 1950s. This approach is strongly related to the first bullet point. Whereas their immanent actions were taken, here only improvements over time could be observed.

Since we cannot provide an absolute proof of the limits of the risk classes in the field of empirical science, and since much information is not available or cannot be obtained at a reasonable cost, all four methods must and are applied equally.

4 Discussion and Summary

Defining risk areas and risk limits within a subject area (e.g. in civil engineering for buildings, bridges, tunnels, earthworks, etc.), within scientific fields (e.g. engineering) and within an entire society remains an extremely challenging task.

According to the previous section, there are four different approaches. Probably the most frequently used procedure of bullet 2 has been used so far (see Sects. 2.1 and 2.3). In recent years, the use of procedure of bullet 3 has increased due to the development of modern procedures such as the LQI. Procedures of bullets 1 and 4 were the most important procedures since the beginning of the development of technical products by humans. In the context of the rapid development of technologies, however, these procedures will continue to be important, as reliable figures are not available for various technologies.

References

1. Arrow, K. J., Cropper, M. L., Eads, G. C., Hahn, R. W., Lave, L. B., Noll, R. G., Portney, P. R., Russell, M., Schmalensee, R., Smith, V. K., & Stavins, R. N. (1996). Is there a role for benefit-cost analysis in environmental, health and safety regulations. *Science, 272*(1996), 221–222.
2. Proske, D. (2008). *Catalogue of risks*. Springer.
3. Faber, M. H., Sorensen, J. D., & Vrouwenvelder, T. A. C. W. M. (2015). On the regulation of life safety risk. In *12th International Conference on Applications of Statistics and Probability in Civil Engineering, ICASP12* (pp. 1–9). Vancouver, Canada, July 12–15, 2015.
4. Maag, T. (2004). *Risikobasierte Beurteilung der Personensicherheit von Wohnbauten im Brandfall unter Verwendung von Bayes'schen Netzen* [Doctorial thesis]. IBK Bericht (Vol. 282). vdf Hochschulverlag AG an der ETH Zürich, Zürich.
5. Fischer, K. (2014). *Societal decision-making for optimal fire safety* [Doctorial thesis 21909]. Zürich.
6. Fischer, K., & Faber, M. H. (2012). The LQI acceptance criterion and human compensation costs for monetary optimization—a discussion note. In *LQI Symposium in Kgs*. Lnynby, Denmark, August 21–23, 2012.
7. De Sanctis, G. (2015). *Generic risk assessment for fire safety: Performance evaluation and optimisation of design provisions performance evaluation and optimisation of design provisions* [Doctorial thesis]. IBK Bericht (Vol. 363). Institut für Baustatik und Konstruktion der ETH Zürich, Zürich.
8. Al-Wazeer, A. A.-R. (2007). *Risk-based bridge maintenance strategies* [Dissertation]. University of Maryland.
9. Vrijling, J. K., van Gelder, P. H. A. J. M., Goossens, L. H. J., Voortman, H. G., & Pandey, M. D. (2001). A framework for risk criteria for critical infrastructures: Fundamentals and case studies in the Netherlands. In *Proceedings of the 5th Conference on Technology, Policy and Innovation, "Critical Infrastructures"*. Uitgeverrij Lemma BV, Delft, The Netherlands, June 26–29, 2001.
10. Baecher, G. B., Abedinisohi, F., & Patev, R. C. (2015). *Societal risk criteria for loss of life, concepts, history, and mathematics*. FN-report draft. University of Maryland, March 7, 2015.
11. Beard, A., & Cope, D. (2007). *Assessment of the safety of tunnels*. IP/A/STOA/SC2005-28/SC29. European Parliament, October 2007.
12. Dukim, N. J. (2009). *Acceptance criteria in Denmark and the EU*. Danmarks Tekniske Universitet, Institut for Planlaegning.
13. Düsgün, H. S. B., & Lacasse, S. (2005). Vulnerability and acceptable risk in integrated risk assessment framework. In *Proceedings of the International Conference on Landslide Risk Management* (pp. 505–515). Vancouver, BC, Canada. Taylor and Francis, May 31–June 3, 2005.
14. Fell, R., Ho, K. K. S., Lacasse, S., & Leroi, E. (2005). State of the art paper 1: A framework for landslide risk assessment and management. In *Proceedings of the International Conference on Landslide Risk Management* (pp. 3–25). Vancouver, BC, Canada. Taylor and Francis, May 31–June 3, 2005.
15. Skjong, R. (2002). *Risk acceptance criteria: Current proposals and IMO position, surface transport technologies for sustainable development*. Valencia, Spain, June 4–6, 2002.

16. Trbojevic, V. M. (2005). *Risk criteria in EU, ESREL 2005* (p. 6), June 27–30, 2005. Poland.
17. Trbojevic, V. M. (2009). Another look at risk and structural reliability criteria. *Structural Safety, 31,* 245–250.
18. van Coile, R., Gernay, T., Hopkin, D., & Khorasani, N. E. (2019). Resilience targets for structural fire design—An exploratory study. In *International Probabilistic Workshop 2019* (pp. 196–201). September 11–13, 2019. Edinburgh, UK.

Dynamic Response Equivalence of a Scaled Bridge Model Due to Vehicular Movement

Paul Cahill and Vikram Pakrashi

Abstract The design of scaled testing is important for establishing equivalence with a full-scale structure but if difficult since the geometry and the material both need to be scaled. For a good, scaled testing, it is important to demonstrate the results of the scaled original structure and the designed scaled testing behave similarly, so that there is control over experimentation. Despite existing guidance around this topic, such equivalence is sometimes not checked appropriately, leading to uncertainties and variations in scaled testing which significantly compromises the usefulness of such experiments. This paper addresses this topic for a bridge-vehicle interaction problem and demonstrates how a scaled testing can show equivalence with respect to its full-scale counterpart. A Buckingham-Pi approach has been taken for scaling and the assumptions around the models and the responses are defined to establish the boundaries of the responses that are intended to be replicated. The non-dimensional parameters are defined and guide the design of future experiments. The conversion of a complex cross-sectional profile to an equivalent beam with made of a different material is dictated by the matching of modelled responses of the scaled responses of the original structure versus the unscaled responses of the experimental structure. The match indicates that establishment of such equivalence is particularly relevant for carrying out future experiments within the laboratory and subsequently linking it to full-scale structures for implementing sensors or carrying our intervention aspects such as repairs. The work also emphasizes on how a well-designed scaled testing should have a numerical benchmark for future interpretation and understanding assumptions around such interpretations when comparing full-scale experiments with controlled laboratory-based experiments, reducing uncertainty around such comparisons. The presented work is expected to be of interest for both researchers and practicing engineers.

Keywords Scaled testing · Bridge-vehicle interaction · Dynamic response · Experimental design

---

P. Cahill · V. Pakrashi (✉)
Dynamical Systems and Risk Laboratory, School of Mechanical and Materials Engineering, University College Dublin, Dublin, Ireland
e-mail: vikram.pakrashi@ucd.ie

© The Author(s), under exclusive license to Springer Nature Switzerland AG 2021
J. C. Matos et al. (eds.), *18th International Probabilistic Workshop*, Lecture Notes in Civil Engineering 153, https://doi.org/10.1007/978-3-030-73616-3_21

1 Introduction

Experimental validation of sensor placement strategies, repair or rehabilitation, or even the assessment for bridges can become complex since it is often difficult, complex and impractical to test the full structure due to inaccessibility, cost and other constraints, including health and safety aspects [1, 2]. Under such circumstances, well defined and designed laboratory experiments can be a surrogate to such tests but the controlled laboratory tests must be linked to the full scale structure in a consistent manner so that the interpretation of the results, the assumptions around the match between the scaled and the full-scale tests, along with their limitations are well understood.

To achieve this, it is important to create numerical benchmarks comparing the responses of the full-scale bridges against the scaled laboratory experiment specimens. Such a benchmark can reduce uncertainty in future experiments and also creates bounds of interpretation. Well designed scaled models can thus be cheap, safe and an effective way to validate the dynamic response of large bridges and have the potential to experimentally determine the applications for integrated sensing [3] or carrying out repairs [4], rehabilitations or other interventions like installation of control systems in future. While full scale experimental results for bridges are often available [5, 6], there is still a paucity in literature in terms of addressing scaling aspects of such bridges to reduce uncertainties around the system. This makes the development of technology and its experimental verification for bridges complex in terms of taking decisions from scaled experimental results in a laboratory.

Under such circumstances, it is important to investigate the abilities of scaled experimental models to replicate the dynamics of a full-scale structure to a reasonably accurate level, depending on the needs and through the use of appropriate scaling methods [7]. Scaled models are effectively used for a wide range of civil applications, including bridges [8, 9], wind turbines [10] and offshore floating platforms [11]. Since scaled experimental models can accurately reflect the response of the structure to external loading, they have the potential to experimentally validate newly developed methods [12].

This paper investigates the use of the scaled model of a bridge and demonstrates how the equivalence in dynamics through a numerical benchmark can guide future experimental design and studies. The quantitative demonstration of the dynamic similitude between the model and the full-scale structure also shows how in future changes to the original bridge can be benchmarked against the model created. The work also informs how such numerical models can reduce uncertainty and problems with interpretation when a scaled experimental model is used in the laboratory to try explaining and understand full-scale bridges.

2 Design of a Scaled Model Bridge

2.1 Scaling Approach

The Buckingham-Pi approach is used for developing the scaled mode, using non-dimensional terms describing relationship between certain physical parameters [13]. Using such terms allows for scaled physical models to replicate a host structure, both statically and dynamically. The use of pi-terms for the creation of a scaled model for a bridge structure has been investigated before [9] and a similar approach is adopted here. Five pi-terms containing the physical parameters are required and are given as

$$\pi_1 = \frac{\delta}{L}, \pi_2 = \frac{\omega^2 L}{g}, \pi_3 = \frac{PL^2}{EI}, \pi_4 = \frac{v^2}{Lg}, \pi_5 = \frac{\rho g L^3}{EI} \tag{1}$$

where L is the length, g is the acceleration due to gravity and the deflection and natural frequency are given as δ and ω respectively, P is the load applied from the vehicle travelling with a constant speed v, while E, I and ρ are the Young's modulus, the second moment of area about the neutral axis and the mass density respectively. These pi-terms can subsequently be used for the creation of scaling factors, from which the relationship between the host structure and the scaled model can be determined [7] and are given as:

$$S_\delta = S_L, S_P = \frac{S_{EI}}{S^2 L}, S_v = \sqrt{S_L}, S_P = \frac{S_{EI}}{S^3 L}, S_\omega = \frac{1}{\sqrt{S_L}} \tag{2}$$

2.2 Properties of the Full-Scale Structure

For this investigation, a host bridge structure was designed (Fig. 1), with the same length and cross-sectional profile as [7] but with differing flexural rigidity and mass

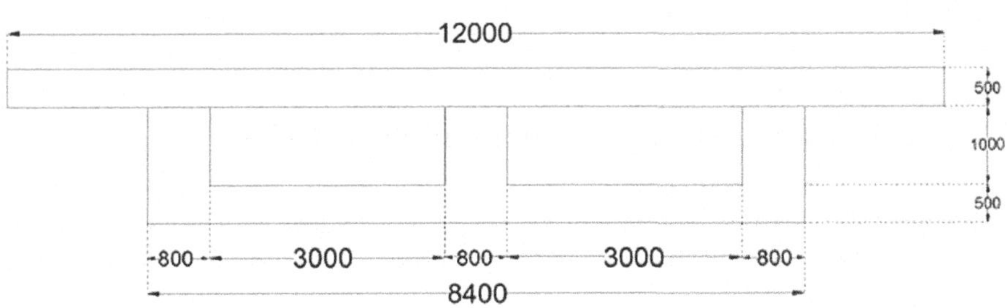

Fig. 1 Cross-sectional geometry of the full-scale structure considered (dimensions in mm)

Table 1 Structural information of the full-scale bridge

Property	Value	Unit
Cross-sectional area	12.6	m^2
Young's modulus	31	GPa
Second moment of area	6.01	m^4
Length	60	m
Width	12	m
Natural frequency	1.081	Hz
Mass density	30,240	kg/m

Table 2 Structural information of the scaled bridge model

Property	Value	Unit
Cross-sectional area	0.0003	m^2
Young's modulus	69	GPa
Second moment of area	6.25×10^{-10}	m^4
Length	1.23	m
Width	0.06	m
Natural frequency	7.58	Hz
Mass density	0.81	kg/m

density. The properties of the host structure as used in this investigation are outlined in Table 1, where the subscript h denotes the full-scale structure.

2.3 Development of a Scaled Model

An equivalent scaled model was designed using the scaling factors presented in this paper. The material used for the model was aluminium and the cross-section of the designed scaled model was rectangular. Considering commercially available materials, the parameters of the model satisfied all five pi-terms. These model parameters are outlined in Table 2.

3 Comparison of Host Structure and Scaled Model

To compare the dynamic response of the full-scale structure with that of the scaled model, two finite element models were created (Fig. 2) with properties as outlined in the previous section.

A moving car, consisting of two axles (as point load) located at a distance of 10% of the length of the model and a width of 0.167, was considered. This results in the

Fig. 2 Finite Element (FE) models of **a** the full-scale bridge and **b** its scaled counterpart for laboratory studies. A dynamical equivalence of the output from the two structures is sought

length and width of the car being 6 m and 3 m for the host structure, and 123 mm and 6 mm for the scaled model respectively.

A constant point load of 200 kN was applied for both axles to the host structure with a scaled load of 0.11 N being applied to the model. The vehicle speed for the host structure was applied at 100 km/h which had a corresponding speed of 14.31 km/h for the scaled model. Using a linear transient solver, the dynamic response of the host structure and scaled model at the midpoint was determined.

Using the calculated scaling factor of 48.7805 for the displacement and 9.08 for the strain, the dynamic response of the scaled model was compared against that of the host (Fig. 3). It was found that for both displacement and strain, the scaled model was accurate in capturing the dynamic response of the host structure. As the strain response of both the host and scaled model match, the use of accurately scaled models for verification of sensor deployment applications is a viable option [14, 15].

4 Conclusions

This paper presented the design for dynamic equivalence between a full-scale bridge and a scaled counterpart of the same for laboratory testing. A Buckingham-Pi approach was taken for establishing this equivalence and both displacement and strain equivalence in terms of dynamic responses was demonstrated for the models through a Finite Element (FE) analysis for a travelling vehicle. The work demonstrates how such designs can reduce uncertainty in scaled experiments and can also be useful in deciding future sensor deployment, repair or other rehabilitation strategies for the structure. The work can also be useful for guiding the development of new sensors and methods of structural health monitoring. The work can be used as a guidance for carrying out inexpensive but insightful decisions around monitoring of bridges.

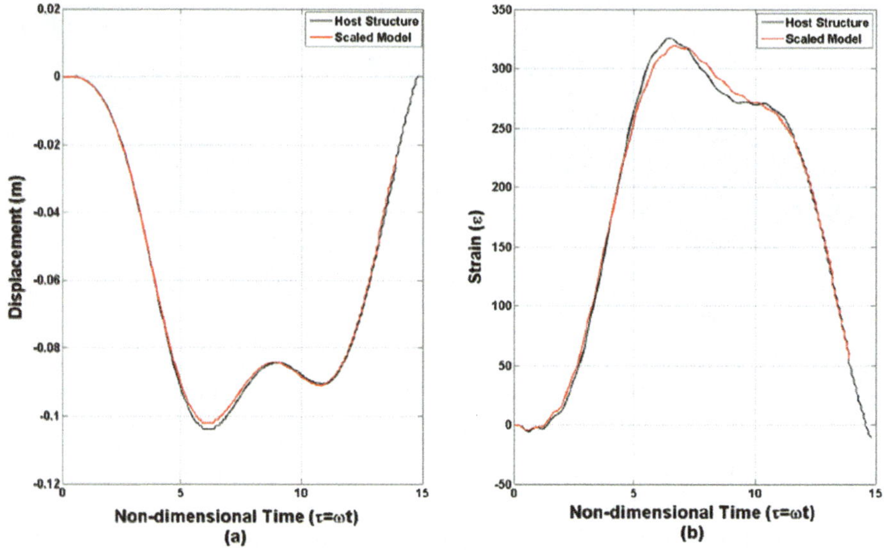

Fig. 3 Dynamical response of the host structure and the scaled model to a moving load for **a** displacement response and **b** strain response

Acknowledgements The authors acknowledge the EU-funded SIRMA (Strengthening Infrastructure Risk Management in the Atlantic Area) project (Grant No. EAPA_826/2018). Vikram Pakrashi would also like to acknowledge the support of SFI MaREI centre under Grant number RC2302_2. For the purpose of Open Access, the author has applied a CC BY public copyright licence to any Author Accepted Manuscript version arising from this submission.

References

1. Chang, P. C., Flatau, A., & Liu, S. C. (2003). Review paper: Health monitoring of civil infrastructure. *Structural Health Monitoring, 2*(3), 257–267.
2. Brownjohn, J. M. W. (2007). Structural health monitoring of civil infrastructure. *Philosophical Transactions Series A, Mathematical, Physical, and Engineering Sciences, 365*(1851), 589–622.
3. Cahill, P., Hazra, B., Karoumi, R., Mathewson, A., & Pakrashi, V. (2018). Vibration energy harvesting based monitoring of an operational bridge undergoing forced vibration and train passage. *Mechanical Systems and Signal Processing, 106,* 265–283.
4. Fitzgerald, F., Malekjafarian, A., Bhowmik, B., Prendergast, L., Cahill, P., Kim, C. W., Hazra, B., Pakrashi, V., & OBrien, E. (2019). Scour damage detection and structural health monitoring of a laboratory-scaled bridge using a vibration energy harvesting device. *Sensors, 19*(11), 2572.
5. Frýba, L., & Pirner, M. (2001). Load tests and modal analysis of bridges. *Engineering Structures, 23*(1), 102–109.
6. Karoumi, R., Wiberg, J., & Liljencrantz, A. (2005). Monitoring traffic loads and dynamic effects using an instrumented railway bridge. *Engineering Structures, 27*(12), 1813–1819.
7. Harris, H. G., & Sabnis, G. M. (1999). *Structural modeling and experimental techniques* (2nd ed.). Boca Raton: CRC Press LLC.

8. Law, S., Chan, T., & Zeng, Q. (1997). Moving force identification: A time domain method. *Journal of Sound and Vibration, 201*(1), 1–22.
9. Bilello, C., Bergman, L. a. & Kuchma, D. (2004). Experimental investigation of a small-scale bridge model under a moving mass. *Journal of Structural Engineering, 130*(5), 799–804.
10. Muir Wood, D., Lombardi, D., & Bhattacharya, S. (2011). Similitude relationships for physical modelling of monopile-supported offshore wind turbines. *International Journal of Physical Modelling in Geotechnics, 11*(2), 58–68.
11. Jaksic, V., O'Shea, R., Cahill, P., Murphy, J., Mandic, D. P., & Pakrashi, V. (2015). Dynamic response signatures of a scaled model platform for floating wind turbines in an ocean wave basin. *Philosophical Transactions of the Royal Society A: Mathematical, Physical and Engineering Sciences, 373*(2035), 20140078:1–18
12. Pakrashi, V., O' Connor, A., & Basu, B. (2010). A bridge—Vehicle interaction based experimental investigation of damage evolution. *Structural Health Monitoring, 9*(4), 285–296.
13. Buckingham, E. (1914). On physically similar systems; Illustrations of the use of dimensional equations. *Physical Review, 4*(4), 345–376.
14. Bhowmik, B., Tripura, T., Hazra, B., & Pakrashi, V. (2019). First order eigen perturbation techniques for real time damage detection of vibrating systems: Theory and applications. *Applied Mechanics Reviews, 71*(6), 060801.
15. Mucchielli, P., Bhowmik, B., Hazra, B., & Pakrashi, V. (2020). Higher-order stabilised perturbation for recursive eigen-decomposition estimation. *ASME Journal of Vibrations and Acoustics, 142*(6), 061010.

Energy Based Model of Vehicle Impacted Reinforced Bridge Piers Accounting for Concrete Contribution to Resilience

Suman Roy and Andrew Sorensen

Abstract The reliability of bridge piers under dynamic loading play an important role in the resilience of bridge structures under high strain rate loading. Vehicle impact is one example of a high strain rate loading condition that has received limited attention compared to other hazard loading types. However, recent studies have shown that vehicular impacted bridge piers have significantly lower capacities and as such as susceptible to secondary hazard loads under multi-hazard loading conditions. Research has shown that dynamic loading criteria require more robust designs; however, most design codes utilized in the United States as well as other international standards do not consider multi-hazard, sequential loading conditions. As such models to determine the resilience of bridge piers to these scenarios are required so they can be accounted for in future code calibrations. In this study, dynamic effect due to vehicular impact has been studied by introducing energy analyses, dissipation and transfer. As such, pier reliability is determined in terms of energy demand. In most vehicle impact models, the concrete is not taken into account as is considered to provide negligible contribution to resist impact. In this study the contribution of the concrete is analyzed using a spring action model. The contribution is evaluated for dynamic impact characterized by energy dissipation, kinetic energy transmitted from vehicle during impact, vehicle material recoiling, and recoiling action from the pier. Monte Carlo simulations are carried out with and without concrete contribution. Results from both the cases are inculcated to predict damage level, failure, and reliability. Concrete's contribution as a part of dynamic behavior instills a good understanding to analyze reinforced concrete (RC) bridge pier in terms of energy concept, and hence will advance design tools for the structural engineers and practitioners enhancing 'Code Calibration' in an optimized way.

Keywords RC pier on impact · Energy dissipation · And limit state equation for reliability

S. Roy (✉) · A. Sorensen
Department of Civil and Environmental Engineering, Utah State University, Logan, USA

© The Author(s), under exclusive license to Springer Nature Switzerland AG 2021

J. C. Matos et al. (eds.), *18th International Probabilistic Workshop*, Lecture Notes in Civil Engineering 153, https://doi.org/10.1007/978-3-030-73616-3_22

1 Introduction

In this research, evaluation for dynamic performance of reinforced concrete (RC) bridge piers has been attempted for quasi-static to plastic state. Bridge piers experience dynamic loading in the form of seismic response, uncontrolled blasting due to accidental as well as purposeful terrorist attack, and vehicular collision.

While the performance of RC bridge piers subject to seismic and blast loading has received significant attention, the response of these piers to vehicular impact has received less attention even though they occur at a much more frequent level [1]. Increased crashworthiness as a result of vehicle collision with the exposed part of the RC bridge piers has led to catastrophic failure followed by collapsing the entire structure. Investigation incorporating the serviceability and age of the structures suggests that a distinguished number of RC bridges are susceptible to failure subject to extreme, as well as sequential multi-hazard loading events. Existing design codes allow for individual hazardous load applications for flexure, axial, shear, and torsional, but the provisions for sequential and simultaneous loading to predict multiple hazards are not included [2, 3]. Unfortunately, little attention with inadequate number of studies has been given on predicting frequently occurring multi-hazard scenario caused by vehicle impacted dynamic load. However, sequential loading for blast and vehicular impact have already been investigated in terms of performance [3, 4].

Satisfactory performance in RC pier bridge require the ductile behavior of the structural member during short duration dynamic response [5]. Dynamic analysis methods to characterize its strength, deformability, and energy method in the flexure-dominated approach is discussed in [6]. The previous research was carried out in a conservative way without considering the contribution of the concrete.

In order to make a holistic approach, a method depicting all possible energy parameters during short duration impact is developed. Aspect ratio and its function during impact for energy dissipation is assessed. Limit state equations incorporating all possible energy parameters to determine the failure possibilities with corresponding reliabilities while experiencing impacts on a RC pier having higher concrete compressive strengths are developed [7]. Vehicle specifications are taken from [8]. Time of impact is considered as non-deterministic to run the Monte Carlo simulations predicting reliability for 11,000 data for each case.

The objective of this study is to examine the effects of vehicular impact, and the structural reliability of circular RC bridge piers in terms of limit state equations governed by energy aspects and dissipation phenomenon.

2 Methodology and Results

A representative RC bridge pier is used in this study to demonstrate the methodology. The pier's unsupported length (L) is 8 feet 6 in. and both ends are restrained from support yielding, rotation and distortion. The cross-section of the pier is circular having an external diameter (h) of 21 in. (53.34 cm). The pier consists of six number #8, ASTM A706 grade 60 steel bars as the main (longitudinal) reinforcement through its foundation bottom. Shear reinforcement is provided with number 4 bar with a pitch of 2.5-in. (63.5 mm) center to center. Using these dimensions, the aspect ratio (L/h) of the pier is calculated at 4.85. The gross cross-sectional area of the pier (A_g) is 346.5 in^2 (2235.48 cm^2) and total area of cross-section of steel (A_s) is 4.74 in^2 (30.58 cm^2). The steel ratio is calculated as ($\rho_l = A_s/A_g$) 0.01368. The nominal axial load $P_{N,Design}$ of the pier is computed to be 687.50 kips (3058.15 kN). Using these dimensions, a numerical model is developed by modifying developed equations from the literatures and journals in order to evaluate the representative pier.

2.1 *Prediction of Plastic Hinge Length (L_p) of Dynamic Impacted Pier*

For reinforced concrete (RC) flexural members, the plastic (permanent) deformation is typically localized in a smaller zone after the yielding of member. The performance of the plastic hinge zone is critical for flexural members as it governs the load carrying capacities of the member [9]. The maximum predicted plastic hinge length (L_p) has been previously defined by [10] and is shown in Eq. 1.

$$L_p = 0.08 * L + 0.15 * f_y * d_b \tag{1}$$

In Eq. 1, L_p and L are the lengths of plastic hinge and unsupported length of pier respectively, d_b is the bar diameter of longitudinal steel reinforcement, and f_y is the yield strength of main reinforcing steel bar taken to be 60 ksi (413.685 MPa) for this study.

The plastic hinge length ratio (L_p/L) can be defined as ψ and using this definition and rearranging Eq. 1 yields Eqs. 2 and 3.

$$\psi = 0.08 + 9.0 * (d_b/h)/(L/h) \tag{2}$$

$$\psi = 0.08 + 9.0 * (\gamma/\eta) \tag{3}$$

In Eq. 3, γ and η are used to replace the ratios d_b/h and L/h (aspect ratio) respectively.

2.2 Rate of Plastic Hinge Formation Considering Strain Rate ($\acute{\varepsilon}$) of Steel

Plastic hinge formation in this research has been illustrated by energy dissipation and the composite behavior of concrete and steel. Concrete is a non-deterministic brittle material composed of aggregate and cementitious matrix. On the other hand, reinforcing steel bar is an isotropic material exhibiting plastic behavior. The overall energy dissipation capacity of the RC member can be evaluated by using energy dissipated by steel rebar while concrete contribution is not typically considered due to its relatively small ductility [11]. Equation 3 can be rewritten in terms of strain (ε) and modulus of elasticity (E_{st}) in steel rebar, yielding Eq. 4.

$$\psi = 0.08 + 0.15 * (E_{st} * \varepsilon) * (d_b/L) \tag{4}$$

In Eq. 4, d_b (diameter of main steel bar) is considered as 1 in (25.4 mm) (#8 steel bar) and L (unsupported length of pier) is 8.5 feet (2.591 m). The modulus of elasticity of steel is taken as 29,000 ksi (2×10^5 MPa).

By taking the partial derivative of Eq. 4 with respect to time and by inserting all values Eq. 5 can be developed.

$$\frac{d\psi}{dt} = 4.35 * 10^3 * (\acute{\epsilon}) \tag{5}$$

In Eq. 9, steel strain rate ($\partial\varepsilon/\partial t$) has been represented as $\acute{\varepsilon}$. This strain rate can therefore be evaluated as a function of the rate of formation of the plastic hinge length ratio ψ. This is represented in Fig. 1. Figure 2 shows the comparative variations of steel strain rate with regards to plastic hinge length (L_p) and plastic hinge length ratio (ψ).

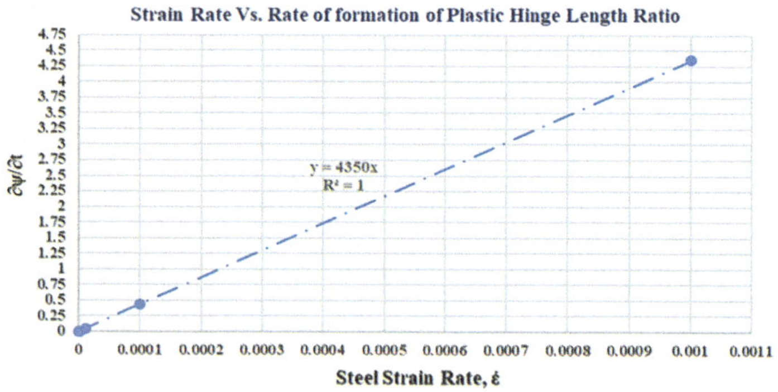

Fig. 1 Steel strain rates with rate of formation of plastic hinge length ratio

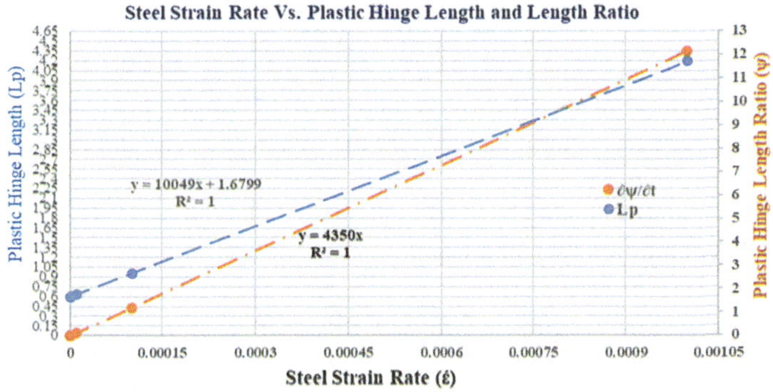

Fig. 2 Steel strain rates with plastic hinge of plastic length ratio and plastic hinge length

2.3 Evaluation of Dissipation of Energy (E_D)

Reinforced concrete (RC) members dissipate energy by experiencing inelastic behavior due to cyclic loading caused by dynamic response. In RC structures, the energy dissipation can be defined by the sum of the energy dissipated by both the concrete and steel. Concrete is a brittle material and hence less ductile. On the other hand, reinforcing steel being an isotropic, dissipates high energy exhibiting plastic behavior compared to concrete dissipating considerable less energy [12]. For that reason, the overall energy dissipation is considered by flexural rebar.

The evaluation of dissipated energy for circular cross-section (e_D) RC pier for static behavior is given in Eq. 6 [10].

$$e_D = 4R_B * \rho * f_y\left(\frac{\pi h^3}{4}\right)_u * [(1-p)\left(\frac{h_s}{2h} - \frac{\varepsilon_y}{d*h}\right) + p(0.5 - \frac{\varepsilon_y}{d*h})^2 \quad (6)$$

In Eq. 6, the dissipated energy (e_D) is considered for steel rebar, Φ_u is the maximum curvature, h is the diameter of the pier, h_s is the distance between the rebar layers located between the boundaries, ε_y is the yield strain of steel rebar (taken as 0.00207), f_y is the yield stress of steel rebar (60 ksi), and p is ratio of lateral steel ratio (ρ_t) over longitudinal steel ratio (ρ_l), which can be calculated as $p = \rho_t/\rho_l = 0.0178/0.01368 = 1.28$. The reduction factor ($R_B$) representing the Bauschinger effect, is considered as 0.75 [13].

The maximum ultimate curvature, Φ_u can be computed from Eq. 7 [13].

$$\Phi_u = 2.45 * (\varepsilon_r/h) \quad (7)$$

The main steel yield strain, ε_r is considered to be 0.06 [13], as such, Eq. 7 yields the Φ_u as 0.007, which is function of pier dimension.

The yield curvature of steel rebar is computed from Eq. 8 [14].

$$\Phi_y = \alpha_{ST} * \left(\frac{\varepsilon_y}{h}\right) \tag{8}$$

In Eq. 8, α_{ST} (modification factor) is taken as 2.12 for pier [14], and ε_y, the yield strain of steel (f_y/E_{st}) is computed as 0.00207, where E_s is modulus of elasticity of steel [13].

Energy dissipation capacity based on kinematic behavior (e_{kh}) is a function of Φ_u and Φ_y, and is computed by using Eq. 9.

$$e_{kh} = 4 * M_r * \left(\Phi_u - \Phi_y\right) \tag{9}$$

In Eq. 9, M_r is the moment carrying capacity of the circular pier cross-section (i.e. moment of resistance), and can be computed as shown in [15].

The total energy to be dissipated at plastic hinge is computed from Eq. 10.

$$E_D = e_D \tag{10}$$

For the representative pier, the moment carrying capacity (M_r) of the circular cross-section RC pier is computed as 6275.30 kip-in (709.013 kN-m) from Eq. 10. Using Eqs. 3, 10 can be further rewritten as Eq. 11.

$$E_D = e_D * (L_p/L) * L = e_D * L * \psi \tag{11}$$

From Eq. 11, E_D can be expressed as shown in Eq. 12, where $\psi = L_p/L$.

$$E_D = (e_D * L) * \left[0.008 + 0.15 * f_y * \left\{\frac{d_b}{\left(\frac{L}{h}\right)} * h\right\}\right] \tag{12}$$

Using Eq. 12 and by simplification, yields Eq. 13 which is a function of aspect ratio (η) and pier dimension ratio (γ). The total energy dissipation (E_D) during formation of plastic hinge caused by impact is illustrated in in terms of γ/η ratio in Fig. 3. Figure 3 shows a linear increment of E_D with the increment of γ/η, whereas Fig. 4 comprises hyperbolic decrement of the aspect ratio (η) and a linear increment of γ/η and E_D. Also Fig. 4 includes a comparative estimate of E_D with η comprising γ/η. As γ/η ratio is a governing function including pier geometry, controls the energy dissipation as a post impact effect.

$$E_D = 0.08 + 86.94 * (\gamma/\eta) \tag{13}$$

In Eq. 13, $\gamma = d_b/h = 0.048$ and aspect ratio, η = L/h which has already been computed as 4.86.

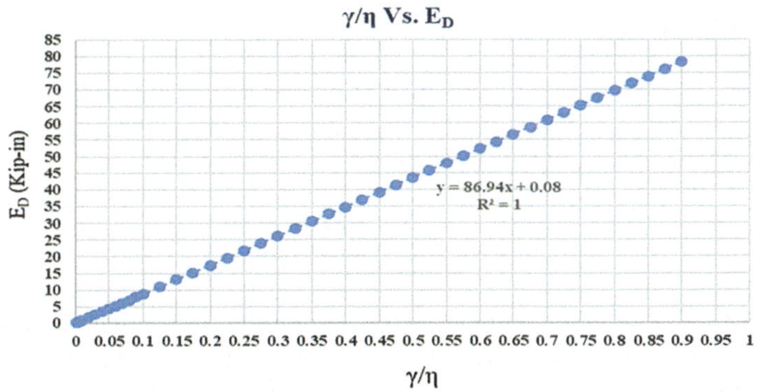

Fig. 3 Total dissipated energy with re-bar diameter, pier diameter and aspect ratio

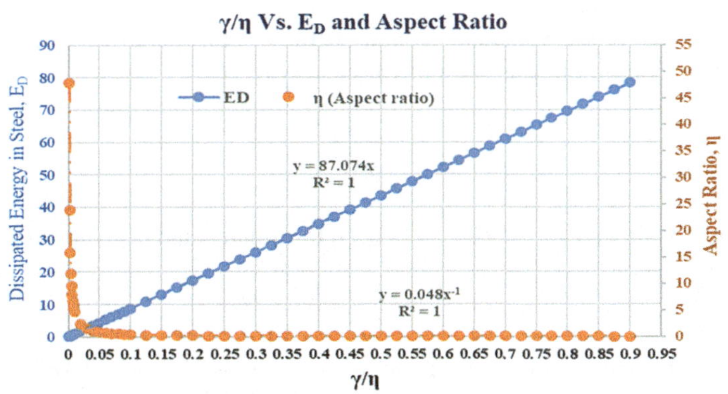

Fig. 4 Comparison of dissipated energy and aspect ratio

2.4 Estimation of Energy During Vehicle Impact

In this research, displacement based approach has been adopted considering flexural response of high-velocity impact of vehicle. Shear criterion considers that the pier might suffer brittle damage, localized spalling followed by shear failure, and deflection [16]. Vehicle data has been taken from published journals [8]. When a heavy mass vehicle moving with high velocity impacts a RC bridge pier, the pier may undergo localized shear deformations at any point of time [17, 18]. In this study, kinetic energy of the moving vehicle before impact, crushing energy causing recoiling of the frontal vehicle material, potential energy of the RC pier, spring action of the concrete pier, and dissipation of energy due to impact are considered.

2.4.1 Crush Energy

Crush energy of vehicle can be described as the amount of kinetic energy required to recoil the frontal part of the vehicle. This value depends on the frontal stiffness of the vehicle. Crush energy (U_{veh}) has been given in Eq. 14 [16].

$$U_{veh} = \frac{1}{2}(F_{max} * u_0) \tag{14}$$

In Eq. 14, F_{max} is the maximum forced inserted by the vehicle on pier during impact, and u_o is the recoiling velocity of the vehicle after impact and can be written as shown in Eq. 15.

$$u_0 = F_{max}/k \tag{15}$$

In Eq. 15, 'k' is the vehicular frontal stiffness, considered as 1.0278 kip/feet (1500 kN/m). Combining Eqs. 14 and 15, yields Eq. 16.

$$U_{veh} = \frac{1}{2}\left(\frac{F_{max}^2}{k}\right) \tag{16}$$

Maximum vehicular force (F_{max}) has been formulated and is given in Eq. 17 [16].

$$F_{max} = M_{veh} * V_{max}/t \tag{17}$$

In Eq. 17, M_{veh} is the mass of the moving vehicle (42.11 kips or 187.315 kN for semi-trailer as maximum load and 26.015 kips or 115.72 kN for most occurrence vehicle load), V_{max} is the maximum velocity of the moving vehicle before impact (44.62 ft./s or 13.6 m/s for maximum and 31.8 ft./s or 9.7 m/s for most occurrence case), and t (time for impact) is the impact duration considered as 10^{-3}–10^{-4} s which governs the strain rates of the steel rebar. In this study, maximum case has been considered for maximum mass and maximum velocity, whereas most frequently occurring case has been taken as most occurring mass with velocity [19]. Equation 17 yields maximum inserted forces 1878.95/t kips and 827.3/t kips for the extreme and most occurring cases, respectively.

2.4.2 Potential Energy of the Pier

Potential energy of the pier (U_{col}) is the energy stored by the pier itself due to the contribution of steel and concrete. In this research, each case (i.e. most vulnerable and frequently occurring) are fragmented into two individual cases. First, the concrete contribution is considered as spring action to recoil and secondly, in a more conservative way without considering the concrete spring action (recoiling action). The potential energy of the RC pier is given in Eq. 18 [16].

$$U_{col} = \int_{0}^{\delta max} F(\delta).d\delta \tag{18}$$

In the Eq. 18, 'F' is represented as the resisting force inserted by the pier and δ = (1.34 in + 0.05 in)/2 = 0.695 in (17.653 mm), which is the average displacement of the pier.

2.4.3 Spring Action of the Pier

To determine the energy responsible for providing the adequate stiffness (E_s) of the RC pier, the shear stiffness has been computed [20]. The pier stiffness has been considered to provide resistance against the shear experienced from horizontal dynamic impact. In this research, the pier displacement concept has been opted to account for the displacement due to rebar slip in flexure deformations [21]. Energy evolved due to pier spring action is introduced in Eq. 19 and as depicted in Fig. 5 [16].

$$E_s = 0.5 * k_{col} * \Delta^2 \tag{19}$$

In Eq. 19, the pier stiffness (k_{col}) can be computed from Eq. 20.

$$k_{col} = 0.8 * E_c * I_g \tag{20}$$

In Eq. 20, E_c is modulus of elasticity of concrete (12,247.45 ksi calculated for 6 ksi compressive strength of concrete), and I_g is the gross moment of inertia of the pier section. Computation of k_{col} using Eq. 20 yields $9.357*10^7$ kip/in pier stiffness while considering concrete contribution in RC pier against impact.

The pier resisting displacement (Δ) can be computed considering the sum of the flexure ($\Delta_{flexure}$) displacement and shear (Δ_{shear}) displacement, and given in Eq. 21

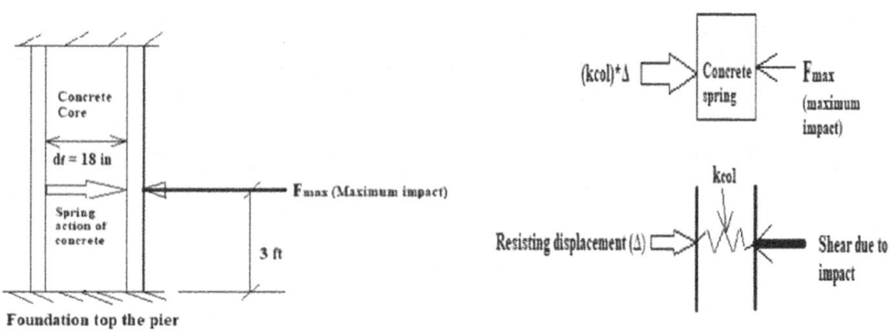

Fig. 5 Concrete spring action and equivalent spring model

[13].

$$\Delta = \Delta_{flexure} + \Delta_{shear} \tag{21}$$

Equation 22 provides the Δ_{flexure}.

$$\Delta_{flexure} = \Phi_y * (\mathrm{L} + L_{sp})^2/6 \tag{22}$$

In Eq. 22, Φ_y is the strain rate of steel at yield (0.00021), and L_{sp} is the strain penetration length and be computed from Eq. 23.

$$L_{sp} = 0.15 * f_y * d_b \tag{23}$$

After computation, Eq. 23 yields L_{sp} as 9.0 in (228.6 mm). Furthermore, from Eq. 22, Δ_{flexure} is computed as 0.05 in (1.27 mm).

Shear displacement (Δ_{shear}) is given in Eq. 24 [13].

$$\Delta_{shear} = \sum_{i=1}^{n} (r_{xy}^{i} + r_{xy}^{i+1})/(2 * h_i) \tag{24}$$

where, r_{xy}^{i} and r_{xy}^{i+1} are the shear strains of lower and upper section of the ith segment, and h_i is considered as height of each segment (Fig. 5).

The shear strain is calculated from Eq. 25,

$$r_{xy} = \tau_f / G \tag{25}$$

where, τ_f is the shear stress and can be computed from Eq. 26 and G is the shear modulus (3×10^3 ksi) (20.7×10^3 MPa) [13].

$$\tau_f = M/\left(h_s * d_f * L\right) \tag{26}$$

In Eq. 26, 'M' is the external maximum moment applied at the base ($F_{max}.3ft. = 2.33 \times 10^8$ kip-in.) (Fig. 5) and hence computed as 2.33×10^8 kip-in (2.633×10^7 kN-m), h_s is the diameter of the pier and d_f is the end to end distance of the longitudinal bar (18 in. or 45.72 cm). Computing Eq. 26, τ_f has been determined as 6042.95 ksi (41.7×10^3 MPa).

Equation 25 yields r_{xy} as 6.043. Combining Eqs. 25 and 26 yield Δ_{shear} as 217.55 in (552.58 cm). Combination of the Eqs. 19–26 result in E_{s} as 2.12×10^{12} kip-in (2.4×10^{11} kN-m).

2.4.4 Prediction of Limit State Equation for Energy Concept

Limit state equations incorporating failure prediction using energy (E) analyses have been developed in this research. Numerical models depicting limit state equations to assess failure during dynamic impact has been determined for a representative pier. Different dynamic cases are considered an intensive assessment. Time (t) during impact for quasi-static to plastic deformation is considered as 10^{-3}–10^{-4} s, hence considered as non-deterministic for short duration impact and high strain rate loading. For that, mean (μ) and co-efficient of variation (V) are considered in this research are 0.00055 and 0.1, respectively [3]. So, standard deviation (σ) has been computed as 0.000055 ($\sigma = \mu*V$), and time for dynamic impact is considered as normally distributed. Steel strain rates are considered to evolve numerical modelling. Finally, Monte Carlo simulations predicting individual cases are carried out for 11,000 data to predict failure (statistical parameters mean and standard deviation converged at about this number of simulations), and determining reliability indices for each cases, considering concrete contribution, and the conservative approach without concrete contribution.

Without Consideration of Concrete Contribution

Spring actions of concrete are not considered in a conservative approach. These are further represented by most frequently occurring and extreme) condition cases (Eqs. 28 and 29). Probability of failure P_f at the limiting case occurs when E ≤ 0. This is further illustrated from Eq. 27 when kinetic energy of the vehicle must not exceed the sum of potential energy of the pier, dissipation of energy due to impact, recoiling of vehicular frontal materials, and spring action incurred by the concrete core. However, being the more conservative analysis, simulations have been undertaken without considering spring action, and consideration of spring action shows more practical results [4] in order for computing P_f (Eq.).27

$$P_f(E < 0) = (\sum_{i=1}^{n} E_i \leq 0) \tag{27}$$

Limit state equations without considering concrete contribution as spring action are illustrated by the Eqs. 28 and 29.

$$E = 614.41 + 86.94 * \frac{\gamma}{\eta} + 3.32 * \frac{10^5}{t^2} - 0.5 * M_{veh} * V_{max}^2 \tag{28}$$

$$E = 3195.81 + 86.94 * \frac{\gamma}{\eta} + 1.72 * \frac{10^6}{t^2} - 4.2 * 10^4 \tag{29}$$

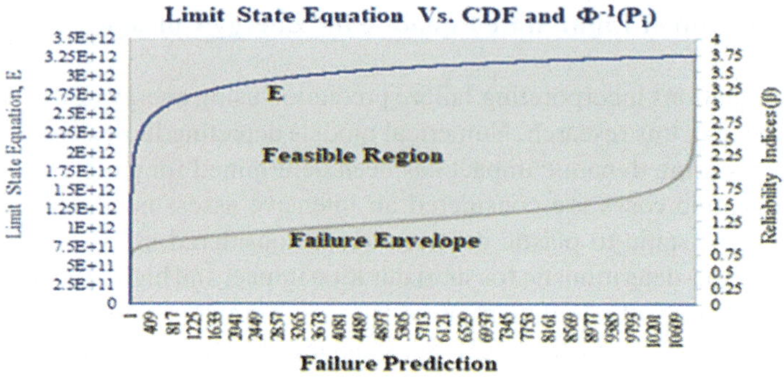

Fig. 6 Results without spring action

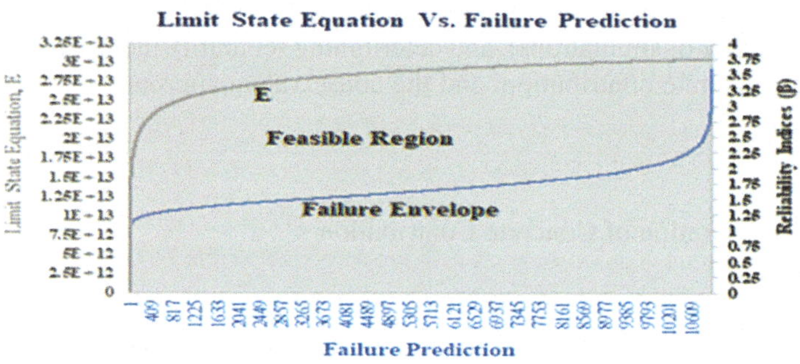

Fig. 7 Results without spring action, most-occurring

Considering Concrete Contribution as Spring Action

Concrete contributions in terms of spring action in limit state equations comprising energy concept are duly considered. This is further represented by the most frequently occurring (Eq. 30) and as extreme (Eq. 31).

$$E = 614.41 + 86.94 * \frac{\gamma}{n} + 0.5 * k_{col} * \Delta^2 + 3.32 * \frac{10^5}{t^2} - 0.5 * M_{veh} * V_{max}^2 \tag{30}$$

$$E = 3195.81 + 86.94 * \frac{\gamma}{n} + 0.5 * k_{col} * \Delta^2 + 1.72 * \frac{10^6}{t^2} - 4.2 \times 10^4 \tag{31}$$

The results of the Monte Carlo for eleven thousand simulations are shown in Figs. 6, 7, 8, and 9.

Probabilities of failure (P_f) of 0.0001 and 0.00012 as little contribution from concrete as a recoiling action for truck and semi-trailer, whereas 0.00005 and 0.000056 considering concrete contributed as a recoiling and spring action. The

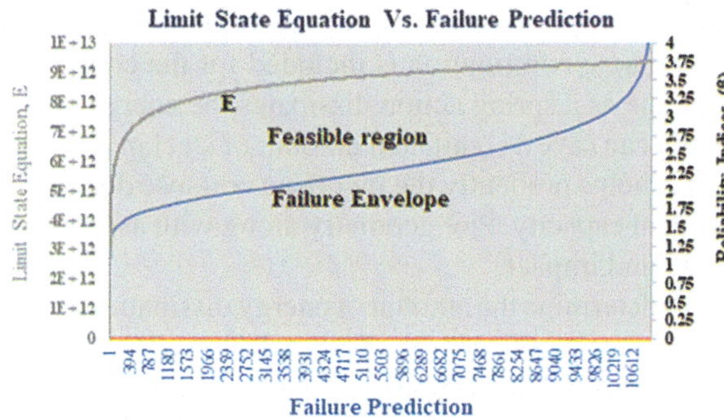

Fig. 8 Results with spring action for most

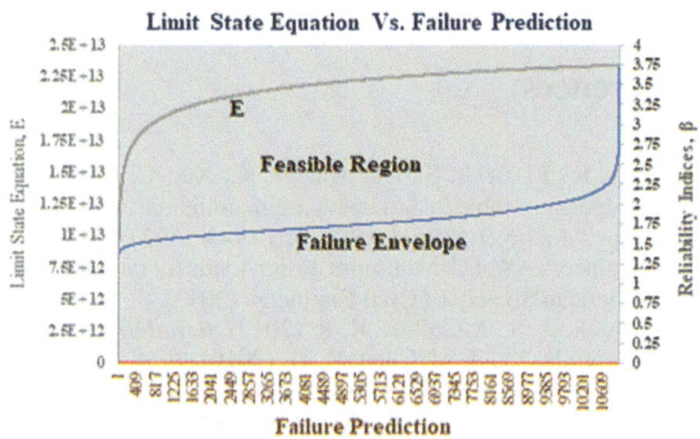

Fig. 9 Results with spring action for extreme most frequently occurring

conservative approach incorporating no concrete contribution in the form of spring action endorses more optimistic result for reliability index. On the other hand, concrete contribution as a spring action resisting localized failure action, yields high reliability index [22]. Boundary conditions assessing reliability indices can be ranged as $2.25 \leq \beta \leq 4.1$ for the respective energy equations where failure predictions are within the range of 10^{-4}–10^{-5}.

Conservative approaches are considered by the feasible regions in which failure does not occur with reliability indices (β) for ranges between 2.25 and 3.75 (Figs. 6 and 8) whereas, Figs. 7 and 9 predicting concrete performance result β lies between 2.75 to 4.1 for the non-conservative approach.

3 Conclusions

In this research, numerical simulation is undertaken to predict the failure of a vehicle impacted RC bridge pier for energy dissipation. The RC pier experiences impact by vehicles at a height of 3 feet from the pier base. Numerical simulations introducing limit state equations are developed to compute energy produced. This is used to assess the failure both for the conservative and non-conservative approaches. Concrete contribution as a spring action is incorporated for the non-conservative approach and no concrete contribution is included for the conservative approach. Contribution of concrete as a spring action dissipates the energy and transmits less load to the steel, which can save a significant amount of steel in order for pursuing the optimal design. It also helps prudently the pier from collapse due to impact, and hence increasing the residual capacity. Pier geometry along with aspect ratio also play significant roles to withstand impact.

To determine the amount of energy dissipation responsible for plastic deformation, γ/η plays a significant role. Failure followed by plastic hinge formation due to impact is only possible if the entire dissipated energy has been governed by steel re-bar in the ranges of quasi-static to plastic strain within 10^{-3}–10^{-4} s time domain.

References

1. Cao, R., El-Tawil, S., Agrawal, A. K., Xu, X., & Wong, W. (2019). Behavior and design of bridge piers subjected to heavy truck collision. *Journal of Bridge Engineering, 24,* 04019057. https://doi.org/10.1061/(ASCE)BE.1943-5592.0001414.
2. Engineers AS of C. Minimum design loads for buildings and other structures (ASCE/SEI 7-10). American Society of Civil Engineers (2013).
3. Nowak, A. S., & Collins, K. R. (2012). *Reliability of structures.* CRC Press.
4. Ayyub, B. M., & McCuen, R. H. (2016). *Probability, statistics, and reliability for engineers and scientists.* CRC Press.
5. Dutta, A., Mander, J. B. (1998). *Capacity design and fatigue analysis of confined concrete columns.* Technical report. Multidisciplinary Center for Earthquake Engineering Research, Buffalo, NY.
6. ATC-40. (1996). *Seismic evaluation and retrofit of concrete buildings.* ATC-40 Report. Report, ATC-40.
7. Pantelides, C. P., Ameli, M. J., Parks, J. E., & Brown, D. N. (2014). *Seismic evaluation of grouted splice sleeve connections for precast RC bridge piers in ABC.* Utah Department of Transportation.
8. Thomas, R. J., Steel, K., & Sorensen, A. D. (2018). Reliability analysis of circular reinforced concrete columns subject to sequential vehicular impact and blast loading. *Engineering Structures, 168,* 838–851.
9. Zhao, X., Wu, Y.-F., Leung, A. Y., & Lam, H. F. (2011). Plastic hinge length in reinforced concrete flexural members. *Procedia Engineering, 14,* 1266–1274.
10. Park, H., & Eom, T. (2004). Energy dissipation capacity of flexure-dominated reinforced concrete members. In *13th World Conference Earthquake Engineering.*
11. Sinha, B. P., Gerstle, K. H., & Tulin, L. G. (1964). Stress-strain relations for concrete under cyclic loading. *Journal Proceedings, 61,* 195–212.
12. Okamoto, S., Yamabe, K. Shiomi, S. (1976). Earthquake resistance of prestressed concrete structures. In *Proceedings of Annual Convention* (pp. 1251–1252). AIJ.
13. Kowalsky, M. J. (2000). Deformation limit states for circular reinforced concrete bridge columns. *Journal of the Structural Engineering. American Society of Civil Engineers, 126,* 869–878.
14. Mander, J. B., Priestley, M. J., & Park, R. (1988). Theoretical stress-strain model for confined concrete. *Journal Structure Engineering (United States), 114,* 1804–1826. https://doi.org/10.1061/(ASCE)0733-9445(1988)114:8(1804).
15. Shi, Y., Hao, H., & Li, Z.-X. (2008). Numerical derivation of pressure–impulse diagrams for prediction of RC column damage to blast loads. *International Journal of Impact Engineering, 35,* 1213–1227.

16. Tsang, H., & Lam, N. T. K. (2008). Collapse of reinforced concrete column by vehicle impact. *Computer Civil Infrastructure Engineering, 23,* 427–436.
17. Crawford, J. E., Holland, T. J., Mendoza, P. J., & Murtha, R. N. (1983). A failure methodology based on shear deformation. In *Recent advances engineering mechanical their impact civil engineering practice* (pp. 985–988). ASCE.
18. Slawson, T. R. (1984). Dynamic shear failure of shallow-buried flat-roofed reinforced concrete structures subjected to blast loading. Army Engineer Waterways Experiment Station Vicksburg MS Structures Lab.
19. Feyerabend, M. (1988). *Hard transverse impacts on steel beams and reinforced concrete beams.* Ger: Univ Karlsruhe (TH).
20. Committee ACI. (2008). Building code requirements for structural concrete (ACI 318–08) and commentary, American Concrete Institute.
21. Li, B. (2012). Initial stiffness of reinforced concrete columns and walls. In *World Conference Earthquake Engineering*. Lisbon.
22. Stewart, M. G. (2019). Reliability-based load factors for airblast and structural reliability of reinforced concrete columns for protective structures. *Structure and Infrastructure Engineering, 15,* 634–646.

Establishment of Suitable General Probabilistic Model for Shear Reliability Analysis

Oladimeji B. Olalusi and Panagiotis Spyridis

Abstract Adequate characterization and quantification of the model uncertainties in shear resistance models are identified as one of the key issues in reliability analysis of shear reinforced concrete beams. Previous studies indicate high model uncertainty in shear prediction. Model uncertainties for various shear predictive models are characterised here, based on a recent and well-vetted database of shear failures. The characterisation includes the estimation of main statistical parameters; and importantly also correlation and regression analysis to assess the consistency of model uncertainties over ranges of design parameters. The aim is to identify models suitable for use as a general probabilistic model (GPM) in future reliability assessments. The Variable Strut Inclination Method (VSIM) displayed high bias, variability and various correlations with shear input parameters which make it an unsuitable choice for GPM. The modified compression field theory (MCFT) showed low bias and variability with consistent model uncertainties over the ranges of shear design parameters and is thus suitable as GPM.

Keywords Shear capacity · Model uncertainty · Variable strut inclination method · Modified compression field theory · Reinforced concrete

1 Introduction

An ideal predictive model is characterised by low bias and variability as well as statistical independence of its model uncertainty concerning the main parameters affecting shear capacity of reinforced concrete beams. Such a model with its corresponding model uncertainty can be used as GPM for beam shear resistance in reliability assessments of shear design formulations [1].

O. B. Olalusi
University of KwaZulu-Natal, Durban, South Africa

P. Spyridis (✉)
Technical University of Dortmund, Dortmund, Germany
e-mail: panagiotis.spyridis@tu-dortmund.de

© The Author(s), under exclusive license to Springer Nature Switzerland AG 2021
J. C. Matos et al. (eds.), *18th International Probabilistic Workshop*, Lecture Notes in Civil Engineering 153, https://doi.org/10.1007/978-3-030-73616-3_23

The purpose of a general probabilistic model (GPM) in reliability assessment is to provide the best estimate of capacity, including expected variability of realisations around such mean value. Design capacity from a standardised design formulation may then be compared to the GPM to assess the probability of failure or adequacy of the safety margin. The GPM would typically be based on the best estimate prediction for the design situation, accounting for material and geometric variability, and adjusted by the predictive model specific model uncertainty to account for bias and scatter in the prediction.

This investigation focuses on the statistical characterization of model uncertainties as a necessary precursor to choosing a suitable GPM for future reliability assessments and allowing its probabilistic description. Quantification of model uncertainty; and evaluation of the consistency of these over design ranges are important objectives since these contribute significantly to reliability performance [1, 2].

The assessment is done for the best estimate shear capacity predictions based on (1) EN 1992-1-1 (EC2) Variable Strut Inclination Method (VSIM) [3] and (2) modified compression field theory (MCFT) as implemented by the analysis program Response 2000 (R2k) [4]. MCFT is a representative of the most advanced level of approximation IV in the fib Model Code, and R2k offers best estimate predictions [5].

2 Description of Model Uncertainties

Models are the means by which we represent and express our understanding of reality. Since it is practically impossible to understand and represent reality in its absolute intricacy, models are a partial representation of reality resulting in a state of uncertainty [6]. In using a model for prediction purposes, there are two different kinds of uncertainties involved, namely aleatory and epistemic uncertainties. Aleatory uncertainties emerge due to inherent variability in a physical process. They are random uncertainties that can only be controlled through efficient design practices. The epistemic uncertainties are systematic uncertainties that emerge from inadequate information and understanding of the model formulation, conservative assumptions and approximations [1, 6].

Epistemic uncertainties can be estimated and calibrated to incorporate adequate conservatism into the design procedure. Model uncertainty can be described as epistemic taking into account the effects forsaken in the model's formulations and simplifications of the mathematical relations [7]. Epistemic uncertainties can be classified as model uncertainty, statistical uncertainty and physical uncertainty [8]. According to them, the physical uncertainties are uncertainties related to the modeling of loading, geometry and material parameters. The statistical uncertainties arise as a result of insufficient statistical data. The model uncertainty is described as the uncertainty related to the idealised mathematical formulation used to approximate the actual performance of the structure. The modern methods of structural reliability and risk analysis are used to estimate these uncertainties to meet acceptable and required levels of safety. For several common cases, model uncertainties for both load and resistance models are recommended by the Probabilistic Model Code [7] and are summarised in Table 1.

In reliability analysis, model uncertainties can be related to resistance models and models for load effects. This contribution is focused only on the uncertainties associated with shear resistance models. Resistance model uncertainties are often found to be significant [1].

2.1 Evaluation of Model Uncertainties

Model uncertainties may be statistically characterised by a probability distribution function with corresponding mean value and variance. In accordance with [7, 9] the model uncertainty MF_x associated with a single experiment, x, is calculated from the ratio of the experimentally observed capacity to the best estimate predictions of the analytical model under consideration Eq. (1).

$$MF_x = V_{\exp,x}/V_{model,x}(X) \tag{1}$$

where, $\mathrm{V}_{\exp,x}$ represents the measured shear failure capacity for a single experimental beam x which is subject to known structural conditions. X is the vector of mean values for basic input variables such as beam depth (d), concrete strength (f_{cm}) amount of longitudinal reinforcement (ρ_l), amount of stirrup reinforcement ($\rho_w f_{ywm}$), shear-span to depth ration (a/d), width (b_w) etc. $\mathrm{V}_{model,x}$ represents for the same experimental beam x the best-estimate shear resistance prediction offered by the predictive model under consideration.

3 Uncertainties Related to Shear Resistance Models

3.1 Model Uncertainty Based on EC2 VSIM (VSIM and VSIM-A)

The model uncertainty derived according to EC2 is based on two variations of VSIM shear design procedure V_{VSIM} (VSIM) and V_{VSIM-A} (VSIM-A), resulting in model uncertainty variables MF_{VSIM} and MF_{VSIM-A} respectively. Considering the VSIM shear design provisions of EC2 [3] shear resistance is taken as the lesser of stirrup capacity (3) or web crushing capacity (4) for an assumed strut angle θ in the range $21.8° \leq \theta \leq 45°$. To obtain best estimate predictions of these, material strength mean values rather than design values are used in (2) and (3) below, while the strut angle is estimated in (4) by setting (2) equal to (3) [10].

$$V_{VSIM-L\theta} = V_{Rm,s} = \frac{A_{sw}}{s} z f_{ywm} \cot \quad (2)$$

$$V_{Rm,max} = \frac{\alpha_{cw} b_w s v_1 \alpha_{cc} f_{cm}}{(\cot\theta + \tan\theta)} \quad (3)$$

$$\theta = \sin^{-1}\sqrt{\frac{A_{sw} f_{ywm}}{\alpha_{cw} b_w s v_1 \alpha_{cc} f_{cm}}} \quad (4)$$

where:

A_{sw} is the cross-sectional area of the shear reinforcement

s denotes the spacing of the stirrups

f_{ywm} and f_{cm} represent the mean values of the steel yield strength and concrete compressive cylinder strength, respectively

θ is the concrete strut angle.

The internal lever arm may be taken as $z = 0.9d$. The value for v_1 maybe taken to be 0.6 for $f_{ck} \leq 60$ MPa, and as $0.9 - \frac{f_{ck}}{200}$ for high-strength concrete beams. The coefficient accounting for long term effects on concrete was assigned a mean value of $\alpha_{cc} = 1.0$ according to [3] corresponding to short-duration loading as would be expected in experimental shear tests. The coefficient accounting for the state of stress in the compression chord is represented by $\alpha_{cw} = 1$ (non-prestressed structures) [3].

MF_{VSIM} represents the model uncertainty variable derived from unbiased EC2 VISM shear resistance model V_{VSIM}. Unbiased V_{VSIM} shear capacity predictions are best estimate predictions obtained based on mean values of material strength (instead of characteristic values) thus partial material safety factors are set equal to one. The strut angle θ is restricted to the range of $21.8° \leq \theta \leq 45°$. The limit placed on the strut angle θ is expected to affect the derived MF_{VSIM} conservatively.

The restrictions imposed on the concrete compressive strut angle (θ) in the operational EC2 VSIM shear design formulation is a form of bias imposed to provide conservatism in design [5]. Mean value predictions may be made with or without the strut angle limits. As a result, alternative function better suited for the best-estimate prediction of EC2 VSIM shear resistance V_{VSIM-A} was investigated. V_{VSIM-A} is the shear resistance prediction of the unbiased EC2 VSIM shear design procedure with no constraint applied to the strut angle (VSIM-A), resulting in model uncertainty denoted MF_{VSIM-A}.

3.2 *Model Uncertainty Based on MCFT-Response 2000 Predictions* (V_{R2k})

The model uncertainty statistics are derived based on the unbiased shear capacity predictions of MCFT based sectional analysis program Response-2000 (V_{R2k}). Unbiased estimates of V_{R2k} were obtained by using the mean values of shear parameters in the sectional analysis program. The characterised model uncertainty is denoted MF_{R2k}.

3.3 Database of Experiments

An extensive database of 511 tests on slender reinforced concrete beams with vertical shear reinforcement was recently compiled by a joint group of ACI-DafStb consisting of ACI Subcommittee 445-D and German Committee for Structural Concrete [Deutscher Ausschub fur Stahlbeton (DafStb)]. In order to develop a set of beam tests that can be used for the purpose of evaluating the accuracy and conservativeness of shear design models, the database was subjected to several control and filtering criteria as presented in [11] resulting in an evaluation database of 160 tests on slender beams used in this study.

The evaluation database consists of simply supported rectangular and flanged beams subjected to point loads. The shear span to depth ratio, (a/d), of the beams, is greater than 2.40 so that they may be considered slender. The beams failed predominantly by diagonal tension, shear-compression and shear tension. No experiments are included that failed in flexure. The evaluation database covers a wide range of beams with low to high concrete strengths, shear reinforcement ratios, and effective depths. Beams with light, moderate and heavy longitudinal reinforcement are included (See Table 1).

Table 1 Range of parameters for the full database (160 experiments)

Parameters	Full database (160 experiments)				
	Minimum value	First quartile (P25)	Median (P50)	Third quartile (P75)	Maximum value
b_w[mm]	75	150	180	250	457
d[mm]	161	263	292.50	451.40	1369
f_{cm}[MPa]	13.40	31.40	50.20	72.43	125.30
ρ_l[%]	0.14	1.10	2.28	3 .01	5.20
$\rho_w f_{ywm}$[MPa]	0.28	0.62	0.87	1.63	9.80
$\boldsymbol{a/d}$	2.40	2.56	3.09	3.52	7.10

4 Statistical Analysis of Model Uncertainty

4.1 Statistical Properties of the Model Uncertainties; Correlation and Regression Analysis

The statistical properties of the model uncertainty (MF) determined for the complete database includes the mean value (μ_{MF}), standard deviation (σ_{MF}) and coefficient of variation (Ω_{MF}). The basic statistical properties of each derived MF are reported in Table 2.

To discern the consistency of each *MF* with variation in important parameters affecting shear strength, a correlation and regression analysis as proposed in [12] is conducted on the calculated model uncertainties. Trends are identified by correlating the model uncertainties with the shear parameters using Pearson's correlation coefficient (r) [12]. The determined r values are presented in Table 2, revealing the level of the correlation between each model uncertainty MF and parameters such as stirrup quantity ($\rho_w f_{ywm}$), concrete strength (f_{cm}), longitudinal reinforcement (ρ_l), beam depth (d), beam width (b_w), and shear span to depth ratio (a/d).

Several of the models revealed a correlation of the model uncertainties with select shear parameters (Table 2). It is worth to reiterate at this stage that a good predictive model is characterised by low bias and variability. Statistical independence of its model uncertainty with respect to the main parameters affecting the shear capacity of reinforced concrete beams would indicate that the model takes proper account of their effect on shear capacity.

Table 2 Statistical properties of model uncertainties for the full experimental database

Statistics of the model uncertainty	MF_{VSIM}	MF_{VSIM-A}	MF_{R2k}
Mean (μ_{MF})	1.57	0.90	1.04
Standard deviation (σ_{MF})	0.48	0.25	0.16
Coefficient of variation	0.31	0.27	0.15
Minimum	0.53	0.43	0.63
Maximum	3.10	1.93	1.62

Table 3 Pearson correlation matrix between model and shear parameters

	MF_{VSIM}	MF_{VSIM-A}	MF_{R2k}
b_w[mm]	0.14	−0.59	−0.12
d[mm]	−0.30	0.21	−0.20
f_{cm} [MPa]	0.12	−0.29	−0.25
ρ_l[%]	0.15	−0.29	−0.16
$\rho_w f_{ywm}$[MPa]	−0.62	0.58	−0.12
a/d	−0.04	0.18	0.06

4.2 Discussion of Model Uncertainty Statistical Results and Trends

MF_{VSIM} has a high mean value of $\mu_{MF} = 1.57$, confirming that the unbiased stirrup resistance function V_{VSIM} generally underpredicts shear capacity, and with substantial variability of $\sigma_{MF} = 0.48$ (see Table 2). A strong correlation value of –0.62 (see Table 3) points to the strong decreasing trend between MF_{VSIM} and $\rho_w f_{ywm}$, which may be observed in Fig. 1a, where V_{VSIM} significantly under-predicts capacity at low $\rho_w f_{ywm}$. The substantial under-prediction of shear capacity at low $\rho_w f_{ywm}$ are due to the neglecting of the concrete contribution in the V_{VSIM} model formulation and to the safety bias introduced by the lower limit on the concrete strut angle.

MF_{VSIM-A} has a mean value of ($\mu_{MF} = 0.90$), indicating that V_{VSIM-A} generally overpredicts shear capacity. MF_{VSIM-A} shows less scatter about the mean with a standard deviation of $\sigma_{MF} = 0.25$. The difference in the mean values of MF_{VSIM-A} ($\mu_{MF} = 0.90$) and $MF_{VSIM-L\theta}$ ($\mu_{MF} = 1.58$) quantifies the conservative influence of the limiting concrete compressive strut angle. A strong, increasing trend exists between MF_{VSIM-A} and increasing $\rho_w f_{ywm}$ with a strong correlation coefficient of 0.58, as indicated in Table 3. From Fig. 1 it can be observed that the model uncertainty was initially unconservative at $\rho_w f_{ywm} < 2$ and started to become conservative at $\rho_w f_{ywm} > 2$ where there is a paucity of data points. The trends, however, between $\rho_w f_{ywm}$ and the model uncertainty variables MF_{VSIM-A} and MF_{VSIM} are opposite. The conservative bias increases for the MF_{VSIM-A} with increasing values of $\rho_w f_{ywm}$ and decrease for MF_{VSIM} with increasing $\rho_w f_{ywm}$. Moreover, a correlation value (r) of -0.59 particularly stated in Table 3 indicates a correlation between V_{VSIM-A} capacity predictions and b_w. The performance of MF_{VSIM-A} in terms of bias ($\mu_{MF} = 0.90$), scatter ($\sigma_{MF} = 0.25$) and strong sensitivity to important shear parameters (shear reinforcement $\rho_w f_{ywm}$ and beam width b_w) makes it evident that the V_{VSIM-A} is not a suitable candidate for the general probabilistic model representation of shear resistance.

MF_{R2k} has a mean value of $\mu_{MF} = 1.04$ and a standard deviation of $\sigma_{MF} = 0.16$, thus sporting the lowest bias and scatter of all models in this study. MF_{R2k} display a mildly negative trend with concrete strength f_{cm} ($r = -0.25$), while other correlations are milder still, see Table 3.

The large bias and scatter displayed by V_{VSIM} and V_{VSIM-A} models and their strong trend with various shear parameters indicate that the models are not well suited for the general probabilistic representation of shear resistance. The high accuracy displayed by MF_{R2k} observations, coupled with their relatively low dispersion and statistical independence with major shear parameters, are features that warrant the consideration of the MCFT (R2k) as suitable to be used as a general probabilistic model. The lognormal distribution is commonly assumed for model uncertainties [7]. This is corroborated by [1, 12] specifically with regards to model uncertainties for shear capacity predictions.

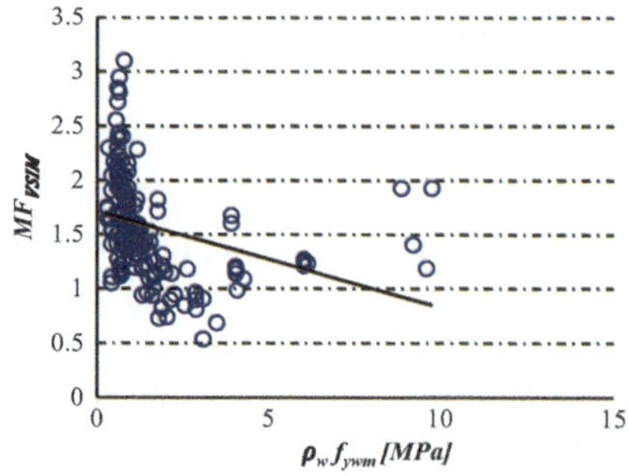

(a) EC2 VSIM

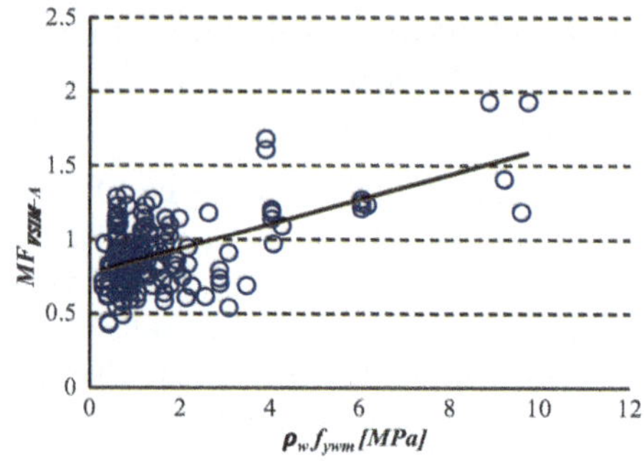

(b) VSIM-A

Fig. 1 Scatter plots, with regression trend lines and correlation statistics of model uncertainty versus shear parameter

5 Conclusion

A general probabilistic model (GPM) in reliability assessment should ideally be based on a predictive model that has low bias and dispersion together with consistent levels of bias and scattering over the range of design parameters. The VSIM and VSIM-A models displayed high bias, variability and various correlations with shear input parameters which makes them an unsuitable choice for GPM. MCFT (R2k) on the other hand, displayed low bias and comparatively lower variability with consistent model uncertainties over the ranges of shear design parameters and are thus suitable as GPM. The VSIM model uncertainty showed a strong correlation with the stirrup reinforcement level $\rho_w f_{ywm}$, significantly underpredicting shear capacity initially and later overpredicting shear capacity as $\rho_w f_{ywm}$ increases. The correlations raise concern that EC2 VSIM may produce designs that are not sufficiently conservative for highly shear reinforced. Reliability assessment is advised to assess the consistency of EC2 reliability performance over the range of practical design situations.

References

1. Holický, M., Retief, J. V., & Sýkora, M. (2016). Assessment of model uncertainties for structural resistance. *Probabilistic Engineering Mechanics, 45,* 188–197. https://doi.org/10.1016/j.probengmech.2015.09.008.
2. Olalusi, O. B., & Viljoen, C. (2020). Model uncertainties and bias in SHEAR strength predictions of slender stirrup reinforced concrete beams. *Structural Concrete, 21*(1), 316–332.
3. EN 1992-1-1. (2004). Eurocode 2: *Design of concrete structures—Part 1-1: General rules and rules for buildings*. Brussels: European Committee for Standardisation (CEN).
4. Bentz, E. C. (2000). *Sectional analysis of reinforced concrete members* (Ph.D. Thesis). Department of Civil Engineering, University of Toronto, Canada.

5. Olalusi, O. B., & Viljoen, C. (2019). Assessment of simplified and advanced models for shear resistance prediction of stirrup-reinforced concrete beams. *Engineering Structures, 186*, 96–109. https://doi.org/10.1016/j.engstruct.2019.01.130.
6. Droguett, E. L., & Mosleh, A. (2008). Bayesian methodology for model uncertainty using model performance data. *Risk Analysis, 28*(5), 1457–1476. https://doi.org/10.1111/j.1539-6924.2008.01117.x.
7. JCSS. (2001). *Probabilistic model code*. Zurich: Joint Committee on Structural Safety. 978-3-909386-79-6.
8. Faber, M., & Sorensen, J. (2002). *Reliability based code calibration*. Zurich, Switzerland: The Joint Committee on Structural Safety.
9. Olalusi, O. B., & Viljoen, C. (2017). Towards effective general probabilistic representation for shear resistance. In *12th International Conference on Structural Safety & Reliability (ICOSSAR)*, TU Wein Vienna, Austria.
10. Beeby, A., & Narayanan, S. (1995). *Designers' handbook to eurocode 2: Part 1.1: design of concrete structures (Eurocode design handbooks) (Pt. 1)*. American Society of Civil Engineers ISBN 10: 0727716689.
11. Reineck, K. H., Bentz, E., Fitik, B., Kuchma, D. A., & Bayrak, O. (2014). ACI-DAfStb databases for shear tests on slender reinforced concrete beams with stirrups. *ACI Structural Journal, 111*(5), 1147.
12. Olalusi, O. B. (2018). *Reliability assessment of shear design provisions for reinforced concrete beams with stirrups* (Doctoral dissertation, Stellenbosch: Stellenbosch University).

Estimation of the Global Health Burden of Structural Collapse

Dirk Proske

Abstract In this article, the global health burden of structural collapse is determined. To this end, mean ratios of structures to inhabitants are determined and applied to the world population, subdivided into industrialized and developing countries. Based on known collapse frequencies of structures, mean annual worldwide collapse numbers of structures are calculated. Furthermore, the average number of victims is estimated and then used to estimate worldwide victim numbers, considering not only fatalities but also injuries. These victim numbers are converted into "lost life years", a parameter often used as a measure of a health risk and compared to some other causes of victims.

Keywords Lost life years · Collapse · Frequency · Building · Bridge

1 Introduction

For some years now, the global burden of disease for the human civilization has been determined in terms of "lost life years" [1]. For some technical products, such as motorized road traffic, the specific global burden is also determined [2]. Therefore, the question also arises for the global burden of structural collapse. This question will be examined in this article.

In contrast to such a global health burden, the advantage of structures was recognized very early in the development of human civilization. Buildings belong to the first technical products in a series of early inventions including wheels and boats. Even today, a large part of the human wealth on earth is tied up in buildings. In Germany, real estate represents 80% of tangible assets [3]. The U.N. counts the right to housing as a fundamental human right [4].

D. Proske (✉)
Department Architecture, Wood and Civil Engineering, Bern University of Applied Sciences, Burgdorf, Switzerland
e-mail: dirk.proske@bfh.ch

© The Author(s), under exclusive license to Springer Nature Switzerland AG 2021
J. C. Matos et al. (eds.), *18th International Probabilistic Workshop*, Lecture Notes in Civil Engineering 153, https://doi.org/10.1007/978-3-030-73616-3_24

2 Work Steps

The global burden caused by structural collapses is determined in the following steps:

- Determination and definition of the construction types to be considered,
- Calculation of a ratio of the number of constructions to the number of inhabitants for countries for which data are available and further extrapolation to the world stock,
- Discussion of the causes of structural collapses,
- Estimation of the frequency of structural collapses,
- Estimation of the average number of victims per collapse and world stock extrapolation,
- Determination of the related "lost life years" as a measure of the burden of disease and
- Plausibility check.

3 Determination and Definition of Construction Types

Basically, structures can be divided into two classes: residential and office buildings as well as sports, leisure and production halls; and infrastructure with bridges, dams, retaining walls and tunnels. This article covers all buildings according to the first class and the five infrastructure objects according to the second class.

A definition of building constructions can be found in [5], different definitions of bridges can be found in [6], definition of dams can be found in [7] and definition of tunnels in [8]. In addition, several large infrastructure operators use their own definitions. The definitions also differ between countries, as shown by different minimum spans for bridges [6].

These different definitions have a considerable impact on the stock figures and thus on the statistics. According to the U.N., approximately one billion people live in slums [4]. The extent to which such housing can be regarded as buildings in the usual sense is at least worth discussing but is not examined in detail in this article. However, this question shows the uncertainty in the estimation of the global stock.

4 Calculation of the Global Stock

There is information on both the mass of the global building stock and the global usable area of buildings. Besides, [9] considers approximately two million settlements on Earth. According to [10], the total mass of all buildings and their equipment is approx. 800 peta-gram and the worldwide usable area of buildings in 2017 was 162.8 billion m^2 [11]. This value is expected to rise to 183.5 billion m^2 by 2026 [11]. In contrast, [12] estimates the total global building floor area with 223.4 billion m^2, which is significantly larger. According to Bilham [13], the number of housing units on earth has doubled between 1950 and 2003. The next doubling will be reached by 2030. Bilham [13] estimates one billion additional housing units from 2009 for the next half century. This dynamic is not considered in this article, i.e. the stock figures are not adjusted to a specific reference point, because it would require a country specific growing function.

However, various countries record and document the building stock development in detail. In Germany, for example, there was a building and housing census [14] during the 2011 census, which will be continued. Various states in the U.S. publish data on the number of buildings and apartments on the Internet, e.g. the Pluto database for New York City [15]. For bridges there are data on the U.S. stock, partly also for Germany [16]. Various infrastructure operators, e.g. the Swiss Railways or the Swiss Roads Office publish figures.

On a global level, both the United Nations (U.N. Statistical Database on Global Housing, U.N. HABITAT Database, see also [9]) and various governmental organizations operate databases on the global building stock [17]. For buildings, the so-called PAGER database of the U.S. Geological Survey [18] is of interest. This database is used to estimate the potential victims and damage of earthquakes very quickly in order to plan suitable rescue measures.

Table 1 shows the buildings and infrastructure stock for different countries. The calculation of the number of buildings in the U.S. was carried out based on numbers in [19]. However, this data results in a significantly higher ratio of buildings to inhabitants of more than 0.41 (see Table 2) than in other industrialized countries such as Switzerland, Germany or Japan. Hence, this value was excluded for the calculation of the average ratio value for industrialized countries. Unfortunately, the author has only data available for China as developing country. Therefore, China was considered as characteristic example for this type of countries. To determine the ratio of the building stock per inhabitants for China, the following approach was applied: In the first step, the function of the ratio of building space per capita gross national product from [19] was used, assuming that this function applies worldwide. This would result in a ratio of floor space per inhabitants between the U.S. and China of approximately 1.56. Besides [19] further information regarding the floor space ratio in China can be found in [20] and [21]. The ratio of 1.56 would further result in a building stock in China of approximately 265 million buildings. The resulting ratio of inhabitants to number of buildings of 0.19 for China was almost confirmed for Kigali in Africa with 0.18 [22].

For bridges and tunnels, data is available for many countries, so that here the calculation of the average ratio values is based on a broader database. However, the density of bridges and tunnels depends on the topography, so that the ratio using the number of inhabitants is only a limited measure. Exact figures for retaining walls and dams are only available for one country. Therefore, the indication of average values in this study contains large uncertainties.

Based on this, albeit incomplete, list, mean ratios of structures to inhabitants are calculated (Table 2). Table 3 lists the worldwide stock of structures thus determined. Some results of Table 3 can be checked: independent estimates of the global bridge stock are between 5 and 6 million [6], estimates for the number of dams are between 800,000 [23] and 850,000 [24] and the number of tunnels is given as approx. 40,000 [25]. These values correspond relatively well to those given in Table 3.

5 Causes of Structural Collapse

This study refers to the risk to the uninvolved public. Collapses during the construction period, on the other hand, are not considered in this document, except for tunnels. Therefore, the calculation of "lost life years" refers to collapses during use. The causes of structural failure during use can be divided into two groups: failure due to accidental loads such as earthquakes, floods and impacts, and failure under normal conditions, such as traffic and payloads. Detailed cause analyses for the collapse of buildings can be found in [43], bridges can be found in [6], dams can be found in [44] and tunnels can be found in [25].

In fact, accidental actions dominate the failure of all types of structures due to the large potential number of collapses per event. For bridges, floods and collisions are the cause of collapse in almost 60% of all cases [6]. For example, almost 300 bridges were destroyed in the 2011 Tohoku earthquake and the following tsunami flood alone. According to [45], more than 190,000 buildings were damaged and more than 45,000 were destroyed. According to [46], even 130,000 buildings were fully destroyed, and 240,000 buildings were partially destroyed during this event. Even greater destruction can be found in the case of severe earthquakes in developing countries, e.g. the 2010 earthquake in Haiti with approximately 250,000 damaged buildings and the highest absolute death toll in terms of magnitude [47].

6 Number of Collapses and the Number of Victims Estimation

6.1 Collapse Frequencies

Various publications have investigated the collapse frequency of structures, for example for buildings and bridges see [6], for dams see [23] and for tunnels see [25]. Based on these studies, Table 4 determines the worldwide absolute number of collapses per type of structure. Rounded collapse frequencies are used for the calculation.

Table 1 Building and infrastructure stock for different types of structures and countries

	Switzerland	Germany	U.S	China	G.B	Japan	Europe
Population in millions (2017/18)	8.42 [26]	82.79 [27]	329 [28]	1 393 [29]	66.04 [30]	126.5 [29]	746 [29]
Buildings in millions	2.5 [31]	24.5 [32][a]	136.6	265.4	–	34 [33]	146.25 [34]
Bridges in thousands	13 [6]	120 [6], 150 [35]	607 [6] 615 [36]	689 [6] 832 [36]	150 [6]	155 [6]	1500 [6]
Retaining walls in thousands	50[b]	–	–	–	–	–	–
Dams in thousands	3[c]	–	–	–	–	–	–
Tunnels in thousands	1.3 [37]	1.5 [25]	0.85 [25]	27 [25]	0.6 [25]	16 [25]	–
Ratio of buildings to inhabitants	0.297	0.290	0.415	0.148–0.189	–	0.268	–
Relationship between bridges and inhabitants	0.0015	0.0014	0.0018	0.0005	0.0023	0.0012	0.0020
Ratio of retaining walls to population[d]	0.0059						
Ratio of dams to population[d]	0.0004						
Ratio of tunnel to population[d]	0.000154	0.000018	0.000003	0.000019	0.000009	0.000126	

The data refer to different years. There was no standardization to a reference year

[a]Calculation of the building stock for Germany with 21.05 million residential buildings and 3.52 million non-residential buildings. This results in 24.52 million buildings [32]. However, there are also other data, e.g. 18.9 million residential buildings according to [38] or 49 million geometric objects, of which 2/3 are main buildings and 1/3 outbuildings, which corresponds to 32 million building ensembles and 22 million postal addresses [39]. Kleist et al. [40] give an average number of inhabitants per residential building of 4.2 which yields to 19.5 million residential buildings

[b]Own estimate based on data provided by ASTRA, SBB

[c]Own estimate based on [41], in Switzerland alone there are 200 dams

[d]Whether the relationship between buildings and inhabitants is a suitable parameter for retaining structures, dams and tunnels is not discussed here. An alternative would be buildings per gross domestic product

Table 2 Mean ratio of buildings to inhabitants for different types of structures

	Industrial countries[a]	Developing countries[a]
Average ratio of buildings to inhabitants	0.264	0.189
Average ratio of bridges to inhabitants	0.0016	0.0005
Mean ratio of retaining walls to inhabitants	0.0059	0.0030[b]
Mean ratio of dams to inhabitants	0.0004	0.00007[c]
Average tunnel to inhabitant ratio	0.000019[d]	0.000003[e]

[a]For a definition of industrialized and developing countries see [42]. For population see [29]
[b]Half the value of industrialized countries (own assumption)
[c]20% of the value of industrialized countries (own assumption)
[d]For tunnels, weighting has been applied to take account of different topographical conditions: 30% of the mean value (own assumption)
[e]For tunnels, weighting has been applied to take account of different topographical conditions: 15% of the value of China (own assumption)

Table 3 Calculation of the global stock for the different types of structures

	Industrialized countries	Developing countries	Worldwide
Population in billions	1.27 [29]	6.3 [29]	7.62
Buildings in millions	334.5	1205.5	1540
Bridges in millions	1.97	3.14	5.11
Retaining walls in millions	7.52	18.87	26.39
Dams in thousands	451	452	903
Tunnels in thousands	23.57	18.5	42.0

6.2 Collapse Victim Numbers

The number of victims of structural collapses can theoretically be given in the form of statistical averages per year for certain areas. Mostly, however, victim figures are given for certain singular events, such as earthquakes or floods, for dam failure see for example [23].

Table 4 Estimated number of collapses per year per building type

	Collapse frequency in industrialized countries	Collapse frequency developing countries	Average number of collapses per year
Buildings	10^{-7} (10^{-4})[a]	10^{-5} (10^{-4})[a]	12,089
Bridges	10^{-5} (10^{-4})	10^{-4}	334
Retaining walls	10^{-4}	10^{-3}	19,621
Dams	10^{-4}	10^{-3}	498
Tunnels	10^{-4}	10^{-3}	21

[a]The values in brackets are informative and result from the consideration of accidental events, such as earthquakes or large-scale flooding according to [6]. However, these values cannot be applied worldwide because they are limited to certain events and regions

Daniell et al. [48] gives a total number of earthquake victims of 2.42 million for the 20th century, corresponding to 24,000 annual fatalities. According to [49] the global average annual number of victims from 1990 to 2011 was about 27,000. Based on the figures in [50] and [51], however, there are also significantly smaller (about 9000 per year) and larger (35,000 per year) average values known. In [52], 2.57 million earthquake deaths are estimated for the 21st century. This corresponds to an annual mean value of 25,700.

In the case of earthquakes, approx. 75% of the fatalities are caused by the collapse of buildings [53]. Floods, fires, landslides and other causes account for only about 25%, with the Tohoku earthquake in Japan in 2011 and the Sumatra Andaman earthquake in 2004 accounting for a significantly higher proportion of flood victims.

Based on the dispersion of the various data, a lower limit value for the worldwide annual victims from building collapses is set at 20,000 (0.75 × 24,000).

The number of worldwide annual fatalities due to flooding is approximately 5000 [51]. In [54], 175,000 deaths from worldwide flooding are mentioned for the period 1975 to 2001. This corresponds to an average annual worldwide death toll of 7000. The author does not know what proportion can be attributed to the failure of dams. In fact, the number depends on the size of the affected population. Considering the maximum annual number of earthquake victims mentioned of approx. 35,000 and the maximum annual number of flood victims of 7000, an upper limit for victims of structural collapses of 50,000 is set.

In addition to large-scale accidental events such as earthquakes and floods, there are also other causes of failure such as impacts. Examples for severe bridge collapses are the collapse of Eschede in Germany in 1998 with 101 fatalities, the collapse of a railway bridge in the Soviet Union in 1983 with 176 fatalities [55] or the collapse of a bridge after a mudslide in China in July 1981 with more than 200 fatalities [56]. However, the average number of bridge collapse victims is below one since many bridge collapses do not include any victims at all. A distribution function of the victims of bridge collapses can be found, for example, in [57].

Building collapses in recent years such as Halstenbeck 1997/87, Bad Reichenhall 2006 and Cologne 2009 in Germany, Marseille 2018 in France, Naples 2017 in Italy,

Mumbai 2017 in India, Accra 2012 in Ghana and Sabhar 2013 in Bangladesh with more than 1000 fatalities or the collapse of the World Trade Center 2001 in New York with more than 3000 fatalities show the range of the number of victims due to building collapse. Even further, they also show the possible maximum victim number of an individual building collapse.

In addition to collapses with a large number of victims, there are also data on building collapses with smaller numbers of victims. Some of these figures also refer to developing countries. Asante and Sasu [58] list 17 collapses of buildings in Kenya and 15 collapses of buildings in Ghana. Based on these figures, the average number of fatalities is about 5 for collapses in Kenya and about 2 for collapses in Ghana. Ayodeij [59] groups 60 collapses of buildings in Nigeria. For these grouped data, an average death toll of about 7 and an injury rate of about 4 are obtained. An evaluation of over 30,000 building collapses during the BAM earthquake in Iran resulted in an average number of fatalities of about 1.5 [60]. Based on this data an average victim number has been estimated for buildings (see Table 5).

A distribution function of the victims of retaining walls can be found in [61]. Based on this study we assume an average victim number of 0.01 for the public. Based on the average number of victims due to dam failure and the conditional mortality which is assumed here with 0.05 we receive an average victim number per dam failure of 5.

6.3 Ratio of Injured to Fatalities

The ratio of injured persons to fatalities is also an important parameter for estimating the severity of accidents. The ratio of injured persons to fatalities for the 17 building collapse cases is approx. 0.28 in Kenya and approx. 0.25 in Ghana. During the BAM earthquake in Iran the ratio of injured persons to fatalities was about 0.5 [60].

Information on the development over time of the ratio of injured persons to fatalities due to earthquakes can be found in [62]. For the period 1986 to 2008, the values changed from 3.6 to 48.

During floods, a factor of 0.1–1% lies between fatalities and affected people [63] and the ratio between injured and affected people lies between 1 and 10% [63]. This yields to a ratio of injured to fatalities of 10.

According to [64], the average ratio of injured persons to fatalities for buildings is 6:1. This number complies with the found range. Therefore, this number will be used for the following computations.

6.4 Estimated Annual Victim Numbers

In fact, not all collapses lead to fatalities. Therefore, conditional death probabilities when buildings collapse can be introduced. Using conditional mortality and the

Table 5 Calculated annual number of fatalities and injuries

Collapse	Average number of victims per collapse	Number of fatalities	Number of injured
Buildings	4.0	48,355	290,130
Bridges	0.20	67	401
Retaining walls	0.01	196	3573
Dams	5.0	2490	14,939
Tunnels	1.0	21	125
Total		51,128	306,772

average number of victims for collapses with fatalities, an average number of victims per collapse is determined. This result is multiplied by the estimated annual number of collapses. In addition, the ratio of injured to fatalities is used to determine the number of people injured. The results are shown in Table 5.

The total number of victims can be compared with the average number of earthquake victims discussed in former sections. Based on the estimates, the number of victims worldwide ranges from approximately 20,000 (lower estimate earthquakes) to 50,000 (upper estimate earthquakes and floods) per year. In Table 5, a value slightly above 50,000 is computed.

7 Estimation of the Lost Life Years

7.1 *Own Estimation*

Assuming a global average life expectancy of about 72 years (WHO 2014) and an average world age distribution as in Europe, the absolute losses in life years estimated in Table 6 result. Considering the ratio of 6:1 for injured persons to fatalities, Table 6 also gives the absolute losses in life years with a permanent loss in quality of health of about 30% [65]. If the sum of "lost life years" from fatalities and injuries is related to the total world population, the related values given in Table 6 are obtained.

Since the annual exposure time in buildings is high with 5500 h after [64] and 5840 h after [66], there is no need to scale to the exposure time. Table 7 allows the comparison of the calculated related "lost life years" in days with other events or technologies. It becomes clear that structures are an extremely safe product.

Table 6 Estimated number of victims and "Lost life years"

Parameter	Lower limit	Computation	Upper limit value
Fatalities due to building collapses	20,000	51,128	50,000
Years of life lost (fatalities)	720,000	1,917,328	1,800,000
Injured by building collapses	120,000	290,130	300,000
Years of life lost (injured)	1,296,000	3,451,191	3,240,000
Total life years	2,016,000	5,368,520	5,040,000
Years of life lost in years per person	0.00026	0.0007	0.00066

Table 7 Examples for lost life years from [55] in days

Activity	Years of lost life in days
Alcoholism (not averaged)	4000
Heart diseases	2000
Cancer	1200
Car crashes	200
Substance abuse	90
Skydiving	25
Flu	2.3
Storms and floods	1
Earthquakes (for exposed areas)	0.2
Buildings (worldwide)	0.19
Poisonous animals and plants	0.08

7.2 *Plausibility Check*

The results from Table 7 will be checked for plausibility. On the one hand, Table 7 gives a value of 0.2 for earthquakes. Since, as described above, on average 75% of earthquake victims are caused by the collapse of buildings, this results in a value of $0.2 \times 0.75 = 0.15$ for building collapse.

Cohen [67] gives an approximate formula for determining "lost life years" (LLY) with LLY $= 1.1 \times 10^6$ days $\times\ r$ with r as additional mortality. For building collapses a mortality rate of 10^{-7} is generally given [6], for bridge collapses 10^{-8} [6]. This gives LLY $= 1.1 \times 10^6$ days $\times\ 10^{-7} = 0.11$ days for buildings and LLY $= 1.1 \times 10^6$ days $\times\ 10^{-8} = 0.011$ days for bridges. These values correspond relatively well with the values in Table 7, bottom row.

Haagsma et al. [68] indicates a loss of approx. 10 million "lost life years" for accidents caused by mechanical effects (without weapons) and approx. 2.2 million "lost life years" for poisonings of all kinds as absolute values. This corresponds to a ratio of approx. 4:1. The absolute death toll in [68] is approx. 200,000 to 100,000 and thus a value of 2:1. In Table 7 there is a ratio of approx. 2.5:1 for structural collapse and poisonous animals and plants. It should be mentioned that the worldwide annual number of accidents requiring medical care is almost one billion.

8 Summary

Based on the estimated worldwide number of structures and the known collapse frequencies, the mean absolute collapses of structures per type are calculated and the number of victims is estimated. From these victim numbers, average "lost life years" are calculated for the population. These values are compared with some other events and technologies. This shows that the values for buildings are significantly lower than for other technologies, such as motor vehicles (car crash). However, the investigation includes large uncertainties regarding the data of the worldwide stock and the number of victims. Therefore, some plausibility checks have been carried out to back the found results.

References

1. Lancet. (2018). *The global burden of disease study 2017*(Vol. 392(10159), pp. 1683–2138, pp. e14–e18).
2. World Bank. (2014). *Transport for health. The global burden of disease from motorized road transport*. Washington D.C: Institute for Health Metrics and Evaluation, University of Washington, The World Bank Group.
3. ZIA. (2017). Zentraler Immobilien Ausschuss e.V.: Immobilienwirtschaft 2017, Berlin.
4. United Nations. (2018). *Tracking progress towards inclusive, safe, resilient and sustainable cities and human settlements* DG 11 Synthesis Report 2018, Nairobi, Kenya.
5. Berner, F., Kochendürfer, B., & Schach, R. (2013). Grundlagen der Baubetriebslehre. Band 1: Baubetriebswirtschaft, Leitfaden des Baubetriebs und der Bauwirtschaft B.G. Teubner Verlag/GWV Fachverlage GmbH, Wiesbaden.
6. Proske, D. (2018a). *Bridge collapse frequencies versus failure probabilities*. Berlin: Springer.
7. ICOLD. (2018). International Commission on Large Dams: Role of Dams.
8. ITA. (2019). International Tunnelling and Underground Space Association, Chatelaine.
9. Wyss, M., Tolis, S., Rosset, P., & Pacchiani, F. (2013). *Approximate model for worldwide building stock in three size categories of settlements*. Geneva: World Agency of Planetary Monitoring & Earthquake Risk Reduction.
10. Krausmann, F., Wiedenhofer, D., Lauk, Chr., Haas, W., Tanikawa, H., Fishman, T., Miatto, A., Schandl, H., & Haberl, H. (2017). Global socioeconomic material stocks rise 23-fold over the 20th century and require half of annual resource use. *PNAS, 114*(8), 1880–1885; first published February 6, https://doi.org/https://doi.org/10.1073/pnas.1613773114.
11. Naviant Research. (2018). *Global building stock database—Commercial and residential building floor space by country and building type* (pp. 2017–2026). Chicago.
12. Zhao, W., Moncaster, A., Reiner, D. M., & Guthrie, P. (2019). Estimation lifetimes and stock turnover dynamics of urban residential buildings in China. *Sustainability, 1*, 3720, 18 pages.
13. Bilham, R. (2009). The seismic future of cities. *Bulletin of Earthquake Engineering*, 49 pages. https://doi.org/10.1007/s10518-009-9147-0.
14. Statistische Ämter des Bundes und der Länder. (2015). *Zensus 2011: Die Gebäude und Wohnungszählung* (p. 2015). Dezember: Hannover.

15. Pluto Database. (2019). Department of City Planning, NYC.
16. Proske, D. (2017). Comparison of computed and observed probabilities of failure and core damage frequencies. In R. Caspeele, L. Taerwe, & D. Proske (Eds.), *14th International Probabilistic Workshop* (pp. 109–122). Cham: Springer.
17. Jaiswal, K. S., & Wald, D. J. (2008). *Creating a global building inventory for earthquake loss assessment and risk management: 2008*, (U.S. Geological Survey Open-File Report 2008-1160), 103 pages
18. U.S. Geotechnical Survey. (2019). Rapid assessment of an earthquakes impact (PAGER).
19. Moura, M. C. P., Smith, S. J., Belzer, D. B. (2015). 120 Years of U.S. Residental housing stock and floor space. *PLoS ONE, 10*, 8: e0134135, 1–18. https://doi.org/10.1371/journal.pone.0134135, August 2015.
20. Hu, M., Bergsdal, H., van der Voet, E., Huppes, G., & Müller, D. B. (2010). Dynamics of urban and rural housing stocks in China. *Building Research and Information, 38*(3), 301–317.
21. Hong, L., Zhou, N., Feng, W., Khanna, N., Fridley, D., Zhao, Y., & Sandholt, K. (2016). Building stock dynamics and its impacts on materials and energy demand in China. *Energy Policy, 94*, 47–55.
22. Bachofer, F., Braun, A., Adamietz, F., Murray, S., d Angelo, P., Kyazze, E., Mumuhire, A. P., & Bower, J. (2019). Build. Stock and building typology of Kigali, Ruanda, 4, 105.
23. Proske, D. (2018b). Comparison of large dam failure frequencies with failure probabilities. 16th International Probabilistic Workshop, Beton- und Stahlbetonbau, 9, pp. 2–6.
24. BUND. (2014). Bund für Umwelt und Naturschutz Deutschland: Weltwassertag 2014.
25. Proske, D., Spyridis, P., & Heinzelmann, L. (2019). Comparison of tunnel failure frequencies and failure probabilities. International Probabilistic Workshop, pp. 177–184
26. BfS (2019): Bundesamt für Statistik: Bevölkerung.
27. Destatis. (2019a). Bevölkerungsstand, 2019.
28. US Census Bureau. (2019). https://www.census.gov/en.html.
29. DSW. (2018). *Deutsche Stiftung Weltbevölkerung, Soziale und demografische Daten weltweit* (DSW-DATENREPORT 2018) Hannover.
30. Office for National Statistics (2019): Population Estimates, 2019
31. HEV Schweiz. (2019). Wohneigentum in Zahlen.
32. Schiller, G., Ortlepp, R., Krauß, N., Steger, S., Schütz, H., Fernández, J. A., et al. (2015). *(2015): Kartierung des anthropogenen Lagers in Deutschland zur Optimierung der Sekundärrohstoffwirtschaft*. Dessau-Roßlau, Juli: Umweltbundesamt.
33. Tanikawa, H., Fishman, T., Okuoka, K., & Sugimoto, K. (2015). The weight of society over time and space—A comprehensive account of the construction material stock of Japan, 1945–2010. *Journal of Industrial Ecology, 9*(5), 778–791. https://doi.org/10.1111/jiec.12284.
34. Fabbri, M. (2012). Clean energy package: Why buildings matter. Presentation Buildings_101_PBIE.pptx.
35. Kroker, H. (2013). Deutschlands Brücken vor dem Kollapse, Die Welt, 3.6.2013, https://www.welt.de/116759711.
36. Alogdianakis, F., Charmpis, D. C., & Balafas, I. (2019). A method to probabilistically estimate the future condition of aging bridges using recorded inspection. Presentation at the 17th International Probabilistic Workshop, 11-13 September 2019, Edinburgh, Heriot-Watt University.
37. Fachgruppe für Untertagebau. (2018). Anzahl der Tunnel und Stollen, https://www.swisstunnel.ch/tunnelbau-schweiz/uebersichtsgrafiken/anzahl-tunnel/.
38. Destatis. (2019b). Anzahl der Wohngebäude in Deutschland in den Jahren 2000 bis 2017 (in 1.000).
39. Behnisch, M., Meinel, G., Burckhardt, M., & Hecht, R. (2012). Auswertungen zum Gebäudebestand in Deutschland auf Grundlage digitaler Geobasisdaten. In G. Meinel, U. Schumacher, M. Behnisch, (Hrsg.) *Flächennutzungsmonitoring IV. Genauere Daten – informierte Akteure – praktisches Handeln* (pp. 151–158). Berlin: IÖR Schriften 60.
40. Kleist, I., Thicken, A. H., Köhler, P., Müller, M., Seifert, I., Borst, D., & Werner, U. (2006). Estimation of the regional stock of residential buildings as a basis for a comparative risk assessment in Germany. *Natural Hazards and Earth Systems Sciences, 6,* 541–552.

41. ICOLD. (2019). World register of dams.
42. Duden. (2016). Wirtschaft von A bis Z: Grundlagenwissen für Schule und Studium, Beruf und Alltag. 6. Aufl. Mannheim: Bibliographisches Institut 2016. Lizenzausgabe Bonn: Bundeszentrale für politische Bildung.
43. Wardhana, A., & Hadipriono, F. C. (2003). Study of recent building failures in the United States. *Journal of Performance of Construction Facilities, ASCE, 8*(2003), 151–158.
44. Zhang, L., Peng, M., Chang, D., & Xu, Y. (2016). *Dam failure mechanisms and risk assessment.* Singapore: Wiley .
45. Norio, O., Ye, T., Kajitani, Y., Shi, P., & Tatano, H. (2011). The 2011 Eastern Japan great earthquake disaster: Overview and comments. *International Journal Disaster Risk Science, 2*(1), 34–42.
46. Kazama, M., & Noda, T. (2012). Damage statistics (Summary of the 2011 off the Pacific Coast of Tohoku earthquake damage). *Soils and Foundations, 52*(5), 780–792.
47. Bilham, R. (2010). Lessons from the Haiti Earthquake. *Nature, 463*(18), 878–879.
48. Daniell, J. E., Khazai, B., Wenzel, F., & Vervaek, A. (2011). The CATDAT damaging earthquakes database. *Natural Hazards and Earth Systems Sciences, 11,* 2235–2251.
49. Guha-Sapir, D., & Vos, F. (2011). Earthquakes, an epidemiological perspective on pattern and trends. In *Human casualties in earthquakes, progress in modelling and mitigation, spence* (pp. 13–24). Heidelberg, London, New York: So and Scawthorn, Springer.
50. Nichols, J. M., & Beavers, J. E. (2008). World earthquake fatalities from the past: Implifications for the present and future. *National Hazards Review, 9*(4), 179–189.
51. Guha-Sapir, D., Hoyois, P., Wallemacq, P., & Below, R. (2016). *Annual disaster statistical review 2016: The numbers and trends.* Brussels: Centre for Research on the Epidemiology of Disasters.
52. Holzer, T. L., & Savage, J. C. (2013). Global earthquake fatalities and population. *Earthquake Spectra, 29*(1), 155–175
53. Coburn, A. W., Spence R. J. S., & Pomonis A. (1992). Factors determining human casualty levels in earthquakes: Mortality prediction in building collapse. In *Tenth World Conference Earthquake Engineering* (pp. 5989–5994). Balkema, Rotterdam.
54. Jonkman, S. N. (2005). Global perspectives on loss of human life caused by floods. *Natural Hazards, 34,* 151–175. https://doi.org/10.1007/s11069-004-8891-3.
55. Proske, D. (2008). *Catalogue of risks.* Berlin, Heidelberg: Springer Verlag.
56. Zhang, S. (1993). A comprehensive approach to the observation and prevention of debris flows in China. *Natural Hazards, 7,* 1–23.
57. Curbach, M., & Proske, D. (2004). Risikountersuchung am Beispiel historischer Brücken unter Schiffsanprall, Beton- und Stahlbetonbau, 99, 12. *Dezember, 2004,* 956–966.
58. Asante, L. A., & Sasu, A. (2018). The challenge of reducing the incidence of building collapse in Ghana: Analyzing the perspectives of building inspectors in Kumasi. *SAGE Open, 2018,* 1–2.
59. Ayodeij, O. (2011). An examination of the causes and effects of building collapse in Nigeria. *Journal of Design and Built Environment, 9,* 37–47.
60. Kuwata, Y., Takada, S., & Bastami, M. (2005). Building damage and human casualties during the Bam-Iran earthquake. *Asian Journal of Civil Engineering, 6*(1–2), 1–9.
61. GEO. (2002). *Geotechnical engineering office: QRA of collapses and excessive displacement of deep excavations* (GEO Report 124) Hong Kong, February 2002.
62. Wyss, M., & Trendafiloski, G. (2009). *Trends in the casualty ration of injured to fatalities in earthquakes.* Second International Workshop on Disaster Casualties, 15–16 June 2009, University of Cambridge, UK, pp. 1–6.
63. Jonkman, S. N. (2003). *Loss of life caused by floods: An overview of mortality statistics for worldwide floods.* Delft Cluster.
64. Rackwitz, R. (1998). *Zuverlässigkeit und Lasten im konstruktiven Ingenieurbau.* Zuverlässigkeitstheoretische Grundlagen: Technische Universität München, Teil I.
65. Lentz, A. (2006). *Acceptability of civil engineering decisions involving human consequences* (Dissertation) TU München, München.

66. Maag, T. (2004). *Risikobasierte Beurteilung der Personensicherheit von Wohnbauten im Brandfall unter Verwendung von Bayes'schen Netzen* (Doctorial Thesis) IBK Bericht, vol. 282, Zürich: vdf Hochschulverlag AG an der ETH Zürich.
67. Cohen, B. L. (1991). Catalogue of risks extended and updated. *Health Physics, 61*(3), 317–335.
68. Haagsma, J. A., Graetz, N., Bolliger, I., et al. (2016). The global burden of injury: Incidence, mortality, disability-adjusted life years and time trends from the global burden of disease study 2013. *Injury Prevention, 22,* 3–18.

Evaluation of Partial Safety Factors for the Structural Assessment of Existings Masonry Buildings

Pietro Croce, Maria L. Beconcini, Paolo Formichi, Filippo Landi, Benedetta Puccini, and Vincenzo Zotti

Abstract The assessment of existing structures and infrastructures is a primary task in modern engineering, both for its key economic significance and for the extent and the significance of the built environment, nonetheless operational rules and standards for existing structures are often missing or insufficient, especially for masonry constructions. Existing masonry buildings, even in limited geographical regions, are characterized by many masonry types, differing in basic material, mortar, block shape, block texture, workmanship, degree of decay and so on. For these reasons, relevant mechanical parameters of masonry are often very uncertain; their rough estimation thus leads to inaccurate conclusions about the reliability of the investigated structure. In this work, a methodology to derive a refined probabilistic description of masonry parameters is first outlined starting from the analysis of a database of in-situ tests results collected by the authors. In particular, material classes, representing low, medium and high-quality masonry, are identified for a given masonry typology by means of the definition of a Gaussian Mixture Model. The probability density functions so obtained are the fundamental basis for the implementation of probabilistic analysis methods. In particular, the study will focus on the evaluation of masonry classes for compressive strength of stone masonry, considering a relevant database of semi-destructive, double flat jacks, in-situ test results. The statistical properties of the identified masonry classes, which can be used for the direct probabilistic assessment of structural performance of masonry walls under vertical loads, are finally considered for the evaluation of suitable partial safety factors, γ_M, to be used in the engineering practice.

Keywords Partial safety factor · Reliability · Existing structures · Masonry · Compressive strength

---

P. Croce · M. L. Beconcini · P. Formichi · F. Landi (✉) · B. Puccini · V. Zotti
Department of Civil and Industrial Engineering, University of Pisa, Pisa, Italy
e-mail: filippo.landi@ing.unipi.it

© The Author(s), under exclusive license to Springer Nature Switzerland AG 2021

J. C. Matos et al. (eds.), *18th International Probabilistic Workshop*, Lecture Notes in Civil Engineering 153, https://doi.org/10.1007/978-3-030-73616-3_25

1 Introduction

The assessment of existing structures and infrastructures is a primary task in modern engineering, both for its key economic significance and for the extent and the importance of the built environment [1], nonetheless operational rules and standards for existing structures are often missing or insufficient, especially for masonry constructions.

In the past centuries, masonry was the main building material; consequently, masonry buildings are a relevant part of existing structures, especially in historical towns [2]. Mostly, they were built according to empirical rules and architectural canons, far away from the modern design approaches. Despite often they successfully perform their functions over time till today, there is a strong need to "measure" their structural performance, especially in seismic-prone areas, mainly in view of prioritization and planning of maintenance and intervention strategies.

As known, many different types of masonry can be identified in existing masonry buildings in Europe and worldwide [3], with significant scatter even in limited geographical regions, differing in basic material, mortar, block shape and texture, workmanship, degree of decay and so on.

Despite existing buildings are commonly declared to be better known than new ones, the relevant mechanical parameters of existing masonry cannot be easily estimated. In fact, mean values and coefficient of variations (COVs) of relevant mechanical properties of a given existing masonry are very scattered, especially in comparison with masonry structures built nowadays. For that reason, the evaluation of masonry parameters cannot overlook the assessment of the related uncertainty that should be properly expressed in probabilistic terms [3, 4].

In practical cases, masonry's mechanical properties are usually evaluated based on limited semi-destructive or non-destructive tests, taking into account relevant uncertainties.

In the assessment, the first step should be the evaluation of the compressive strength of masonry walls and pillars. Despite a relatively extensive research into masonry structures, the issue of a reliable determination of the load-bearing capacity of existing, mainly historic, stone masonry structures, is still waiting for a satisfactory solution.

A probabilistic description of compressive strength of regular masonry types can be found in [3, 5], based on the EN1996-1-1 model [6], considering tests and the related probabilistic models for masonry units and mortar. But, also considering that the extraction of an appropriate number of samples from the investigated walls is often impossible, this approach cannot be applied to irregular stone masonry. In the following, a procedure to identify masonry classes and their main statistical parameters is proposed, based on masonry compressive strength derived from double flat-jacks tests [7]. Results are presented for stone masonry, for which a considerable wide database of in situ compression tests on masonry walls was collected by the authors [8, 9] in the framework of the in-situ experimental campaign for the assessment of seismic vulnerability of masonry school buildings in the Municipality of Florence.

These probabilistic models for compressive strength are the basis for a reliability analysis devoted to assess, depending on the required target reliability level, partial safety factor γ_M for existing masonry, to be used in the partial factor method implemented in the Eurocodes [10].

1.1 Experimental Tests for the Evaluation of Masonry Mechanical Parameters

As anticipated earlier, assessing the relevant mechanical properties of existing masonry walls should need an ad-hoc experimental test campaign, both in the laboratory and in-situ.

Laboratory tests may be, for example, direct compressive tests on masonry samples extracted from the structure, as well as compressive tests on single bricks or blocks associated with tests on mortar samples, such as Darmstadt test [11], PNT-G [12] or direct compression.

In situ tests are, instead, carried out, for example, by means of single and double flat jacks [7]. The idea of the test is similar to a standard compressive test, with the difference that it is carried out directly onto the investigated panel, to which the load is applied via two flat jacks, inserted in horizontal cuts, within the panel's thickness. During the tests, four inductive deformation transducers, three vertical and one horizontal (Fig. 1), are used to measure vertical and horizontal displacement in the area between the two cuts, approximately 500 mm from one another.

During the test, it is possible to detect the first cracks occurring in the compressed portion of wall, and to check if they affect the mortar or the stones. The pressure value, which causes the crack formation, is used to estimate the compressive strength of the masonry, while the elastic modulus E and the apparent Poisson ratio, ν, are derived from measurements of vertical and horizontal deformations.

Flat jack tests represent one of the less intrusive method for masonry testing and provide, as shown in [8], useful data to obtain a complete mechanical characterizations of masonry walls. Indeed, the compressive strength, f_m, and the elastic modulus, E, are directly estimated by the test, but also shear modulus, G, and shear strength, τ_0, can be derived by means of appropriate experimental relationships [8].

It must be remarked once again that safeguard of the structural integrity calls for a severe limitation of the number of destructive or semi-destructive tests to be carried out on a single structure. As a consequence, even in the most favorable situation, test results only allow to broadly assess the mean value of mechanical parameters and the material's degree of homogeneity throughout the structure, being generally not sufficient to derive the appropriate statistical description of mechanical parameters, which are needed for reliability assessment.

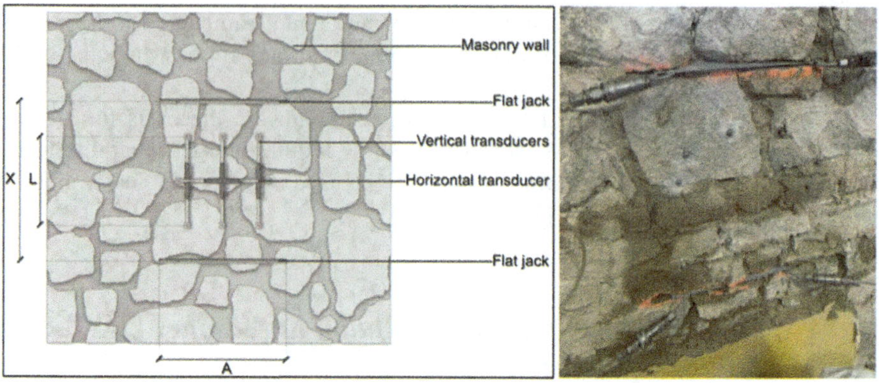

Fig. 1 Double flat jacks test on a stone masonry wall

To overcome the lack of information about mechanical properties in terms of probability density functions (*pdfs*) and relevant statistical parameters it is possible to carry out analysis based upon valid databases of tests results carried out on similar masonry panels [13].

During the last years, a wide experimental campaign has been carried out by the authors on rather homogenous sets of brick and stone masonry buildings located in the same geographical area (the Municipality of Florence). The results will be discussed in the following.

1.2 Database of Test Results

In the framework of static and seismic vulnerability assessments carried by the authors during the last three years on about 80 masonry school buildings of the Municipality of Florence, a large and consistent database of masonry mechanical parameters has been set up, collecting the results of ad hoc in situ and laboratory tests carried out on several masonry typologies characterizing these buildings. The experimental results, supplemented with literature data made it possible to set up a rational classification for various types of masonry, providing, at the same time, sound information about statistical properties of relevant investigated mechanical parameters.

The buildings differ in size and historical-artistic importance. Most of them date back to the end of 1800 and the beginning of 1900, but more ancient buildings, built before 1700, as well as modern buildings, built after the Second World War, have been also investigated.

The values collected in the database are critically discussed, also referring to the values recommended in the Guidelines for the application of the Italian Building Code [14] for the different existing masonry typologies. Moreover, an estimation

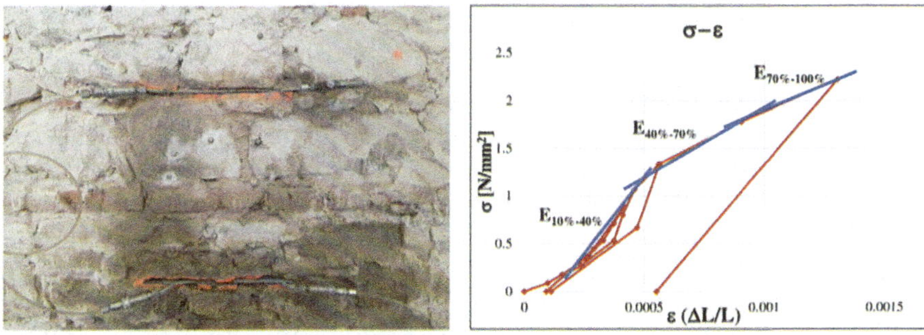

DOUBLE FLAT JACK TEST									ITALIAN GUIDELINES			MQI
σ_0 (N/mm²)	σ (N/mm²)	$E_{10\%-40\%}$ (N/mm²)	$E_{40\%-70\%}$ (N/mm²)	$E_{70\%-100\%}$ (N/mm²)	ν	$G_{10\%-40\%}$ (N/mm²)	$G_{40\%-70\%}$ (N/mm²)	$G_{70\%-100\%}$ (N/mm²)	f_m (N/mm²)	E_m (N/mm²)	G_m (N/mm²)	MQI_v
0.76	2.25	2767	1829	1113	0.5÷1.9	943	376	192	2.00	1230	410	3

Fig. 2 In-situ double flat jack test, stress-deformation diagram resulting and test data processing

of the masonry quality index (MQI) [15] is provided considering information on masonry quality obtained by visual inspection.

The present study is mainly focused on the analysis of the results of 95 double flat jack tests, performed by three different Laboratories according to ASTM standards [7]. Among the 95 tests, 67 concern stone masonry walls, 25 solid brick masonry and 3 other masonry types.

For the sake of the consistency and homogeneity of the results, all the information obtained from the in-situ tests have been analysed, processed and evaluated according to a unique procedure. In Fig. 2 the synthesis report is reported, as an example, for a stone masonry wall.

The main parameters considered are: the normal stress in the masonry panel due to permanent loads, σ_0, the masonry compressive strength, f_m, the elastic modulus, E, and, as ratio between horizontal and longitudinal displacements, the apparent value of the Poisson modulus, ν, which is often outside the limits for isotropic and homogenous materials, since determined in a post-cracking state.

The apparent value of the shear modulus, G, has been thus estimated adopting the usual relationship for isotropic and homogenous materials, again disregarding cracks.

The elastic modulus E and the shear modulus G have been evaluated linearizing three different parts of the $\sigma - \varepsilon$ diagram, to reproduce the masonry behaviour in the mainly elastic and plastic phase. In fact, in the $\sigma - \varepsilon$ diagram, three interval in terms of normal stresses have been considered, ranging from 10 to 40%, from 40 to 70% and from 70 to 100% of the compressive strength respectively, as summarized, for example, in Fig. 2. The first interval corresponds to the quasi-elastic section of the diagram, the intermediate interval refers to the cracked condition, while the higher interval reflects the plastic section.

2 Analysis of Test Results

Starting from the test results collected in the database, the statistical parameters, i.e. mean and coefficient of variation, of the relevant masonry mechanical parameters have been derived. The results for stone masonry compressive strength, f_m, the elastic modulus, *E*, and the shear modulus *G* in different conditions, are summarized in Table 1.

As expected, data are characterized by high coefficient of variation, especially concerning elastic and shear modulus [8], due to the wide variability of masonry properties, even between those belonging to the same typology, depending not only on the quality of the original raw materials, but also on the texture, on the workmanship and on the degradation. In fact, the quality of the mortar, the presence of irregular or dressed stones, as well as the different shape and size of the stones, well justify the existence of different classes within the same masonry typology.

A further analysis is then needed to identify homogenous statistical populations for masonry mechanical parameters. In particular, the general procedure already applied for the identification of concrete classes in [16] or rebar classes in [17], can be used as previously illustrated in [13].

The basic idea of the method is to subdivide mechanical tests results by means of a cluster analysis based on Gaussian Mixture Models (GMM), in such a way that homogenous statistical populations for masonry mechanical parameters can be identified. GMM is a cluster algorithm which provides a mixture of Gaussian distributions of vectors processing the subpopulations which are part of the whole Gaussian distribution.

2.1 *Identification of Masonry Classes*

The cluster analysis has been carried out by means of Gaussian Mixture Model (GMM), which is an algorithm, of unsupervised learning, able to identify subpopulations in of a whole population made up by a mixture of several unknown Gaussian distributions [18]. Data are analysed with the aim to find a better probabilistic model consisting of different distributions, following the experience-based awareness of a priori existence of different sub-population in the whole dataset. Each identified subpopulation represents a masonry stone class, characterized by its Coefficient of Variation (COV).

Table 1 Statistical parameters for stone masonry properties

Variable	Mean (N/mm^2)	COV
f_{m}	1.88	0.33
$E_{10\text{–}40}$	1789	0.47
$G_{10\text{–}40}$	659	0.57

Let $X_1, \ldots, X_n$ a random sample of size n, X_i is the p-dimensional random vector with *pdf* $f(x_i)$ on $\mathbb{R}^p$. In a mixture models (MM), with k components the distribution $f(x_i)$ is associated with the following density [18]:

$$f(x_i) = \sum_{j=1}^{k} w_i f_j(x_i), \tag{1.}$$

where $f_j(x_i)$ are the component densities of the mixture and the quantities $w_1, \ldots, w_k$ are the mixing proportions (or weights) with

$$\sum_{i=1}^{k} w_i = 1 \tag{2.}$$

To speed up the process, an engineering evaluation of a priori value of k is needed. In the present study, k is set equal to 3, considering low, medium and high-quality stone masonry. Then, the mixture model has been fitted by means of the Expectation Maximization (EM) algorithm, which is a tool able to simplify maximum likelihood problems, starting with an *Expectation* (E) step where a first assignment of each observation to each model is performed, then a *Maximization* (M) step computes the weights, the variance and the mixing probability, and finally the E and M steps are iterated until convergence [19].

The results are illustrated in Fig. 3, focusing on the compressive strength on masonry. In particular, the frequency histogram is plotted together with the probability density function obtained by fitting the whole dataset with a Normal distribution

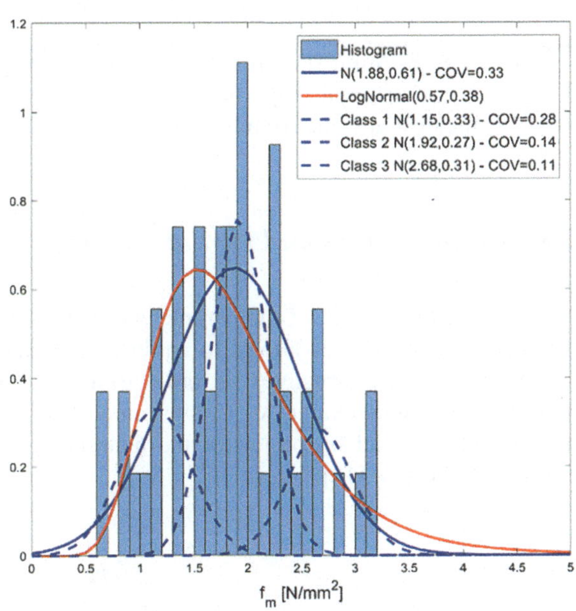

Fig. 3 Identification of stone masonry classes for the compressive strength f_m

(in blue) and a Lognormal distribution (in red), while the GMM is shown with blue dashed lines.

As already noted for concrete in [16] and rebars in [17], the cluster analysis leads to a better evaluation of statistical parameters for masonry compressive strength rather than the analysis of the whole dataset. Indeed, the proper identification of sub-classes allows to significantly improve the estimate of the coefficient of variation to be associated with each class: the COVs, which results to be 28% for the lower class, 14% for the intermediate class, and 11% for the upper class, are significantly smaller than that resulting from the analysis of the whole dataset (COV = 33%).

3 Structural Assessment of Existing Masonry Structures

The assessment of existing structures should be based on the principles of limit states [20], selecting the relevant situations (equivalent to those used for design of new structures) and taking into account the updated information on the actual conditions and circumstances under which the structure is required to fulfill its function during its design working life.

In particular, the structural assessment aims to determine the reliability of a structure as a whole or in terms of individual members, with respect to prescribed limit states and for a notional time period. In the assessment of actual reliability, the verifications are mostly based on the partial factor method [10, 21], but probabilistic methods can also be applied in special cases. In mathematical terms, the following condition should be fulfilled for each relevant limit state:

$$g(F_d, X_d, a_d, \theta_d) > 0 \tag{3}$$

where g is the limit state function, F_d is the design value of actions, X_d is the design value of material properties, a_d is the design value of geometrical quantities and θ_d is the design value for model uncertainty.

3.1 Design Values and Partial Factors

The suitable knowledge of the parameters of the statistical distribution describing the materials' properties allows to calibrate partial factors to be adopted for the assessment existing structures, in order to achieve a given target reliability. The design or assessment value X_d is determined from the characteristic value X_k, by means of the partial factor γ_m for the material resistance, and, in some cases, a conversion factor η [10]

$$X_d = \eta \frac{X_k}{\gamma_m}. \tag{4}$$

From Eq. (4), the partial factor γ_m can be derived, assuming a unit conversion factor η, as follows

$$\gamma_m = \frac{X_k}{X_d}. \tag{5}$$

The characteristic values X_k is generally evaluated, according to EN1990 [10], as the 5%-fractile of X, while the design value X_d can be evaluated based on the value of X at the FORM design point, i.e. the point on the failure surface ($g = 0$) closest to the average point in the space of normalised variables [10]. In this way, the partial factor on the material strength can be determined in case X is described by a normal distribution as

$$\gamma_m = \frac{X_k}{X_d} = \frac{\mu_X(1 - 1.645V_X)}{\mu_X(1 - \alpha_R\beta_t V_X)} = \frac{(1 - 1.645V_X)}{(1 - \alpha_R\beta_t V_X)} \tag{6}$$

and in case of Lognormal distribution as

$$\gamma_m = \frac{X_k}{X_d} = \frac{\mu_X \exp\left(-0.5\ln\left(1 + V_X^2\right) - 1.645\sqrt{\ln\left(1 + V_X^2\right)}\right)}{\mu_X \exp\left(-0.5\ln\left(1 + V_X^2\right) - \alpha_R\beta_t\sqrt{\ln\left(1 + V_X^2\right)}\right)}$$
$$= \exp\left((\alpha_R\beta_t - 1.645)\sqrt{\ln\left(1 + V_X^2\right)}\right) \tag{7}$$

where V_X is the COV of the material properties, α_R is the sensitivity factor for resistance in the FORM analysis, which can be assumed equal to 0.8 [10], and β_t is the target reliability index.

Obviously, the verification consists in checking that the design value of the resistance R_d is not less of the corresponding design value of the action effects E_d:

$$R_d \geq E_d. \tag{8}$$

The design value of the resistance, R_d, should be estimated considering the design value of the material properties, X_d, the geometry a_d and the model uncertainty θ_d. In particular, a model uncertainty factor for the resistance γ_{Rd} is defined, which takes into account the uncertainties in the resisting model and geometrical deviations if these are not modelled explicitly, so that the design value of the resistance, R_d results.

$$R_d = R\left(\frac{X_k}{\gamma_M}; a_d\right), \quad \text{where } \gamma_M = \gamma_{Rd}\gamma_m \tag{9}$$

The partial factor for model uncertainty, γ_{Rd}, can be obtained, in case of normal distribution, as the ratio

$$\gamma_{Rd} = \frac{1}{\mu_\theta (1 - \alpha_R \beta_t V_\theta)}, \tag{10}$$

and in case of Lognormal distribution as

$$\gamma_{Rd} = \frac{1}{\mu_\theta \exp(1 - \alpha_R \beta_t V_\theta)}, \tag{11}$$

where μ_θ is the mean value, V_θ is the coefficient of variation, and α_R is the sensitivity factor in the FORM analysis, assumed equal to 0.32 ("non-dominant resistance variable"). γ_{Rd} is generally set equal to 1.1 for new structures [2], but higher values are proposed for existing masonry structures, which detailed identification is limited [2]. In the following calculations a Lognormal distribution with $\mu_\theta = 1$ and $V_\theta = 0.18$ is assumed.

Adopting the distributions previously obtained for the masonry compressive strength, f_m, the partial factor γ_M can be evaluated depending on the target reliability, β_t, combining Eqs. (6) and (11). In Fig. 4, the results are reported for the whole population of the dataset and for the three identified sub-classes of stone masonry, depending on the adopted β_t.

Target values for the reliability index are given for new structures in the Annex C of EN1990 [10] depending on the consequences classes (CC) of the building, i.e. the "*categorization of the consequences of structural failure in terms of loss of human lives or personal injury and of economic, social, or environmental losses*" [10]. Three consequence classes are defined, CC1 (low) CC2 (medium) and CC3 (high), and the corresponding reliability levels are 4.2, 4.7 and 5.3 for one-year reference period,

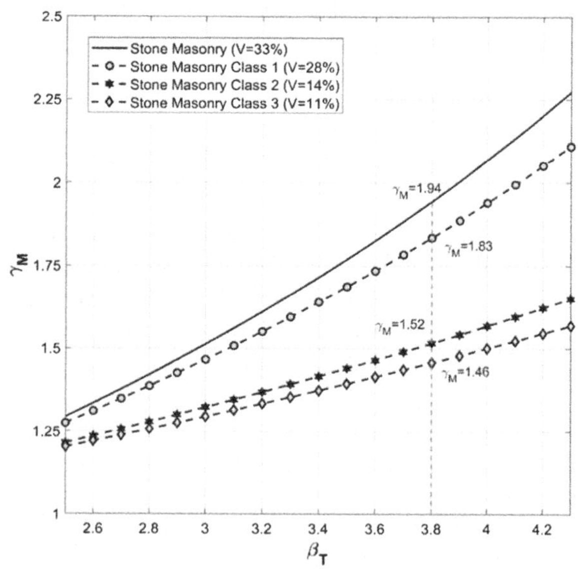

Fig. 4 Partial factor γ_M for existing stone masonry compression strength, variation with target reliability

Table 2 Characteristic and design values of masonry strength and partial factors

Masonry	f_m(N/mm^2)	$f_{m,k}$(N/mm^2)	γ_M	f_d(N/mm^2)
Stone masonry ($\beta_t = 3.8$)	1.88	0.86	1.94	0.45
Stone masonry class 1 ($\beta_t = 3.8$)	1.14	0.61	1.83	0.33
Stone masonry class 2 ($\beta_t = 3.8$)	1.92	1.47	1.52	0.97
Stone masonry class 3 ($\beta_t = 3.8$)	2.68	2.18	1.46	1.50
Rubble stone [14]	1.00–2.00		2.00–3.00	0.33–1.00
Undressed stone [14]	2.00		2.00–3.00	0.67–1.00

while they are set equal to 3.3, 3.8 and 4.3 for a 50 year reference period. Most existing masonry buildings can be classified to consequences class of failure CC2 or CC3 in terms of loss of human life, but it must be highlighted, in case of ancient building, that the consequence of failure should be considered also for the unrecoverable loss of the historical value of the building. For existing structures, some reduction in the reliability index is often acceptable, a discussion about appropriate target reliability levels can be found, for example, in [22] and [23], but it must be highlighted that this concept is often not correctly applied.

Assuming a target value $\beta_t = 3.8$, the characteristic values, the partial factors and the design values for masonry compressive strength are reported in Table 2. These values are finally compared with those provided by the Italian Guidelines [14] in terms of mean values, also reported in the table.

4 Conclusions

The assessment of the structural performance of existing masonry buildings is still a critical issue due to the significant uncertainties characterizing the definition of masonry mechanical parameters. In the paper, a methodology based on Gaussian Mixture Model is outlined to identify masonry classes and their main statistical parameters, mean and coefficient of variation, starting from the analysis of a wide database of in-situ tests results collected by the authors.

In particular, material classes, representing low, medium and high quality stone masonry, are presented focusing on masonry compressive strength, together with the corresponding partial safety factors to be used for the structural verification of masonry walls under vertical loads according to the partial factor method implemented in the Eurocodes.

The obtained probability density functions for the masonry classes provide also a sound basis to perform reliability assessment of existing masonry buildings.

References

1. Croce, P., & Holicky, M. (2015). *Operational methods for the assessment and management of aging infrastructure*. TEP: Pisa, Italy.
2. Domański, T., & Matysek, P. (2018). The reliability of masonry structures—Evaluation methods for historical buildings. *Technical Transactions, 9,* 91–108.
3. Sykora, M., et al. (2013). Probabilistic model for compressive strength of historic masonry. In *Safety, reliability and risk analysis beyond the horizon* (ESREL 2013).
4. Marsili, F., et al. (2017). A Bayesian network for the definition of probability models for masonry mechanical parameters. In *14th IPW* (pp. 253–268). Springer.
5. Witzany, J., Čejka, T., Sykora, M., & Holický, M. (2015). Strength assessment of historic brick masonry. *Journal of Civil Engineering and Management*.
6. EN 1996-1-1:2005. (2005). *Eurocode 6—Design of masonry structures*. Brussels: CEN.
7. ASTM, American Society for Testing and Materials. (1991). Standard test method for in-situ measurement of masonry deformability properties using flat jack method. C 1197-91.
8. Croce, P., et al. (2018). Shear modulus of masonry walls: A critical review. *Procedia Structure Integrity, 11,* 339–346.
9. Croce, P., et al. (2020). In situ tests procedures for the evaluation of masonry mechanical parameters. In *4th International Conference on Protection of Historical Constructions, PROHITECH 2020*. Athens, Greece (Accepted for publication).
10. EN 1990:2002. (2002). *Eurocode—Basis of structural design*. Brussels: CEN.
11. Henzel, J., & Karl, S. (1987). Determination of strength of mortar in the joints of masonry by compression tests on small specimens. *Darmstadt Concrete, 2,* 123–136.
12. Gucci, N., & Barsotti, R. (1995). A non-destructive technique for the determination of mortar load capacity in situ. *Materials and Structures, 28,* 276–283.
13. Croce, P., et al. (2021). Bayesian methodology for probabilistic description of mechanical parameters of masonry walls. *ASCE-ASME Journal of Risk and Uncertainty in Engineering Systems, Part A: Civil Engineering, 7*(2), 04021008.
14. Council, I. P. W. (2019). *Guidelines for application of Italian building code*. Roma, Italy: Istituto Poligrafico e Zecca dello Stato (In Italian).
15. Borri, A., et al. (2015). A method for the analysis and classification of historic masonry. *Bulletin of Earthquake Engineering, 13*(9), 2647–2665.
16. Croce, P., Marsili, F., Klawonn, F., Formichi, P., & Landi, F. (2018). (2018) Evaluation of statistical parameters of concrete strength from secondary experimental test data. *Construction and Building Materials, 163,* 343–359.
17. Croce, P., et al. (2020). Statistical parameters of steel rebars of reinforced concrete existing structures. In *Proceedings of the 30th European Safety and Reliability Conference and the 15th Probabilistic Safety Assessment and Management Conference*, 4751–4757, Research Publishing: Singapore.
18. Press, W. H., Tevkolsky, S. A., Vetterling, W. T., & Flannery, B. P. (2007). *Numerical recipes, the art of scientific computing* (3rd ed.). NY: Cambridge University Press.
19. MacLachlan, G., & Peel, D. (2000). *Finite mixture models*. New York: Wiley.
20. Luechinger, P., et al. (2015). *New European technical rules for the assessment and retrofitting of existing structures* (Joint Research Center (JRC) 94918 Report).
21. Fédération internationale du béton (fib) Bulletin 80. (2016). *Partial factor methods for existing concrete structures*. Germany: DCC, Siegmar Kästl e.K.
22. Vrouwenvelder, T., & Scholten, N. (2010). Assessment criteria for existing structures. *Structural Engineering International, 20*(1), 62–65.
23. Sykora, M., & Holicky, M. (2012). Target reliability levels for the assessment of existing structures—Case study. In *Proceedings of IALCCE 2012*, Vienna, Leiden: CRC Press/Balkema.

FORM/SORM, SS and MCMC: A Mathematical Analysis of Methods for Calculating Failure Probabilities

Karl Breitung

Abstract A basic problem in structural reliability is the calculation of failure probabilities in high dimensional spaces. FORM/SORM concepts are based on the Laplace method for the pdf of the failure domain at its modes. With increasing dimensions the quality of SORM decreases considerably. The straightforward solution would have been to improve the SORM approximations. However, instead of this, a new approach, subset simulation (SS) was championed by many researchers. By the proponents of SS it is maintained that SS does not suffer from the deficiencies of SORM and can solve high-dimensional reliability problems for very small probabilities easily. However by the author in numerous examples the shortcomings of SS were outlined and it was finally shown that SS is in fact a disguised Monte Carlo copy of asymptotic SORM. The points computed by SS are converging towards the beta points as seen for example in the diagrams in many SS papers. One way to improve FORM/SORM one runs, starting near the modes i.e. beta points, MCMC's which move through the failure domain $F = \{\mathbf{x}; g(\mathbf{x}) < 0\}$ with $g(\mathbf{x})$ the LSF. With MCMC one can calculate integrals over F with the pdf $\phi(\mathbf{x})$, but not the normalizing constant $P(F)$. However, a little artifice helps. Comparing the failure domain with another having a known probability content; not $P(F)$ has to be estimated, but the quotient of these two probabilities. A good choice for this is $F_L = \{x; g_L(\mathbf{x}) < 0\}$ given by the linearized LSF $g_L(\mathbf{x})$, so $P(F_L) = \Phi(-|\mathbf{x}^*|)$ with $\mathbf{x}^*$ the beta point. Running two MCMC's, one on F and one on F_L by comparing them it is possible to obtain an estimate for the failure probability $P(F)$. Another way is to use a modified line sampling method. For each design point for a random set of points on the tangential plane the distance of the plane to the limit state surface on the ray normal to the tangential space is determined and the corresponding normal line integral. Improving FORM/SORM by MCMC adds the advantages of analytic methods to the flexibility of the Monte Carlo approach.

Keywords FORM/SORM · MCMC · Asymptotic approximations · Subset simulation

---

K. Breitung (✉)
TU Munich, Munich, Germany
e-mail: breitu@aol.com

© The Author(s), under exclusive license to Springer Nature Switzerland AG 2021
J. C. Matos et al. (eds.), *18th International Probabilistic Workshop*, Lecture Notes in Civil Engineering 153, https://doi.org/10.1007/978-3-030-73616-3_26

1 Introduction

One of the central problems in structural reliability is the calculation of failure probabilities. In the beginnings in the seventies last century the mechanical/mathematical models consisted in most cases of less than ten variables. And then this was already computational challenging. The computational capacities have increased since then, but also the variables of which models are composed. So maybe for structures which were studied forty years ago it would be possible nowadays to find the relevant probabilities by crude Monte Carlo, but due to the much larger complexity of the models today pure brute force approaches again will not work for systems studied now.

An important aspect overlooked in most publications is: what is the purpose of these computations? R. Hamming said: *The purpose of scientific computing is insight, not numbers*. However in most of the algorithms for failure probability computations one sees only a lot of number crunching and nothing else. Is this really the purpose?

To understand the problem one has to look at the development of structural reliability in the last forty years. The structures studied became more complicated. In ye-olde times one could sketch with pencil and paper the limit state surface (here usually a curve). One needed for a known model the failure probability as additional information. Nowadays in most cases there is no intuitive understanding of the structure of the limit state surface anymore. So the data obtained by failure probability estimation methods should be used also to grasp this structure in some way.

Today there is a jungle of innumerable algorithms almost all based on imprecise heuristic arguments. The efficiency of these is often shown by three or four examples, the more complex of those are irreproducible due to the use of some FEM programs. And all want only to produce numbers.

In the opinion of the author one should look at these problems more from the viewpoint of structuralist concepts. In this view mathematical methods are instruments to find structures. Introductions to this philosophy can be found in [20, 23]. Some aspects are treated in [8].

To see what structures there are to find, one should remember the α-factors. They give a crude idea about the importance of the various random variables on the failure probability. So they are an example of seeing the failure probability computation problem not only as a forward but also as an inverse problem, i.e. to find out causes for the observed data. The differences between forward and inverse problems are outlined for example in [24]. Today where there many problems cannot be understood in an intuitive way, it is essential to try to grok what variables have influence on the failure. And here again the geometry of the limit state surface comes into the play. Studying its properties such information can be obtained.

2 FORM/SORM Approximations

These approximations seem to be forgotten since some years they had being discredited by various authors as inaccurate, see [14, 27]. The first example [14] does not make sense since the point which the authors identify as design or beta point is not the design point of the failure domain they investigate, the interior of the parabolic region. In the second paper it is only shown that for increasing dimensions the quality of SORM decreases. The problem with these arguments is that the authors do not see the essential role of the beta points as indicators for the important domains where the main mass of the probability in the failure domain lies. In [4] it is explained that the probability mass in the failure domain moves away from the beta point since with increasing dimensions the volume of the domain around it increases much faster. In [21] it is outlined that for starting MCMC algorithms one needs suitable starting points. In the case of multimodal PDF's these are the modes. Translated into structural reliability lingo, in the case of several disjoint failure domains, these are the beta points. So the objections towards the use of beta points are quite moot and not shared by the experts in MCMC. Certainly the quality decreases with increasing dimensions, but this can be remedied by using suitable MCMC methods as will be described in this paper.

The basic idea of SORM is to use asymptotic analysis concepts for finding simple approximations [6, 12]. In asymptotic analysis by a simple inverse transformation the large failure domain outside is transformed into a small domain near the origin. Here one can derive approximations by Taylor expansions at the origin (Fig. 1).

Given a failure domain $F = \{\mathbf{x}; g(\mathbf{x}) < 0\}$ who has beta points $\mathbf{x}_1, \ldots, \mathbf{x}_k$ with $|\mathbf{x}_i| = \beta$, a new failure domain F^* is defined by $F^* = \beta^{-1}F$. This failure domain has distance unity to the origin.

$$\mathrm{P}(F) \sim \Phi(-\beta) \sum_{j=1}^{k} \left[\prod_{i=1}^{n-1} \left(1 - \beta\kappa_i^{(j)}\right)^{-1/2} \right], \; \beta \to \infty \tag{1}$$

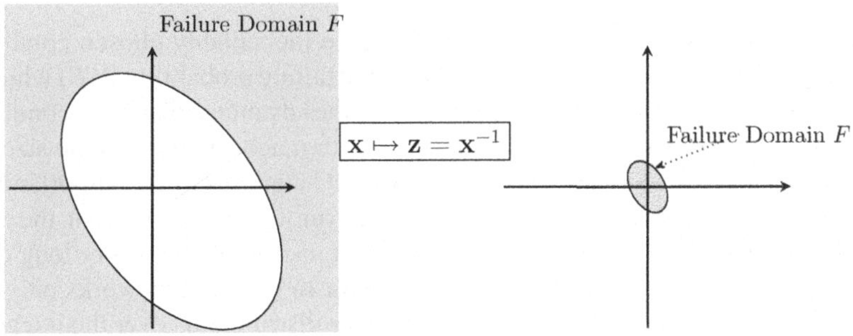

Fig. 1 Transformation for asymptotic analysis

Here the $\kappa_i^{(j)}$'s are the main curvatures of the limit state surface at the beta point $\mathbf{x}_j$. It is important to notice that the relation is valid if several beta points are present. The formula above gives an intuitive understanding about the influence of the structure of the limit state surface on the failure probability, however it requires an eigenvalue analysis of the Weingarten map (for details see [25], Chap. 12). An equivalent formula which requires only the Hessian of the LSF's at these points and no eigenvalue analysis was derived in [7].

How to use these FORM/SORM approximations as starting point for obtaining better approximations for the failure probability $\mathrm{P}(F)$ will be outlined in Sect. 4.2. Before the sequential methods which are used now often will be described and their deficiencies explained.

3 Subset Simulation (SS), Sequential Importance Sampling and Variants

The subset simulation concept is a variant of Monte Carlo methods; here it is tried to avoid the large number of data points needed in the usual Monte Carlo. This is done by using an iterative procedure. The algorithm can be seen as a sort of a global stochastic optimization procedure combined with MC integration.

While importance sampling methods try to improve the efficiency of Monte Carlo by identifying regions with high probability content and moving more date points there, SS starts from an enlarged failure domain whose beta points are much nearer to the origin and then moves step by step towards the original failure domain. These intermediate regions are defined in the form $F_i = \{g(\mathbf{u}) < a_i\}$ where the a_i's are positive and $a_i \to 0$. The basic idea of the method (see [1, 2]) is to write the failure probability $\mathrm{P}(F)$ as a product of conditional probabilities

$$\mathrm{P}(F_n) = \mathrm{P}(F_1|F_0)\cdot\mathrm{P}(F_2|F_1)\ldots\mathrm{P}(F_n|F_{n-1}) = \prod_{k=0}^{n-1}\mathrm{P}(F_{k+1}|F_k)$$

Here $\mathbf{R}^n = F_0 \supset F_1 \supset F_2 \supset \ldots \supset F_n = F$. Since the suitably chosen conditional probabilities are relatively large compared with the failure probability $\mathrm{P}(F)$ which has to be estimated, such an access to the problem has the advantage that these conditional probabilities can be estimated more efficiently with much smaller sample sizes.

The proponents of SS claim that this is a MCMC algorithm. One important point in MCMC methods is that these chains have to run quite long to visit the whole integration domain. it is not correct to assume that instead of some very long chains one can replace them with very many short chains. In general this works only if the starting points of the short chains have a stationary distributions over the integration domain. However this is what should be obtained by the MCMC algorithm. This problem is a little bit tricky, it is a sort of *catch 22*. The justification that SS methods do this is usually the claim that the seeds of the SS sequence in the next domain F_{j+1}

have already a stationary distribution in this domain. However, this in the subset community often repeated claim is wrong. This has been shown in [8], based on earlier work by [5]. From this follows that the data points of the SS method have no stationary distribution in these failure domains F_i, i.e. their PDF is not equal to the standard normal density constrained to F_i. Therefore then the data of the algorithm cluster around the seed points since the chain length is very short, in general it is chosen as ten. In Fig. 2 this is shown.

A further clear disadvantage of the method is that with decreasing failure probability the corresponding error variances for the estimator are increasing. The problems in SS methods are:

1. The calculated data sets can converge to local minima and not to the beta points (see [10, 11]),
2. The variance of the estimators increases with decreasing failure probabilities,
3. The information about the location of the beta points obtained from the convergence of the data points is ignored (see [9]).

In the published examples the problem under (1) never shows up; since all examples have a well-behaved LSF which is a linear or homogeneous function. However if one has a complicated black-box algorithm which produces the values of the LSF, how can one verify that the LSF is well-behaved?

The examples the author studied in [10] show that already in simple cases all these methods can lead to wrong results. The trust that the SS-community sets into theirs methods is psychologically understandable. However, is it justified? The counterexamples show that there cannot be a general proof of convergence for these approaches. Maybe one can give a proof for special cases of well-behaved LSF's. This would not be sufficient for the claim that the SS approach works also for complex high-dimensional cases. The proponents of SS refrain from a precise mathematical explanation what the algorithm is doing and why it is working. One problem which makes it difficult to understand the interior machinery is certainly that it is a twisted up combination of stochastic minimization and integration. Everybody who looks at the diagrams of SS calculations can see that the point clouds converge towards the beta points [13, 18]. However, in the vast SS literature the concept of beta points is

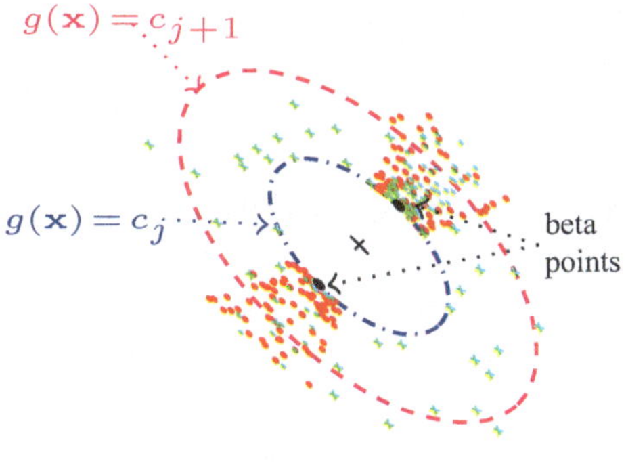

Fig. 2 The distribution of points for stationary MCMC (crosses) and for SS (dots)

practically never mentioned and is more or less subject to a *damnatio memoriae*. Only in analyzing the algorithm in context with FORM/SORM concepts and asymptotic analysis methods one is able to understand what is going on in the algorithm. IN [19] it si claimed that SS is a form of sequential importance sampling and the convergence proofs for these methods can be applied therefore. The problem with these proofs is that practically the show that everything converges to almost everything; in reality it is essential that the algorithms works in an acceptable time frame.

The gist of the arguments for SS is that no information about the structure of the problem is necessary. The user starts the algorithm and automatically a failure probability estimate pops up as result. This may be very nice for the people applying the method, but doing it this way any structural information is lost completely. And there is the clear danger that due to a lack of understanding the structure wrong results are not recognized. In [17], p. 394, it is written about the indiscriminate application of MCMC algorithms:

For MCMC, an extremely naive user can generate a lot of output without even understanding the problem. The lack of discipline of learning about the problem that other methods require can lead to unfounded optimism and confidence in the results.

The important information about the location of the beta points is thrown away. This seems to follow from the ideology of SS which claims to be basically different from the FORM/SORM approach, so any mention of these points is considered almost as an anathema.

The philosophical view here seems to be that the goal of structural reliability methods for failure probability calculation is only to produce numbers, i.e. probabilities. This contradicts the opinion of the author that failure probability calculations should be seen more under a structuralistic point of view. The efficiency of the SS/SiS approach has been illustrated by a deluge of examples. However all these examples have the same deficiency, the structure of the graph of the LSF's. The LSF' are always of a simple structure, homogeneous functions or similar types.

The bold conclusion made by the proponents of these concepts that one can conclude from the good results obtained for simple examples that they work also for more complex structured cases is a fallacy. Examples are an important tool in mathematics if they are used in the right way. Especially counterexamples help to understand the limitations of mathematical results and they show the way to possible generalizations. In no way one can give a general proof that a method works by examples only.

4 Methods for Improving Form Estimates

Now if FORM/SORM is not exact enough and SS has a number of serious deficiencies, what can be done to get good estimates? Here two approaches will be considered, Bennett's accceptance ratio and line sampling.

4.1 Bennett's Acceptance Ratio

The method outlined here for calculating normalizing constants will be applied in the next section to approximate failure probabilities.

A common case in MCMC is that a non-normalized PDF $q(\mathbf{x})$ is given, i.e. a function $q(\mathbf{x}) = p(\mathbf{x})/c$ over a domain D, where the value of p is unknown and $\int_D p(\mathbf{x})\, d\mathbf{x} = 1$. Using MCMC one can calculate approximations for integrals:

$$E = \int_D h(\mathbf{x}) p(\mathbf{x})\, d\mathbf{x} \tag{2}$$

without determining the constant c. The algorithms produce a sequence of random points $\mathbf{x}_i$ whose (one-dimensional) distribution converges towards the distribution with the PDF $p(\mathbf{x})$. In general, these algorithms need some burn-in time, before the distribution comes near the desired distribution.

To calculate the normalizing constant many methods have been proposed. One of these is Bennett's acceptance ratio abbreviated BAR. This method was invented by Bennett [3] and it received his name afterwards. Originally it was used for calculating free energy differences. This approach seems to be well suited for failure probability calculations. Here only the most simple form will be described. A number of ramifications is possible taking this as starting point. The following exposition is more or less a paraphrase of Sect. 2 in [16].

Given are two densities $p_i(\mathbf{x}),\ i = 1, 2$ with respect to the usual Lebesgue measure in the n-dimensional space. We know these densities only up to a normalizing constant, i.e.

$$p_i(\mathbf{x}) = \frac{q_i(\mathbf{x})}{c_i}. \tag{3}$$

Here D_i is the support of $p_i(\mathbf{x})$. It is assumed that the unnormalized PDF's $q_i(\mathbf{x})$ can be evaluated at each point $\mathbf{x} \in D_i$. Further it is assumed that the first normalizing constant c_1 is known. The algorithm will give a possibility to estimate the unknown c_2 comparing it to the known c_1.

Let now $h(\mathbf{x})$ be an arbitrary function which is defined on $D_1 \cap D_2$ the intersection of the domains such that

$$0 < \int\limits_{D_1 \cap D_2} |h(\mathbf{x}) p_1(\mathbf{x}) p_2(\mathbf{x}) d\mathbf{x}|\, d\mathbf{x} < \infty. \tag{4}$$

Such a function exists if and only if one has $\int_{D_1 \cap D_2} p_1(\mathbf{x}) p_2(\mathbf{x}) d\mathbf{x} > 0$. This quantity is measuring the overlap between the support of the both PDF's. Given now such a function $h(\mathbf{x})$, one can write the identity:

$$\frac{\int_{D_1 \cap D_2} h(\mathbf{x}) p_1(\mathbf{x}) p_2(\mathbf{x}) d\mathbf{x}}{\int_{D_1 \cap D_2} h(\mathbf{x}) p_1(\mathbf{x}) p_2(\mathbf{x}) d\mathbf{x}} = 1 \tag{5}$$

Replacing in the nominator and the denominator the functions $p_2(\mathbf{x})$ and $p_1(\mathbf{x})$ by the unnormalized densities, one obtains

$$
\begin{aligned}
1 &= \frac{\int_{D_1\cap D_2} h(\mathbf{x}) \left[\frac{q_2(\mathbf{x})}{c_2}\right] p_1(\mathbf{x})d\mathbf{x}}{\int_{D_1\cap D_2} h(\mathbf{x}) \left[\frac{q_1(\mathbf{x})}{c_1}\right] p_2(\mathbf{x})d\mathbf{x}} \\
&= \frac{c_1}{c_2} \times \frac{\int_{D_1\cap D_2} h(\mathbf{x})q_2(\mathbf{x})p_1(\mathbf{x})d\mathbf{x}}{\int_{D_1\cap D_2} h(\mathbf{x})q_1(\mathbf{x})p_2(\mathbf{x})d\mathbf{x}} \\
&= c_1 \times \underbrace{\frac{I\!E_1(h(\mathbf{x})q_2(\mathbf{x}))}{I\!E_2(h(\mathbf{x})q_1(\mathbf{x}))}}_{=r}
\end{aligned}
\tag{6}
$$

Here $I\!E_i(.)$ denotes the expected value with respect to the probability measure with PDF $p_i(\mathbf{x})$. Since the unnormalized densities q_i are zero outside of D_i, one can replace the integral over $D_1 \cap D_2$ by the integral over D_i in the nominator and denominator, respectively. So now, if one knows the quantities on the right side or can estimate them, one obtains an estimator for the normalizing constant c_2. Important here is that the function $h(\mathbf{x})$ can be chosen freely as long as Eq. (4) is satisfied. So it can be adjusted to simplify the function or to decrease the variance of the estimation. Various choices for it are discussed in [16].

The expected values in the fraction can be estimated using MCMC methods. Assume now that one has two MCMC chains, one producing points $\mathbf{x}_i$ with the PDF p_1 as target distribution and the other points $\mathbf{y}_j$ with target distribution PDF p_2, each with run length n. Then an estimator of r is given by:

$$
\hat{r} = \frac{\sum_{i=1}^{n} h(\mathbf{x}_i)q_2(\mathbf{x}_i)}{\sum_{j=1}^{n} h(\mathbf{y}_j)q_1(\mathbf{y}_j)}
\tag{7}
$$

4.2 Failure Probability Calculation Using Bennett's Acceptance Ratio

Bennett's acceptance ratio can be used for calculating failure probabilities. In this case the difference between the both integrals is only the integration domain. The integrand is always the PDF. In the standard normal case it is $\varphi(\mathbf{x})$. Again the idea is to compare a failure domain whose probability content is to be estimated with another one whose probability content is known. Obvious choices for domains with known contents are the domains defined in the FORM/SORM algorithms.

The failure domains for the FORM/SORM approximation functions $F_L = \{g_L(\mathbf{x}) < 0\}$ and $F_Q = \{g_Q(\mathbf{x}) < 0\}$ are approximating the original failure domain $F = \{g(\mathbf{x}) < 0\}$. Certainly these domains might be not so accurate approximations for the failure domain if the dimension is high and/or the shape is complex.

However, in general these domains will be nearer to F than the domains $\{g(\mathbf{x}) < c_1\}$ with $c_1 > 0$ usually chosen a starting point in SS/SuS/SiS methods. Further one can assume that in the most cases the probabilities $\mathrm{P}(F)$, $\mathrm{P}(F_L)$ and $\mathrm{P}(F_Q)$ are in the same order of magnitude. In the case of SS and similar methods the first probability $\mathrm{P}(F_1)$ is usually taken as 10^{-2} which is several orders of magnitude away from the true failure probability in most examples.

Let now be given two domains F_1 and F_2. For F_1 one knows the probability $\mathrm{P}(F_1)$, and $\mathrm{P}(F_2)$ has to be estimated. If one takes as F_1 either the linearized domain or the quadratic approximation domain, one can be certain that the domains overlap, since a neighborhood of the beta point is contained in the approximating domain and the original one, i.e. the sets D_1 and D_2 do overlap is satisfied.

For the failure probability estimation, one considers the two PDF's p_i and its unnormalized densities q_i

$$p_i(\mathbf{x}) = \frac{\varphi_n(\mathbf{x})\mathbf{1}_{F_i}(\mathbf{x})}{\mathrm{P}(F_i)}, \; q_i(\mathbf{x}) = \mathbf{1}_{F_i}(\mathbf{x})\varphi_n(\mathbf{x}), \; i = 1, 2 \tag{8}$$

with $\varphi_n(\mathbf{x})$ the n-dimensional standard normal density. Taking as the function $h(\mathbf{x})$ the inverse of the standard normal PDF $\varphi_n(\mathbf{x})$, i.e. $h(\mathbf{x}) = 1/\varphi_n(\mathbf{x})$ gives then

$$h(\mathbf{x})q_i(\mathbf{x}) = \mathbf{1}_{F_i} \text{ on } F_1 \cap F_2 \tag{9}$$

Inserting this into Eq.(6) one obtains

$$\mathrm{P}(F_2) = \mathrm{P}(F_1) \times \underbrace{\frac{\mathrm{P}_1(F_1 \cap F_2)}{\mathrm{P}_2(F_1 \cap F_2)}}_{=r_p} \tag{10}$$

Here analogously one writes $\mathrm{P}_i(A) = I\!E_i(\mathbf{1}_A)$.

The quotient on the right side one can estimate running two MCMC's, one on F_1 and the other on F_2. Denoting the points of the first chain by $\mathbf{x}_i$ and those of the second chain by $\mathbf{y}_j$, one obtains using Eq. (7) the following estimator for

$$\widehat{r_p} = \frac{\sum_{i=1}^{n} \mathbf{1}_{F_2}(\mathbf{x}_i)}{\sum_{j=1}^{n} \mathbf{1}_{F_1}(\mathbf{y}_j)} \tag{11}$$

or written with the number sign # as

$$\widehat{r_p} = \frac{\#\{\mathbf{x}_i;\, i = 1 \ldots n,\ \mathbf{x}_i \in F_2\}}{\#\{\mathbf{y}_j;\, j = 1 \ldots n,\ \mathbf{y}_j \in F_1\}} \tag{12}$$

So the quotient r_p can be estimated by counting the points in the Markov chains over the sets $F_i,\ i = 1, 2$ which are in the other set.

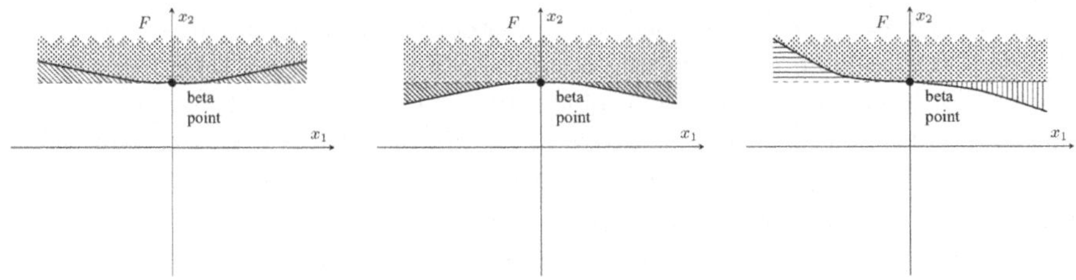

Fig. 3 Bennett's acceptance ratio for failure probabilities

Given a failure domain F_1 whose probability content $\mathrm{P}(F_1)$ is known, one obtains as estimate for $\mathrm{P}(F_2)$

$$\widehat{\mathrm{P}(F_2)} = \mathrm{P}(F_1) \times \widehat{r_p} \tag{13}$$

As starting point one can take the linear FORM approximation. The probability $P_L = \{g_L(\mathbf{x}) < 0\}$ is compared with the true failure probability $\mathrm{P}(F)$.

In Fig. 3 three cases are illustrated. As reference set with known probability the linear approximation is taken, its limit is denoted by the dashed horizontal line. In the first case $F_2 \subset F_1$, one has to estimate the probability of the set $F_1 \setminus F_2$. In the second case $F_1 \subset F_2$, here the probability of $F_2 \setminus F_1$ has to be found. In the third case there are two sets $F_1 \setminus F_2$ (horizontal lines) and $F_2 \setminus F_1$ (vertical lines) whose joint probability has to be determined.

4.3 *Line Sampling Using FORM*

Here the linear FORM approximation is used as initial failure probability estimate. Assume that the design point is rotated into the x_n-axis, i.e. it is $(0, \ldots, 0, \beta)$. Then a sample of N $(n-1)$-dimensional standard normal random vectors $\mathbf{y}_i$ are created and in the usual way an importance sampling estimate for the ratio $\mathrm{P}(F)/\mathrm{P}(F_L)$ is made by

$$\frac{\mathrm{P}(F)}{\mathrm{P}(F_L)} \approx \sum_{i=1}^{N} \frac{\Phi(-\beta - k(\mathbf{x}))}{\Phi(-\beta)} \tag{14}$$

and

$$\mathrm{P}(F) \approx \mathrm{P}(F_L) \cdot \left[\sum_{i=1}^{N} \frac{\Phi(-\beta - k(\mathbf{x}))}{\Phi(-\beta)} \right] \tag{15}$$

Fig. 4 Definition of $h(\mathbf{u})$

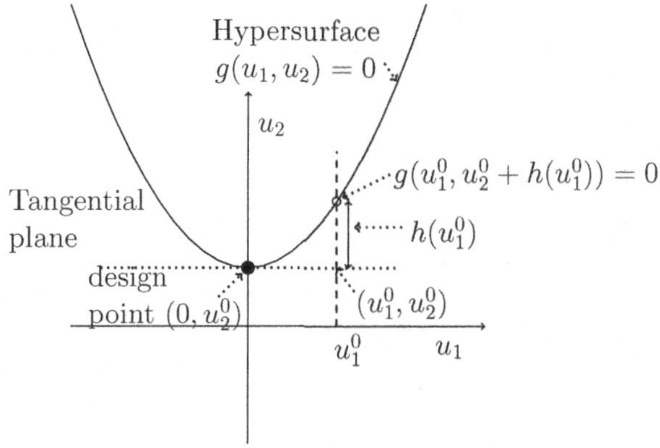

Here the function $k(\mathbf{x})$ defined as the root of the equation $g(\mathbf{y} + \lambda \mathbf{e}_n) = 0$ with $\mathbf{y} = (y_1, \ldots, y_{n-1}, \beta)$ and $\mathbf{e}_n = (0, \ldots, 0, 1)$, i.e. the x_n-coordinate minus β of the point where the line $\mathbf{y} + \lambda \mathbf{e}_n$ meets the limit state surface, see Fig. 4.

This is similar to the line sampling approach in [28]; the difference is that there no design points are used and the orientation of the hyperplane from which it is sampled, it determined by an iterative process. If there are several design points it is unclear where the line sampling method will find the "important direction".

5 Examples

To give an idea of the quality of the approximations the concept of the *efficiency* of statistical estimators is used (see [15]). The efficiency eff of an MC estimator $\hat{p}$ is given by

$$\text{eff} = \left[n \cdot \text{MSE}(\hat{p})\right]^{-1} \tag{16}$$

with the mean square error defined by $\text{MSE}(\hat{p}) = I\!E((\hat{p} - p)^2) + \text{var}(\hat{p})$ and n the number of samples. This allows to compare estimators with different sample sizes.

5.1 Approximation of a Parabolic Failure Domain

A very simple example to illustrate the methods. Given is a parabola:

$$g(x_1, x_2) = \beta - x_2 - \frac{\kappa}{2} x_1^2 \tag{17}$$

Table 1 Bennett's acceptance ratio

Log of failure probability	Mean of estimate	Standard deviation	Mean square error	Efficiency
−4.9935	−4.9985	0.12717	0.12726	0.0039290

Table 2 Design point line sampling

Log of failure probability	Mean of estimate	Standard deviation	Mean square error	Efficiency
−4.9935	−4.9924	0.019783	0.00039261	1.2736

As approximating failure domain the linear approximation $F = \{\beta - x_2 < 0\}$ is taken with probability content $\Phi(-\beta)$. As parameters were taken $\beta = 4$ and $\kappa = -2$. The exact failure probability is $1.0150e - 5$. For the data sets with $n = 2000$ the following results were obtained. The logarithms are to the base 10 (Tables 1 and 2).

The example shows that the line sampling approach is much more efficient that the Bennett' acceptance ratio method.

5.2 *Sum of Exponential Random Variables*

This example is from [6]. It appears also in [18] as an example for SS algorithms. Unfortunately a comparison of the results is not possible, since in this paper important information about the performance of the method is missing, i.e. the number of SS steps, the bias of the estimates and the moments of the logarithm of the estimator; therefore here the SS algorithm in [26] is used for comparisons. The coefficient of variation of the estimates given there is no good indicator of its variation, since the histogram of the data is quite skewed to the right (see [22] for an explanation). So the error of the estimator of the logarithm is taken as quality measure.

Given are k independent random variables $Y_1, \ldots, Y_k$, each with a standard exponential distribution, i.e. with PDF $f(x) = \exp(-x)$ for $x \geq 0$. The sum $Y = \sum_{j=1}^{k} Y_j$ has then an Erlang distribution with shape parameter n and rate 1. The LSF is given by $g(\mathbf{y}) = n + \alpha\sqrt{k} - \sum_{i=1}^{K} y_i$. Now this is transformed into the standard normal space.

There is a unique beta point at:

$$\mathbf{z} = (z, \ldots, z) \text{ with } z = -\Phi^{-1}\left(\exp\left(-\frac{\alpha}{\sqrt{k}} - 1\right)\right) \tag{18}$$

In the original paper there are typos in the equations (19) and (20a, b). The corrections are: Equation (19) corrected:

$$z = -\Phi^{-1}\left(\exp\left(-\frac{\alpha}{\sqrt{k}} - 1\right)\right)$$

Equation (20a) corrected:

$$J_1 = \left\{1 - z\left[\frac{\varphi(z)}{\Phi(-z)} - z\right]\right\}^{k-1}$$

Equation (20b) corrected:

$$P(\tilde{g}(\underline{X}) < 0) \sim \Phi(-\sqrt{k}\,z) \cdot \left\{1 - z\left[\frac{\varphi(z)}{\Phi(-z)} - z\right]\right\}^{-(k-1)/2}$$

The curvature of the limit state surface at the beta point is constant

$$\kappa = \frac{z}{\sqrt{k}}\left[\frac{\varphi(z)}{\Phi(-z)} - z\right] \quad (19)$$

Now, for $k = 100$ SS and the design point line sampling approach methods are compared for $n = 500$ and $\alpha = 5,\ 6$. For SS the usual parameters are used, i.e. $p = 0.1$ (Table 3 and 4).

To be not too braggodocian, the comparison is not quite fair, since to the design point line sampling the costs fro finding the design point have to be added. However, this shows that the proposed line sampling method is at least competitive.

Table 3 Subset simulation

Log of failure probability	Mean of estimate	Standard deviation	Mean square error	Efficiency
−5.2274	−5.2506	0.10264	0.011073	0.025307
−6.8410	−6.9061	0.13661	0.022896	0.012311

Table 4 Design point line sampling

Log of failure probability	Mean of estimate	Standard deviation	Mean square error	Efficiency
−5.2274	−5.2640	0.16668	0.029124	0.0049476
−6.8410	−6.8783	0.18850	0.036924	0.0032867

6 Conclusions

Here methods for improving FORM/SORM estimates were outlined. The method starts from the regions where the PDF in the failure domain is maximal and the difference between the approximation at the start and the true failure probability is much smaller. This is based on the line sampling method but using the design points. With this concept one has again a meaning for the geometrical structure of the failure domain and the beta points for high dimensional problems. This approach allows a number of generalizations and variations which can be tailored to suit complex problems.

References

1. Au, S. K., & Beck, J. L. (2001). Estimation of small failure probabilities in high dimensions by subset simulation. *Probabilistic Engineering Mechanics, 16*, 263–277.
2. Au, S.-K., & Wang, Y. (2014). *Engineering risk assessment with subset simulation*. New York: Wiley.
3. Bennett, C. H. (1976). Efficient estimation of free energy differences from Monte Carlo data. *Journal of Computational Physics, 22*, 245–268.
4. Betancourt, M. (2017). A conceptual introduction to Hamiltonian Monte Carlo. In: *arXiv e-prints*, S. 1701.02434.
5. Botev, Z., & Kroese, D. (2012). Efficient Monte Carlo simulation via the generalized splitting method. *Statistics and Computing, 22*, 1–16.
6. Breitung, K. (1984). Asymptotic approximations for multinormal integrals. *Journal of the Engineering Mechanics Division ASCE, 110*, Nr. 3, S. 357–366.
7. Breitung, K. (2015). 40 years FORM: Some new aspects? *Probabilistic Engineering Mechanics, 42*, 71–77. https://doi.org/10.1016/j.probengmech.2015.09.012. ISSN 0266–8920
8. Breitung, K. (2018). On subsets and onions: Lost in outer space. In *Proceedings of the Joint ICVRAM ISUMA UNCERTAINTIES Conference 2018*. Available on researchgate.net.
9. Breitung, K. (2019). SORM, subset simulation and simulated annealing. In *IPW2019, 17th International Probabilistic Workshop*, 11–13 September, Edinburgh, UK, S. 57–61.
10. Breitung, K. (2019). The geometry of limit state function graphs and subset simulation. *Reliability Engineering and System Safety, 182*, 98–106.
11. Breitung, K. (2020). Improvement of FORM/SORM estimates by Markov Chain Monte Carlo. In E. Zio, (Hrsg.): *Proceedings of ESREL 2020* to appear.
12. Breitung, K., & Hohenbichler, M. (1989). Asymptotic approximations for multivariate integrals with an application to multinormal probabilities. *Journal of Multivariate Analysis, 30*, 80–97.
13. Cui, F., & Ghosn, M. (2019). Implementation of machine learning techniques into the Subset Simulation method. Structural Safety, S. 12–25.
14. Katafygiotis, L., & Zuev, K. (2007). Geometric insight into the challenges of solving high-dimensional reliability problems. In: *Probabilistic Engineering Mechanics, 23*, Nr. 2, S. 208–218.
15. Lemieux, C. (2009). *Monte Carlo and Quasi-Monte Carlo sampling*. Dordrecht : Springer (Springer Series in Statistics).
16. Meng, X.-L., & Wong, W. (1996). Simulating ratios of normalizing constants via a simple identity: A theoretical exploration. *Statistica Sinica, 6*, 831–860.
17. Monahan, J. (2011). *Numerical methods in statistics*. Cambridge: Cambridge University Press.
18. Papaioannou, I., Betz, W., Zwirglmaier, K., & Straub, D. (2015). MCMC algorithms for subset simulation. *Probabilistic Engineering Mechanics, 41*, 89 –103. https://doi.org/10.1016/j.probengmech.2015.06.006. ISSN 0266–8920.

19. Papaioannou, I., Papadimitriou, C., & Straub, D. (2016). Sequential importance sampling for structural reliability analysis. *Structural Safety, 62*, 66 –75. https://doi.org/10.1016/j.strusafe.2016.06.002. ISSN 0167–4730.
20. Piaget, J. (1971). *Structuralism*. London, UK: Routledge & Kegan Paul PLC, 1971. Translated from the French.
21. Pompe, E., Holmes, C., & Łatuszyński, K. (2018). A framework for adaptive MCMC targeting multimodal distributions. In: *arXiv e-prints*, Dec, S. arXiv:1812.02609.
22. Ramsey, F., & Schafer, D. (2012). *The statistical sleuth: A course in methods of data analysis*. Boston: Cengage Learning.
23. Rickart, C. (1995). *Structuralism and structures: A mathematical perspective*. Singapore: World Scientific.
24. Tarantola, A. (2005). *Inverse problem theory*. Philadelphia, PA: SIAM.
25. Thorpe, J. (1979). *Elementary topics in differential geometry*. New York: Springer.
26. Uribe, F. (2018). *Subset simulation*. Engineering Risk Analysis Group Technische Universitat Munchen.
27. Valdebenito, M. A., Pradlwater, H., & Schuëller, G. : The role of the design point for calculating failure probabilities in view of dimensionality and structural nonlinearities. *Structural Safety, 32*, Nr. 2, S. 101–111.
28. Zio, E. (2013). *The Monte Carlo simulation method for system reliability and risk analysis*. London, UK: Springer.

Fractile Based Sampling Procedure for the Effective Analysis of Engineering Structures

Alfred Strauss, Beatrice Belletti, and Thomas Zimmermann

Abstract The non-linear analysis of the performance of engineering structures requires in general a huge computational effort. Moreover, in some cases a model updating procedure is needed. In this contribution, a model updating procedure has been applied for the simulation of pre-stressed reinforced concrete (RC) beams. The combined ultimate shear and flexure capacity of the beams is affected by many complex phenomena, such as the multi-axial state of stress, the anisotropy induced by diagonal concrete cracking, the interaction between concrete and reinforcement (bond), and the brittleness of the failure mode. Spatial distribution of material properties may be considered by random fields. Furthermore, statistical and energetic size effects may influence the analysis. To incorporate all the mentioned affects within a probabilistic analysis by using Monte Carlo simulation, feasibility limits are achieved quickly. Therefore, the aim was to improve the sampling technique for the generation of the realizations of the basic variables for, a general, computationally complex analysis tasks. The target was to develop a method similar to a simplified probabilistic method e.g. Estimation of Coefficient of Variation (ECoV). Therefore the so-called fractile based sampling procedure (FBSP) by using Latin Hypercube Sampling (LHS) has been developed. It allows a drastic reduction in the computational effort and allows the consideration of correlations between the individual basic variables (BV). However, fundamental aspect of the presented procedure is the appropriate selection of a leading basic variable (LBV). The appropriate choice of the LBV among the defined BVs is essential for mapping the correct correlation. Three methods for the determination of the LBV were investigated in this paper.

A. Strauss (✉)
Institute of Structural Engineering, University of Natural Resources and Life Sciences, Vienna, Austria
e-mail: alfred.strauss@boku.ac.at

B. Belletti
Department of Engineering and Architecture, University of Parma, Parma, Italy

T. Zimmermann
Building Inspection (MA37), City of Vienna, Vienna, Austria

© The Author(s), under exclusive license to Springer Nature Switzerland AG 2021

J. C. Matos et al. (eds.), *18th International Probabilistic Workshop*, Lecture Notes in Civil Engineering 153, https://doi.org/10.1007/978-3-030-73616-3_27

Keywords Fractile based sampling · Non-linear finite element analysis · Probabilistic analysis · Sensitivity and reliability

1 Introduction

The development of a non-linear numerical computational model for the representation of experimental data and further for reliability assessment tasks is a complex challenge and generally an updating procedure is needed. The load bearing capacity of a reinforced concrete beam is affected by different influences. These are e.g. multi-axial stress states, diagonal cracking causes by anisotropy of the material matrix, description of the bond interaction between concrete and reinforcement or brittle failure of the structural system. However, material properties vary in a spatial manner and might be considered by random fields. Furthermore, statistical and energetic size effects may influence the analysis. In order to incorporate all these influences on the bearing capacity, a non-linear finite element analysis (NLFEA) can be used.

For the analysis of the load bearing capacity in terms of shear and bending interaction limit state functions are needed, whereby the afore mentioned phenomena can be incorporated in a more or less detailed way. However, these parameters are generally not deterministic but rather probabilistic in order to incorporate uncertainties. Code bases traditional approaches simplify the problem by considering the uncertain parameters to be deterministic and to use partial safety factors to account for the uncertainties. Such an approach does not absolutely guarantee the required reliability level and the influence of individual parameters on the reliability is not determinable. Compared to that the application of a fully probabilistic (FA) approach can be used instead [1]. Verification of a structure with respect to a particular limit state is carried out via a model describing the limit state in terms of a function, whose value depends on all relevant design parameters. Verification of the limit states shall be realized by a probability-based method. The Model Code 2010 recommends to use different safety formats for verification of the limit state, see [2, 3]. A review of these safety formats can be found e.g. in [4–6]. The most common are the following:

- Semi-probabilistic approach: Computational requirements are significantly reduced, whereby the design value of response R is evaluated instead of the probability of failure.
- Global safety factor approach: It is defined in EN 1992-2 and allows only compressive type of failure. However, the study presented in [5] extended the application also to shear failure modes.
- The Estimation of Coefficient of Variation (ECoV) method: This is based on the semi-probabilistic approach, the difference among them consists in the procedure adopted to estimate the coefficient of variation and mean value of the response [7]. Only two simulations of NLFEA are required, the first one is carried out with mean values of basic variables (BV) and the second simulation with characteristic values.

- The extended ECoV method proposed in [4, 8] allows to evaluate not only the material uncertainties but also the model and geometrical uncertainties.

The probabilistic safety format, sometimes referred to as fully probabilistic method, allows explicitly including the reliability requirements in terms of the reliability index β and the reference period. This latter safety format may be used for structures to be designed and for existing structures in cases where such an increased effort is economically justified. However, the FP approach is less often used for the design of new structures due to lack of statistical data. It is often used in the assessment of existing structures in order to determine the residual service life. For the FP approach different simulation procedures, e.g. Monte Carlo (MC) or Latin Hyper Cube Sampling (LHS) can be used to generate realizations of the probabilistic variables. In case of MC a large number of simulations is needed in order to receive an accurate result. In combination with complex calculations this method is often not applicable [9]. Remedial measure is to use LHS instead. This technique is an advanced MC sampling procedure, firstly described in [10]. LHS allows a significant reduction of the required number of realizations and it allows also the incorporation of correlation between the BVs [11].

In addition to these classical methods, the objective within this research was to develop a more efficient sampling method which, approximates the ECoV method in terms of the computational effort but includes the features of the well-established LHS. Therefore the so-called Fractile-Based Sampling Procedure (FBSP) has been developed.

2 Experimental Data

In a comprehensive research project the behavior of pre-stressed reinforced T-shaped concrete beams were investigated. Thereby experimental investigations to characterize the material properties were carried out as well as large scale tests with proof-loading on the concrete beams with a comprehensive monitoring program were carried out. Based on this research, Table 1 shows the material characteristics in terms of mean value, coefficient of variation (cov) and an appropriate probability density function (PDF) obtained from experiments or code information. Further details on the research project can be found in [12, 13].

3 Probabilistic Sampling Procedure

The failure probability is depending mainly on the proper characterization of input BVs, computational models and sampling techniques which are needed to create the input samples from the BVs and to be used in the NLFEA.

Table 1 Input random variables, obtained from [12, 13]

Symbol	Mean	cov(%)	PDF	Unit
Concrete mix (C50/60 B4—28 Days), 3D non-linear cementitious				
f_c	77	16.4	GMB min EVI	MPa
f_{ct}	3.9	20.6	GMB max EVI	MPa
Ec	34.8	20.6	WBL min(3par)	GPa
G_F	219.8	32.8	GMB max EVI	J m^{-2}
Steel reinforcement (BSt 550B), multilinear diagram				
E_s	200	2	Normal	GPa
f_{ys}	550	4	Normal	MPa
Tendons (Cables—St1570/1770), bilinear diagram with hardening				
E_t	195	2.5	Normal	GPa
f_{yt}	1670	2	Normal	MPa
Pre-stressing force				
P	0.0418	6	Normal	MN

For structural engineering systems, typical BVs for capturing uncertainties are (a) material properties such as the elastic modulus of concrete or steel, (b) geometrical properties such as the cross section dimensions or the concrete cover of the reinforcement and (c) the environmental properties such as the chloride content in the air or the humidity.

For the probabilistic sampling procedure itself MC or LHS technique can be used. Whereby, LHS allows a significant reduction of the required number of realizations of the BVs due to a "controlled" random generator process [14]. The multipurpose probabilistic software for statistical, sensitivity and reliability analyses of engineering problems (FReET) is based on the LHS technique. FReET can be used for the estimation of statistical parameters of the structural response, the estimation of the theoretical failure probability, the sensitivity analysis, the response approximation and reliability-based optimization.

Now, in order to combine the low computational effort of ECoV method with the significance of a FP approach with respect to sensitivity and reliability the FBSP is proposed. Thereby the BVs are defined by PDFs and according to the LHS strategy the representative parameters of variables are selected randomly, being random permutations of integers $k = 1, 2, \ldots$ N and the representative value of each interval is the mean value [15, 16].

$$x_{i,k} = N \cdot \int_{y_{i,k-1}}^{y_{i,k}} x \cdot f_i(x) \cdot dx \quad with\ y_{i,k} = F_i^{-1}\left(\frac{k}{N}\right) \tag{1}$$

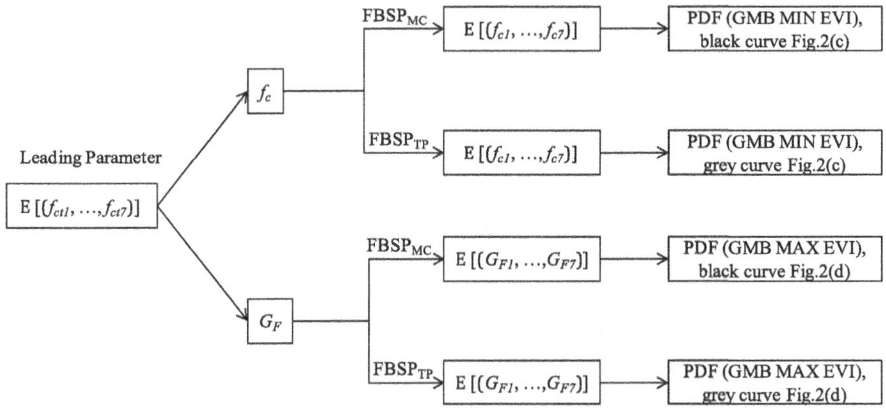

Fig. 1 Flow-chart explanation for the mechanical parameters estimation by using FBSP based on Model Code 2010 correlations ($FBSP_{MC}$) and based on experimental information ($FBSP_{TP}$), with f_{ct} as LBV

In the FBSP as a first step a set of 300 realizations of the BVs were generated by using LHS and as a second step sub-sample fields were extracted. The extracted sub-sample field were chosen by a so-called leading-basic-variable (LBV). By selecting a LBV X_i the randomization sets k for which X_{ik} of the LBV are closest to the predefined fractile $X_{i,p\%} = \{X_{i,5\%}X_{i,15\%}X_{i,30\%}X_{i,70\%}X_{i,85\%}X_{i,95\%}\}$ are extracted. The selection procedure showed that already for a small number of simulation sets k a very good mapping with the target correlation of the BVs could be achieved. The values of the BVs are differently coupled with respect to the LBV, while maintaining the correlations, e.g. the LBV $X_{i,5\%}$ is not necessarily associated with the LBV $X_{i,15\%}$ and it significantly influences the sample sets k, and the LBV also maybe significantly influences the simulation process e.g. the crack initiation and the crack pattern development associated with the NLFEA of concrete structures. In the following there are three approaches suggested for the appropriate determination of the LBV.

The whole FBSP was carried out by using two different resources (1) $FBSP_{MC}$ based on Model Code 2010 information and (2) $FBSP_{TP}$ based on information gathered from experimental data (proof loading of reinforced concrete beam). Figure 1 shows the extraction process of the FBSP specific sub-sample fields.

3.1 Determination of the LBV Using the "FBSP Based LBV" Procedure

In this approach, the reference is the structural response obtained by the application of a FP analysis. Further, simulations are performed for n FBSP sub-sample fields according to the n-selected LBVs. Consequently, the appropriate LBV can be determined from the consistency of the FBSP generated structural response with the

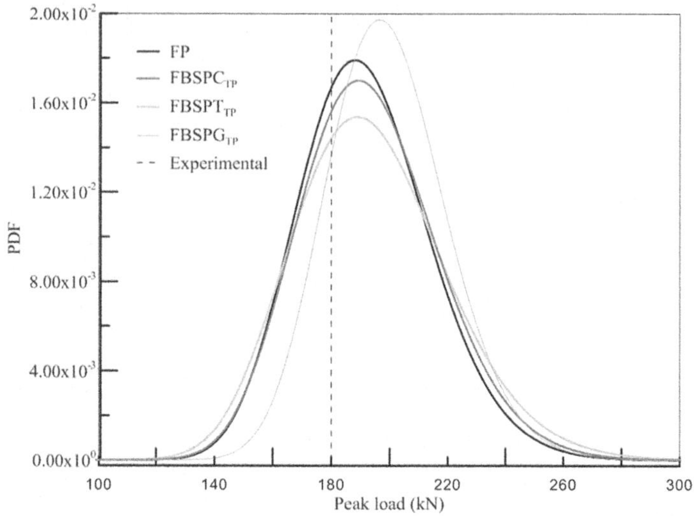

Fig. 2 Different log-normal responses curves by using FBSP adopting different LBVs: f_c (FBSPC), f_{ct} (FBSPT) and G_f (FBSPG) and response curve obtained by using FP approach

structural response obtained by FP analysis method. Within the investigation on the pre-stressed reinforced T-shaped concrete beam, in particular predefined fractiles (a) of the compressive strength f_c as LBV, (b) of the tensile strength f_{ct} as LBV and (c) of the specific fracture energy G_f as LBV and its accompanying parameters are extracted from the basic sample field. The computed PDFs of the structural responses of are shown in Fig. 2. The comparison with FP results using lognormal PDFs showed that f_c is an appropriate LBV. The major disadvantage of the "FBSP based LBV" procedure for determining the LBV is the necessity of processing the complete FP analysis.

3.2 Determination of the LBV Using the "Target Correlation Matrix" Procedure

In this approach, the target correlation matrix, defined for the BV of the FP analysis serves as a reference for the comparison between the correlation matrix obtained from the considered FBSP sub-sample fields (e.g. for n-LBV the correlation matrixes extracted from the n-FBSP sub-sample fields are analyzed and compared with the reference one).

The computation of the correlation matrix coefficients of the basic sample field as well as for the FBSP specific sub-sample fields can be computed by using the Pearson methodology. The big advantage of this procedure is that there is no analysis of the structural response for the determination of the LBV necessary, since only the BVs are used for the comparison of the correlations. The Pearson correlation coefficient is evaluated for each couple of BVs and the correlation matrix of FBSP with a specific LBV is setup. Hence, an error matrix can be derived between the exact correlation matrix $\mathbf{T_x}$ and the target correlation matrix $\mathbf{T}$ by using Eq. 2.

$$\mathbf{E} = \mathbf{T} - \mathbf{T_x} \tag{2}$$

To evaluate the error with respect of changing the LBV, the second-order-norm of matrix $\mathbf{E}$ is calculated. Denoting $\mathbf{A}$ as a generic matrix, $n_\mathbf{A}$ is the second-order-norm of $\mathbf{A}$, and is derived according to Eq. 3.

$$n_\mathbf{A} = \left[f\left(\mathbf{A}^\mathrm{T}.\mathbf{A}\right)\right]^{1/2} \tag{3}$$

Thereby $f(.)$ is a function that provides the maximum eigenvalue between all eigenvalues of the matrix in square brackets, which is equal to the scalar product between $\mathbf{A}$ and its transposed matrix.

3.3 Determination of the LBV Using the "Sensitivity Analyses" Procedure

In this approach, the simulation of the structural response with the basic sample field or with the FBSP specific sub-sample field serves as a basis for the sensitivity analyses considerations. For instance, the sensitivity analysis is processed for the maximum bearing capacity, as LBV those BVs are defined which have the largest impact on the bearing capacity. The big advantage of this procedure is that a FP analysis of the structural response for the determination of the LBV is not necessary, because the sensitivity analyses can also be performed on the FBSP specific sub-sample fields.

4 Comparison of Basic Sampling and FBSP

The LBV plays an important role in the context of FBSP. The LBV defines the composition of the sample set in the FBSP sub-sample field on the basis of the predefined fractile values. Figure 3a shows a comparison of the PDFs of f_{ct} and Fig. 3b shows the comparison of the PDFs of G_f. In both cases f_c was considered as LBV and the PDFs are generated from (a) basic sample field, (b) the $\mathrm{FBSP_{TP}}$ sub-sample field, and (c) the $\mathrm{FBSP_{MC}}$ sub-sample field. Figure 3c shows a comparison of the PDFs of f_c and Fig. 3d shows the comparison of the PDFs of G_f. In both cases, f_{ct} was considered as LBV and the PDFs are generated as before.

According to [1, 13], a Gumbel Maximum distribution (GMB MAX EVI) has been used for f_{ct} and G_f for the comparison of the PDFs. The comparison of the PDFs

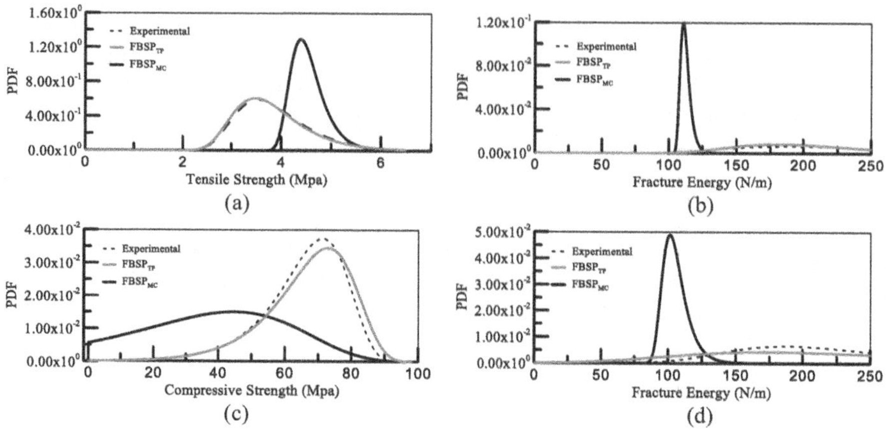

Fig. 3 PDFs for different mechanical parameters in which the probabilistic distribution parameters were obtained from the basic sample field (experimental in the legend), FBSP$_{TP}$ (by using target correlation matrix) and FBSP$_{MC}$ (by using Model Code 2010 correlations): (**a**) and (**b**) show the results if f_c is the LBV, (**c**) and (**d**) show the results if f_{ct} is the LBV

of f_{ct} and G_f generated from the FBSP$_{TP}$ sub-sample field using f_c as LBV with the PDF based on the basic sample field shows a very good agreement and provides a very strong argument for the proposed FBSP. On the other hand, the comparison of the PDFs generated based on the FBSP$_{MC}$ sub-sample fields shows a significant deviation in the mean as well as the standard deviation with respect to the PDFs of the basic sample field.

As can be seen from Fig. 3a, the mean value of f_{ct} is significantly higher and the standard deviation of the FBSP$_{MC}$-PDF is significantly smaller than the FBSP$_{TP}$-PDF or the experimental based PDF. Figure 3b shows a smaller mean and standard deviation for G_f by using FBSP$_{MC}$ with respect to the experimental based PDFs. It is evident, that the Model Code 2010 formulations produces a higher value of f_{ct} and a lower value of G_f compared to the experimentally derived PDFs. As can be seen from Fig. 3c the mean value of f_c is significantly lower and the standard deviation is significantly higher in case of FBSP$_{MC}$ compared to the FBSP$_{TP}$ or to the experimental based PDF. Figure 3d shows significant smaller mean and standard deviation values of G_f in case of FBSP$_{MC}$ with respect to the experimentally based PDFs.

5 Conclusions

Numerous simulation methods, which originated from the MC method and subsequently adapted for advanced probabilistic analyses, reliability and safety considerations, have been already established in the scientific community. These methods include e.g. the LHS method as well as the ECoV method. For time-consuming analyses, e.g. NLFEA of engineering structural components, including shear and normal force interaction as well as pre-stressed reinforcement, an average of up to 300 LHS simulations are required for a serious probabilistic statement. Conversely, if the ECoV method is adopted only 2 simulations are necessary, but this method cannot map correlations among the BVs and the failure modes.

Therefore, the FBSP method as presented can be located between the well-established LHS and the ECoV method. It could be shown, that a proper selection of the LBV results in a decrease of necessary simulations but still provides reliable predictions. However, the appropriate choice of the LBV among the defined BVs is essential for mapping the correct correlation.

Three methods for the determination of the LBV were investigated, in which the sensitivity-based correlation analysis provided the best results.

Acknowledgments Austrian Research Promotion Agency (FFG) and the National Foundation for Research, Technology and Development supported this work by the project OMZIN [FFG-N° 836472]. This paper also reports on the scientific results obtained by the University of Parma within the project PRIN (Project of Prominent National Interest, Italian Research) and financially co-supported by MIUR (the Italian Ministry of Education, University and Research). Finally, the support of DASSAULT SYSTEMES is also strongly acknowledged.

References

1. Vořechovská, D., Teplý, B., & Chromá, M. (2010). Probabilistic assessment of concrete structure durability under reinforcement corrosion attack. *Journal of Performance of Constructed Facilities, 24*(6), 571–579.
2. Belletti, B., Damoni, C., Hendriks, M. A. N., & De Boer, A. (2014). Analytical and numerical evaluation of the design shear resistance of reinforced concrete slabs. *Structural Concrete, 15*(3), 317–330.
3. Belletti, B., Pimentel, M., Scolari, M., & Walraven, J. C. (2015). Safety assessment of punching shear failure according to the level of approximation approach. *Structural Concrete, 16*(3), 366–380.
4. Schlune, H., Plos, M., & Gylltoft, K. (2011). Comparative study of safety formats for nonlinear finite element analysis of concrete structures. In *Applications of Statistics and Probability in Civil Engineering, Proceedings of the 11th International Conference on Applications of Statistics and Probability in Civil Engineering* (pp. 2542–2548).
5. Cervenka, V. (2013). Reliability-based non-linear analysis according to fib Model Code 2010. *Structural Concrete, 14*(1), 19–28.
6. Holický, M., & Sykora, M. (2010). Global resistance factors for reinforced concrete structures. In *Advances and Trends in Structural Engineering, Mechanics and Computation - Proceedings of the 4th International Conference on Structural Engineering, Mechanics and Computation, SEMC* (pp. 771–774).
7. Holický, M. (2006). Global resistance factors for reinforced concrete members. Presented at the ACTA POLYTECHNICA.
8. Schlune, H., Plos, M., & Gylltoft, K. (2011). Safety formats for nonlinear analysis tested on concrete beams subjected to shear forces and bending moments. *Engineering Structures, 33*(8), 2350–2356.

9. Hurd, C. C. (1985). A note on early Monte Carlo computations and scientific meetings. *Annual History Computing, 7*(2), 141–155.
10. McKay, M. D., Beckman, R. J., & Conover, W. J. (1979). Comparison of three methods for selecting values of input variables in the analysis of output from a computer code. *Technometrics, 21*(2), 239–245.
11. Vořechovský, M., & Novák, D. (2003). Efficient random fields simulation for stochastic FEM analyses. *Computational Fluid and Solid Mechanics, 2003,* 2383–2386.
12. Strauss, A., Krug, B., Slowik, O., & Novák, D. (2017). Combined shear and flexure performance of prestressing concrete T-shaped beams: Experiment and deterministic modeling. *Structural Concrete, 19*(1), 16–35.
13. Zimmermann, T., & Lehký, D. (2015). Fracture parameters of concrete C40/50 and C50/60 determined by experimental testing and numerical simulation via inverse analysis. *Journal of Fractional, 192*(2), 179–189.
14. Novák, D., Vořechovský, M., & Teplý, B. (2014). FReET: Software for the statistical and reliability analysis of engineering problems and FReET-D: Degradation module. *Advances in Engineering Software, 72,* 179–192.
15. Keramat, M., & Kielbasa, R. (1997). Efficient average quality index estimation of integrated circuits by modified Latin hypercube sampling Monte Carlo (MLHSMC). In *Proceedings—IEEE International Symposium on Circuits and Systems* (Vol. 3, pp. 1648–1651).
16. Huntington, D. E., & Lyrintzis, C. S. (1998). Improvements to and limitations of Latin hypercube sampling. *Probabilistic Engineering Mechanics, 13*(4), 245–253.

Fragility Curves for Fire Exposed Structural Elements Through Application of Regression Techniques

Ranjit K. Chaudhary, Ruben Van Coile, and Thomas Gernay

Abstract The structural fire engineering community has demonstrated a growing interest in probabilistic methods in recent years. The trend towards consideration of probability is, amongst others, driven by an understanding that further advances in detailed numerical models are potentially offset by the basic uncertainty in the input parameters. Consequently, there has been a call for the development of fragility curves for fire-exposed structural elements, to support the application of probabilistic methods both in design as well as in standardization. State-of-the-art structural fire engineering models are, however, commonly very computationally expensive, even for simple cases such as isolated structural elements. This can be attributed to the requirement of coupling thermal and mechanical analyses, and to the large non-linearity in both the heating of structural elements and the resulting mechanical effects of temperature-induced degradation and strains. This severely hinders the development of fragility curves beyond very specific cases, especially when including a stochastic description of the (natural) fire exposure. In the current contribution the application of regression techniques to structural fire engineering modeling is explored, as a stepping stone towards establishing a methodology for the efficient development of fragility curves for fire-exposed structural members. A simplified model with limited computational expense is applied to allow for validation of the proof-of-concept.

Keywords Structural fire engineering · Probabilistic method · Fragility curves · Regression techniques

---

R. K. Chaudhary (✉) · R. Van Coile
Ghent University, Ghent, Belgium
e-mail: RanjitKumar.Chaudhary@Ugent.be

T. Gernay
Johns Hopkins University, Baltimore, USA

© The Author(s), under exclusive license to Springer Nature Switzerland AG 2021

J. C. Matos et al. (eds.), *18th International Probabilistic Workshop*, Lecture Notes in Civil Engineering 153, https://doi.org/10.1007/978-3-030-73616-3_28

1 Introduction

The interest in probabilistic studies of structures under fire hazard has increased significantly in recent years. This is mainly because the available prescriptive design approaches are at times insufficient for the design of structures under fire hazard [1], due to the uncertain behavior of structural materials and the complexity of system behavior under variable fire hazard scenarios. Several investigations have been conducted till date to encompass the uncertainties of fire load and structural materials [1–5]. These investigations have shown the significant effect of randomness on structural behavior under fire hazard. Although these studies by themselves can provide adequate procedures to attain safe and reliable structural design, their implementation is arduous because of substantial computational expenses [5]. This huge computational expense for the development of structural fragility functions can be attributed to a wide range of factors, which include: structural analysis technique, number of realizations, nonlinear structural behavior, the need for a two-step analysis of the thermal response followed by the mechanical analysis, the need for modeling of structural assemblies given the effects of thermally-induced deformations and forces, etc. Therefore, the development of fragility curves through sampling schemes such as Monte-Carlo Simulations (MCS) or Latin-Hypercube Sampling (LHS) is impractical for any but the simplest cases [5].

The fragility curves derived through the above methodologies have a further drawback as they apply only to a specific case, i.e. for a particular geometrical condition or a certain fire scenario (e.g. single compartment with known nominal fire load density and ventilation characteristics undergoing flashover). This makes design iterations most cumbersome, as a change in compartment geometry or structural configuration requires a new fragility evaluation. In effect, it is therefore currently rarely worthwhile to expend large computational effort to quantify the structural vulnerability for a specific design iteration. To allow for probabilistic approaches to be integrated in structural fire engineering design, there is a need of a technique which can both simulate the entire complex non-linear system behavior, taking into account design and exposure uncertainties, and at the same time allows for fast design iterations. These requirements imply that the structural responses should be predicted quasi-instantly, so that fragility curves can be developed easily even if the structural parameters are varied. Consequently, machine learning techniques such as surrogate modeling by regression can be helpful in this regard [6]. Given adequate 'training' of the model, the surrogate model allows to approximate the output of a full non-linear coupled thermal and structural calculation, which would otherwise require large computation effort. Such surrogate model can then be efficiently used to quasi-instantaneously develop fragility curves for the structure for a wide range of scenarios.

The present study intends to explore the application of regression techniques to the field of structural fire engineering and explore a methodology for the development of fragility curves for structures subjected to fire hazard. Herein, a surrogate model is developed for a non-linear model of a reinforced concrete slab, which is available in the literature [7] and requires limited computational expense, allowing to readily compare and validate the response and obtain fragility curves, thereby validating the proof-of-concept. The paper describes the methodology for the probabilistic analysis of fire exposed structural elements, and then focusses on the efficacy and accuracy of surrogate models in reducing the computational expense for the probabilistic analysis.

2 Methodology: Applied Polynomial Regression

Machine learning techniques have been gaining wide popularity in almost every arena, owing to the development of new efficient computational technologies. Polynomial regression is a type of machine learning technique, in which polynomial equations are used as approximate (i.e. surrogate) models to predict the result of a physical system. The surrogate models developed through polynomial regression are quite flexible, straightforward and intuitive, and can be capable of predicting the result of complex non-linear systems. The first step involved in the application of polynomial regression is the identification of the independent variables, on which the physical model behavior depends. The identified independent variables are commonly denoted as the 'features' of the data in machine learning terms. If X represents the vector of independent variables ($x_1, x_2, \ldots, x_r$) of the physical system, then the response (Y) of the physical system is given by:

$$Y = g(X) \tag{1}$$

where, $g(X)$ refers to the function characterizing the physical system. Similarly, if $f(X)$ represents a surrogate model for the physical system, then the predicted response is given by:

$$Y_{\text{pred}} = f(X) \tag{2}$$

A first order polynomial equation for the polynomial regression, relating the features with the model output can be written as

$$f(X) = \theta_0 + \sum_{i=1}^{r} \theta_i x_i \tag{3}$$

where, $\theta_0, \theta_1, \ldots, \theta_r$ are referred to as regression coefficients, in which the first term, θ_0 is the bias term and θ_1 to θ_r are the weights associated with the polynomial terms in the function.

Assuming the independent variables have been properly selected, the accuracy of the polynomial model depends on the appropriateness of the chosen regression model (polynomial equation, and its degree), and how precisely it has been trained, which in turn depends on how well the independent variables are sampled. In other words, an appropriate sampling scheme needs to be adopted, which is effective in encompassing the entire sample space of the independent variables at reasonable computational cost. In this regard, LHS has been found to be an effective sampling method and has been adopted in this study [8].

To confirm the accuracy of the fitted surrogate model, three different sample sets are considered: a training set, a cross-validation (CV) set, and a test set. The training set comprises of a single LHS set, while cross-validation set and test set are randomly developed. The surrogate model considered is trained only for the training data set, while the cross-validation set and test set are used to adjust the learning parameters and assess the accuracy of the trained model, respectively. The learning parameters for the polynomial regression refer to the polynomial degree of surrogate model (*D*), and the regularization parameter (λ), while the accuracy is assessed in relation to the number of training data set samples (*n*). The regularization parameter prevents the overfitting of the surrogate model to the training data set.

The regression coefficients of the surrogate model ($\theta_0, \theta_1, \ldots, \theta_r$) are obtained through minimization of the 'cost function'. In case of polynomial regression, the cost function for training is given by Eq. (4), where *n* is the number of training samples considered and λ is the regularization parameter, which governs the magnitude of the penalty 'regularization' term to avoid overfitting during training.

$$J_{\text{learn}}(\theta) = \frac{1}{2n}\sum_{i=1}^{n}(f(X^{i}) - Y^{i})^2 + \frac{\lambda}{2n}\sum_{j=1}^{r}\theta_j^2 \tag{4}$$

To minimize the cost function, an optimization algorithm is needed. The 'gradient descent' algorithm is adopted in the current study [10]. For given λ and *D*, the regression coefficients are thus readily estimated through the minimization of the cost function. In order to determine the appropriate learning parameters, non-regularized costs are considered for the training and cross-validation data, i.e.:

$$J_{\text{train}}(\theta) = \frac{1}{2n_{\text{train}}}\sum_{i=1}^{n_{\text{train}}}(f(X^{i}_{train}) - Y^{i}_{train})^2 \tag{5}$$

$$J_{cv}(\theta) = \frac{1}{2n_{\text{cv}}}\sum_{i=1}^{n_{\text{cv}}}(f(X^{i}_{cv}) - Y^{i}_{cv})^2 \tag{6}$$

where, J_{train} and J_{cv} are the training and cross-validation errors, respectively. To train the surrogate model for the physical system, the LHS samples are scaled first as the independent variables have different units and range of magnitudes, thereby making it difficult in the convergence of the objective function. Thus in this study, the LHS samples are scaled by standardization technique, where the independent variables are scaled to have zero mean and unit variance, enabling faster convergence [9].

The accuracy of the developed surrogate model is assessed by estimating the value for the coefficient of determination (R^2) based on the test set data, which is given by

Draper and Smith [11]:

$$R^2 = 1 - R_{res}/R_{tot} \tag{7}$$

where, $R_{res} = \sum_{i=1}^{n_{test}} (Y_{test}^i - f(X_{test}^i))^2$ and $R_{tot} = \sum_{i=1}^{n_{test}} (Y_{test}^i - \overline{Y}_{test})^2$ refers to the explained variance and total variance for the model, respectively. The R^2 value ranges from 0 to 100%, with 100% indicating a perfectly fitted surrogate model. Once the desired accuracy is achieved (i.e. R^2 exceeds a predefined threshold), the surrogate model can be used as a predictive model, to simulate the response of the system and perform fragility analysis of the fire-exposed structure.

3 Fragility Curve for the Fire Exposed Concrete Slab

To validate the application of regression techniques as a tool to develop fragility curves for fire exposed structures, a simple test case is considered. Specifically, a simple model for the resisting moment of a fire-exposed reinforced concrete slab is applied, considering [7]. The resisting moment is given by:

$$M_{\mathrm{R}} = A_{\mathrm{s}} k_{f_{\mathrm{y}}(\mathrm{T})} f_{\mathrm{y},20\,^{\circ}\mathrm{C}} \left(h - c - \frac{\phi}{2} \right) - 0.5 \frac{(A_{\mathrm{s}} k_{f_{\mathrm{y}}(\mathrm{T})} f_{\mathrm{y},20^{\circ}\mathrm{C}})^2}{b f_{\mathrm{c},20^{\circ}\mathrm{C}}} \tag{8}$$

where, A_{s} is the tensile reinforcement in the slab, ϕ is the diameter of the reinforcing bars in the slab, h is the depth of the slab, b is the width of the slab, c is the concrete cover to the reinforcement, $f_{\mathrm{c},20\,^{\circ}\mathrm{C}}$ is the compressive strength of the concrete, and $f_{\mathrm{y},20\,^{\circ}\mathrm{C}}$ and $k_{\mathrm{fy(T)}}$ are respectively the 20 °C reinforcement yield strength and the yield strength retention factor at temperature T. The moment capacity of the concrete slab in the expression above depends on several independent factors and is highly non-linear. The temperature field in the slab is considered deterministic in the current test study. The slab is exposed to the ISO 834 temperature curve at the bottom, and the unexposed surface of slab is cooled by convection. The temperature profile developed across the slab depth is evaluated with a validated 1-D temperature model, as specified in [12], considering the Eurocode EN 1992-1-2:2004 nominal thermal properties for concrete. Therefore, the objective of the test case is to explore the effect of uncertainties in capacity for a concrete slab, while the thermal action is deterministic and can be understood as representative of a standard fire rating requirement.

3.1 Surrogate Modeling

In Eq. (8), the independent variables, on which the moment capacity of the slab depends, are identified as $f_{\mathrm{c},20\,^{\circ}\mathrm{C}}, f_{\mathrm{y,T}} = (k_{\mathrm{fy(T)}} \cdot f_{\mathrm{y},20\,^{\circ}\mathrm{C}})$, c, h and A_{s}. The parameters b

Table 1 Independent variables considered for regression

Independent variables	Lower limit	Upper limit	Unit	Distribution
Concrete strength ($f_{c,20\,°C}$)	15	80	MPa	Uniform
Rebar strength ($f_{y,T}$)	100	1000	MPa	
Concrete cover (c)	20	70	mm	
Slab thickness (h)	100	300	mm	
Reinforcement area (A_s)	0.10	0.25	% (section area)	

and ϕ are considered constant, with magnitude of 1 m (unit width) and 12 mm, respectively. The variable $f_{y,T}$ refers to the yield strength of reinforcing bars at elevated temperature (T), where T is evaluated considering the ISO 834 standard fire exposure duration, as specified above, and considering the axis position of the rebar [12]. In order to obtain a regression model which is capable of being used for design iterations, the features are specified by lower and upper limits. These lower and upper limits constitute boundaries for the regression model. As the regression model is intended to provide a fast and accurate prediction within parameter space defined by the limits in Table 1, a dispersed sampling scheme within this parameter space is required for training the surrogate model. An LHS scheme is adopted in sampling of the considered features, applying a uniform distribution over the respective parameter spaces. Table 1 summarizes the range and distribution of the independent variables adopted for the sampling scheme.

The sample space is standardized, as discussed earlier, to enable faster minimization of cost function during the development of surrogate model. The scaling is furthermore a necessity for an unbiased application of the regularization. The minimization of cost function is carried out using a gradient descent algorithm.

In the current study, the methodology for the development of surrogate model is demonstrated first. In this regard, a surrogate model to estimate moment capacity of the concrete slab is developed considering the assumed independent variables. Figure 1a shows the J_{train} and J_{cv}, calculated through Eqs. (5) and (6), for the surrogate models fitted with different degree of polynomials, considering parameters $\lambda = 0$, $n_{train} = 2500$ and $n_{cv} = 500$. It is observed that models with 2nd degree polynomial have considerably higher accuracy of prediction over the 1st degree of polynomial, while further improvement of accuracy over 2nd degree polynomial for higher degree polynomials is not substantial. Also, the surrogate model with 3rd and higher degree polynomial is considerably slower in optimization ('training'). Therefore, a surrogate model of 2nd degree polynomial is adopted for the further calculations in the study. Similarly, Fig. 1b allows to determine the optimum number of LHS training sample points to be considered for the regression, as J_{train} and J_{cv} are plotted against the number of training samples, considering a cross-validation set of a fixed size ($n_{cv} = 500$). To avoid stochastic variation to influence the estimation of J_{train} and J_{cv}, a repeated sampling approach is adopted. In effect, errors are estimated as the average result from 10^3 repetitions for the LHS samples. Based on the figure, a training sample size comprising of 2000 sample points ($N_{optimum}$) can be recommended for the considered fire exposed concrete slab and independent variables (with their ranges), while a model based on 100 sample points ($N_{\text{Insufficient}}$) is found to have higher errors and is considered insufficient for developing the surrogate model for the considered scenario.

Figure 1c shows the variation of the errors (J_{train} and J_{cv}) with the regularization parameter (λ) for the considered training sample size of 2000. The optimum value of λ refers to its value at minimum J_{cv}, which in the figure is almost equal to zero. Therefore, the regularization parameter can be neglected for the present study. In case of a higher degree hypothesis for the polynomial, however, regularization may be crucial. Finally, Fig. 1d shows the accuracy of the fit obtained for the surrogate model of moment resistance of the slab based on test set data ($n_{\text{test}} = 500$) for n_{train} of 2000 (as recommended), without regularization parameter ($\lambda = 0$) as concluded previously. In the figure, moment capacity of the slab estimated from the actual model i.e., Eq. (8) is compared with the predicted capacity based on surrogate model. The coefficient of determination (R^2) as obtained from the fitted test data is found to be 99.75%, which is considered adequate enough for validation of the model.

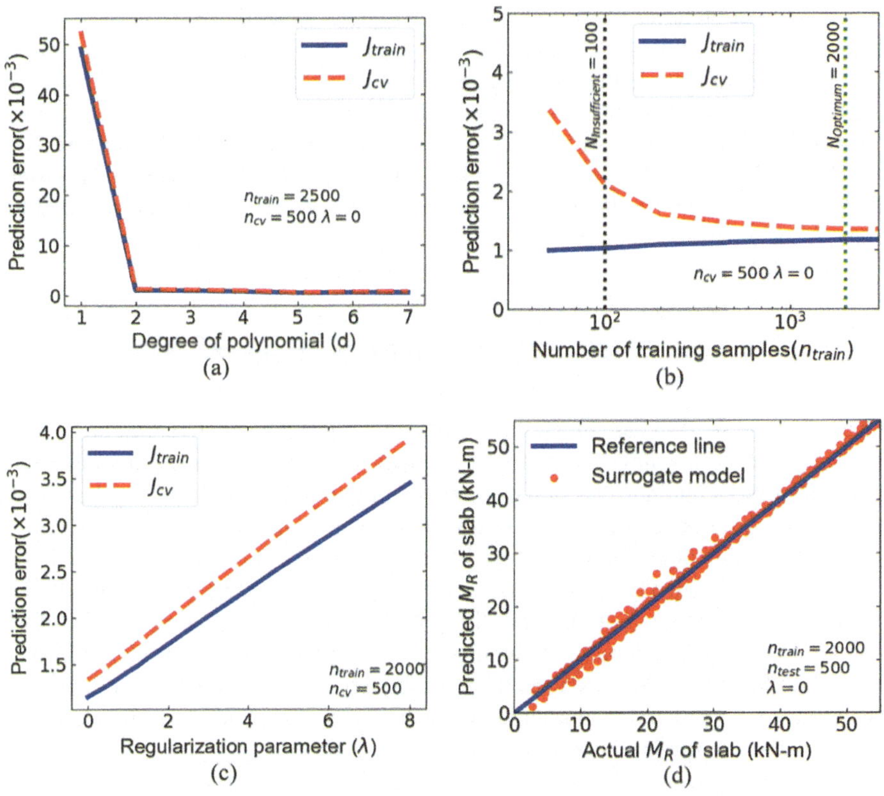

Fig. 1 Learning curves for the polynomial regression

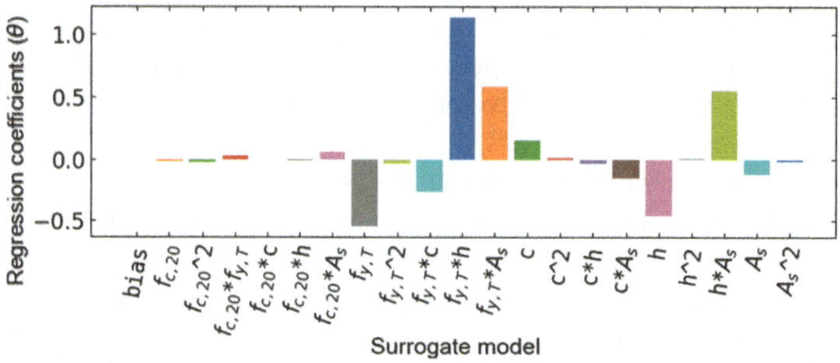

Fig. 2 Regression coefficients of the second degree surrogate model

Figure 2 represents the standardized regression coefficients for the 2nd degree surrogate model trained with 2000 LHS samples. In the figure, the independent variable, $f_{y,T}$ is found to have the highest regression coefficient, and thus has the highest influence on the response of the slab in comparison to all other independent variables. The 2nd degree surrogate model also includes the regression coefficients for the interaction terms of the independent variables. Overall, the term ($f_{y,T} \times$ h) is found to have the highest absolute value, being the most influential parameter. These interaction terms refer to the combined effect of different independent variables, i.e. where the multiplication of two terms influences the model output. The bias term and, for example, the concrete strength have low values for the standardized regression coefficients, and therefore have negligible contribution to the response of the slab. The surrogate model can be used to compare the influence of the different variables to the response of the system.

3.2 *Application for Fragility Curve Evaluation*

The preceding section has shown that surrogate models are able to predict the structural response of a simple fire-exposed member, demonstrating the potential effectiveness of the technique. As noted in the introduction, probabilistic studies could take advantage of surrogate models to substitute complex non-linear models, allowing for fast and efficient evaluation of fragility curves in case of design iterations. In the current study, fragility curves determined through the trained surrogate model are compared with fragility curves using the actual model for moment resistance of the slab, i.e. Equation (8), demonstrating the applicability of surrogate model for probabilistic studies of fire exposed structures. In this regard, the stochastic models listed in Table 2 are considered, being standard uncertainties for the input parameters; these are similar to the models adopted in [12].

Considering the stochastic models in Table 2, 10^4 LHS samples are developed. The moment capacity of the slab is then evaluated for each of the LHS samples using

Table 2 Probabilistic models for the stochastic variables, with μ the mean value and σ the standard deviation

Stochastic variables	Distribution	Mean	COV
Concrete strength, $f_{c,20\,°C}$ ($f_{ck,20} = 30$ MPa)	Lognormal	42.9 MPa	0.15
Retention factor for the steel bars, $f_{y,T}$ ($f_{yk,20} = 500$ MPa; $\mu_{fy,20} =$ 584 MPa)	Logistic model [4]	Eurocode 3	Temperature-dependent
Concrete cover (c)	Beta [μ-3σ; μ + 3σ]	35 mm	0.14 (σ = 5 mm)
Slab thickness (h)	Normal	200 mm	0.025
Bottom reinforcement area (A_s)	Normal	1.02 (0.1965%)	0.02

the surrogate model and the actual model. Finally, fragility curves are defined by the estimated response of the slab.

Figure 3 compares the observed probability density function (PDF) for the moment of resistance of the slab for 120 min of ISO 834 standard exposure, developed based on the surrogate and actual models. The figure shows that the PDF developed from the surrogate model ($N_{\text{Optimum}} = 2000$) approximately coincides with the PDF developed from the actual model. Meanwhile, although the PDF from the surrogate model with only 100 LHS ($N_{\text{Insufficient}}$) samples deviates from the actual model, the difference is small. The mean value for the moment capacity of slab based on actual model is 32.61 kN m, whereas it is 32.58 kN m and 32.77 kN m for the surrogate models developed from 2000 and 100 LHS samples, indicating a deviation on the mean value of 0.09% and 0.5%, respectively.

Figure 4a shows the comparison of the fragility curves (complementary CDF, or cCDF and CDF) observed for both the actual model and surrogate model. Fragility curves provide an easy way to estimate the capacity with a target probability of exceedance. In the figure, the surrogate models developed based on 100 and 2000

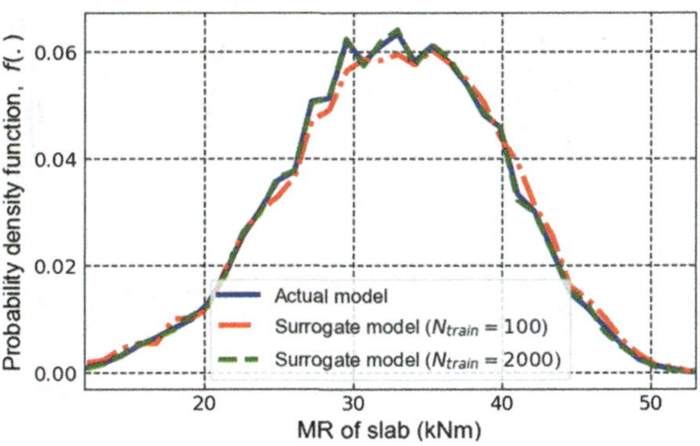

Fig. 3 Comparison of probability density function based on actual and surrogate model

Table 3 Comparison of predicted and actual moment of resistance of slab

S. No.	CDF (.)	M_R of slab (kN) for ISO fire		
		$n_{train} = 100$	$n_{train} = 2000$	Actual model
1	10^{-1}	23.21	23.39	23.40
2	10^{-2}	14.91	15.83	15.89
3	10^{-3}	10.86	12.08	12.04
4	10^{-4}	8.45	9.45	9.34

LHS samples predict the 10^{-2} capacity quantile (i.e. capacity with 99% probability of exceedance) as 14.91 and 15.83 kN, which is very close to the actual value of 15.89 kN based on the actual model as shown in Table 3. Thus, the error in prediction of 10^{-2} capacity quantile based on surrogate model for 2000 LHS samples ($N_{optimum}$) is approximately 0.37%. Table 3 also shows the comparison of the predicted response of the moment capacity of slab based on actual and surrogate models for different capacity quantiles. Note that the fragility curves are very accurate also for the 10^{-4} capacity quantile.

As the surrogate model relies on 2000 model evaluations (over a wide parameter range), this indicates a reduction in the number of model evaluations relative to direct fragility evaluation through 10^4 Monte Carlo realizations Therefore, it can be concluded that surrogate models can be effectively used for the probabilistic studies of the fire-exposed concrete slab.

Furthermore, the regression model is immediately capable of accommodating design iterations, or, in this case, different exposure scenarios. Figure 4b shows the fragility curve for the considered concrete slab exposed to an EN 1991-1-2:2002

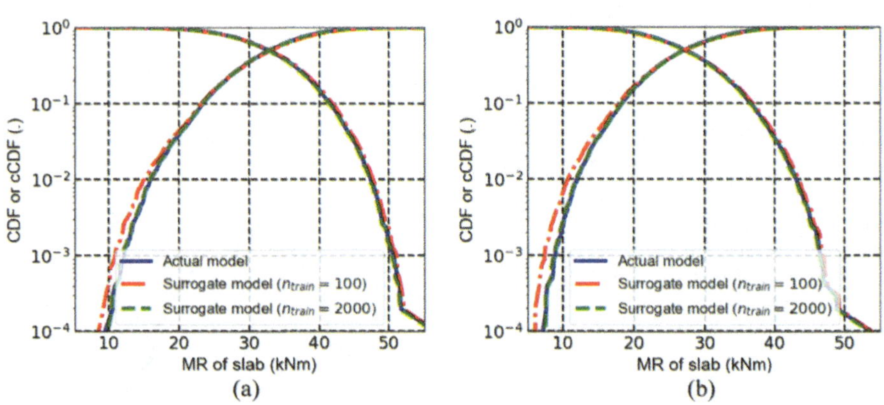

Fig. 4 Comparison of fragility curve based on actual model and surrogate model for **a** 120 min ISO fire and **b** parametric fire ($\Gamma = 1$, $t_{max} = 120$ min)

Eurocode parametric fire with $\Gamma = 1$, $t_{max} = 120$ min, based on the actual and surrogate model. The temperature of the reinforcement is calculated considering a (regression) relation proposed by Thienpont et al. (2019), giving reinforcement temperature in function of the heating phase duration and concrete cover. The obtained mean value for the (minimum or 'burnout' [13]) moment capacity of the slab based on surrogate model with 100 and 2000 LHS samples is approximately 27.27 kN m, which is close to the actual mean capacity estimated as 27.28 kN m. Thus, the fragility curve developed from the surrogate model based on LHS samples are quite in agreement with the actual result, thereby demonstrating the potential of surrogate models for structural fire engineering fragility curve evaluation.

4 Conclusions

The available prescriptive design guidelines are at times inadequate for structural design under fire loading. In such cases, there is a need for an explicit evaluation of the safety level through probabilistic studies which take into account the effects of uncertainty associated with the structural system. Probabilistic studies can however be complicated and commonly demand substantial computational expense. This makes the implementation of probabilistic methods in structural fire engineering design difficult. To alleviate these difficulties, probabilistic studies can be carried out by developing surrogate models, which are capable of approximating the output of complex, coupled thermal and mechanical calculations. When applied appropriately, these surrogate models are flexible and can be applied in general over a broad area of problems to estimate the structural response for a particular scenario. Once an adequate surrogate model is developed, fragility curves can be efficiently determined to represent the structural vulnerability to fire hazard. From the current investigation, focusing on a simple example of a slab in uniaxial bending subjected to fire on its lower face, it is found that the surrogate models are capable of predicting the capacity with an error of less than 1% and thus can precisely represent the non-linear response. The fragility curve developed for the slab indicates a moment capacity at the 10^{-2} quantile of 15.89 and 15.83 kN m, considering the actual and surrogate models respectively, which validates the sufficiency of regression technique for probabilistic studies of this fire exposed structure. Moreover, the surrogate models also give an idea of the most influential parameter in the analysis of structural systems. As a result, the current study gives a starting point for the further development of a framework for carrying out probabilistic studies on structures subjected to fire hazard by using regression techniques. The next step will explore the applicability to structural members exhibiting more complex fire behavior.

References

1. Gernay, T., Khorasani, N. E., & Garlock, M. (2019). Fire fragility functions for steel frame buildings: Sensitivity analysis and reliability framework. *Fire Technology, 55*(4), 1175–1210.
2. Iqbal, S., & Harichandran, R. S. (2010). Capacity reduction and fire load factors for design of steel members exposed to fire. *Journal of structural engineering, 136*(12), 1554–1562.
3. Khorasani, N. E., Garlock, M., & Gardoni, P. (2014). Fire load: Survey data, recent standards, and probabilistic models for office buildings. *Engineering Structures, 58,* 152–165.
4. Qureshi, R., Ni, S., Elhami Khorasani, N., Van Coile, R., Hopkin, D., & Gernay, T. (2020). Probabilistic models for temperature-dependent strength of steel and concrete. *Journal of Structural Engineering, 146*(6), 04020102.
5. Gernay, T., Khorasani, N. E., & Garlock, M. (2016). Fire fragility curves for steel buildings in a community context: A methodology. *Engineering Structures, 113,* 259–276.
6. Naser, M. Z. (2019). Can past failures help identify vulnerable bridges to extreme events? A biomimetical machine learning approach. *Engineering with Computers*, pp. 1–33.
7. Van Coile, R., Caspeele, R., & Taerwe, L. (2013). The mixed lognormal distribution for a more precise assessment of the reliability of concrete slabs exposed to fire. In *Proceedings of ESREL*, (Vol. 2013, No. 29/09, pp. 02–10.).
8. Burhenne, S., Jacob, D., & Henze, G. P. (2011). Sampling based on Sobol' sequences for Monte Carlo techniques applied to building simulations. In *Proceedings international building performance simulation association* (pp. 1816–1823).
9. Ioffe, S., & Szegedy, C. (2015). Batch normalization: Accelerating deep network training by reducing internal covariate shift, 1502.03167.
10. Forrester, A. I., Bressloff, N. W., & Keane, A. J. (2006). Optimization using surrogate models and partially converged computational fluid dynamics simulations. *Proceedings of the Royal Society A: Mathematical, Physical and Engineering Sciences, 462*(2071), 2177–2204.
11. Draper, N. R., & Smith, H. (1998). *Applied regression analysis* (Vol. 326). Wiley-Interscience.
12. Thienpont, T., Van Coile, R., Caspeele, R., & De Corte, W. (2019). Comparison of fire resistance and burnout resistance of simply supported reinforced concrete slabs exposed to parametric fires. In *3rd International Conference on Structural Safety under Fire and Blast.*
13. Gernay, T. (2019). Fire resistance and burnout resistance of reinforced concrete columns. *Fire Safety Journal, 104,* 67–78.

Identification of Risk Management Models and Parameters for Critical Infrastructures

Oscar J. Urbina, Elisabete R. Teixeira, and José C. Matos

Abstract The resilience of an area/region/country or society is directly related to the performance of its Critical infrastructures (CI), especially when it is affected by extreme events. The increasing number of catastrophic events, such as terrorist attacks or natural disasters (tsunamis, fires, floods), alerted Europe and other nations worldwide to take measures for preventing or reducing possible consequences against these situations. CI are commonly defined as facilities, systems and assets, essential for the maintenance of vital social functions, and their disruption or destruction may significantly impact the well-being of society. It is mandatory for any nation to identify which Infrastructures must be defined as critical, by analyzing the impacts provoked by an extreme event and the society's dependence towards this Infrastructure. For this purpose, European Commission established a procedure for the identification and designation of European CI ensuring to avoid different approaches within the EU. Three cross-cutting criteria where defined: (a) Casualties; (b) Economic-effect; (c) Public effect. This paper aims to introduce different risk management models for CI and the parameters necessary for quantification of these Methodologies. There are several models for risk management, the ones studied and introduced in this paper were applied in different countries and types of CI, these vary from deterministic approaches to probabilistic methods. The critically parameters are related in governmental, economical, security and welfare terms, these parameters are important for two main reasons: (1) to keep updated the critical index and the maps of risks and vulnerability that predictive models may use; (2) Current tools are essentially based on models weighed by qualitative weights, not allowing the complete analysis of one-off events.

Keywords Critical infrastructures · Extreme events · Risk management models · Predictive models

O. J. Urbina (✉) · E. R. Teixeira · J. C. Matos
Institute for Sustainability and Innovation in Structural Engineering (ISISE), University of Minho, Campus de Azurém, 4800-058 Guimarães, Portugal

© The Author(s), under exclusive license to Springer Nature Switzerland AG 2021

J. C. Matos et al. (eds.), *18th International Probabilistic Workshop*, Lecture Notes in Civil Engineering 153, https://doi.org/10.1007/978-3-030-73616-3_29

1 Introduction

Critical infrastructures (CIs), play a vital role in today's societies, enabling many of the functions and services of modern societies. From financial services and emergency services, to energy production and water supply, these infrastructures fundamentally impact and continue to improve the quality of life of societies. Today, CI systems face several potential hazards, such as natural (earthquakes, floods, fires), technological (operational failures in systems) and human (fires, cyber-attacks or malicious activity) that can intervene in the functionality of these systems [1, 2]. The malfunctioning of these systems causes a cascading effect through the community and causes social, economic and functional disruption.

Therefore, it is important to understand the potential hazards/risks affecting the use of infrastructures and their methods of analysis for the development of these systems. Risk management models are the best solution to assess and find planning, mitigation and recovery solutions, where the consequences of damage and losses are quantified for decision making [2]. In this context, risk analysis plays an important role as it provides information and helps the development of risk mitigation plans and strategy by decision makers.

There are a significant number of risk assessment models for CI. In general, the approach used in these models is a common and linear approach, consisting of only a few elements: (i) identification and classification of threats; and (ii) identification of vulnerabilities and impact assessment [3]. Nonetheless, there is a big difference between these methodologies and the target audience of their application (policy makers, managers, research institutes), as well as the field of applicability (asset level, system/infrastructure level). Resulting in the need to study and develop methodologies that attempt to assess the issue of several simultaneous risks and the interdependencies of infrastructures.

The matter of interdependencies in the critical infrastructure system is very important, and according to Rinaldi et al. [4] four interdependencies: (i) physical (the operation of one infrastructure depends on the material output of the other); (ii) cyber (dependency on the information transmitted through the information in the infrastructure); (iii) geographic (dependency on the effect of the local environment that simultaneously affects several infrastructures); (iv) logical (all those that do not fit into the previous points). Continuously, methodologies take interdependencies into account and this reflects the natural evolution of risk assessment models.

In the following sections several risk management models for critical infrastructures obtained by a detailed state of the art review will be presented and discussed.

2 State of the Art on Risk Management Models for Critical Infrastructure

2.1 Critical Infrastructure: Concepts and Definitions

It is important to know the specific definition of CI, however, there is no Universal definition, despite the similarities, the definition differs from one country to another due to its own specifications and socio-cultural characteristics [5]. In Australia, for example, a CI is defined as "*those physical facilities, supply chains, information technologies and communication networks which, if destroyed, degraded or made unavailable for an extended period of time, would have a significant impact on the social or economic well-being of the nation, or would affect Australia's ability to conduct national defense and ensure national security*" [6]. On the other hand, the US defines CI as "*systems and assets, physical or virtual, so vital to the United States that the inability or destruction of such systems and assets would have a debilitating impact on security, national economic security, public health or national security, or any combination thereof*" [6]. The official definition given by the European Union Council determines a CI as "*an asset, system or part thereof situated in Member States which is essential for the maintenance of vital functions in society, health, safety, security, economic or social well-being of persons, and the disruption or destruction of which would have a significant impact in a Member State as a result of the failure to maintain those functions*" [7].

In addition to differences in definition, there are more substantial differences when defining a CI, for example: in Germany the CI is divided into vital technical infrastructure and vital socio-economic services infrastructure; on the other hand, in Great Britain the CI is divided into national critical infrastructure and other critical infrastructure [8]. Moreover, it is important to know how any infrastructure can be designated as critical. For Alexander Fekete [9], there are two main questions that need to be considered: (1) On what are we dependent? That is, whether the infrastructure under study, as referred to in the Council's definition of the European Union, is essential for the maintenance of vital societal functions, and (2) What would be the impacts of failure? in terms of potential number of victims (victims or injured) and socio-economic impacts.

The European Commission has established a procedure for the identification and designation of European Critical Infrastructure (ECI) in order to have a common approach to the protection of these CI by defining three cross-cutting criteria: (a) Victims criterion (assessed in terms of the potential number of fatalities or injuries); (b) Economic effects criterion (assessed in terms of the significance of economic loss and/or degradation of products or services; including potential environmental effects); and (c) Public effects criterion (assessed in terms of the impact on public confidence, physical suffering and disruption of daily life, including the loss of essential services). While the limits for each of the cross-cutting criteria are determined on a case-by-case basis by the Member States concerned for a particular critical infrastructure [7].

According to Gritzalis, Theocharidou and Stergiopoulos, based on the relevant literature and international practices used, a precise process to identify and designate the national CIs should be applied in four steps [8]:

- Identification of critical sectors/sub-sectors. At this stage, sectors and/or sub-sectors that are considered important for national interests are identified.
- This means that each EU member should declare a list of its national critical sectors because not all sectors are equally critical, resulting in some sectors more critical than others. However, in terms of defining a common EU framework for Critical Infrastructure Protection (EPCIP), a list of common critical sectors and sub-sectors within the related critical services is provided [6].
- Identification of critical services. Critical (or vital) sector/sub-sector services are identified and designated for each critical sector.
- CI designation. For each critical service, the critical assets/components that make up the CI are identified and designated.
- CI protection. Protection and security procedures are implemented for each CI.

2.2 Criteria for the Assessment of Risk Management Models

Once the CI is identified, the first step to protect it involves identifying and evaluating the factors that may negatively influence its operations, defining a systematic and analytical approach to prioritize resilience measures for CI. This analysis must include an assessment of the impacts of the CI breakdown by pre-established criteria. Several approaches are used in OECD countries [10], for example, in Switzerland a first contrast is made between different sectors and subsectors with three categories of criticality (very high, high and normal criticality). In the Netherlands, economic, physical and social criteria allow different critical infrastructure processes to be defined, but then a distinction is made between the intensity of the effects, stating two categories, the first one, category A, when disturbances produce large impacts and cascading effects, and, the second category, category B, when impacts caused are smaller, in order to reflect diversity within CI and to establish priorities. In terms of criteria, the European Commission defines a minimum set for the assessment of CI, including public, economic, environmental, policy and psychologic impacts, and interdependencies [7, 10].

Identifying the weak points on a CI makes possible to prioritize and focus the resilience efforts and investments on existing infrastructure systems: on points of failure that would have the most serious consequences. This prioritization can be a decisive variable in decision making, such as which infrastructure should be hardened or relocated, or which CI should receive priority restoration after a disaster to ensure rapid recovery [10]. This study attempts to present a structured review of existing methodologies/models at both national and international levels, identifying which models have already been developed and which failures still exist. Clearly, there is a huge list of models, but only a few will be presented in this report, which have been chosen according to the following criteria:

- Scope of the model: sector and end-user;
- Model Objectives;
- Applied methodology and standards;
- The study of interdependencies;
- Consideration of infrastructure resilience;
- Risks analysis in different sectors.

In the following points different methodologies will be presented, applied in different countries, showing different methods and approaches, depending on the type of CI and its risks.

2.3 *Risk Management Models*

Despite the existence of numerous risk assessment methods, this article will present methods ranging from deterministic approaches to probabilistic methods. These deterministic approaches analyze and interpret historical events of disasters and available back data considering new developments, scenarios and simulations that expand on back analysis [6, 8]. Below it is presented several risk managements models, describing their most important features and parameters for the risk assessment:

2.3.1 The Dynamic Inoperability Input-Output Model (DIIM)

A methodology was proposed to measure network resilience considering physical and cyber-threats dependencies between facilities in critical civil infrastructure systems by modifying the DIIM (Dynamic Inoperability Input-output Model) and combining it with graphics theory. Additionally, recovery coefficients for each type of installation were also studied and used to model the operability of each network installation over time [11]. It is based on the Input-Output Dynamic Inoperability Model to evaluate the recovery of civil infrastructure facilities, considering the dependencies at the infrastructure level.

2.3.2 GIS-Based High-Level Approach

This model evaluates the impacts of climate change on CI. This approach is based on the application of high-resolution climate change forecast maps and provides the recognition of critical high-risk zones. This GIS tool aims to provide stakeholders information necessary to take decisions regarding potential impacts and opportunities of climate change impacts and highlights the critical points of climate change for more detailed analysis. The results of this model are: (1) Matrices with information regions that highlight vital connections between infrastructure assets and climate threats; (2) Sectorized maps that show the vulnerabilities of CI networks to current and future climate extreme events; (3) Cross-sectoral geospatial criticality maps for several climate threats [12].

2.3.3 BIRR—Better Infrastructure Risk and Resilience

This methodology was created by the Argonne National Laboratory (United States) and includes 18 critical infrastructures (energy, critical production, etc.). This methodology has a sector-wide approach that goes down to the asset level and gives priority to protection measures and is applied primarily to terrorist threats. The interesting features of this methodology are the concepts of vulnerability index (IV), protection measures index (IMP) and resilience index (IR). The concept behind the IR, is to have a common metric and facilitate the comparison between the various infrastructure sectors that are covered by this methodology. The procedure for establishing the IV part of the IMP is designed to reflect the increased protection of certain assets as new measures are implemented. It should be noted that in determining the LMI, the question of dependencies is considered. For each asset analyzed, it is possible to define in which main sectors (electricity, gas, ICT, etc.) its operation is dependent and quantifies this by means of redundancy, resilience and impact indices. Essentially, this methodology allows policy makers to create tools that help analyze and identify the vulnerabilities of different sectors and prepare risk reports [3].

2.3.4 Damage Estimation Model (DEM)

This model predicts and analyzes the effects that a hurricane can have on the performance of interdependent infrastructure systems by producing scenarios of damage to a set of CIs. It uses a Monte Carlo simulation and statistical method to predict the damage caused by a hurricane to the systems under study with the objective to provide the stakeholders a variety of options in order to choose a scenario based on the user's needs. The DEM is capable to employ forecasted, historic or customized risk scenarios to produce its predictions [13].

2.3.5 Infrastructure Disruption Model (IDM)

This model applies optimization techniques to determine the cascading effects that an event can cause on all the Infrastructure Systems considered, in order to anticipate and analyze the outcomes that a hurricane may produce on the performance of interdependent infrastructure systems. The IDM applies optimization techniques to distinguish which components receive the critical services and which components do not. The model performs the analysis employing the DEM results as an input, providing maps of Infrastructure damage and service disruptions [13].

2.3.6 BMI—Baseline Protection Concept

The German Federal Ministry of the Interior has created a basic protection plan, which contains a risk assessment methodology. It is essentially aimed at industries in the area of infrastructure and aims at human protection [14]. This plan contains a broad list of possible risks ranging from natural hazards to terrorism and criminal acts. It presents the potential points of vulnerability for each category of risk, and the mitigation/protection measures against the risks.

2.3.7 DIESIS—Design of an Interoperable European Federated Simulation Network for CI

DIESIS is a study that seeks the implementation of a Unified European Centre for Infrastructure Simulation and Analysis. DIESIS includes the CI simulators SINCAL (electrical network), NS2 (telecommunications network), Open Track (rail traffic) and a flood simulator (Aqua) for the simulation of external events. This assemblage of models enable the assessment of the scientific, technical and financial viability of the proposed e-Infrastructure towards the possible impacts it may be exposed to [15].

2.3.8 Carver2

CARVER2 is a tool that has been developed to address the analysis needs of critical infrastructure primarily from the point of view of the policy maker [3]. CARVER takes into account natural risks but also terrorist threats, the following aspects are the criteria considered for the risk assessment [3]:

- Criticality: This is in fact part of the methodology impact assessment.
- Accessibility: Refers to the possibility of terrorists being able to enter the infrastructure and cause its destruction, being essentially evaluated by the vulnerability of the infrastructure in terms of physical security.
- Recoverability: Partially analyzes resilience since it refers to the infrastructure's ability to recover its original state after failure.
- Vulnerability: Analysis of infrastructure vulnerability (in this methodology, vulnerability to terrorist attacks is very well defined, but little in relation to natural risks).
- Notoriety: Assessment of infrastructure as an icon (e.g. cultural site) with indirect impact.
- Redundancy: Presentation of the alternatives that result from the evaluation.

In this methodology the interdependencies are considered, users have a list of sectors that are affected by the loss of another one, the level at which these interdependencies were defined and which interdependencies were included in the tool (cyber, physical, functional, geographic) are still to be defined. At the end of the evaluation, a report is generated with a classification, which allows the comparison of completely different infrastructures, as it presents a standardized metric.

2.3.9 CIP/DSS—Critical Infrastructure Protection Decision Support System

The Decision Support System for Critical Infrastructure Protection (CIP/DSS) simulates the dynamics of each infrastructure and links one infrastructure to another according to their interdependencies [16]. For example, repairing damage to a city's electrical grid requires transportation to the fault sites and delivery of repair equipment, fuel for the vehicles/equipment needed for repair, telecommunications for problem diagnosis and coordination of repairs and availability of work. The repair itself involves diagnosis, ordering parts, dispatching teams and performing work. The electrical network responds to the initial damage and to complete the repair needs to make changes in its operational characteristics. Dynamic processes like this are represented in the simulations generated by CIP/DSS, through differential equations, discrete events and operation coding rules. CIP/DSS develops a risk-based decision support system that provides information for decision making on infrastructure protection, considering critical infrastructures and their primary interdependencies [16].

2.3.10 Multicriteria Identification and Prioritization Methodology

The Local Disaster Index (LDI) identifies the social and environmental risks that arise from more recurrent lower-level events that are often chronic at local and subnational levels. These particularly affect the most socially and economically fragile population and generate a highly detrimental impact on countries' development [6].

LDI is calculated by adding three sub-indicators (K, A, and L). According to the author of the study, the calculation formula results from the following mathematical equation: LDI = LDI_K + LDI_A + LDI_L, where LDI_K represents the Local Sub-indicator of the Number of Deaths, LDI_A represents the Sub-indicator of the Number of People Affected, and LDI_LO Sub-indicator of Losses in four varieties of events, such as landslides, earthquakes, floods and storms, among others. The prevailing Vulnerability Index (PVI) which is composed of a series of indicators that characterize the prevailing vulnerability conditions reflected in exposure in prone areas, socio-economic fragility and general lack of social resilience [6].

The Risk Management Index (RMI), which gathers a set of indicators related to the country's risk management performance. These reflect organizational, development, capacity and institutional actions taken to reduce vulnerability and losses, to prepare for the crisis and for efficient recovery [6]. Their calculation formula is identical to that of the PVI index. Obtaining the RMI index results from averaging the four subindicators RI, RR, DM and FP, as shown in Eq. (1), where, RI: Risk Identification Indicator, RR is the Risk Reduction Indicator, DM: Risk Management Indicator, and

FP is the Financial Protection Indicator.

$$RMI = \frac{RMI_{RI} + RMI_{RR} + RMI_{DM} + RMI_{FP}}{4} \quad (1)$$

2.4 *Predictive Models' Methodology*

This section presents some methodologies used to predict possible events that put the critical infrastructure at risk.

2.4.1 ASOR—Evaluation and Strengthening of Organizational Resilience in the Critical Infrastructure System

The ASOR (Assessing and Strengthening Organizational Resilience in Critical Infrastructure System) method is based on the principle of assessing the factors that determine organizational resilience, identifying weaknesses, and proposing measures to strengthen organizational resilience in a critical infrastructure entity. The core of this method is the process of assessing and strengthening organizational resilience in a critical infrastructure system. This procedure is based on available resources, focusing essentially on CI resilience factors [17].

2.4.2 SafeCity—Geographic Information System for Critical Infrastructure Protection

The SafeCity is built on a web-based Geographic Information System for Critical Infrastructure Assessment and Visualization and Hazards for the Civil services. The system allows operators to identify Critical Infrastructure in the context of the multi-layer maps of a city, perform analysis and simulate different hazard scenarios. Once a degree of criticality has been assigned to each identified Critical Infrastructure, the system allows a spatial analysis of the density of vulnerable structures in the different areas of the city [18].

3 Conclusions and Future Research

Based on the analysis of the different risk management and predictive models, the creation of a matrix was initiated, that ease the understanding of the focus of each Methodology and the inputs and outputs that each methodology needs and provides during the risk management and assessment (Table 1).

Table 1 Descriptive matrix of risk management models

Methodology	Type of risk	Necessary parameters to insert (inputs)	Parameters obtained (outputs)	Focus of the study
Dynamic inoperability input-output model (DIIM) [11]	Natural disasters: hurricanes	– Inputs of hazard information, types of structure, fragility curves, minimum operability requirements, recovery coefficients, dependency matrix and weights for all systems are required – Loss ratio	– Coefficients of recovery and operability for different types of infrastructure against extreme events	Evaluates the recovery of civil infrastructure facilities, considering the dependencies at the infrastructure level
GIS-based high-level approach [12]	Coastal floods and erosion, river and rain floods, bridge runoff, extreme storms, cold, heat and landslides	Change in: -mean seasonal precipitation – Change in maximum daily seasonal temperature and number of consecutive dry days – Number of ice and frost days – Number of extreme wind speeds – Number of very wet days – Precipitation intensity and number of very wet days – Number of very wet days and seasonal variation in precipitation	– Sector information matrices highlighting the main links between infrastructure assets and climate threats – Sector letters showing the exposure of current and future plots of vulnerable CI networks – Maps for classifying cross-cutting geo-sectoral risks to the various climate threats	Aims to provide decision-makers with the information they need to make their decisions on the potential impacts and opportunities of climate change and highlights the critical points of climate change for more detailed analysis

(continued)

Table 1 (continued)

Methodology	Type of risk	Necessary parameters to insert (inputs)	Parameters obtained (outputs)	Focus of the study
Damage estimation model (DEM) and infrastructure disruption model (IDM) [13]	Hurricanes	– Wind speed – Radio to maximum wind – Storm track – Maximum wind speed in the region for the chosen hurricane scenario – Infrastructure components and the interdependencies under study	It provides three wind damage maps and the flood damage map, including a list of all nodes (components of the Infrastructure studied), whether they have been damaged (from damage estimate model), and types of interruptions are occurring	Analysis of hurricanes to estimate the CIs damage and service disruptions. Study how a community is affected by the CI interdependencies and prepare it against extreme events
The safe city [18]	Natural disasters and terrorist attacks	– Digital city terrain model and layers representing existing water bodies – Other parameters must be defined by the end user individually for each analysis performed. These include the amount by which the simulated water level is increased over its default value and the extent of the area for which the simulation will be performed	The creation of hazard scenarios during the preparedness phase, mapping of threats identified in the response phase, and the simulation of the geographic impact of threats during the mitigation phase	It presents a municipal CI analysis system, that offers integrated tools for target analysis, risk scenario simulations and spatial analysis within a web-based and remotely accessible geographic information system

(continued)

Table 1 (continued)

Methodology	Type of risk	Necessary parameters to insert (inputs)	Parameters obtained (outputs)	Focus of the study
DIESIS [15]	Floods, additional risks generated from models	– Electrical networks: voltage and load level of the electrical networks – Telecommunication network: operational level of connection – Railroad Networks: maps with routes and timetables, list of rules and equipment – Water Networks: demographic map, initial water levels, flood configurations and points of interest	The results of the federated simulation are viewed with Google earth software. This visualization tool is based on keyhole markup language (KML). The visualization module also allows a graphical representation for a flood	The design of an Interoperable European Federative simulation network for critical infrastructure projects (DIESIS), studied four types of infrastructures modeling them in order to avoid possible extreme events

This paper provided a critical analysis of the capabilities of different strategies, applications and methodologies for the identification and evaluation of risks in critical infrastructure, as an opening towards developing an integrated and multidisciplinary methodology that allows to assess various risks in different types of Critical Infrastructures. Considering the importance of infrastructure interdependencies and the complex network modeling techniques it requires.

It was observed from the most relevant articles related to risk management and predictive models found that whether for risk assessment or predictive models, these were created according to a specific necessity. This led to the existence of non-multidisciplinary methodologies and assessments, varying them from types of risks, to the focus of study and groups of CIs, creating a gap of approaches consider various types of Critical Infrastructures, risks and objectives of the assessment.

In general, the studies of risks and vulnerabilities in Critical Infrastructures show two noticeable aims in the development of these methodologies:

- The first one relates to the identification of methods, techniques, tools and diagrams to describe the current state of infrastructure. Hazards and extreme events are used to obtain a survey of the infrastructure performance and its response to these events.
- The second one aims to understand the performance of the CI in different scenarios under diverse simulations, resulting in the identification of weaknesses on the Critical Infrastructure.

Finally, it is essential to continue investigating and improving the way of collecting and processing the methodologies about critical infrastructure's resilience and risk assessments. Hence, future studies are needed to conduct to validate the usefulness and reliability of these methodologies described to evaluate the resilience of CI exposed to extreme events, natural or manmade. Additionally, it is vital to realize empirical applications applied on case studies embracing different scenarios for each dimension of resilience.

Acknowledgments This work was partly financed by FEDER funds through the Competitivity Factors Operational Programme—COMPETE and by national funds through FCT (Foundation for Science and Technology) within the scope of the project POCI-01-0247-FEDER-039555).

References

1. Ficco, M., Choraś, M., & Kozik, R. (2017). Simulation platform for cyber-security and vulnerability analysis of critical infrastructures. *Journal of Computer Science, 22,* 179–186. https://doi.org/10.1016/j.jocs.2017.03.025.
2. Ongkowijoyo, C. S., Doloi, H., & Gurmu, A. T. (2020). Hybrid risk analysis model for analyzing the urban infrastructure risk. *International Journal Disaster Risk Reduction, 48*, 101600. https://doi.org/10.1016/j.ijdrr.2020.101600.
3. Giannopoulos, G., Filippini, R., & Schimmer, M. (2012). Risk assessment methodologies for critical infrastructure protection. Part I: A state of the art.

4. Rinaldi, S. M. (2004). Modeling and simulating critical infrastructures and their interdependencies. In *Proceedings of Hawaii International Conference System Science* (Vol. 37, pp. 873–880). https://doi.org/https://doi.org/10.1109/hicss.2004.1265180.
5. Almeida, A., & Técnico, I. S. (2008). *A multi-criteria methodology for the identification & ranking of critical infrastructures* (p. 10).
6. Ramos, A. (2011). Metodologia MULTICRITÉRIO DE Identificação e Priorização de António Orlando de Novais Ramos Henriques de Almeida Dissertação para a atribuição do Grau de Mestre em Engenharia e Gestão Industrial Maio 2011.
7. EU Council. (2008). The Council of the European Union 2008/114/EC of 8 December 2008 on the identification and designation of European critical infrastructures and the assessment of the need to improve their protection. In *Human response to vibration* (pp. 187–202). https://doi.org/10.1201/b12481-8.
8. Dimitris Gritzalis, G. S., & Theocharidou, M. (2019). *Critical infrastructure security and resilience.*
9. Fekete, A. (2011). Common criteria for the assessment of critical infrastructures. *International Journal Disaster Risk Science, 2*(1), 15–24. https://doi.org/10.1007/s13753-011-0002-y.
10. OECD. (2019). *Good governance for critical infrastructure resilience*, OECD Revie. Paris: OECD Publishing.
11. He, X., & Cha, E. J. (2018). Modeling the damage and recovery of interdependent critical infrastructure systems from natural hazards. *Reliability Engineering System Safety* (Vol. 177, pp. 162–175). https://doi.org/10.1016/j.ress.2018.04.029
12. Hawchar, L., Naughton, O., Nolan, P., Stewart, M. G., & Ryan, P. C. (2020). A GIS-based framework for high-level climate change risk assessment of critical infrastructure. *Climate Risk Management, 29*, 100235. https://doi.org/10.1016/j.crm.2020.100235
13. Loggins, R. A., & Wallace, W. A. (2015). Rapid assessment of hurricane damage and disruption to interdependent civil infrastructure systems. *Journal of Infrastructure Systems, 21*(4), 1–2. https://doi.org/10.1061/(ASCE)IS.1943-555X.0000249.
14. Interior Ministry Germany. (2009). *National strategy for critical infrastructure protection (CIP Strategy)* (p. 18). https://doi.org/10.2307/3433134
15. Tofani, A., Usov, A., Castorini, E., & Rome, E. (2009). DIESIS-design of an interoperable European federated simulation network for critical infrastructures. https://doi.org/10.13140/RG.2.1.4001.2000.
16. Bush, B., Dauelsberg, L., LeClaire, R., Powell, D. (2005). *Critical infrastructure protection decision support system (CIP/DSS) project overview* (p. 13).
17. Rehak, D. (2019). Assessing and strengthening organisational resilience in a critical infrastructure system: Case study of the Slovak Republic. *Safety Science, 123*(November), 2020. https://doi.org/10.1016/j.ssci.2019.104573.
18. Kulawiak, M., & Lubniewski, Z. (2014). SafeCity—A GIS-based tool profiled for supporting decision making in urban development and infrastructure protection. *Technology Forecasting Social Change, 89*, 174–187. https://doi.org/10.1016/j.techfore.2013.08.031.

Implementation of Reliability Methods in a New Developed Open-Source Software Library

Jan Philip Schulze-Ardey, Tânia Feiri, Josef Hegger, and Marcus Ricker

Abstract Structural reliability methods aim at the computation of failure probabilities of structural systems with methods of statistical analysis due to varied uncertainties occurring during their design, building or even operating conditions. However, in the field of civil engineering, the use of structural reliability methods unfortunately remains limited to specific cases. Most of the software available has still a limited range concerning wide parametric studies for analysis with reliability methods in civil engineering. This paper describes a new open-source software library as an effective tool for reliability analysis in civil engineering. The goal is to facilitate the adoption of reliability methods among engineers in practice as well as to provide an open platform for further scientific collaboration. The new library is being developed as a so-called "R package" in open-source programming software "R". The package is capable of carrying out systematic parameter studies using different probabilistic reliability methods, as FORM, SORM, Monte Carlo Simulation. Based on this, an overview on the probabilistic reliability methods implemented in the package as well as results of parametric studies is given. The performance of the package will be shown with a parametric study on a practical example. Most important results of the parametric study as well as the correctness of different reliability methods will be described in the paper. By describing probabilistic methods using an example in practice, engineers can get a basic understanding behind the ideas of probability theories. Further work will result in large parameter studies, which will support the development of a new guideline for reliability in civil engineering. This guideline describes techniques of code calibration as well as to determine new partial safety factors (e.g. for non-metallic reinforced concrete, fixing anchors, etc.). Furthermore, advanced reliability methods (e.g. Monte Carlo with Subset Sampling) will be implemented in the new R package.

J. P. Schulze-Ardey (✉) · J. Hegger
Institute of Structural Concrete, RWTH Aachen University, Aachen, Germany
e-mail: jschulze@imb.rwth-aachen.de

T. Feiri · M. Ricker
Institute of Structural Engineering, Hochschule Biberach University of Applied Sciences, Biberach, Germany

© The Author(s), under exclusive license to Springer Nature Switzerland AG 2021
J. C. Matos et al. (eds.), *18th International Probabilistic Workshop*, Lecture Notes in Civil Engineering 153, https://doi.org/10.1007/978-3-030-73616-3_30

Keywords Probability software · Reliability methods · Parameter studies · Statistical analysis

1 Introduction

By accounting uncertainty of load and material properties, civil engineering researchers like Freudenthal [1] changed the classic deterministic perspective of structural design towards a more scientific approach. Since 1950, substantial research has been done and published, e.g. refer to the CEB/FIP Bulletin No. 112 [2] or the "Probabilistic Model Code" documents [3], developed by the Joint Committee on Structural Safety. Furthermore, scientific committees provided the international code ISO 2394 "General principles on reliability for structures" [4] as a step for international standardisation of safety elements.

The goal of reliability analysis is to determinate the probability of failure with statistical methods. Safety elements can be derived by deterministic, the so-called "semi-probabilistic", and probabilistic methods. Eurocode 0 [5] categorises such probabilistic methods into Level I (partial safety factors are assumed to achieve a certain failure probability), Level II (approximated calculation of the failure probability) and Level III (exact determination of the failure probability). The Eurocode itself uses Level I methods in the design equations and offers a generic description of Level II and Level III methods. Eurocode 0 [5] gives only a detailed view on the Mean Value First Order Second Moment Method (MVFOSM), which can be considered inconsistent regarding the reliability, as it is shown e.g. by RICKER [6].

In mathematical terms, the determination of the reliability index β is easier than the calculation of the probability of failure. Current international codes, as Eurocode 0 [5], provide different target values for the probability of failure and the respective reliability indices β, depending on certain boundary conditions, e.g. $\beta = 3.8$ is defined for a 50-year reference period. To calculate the probability of failure with Level II and Level III reliability methods, it is needed an algorithm to solve the multidimensional probability integral. In most cases, it is not possible to use analytic mathematic methods for joint density functions, depending on an arbitrary number of random variables with different distribution functions, and sophisticated limit state functions.

So far, there are few commercial software tools as well as non-commercial and open-source software tools for reliability analysis available. An example is the software tool "mistral" (Methods in Structural Reliability Analysis) that is written as a R-package [7]. The new software tool, which is described in this paper, has several more features (e.g. an algorithm for the automation of parametric studies) and more probabilistic methods are available.

2 Reliability Methods

The fundamental mathematical problem of reliability analysis is based on the assessment of the probability of failure p_f by solving the following high-dimensional convolution integral:

$$p_f = P(g(\vec{x})) \leq 0 = \int_{g(\vec{x}) \leq 0} F_X(\vec{x})\mathrm{d}x \quad (1)$$

where $g(\vec{x}) \leq 0$ is denoted the failure domain and $f_x(\vec{x})$ is the joint probability density function of the basic random variables in a resistance or load function. In many cases, no analytical mathematical solution exists. Thus, only numerical methods give acceptable (or satisfactory) results. There are several reliability analysis techniques to calculate a reliability index and the respective probability of failure. Table 1 gives an overview on some common methods and their respective accuracy.

In the field reliability analysis in structural concrete members, FORM, SORM, and Monte Carlo simulation methods are the most relevant techniques. The solution of a high dimensional integral, which is the probability of failure, can be described as a (non-linear) optimisation problem with boundary conditions. Figure 1 illustrates

Table 1 Reliability methods

Method	Accuracy
First order reliability method (FORM)	Approximation (Level II)
Second order reliability method (SORM)	
Crude Monte Carlo simulation	Exact solution (Level III)
Monte Carlo method importance sampling	
Monte Carlo method subset sampling	

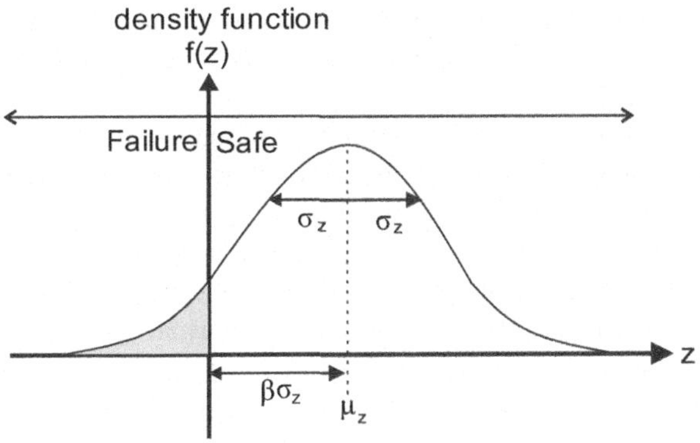

Fig. 1 Fundamental mathematical problem of reliability analysis

the geometrical interpretation of the reliability index β in relation to the probability of failure and respective the safe region or the unsafe region (failure).

2.1 FORM Algorithm

The solution of a high dimensional integral can be described as a (non-linear) optimisation problem with side conditions. This optimisation problem is not simple and, therefore, leads to the development of several algorithms. One of the most relevant approximation methods, the so-called "First Order Reliability Methods" (FORM), were developed 40 years ago and are still considered as robust algorithms for the safety level assessment. In fact, the FORM methods have a great importance in civil engineering regarding code calibrating and reliability in general [8]. For an almost linear limit state function, the FORM algorithm provides satisfactory results that are comparable with the results attained with Level III methods.

The FORM algorithm is an iterative procedure and non-normal distributed random variables are approximated by the so called "Tail Approximation" whereby the density function $f_X(\vec{x})$ and probability function $F_X(\vec{x})$ in the point $\vec{x}_i^*$ from the original distribution and the standard normal distribution are equalised. The Starting vector $\vec{x}_{i=1}^*$ is of great importance because it is possible that the algorithm finds local minimas.

Figure 2 shown the procedure of a common FORM algorithm.

2.2 SORM Algorithm

The second-order reliability method (SORM) has been established as an attempt to improve the accuracy of the first-order methods, as the FORM. In first-order methods, since the limit state function is approximated by a linear function, accuracy problems can occur when the performance function is strongly nonlinear [9]. As opposed to the first-order methods, in the SORM, the integration boundary $g(\vec{x}) = 0$, denoted the limit-state surface, is no longer approximated by a hyperplane; instead, the boundary $g(\vec{x}) = 0$ is replaced by a paraboloid in a transformed standard normal space [10–12].

The requirements for this approximation are, however, that the limit state function is continuous near the approximation point and can be differentiated at least twice. Fundamentally, for convex functions $g(\vec{x}) = 0$ an approximation as a hypersphere and an approximation as a linear hyperplane represent an upper limit and a lower limit for the failure probability p_f (Fig. 3).

It is assumed that, in the standard normal space, the reliability index β corresponds to the minimum distance from the origin of the axes to the limit state surface. The minimum distance point on the limit state surface is denoted the design point $\vec{x}^*$.

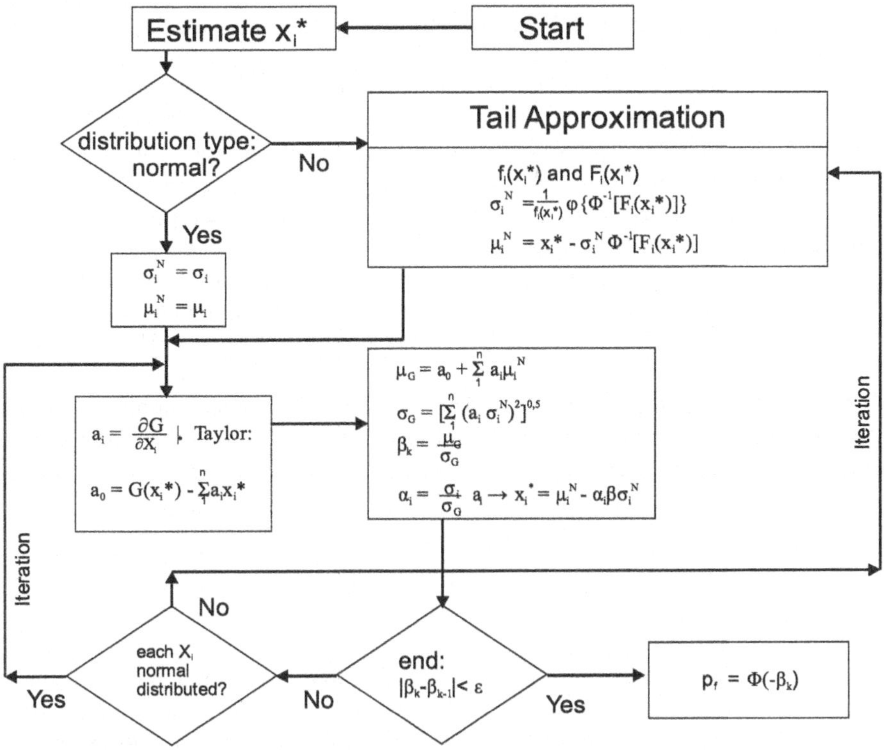

Fig. 2 Procedure of a FORM algorithm (adapted from [17])

Fig. 3 Schematic representation of the integration areas (adapted from [18])

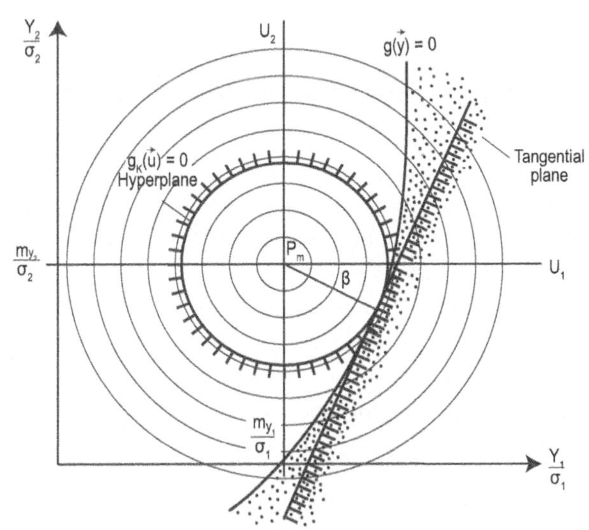

In the curvature-fitting SORM, the paraboloid is defined by matching its principal curvatures to the principal curvatures of the limit state surface at the design-point $\vec{x}^*$ [13]. To this end, Eq. (1) is transformed into a so-called quadric function. A quadric function depends on the number of variables and can be a curve, surface or hyper surface of second order. The basic variables X_i are converted into standard normal distributed variables U_i. The coordinate system $(u_1, u_2, \ldots, u_n)$ is rotated around its origin so that one of the coordinate axes coincides with the design point. In the new coordinate system, the design point has the coordinates $(0, \ldots, \beta)$. This rotation is carried out through an orthogonal transformation matrix by using, for example, the Gram-Schmidt orthogonalization algorithm. Then, at the design point, the principal curvatures of the limit-state surface are obtained as the eigenvalues of Hessian matrix [13].

The exact calculation of the probability of failure can be rather complex. Breitung [10], for example, has derived an asymptotic approximate equation that provides insight into the nature of the contribution of each curvature, where the probability of failure is expressed as:

$$p_f \approx \Phi(-\beta) \prod_{i=1}^{n-1} (1 - \beta\kappa_i)^{-1/2} \tag{2}$$

in which $\Phi(.)$ is the standard normal cumulative distribution function, β is the distance from the coordinate origin (i.e. reliability index). The first term in Eq. (1) represents the first-order approximation of the failure probabilities p_f, and the product term involving the quantities $(1 - \beta\, k_i)$, with $\beta\, k_i$ being the main curvatures at the design point, represents the second-order correction [13].

2.3 *Monte Carlo Simulation*

The Monte Carlo simulation method uses techniques of statistical calculation by generating uniform distributed (pseudo) random numbers. By generating a stochastically independent, high number of those random variables, the probability of failure can be calculated using Eq. (3):

$$p_F = \frac{n_F}{N} \tag{3}$$

where N is the total number of realisations (or number of simulations) and n_F is the number of simulations, for which the performance function is less or equal to zero: $g \leq 0$).

If the number of realizations increases, the accuracy of the simulation will also increase, whereas the coefficient of variation will decrease.

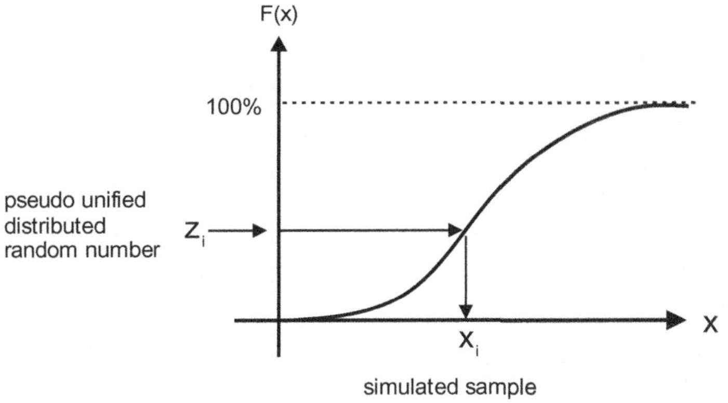

Fig. 4 Principle of Monte Carlo simulation

For arbitrary types of distributions (e.g. lognormal, gamma, …), the generated uniformly distributed random variables have to be transformed with the probability function $F_X(\vec{x})$, applicable for the certain distribution type (Fig. 4).

3 Implementation of Reliability Methods

Chapter 3 describes a new open-source software library for reliability analysis in civil engineering. The goal is to facilitate the adoption of reliability methods among engineers in practice as well as to provide an open platform for further scientific collaboration in software language "R" [14].

3.1 Description of Software Tool

The new library is being developed as a so-called "R package" in open-source programming software "R". The package is capable of carrying out systematic parameter studies using different probabilistic reliability methods (e.g. FORM, SORM, Monte Carlo Simulation). Based on this, an overview on the probabilistic reliability methods implemented in the package as well as results of first parametric studies are given. The developed package allows to perform systematic and large parameter studies and provides different algorithms of reliability analysis in an effective way. The structure of the software tool is shown in Fig. 5.

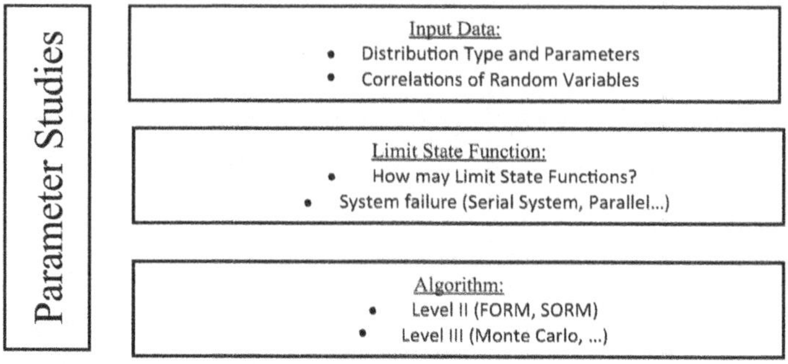

Fig. 5 Structure of the software tool

3.2 *Example of Parameter Study*

As a practical example, the limit state function of the bending problem for steel reinforced concrete members is chosen. Equations (5) and (6) shows a formula of the state function g which is used for the parameter study.

$$g = M_R - M_E \tag{4}$$

$$M_R = \theta_R \cdot A_s \cdot f_y \cdot d \cdot \left(1 - \frac{A_s \cdot f_y}{2 \cdot d \cdot f_c}\right) \tag{5}$$

$$M_E = \theta_E \cdot (g + q) \cdot \frac{6^2}{8} \tag{6}$$

Table 2 shows the statistical parameters of the basic variables (mean and standard deviation) and their distribution types.

In Fig. 6, the resulting reliability indices β are presented in dependence of the varied effective depth. For the parametric study, three different reliability methods (FORM, SORM, Monte Carlo) were used.

It can be seen that the curvature of the limit state function shows the non-linear effect of the limit state function (Fig. 6). The results of the SORM algorithm (Level II) and the Crude Monte Carlo method (Level III) are almost the same, and therefore, it gives a first indication that the software is suitable for parametric studies with the reliability methods described in Chap. 2. The new software code is working well and this first example shows the effectiveness of the new software library, especially for large parameter studies.

Table 2 Limit state function 1

No.	RV	Name	Dist. Typ	Mean	Std. deviation	References
X1	θ_R	Uncertainty resistance	LogNormal	1.0	0.05	[3] Table 3.9.1
X2	A_s	Area of steel	Normal	A_s	$0.02 \times A_s$	[3] Chapter 3.2.2
X3	f_y	Yield strength steel	Normal	560 N/mm^2	30 N/mm^2	[15] Page 109
X4	d	Effective depth	Normal	d	10 mm	[3] Chapter 3.10.3
X5	f_c	Concrete strength	LogNormal	33 N/mm^2	5 N/mm^2	[16] Table 3.1
X6	θ_E	Uncertainty load	LogNormal	1.0	0.10	[3] Table 3.9.1
X7	g	Self-load	Normal	g_k	$0.05g_k$	[5] Chapter 4.1.2
X8	q	Live-load	Gumbel	$0.7q_k$	$0.2q_k$	[15] Page 109

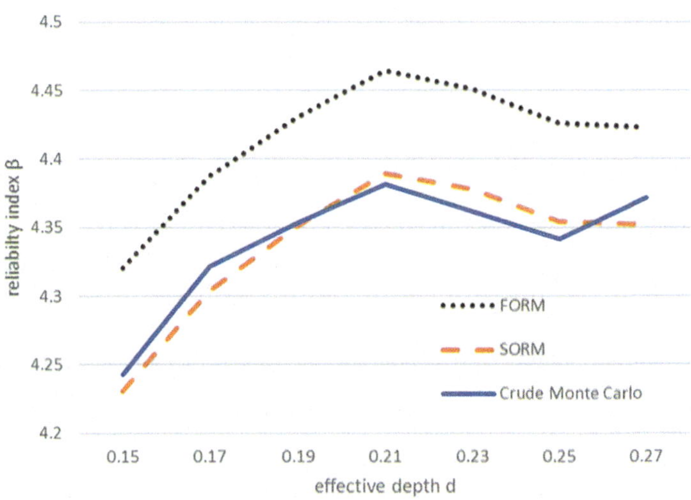

Fig. 6 Results parameter study of bending problem

4 Conclusions and Outlook

It is shown in this paper how different reliability methods can be implemented in program code. In addition, the results of a first parameter studies are presented to illustrate the correctness and functionality of the new software package.

The parametric study highlighted two important aspects. On the one hand, the implementation of reliability methods in civil engineering is an important step towards a wider application of statistical methods, to which this contribution should motivate. On the other hand, the parameter study presented shows the application of the probabilistic methods using a practical example with a nonlinear limit state function and non-normally distributed basic variables.

Furthermore, advanced reliability methods (e.g. Monte Carlo with Subset Sampling Simulation) will be implemented in the new R package. Further work will result in larger parameter studies, which will support the development of a new guideline for the application of reliability methods in civil engineering, and will continue the progress of reliability research mentioned in Chap. 1. In the project "TesiproV", the authors will provide a new guideline, which describes certain techniques of code calibration, based on reliability methods, as well as the assessment of new partial safety factors.

Acknowledgements The authors would like to thank the German Federal Ministry for Economic Affairs and Energy (BMWI) for funding the project TesiproV "Allgemeingültiges Verfahren zur Herleitung von Teilsicherheitsbeiwerten im Massivbau auf Basis probabilistischer Verfahren anhand ausgewählter Versagensarten – Erstellung eines Richtlinienentwurfs" within the research funding program "WIPANO" (No. 03TNK003).

References

1. Freudenthal, A. (1947). The safety of structure. *Transactions of the American Society of Civil Engineers, 112*(1), 125–159.
2. Joint-Committee CEB/CECM/CIB/FIB/IABSE Structural Safety. (1974). *First order reliability concepts for design codes (documentation).* Paris: Comite Europeen du Beton (Bulletin d'Information, No. 112).
3. Joint Committee on Structural Safety. (2001). *Probabilistic model code (12th draft).* www.jcss.ethc.ch.
4. ISO 2394:2015(E). (2015). *General principles on reliability for structures.* Geneva: International Organization for Standardization (ISO).
5. DIN EN 1990. (2010). *Grundlagen der Tragwerksplanung; Deutsche Fassung EN 1990:2002.* Berlin: Beuth Verlag.
6. Ricker, M., Feiri, T., Nille-Hauf, K., Adam, V., & Hegger, J. (2020). *Enhanced reliability assessment of punching shear resistance models for flat slabs without shear reinforcement.* Preprint submitted to Engineering Structures.
7. Walter, C. et al (2016). *Mistral: Methods in structural reliability analysis.* http://cran.r-project.org/package=mistral.
8. Melchers, R., & Becks, A. (2018). *Structural analysis and prediction.* Hoboken: Wiley.
9. Zhao, Y. G., & Ono, T. (1999). A general procedure for first/second-order reliability method (FORM/SORM). *Structural Safety, 21*(2), 95–112.
10. Breitung, K. (1984). Asymptotic approximations for multinormal integrals. *Journal of Engineering Mechanics, 110*(3), 357–366.
11. Der Kiureghian, A., Lin, H. Z., & Hwang, S. J. (1987). Second-order reliability approximations. *Journal of Engineering Mechanics, 13*(8), 1208–1225.
12. Tvedt, L. (1990). Distribution of quadratic form in normal space—Application to structural reliability. *Journal of Engineering Mechanic, 116*(6), 1183–1197.
13. Kirugehian, A. D., & Stefano, M. D. (1991). Efficient algorithm for second-order reliability analysis. *Journal of Enigneering Mechanics, 117*(12), 2904–2923.
14. R Project. (2020). www.r-project.org.

15. Ricker, M. (2009). *Zur Zuverlässigkeit der Bemessung gegen Durchstanzen bei Einzelfundamenten*, Dissertation, RWTH Aachen University.
16. DIN EN 1992-1-1. (2011). *Bemessung und Konstruktion von Stahlbeton- und Spannbetontragwerken – Teil 1-1: Allgemeine Bemessungsregeln und Regeln für den Hochbau; Deutsche Fassung EN 1992-1-1:2004 + AC:2010*. Berlin: Beuth Verlag.
17. Faber, M. H. (2009). *Risk and safety in engineering. Lectures Notes*. Zurich: ETH Zurich.
18. Neumann, H. J., Fiessler, B., & Rackwitz, R. (1977). *Die genäherte Berechnung der Versagenswahrscheinlichkeit mit Hilfe rotationssymmetrischer Grenzzustandsflächen* 2. [zweiter] Ordnung. Sonderforschungsbereich 96, Techn. Univ.
19. Ditlevsen, O., & Madsen, H.O. (2007). *Structural reliability methods*.
20. Bronstein, I. N., & Semendjajew, K. A. (1988). *Ergänzende Kapitel zum Taschenbuch der Matehmatik, Bd 1, 5 bearbeitete und erweitert Auflage*. Frankfurt: Verlag Harri Deutsch.

Influence of an In-Situ Inspection on the Reliability Analysis of an Ancient Timber Roof

Leonardo G. Rodrigues and Hélder S. Sousa

Abstract Despite the durability of timber and its efficient performance seen in the built heritage, it has become a common practice, in Portugal, to replace ancient timber roof structures by concrete or steel roof ones. The main reason may be attributed to the difficulty in assessing the real condition of timber structures with respect to its actual level of conservation. In this work a reliability analysis of an ancient timber roof, from a Portuguese neoclassic building of the eighteenth century, is made to evidence the importance of different levels of information taking into account visual and geometric inspections. The impact of posterior knowledge obtained from non-destructive tests is evaluated by comparing the probability of failure and the reliability index on two distinct scenarios. The first scenario considers only prior information for the mechanical properties of timber elements and apparent cross-sections for the structural members. On the other hand, the second scenario considers the results of an in-situ inspection that provides the residual cross-section of the principal members, as well as the updating of the modulus of elasticity and density, based on a Bayesian Updating procedure that takes advantage of the results of a database of non-destructive tests. Latin Hypercube Sampling (LHS) was used in this study to generate a set of structural models, in which each model corresponds to a realization of the assumed random variables. Apart from the mechanical properties, the uncertainties related to permanent, snow and wind loads, are included according to the provisions of the Joint Committee on Structural Safety (JCSS). The presented results indicate that in-situ inspections have to be a priority on the assessment of ancient timber structures. The absence of a careful assessment of deteriorated elements can lead to incorrect conclusions about the structural safety. Additionally, the use of a probabilistic framework allows to a better definition of intervention plans by providing the reliability of distinct critical elements.

Keywords Bayesian updating · Non-destructive tests · Reliability

---

L. G. Rodrigues (✉) · H. S. Sousa
ISISE, Department of Civil Engineering, University of Minho, Guimarães, Portugal

© The Author(s), under exclusive license to Springer Nature Switzerland AG 2021

J. C. Matos et al. (eds.), *18th International Probabilistic Workshop*, Lecture Notes in Civil Engineering 153, https://doi.org/10.1007/978-3-030-73616-3_31

1 Introduction

The failure of several timber roof structures under expected snow loads has shown that design and/or construction errors, unexpected deterioration, and insufficient maintenance can lead to consequences far greater than the initial event [1]. On the other hand, when each structural member is properly designed and an adequate maintenance plan is put into practice, timber structures are durable and can remain in service for periods greater than the design life-span [2]. Despite the durability of ancient timber, it has become a usual practice, in Portugal, to replace existing timber roofs structures by concrete or steel roof structures, supposedly due to lower construction costs. This is mainly due to the difficulty in assessing the real condition of timber structures, with respect to its actual level of conservation. In fact, the geometric assessment and visual inspection of existing timber structures are two of the most important tasks in evaluating their integrity. The inherent variability of timber and its susceptibility to decay emphasize the need to carry out a characterization of the constitutive timber elements [3]. The information on the geometry and the state of degradation can be measured by means of non-destructive tests (NDT), mechanical tests, and other means [4]. However, the uncertainties related to the actual mechanical properties constitute a drawback for practitioners, which in turn may design ultra-conservative solutions or decide for a complete replacement of the structure. The existence of a database with correlations between NDT results and mechanical properties of wood from different species would constitute an important achievement. On the other hand, engineers can perform reliability-based analysis, where they can assume that mechanical properties, as well as the external loads, as random variables, thus accounting to different sources of uncertainty. In this regard, Köhler et al. [5] proposed distribution functions for the main mechanical properties of timber members, based on test programs and investigations considering European and North American softwoods, whereas the probabilistic models for loads can be derived based on the Probabilistic Model Code and other standards. Additionally, the parameters of these probabilistic models can be updated through the use of Bayesian methods, when new empirical or monitoring data is available.

The main objective of this paper is to implement a probabilistic methodology to evaluate the safety of an ancient timber roof, from a Portuguese neo-classic building of the eighteenth century, for partial collapse prevention limit state. The evaluation includes the inherent uncertainties of timber, as well as the uncertainties related to external loading. Distinct damage scenarios are considered, in order to stress the importance of performing adequate in-situ inspections before the design of solutions to retrofit the structure. The first assumed scenario (Scenario 1) considers the cross-sections with their apparent dimensions, and the mechanical properties as random variables represented through the probabilistic distributions proposed in [5] and considering the mean values presented in national standards [6]. The second reliability analysis takes advantage of various NDT, such as impact penetration, drilling resistance and ultrasounds tests, where effective cross-sections are considered for the different structural members, whereas a Bayesian Updating procedure was performed to obtain new model parameters for density and modulus of elasticity (Scenario 2).

2 Description of the Structure

The two pitched roof structure is composed by four Maritime Pine (*Pinus pinaster*) collar beam trusses, spaced 3 m from each other, as illustrated in Fig. 1. The disposition of elements was based on the structural configuration of the Chimico Laboratory, a Portuguese neoclassic building from the eighteenth century.

In the case of the collar beam trusses considered in the current study, a geometrical assessment was performed at intervals of 40 cm for each individual member. The assessed sections were marked and identified so that the dimensions of these same sections could be measured using non-destructive tests in order to obtain the effective cross section. Information on the density, the presence of voids and their location, as well as the modulus of elasticity were obtained using impact penetration, drilling resistance and ultrasounds tests. Figure 2 illustrates a mapped diagram with measured sections of each member of one collar truss. Due to the biological attack found on the surface of the elements, significant coefficients of variation were found pertaining to the cross-section dimensions. Moreover, extensive wane was found affecting the rectangular section of the elements. This was especially found on the rafters and collar beams. as presented in Table 1.

One of the collar trusses was tested at the Structural Laboratory of the University of Minho, Portugal. The truss was subjected to two downward loads, applied on the rafters, at a displacement rate of 0.05 mm/s until failure. Further information is available in [9]. Three main failures were found during the mechanical test, namely, failure of both rafters in the sections near the connection with the tie-beam and failure near the connection of one of the rafters with the post. Those failures coincided with sections having lower visual grades due to the presence of significant knots.

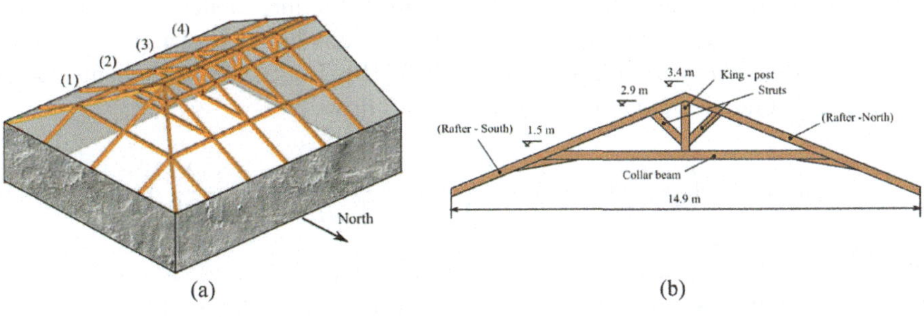

Fig. 1 Timber roof: **a** three-dimensional perspective and **b** planar collar beam truss [7]

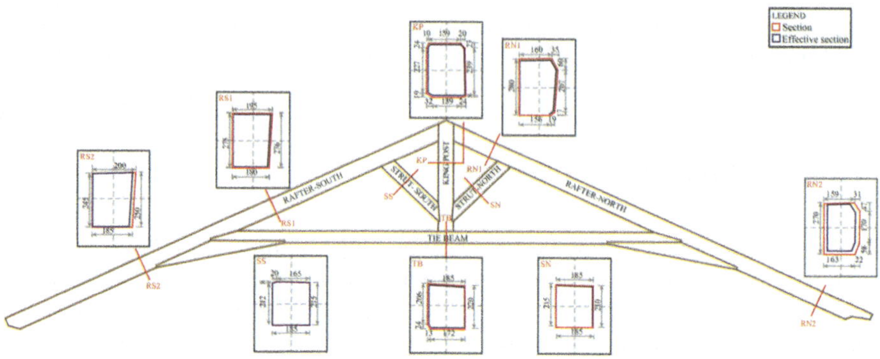

Fig. 2 The collar truss with representative sections obtained from the geometrical survey mapped [8]

3 Probabilistic Model

3.1 Structural Model, Limit States and Variables

For the structural analysis, 2D linear finite element models were developed within the OpenSees framework. The trusses were modelled using linear elastic frame elements connected with zero-length elements. Each structural element is divided in frame segments of in 40 cm length as per the locations of the non-destructive tests carried out. The stiffness between the collar beam and the rafters is considered as rigid for all the three degrees of freedom (translational and rotational), whereas the remaining ones were considered as hinged connections. The supports of the truss on the walls were also modelled by hinges. The connection between the purlins and the trusses is considered as weak. In this scenario, in case of failure of one of the trusses, the loads are not redistributed to another undamaged parts of the structure. Additionally, one can evaluate separately the reliability of each truss, while considering uniform distribution of loads within the roof. The considered failure modes are related with the ultimate limit state verifications for bending and combined axial and shear forces in the timber elements according to Eurocode 5 [10]. The failure of elements due to bending and combined axial forces, is given as follows:

$$g_i = X_R - \left(\frac{a_i \cdot G + b_i \cdot Q}{k_{mod} \cdot k_{c,z} \cdot R_{c,0} \cdot A} + k_m \cdot \frac{c_i \cdot G + d_i \cdot Q}{k_{mod} \cdot k_h \cdot R_m \cdot W} \right) = 0 \quad (1)$$

where A is the cross-section area, W is the section modulus, and X_R is the model uncertainty, which is modelled through a lognormal distribution with an expected value of 1.0 and a coefficient of variation of 10% [11]. The internal bending moment, as well as the normal force, are given by linear combinations the permanent loads G, and the variable loads Q, which englobes imposed, wind, and snow loads. R_m is the bending strength parallel to the grain and $R_{c,0}$ is the compression strength parallel to

Table 1 Apparent and residual (in parenthesis) dimension of the structural elements cross-sections

Truss	Element	Height (mm)			Width (mm)		
		Min	Mean	Max	Min	Mean	Max
(1)	North rafter	255.0 (170.0)	271.3 (256.1)	280.0 (280.0)	180.0 (60.0)	189.4 (90.6)	200.0 (165.0)
	South rafter	260.0 (138.0)	280.0 (209.8)	290.0 (285.0)	165.0 (50.0)	183.1 (101.9)	190.0 (190.0)
	Collar beam	180.0 (90.0)	217.2 (170.2)	240.0 (214.0)	190.0 (130.0)	195.0 (176.7)	200.0 (195.0)
	King-Post	265.0 (227.0)	265.0 (252.3)	265.0 (265.0)	200.0 (200.0)	260.0 (260.0)	380.0 (380.0)
	North strut		215.0 (205.0)			190.0 (185.0)	
	South strut		210.0 (210.0)			185.0 (185.0)	
(2)	North rafter	215.0 (90.0)	240.0 (179.9)	250.0 (235.0)	180.0 (180.0)	186.7 (185.7)	190.0 (190.0)
	South rafter	225.0 (123.0)	243.3 (150.3)	260.0 (190.0)	195.0 (100.0)	202.9 (155.0)	220.0 (220.0)
	Collar beam	200.0 (70.0)	215.0 (184.5)	225.0 (225.0)	190.0 (115.0)	199.6 (186.4)	205.0 (205.0)
	King-Post		260.0 (230.0)			425.0 (425.0)	
	North strut		195.0 (195.0)			190.0 (100.0)	
	South strut		225.0 (125.0)			200.0 (200.0)	
(3)	North rafter	245.0 (165.0)	262.8 (230.9)	270.0 (270.0)	190.0 (110.0)	195.0 (163.9)	200.0 (200.0)
	South rafter	240.0 (200.0)	270.0 (242.5)	290.0 (262.0)	180.0 (140.0)	190.8 (176.7)	200.0 (190.0)
	Collar beam	215.0 (163.0)	231.7 (216.0)	240.0 (237.0)	190.0 (185.0)	195.8 (194.2)	200.0 (200.0)
	King-Post	260.0 (230.0)	260.0 (245.0)	260.0 (260.0)	195.0 (195.0)	250.0 (201.3)	415.0 (220.0)
	North strut	235.0 (199.0)	237.5 (211.5)	240.0 (224.0)	195.0 (195.0)	195.0 (195.0)	195.0 (195.0)
	South strut		230.0 (230.0)			200.0 (160.0)	
(4)	North rafter	260.0 (175.0)	267.5 (230.5)	280.0 (268.0)	190.0 (125.0)	200.4 (195.0)	205.0 (205.0)
	South rafter	250.0 (130.0)	265.0 (206.9)	270.0 (236.0)	190.0 (95.0)	195.0 (159.5)	205.0 (195.0)

(continued)

Table 1 (continued)

Truss	Element	Height (mm)			Width (mm)		
		Min	Mean	Max	Min	Mean	Max
	Collar beam	180.0 (175.0)	207.9 (203.7)	220.0 (220.0)	190.0 (100.0)	196.4 (167.9)	200.0 (200.0)
	King-Post	265.0 (205.0)	267.5 (205.0)	270.0 (205.0)	195.0 (190.0)	312.5 (190.0)	430.0 (190.0)
	North strut		220.0 (170.0)			190.0 (190.0)	
	South strut		230.0 (150.0)			190.0 (190.0)	

the grain. k_{mod} (equal to 0.9) is a modification factor taking into account the effect of the duration of load and moisture content. $k_{c,z}$ is an instability factor, which is dependent of the slenderness of each structural element, and k_h is a size effect factor [10]. On the other hand, the shear failure of elements is modelled with the following short-term ultimate limit state function:

$$g_i = X_R - \left(\frac{e_i \cdot G + f_i \cdot Q}{k_{mod} \cdot R_v \cdot A} \right) = 0 \qquad (2)$$

where R_v is the shear strength, and the internal shear force is given by linear combinations of the variable loads Q and the permanent loads G. As presented in Table 2, seven random variables are evaluated for each timber element. The distribution parameters of the mechanical properties (bending strength, bending modulus of elasticity, and density), used for the first scenario assumed, are computed based on characteristic values and mean values, defined in NP 4305:1995 [6] for structural Maritime

Table 2 Random variables for timber material properties

	Distribution parameters				Correlation matrix						
X	Dist.	$E[X]$	$CoV[X]$		R_m	E_m	ρ_{den}	$R_{t,0}$	$R_{c,0}$	G_v	R_v
R_m	LN	27.8	0.25	R_m	1.0	0.8	0.6	0.8	0.8	0.4	0.4
E_m	LN	7500	0.13	E_m		1.0	0.6	0.6	0.6	0.6	0.4
ρ_{den}	N	580	0.13	ρ_{den}			1.0	0.4	0.8	0.6	0.6
$R_{t,0}$	LN	18.6	0.30	$R_{t,0}$				1.0	0.5	0.4	0.6
$R_{c,0}$	LN	25.4	0.20	$R_{c,0}$					1.0	0.4	0.4
G_v	LN	470.0	0.13	G_v						1.0	0.6
R_v	LN	3.1	0.25	R_v							1.0

Description: R_m—Bending strength to the grain (N/mm^2); E_m—Bending modulus of elasticity (N/mm^2); ρ_{den}—Density (kg/m^3); $R_{t,0}$—Tension strength to the grain (N/mm^2); $R_{c,0}$—Compression strength to the grain (N/mm^2); G_v—Shear modulus (N/mm^2); R_v—Shear strength (N/mm^2)

Table 3 Random variables for timber material properties

Variable	Distribution	$E[X]$ (kN/m^2)	$CoV[X]$
Permanent load, P	Normal	1.0	0.25
Live load, LL	Gumbel	0.32	0.40
Wind load—upwind, W_{up}	Gumbel	0.23	0.35
Wind load—downwind, W_{down}	Gumbel	0.32	0.35
Snow load, S	Gumbel	0.20	0.40

Pine swan timber. Table 2 also presents the intra-element correlation coefficients considered, which were also taken from the work developed by Köhler [6].

The loads considered are the permanent (i.e., weight of the trusses, roof tiles and sheeting), imposed (i.e. live load), wind loads (i.e., upwind and downwind), and snow load. The expected values (mean values), coefficient of variation (CoV), and probabilistic distributions used for each load type are provided in Table 3.

3.2 Bayesian Updating of Mechanical Properties

The updating procedure was detailed for the case of existing timber structures in [12], following a brief description is provided. When samples or measurements of a stochastic variable X are provided, the probabilistic model may be updated and, thus, also the probability of failure. Considering a stochastic variable X with density function $f_X(x)$, and if q denotes a vector of parameters defining the distribution for X, the density function of the stochastic variable X can be written as $f_X(x,q)$. When the parameters q are uncertain then $f_X(x,q)$ can be considered as a conditional density function: $f_X(x|Q=q)$ where q denotes a realization of Q, therefore q is a vector of distribution parameters (e.g. mean μ, and standard deviation σ). The initial density function for the parameters Q is denoted $f_Q'(q)$ and is termed the *prior* density function.

Taking into account new information, it is assumed that n observations or measurements of the stochastic variable X are available making up a sample $\hat{x} = (\hat{x}_1, \hat{x}_2, \ldots, \hat{x}_N)$. The measurements are assumed to be statistically independent. The updated density function $f_Q''(q|\hat{x})$ of the uncertain parameters Q given the realizations is denoted the *posterior* density function and is given by:

$$f_Q''(q|\hat{x}) = \frac{f_N(\hat{x}|q) f_Q'(q)}{\int f_N(\hat{x}|q) f_Q'(q) dq} \tag{3}$$

where the *Likelihood function* $f_N(\hat{x}|q) = \prod_{i=1}^{N} f_X(\hat{x}_i|q)$ is the probability density of the given observations assuming that the distribution parameters are q. Then the updated density function of the stochastic variable X given the realization $\hat{x}$ is denoted the *predictive* density function and is defined by:

$$f_X(x|\hat{x}) = \int f_X(x|q) f''_Q(q|\hat{x}) dq \tag{4}$$

The *prior* and *posterior* distributions are often chosen according to the available data and to the importance of the analysis. Normal distributions are often used for that purpose. Assuming that X is Normal distributed and both mean value, μ and standard deviation, σ are uncertain then the *prior* distribution of the resistance function R is denoted $f_Q{}'(q) = f_R{}'(\mu, \sigma)$ and can be defined as Eq. (5):

$$f'_R(\mu, \sigma) = k\sigma^{-(\nu'+\delta(n')+1)} \exp\left(-\frac{1}{2\sigma^2}\left(\upsilon'(s')^2 + n'(\mu - m')^2\right)\right) \tag{5}$$

with $\delta(n') = 0$ for $n' = 0$ and $\delta(n') = 1$ for $n' > 0$. The *prior* information about the standard deviation σ is given by parameters s' and ν'. The expected value $E(\sigma)$ and coefficient of variation $COV(\sigma)$ of σ can asymptotically (for large ν') be expressed as:

$$E(\sigma) = s' \tag{6}$$

$$COV(\sigma) = \frac{1}{\sqrt{2\upsilon'}} \tag{7}$$

The *prior* information about the mean μ is given by parameters m', n' and s'. The expected value $E(\mu)$ and coefficient of variation $COV(\mu)$ of μ can asymptotically (for large ν') be expressed as:

$$E(\mu) = m' \tag{8}$$

$$COV(\mu) = \frac{s'}{m'\sqrt{n'}} \tag{9}$$

Another possible way to interpret the *prior* information is to consider the results of a hypothetical prior test series, for mean and standard deviation analysis. For that case the standard deviation is characterized by s' (hypothetical sample value) and ν' (hypothetical number of degrees of freedom for s'). The information about the mean is given by: m' (hypothetical sample average) and n' (hypothetical number of observations for m').

Usually the degrees of freedom for the number of observations n is given by $\nu = n - 1$, but the *prior* parameters n' and ν' are independent from each other. When new information is available, the resistant model given by the *prior* distribution $f_R{}'(\mu,$

σ) may be updated according to Eq. (3), with the parameters:

$$n'' = n' + n \tag{10}$$

$$\nu'' = \nu' + \nu + \delta(n') \tag{11}$$

$$m''n'' = n'm' + nm \tag{12}$$

$$\nu''(s'')^2 + n''(m'')^2 = \nu'(s')^2 + n'(m')^2 + \nu s^2 + nm^2 \tag{13}$$

With this procedure the predictive value of the resistance R is given by:

$$f_R = m'' - t_{\nu''} s'' \sqrt{\left(1 + \frac{1}{n''}\right)} \tag{14}$$

where $t_{\nu''}$ has a central t-distribution [11]. The new information may be considered from the data gathered in non-destructive tests and with that data reliability may be updated.

3.3 Data for Updating

In this work, Bayesian updating methods are used to update two key mechanical properties of the material, namely bending modulus of elasticity and density. The updating data was obtained through NDT's results collected using pin penetration tests, conducted on one of the trusses, as well as from tests allowing to estimate the correlations between those results and the mechanical properties. Further detail regarding the experimental results and linear correlation analysis can be found in [9, 13, 14].

3.4 Updated Values for the Key Mechanical Properties (Scenario 2)

Information is considered by vague prior information on both mean value and standard deviation. Therefore, the prior information parameters can be presented as hypothetical sample average, m', and sample standard deviation, s', are not relevant; hypothetical number of observations for m', $n' = 0$; hypothetical number of degrees of freedom for s', $\nu' = 0$. Thus, the posterior parameters become: $n'' = n$; $\nu'' = n - 1$; $m'' = m$ and $(s'')^2 = s^2$. The predictive value of r is given by:

Table 4 Updated values for bending modulus of elasticity and density for the tested truss, assuming vague information on both mean and standard deviation

Element	MOE		Density	
	Mean (kN/mm^2)	CoV (%)	Mean (kg/m^3)	CoV (%)
North rafter	5596	7.8	575	1.0
South rafter	8218	1.9	601	0.5
Collar beam	8992	1.9	613	0.5
Post	8672	1.6	608	0.47
North diagonal	10,004	0.9	626	0.3
South diagonal	11,344	0.6	645	0.2
Truss (global)	7830	4.8	601	0.8

$$r_d = \exp(m(Y)) \cdot \exp\left(-t_{vd} \cdot s(Y) \cdot \sqrt{1 + \frac{1}{n}}\right) \tag{15}$$

where $t_{\nu d}$ has a central t-distribution.

The results of the updated values for bending modulus of elasticity and density are provided for the tested truss, for each element and globally on Table 4. By analysis of the results it is seen that a low variation is found within elements with exception of the North rafter. This is consistent with the visual inspection carried out to the trusses which showed that this element was severely decayed on localized segments (near the wall support).

From the results obtained, one concluded that for Scenario 2, the modulus of elasticity and the density should be modelled considering the global updated parameters (expected value and coefficient of variation), presented in Table 4, whereas the remaining random variables are represented with the same parameters and distributions, already presented in Table 2.

4 Reliability Analysis

The Latin Hypercube Sampling (LHS) was used to generate 100,000 structural models for each scenario considered in this analysis. As mentioned above, the four trusses where analyzed separately through 2D linear finite element models. Thus, the reliability analysis implied the performance of 800,000 numerical analyses. A load controlled method was applied with a load factor increment $\Delta\lambda_L = 0.1$. For each structural realization, the analysis finished when the short-term ultimate limit state function (Eqs. 1–2) did not hold, which is associated to the partial collapse limit state. During the analysis, a structural failure is considered when the load factor is lower or equal to 1.0, or when the structure is not able to sustain the applied loads. The probability of failure is then given by the ratio between the number of failures

(z_0) and the number of realizations (z):

$$p_f = \frac{z_0}{z} \tag{16}$$

Given the probability of failure (p_f), it is possible to determine the structural reliability index (β) through the inverse cumulative distribution function (Φ^{-1}) of standard normal distribution:

$$\beta = \Phi^{-1}(1 - p_f) \tag{17}$$

Steenbergen et al. [15] proposed a target reliability β_0, which defines the minimum reliability for which not to upgrade the structure is assumed as acceptable. The index β_0 is dependent on the lifetime period, collapsed area, and consequence class. If one considers a lifetime of 5 years and a high consequence class, the structure under study needs an intervention when the reliability index is lower than 3.6. From the results presented in Table 5, one can conclude that in-situ inspections have to be a priority on the assessment of ancient timber structures. The use of apparent cross-sections and timber mechanical properties given by NP 4305:1995 [6], could lead to the decision of postponing an intervention. However, when the analysis considers the effective cross-section, as well as the updated mechanical properties, one can conclude that the structure needs urgent interventions. The impact of assessing the deterioration of timber elements can be also evaluated through the cumulative distribution functions presented in Fig. 3, which were obtain by fitting a lognormal distribution to the values of load factor measured at failure for each truss. The reduction of the effective

Table 5 Reliability indices and probabilities of failure for distinct Scenarios

Scenario	p_f	β
(1) Without in-situ inspection	9×10^{-5}	3.75
(2) After an in-situ inspection	0.96	−1.72

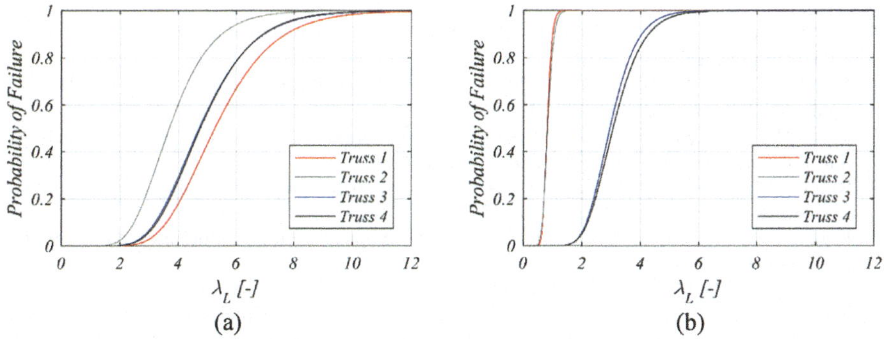

Fig. 3 Cumulative distribution functions for load factors obtained at the partial collapse limit state: **a** Scenario 1 and **b** Scenario 2

cross-section of the rafters of trusses 1 and 2 (see Fig. 1), are compromising their strength, especially for bending and compression forces. On the other hand, one can conclude that trusses 3 and 4 are still able to sustain the applied loads. Thus, a reliability analysis can be a useful tool to plan future interventions in ancient timber structures.

5 Conclusions

A reliability analysis was presented for an ancient timber roof. The evaluation was based on the results of an in-situ inspection including non-destructive tests. The inspection allowed to gather the dimensions of apparent and effective cross-sections, while the results of impact penetration, drilling resistance and ultrasounds tests, permitted the Bayesian update of mechanical properties, such as the elastic modulus and wood's density. The presented results are indicative that in-situ inspections have to be a priority on the assessment of ancient timber structures in order to have an optimized intervention plan. The absence of a careful assessment of deteriorated elements can lead incorrect conclusion about the global structural safety. For the presented study case, the use of a probabilistic framework would allow to define the intervention plan by providing the reliability of distinct primary trusses. Future studies shall assess the impact of defect on the reliability of ancient structures, given that it is difficult to measure in-situ the size and dispersion of knots.

Acknowledgements This work was partly financed by FCT/ MCTES through national funds (PIDDAC) under the R&D Unit Institute for Sustainability and Innovation in Structural Engineering (ISISE), under reference UIDB/ 04029/2020.

References

1. Munch-Andersen, J., & Dietsch, P. (2011). Robustness of large-span timber roof structures—Two examples. *Engineering Structures*.
2. Woodard, A. C., & Milner, H. R. (2016). *Sustainability of timber and wood in construction* (2nd ed.). Amsterdam: Elsevier Ltd.
3. Cruz, H., et al. (2015). Guidelines for on-site assessment of historic timber structures. *International Journal of Architectural Heritage, 9*(3).
4. Riggio, M., et al. (2014). In situ assessment of structural timber using non-destructive techniques. *Materials and Structures, 47*(5), 749–766.
5. Köhler, J. (2006). *Reliability of timber structures*. Swiss Federal Institute of Technology, no. DISS. ETH NO. 16378.
6. LNEC, NP 4305:1995. (1995). Structural maritime pine swan timber—Visual grading, CT 14.
7. Sousa, H. S., & Neves, L. C. (2018). Reliability-based design of interventions in deteriorated timber structures. *International Journal of Architectural Heritage, 12*(4), 507–515.
8. Rodrigues, L. G., Aranha, C. A., & Sousa, H. S. (2016). Robustness assessment of an ancient timber roof structure. In *Historical earthquake-resistant timber framing in the mediterranean area* (pp. 447–457). Cham: Springer.

9. Branco, J. M., Sousa, H. S., & Tsakanika, E. (2017). Non-destructive assessment, full-scale load-carrying tests and local interventions on two historic timber collar roof trusses. *Engineering Structures, 140,* 209–224.
10. CEN, EN 1995-1-1:2004. (2004). *Eurocode 5: Design of timber structures—Part 1-1: General—Common rules and rules for buildings* (vol. 1).
11. Köhler, J., Sørensen, J. D., & Faber, M. H. (2007). Probabilistic modeling of timber structures. *Structural Safety, 29*(4), 255–267.
12. Sousa, H. S., Sørensen, J. D., Kirkegaard, P. H., Branco, J. M., & Lourenço, P. B. (2013). On the use of NDT data for reliability-based assessment of existing timber structures. *Engineering Structures, 56,* 298–311.
13. Gomes, I. D., Kondis, F., Sousa, H. S., Branco, J. M., & Lourenço, P. B. (2016). Assessment and diagnosis of two collar timber trusses by means of visual grading and non-destructive tests. In *Historical earthquake-resistant timber framing in the Mediterranean area* (pp. 311–320). Berlin: Springer.
14. Yurrita, M., & Cabrero, J. M. (2018). New criteria for the determination of the parallel-to-grain embedment strength of wood. *Construction and Building Materials, 173,* 238–250.
15. Steenbergen, R. D. J. M., Sýkora, M., Diamantidis, D., Holický, M., & Vrouwenvelder, T. (2015). Economic and human safety reliability levels for existing structures. *Structural Concrete, 16*(3), 323–332.

Inherent Variability of Geotechnical Properties for Finnish Clay Soils

Monica S. Löfman and Leena K. Korkiala-Tanttu

Abstract Compared to manufactured materials, soil properties often exhibit significant variability even within a seemingly homogeneous soil layer. The uncertainty related to this variability can be dealt in a robust manner by means of reliability-based methods. Hence, effort has been made to collect soil statistics in order to provide approximate guidelines for selecting the value of coefficient of variation (COV) of inherent variability. It has been observed that the COV value for the same physical property tends to vary within a relatively narrow range, meaning that the literature COV ranges could be utilized with some confidence on sites which lack sufficient soil data. However, it is not certain whether these prior COV values can be used in Finland since many Finnish clay soils are soft and sensitive due to their unique geological history shaped by the last post-glacial processes. Hence, this paper evaluates the inherent variability of various geotechnical properties (index, strength and consolidation properties) in four clay soil sites located in Southern Finland. Besides prior ranges of COV, this paper provides prior ranges of the mean soil property, applicable for soft post-glacial clays and clayey gyttjas. Furthermore, the shape of the probability distribution is evaluated for various soil properties at one clay site by means of normality tests and visual assessment. It is concluded that the derived COV values are in accordance with literature ranges, but for more reliable estimates, soil statistics derived for Finnish clay soils should be preferred when possible. Nonetheless, no literature range can replace extensive site-specific soil statistics. Finally, it is confirmed that nearly all the soil properties at the studied Finnish clay site can be modelled as normal or lognormal distribution.

Keywords Variability · Coefficient of variation · Clay · Soil statistics · Soil property

M. S. Löfman (✉) · L. K. Korkiala-Tanttu
Department of Civil Engineering, Aalto University, Espoo, Finland
e-mail: monica.lofman@aalto.fi

© The Author(s), under exclusive license to Springer Nature Switzerland AG 2021

J. C. Matos et al. (eds.), *18th International Probabilistic Workshop*, Lecture Notes in Civil Engineering 153, https://doi.org/10.1007/978-3-030-73616-3_32

1 Introduction

Compared to manufactured materials such as steel, soil properties often exhibit much greater variability since their formation has been affected by complex geological processes. Hence, the uncertainty in soil parameters used in geotechnical design calculations can be quite significant. According to Phoon and Kulhawy [1], the sources of uncertainty in estimating design soil properties include inherent variability, measurement error, statistical estimation error and transformation uncertainty. Accordingly, probabilistic approaches and reliability-based design can be powerful in dealing with these underlying uncertainties in a robust manner (e.g., Phoon [2]).

However, in typical geotechnical design situations, the designer does not have the needed amount of site-specific data for estimating the variability statistics. Due to this "curse of small samples" [2], great effort has been made to collect existing soil statistics and to develop methods like Bayesian approaches to combine the limited site-specific data to existing prior knowledge on inherent variability. It has been observed that the coefficient of variation (COV), defined as the standard deviation divided by the mean, tends to vary within a relatively narrow range for the same physical soil properties (e.g., [1, 3, 4]). Furthermore, the COV values are considered to be universal, i.e., not dependent on the geological age of the soil (e.g., Uzielli et al. [4]) Consequently, the literature COV ranges could be utilized with some confidence on sites which lack sufficient soil data [1] or as prior distributions in Bayesian framework for geotechnical characterization.

Even though many COV ranges for inherent variability have been presented in the literature, the question remains how applicable these prior COV values are to be used in Finland. In particular, many Finnish clay soils are highly sensitive and soft due to their unique geological history shaped by the last post-glacial processes. The inherent variability of Finnish clays has been studied very little, which in turn has hindered the uptake of reliability-based design. Hence, more studies are needed to verify the suitable prior range of COV values. Therefore, this paper evaluates the inherent variability of various geotechnical properties (index, strength and consolidation properties) in four extensively studied clay soil sites located in Southern Finland. Besides prior ranges of COV, this paper will provide prior ranges of the mean soil property, applicable for soft post-glacial clays and clayey gyttjas. Lastly, the shape of the probability distribution is evaluated by means of normality tests and visual assessment for various soil properties at one clay site.

2 Estimation of Inherent Variability of Soil Properties

2.1 Variability of Measured Soil Properties

The observed variability of measured soil properties is affected by both the inherent variability and measurement error. The extent of measurement error can be minimized by limiting the data to a certain measurement test type and my maximizing the equipment and procedural controls [1]. Alternatively, if the measurement error (COV_{meas}) is known, the inherent (spatial) variability COV_{inh} can be estimated by Orchant et al. [5]:

$$COV_{inh}^2 = COV_{obs}^2 - COV_{meas}^2 \tag{1}$$

where COV_{obs} is the observed variability, i.e., the coefficient of variation calculated using the measured soil properties. However, since the measurement error is available only for some of the soil properties studied in this paper, the COV values presented in the results represent total (observed) variability COV_{obs}. In the discussion section however, the actual inherent variability (COV_{inh}) is estimated for the undrained shear strength and unit weight.

2.2 De-trending of Soil Properties with Spatial Trend

Many geotechnical properties, such as undrained shear strength, exhibit a trend with respect to depth. In such situations, it is important to apply de-trending to the measurements before estimating the statistics of inherent variability [1]. For linear dependence with depth z, the trend function for the soil property y can be defined as:

$$\hat{y} = \beta_0 + \beta_1 \cdot z \tag{2}$$

where β_0 is the intercept and β_1 is the slope. According to Phoon and Kulhawy [1], the de-trended coefficient of variation for a soil property with trend can be defined with:

$$COV_{\text{de - trended}} = \frac{\sqrt{\frac{1}{n-1}\sum_{i=1}^{n}[w(z_i)]^2}}{t} \tag{3}$$

where $w(z_i)$ is the fluctuation at depth z_i, n is the number of samples within the studied soil layer, and t is the mean soil property trend. In this study, $w(z_i)$ was defined as ($y_i - \hat{y}_i$), in which $\hat{y}_i$ is the soil property value predicted by the trend function (Eq. 2) and y_i is the actual soil property value at depth z_i. The mean soil property trend t was defined as the average of all trend values $\hat{y}_i$ within the studied soil layer.

3 Inherent Variability Within Clay Soil Layers

3.1 Site Descriptions

Four extensively studied clay soil sites from Southern Finland were selected for the variability analysis. Haarajoki site in Järvenpää was well studied as a part of test embankment construction and settlement calculation competition. The behavior of Haarajoki test embankment has been studied by many (e.g., Yildiz et al. [6]). The properties of Suurpelto site in Espoo are reported by e.g., Ojala et al. [7] and Pätsi [8]. POKO site (located near the City of Porvoo), has been studied by Koskinen [9]. Lastly, Tolsa site in Kirkkonummi was studied as part of a railway development project. The dominant soil types and classification properties of each site are presented in Table 1. Most of the studied soil profiles are located within 20-m distance from each other; however, the Suurpelto soil layers include three sampling profiles within a 400-m study line (as result, the layer depths somewhat overlap as the deposit thicknesses vary).

The soil properties with respect to depth are shown in Figs. 1, 2 and 3. The undrained shear strength (s_u) was defined using the fall cone test. An example of the de-trending of the data is shown in Fig. 1 (right). All specimens expect for dry crust

Table 1 Properties of the studied clay soil layers

Site (layer)	Dominant soil types (Cl = clay)	Layer depth (m)	Clay fraction (%)	Organic content (%)	Plasticity index (%)	Sensitivity (–)	OCR[a] (–)
Haarajoki (A)	Dry crust Cl	0.3–1.8	49–69	0.4–2.0	56–61	1–10	2.0–27.5
Haarajoki (B)	Fat Cl and gyttja Cl	2.0–7.1	66–89	1.4–2.2	58–75	23–30	1.5–4.4
Haarajoki (C)	Fat Cl	6.9–9.9	68–74	1.8–2.0	47–53	22–46	1.3–2.3
Haarajoki (D)	Fat Cl	9.9–17.6	66–81	1.6–1.8	46–54	46–53	1.6–2.3
Suurpelto (A)	Clayey gyttja	4.1–9.5	45–47	7.7–7.8	≈130	7–9	1.0–1.6
Suurpelto (B)	Cl and gyttja Cl	8.0–13.3	45–58	0.6–2.5	46–49	8–13	1.0–1.6
POKO	Fat Cl and gyttja Cl	4.5–10.9	40–79	0.5–2.1	35–63	9–47	0.9–2.7
Tolsa	Lean Cl and fat Cl	1.9–3.8	50–58	0.0–0.0	24–27	11–30	1.4–3.0

[a]OCR = Overconsolidation ratio (i.e., preconsolidation pressure divided by the effective in situ stress)

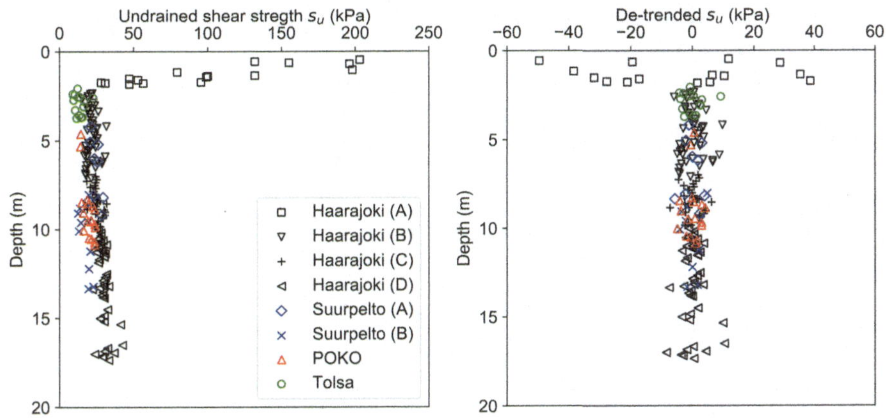

Fig. 1 Undrained shear strength before de-trending (left) and after de-trending (right)

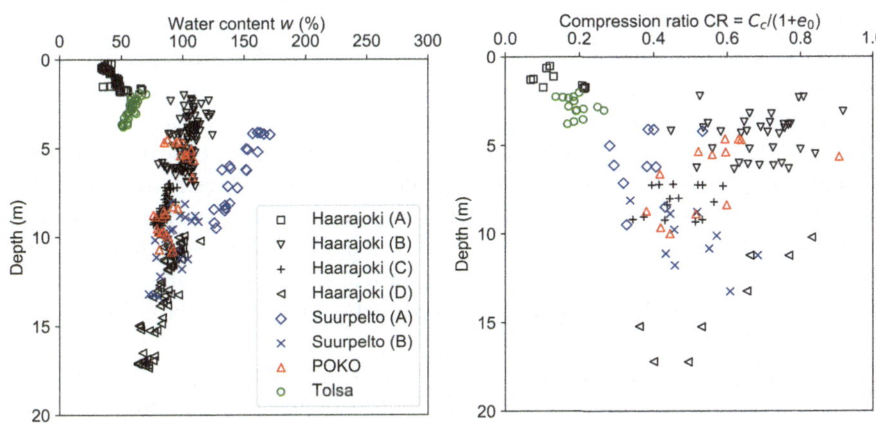

Fig. 2 Water content (left) and compression ratio (right)

clay of Haarajoki (A) are taken below the groundwater level and can be considered fully saturated.

The compressibility and consolidation properties were defined from an oedometer test performed on undisturbed clay specimen. Almost half of the oedometer tests on Haarajoki clay were constant rate of strain (CRS) tests; the rest represent incrementally loaded oedometer tests. In the case of CRS test, the compression index (C_c) was approximated from alternative compressibility parameters as described by Löfman and Korkiala-Tanttu [10]. For most of the CRS tests, the preconsolidation pressure σ_p' was rate-corrected using the Sällfors method [11]. In addition, in Haarajoki and Tolsa sites, approximately one fifth of the oedometer tests were performed on horizontally oriented specimens to study the possible anisotropy of consolidation properties. However, these results did not deviate significantly from the vertical

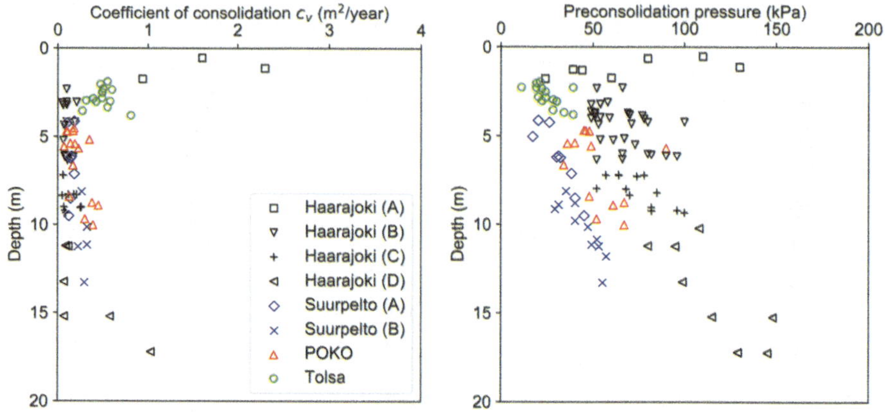

Fig. 3 Coefficient of consolidation in normally consolidated state (left) and preconsolidation pressure (right)

specimens, and hence they were included in the analysis. Nevertheless, the differences between the oedometer test types may contribute to the observed variability of consolidation properties.

The coefficient of consolidation (c_v) was defined using the log-time method. Since the value of c_v tends to be stress-state-dependent, only the values corresponding to normally consolidated stress states adjacent to preconsolidation pressure σ_p' were considered in this study. That is, the values shown in Fig. 3 (left) represent c_v values from load ratios $(\sigma_v'/\sigma_p') = 1.5$–$3$ (σ_v' is the vertical stress applied in the oedometer test). Accordingly, for the coefficient of creep (coefficient of secondary compression defined from log-time graph, $C_{\alpha\varepsilon}$) ratios $(\sigma_v'/\sigma_p') = 1.5$–$5$ were selected. A higher upper bound was selected because the decrease in $C_{\alpha\varepsilon}$ with respect to increasing load ratio was observed to be less prominent. Lastly, the initial permeability (k_1) corresponds to a permeability value k at zero strain, extrapolated from a fitted linear trend between the strain and the logarithm of permeability k. (e.g., [12]).

3.2 *Second-Moment Statistics of Inherent Variability*

The second-moment statistics (arithmetic mean and sample COV with and without de-trending) are collected to Table 2. In addition, the COV values with respect to mean values are shown in Fig. 4 for the selected soil properties. When deriving the statistics, samples with less than $n = 10$ measurements were not considered (Table 3).

As expected, the amount of decrease in the observed variability was dependent on the prominence of the spatial trend (e.g., s_u of the dry crust clay versus the other layers). The COV values are quite consistent, except for the undrained shear strength in the dry crust clay which exhibited greater variability. Nevertheless, no clear pattern with respect to soil type was found.

Table 2 Second moment statistics of observed variability within a clay soil layer

Soil property[a]	Soil type (Cl = clay)	Number of data groups (ID[c])	Number of samples n		Soil property, mean value		Coefficient of variation, COV_{obs}	
			Range	Mean	Range	Mean	Range	Mean[b]
s_u (kPa)	Dry crust Cl	1 (a)	–	16	–	103	–	0.32 (0.58)
s_u (kPa)	Cl and gyttja Cl	6 (b)	11–46	27	13.7–29.5	21.4	0.11–0.26	0.16 (0.18)
w (%)	Cl and clayey gyttja	8 (c)	22–98	46	44.9–145.3	88.2	0.03–0.11	0.07 (0.11)
e_0 (–)	Cl and clayey gyttja	7 (d)	10–42	18	1.35–3.85	2.47	0.02–0.12	0.06 (0.10)
γ (kN/m^3)	Cl and clayey gyttja	8 (c)	14–79	34	12.9–17.1	15.1	0.01–0.03	0.02 (0.03)
$\sigma_p{}'$ (kPa)	Cl and clayey gyttja	6 (d)	10–32	16	25.6–74.8	48.8	0.11–0.27	0.18 (0.26)
OCR	Cl and clayey gyttja	6 (d)	10–32	16	1.3–2.6	1.8	0.13–0.30	0.19 (0.21)
CR	Cl and clayey gyttja	6 (d)	10–34	16	0.192–0.686	0.463	0.15–0.20	0.17 (0.19)
C_c	Cl and clayey gyttja	6 (d)	10–34	16	0.493–2.704	1.729	0.15–0.21	0.18 (0.21)
c_v (m^2/yr)	Cl and gyttja Cl	4 (e)	12–19	15	0.102–0.495	0.236	0.28–0.61	0.43 (0.47)
k_I (10^{-9} m/s)	Cl and clayey gyttja	3 (f)	10–13	12	0.450–1.396	0.902	0.29–0.56	0.46 (0.49)
$C_{\alpha\varepsilon}$ (%)	Cl and clayey gyttja	4 (g)	12–25	19	0.674–2.472	1.457	0.26–0.52	0.36 (0.37)

[a]s_u = undrained shear strength (fall cone); w = water content; e_0 = initial void ratio; γ = unit weight; $\sigma_p{}'$= preconsolidation pressure; OCR = overconsolidation ratio; CR = compression ratio (= $C_c/(1 + e_0)$); C_c = compression index; c_v = coefficient of consolidation; k_I = initial permeability; $C_{\alpha\varepsilon}$ = coefficient of creep

[b]COV in parenthesis is defined for the data without de-trending

[c]See Table 3

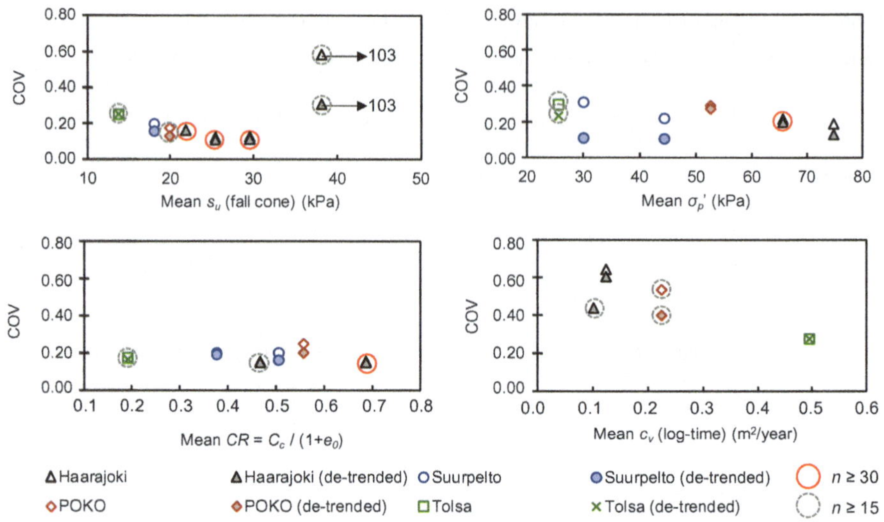

Fig. 4 COV and mean values for the selected soil properties ($n \geq 10$ in all datasets)

Table 3 Identifications for different datasets

ID	Layers included in the dataset
a	Only Haarajoki (A) included
b	Haarajoki (A) and Suurpelto (A) *not included*
c	All layers included
d	Haarajoki (D) *not included*
e	Haarajoki (A, D) and Suurpelto (A, B) *not included*
f	Haarajoki (B), Suurpelto (A) and POKO included
g	Haarajoki (A, C–D) and Suurpelto (B) *not included*

In Table 4, the ranges of COV_{obs} for each soil property category are compared to the literature ranges. It can be observed that the variability in the studied clay layers is in accordance with the approximate guideline ranges. However, the COV values for the index properties (unit weight, water content and void ratio) seem to be at the lower end of the reported range. This smaller variability could be due to differences in datasets: most specimens in this study are very homogeneous clays.

3.3 Probability Distributions for the Inherent Variability of Haarajoki Clay

The depth profiles presented above (Figs. 1, 2 and 3) show that if the trend functions are considered, the clay layers B-D in Haarajoki could also be modeled as a single

Table 4 Comparison of the observed variability in clay soil properties to literature ranges

Soil property category	Soil properties (this study)	COV_{obs} range (this study)	Approximate guideline[a]
Laboratory strength	Undrained shear strength s_u	0.11–0.32	0.10–0.55
Density	Unit weight (γ)	0.01–0.03	<0.10
Laboratory index	Water content w and void ratio e_0	0.02–0.12	0.07–0.30
Compressibility	Compression index C_c and ratio CR, OCR, preconsolidation pressure $\sigma_p{}'$	0.11–0.30	0.10–0.37
Time-dependent compression and hydraulic permeability	Permeability k_l, coefficient of consolidation c_v, coefficient of creep $C_{\alpha\varepsilon}$	0.26–0.61	0.33–0.90

[a]Table 7 in Uzielli et al. [4]

geotechnical layer regarding undrained shear strength and preconsolidaiton pressure. For the remaining soil properties, layers B-C were merged together. As a result, the number of samples n per dataset became greater, hence allowing the calculation of skew and kurtosis to study the shape of the distribution. For testing the normality of the data, Shapiro-Wilk test [13] is often recommended (e.g., [4, 14]) and thus selected for this study with significance level was set to $\alpha = 0.05$. IN other words, a p-value greater than 0.05 indicated normality. The suitability of lognormal distribution was evaluated by performing the normality test for natural logarithms of the soil property values. The coefficient of variation for the data after natural logarithm transformation is calculated with:

$$COV_{\ln(y)} = \sqrt{\exp\left(SD^2_{\ln(y)}\right) - 1} \tag{4}$$

where $SD_{\ln(y)}$ is the sample standard deviation calculated for natural logarithms of the soil property y.

The statistics and the results of the normality test are shown in Table 5. De-trending of the data was applied to all soil properties that exhibited any noticeable spatial trend.

In Table 5, the kurtosis values represent excess kurtosis, i.e., normal distribution would have a kurtosis equal to zero. All the index properties and the undrained shear strength have a positive excess kurtosis, which implies that the distributions have heavier tales than the normal. On the other hand, negative excess kurtosis was more common in the compressibility and consolidation properties (implying thinner tails). However, the latter also had much smaller sample sizes (n).

Figures 5 and 6 show the fitted probability distributions for the selected soil properties. For the majority of the soil properties, both normal and lognormal distributions

Table 5 Observed variability of the de-trended soil properties (Haarajoki clay)

Soil property (unit)	n	Range, values	Mean	SD	COV[a]	Skew[a]	Excess kurtosis[a]	Normality test, p-value[a]	Probability distribution ($\alpha = 0.05$)
s_u (kPa)	121[b]	15.6–42.4	25.6	3.44	0.13 (0.14)	0.283 (−0.251)	0.825 (0.662)	0.058 (0.083)	N/LN
w (%)	155	76.3–124	97.1	6.75	0.07 (0.07)	−0.081 (−0.386)	0.788 (1.161)	0.211 (0.033)	N
e_0 (–)	59	2.176–3.413	2.787	0.170	0.06 (0.06)	−0.317 (−0.524))	0.320 (0.555)	0.518 (0.205)	N/LN
γ (kN/m^3)	121	13.7–15.6	14.6	0.272	0.02 (0.02)	−0.406 (−0.486)	0.985 (1.136)	0.085 (0.041)	N
$\sigma_p{}'$(kPa)	53[b]	49–148	75.3	13.5	0.18 (0.18)	0.452 (−0.048)	0.048 (−0.016)	0.362 (0.847)	N/LN
OCR (–)	45	1.3–4.4	2.4	0.51	0.21 (0.21)	0.670 (0.264)	0.038 (−0.725)	0.017 (0.124)	LN
CR (–)	49	0.343–0.918	0.619	0.104	0.17 (0.18)	−0.183 (−0.676)	−0.181 (0.439)	0.766 (0.101)	N/LN
C_c (–)	49	1.091–4.050	2.366	0.446	0.19 (0.20)	−0.087 (−0.839)	0.433 (1.371)	0.344 (0.017)	N
c_v (m^2/yr)	31	0.110–0.548	0.224	0.060	0.55 (0.53)	1.161 (0.501)	0.224 (−0.760)	0.000 (0.045)	–
$C_{\alpha\varepsilon}$ (%)	28	0.6–2.71	1.43	0.551	0.38 (0.41)	0.675 (−0.197)	−0.098 (−0.315)	0.114 (0.501)	N/LN

[a]The value in parenthesis is calculated for the natural logarithm transformed values (Eq. 4)
[b]Layer D included

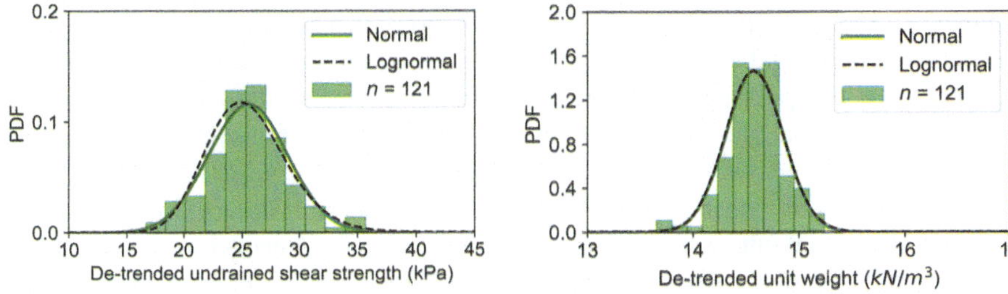

Fig. 5 Probability distributions for undrained shear strength (left) and unit weight (right)

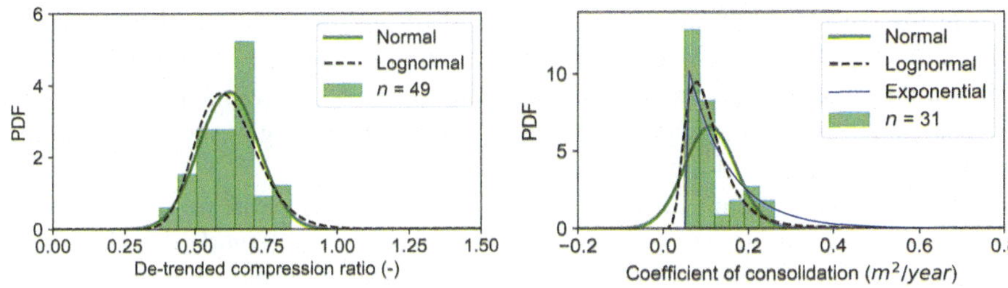

Fig. 6 Probability distributions for compression ratio (left) and coefficient of consolidation (right)

were suitable according to the normality test (Table 5). The only exceptions were the water content, unit weight and compression index (normal only) and the OCR (lognormal only). The findings are mostly in accordance with the existing literature [4, 15]: normal distribution is usually recommended for index properties, and for undrained shear strength and OCR, either normal or lognormal can be suitable. If the COV value is very small (such as $COV_{obs} = 0.02$ for unit weight), there is no visible difference between normal and lognormal (see Fig. 5 right). Nevertheless, lognormal is often preferred over normal due to its non-negativity.

For the coefficient of consolidation (c_v), neither normal nor lognormal passed the normality test. Indeed, c_v exhibited a significant positive skew (1.161). Figure 6 (right) shows that the lognormal fit is more or less adequate despite the result of the normality test. Further analysis showed that a better fit might be provided by a shifted exponential distribution.

4 Discussion

As pointed out earlier, the observed variability is an upper bound for the inherent variability due to other uncertainties, measurement error especially, included in the observations. By applying Eq. 1, the inherent variability can be estimated if the measurement error COV_{meas} related to the soil property in question is known. For instance, COV_{meas} in defining undrained shear strength with fall cone test has been studied by Knuuti and Länsivaara [16]. They found an average value of $COV_{meas} = 0.05$. Thus, by applying Eq. 1, the actual inherent variability of undrained shear strength was estimated to be approximately one (percent-)point smaller than the observed variability COV_{obs} collected in Table 2; for instance, COV_{inh} the fall cone undrained shear strength of clays soils is 0.15 in average (while $COV_{obs} = 0.16$). Similarly, the actual inherent variability of unit weight (i.e. density) can be estimated using the measurement errors summarized by Phoon and Kulhawy [1]: COV_{meas} for unit weight of fine-grained soils was found to vary between 0.01–0.02 (three data groups). The range of COV_{obs} values was 0.01–0.03, hence resulting in COV_{inh} varying between 0–0.028. Indeed, Eq. 1 implies that the inherent variability can be zero in the case of high enough measurement uncertainty compared to observed variability. In the case of soil properties with small observed variability, such as unit weight of homogeneous clay, inherent variability being practically zero seems reasonable. However, for properties like undrained shear strength, such result should prompt a re-evaluation of the assumptions made.

Finally, it should be noted that the provided ranges of coefficient of variability COV_{obs} are subject to statistical uncertainty. Both the mean and standard deviation are evaluated from a limited sample of measurements instead of the whole population. Kelley [17] has proposed a method to evaluate the required number of samples n to reach a desired width ω of the confidence interval for the population COV. The required n depends on the estimated population COV (higher COV implies higher required n) and the level of assurance (e.g., 99% assurance that the obtained 99% confidence interval for COV will not be wider than ω units). Therefore, one can assess the confidence in the given COV_{obs} range (Table 4) in a following manner: The width ω is taken as the COV_{obs} range, and the population COV is assumed to be in the same scale as the mean COV_{obs} (Table 2). Therefore, in the case of density and laboratory index properties, 99% assurance that the population COV is within the given range requires 31–32 samples [17]. The mean n was 34 in unit weight and 46 in water content data, thus implying quite high confidence in the given COV_{obs} range for these physical properties. For undrained shear strength and compressibility, the required n for similar confidence would be 24–37. The mean n in undrained shear strength data was high enough, but the mean n in compressibility was only 16. Hence, due to a smaller sample size ($n = 10$ being the smallest dataset), the COV_{obs} range for compressibility corresponds to a lower level of confidence: 95% confidence interval with 80% assurance (where required n would be 11–16).

5 Conclusions

This paper evaluated the inherent variability of various soil properties at four clay soil sites in Southern Finland. The derived ranges of coefficient of variation (COV) verified that the approximate guidelines found from the literature can be applied to Finnish clays; however, for more accurate (and less conservative) estimates, soil statistics derived for Finnish clay soils should be preferred over global literature values when possible. However, it should be noted that no literature range can replace extensive site-specific soil statistics.

Lastly, the suitability of normal or lognormal distribution was evaluated using the Haarajoki clay data. It was observed the existing guidelines regarding the soil property distributions were mostly in accordance with the normality test results, that is, most soil properties can be modelled as normal or lognormal distributions.

Acknowledgements This study was funded by the Finnish Transport Infrastructure Agency.

References

1. Phoon, K.-K., & Kulhawy, F. H. (1999). Characterization of geotechnical variability. *Canadian Geotechnical Journal, 36,* 612–624.
2. Phoon, K.-K. (2017). Role of reliability calculations in geotechnical design. *Georisk, 11,* 4–21.
3. Lacasse, S., Nadim, F., Rahim, A., & Guttormsen, T. R. (2007). Statistical description of characteristic soil properties. In *Offshore Technology Conference.*
4. Uzielli, M., Lacasse, S., Nadim, F., & Phoon K.-K. (2006). Soil variability analysis for geotechnical practice. In *Proceedings of the 2nd International Workshop on Characterisation and Engineering Properties of Natural Soils*, Singapore.
5. Orchant, C. J., Kulhawy, F. H., & Trautmann, C. H. (1988). *Critical evaluation of in-situ test methods and their variability.* Report EL-5507(2), Palo Alto.
6. Yildiz, A., Karstunen, M., & Krenn, H. (2009). Effect of Anisotropy and Destructuration on Behavior of Haarajoki Test Embankment. *International Journal of Geomechanics, 9,* 153–168.
7. Ojala, A. E. K., Ikävalko, O., Palmu, J.-P., Vanhala, H., et al. (2007). *Espoon Suurpellon alueen maaperän ominaispiirteet.* Report, Geological Survey of Finland, Espoo.
8. Pätsi, K. (2009). *Suurpellon syvästabiloidun koepenkereen analysointi.* Master thesis, Helsinki University of Technology, Espoo.
9. Koskinen, M. (2014). *Plastic anisotropy and destructuration of soft Finnish clays.* Doctoral dissertation, Aalto University, Espoo.
10. Löfman, M. S., & Korkiala-Tanttu, L. K. (2019). Variability and typical value distributions of compressibility properties of fine-grained sediments in Finland. In *Proceedings of the 7th International Symposium on Geotechnical Safety and Risk (ISGSR 2019).*
11. Sällfors, G. (1975). *Preconsolidation pressure of soft, high-plastic clays.* Doctoral dissertation, Chalmers University of Technology.
12. Larsson, R., Bengtsson, P.-E., & Eriksson, L. (1997). Prediction of settlements of embankments on soft, fine-grained soils. In *Calculation of settlements and their course with time. Information 13E*, Swedish Geotechnical Institute.
13. Shapiro, S. S., & Wilk, M. B. (1965). An analysis of variance test for normality (complete samples). *Biometrika, 52,* 591–611.
14. Thode, H. C. (2002). *Testing for normality.* New York: Marcel Dekker.
15. Lacasse, S., & Nadim, F. (1996). Uncertainties in characterising soil properties. In *Uncertainty in the geologic environment: from theory to practice.* New York: ASCE.
16. Knuuti, M., & Länsivaara, T. (In press). Measurement uncertainty of the fall cone (FC) test. In *Proceedings of the 18th Nordic Geotechnical Meeting*, 2021, Helsinki.
17. Kelley, K. (2007). Sample size planning for the coefficient of variation from the accuracy in parameter estimation approach. *Behavior Research Methods, 39,* 755–766.

Integration of the Analysis of the Error of Geometric Dimensions Modeled with a Probabilistic Approach

Marc Gille, Pierre Beaurepaire, Fabien Taghon, Antoine Dumas, Nicolas Gayton, and Thierry Yalamas

Abstract Metrology is extensively used in the manufacturing industry to determine whether the dimensions of parts are within their tolerance interval. However, errors cannot be avoided. If the metrology experts are actually aware of it, and currently able to identify the different sources that contribute to making errors, very few research has been made in this area to develop metrology methods accounting for such errors. The probability density function of the error is here assumed to be given as an input. This work deals with a batch of measures and its statistical properties. The first proposed method aims to correct the effects of the errors to the distribution that characterize the entire batch. Then a second method tries to estimate for each single measure, the dimension that is being the most likely given by a measure, after the error is deducted. It is based on the output knowledge of the first method and integrates it with Bayesian statistics. Only Gaussian distributions are considered in the paper. Their relevance is shown through one example applied on simulated data.

Keywords Metrology · Uncertainties · Statistics · Bayesian

1 Introduction

Metrology is a key stage in industry as it validates the quality requirements at different steps of the production process. Designers prepare a Computer-Aided Design (CAD) model of the parts; a nominal value and tolerance interval are associated with each dimension (such as for instance the thickness of the part, the diameter of a hole, the radius of a fillet, etc.). Off-tolerance dimensions may have detrimental effects, for instance two mating parts need to be geometrically compatible to perform assembly. These dimensions are measured when the parts are manufactured. However, as every

M. Gille (✉) · P. Beaurepaire · N. Gayton
SIGMA Clermont-Ferrand, Aubière, France
e-mail: gille@phimeca.com

M. Gille · F. Taghon · A. Dumas · T. Yalamas
PHIMECA Engineering, Cournon d'Auvergne Paris, France

© The Author(s), under exclusive license to Springer Nature Switzerland AG 2021

J. C. Matos et al. (eds.), *18th International Probabilistic Workshop*, Lecture Notes in Civil Engineering 153, https://doi.org/10.1007/978-3-030-73616-3_33

measure is always tarnished with errors, the issue of the relevance of such measures arises.

The consequence of an improper measure gives a wrong idea of the part conformity. It may lead to additional machining, scraping of compliant parts, or delivery of non-compliant parts to a customer. When studying a batch of parts, all of these dimensions are used to compute some quality metrics—such as the well-known *Cp* and *Cpk*, thus also subject to errors. Therefore, the errors also propagate to these quality metrics.

To the authors' knowledge, the measure process currently takes into account the measurement tool accuracy. A common practice consists in applying the maximum accuracy error on both side of the measure to create the tolerance interval. However there is actually no reason to center the error interval on the measure. Moreover, only the bounds of the error interval are retained in the end, suggesting that all errors inside this interval are equiprobable. This actual lack of knowledge can be seen as a uniform probability density function modeling of the error.

If the measurement tool accuracy is mistaken, many other error sources are currently missing. This can be studied with the Ishikawa diagram which analyzes the method, environment, material, process and people sources of uncertainties [1]. The temperature, pressure, dust, altitude, operator repeatability and inter-operability are all example of errors that are being ignored. If metrologists are actually aware of these input errors, there is nowadays no mean to integrate them. By default, a certain amount of skepticism is introduced into the safety factors to prevent any risks, and the cost of these coefficients has already been widely stated in the literature.

This paper aims to use all the available information. The *error* is modeled as a random variable and its probability density function is considered as a given input. This research is part of the European project STAM [Acknowledgments] leading by an expert metrology company where engineers assume today to be able to identify it. The *measure* is the value that can be read on the measurement tool. This work aims at correcting (at least partially) the effects of the error on the measure, and tries to estimate the *true dimension*. The search for the true dimension is based on this fundamental formula [1]:

$$m = v + \mathcal{E} \tag{1}$$

where $m \in \mathbb{R}$ is the measure of the true dimension v tarnished with the error $\mathcal{E}$. The true dimension is a notion that needs to be carefully manipulated since it is actually impossible to know its exact value; many metrologists believe it is very complex to define it. The uncertainties around the manufacturing process cannot be avoided, therefore, the made parts have not exactly the same dimension. That is why the true dimension is considered as a random variable.

The paper finally aims to solve two main issues: (i) the correction of the density associated with a batch of measures; (ii) the correction of the measure associated with one part. Two methods that complement each other are developed and illustrated with an example. This study assumes that the error and measure input probability density functions are Gaussian as it is widely used and covers most of the practical cases. Note that this is particularly relevant for the error as it is the sum of numerous independent variables. The use of Gaussian distribution is therefore in good agreement with the central limit theorem. This paper develops the methodology in Sect. 2, Sect. 3 presents an example of their application and the paper closes with the conclusions.

2 Proposed Methods of Analysis

2.1 *Correction of the Density Associated with a Batch of Measures*

A manufactured part is meant to verify some key dimension expectations. To check the quality of a batch of parts, these essential dimensions are measured. The distributions obtained are all the more important that they are used to establish the quality metrics, such as process capability indexes [2, 3]. However, just as a measure is always tarnished with an error, the distribution of a set of measures is as well tarnished with errors. To estimate the true dimension distribution, the idea is to exploit the error distribution we assumed to know. The measure dimension is expressed as the sum of two random variables and its probability density function is expressed as:

$$f_M(x) = \int_{-\infty}^{+\infty} f_V(t) f_\varepsilon(x - t)\mathrm{d}t \tag{2}$$

where $f_{M,}$ is the probability density function of the measure, f_V and f_ε are respectively the probability density functions of the two input random variables: the true dimension V and the error ε. This formula corresponds to the measure distribution and it happens to be a convolution product. Convolution is a well-documented issue, notably resolvable with a Monte Carlo simulation or Gauss' integration scheme [4]. In our context, the convolution product is known—the results are available (measures)—and one term of the product is also known the probability density function of the error. The identification of the second term of the convolution product is referred to as *deconvolution.*

Deconvolution methods to deal with continuous signals, either temporal or spatial, are mainly used and developed in image processing [5] and signals processing [6, 7]. The metrological approach is not about filtering: a sample set is available, it is not time-dependent and each measure is a unique phenomenon. Therefore, these well documented methods are not applicable to the deconvolution of probabilistic distributions.

We recall that we introduced the assumption of a Gaussian modeling. In this context the most relevant method in our context is the use of characteristic functions, as they provide an analytical solution. If a random variable has a probability density function, then the characteristic function is its inverse Fourier transform. As a Gaussian distribution satisfies the conditions of Bochner's theorem [8], its characteristic function exists and its expression is [9]:

$$\phi_X(t) = e^{it\mu_X - \frac{1}{2}t^2\sigma_X^2} \tag{3}$$

with μ_X and σ_X respectively the mean and the standard deviation of the normal distribution which characterize the random variable X. Then is used the interesting property that the characteristic function of the sum of two independent random variables is equal to the product of each of their characteristic function [10]. As we can assume the independence between the true dimension V and the error ε, the relation obtained is:

$$\phi_M(t) = \phi_{V+\varepsilon}(t) = \phi_V(t)\phi_\varepsilon(t) \tag{4}$$

This last expression makes the deconvolution straightforward: the characteristic function of the true dimension density is obtained by a division. The division of the two exponential functions gives an exponential function with the same shape, from which it is very easy to identify the parameters. The true dimension distribution that we finally get is Gaussian; its mean is the measure mean minus the error mean; its variance is the measure variance minus the error variance.

$$V \sim \mathcal{N}\left(\mu_M - \mu_\varepsilon, \sigma_M^2 - \sigma_\varepsilon^2\right) \tag{5}$$

2.2 *Correction of the Measure Associated with One Part*

The first method deals with the distribution of the measure, as we considered a batch of parts. This subsection describes another method which is focused on the measure of each part of the batch. The main objective is to determine if the parts respect the tolerance specifications, and more generally for each part the probability that the dimension lays within its tolerance interval. It is impossible to know the true dimension density, so this method aims to get an approximation of it, at least better than the measure which is always tarnished with errors.

Recall that the measure has been introduced as the sum of a true dimension and an error. So, each measure can be associated with an infinite number of combinations of true dimension and error. However, the combinations are not all equiprobable. The research consists in finding the most probable couple of terms of which the sum is equal to the measure. On one side, the error measurement probability function is obviously used to weigh the plausibility of a supposed error. On the other side, it is necessary to provide some other new information for the *a priori* to estimate the plausibility of a supposed true dimension.

The production team possesses a specific expertise, which includes some theories and a long time experiences for operators and engineers, as long as the machine in the manufacturing process is maintained and keeps working to make the investment the most profitable. This non negligible knowledge exists and just waits to be picked up. Nevertheless, it must be noticed that it remains very complex to quantify the truthfulness and integrity of this source. Another interesting alternative is to use the deconvoluted distribution from the last application as the *a priori* probability density function. Like the production team's knowledge, this shall identify the production behavior, with an experience based on a previous huge batch of parts and being "conscious" of the error probability density function. This is the *a priori* information chosen in this paper. Remark that the batch being corrected must be independent from the one used for the deconvolution [11]. In order to assimilate this knowledge, we use the Bayesian statistics. From these knowledge sources, the Bayesian theorem [11] highlights the probability for a given value x to be the true dimension knowing the measure m is given by:

$$P(x|m) = \frac{1}{C}\mathcal{L}(m - x)\pi(x) \tag{6}$$

where $\mathcal{L}$ is the *likelihood*, P and π are respectively the *a posteriori* (or *posterior*) and the *a priori* (or *prior*) probability density functions, and C is the evidence which guarantees that the *a posteriori* distribution integrates to one. The *a priori* distribution corresponds to the expert knowledge which weighs the plausibility of a supposed true dimension. It would be the distribution that characterizes the manufacturing process if no measure were made. The likelihood gives the probability for m to have been measured knowing that the true dimension is x. It is based on the error density, which is known, we get: $\mathcal{L}(m - x) = \mathcal{E}(x - m)$. The likelihood would be used to compute the probability of being out-of-tolerance if there was no *a priori*.

The outcome of the procedure for each measure is an *a posteriori* distribution, which contains much more information. In particular, the highest density of the a posteriori is called the *revised dimension*: this is the most probable true dimension. While the traditional use of the Bayesian theorem is meant to update the *a priori* with the observations, we may say that we update our observations with our a priori. This Bayesian approach can easily be extended to any arbitrary probability density function. In our context, the input distributions are supposed to be Gaussian. So, the product between the likelihood and the *a priori* is a density product of two Gaussian distributions which can be solved analytically. The posterior distribution obtained is also Gaussian and, incorporating the previous expression of the likelihood and that the *a priori* input is set as a deconvoluted distribution, it is expressed as:

$$R_m \sim \mathcal{N}\left(m\left(1 - \frac{\sigma_\varepsilon^2}{\sigma_M^2}\right) + \frac{\sigma_\varepsilon^2}{\sigma_M^2}\mu_M - \mu_\varepsilon, \sigma_\varepsilon^2\left(1 - \frac{\sigma_\varepsilon^2}{\sigma_M^2}\right)\right) \tag{7}$$

where R_m is the random variable of the revised dimension. It is worth noting that there is a linear relationship between the mean of the revised distribution and the measure. Also, the standard deviation is constant and does not depend on the input measure.

3 Examples

3.1 Presentation of the Example

To illustrate the effect of these two methods we consider the following analytical examples; the data (measure) is generated using a Monte Carlo simulation, which allows us to compare the results to the known solution, thus to quantify their quality.

The first major input is the error probability density function. Recall that it is assumed to be known using the metrologist expertise. It is important to remind that this information is actually impossible to identify perfectly. Yet to estimate the extent of the method potential, the error probability density function is supposed to describe perfectly the mistakes made in reality. This probability density function used here is unbiased, in the example its mean is equal to zero. Bias correction is straight-forward and already applied in industry, this is therefore not considered here.

The second major input is the true dimension probability function which is also set arbitrarily. Its mean value is not discussed in detail as it does not influence the study. The standard deviation is set in order to be twice the standard deviation of the error. If the measurement tools have the same dispersion than the production—which is very unlikely—then they are useless. On the contrary, if they are too precise then there may be no need for the application of these methods.

This sample size is set to 1000. It is moderately large from a statistical perspective and we assume that the uncertainties are fairly well represented. On the other hand, collecting and measuring 1000 dimensions requires quite some efforts for a real-life manufactured component. We therefore assume that our sample size is a fair trade-off between the statistical relevance and industrial constraints.

To get a measure sample, two samples are drawn from the error and the true dimension probability density function and added together as a whole. The limits of the tolerance interval are set to 14.0 and 14.4. Recall that the input distributions are assumed to be Gaussian. Table 1 summarizes the input distributions:

3.2 Deconvolution of a Batch of Parts

The methods described in Sect. 2.1 are applied to remove the effects of the errors to the distribution identified from a set of measures. Figure 1 shows the deconvolution result. The deconvoluted line is very close to the solution—the true dimension density. Despite an analytical solving of the problem, the deconvolution probability density

Table 1 Parameters of the input distributions

Probability density function	Mean	Standard deviation
True dimension	14.2	0.1
Error	0.0	0.05

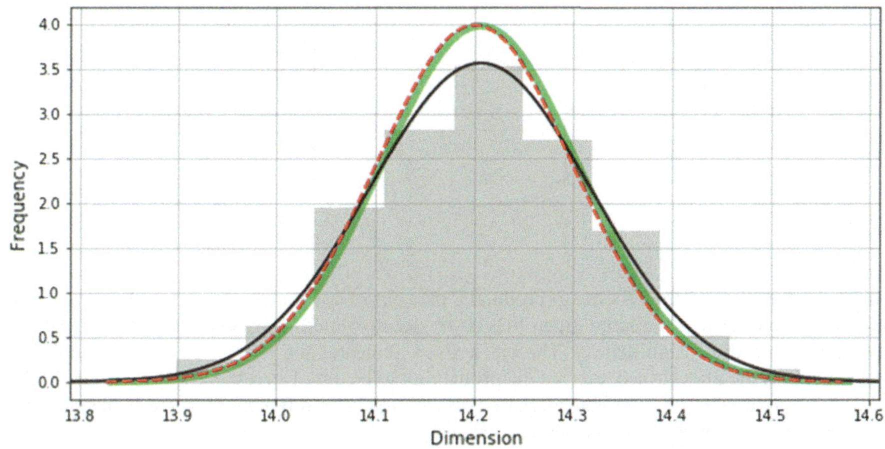

Fig. 1 Deconvolution of the density associated with a batch of parts. The dashed red line is the true dimension distribution. The black line is the Gaussian distribution inferred on the measure sample represented by the transparent grey histogram. The thick green line is the deconvoluted distribution

function does not exactly match the solution. This difference is only due to the sample size, as the mean and variance of the measure sample differ slightly from those of the law from which it is extracted. Both the mean and variance of the deconvoluted distribution are still very satisfying: the error made for these parameters is respectively about 0.059 and 0.001 times the standard deviation. Nevertheless, the high quality of the results obtained must be put in parallel with the Gaussian modeling hypothesis and the perfect identification of the error probability density function that have been assumed.

The deconvoluted distribution can be subsequently used as the reference to characterize the production process. The re-estimation of the parameters directly impacts the percentage for parts to be out-of-tolerance. This is explicitly shown in Fig. 2 where the probability goes from 0.0736 with the measures to 0.0462 with the deconvolution.

3.3 Revision of the Measure Associated with a Part

The methods described in Sect. 3.2 are applied to remove the effects of the errors to each measure of a simulated set. The deconvoluted distribution from the last application is set as the *a priori* probability density function. The scatter plot of the measure residues in Fig. 3a shows a concentric dispersion which confirms the fair construction of the sample, based on two Gaussian probability density functions. The scatter of the revised dimension shown in Fig. 3b presents a dispersion smaller than the measured ones: the mean squared error goes from 1.555 to 1.431. It also presents a linear decreasing tendency, pointing out the dependence between the revised dimension and the true dimension value. This is clearly visible for the extreme values of

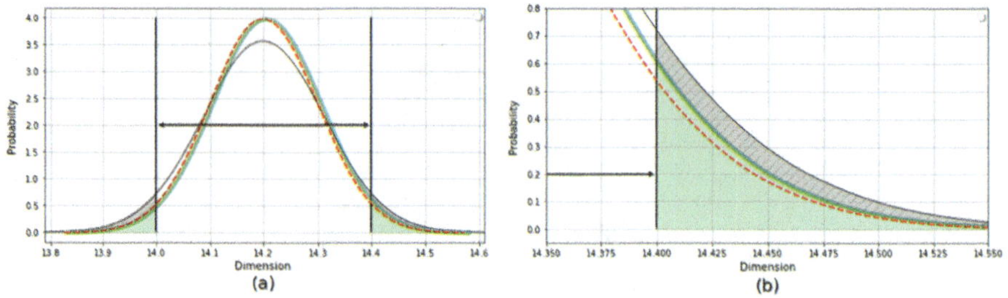

Fig. 2 Probability for the process to produce a part out-of-tolerance. The red dashed line is the true dimension density, the large green line is the deconvoluted density and the solid black line is the measured dimension density. The black arrow indicates the tolerance interval bounds, the green area and the grey striped area are the probability to be out of it, respectively according to the deconvoluted distribution and to the measured distribution. Graph **b** focuses on the right side of the graph **a**, to explicit the probability failure reduction

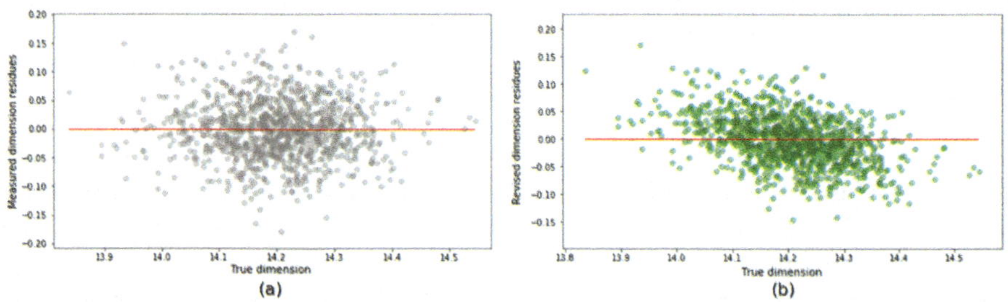

Fig. 3 Measures and revised dimension residues. **a** The measures residues with respect to the true dimension. **b** The revised dimensions residues with respect to the true dimension

the true dimension. The introduction of a bias is the price of an improvement of the measure.

The coefficient of determination is used to quantify the improvement introduced by the proposed method. This metrics is frequently used in the context of (non-)linear regression, as it can be considered as the score of the predictive model. The idea in this application is to consider that the measure and revised dimension can be seen as predictions of the true dimension:

$$R^2 = \frac{\sum_{i=1}^{n} \left(x^{(i)} - \mu_V^2\right)^2}{\sum_{i=1}^{n} \left(v^{(i)} - \mu_V^2\right)^2} \tag{8}$$

where x is either the measure or the revised dimension. If one of these two predictions is perfect then its predicted value is equal to the true dimension and the score is equal to 1. The uncertainties present in this model make the search of almost perfect values irrelevant. The coefficient of determination computed in this example on the measured and revised dimensions are respectively equal to 0.776 and 0.811. This result is satisfying as it endorses the general increase in quality.

The mean squared errors and the coefficients of determination confirm that the revised dimensions obtained through Bayesian statistics are generally better than the measure. But no conclusion regarding the conformity of each parts in the batch can be drawn. The confusion matrix is a relevant tool to synthesize all this information. It is also based on the idea that both the measured and the revised dimension are predictions of the truth. The two confusion matrices are shown in the Fig. 4. In the top left corner is presented the number of right predictions about the actual conformity of the true dimension (*True compliant*); in the bottom right corner, the number of the right predictions about the actual non-conformity of the true dimension (*True non- compliant*). These right predictions constitute a diagonal in the matrix. The other diagonal groups all the errors: either when it is mistakenly believed the part is acceptable—in the bottom left corner (*False compliant*)—or when it is mistakenly believed the part is not acceptable—in the top right corner (*False non-compliant*).

The sum of True compliant and True non-compliant is higher using the revised dimensions and this is an encouraging result (960 against 942). However, these are many more False compliant (30 against 13) which leads to the conclusion that this method is "optimistic" as it tends to report more conformed value. To evaluate the true benefit with the confusion matrix, it is necessary for the industrialist to associate a cost for each of the four cases; or at least for the False compliant and False non-compliant which are respectively known as client risk and supplier risk.

The optimistic behavior identified can be explained by the linear relation between the revised dimension and the measure that has been given in Eq. (7). Indeed, applying the tolerance interval to the linearly transformed measure is equivalent to apply the linearly transformed tolerance interval to the measure. The new tolerance interval obtained after the Bayesian revision is called the *acceptance interval*. In our application its bounds are equal to 13.96 and 14.44 (against 14.0 and 14.2 originally).

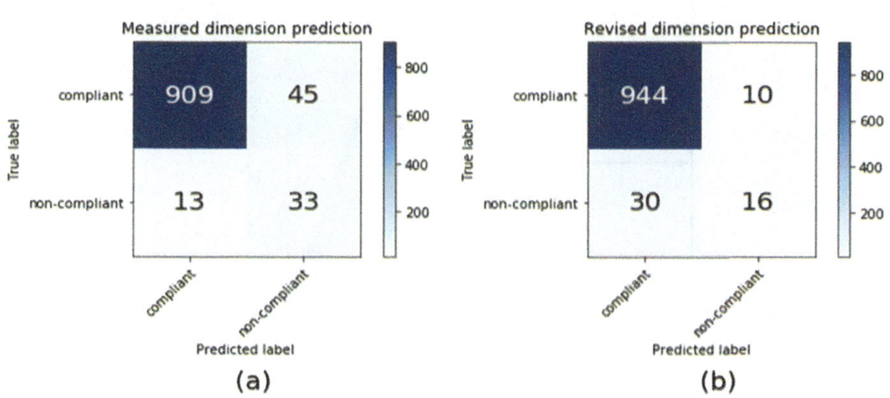

Fig. 4 Confusion matrix of the different predictions. The confusion matrix of the **a** measures and **b** revised dimensions

The width of the acceptance interval is increased; therefore the method reduces the number of rejects.

The Bayesian revision explains this enlargement as the measures are attracted by the *a priori*. In most of the common cases, it is in the tolerance interval and so, the small dimensions are increased and the large ones are reduced. The consequence of this optimistic attraction into the tolerance interval is an increase of the risk. Recall that the criterion to choose the revised dimension is to maximize the probability of the *a posteriori* distribution. However, this definition does not consider the probability to be out-of-tolerance which is a risk that the user should also be aware of. In the worst case shown in Fig. 5, the revised dimension falls on an extreme bound of the tolerance interval and, with Gaussian probability density function only, the probability to be out-of-it is 0.5. To summarize, by considering the revised value or by applying the acceptance interval, industrialists are setting the *admissible risk* to mispredict the conformity up to 0.5.

The Bayesian revision ends with an *a posteriori* probability density function which makes possible to compute the probability to be out-of-tolerance for the revised dimension. In this Gaussian modeling context, the acceptance interval has been determined analytically with the analytical expression of the mean. By proceeding experimentally, if the admissible risk is set to 0.5 then it is possible to identify all the revised value that are conformed without exceeding this probability: therefore, to identify the acceptance interval which must be applied on the measure to obtain the set of revised values which satisfy this conformity status. This approach may be used with any admissible risk threshold. Figure 6 shows an abacus to read the acceptance interval for any admissible risk threshold for the numerical values of this example.

Figure 6 is symmetric as the tolerance interval is centered on the true dimension. The basic results can be confirmed: if the wanted risk is null, then none of the

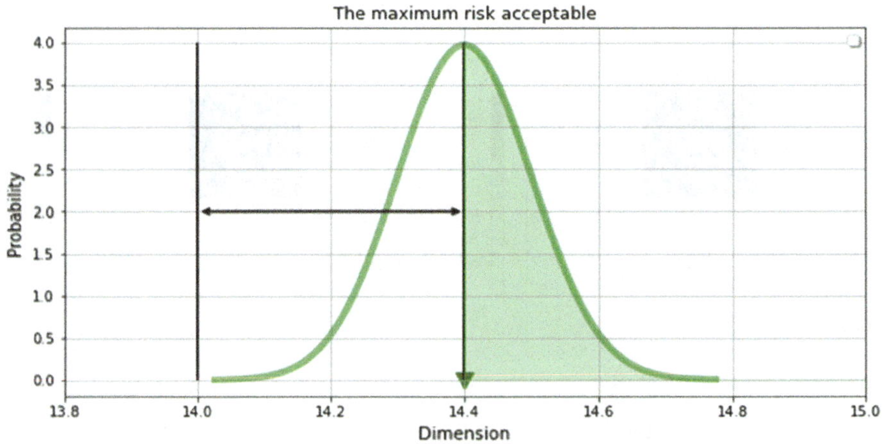

Fig. 5 The maximum acceptable risk. The large green curve is the *a posteriori* density. The green triangle is the revised dimension and the green area is the risk associated with this chosen value

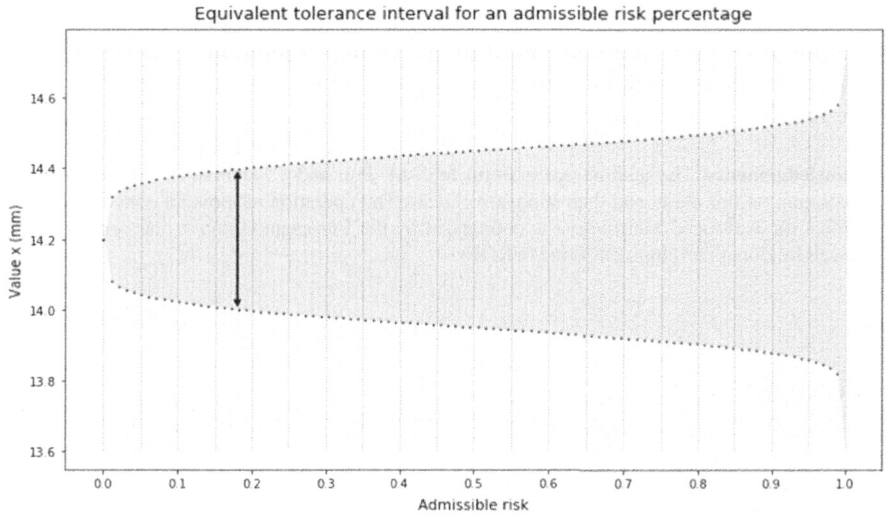

Fig. 6 Acceptance interval for an admissible risk. The black thick arrow reminds the original tolerance interval bounds at its equivalent risk. The black dots define the bounds of the acceptance interval included in the shaded area

measure meet the requirement, therefore none is compliant. If there is no limit to the acceptable risk then all of the measures are considered as compliant. The acceptance interval calculated before for a 0.5 risk with bounds equal to 13.96 and 14.44 can be found. An interesting result, read in reverse, is the revelation of the risk taking with the measurement process: in our application it is 0.18.

4 Conclusions

In this paper, we present two methods that complement each other, which can be used by metrologists to consider the error random variable. The first deals with a batch of measures. It aims to correct the distribution which characterizes this batch from the error disturbances. The output is the new reference distribution to compute and estimate the usual quality metrics. The second method deals with measures one at a time. The output of the first method and the probability density function of the error are combined with the Bayesian statistics. It finally gives a new distribution to characterize the measure. The most probable dimension can be identified from this distribution, as well as the probability of being out-of-tolerance, for a given tolerance interval.

Both methods are applied to the same example; the data are generated using Monte Carlo simulation and it is observed that errors can be partially corrected. It also provides a diagram of the acceptance interval for a better control of the production risk. The concrete benefit is relative to the industrial application context, as it involves the actual costs of the client and the supplier risk. Remark that, if the Bayesian correction does not require the Gaussian model assumption, the deconvolution of non-parametric distribution is more challenging and will be studied in more detail in the future.

Acknowledgements The authors are grateful to J.-M. Pou and C. Dubois (DeltaMu) for their collaboration on this study and their innovative ideas. The operation «Proove of concept Services and Tools for Advanced Metrology» is cofinanced by the European Union in the context of the European Regional Development Fund (ERDF).

References

1. Les guides techniques du collège français de métrologie, Processus de mesure: évaluer les incertitudes 20 exemples. Afnor edition (2018).
2. Kane, V. (1986). Process capability indices. *Journal of Quality Technology*, 41–52.
3. Statistical Methods in Process Management. (2014). Capability and performance. *Part 1: General principles and concepts*. NF ISO 22514-1.
4. Bailon, M. L., & Hornrop, D. J. (2006). *On the calculation of convolutions with Gaussian Kernels CAMS Report 0506-46*.
5. Caballero-Gaudes, C., et al. (2019). A deconvolution algorithm for multi-echo functional MRI: Multi-echo sparse: Paradigm free mapping. *NeuroImage, 202*, 116081.
6. Mahata, K. (2020). Sparse deconvolution via off-grid T.V minimization. *Signal Processing, 170*, 107406.
7. Mirel, M., & Cohen, I. (2017). Multichannel semi-blind deconvolution (MSBD) of seismic signals. *Signal Processing, 135*, 253–262.
8. Dugué, D. (1998). «Calcul des probabilités». Encyclopædia Universalis, §4.
9. Université de Reims Champagne Ardenne, MA 0804-(2013–2014) Master 1, 2 Loi usuelles.
10. Bromiley, P. A. (2018). *Products and convolutions of gaussian probability density functions*. The Product of Two Univariate Gaussian PDFs, p. 1.
11. Boreux, J.-J., Parent, E., & Bernier, J. (2010). Pratique du calcul bayesien. Berlin: Springer.

International Codes in the Prediction of Load-Bearing Capacity of Slender Columns

Alfred Strauss, Neryvaldo Galvão, José C. Matos, Michael Hauser, Benjamin Täubling, Mohamed Soliman, Mohammad Tamini, Xin Ruan, Lingfeng Zhu, and Hiroki Ishibashi

Abstract The bearing-buckling capacity of slender columns is carefully addressed in modern codes by different analytical methods and non-linear advanced numerical methods for structural analysis. This work summarizes an experimental campaign and the safety formats of international codes available for the prediction of slender columns load-bearing capacity. This work is motivated on a round-robin investigation considering numerical simulations supported on non-linear material models and second-order effects (one of the methods suggested in the Eurocode), that proved an expressive overestimation of the load capacity of single slender columns when comparing the numerical results with the experimental ones. Consequently, a high interest in the safety formats of international codes arises, aiming the comparison of the predictions provided by different methods described in design codes from Canada, China, Japan and USA with the European one (namely, the nominal curvature method), as well as the identification of strengths and weaknesses of such methods.

Keywords Slender columns · International safety formats · Load-bearing capacity · Buckling capacity · Round-robin tests

---

A. Strauss (✉) · M. Hauser · B. Täubling
University of Natural Resources and Life Sciences, Vienna, Austria
e-mail: alfred.strauss@boku.ac.at

N. Galvão · J. C. Matos
University of Minho, Guimaraes, Portugal

M. Soliman · M. Tamini
Oklahoma State University, Oklahoma, USA

X. Ruan · L. Zhu
Tongji University, Shanghai, China

H. Ishibashi
Waseda University, Tokyo, Japan

© The Author(s), under exclusive license to Springer Nature Switzerland AG 2021
J. C. Matos et al. (eds.), *18th International Probabilistic Workshop*, Lecture Notes in Civil Engineering 153, https://doi.org/10.1007/978-3-030-73616-3_34

1 Introduction

The Eurocode is a semi-probabilistic based design code that takes into account geometric, material, model and load uncertainties, by means of partial safety factors. Thus, the reliability of the structures designed accordingly is assured by partial safety factors determined based on reliability classes, characterized by a target probability of failure or reliability index and expected levels of consequences given the violation of a certain limit states [1]. The partial safety factors are usually separated into two main groups, one for the material strength and the other one for load effects. The material strength is reduced by a partial safety factor γ_M and load effects are amplified by a partial safety factor γ_F. The design of slender columns according to Eurocode can be achieved by three different methods, aiming the quantification of the buckling load, namely, the general method non-linear second-order analysis, the nominal stiffness and the nominal curvature method [2]. The buckling load of slender compressed elements must be determined since the failure of such an element can occur before the material strength is reached in its critical cross-section. Hence, a partial safety factor for buckling failure is a must for the assurance of adequate reliability levels [3].

The design of slender columns is still nowadays a matter of controversy because of known contradictions of the design concepts. Concerning the non-linear method suggested in EN 1992-1-1 [2], there is still a need for research given the overestimation of the buckling load by such method as reported in the preceding paper [4] of the paper series that this paper is part of. The work developed in Strauss [4] is focussed in a collaborative round-robin investigation on numerical simulation aiming the prediction of the bulking load of a single slender element with different finite element software and engineers from several countries of Europe. The accuracy of the predictions performed according to non-linear methods is afterwards assessed according to the laboratory experimental results.

The IABSE Commission 1—Task Group TG1.4 was inspired to participate in the investigation due to the following reasons:

- Recently engineers have shown some interest in the non-linear analysis given the method ability to better duplicate real structural behaviour and enhance the efficiency of material used for the overall safety of the structure. In times that sustainability is significantly addressed such method can prove very useful.
- Nevertheless, the non-linear method demands some extra efforts from the users since it requires careful selection of the appropriate incremental and iterative solution procedure and more complex material and geometrical models.
- The assessment of existing structures and design of new structures can be further improved through probabilistic methods by direct consideration of uncertainties (e.g. material, model, geometric and modeling uncertainties) into the structural non-linear analysis and response prediction, addressing also failure modes sensitivity to such uncertainties.

Committed to addressing the previously enumerated concerns, the IABSE Commission 1 has documented some findings in [4] concerning different approaches available i.e. analytical methods and non-linear finite element methods. Furthermore, high interest in international safety formats and their differences arise. Hence, safety formats from Canada, China, Europe, Japan and the USA, are addressed in this second paper of the IABSE paper series concerning carrying capacity of isolated slender columns.

2 Experimental Investigation

To validate and assess the accuracy of the analytical and non-linear finite element models (NLFEM) outputs, an experimental campaign was conducted aiming the quantification of the slender column true load-carrying capacity. The slender column geometry (cross-section and height) as well as its reinforcement layout (see Fig. 1),

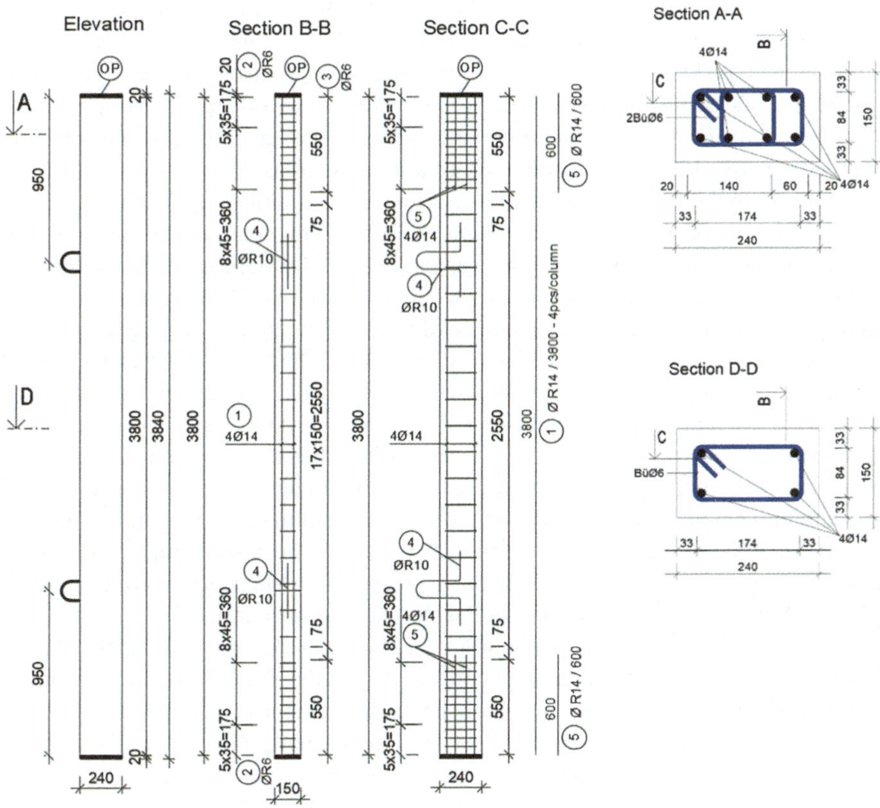

Fig. 1 Investigated slender column [4]

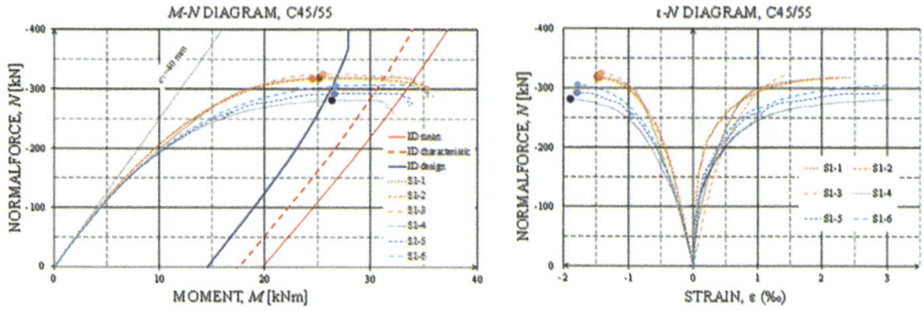

Fig. 2 Experimental results of the specimens S1-1 to S1-6 in the fracture-prone cross-section at the middle height of the column [4]

was carefully designed so it could fail due the loss of stability considering an initial eccentricity of 40 mm for the axial force applied. The column was designed considering a C45/55 concrete and B 500B steel reinforcement. A total number of 6 columns (see [4] for more details on NLFEM prediction) were tested in the laboratory. The column geometry and further information are carefully detailed in Fig. 1.

The experimental results are here displayed in Fig. 2 through two curves. The figure on the left side displays the axial force variation versus the moment generated by the force due to the initial eccentricity and second-order effects or, in other words, geometric non-linearity; and the other shows how the axial force plus the bending moment generated by second other deformation is increasing the compressive and tensile strain in the fracture-prone cross-section at the middle height of the column. The blue and orange dots in Fig. 2 features the verified maximum axial force of each column i.e. the column load-carrying capacity $N_{\max}$. The failure of the column by stability loss is evident since the failure occurs before the N-M interaction threshold curve given by the EN1992-1-1. Three N-M interaction curve as displayed in Fig. 2 based on design, characteristic and mean values. Given the nature of the experimental campaign, the N-M interaction curve given by mean values should be considered as the reference to determine if a stability failure or not have taken place. Additionally, the documented compressive strain was somewhere between 1.4 and 1.8‰, below the maximum compressive strain allowed by the Eurocode i.e. 3.5 ‰. The results are also summarized in Table 1 for each of the specimens of the experimental campaign, namely, the maximum axial force (Nmax) as well as the corresponding second-order eccentricity in the critical cross-section (e_2), and the associated bending moment (Mmax). Some statistics regarding the results are also summarized.

Employing the method of estimation of a coefficient of variation of resistance, also known as the ECOV method, according to [5], the slender column resistance uncertainty can be summarized by a global resistance factor γ_R given by Eq. 1. Here, the resistance of the column is modelled by a lognormal probabilistic distribution. The ECOV method is summarized in Eqs. 1, 2 and 3:

$$\gamma_R = \exp(\alpha_R \beta V_R) \tag{1}$$

Table 1 Statistic of the experiments

Test	Nmax (kN)	e_2 (mm)	Mmax (kNm)
S1-1	324.4	57.6	31.7
S1-2	323.4	42.7	26.8
S1-3	332.6	38.3	26.0
S1-4	271.2	58.4	26.7
S1-5	296.0	59.4	29.4
S1-6	311.4	55.0	29.6
S1-1 bis S1-3	326.8 (0.02)	46.2(0.22)	28.2 (0.11)
S1-4 bis S1-6	292.9 (0.07)	57.6 (0.04)	28.6 (0.06)
S1-1 bis S1-6	309.8 (0.07)	51.9 (0.17)	28.4 (0.08)

where:

α_R is a sensitivity factor for the reliability of resistance.

β is the target reliability index.

V_R is the coefficient of variation of resistance determined according to Eq. 2:

$$V_R = \frac{1}{1.65}\ln\frac{R_m}{R_k} \tag{2}$$

where:

R_m is the mean resistance value.

R_k is the characteristic value of resistance correspond to 95% exceedance probability (or 5% fractile).

Considering a target reliability index of 3.8 ($P_f \approx 10^{-4}$) for a 50-year reference period and a sensitivity factor of 0.8 the global resistance factor of the column maximum axial force should be 1.24 and 1.28 for the corresponding bending moment (M_{Nd}). If a well-validated mathematical model (Analytical or numerical) is used for the design, a partial factor of 1.06 targeting the model uncertainty in resistance (a generous value) can be used, aiming to cover the gap between the model (not the geometric and material uncertainty) and the real slender column performance. Thus, the design buckling axial force can be computed according to Eq. 3. In summary, for a reliable design of the slender column, the maximum axial load that the column should be exposed to is 227.98 kN, in order to comply with a 3.8 target reliability index.

$$N_d = \frac{N_m}{\gamma_{Rd}\gamma_R} = 227.98\,\text{kN and } M_d = \frac{M_m}{\gamma_{Rd}\gamma_R} = 20.89\,\text{kNm} \tag{3}$$

3 European Design Format

The European standardized rules for the design of engineering structures are summarized in the Eurocode. The Eurocode was developed by the European Committee for Standardization (CEN) which is constituted by several researchers, engineers, and users from different member states. In Germany, the Eurocode was introduced by the authorities since July 2012 as applicable law, as well as in Austria with their national annexe since July 2009.

As mentioned in the introduction, the Eurocode provides three methods for the design of slender columns, that is [2]: (i) a general method based on non-linear analysis considering geometric nonlinearity of the structure i.e. second-order effects and material nonlinearity; (ii) the method based on nominal stiffness where the second-order analysis is based on stiffness, nominal values of the flexural stiffness, considering cracking, creep and material non-linearity on the performance of the column. This method also allows the consideration of the stiffness of adjacent members such as beams, slabs or foundations. Soil structure interaction should be considered when relevant. Here the total design considering the second-order effect is obtained by a moment magnification factor applied to the moment obtained from the first-order analysis. (iii) the method based on nominal curvature which is mostly suitable for isolated elements with constant axial force and a defined effective length l_0 conditioned by the column supports and other factors. Here the nominal second-order moment is dependent on the column deflection which turns to be dependent on the effective length and an estimated maximum curvature.

In Fig. 3 is summarized the basics of the Eurocode's nominal stiffness and nominal curvature method based on the M–N interaction threshold of the cross-section capacity considering the second-order effects by means of a magnification factor or rotation. Aiming the assurance of failure by material damage and not stability loss, the buckling load (N_B) must be greater than the cross-section resistance (N_{Rd}).

According to the nominal stiffness method, the eccentrically compressed column ultimate bearing capacity is $N_d = 205$ kN, below the design value (227.98 kN) computed according to the global resistance factor obtained from the ECOV method

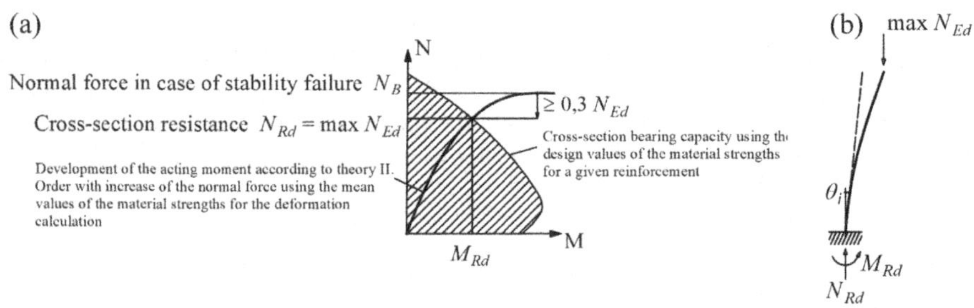

Fig. 3 The Eurocode's 'nominal stiffness' and the 'nominal curvature' approach basics: **a** based on *M–N* interaction threshold diagram of the cross-section capacity, **b** including second-order moment effects, using magnification factor ψ_I or (θi), [6]

applied to the experimental results in Eq. 3 (see Sect. 3). The nominal curvature method, however, provided an ultimate bearing capacity of $N_d = 222$ kN ($M_{Nd} =$ 24.28 kNm), closer to the design value given by the ECOV method. For practical applications, the nominal stiffness method may be used for both isolated members and whole structures if the nominal stiffness of the adjacent elements are estimated properly since it takes into account the stiffness contribution of other elements connected to the main element. On the other hand, the nominal curvature method does not consider such information and is considered mainly suitable for isolated elements. Thus, here the nominal curvature proves again to be most suited for isolated elements. However, when efficiency is concerned, which by the way is the aim of engineering activities, the nominal curvature method is the approach one should consider for isolated elements.

4 American Design Format

4.1 USA Design Format

The American Concrete Institute (ACI) in ACI 318-14 [7] describes three methods to assess or predict the carrying capacity of slender columns taking into account second orders deformations, namely: (a) nonlinear second-order analysis: the method should take into account material a geometric non-linearity, lateral drifting, creep and shrinkage. Considered as the optimum approach, but the computational complexity turns it not very attractive; (b) elastic second-order analysis: this method considers the cracks down the length of the column and creep effects, and; (c) moment magnification procedure: here the first-order elastic effects are quantified and then scaled to consider second-order effects. The member stiffness and cracking effects are quantified through a set of equations based on studies of Macgregor et al. [8].

To maintain reliable design levels ACI employee's resistance and load factor accounting for uncertainties of simplification assumptions, material properties variation and demanded reliability levels. An N-M interaction diagram is usually generated and divided into two regions where two different safety factors are considered. For potential brittle failure (above the balance point) a 0.65 factor is used and for ductile failure mechanism (below the balance point) a 0.90 factor is considered.

Considering the moment magnification factor, the design capacity of the slender column was estimated to be $N_d = 280.00$ kN for a correspondent moment of $M_{Nd} =$ 17.5 kNm.

4.2 Canada Design Format

The Canadian standards approach for the design of reinforced concrete slender columns as several resemblances to ACI 318-14, considering column slenderness ratio, interaction diagrams, second-order effects through a moment magnification factor. Thus, the obtained design axial force is $N_d = 280.00$ kN for a correspondent moment of $M_{Nd} = 16.6$ kNm. The Canadian design format can be found detailed reported in a handbook from the Cement Association of Canada [9].

5 Asia Design Format

5.1 China Design Format

Concerning design and prediction of carrying capacity of eccentric compressed components, an investigation was initiated in 1980 by the Ministry of Housing and Urban-Rural Development of the People's Republic of China. The research was supported by experience and theoretical formulations on eccentric compressed elements from Germany, USA and the former Soviet Union [10]. Several experimental studies were carried out, nevertheless, the investigation was only restricted to material damage considering second-order effects, disregarding true instability problems. Furthermore, the investigation considered beam and columns of ordinary houses, generating several uncertainties or unexpected behaviour with regarding columns of big size. Therefore, later on in Specifications for Design of Highway Reinforced Concrete and Prestressed Concrete Bridges and Culverts [11] a method considering the P-Δ effect is provided, introducing the instability of slender columns. The concept described in such specification is similar to the one implemented in ACI318-08 [12], but the limit curvature expression still adopts the Chinese customary. Here the eccentricity of unfavourable direction is considered.

The Chinese approach can be summarized into the following steps: initially, the compression bearing capacity of the column is quantified, followed by the consideration of eccentricity effects. The height of the pressure zone is computed to determine if the section is under small eccentricity or large eccentricity followed by the quantification of a bending moment increasing coefficient to take into account secondary effects and stiffness of the element. Finally, the column ultimate bearing capacity is obtained, and the column's performance is identified in an N-M interaction diagram to highlight its theoretical performance. According to china design format the column capacity is summarized in $N_d = 282.32$ kN and $M_{Nd} = 13.09$ kNm.

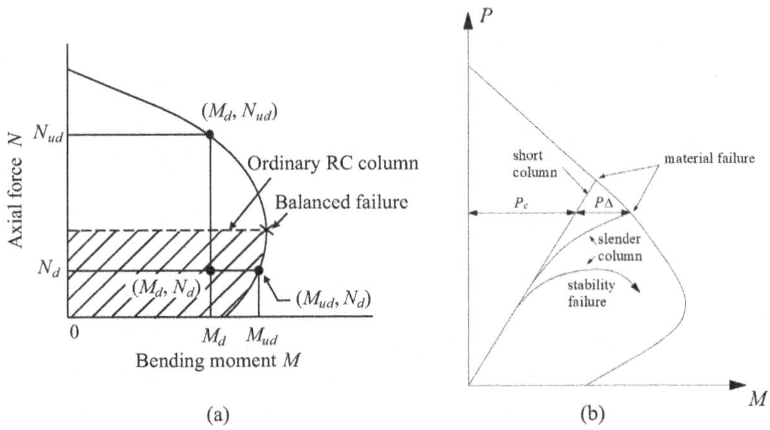

Fig. 4 **a** *N-M* interaction diagram (adopted from JRA [13]. **b** N-M interaction diagram of short and slender columns

5.2 *Japan Design Format*

In Japan, the research on eccentrically compressed columns in civil engineering structures focused mainly on the bending performance of short column (see Fig. 4b), in opposition to the extensive studies on slender columns carried out by countries where strong earthquakes are not so frequent as in Japan. The effects of slenderness ratio on the carrying capacity of reinforced concrete columns are not considered in japan since it is demanded that every highway bridge in japan must be designed taking into consideration earthquake forces of high intensity. Thus, the shear span ratio of reinforced concrete columns allowed is lesser than the acceptable in non-susceptible earthquake regions. In other words, the slender columns analysed here in this paper is not in agreement with safety formats of the Japan Road Association (JRA). Nevertheless, the carrying capacity of a short-column can be computed based on the cross-sectional resistance design procedure given in JRA [13]. Here, the N-M interaction threshold displayed in Fig. 4a is used as the flexural and axial capacity of the column cross-section. Accordingly, the required flexural (M_d) and axial (N_d) capacity due to the loading conditions must be beneath the balanced failure point and inside the hatched area so the failure can be ductile.

6 Conclusions and Final Discussion

The design formats from each country were further investigated considering geometric and material uncertainties, disregarding safety factors, to produce the resistance probabilistic distribution given a random consideration of such uncertainties. The column resistance is later on modelled by a lognormal probabilistic distribution.

The ECOV method (as described in Sect. 2) is also employed to compute the global resistance factor that is used as a measure of the resistance deviation due to material and geometric uncertainties. The safety formats outputs are also plotted against the experimental result to provide a clear view of the agreement and/or disagreement of the safety formats and the real performance of the slender column (experiment). The results are summarized in Fig. 5.

The European design method produced a histogram of the axial resistance of column close to the experimental results with a mean value of 300 kN and standard deviation of 16.4 kN in opposition to 309.8 and 21.68 kN produced by the experimental results, respectively. According to the ECOV method, one produced a global resistance factor equal to 1.18 (Eurocode) and the other 1.24 (Experiment). In summary, the nominal curvature method exhibits minor influence from material and geometric uncertainties, when compared to experimental results. Concerning the mean axial resistance, a deviation of [−3.1%] is observed with respect to the experimental results. On the other hand, the mean bending moment caused by the maximum mean normal force given geometric imperfections and second-order deformations exhibits a deviation of [+15.5%] with respect to the experimental results.

The European design format considered here (i.e. nominal curvature), according to Fig. 5, match more or less the experimental results and provides the lowest global safety factor according to ECOV method, while the others safety formats from other continents clearly overestimate the column axial force required for stability failure. Nevertheless, the general method based on non-linear analysis also suggested in the Eurocode [2] leads to the overestimation of the load-bearing capacity of slender columns, as demonstrated in [4].

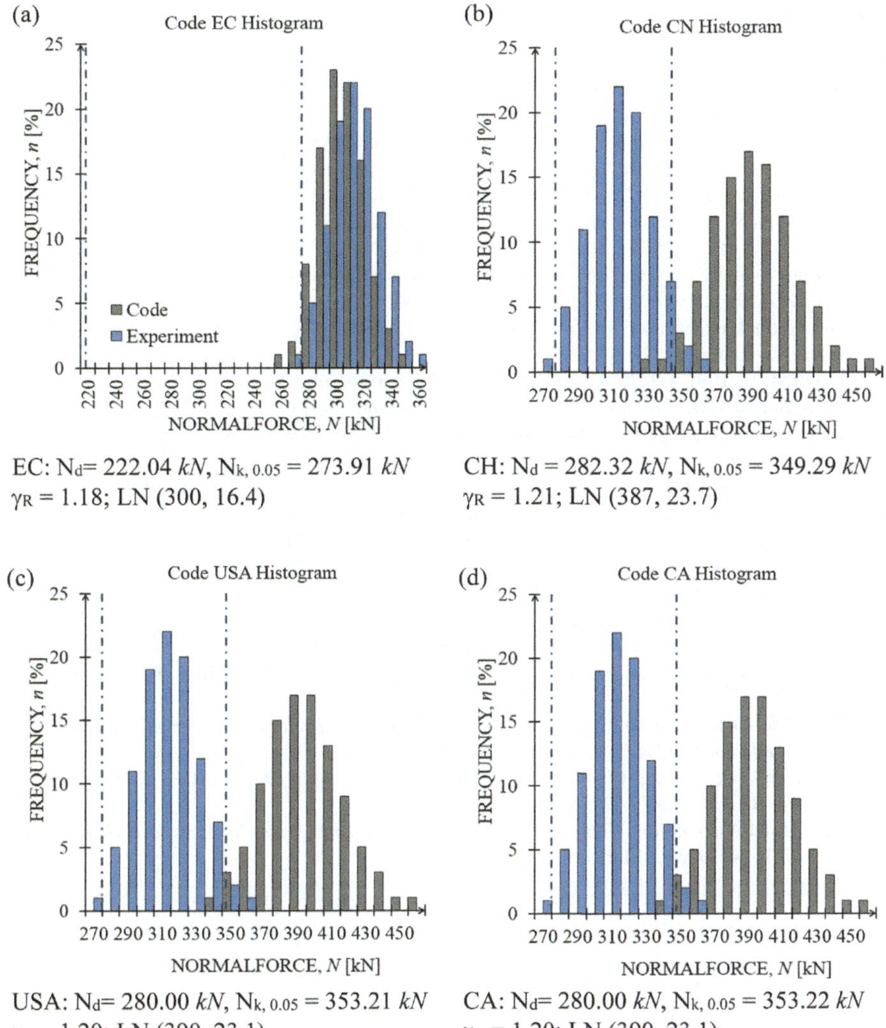

EC: $N_d = 222.04\ kN$, $N_{k,\,0.05} = 273.91\ kN$
$\gamma_R = 1.18$; LN (300, 16.4)

CH: $N_d = 282.32\ kN$, $N_{k,\,0.05} = 349.29\ kN$
$\gamma_R = 1.21$; LN (387, 23.7)

USA: $N_d = 280.00\ kN$, $N_{k,\,0.05} = 353.21\ kN$
$\gamma_R = 1.20$; LN (390, 23.1)

CA: $N_d = 280.00\ kN$, $N_{k,\,0.05} = 353.22\ kN$
$\gamma_R = 1.20$; LN (390, 23.1)

Fig. 5 International design format sensitivity to geometric and material parameters uncertainties

Acknowledgments The authors would like to acknowledge IABSE COM1 and the IABSE TG1.4 members and international partners for supporting this project.

References

1. European Committee for Standardization (CEN). (2005). *EN 1990, Eurocode 0: Basis of structural design*. Belgium: Brussels.
2. European Committee for Standardization (CEN). (2004). *EN 1992-1-1, Eurocode 2: Design of concrete structures—Part 1-1: General rules and rules for buildings*. Belgium: Brussels
3. Benko, V., Dobrý, J., & Čuhák, M. (2019). Failure of slender concrete columns due to a loss of stability. *Slovak Journal* of *Civil Engineering, 27*, 45–51.
4. Strauss, A., et al. (2020). Round-Robin modelling of the load-bearing capacity of slender columns by using classical and advanced non-linear numerical and analytical prediction tools. *Structural Engineering International, 31*, 118–135.
5. International Federation for Structural Concrete (FIB). (2010). *Model code for concrete structures*. Lausanne, Switzerland: Ernst&Sohn.
6. ÖNORM B 1992-1-1:20xx, *Eurocode 2—Bemessung und Konstruktion von Stahlbeton- und Spannbetontragwerken Teil 1-1: Grundlagen und Anwendungsregeln für den Hochbau. Nationale Festlegungen zu ÖNORM EN 1992-1-1, nationale Erläuterungen und nationale Erg+nzungen*.
7. ACI committee 318. (2014). *Building Code Requirements for Structural Concrete and Commentary (ACI 318–14)*. Michigan: Farmington Hills.
8. MacGregor, J. G., Breen, J. E., & Pfrang, E. O. (1970). Design of slender columns. *ACI Structural Journal, 67*, 6–28.
9. Cement Association of Canada and Canadian Standards Association. (2006). *Concrete design handbook*. Ottawa, Canada: Cement Association of Canada.
10. Ministry of Housing and Urban-Rural Development of the People's Republic of China. (2010). *Code for design of concrete structures (GB 50010–2010)[S]*. Beijing, China: Building Industry Press.
11. JTG 3362-2018. (2018). *Specifications for design of highway reinforced concrete and prestressed concrete bridges and culverts*. Beijing, China.
12. American Concrete Institute (ACI). *Reinforced Concrete Beam Design ACI 318-08 Reinforced Concrete Beam Design ACI 318-08*.
13. Japan Road Association (JRA). (2017). *Specifications for highway bridges. Part III: Concrete bridges*. Tokyo, Japan: Maruzen.

Investigation of Parameter Uncertainties Inherent to the Geotechnical Design of Bank Revetments at Inland Waterways

Julia Sorgatz and Jan Kayser

Abstract This paper describes the effects of uncertainty inherent to the choice of hydraulic load and soil parameters on the geotechnical revetment design. As for the practitioner, the effect of uncertainties on the required armour layer thickness is studied. Uncertainties inherent to revetment design mainly result from the load and resistance parameters employed in the design. At present, design loads are obtained from empirical equations and worst-case 'design vessel passages'. Characteristic soil parameters are defined on the basis of a limited number of field and laboratory tests. Thus, uncertainties arise with regard to the choice of characteristic values. In order to investigate the effects of parameter uncertainty on the revetment design, distributions and correlations of loads are assessed using vessel passages observed in the field. In ensuing uncertainty analyses it is found that at present available data does not allow approximating loads by means of probability functions, whereas for the soil parameters the results indicate that the minima of the soil parameters govern the design. However, it is also found that when considering more than one soil parameter as random variable, a less conservative design can be achieved as with the individual minima. As a conclusion, recommendations regarding parameter choice and design procedure are provided.

Keywords Uncertainty analysis · Revetment design · Slope stability under rapid drawdown · Characteristic values

1 Introduction

Bank revetments at German inland waterways are mainly secured by loose or grouted armour stones on a filter layer. Their design according to BAW Code of Practice: Principles for the Design of Bank and Bottom Protection for Inland Waterways (GBB) [1] encompasses a hydraulic and a geotechnical design, where the former defines the armour stone diameter necessary to withstand waves and currents and the

---

J. Sorgatz (✉) · J. Kayser
Bundesanstalt Für Wasserbau (BAW), Karlsruhe, Germany
e-mail: julia.sorgatz@baw.de

© The Author(s), under exclusive license to Springer Nature Switzerland AG 2021
J. C. Matos et al. (eds.), *18th International Probabilistic Workshop*, Lecture Notes in Civil Engineering 153, https://doi.org/10.1007/978-3-030-73616-3_35

latter evaluates the embankment stability under a rapid water level drawdown. The presented study focuses on the geotechnical design.

The National Research Council [2, p. 2] states that “probabilistic methods, while not a substitute for traditional deterministic design methods, do offer a systematic and quantitative way of accounting for uncertainties encountered by geotechnical engineers, and they are most effective when used to organize and quantify these uncertainties for engineering designs and decisions.”

First investigations of uncertainties inherent to shore protection structures can be found in the Netherlands. Triggered by a severe storm event in 1953, van Danzig [3] presents a probabilistic approach for the geotechnical design of flood defence systems. Since then, numerous concepts for the design of sea defence structures, i. e. dikes, dunes and breakwaters, have been published, i. e. [4–9]. Approximately ten years ago, first studies of the hydraulic [10] and geotechnical stability [11, 12] of river and canal embankments were published, which focus on flood events and natural flow. So far, revetment stability under ship-induced loads has not been addressed.

The uncertainty inherent to the design of hydraulic structures is a result of aleatory and epistemic uncertainties. A third category of uncertainties are the so-called ‘unknown unknowns’ which refer to unidentified aleatory and epistemic uncertainties. This work primarily deals with aleatory uncertainty by introducing load and resistance parameters as random variables.

PIANC [13] states that the effects of ship or wind-induced water motion have a random character, however, to the knowledge of the authors, the uncertainty of ship-induced drawdowns and their effect on revetment design has not been investigated yet.

Sources of uncertainty inherent to resistance parameters are natural (inherent) variability, measurement error, transformation error and statistical uncertainty [14]. The natural variability is a result of the genesis of soil. Transformation uncertainty is “related to the accuracy of physical or statistical models” [14] and a result of empirical or other correlation models. For instance, the friction angle is usually determined by direct shear tests based on the relationship between measured shear stress failure and normal stress (Mohr-Coulomb failure criterion). Statistical uncertainty arises from the choice of probability function and its parameters. For the probability function of the effective friction angle φ', Lacasse and Nadim [15], Wolff et al. [16], Lumb [17] and JCSS [18] suggest a Gaussian distribution, whereas Schultze [19] recommends a Lognormal distribution. The hydraulic conductivity k is commonly considered as lognormally distributed [20–22]. This paper focuses on statistical uncertainty. So far, the effects of different probability functions and parameters on revetment design have not been studied.

2 Reliability-Based Revetment Design: Theory and Methods

2.1 Principles of the Revetment Design

In simple terms, a vessel passage in a confined waterway cross-section, i. e. canal or river, leads to changed discharge conditions, which trigger a flow around the vessel and, thereby, cause a lowering of the water level next to the vessel, referred to as drawdown [1]. If the water level is lowered faster than the hydrostatic pore pressure in the soil can adapt to, excess pore pressure may develop (Fig. 1). This process is caused by a delayed pressure equalisation due to gas bubbles in the water at shallower depth [23, 24] and influenced by the compressibility of the water-gas-mixture [25, 26]. The excess pore pressure leads to reduced effective stress, which lowers the shear strength of the soil. This may result in local slope sliding or liquefaction [1].

Wave-induced drawdowns can be simplified by a uniformly decreasing water level with constant drawdown rate [23, 24]. The excess pore pressure attains a maximum at the end of the drawdown z_a, which allows to assess the acting forces as a steady-state problem. A depth-depending excess pore pressure $\Delta p\ (z)$ may develop:

$$\Delta p(z) = \gamma_w z_a \left(1 - a\mathrm{e}^{-bz}\right) \tag{1}$$

where γ_w is the unit weight of water and $a = 1$ and b are pore pressure parameters; b describes the shape of $\Delta p\ (z)$ as response to k and the ratio of design drawdown time $t_a^* = 5$ s and drawdown time t_a, Eqs. (2) and (3), for a gas content of 5–15% in the pore fluid [25].

$$\mathrm{b}^* = 0.166 \cdot \mathrm{k}^{-0.327} \tag{2}$$

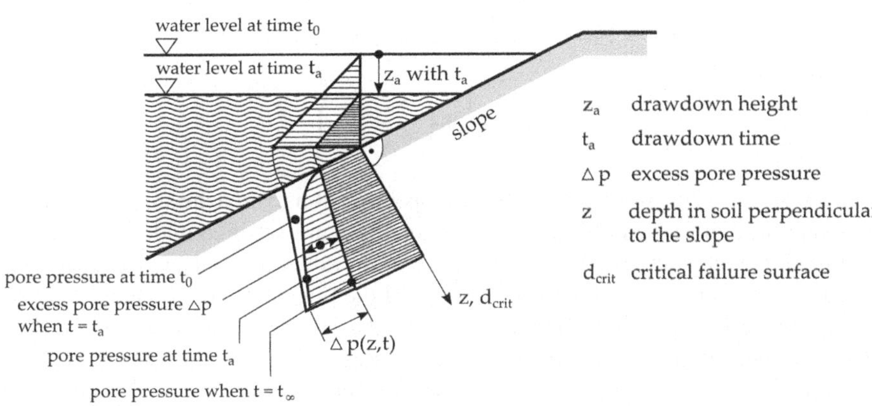

Fig. 1 Hydrostatic pore water pressure and excess pore pressure during rapid drawdown [1]

$$b = b^* \sqrt{\frac{t_a^*}{t_a}} \tag{3}$$

The excess pore pressure may cause driving forces to exceed resisting forces at the vertical slice of an infinite slope leading to local slope failure. The resisting forces are a function of buoyant unit weight of the soil γ_B', slope inclination β and φ′, the driving forces are governed by z_a. The difference of resisting and driving forces reaches a minimum at a critical depth d_{crit} as follows:

$$d_{crit} = \frac{1}{b}\ln\frac{\tan\varphi'\gamma_w z_a b}{\cos\beta\gamma_B'(\tan\varphi' - \tan\beta)}\ for\ \varphi' > \beta \tag{4}$$

If $d_{crit} > 0$, the equilibrium condition stated in Eq. (5) is used to determine the required unit weight of armour stones under buoyancy g', which is the product of required armour layer thickness d_D and buoyant unit weight of armour stones γ_D'. Furthermore, Eq. (5) encompasses the filter layer thickness d_F, the buoyant unit weight of the filter layer γ_F', the effective cohesion c' and the shear strength τ, e.g. as a result of a toe support.

$$g' = \gamma_D' d_D = \frac{\Delta p \tan\varphi' - c' - \tau}{\cos\beta\tan\varphi' - \sin\beta} - \left(\gamma_F' d_F + \gamma_B' d_{crit}\right) \tag{5}$$

Waves and drawdown generate flow in permeable soils, which may lead to considerable vertical hydraulic gradients at bottom and bank and, thereby, liquefaction of near-surface soil layers. In the case of a toe support or a moderate slope inclination, the revetment dimensions determined in Eq. (5) may not satisfy the equilibrium of liquefaction given in Eq. (6) with the corresponding critical depth d_{critB} stated in Eq. (7).

$$g' = \gamma_D' d_D \geq \frac{\Delta p}{\cos\beta} - \left(\gamma_F' d_F + \gamma_B' d_{critB}\right) \tag{6}$$

$$d_{critB} = \frac{1}{b}\ln\left(\frac{\gamma_w z_a b}{\gamma_B'\cos\beta}\right) \tag{7}$$

A design according to GBB [1] fulfils the specifications, when the analyses demonstrate that the limiting equilibrium states (Eqs. 5, 6) are satisfied under the relevant combination of characteristic values. Characteristic values are "selected as a cautious estimate of the value affecting the occurrence of the limit state" [27]. Their selection is either based on expert knowledge or statistical methods. GBB [1] recommends a value at the lower end of the range of possible values for *k*; for the choice of φ′ no recommendations are provided; z_a and t_a are derived from worst-case design vessel passages [28].

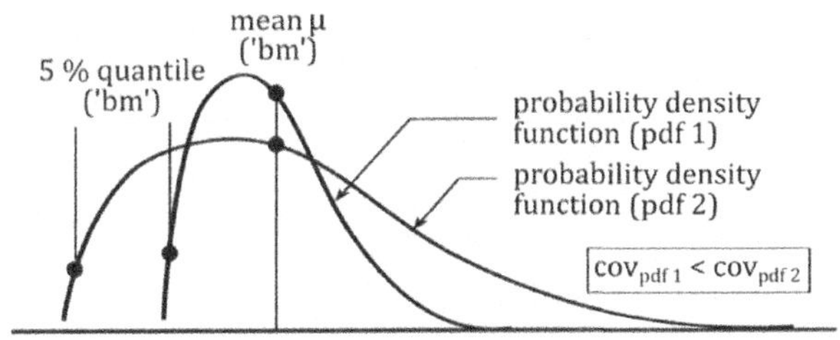

Fig. 2 Parameter definition to compare the random analyses to deterministic benchmark solutions

2.2 Reliability Analyses and Parameter Combinations

The theory of reliability-based methods and their application in geotechnical engineering is well-known. Thus, for mathematical basics reference is made to literature [29, 30]. This paper directly presents the results of the uncertainty analyses, which are conducted with the Python package OpenTURNS [30]. A minimum of 10,000 Monte Carlo simulations is run to obtain a range of possible armour layer thicknesses. Subsequently, the 95% quantiles of the random output are computed to compare the probabilistic results to deterministic benchmark solutions.

The probability functions of the soil parameters are defined as follows: The mean value is constant, while the variability of the soil properties relative to the mean is governed by the coefficient of variation (cov). Different cov are investigated as indicated by the two probability density functions shown in Fig. 2. Since GBB [1] does not states particular bounds that represent characteristic values, a range of φ' and k are considered for a deterministic benchmark solution. The lower bound of the benchmark solution is defined as the 5% quantile of the respective distribution. The upper bound of the deterministic benchmark solution is the mean of the respective distribution. For illustrative purposes, case studies with permeable sand (SW) and with silty sand (SU) are conducted. The variability of the soil is expressed via mean and cov. The soil parameters originate from the German design standard EAU [31].

As it will be shown in Sect. 3.1, currently available data does not support an uncertainty representation of drawdown parameters. The drawdown combinations used beyond Sect. 3.1 are thus based on literature and valid for a standardised rectangular trapezoidal waterway cross-section [28]. Combining loads and soil types, four case studies are investigated (see Table 1).

3 Results

3.1 Uncertainty of Loads

Two field campaigns conducted by the Bundesanstalt für Wasserbau (BAW) at Rhine river and Küsten Canal were used to investigate distributions and correlations of vessel-induced drawdowns. A campaign commonly lasted between one to two

Table 1: Combinations of loads and soil types with their physical properties based on literature [28, 31]

Soil type		Friction angle φ'	$cov_{\varphi'}$	Hydraulic conductivity k	cov_k	Unit weight γ'_B	Time t_a	Height z_a
–		°	–	m/s	–	kN/m³	s	m
SW1	Sand, widely graded	32.5–37.5	0.01–0.10	1×10^{-4}–1×10^{-5}	0.1–1.00	11.5	4.5	0.63
SW2	Sand, widely graded	32.5–37.5	0.01–0.10	1×10^{-4}–1×10^{-5}	0.1–1.00	11.5	27.6	0.83
SU1	Silty sand	32.5–37.5	0.01–0.10	1×10^{-5}–1×10^{-6}	0.1–1.00	9.5	4.5	0.63
SU2	Silty sand	32.5–37.5	0.01–0.10	1×10^{-5}–1×10^{-6}	0.1–1.00	9.5	27.6	0.83

weeks. The measured values encompass dimensions and draught of the vessel, vessel velocity, passing distance and resulting water level fluctuations and flow velocities. Within the scope of this study only drawdowns are evaluated, which were recorded by absolute pressure probes at a minimum of two different heights.

It was found that based on the existing data drawdown parameters (1) are difficult to describe by generally valid probability functions; a Lognormal distribution tends to fit the data best (see Fig. 3), although not all datasets confirm this result. (2) With a Pearson coefficient $\rho_P = 0.2$ there is no significant correlation between z_a and t_a.

Commonly, the armour layer thickness required to protect an embankment against slope sliding ranges between 0.60 and 0.80 m. The current analysis, however, results in armour layer thicknesses greater 1.00 m (see Fig. 4). It is assumed that, although, data analyses indicate a negligible correlation of z_a and t_a, random parameter combination and large uncertainty inherent to distribution fitting results in overly large revetment dimensions. Despite this fact, the observations raise the question whether

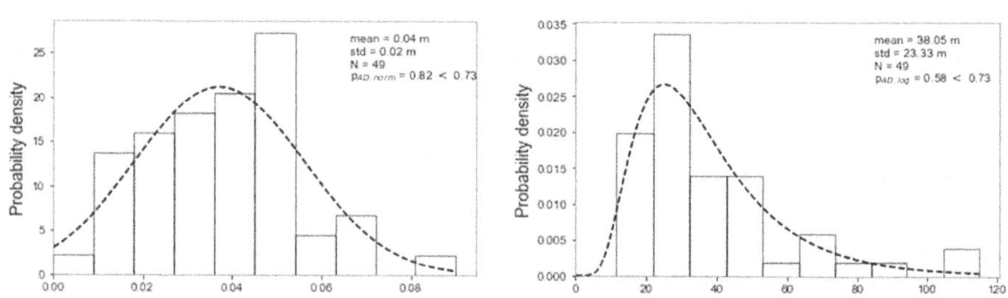

Fig. 3 Examples of probability density functions of sternal drawdown height z_a and drawdown time t_a. Probability density functions of bow drawdowns are similar

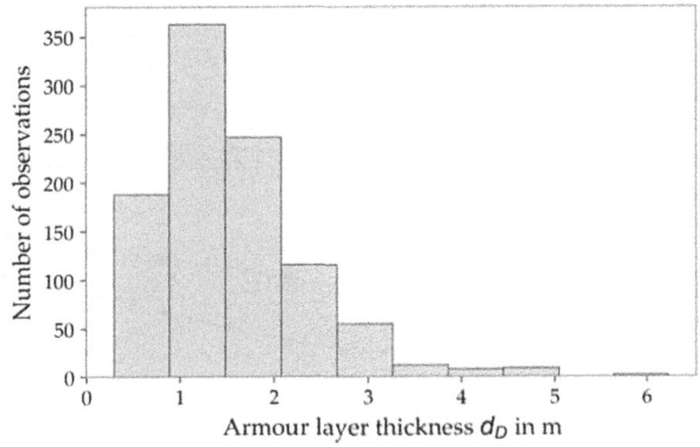

Fig. 4 Armour layer thickness obtained from the uncertainty analysis with random drawdown z_a and drawdown time t_a. It is obvious that the analysis results in significantly larger revetment dimensions than usual

current design drawdowns represent the most critical cases. In the future, long-term observations may assist in determining distributions and correlations of the drawdown parameters as well as critical drawdown combinations.

For now, deterministic loads as proposed in MAR [28], which are obtained from conservative equations in combination with worst case design scenarios, are used for further analyses. From the available load combinations, the most unfavourable are chosen (see Sect. 2.2, Table 1).

3.2 *Effect of Parameter Uncertainty of Random Friction Angle or Hydraulic Conductivity*

Figure 5 illustrates the effects of distribution uncertainty of φ' on the armour layer thickness for two distribution types. It is observed that neither the permeable sand (SW) nor the silty sand (SU) requires a thicker armour layer as a result of the

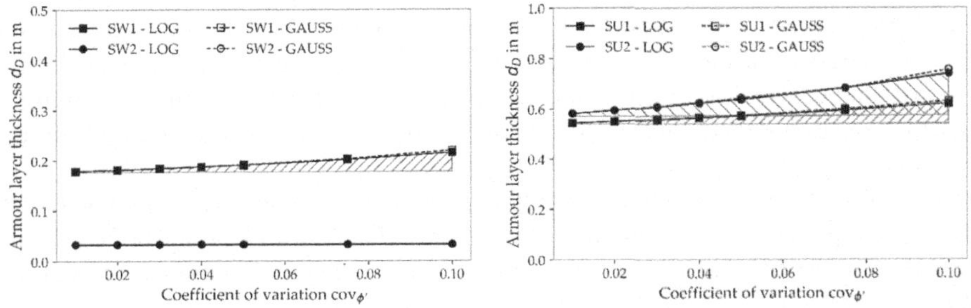

Fig. 5 Armour layer thickness as a function of $cov_{\varphi'}$ for permeable sand (left, SW) and silty sand (right, SU). The effective friction angle is approximated by a Lognormal (LOG) and a Gaussian (GAUSS) distribution. The hatched areas indicate the deterministic benchmark solutions obtained with the 5 and 50% quantiles of φ'. The black graphs depict the 95% quantiles obtained from the uncertainty analysis with random φ'

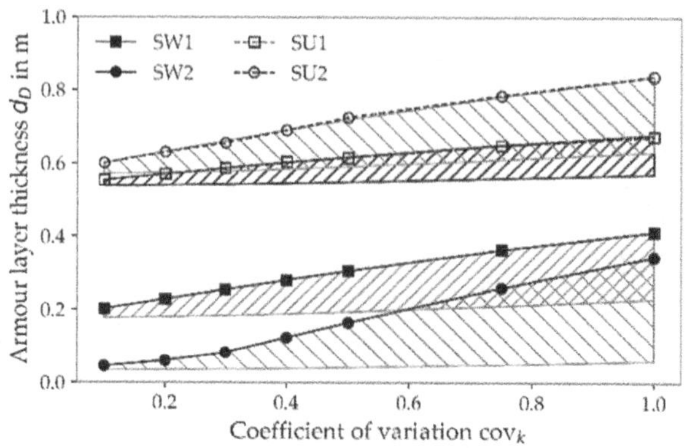

Fig. 6 Armour layer thickness as a function of cov_k for permeable sand (SW) and silty sand (SU). The hydraulic conductivity is approximated by a Lognormal distribution. The hatched areas indicate the deterministic benchmark solutions obtained with the 5 and 50%quantiles of *k*. The black graphs depict the 95% quantiles obtained from the uncertainty analysis with random *k*.

different distribution types. Eventually, with $cov_{\varphi'} > 0.1$, the Gaussian distribution may lead to larger armour layer thicknesses. This observation agrees with Schneider and Schneider [32], who recommend the use of Lognormal distributions in particular for larger coefficients of variation ($cov_{total} > 0.3$).

An increasing $cov_{\varphi'}$ leads to an increasing armour layer thickness. Depending on soil type and drawdown combination a maximum increase of 20 cm is observed for the investigated parameter range. In short, the uncertainty of $cov_{\varphi'}$ affects the uncertainty of the required armour layer thickness more than the choice of distribution. The design of less permeable soils is more strongly affected than the design of permeable soils.

The effects of the uncertainty of *k* on the required armour layer thickness are shown in Fig. 6. In contrast to φ', literature clearly states that a Lognormal distribution is the most suitable choice for *k*. Compared to the results with random φ', an uncertain *k* contributes more strongly to the uncertainty of the required armour layer thickness due to the larger variability of *k*. However, the required armour layer thickness does not rise linearly; the larger cov_k the smaller the increase of the armour layer thickness. Consequently, the variability of *k* affects the armour layer thickness less with increasing cov_k. In conclusion, the results confirm the recommendations of GBB [1] regarding the choice of the characteristic value of *k* as minimum observed in field or laboratory tests.

Finally, it is pointed out that for random φ' and *k* the armour layer thickness in the SW cases is governed by the small t_a at moderate z_a (SW2), whereas the armour layer thickness in the SU cases is governed by the large t_a in combination with large z_a (SU1). This behaviour is explained by the time to reach a quasi-stationary state and, thus, the maximum excess pore pressure. In soils of smaller hydraulic conductivity it takes longer to reach a quasi-stationary state, while in permeable soils the quasi-stationary state is reached faster. In less permeable soil the maximum excess pore pressure is thus reached with large t_a, whereas small t_a do not allow the excess pore pressure to fully build up. As a result, the SU cases require larger armour layer thicknesses with larger, but slower drawdowns, whereas the SW cases require more armour stones with smaller, but faster drawdowns. The observation emphasises the importance of investigating different drawdown combinations to identify the most critical combination.

3.3 *Effect of Parameter Uncertainty of Random Friction Angle and Hydraulic Conductivity*

The armour layer thickness obtained when considering combinations of random φ' and k is shown in Fig. 7. Again, the choice of the probability function does not significantly affect the result. As the minima of φ' and k govern the design, there is hardly any difference between the results obtained with Lognormal (LOG) and Gaussian (GAUSS) φ'.

Compared to analyses with either random φ' or k, the uncertainty inherent to the design increases when considering both, φ' and k, as random. At the same time, it is observed that the random combination of φ' and k results in smaller armour layer thicknesses than the deterministic 5% quantile result. Depending on the soil type, a maximum difference of approximately 15 cm is observed between deterministic benchmark solution and uncertainty analysis.

The analyses assume uncorrelated φ' and k. However, in the case of the investigated parameters the correlation is slightly negative [32]. A zone that is characterised by small φ' is more likely to be associated with high k and vice versa. Using the minima of φ' and k as characteristic values may therefore result in a conservative design.

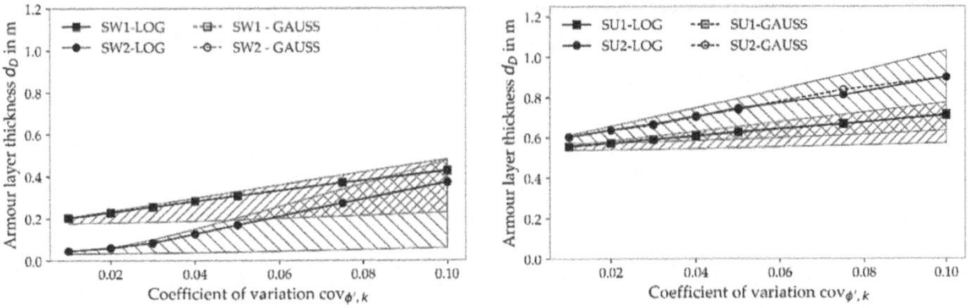

Fig. 7. Armour layer thickness as a function of $cov_{\varphi',k}$ for permeable sand (left, SW) and silty sand (right, SU). The effective friction angle is approximated by a Lognormal (LOG) and a Gaussian (GAUSS) distribution. The hatched areas indicate the deterministic benchmark solutions obtained with the 5% and 50% quantiles. The black graphs depict the 95% quantiles obtained from the uncertainty analysis with random φ' and k.

4 Discussion and Conclusions

The analyses demonstrate that the majority of uncertainty inherent to the geotechnical revetment design results from the employed load combinations. At present available data does not allow approximating drawdowns by means of probability functions. Long-term observations may assist in determining adequate distributions and correlations. With an increased observation period it may also be conceivable to apply extreme value theory to determine characteristic values. At present, it is recommended that a design or assessment of the geotechnical design should explore a number of drawdown combinations, e.g. obtained from field observations or from MAR [28], to identify the most relevant drawdown combination.

The analysis with random soil parameters indicates that the choice of distribution type does not affect the armour layer thickness significantly. In contrast to that, the variance of soil parameters affects the required armour layer thickness significantly; if considered separately, the minima of φ' and k govern the design.

Considering uncorrelated or negatively correlated soil parameters, the minima φ' and k as characteristic values may result in an overly conservative design. The presented uncorrelated reliability analyses indicate that the armour layer thickness may be reduced by a maximum of 15 cm if considering both, φ' and k, as random variables. For practical design purposes, this means that using reliability-based methods with random soil parameters and the combined 95%-exceedance value may allow for a more economic design than using the 5%-characteristic values of φ' and k. Further investigations regarding the required target reliability and a corresponding semi-probabilistic design approach with partial factors are required.

To conclude with, it is important to note that the stability of a slope in rapid drawdown situations depends on the local excess pore pressure and shear strength. Therefore, it may not be sufficient to examine the statistical variability of the soil parameters. Future investigations that account for the spatial variability of soil, e.g. by means of random fields, may supplement current knowledge regarding the choice of characteristic soil parameters.

References

1. GBB. (2010). *Grundlagen zur Bemessung von Böschungs- und Sohlensicherungen an Binnenwasserstraßen*. Bundesanstalt für Wasserbau: BAWMerkblatt.
2. National Research Council. (1995). *Probabilistic methods in Geotechnical Engineering*. Washington, D.C.: National Academies Press.
3. van Dantzig, D. (1956). Economic decision problems for flood prevention. *Econometrica, 24*(3), 276.
4. Bakker, K. J., & Vrijling, J. K. (1980). *Probabilistic design of sea defences*. Delft: Delft University of Technology.
5. Mol, A., Groeneveld, R. L., & Waanders, A. J. (1984). Safety and reliability of breakwaters. *Coastal Engineering Proceedings, 1*(19), 2451–2466.
6. Barends, F. B. J., & van Dijk, J. J. (1985). *Computer aided evaluation of the reliability of a breakwater design*. Zoetermeer: Final report CIAD.

7. van der Meer, J. W. (1988). *Rock Slopes and Gravel Beaches under Wave Attack*. Thesis. Delft Hydraulics Laboratory.
8. Vrijling, J. K. (2001). Probabilistic design of water defense systems in The Netherlands. *Reliability Engineering & System Safety, 74*(3), 337–344.
9. Kortenhaus, A. (2003). *Probabilistische Methoden für Nordseedeiche*. PhD thesis. Technische Universität Braunschweig.
10. Jafarnejad, M., Pfister, M., Brühwiler, E., & Schleiss, A. J. (2017). Probabilistic failure analysis of riprap as riverbank protection under flood uncertainties. *Stochastic Environmental Research and Risk Assessment, 3*(4), 1839–1851.
11. Möllmann, A. (2009). *Probabilistische Untersuchung von Hochwasserschutzdeichen mit analytischen Verfahren und der Finite-Elemente-Methode*. PhD thesis. Universität Stuttgart.
12. Weißmann, R. (2014). *Probabilistische Bewertung der Zuverlässigkeit von Flussdeichen unter hydraulischen und geotechnischen Gesichtspunkten*. PhD thesis. Karlsruher Institut für Technologie.
13. PIANC. (1987). Risk consideration when determining bank protection requirements. *Supplement to PIANC Bulletin, 58*, Permanent International Association of Navigation Congresses (PIANC).
14. ISO 2394 2015. *General principles on reliability for structures*.
15. Lacasse, S., & Nadim, F. (1996). Uncertainties in characterising soil properties (Plenary). In: C.D. Shackelford, P.P. Nelson, & M.J.S. Roth (Eds.), *Uncertainty in the geologic environment—From theory to practice. Proceedings of Uncertainty '96* (pp. 49–75). New York: American Society of Civil Engineers (ASCE).
16. Wolff, T.F., Demsky, E.C., Schauer, J., & Perry, E. (1996). Reliability assessment of dike and levee embankment. In: C. D. Shackelford, P. P. Nelson, & M. J. S. Roth (Eds.), *Uncertainty in the geologic environment—From theory to practice. Proceedings of Uncertainty '96* (pp. 636–650). New York: American Society of Civil Engineers (ASCE).
17. Lumb, P. (1966). The variability of natural soils. *Canadian Geotechnical Journal, 3*(2), 74–97.
18. JCSS. (2006). *Probabilistic model code. Section 3.7: Soil properties*. Joint Committee on Structural Safety.
19. Schultze, E. (1972). Frequency distributions and correlations of soil properties. In: Hong Kong University Press (Ed.), *Statistics and Probability in Civil Engineering*.
20. Carsel, R. F., & Parrish, R. S. (1988). Developing joint probability distributions of soil water retention characteristics. *Water Resources Research, 24*(5), 755–769.
21. Mallants, D., Mohanty, B. P., Vervoort, A., & Feyen, J. (1997). Spatial analysis of saturated hydraulic conductivity in a soil with macropores. *Soil Technology, 10*(2), 115–131.
22. de Rooij, G. H., Kasteel, R. T. A., Papritz, A., & Flühler, H. (2004). Joint distributions of the unsaturated soil hydraulic parameters and their effect on other variates. *Vadose Zone Journal, 3*(3), 947–955.
23. Köhler, H. J. (1993). The influence of hydraulic head and hydraulic gradient on the filtration process. In J. Brauns (Ed.), *Filters in geotechnical and hydraulic engineering* (pp. 225–240). Rotterdam: Balkema.
24. Köhler, H. J. (1997). Boden und Wasser—Druck und Strömung. *Mitteilungsblatt der Bundesanstalt für Wasserbau, 76*, 15–33.
25. Montenegro, H. (2016). *FuE-Abschlussbericht: Infiltrationsdynamik in Erdbauwerken*.
26. Ewers, J., Sorgatz, J., & Montenegro, H. (2017). Laborversuche und gekoppelte Berechnungen zur Untersuchung von Porenwasserüberdrücken infolge schneller Wasserstandsabsenkungen. In: Deutsche Gesellschaft für Geotechnik e.V. (DGGT) (Ed.), *Fachsektionstage Geotechnik der Deutschen Gesellschaft für Geotechnik*.
27. DIN EC7. (2015). Handbuch Eurocode 7—Geotechnische Bemessung. *Band 1, Deutsches Institut für Normung e. V. (DIN)*. Berlin: Beuth.
28. MAR. (2008). *Anwendung von Regelbauweisen für Böschungs- und Sohlensicherungen an Binnenwasserstraßen*. BAWMerkblatt, Bundesanstalt für Wasserbau.
29. Phoon, K.-K., & Ching, J. (2015). *Risk and reliability in geotechnical engineering*. Hoboken: Taylor and Francis.

30. Baudin, M., Dutfoy, A., Iooss, B., & Popelin, A.-L. (2015). *OpenTURNS. An industrial software for uncertainty quantification in simulation.*
31. EAU. (2012). *Empfehlungen des Arbeitsausschusses "Ufereinfassungen". Häfen und Wasserstraßen.* Deutsche Gesellschaft für Geotechnik e.V. (DGGT), Ernst, Berlin.
32. Vardon, P.J., Liu, K., & Hicks, M.A. (2016). Reduction of slope stability uncertainty based on hydraulic measurement via inverse analysis. *Georisk: Assessment and Management of Risk for Engineered Systems and Geohazards, 10*(3), 223–240.
33. Baecher, G. B., & Christian, J. T. (2003). *Reliability and statistics in geotechnical engineering.* Chichester: Wiley.
34. Schneider, H. R., & Schneider, M. A. (2013). Dealing with Uncertainties in EC7. In: P. Arnold, G.A. Fenton, M.A. Hicks, & T. Schweckendiek (Eds.), *Modern geotechnical design codes of practice* (pp. 87–101). Amsterdam: IOS Press.

Life-Cycle Cost Analysis of a Viaduct Considering Uncertainties on the Interventions Plan

Carlos Santos, Mário Coelho, Monica Santamaria, José C. Matos, and Mauricio Sanchéz-Silva

Abstract Life-Cycle Analysis is usually referred to as the assessment of a system that includes the three pillars of sustainability, i.e. economic, environmental, and social aspects. Many works have been developed presenting and discussing methodologies and frameworks to include the evaluation of these aspects. In general, economic performance is the most addressed, followed by the environmental aspect. A highly generalized formulation is usually used regarding social aspects since it corresponds to the society at large. The present work is focused on the economic aspects of the assessment, implementing a life-cycle cost analysis methodology to an infrastructure system (viaduct) covering all direct costs for agency/owners. Furthermore, to indirectly account for climate changes, and the requirements of the quality control plan (QCP) for operational issues, a semi-probabilistic approach is carried out. The life cycle cost analysis assumes a uniform probabilistic distribution for the intervention time of maintenance and rehabilitation activities. With this, variations in the degradation processes due to climate changes and/or intervention needs to fulfill requirements from QCP, are considered. For this purpose, the time at which each intervention might occur in the future is assumed as a random variable. Monte Carlo simulation is then used to compute the cost of several different scenarios. On other hand, the present work provides another approach referred as the deterministic approach. It estimates the life cycle costs deterministically, and considers at the end of the analysis the uncertainties associated with the life-cycle process in a rough way assuming a general coefficient of variation of ±20%. The main goal of this work is to understand the impact that uncertainties in the intervention schedule might have on the final life-cycle cost.

Keywords LCCA · Direct costs · Infrastructures · Uncertainties · Probabilistic assessment

C. Santos · M. Coelho (✉) · M. Santamaria · J. C. Matos
University of Minho, Guimarães, Portugal
e-mail: mcoelho@civil.uminho.pt

M. Sanchéz-Silva
University of Los Andes, Bogota, Colombia

© The Author(s), under exclusive license to Springer Nature Switzerland AG 2021
J. C. Matos et al. (eds.), *18th International Probabilistic Workshop*, Lecture Notes in Civil Engineering 153, https://doi.org/10.1007/978-3-030-73616-3_36

1 Introduction

Within the scope of transportation infrastructures system's management in coastal zones, decision-making should be based on the Life-Cycle Analysis (LCA) covering a set of key performance indicators (KPI) associated with each pillar of sustainability (environmental, economic, and social).

Life-Cycle Analysis corresponds to an assessment of the infrastructure system performance within the whole life cycle. This holistic methodology considers a set of (simultaneous or interactive) different aspects such as: (*i*) sustainability requirements (environmental, economic, and social criteria); (*ii*) technical and functional requirements (KPI´s are used to measure infrastructure's performance—reliability, availability, durability, see more details in [1]); (*iii*) different life-cycle phases; and (*iv*) various functional units (material, component or structure) [2]. Due to the wide range of the analysis and its numerous issues in the quantification of involved parameters, the goal and scope of an LCA must be clearly defined and be consistent with its intended application. Furthermore, relevant aspects such as the safety features, condition assessment, among others, should be introduced in LCA of existing structures.

Regarding sustainability requirements, large infrastructure systems with long life cycles (e.g., dams, bridges, roadways) usually have an impact on the long-term socio-economic development of a country. Sustainable development is widely accepted as "meeting the needs of the present without compromising the ability of future generations to meet their needs". Environmental requirements can be determined from several variables such as toxicity, global warming, among others. Economic requirements are evaluated from the life-cycle cost analysis (LCCA) which takes into account all costs emerging from the existence of the infrastructure. Lastly, the social pillar is associated with the infrastructure cultural value and it can be measured with aesthetical factors or other [3].

LCA arose as a new paradigm based on the design practitioners awareness that all actions and decisions to be made throughout the service life should be part of the sustainable structural design. Thus, this vision replaces the traditional idea that gives significant importance to the conception and design, since it is directly linked to the necessary initial investment, namely construction costs. Once maintenance becomes an integral part of the design, a life cycle plan should be carried out by the infrastructure designer, i.e. the designer should plan the maintenance actions during the infrastructure´s service life. This maintenance includes what activities can be done, when they should be done and how much traffic disturbance will the maintenance works introduce, among others [3].

LCA has also emerged over the last years as a valuable decision-making support tool in assessing lifetime impacts of the built environment, for both policymakers and industry [4]. This has been reflected in the time invested by the scientific community on the topic, namely from European projects like Bridge Life Cycle Optimisation (ETSI), Sustainable steel-composite bridges in the built environment

(SBRI), Life Cycle Management of concrete infrastructures for improved sustainability (LIFECON), among others [3, 9]. Furthermore, International and European standards for sustainable construction have also been developed as an effort to balance performance targets between conflictive criteria such as environmental and economic indicators.

The present work provides two approaches to determine the life-cycle costs emerging throughout an infrastructure´s service life. The first one follows a deterministic formulation, which is based on a standard interventions plan with fixed intervention times (single scenario). The second approach comprises a semi-probabilistic methodology for considering uncertainties, i.e. the service life prediction of bridge components and its intervention times follow a probabilistic uniform distribution. Both approaches can be useful to support the infrastructure managers in decision-making.

2 Life-Cycle Cost Analysis (LCCA)

As previously mentioned, Life-Cycle Cost (LCC) is one of the three pillars of sustainability included in LCA. LCC is related to the economic performance and is the one that is specifically addressed in the present work.

2.1 *Basic Principles*

Optimal use of financial resources is an essential part of managing infrastructures. In the past decades, it has become very clear that management and decision-making based only on the lowest cost does not meet modern sustainability requirements [5]. Recent tools, guidelines, and standards for assessment of the LCC of new and existing assets (e.g., buildings, bridges, roadways, built environment in general) have confirmed the effort made in this sense in order to achieve sustainability goals [6]. Its implementation from agencies/owners to manage infrastructures allows optimization of operation and maintenance costs through cost-effective intervention plans where the intervention time and frequency are described. Moreover, the operation and maintenance of infrastructure systems highly depend on their performance throughout service life. The costs associated with the referred activities constitute a significant part of an often limited budget.

The continuous increase of sustainability requirements has led to the ambition of assessing the life-cycle costs for longer periods (e.g. during service life). However, in many cases it is not feasible to perform it due to the uncertainties inherent to all stages of the asset´s entire life, which increase the complexity of LCCA. Moreover, this complexity goes beyond the mathematical models, requiring a thorough understanding of the relationship between different stages (processes related and executed by different actors) [7].

Basically, in LCCA main uncertainties come from:

- Structural system: (*i*) System state (condition) at a given time t, and the corresponding need for intervention; (*ii*) Time and level of interventions (when and to what extent);
- Finance: (*i*) Discounting function modeling; and (*ii*) Financial costs.

For models considering uncertainties, LCCA can only be evaluated in terms of the expected value of the costs. For these cases, the mathematical model is quite complex and it requires important simplifications (assumptions). Most models focus on the problem related to the variability of financial costs, e.g. considering probabilistic distributions to each unit cost [8].

2.2 *Methodology of LCCA*

The definition of an LCC model begins with the cost classification structure that aims to define all costs at a satisfactory level. Basically, it includes all costs the infrastructure will incur at different phases. The two main cost categories are direct and indirect costs, and the total cost of infrastructure corresponds to the sum of them.

Direct costs are those directly borne by the owner during the entire life of an asset but can include also other costs such as those associated with the planning and pre-construction processes. Normally, these are divided into three subcategories: (*i*) design and construction costs; (*ii*) operation costs (namely, inspection, maintenance, replacement, and rehabilitation); and (*iii*) decommissioning (end of service life costs due to removing the system form service). Following a deterministic approach considering a fixed time interventions plan, and also fixing the end of service lifetime, the total discounted direct cost (C_{Td}) is given as:

$$C_{T_d}(\boldsymbol{p}) = C_0(\boldsymbol{p}) + \sum_{n=1}^{N(t_D)} C_{m,n}(\boldsymbol{p})\delta(t_n) + C_D\delta(t_D) \tag{1}$$

Where: C_0 = Design and construction costs (monetary unit); $C_{m,n}$ = Cost of the nth preventive intervention (monetary unit); C_D = Decommissioning costs (monetary unit); $\delta(t)$= Discounting function; t_n= Time at which the n-th intervention occurs; N(t_D) = the total number of interventions within the time frame t_D; t_D= end of the service life of the system; and $\boldsymbol{p}$ = vector parameter of system properties.

Indirect costs are those which society incurs as a result of the construction, operation, interventions, or decommissioning of the asset. In some cases, the evaluation of the three pillars of sustainability can be grouped in economic parameters converting environmental and social aspects into monetary units. Thus, indirect costs depend on the type and extent of the evaluation but the most commonly used are user costs, environmental costs, and societal costs.

In order to compute the Net Present Value (NPV) of future investments, costs should be discounted to time t = 0 (corresponding to the construction year for new assets or to the current instant for existing structures). This allows having a standard value representation for comparison purposes. The most widely used model for computing NPV is given by [7]:

$$\delta(t) = \frac{1}{(1+\gamma)^t} \approx \exp(-\gamma t) \text{ for } \gamma \ll 1 \quad (2)$$

Where γ is called a discount rate.

The values of the discount rate vary depending upon the location of the infrastructure system. Typical annual discounting rates in Europe and the US are between 2 and 4%, in middle-income countries the rate varies between 5 and 12%, and larger values are only observed in other specific locations [7].

Normally, the future values are discounted using a time-independent discount rate (constant) in order to enable comparison between scenarios. Present values are determined by employing equation Eq. 2 so that the life-cycle costs can be referred to the same point in time:

$$LCC^{PV} = \sum_{i=1}^{Z} \frac{LCC(t)_i^{FV}}{(1+\frac{\gamma}{100})^m} \quad (3)$$

Where: LCC^{PV} represents the present value of the life-cycle cost; LCC is the future cost; t corresponds to the time (years); m is the difference between the year of observation and the reference year; and i is the running index.

3 Implementation on a Case Study

The implementation of the methodology on a case study covers all direct costs for agency/owners. In the next section a summary of the work conducted is presented.

3.1 General Description

A reinforced concrete viaduct with the dimensions shown in Table 1 was analysed. The corresponding life-cycle costs for a period of 100 years were determined based on the construction costs, interventions costs, and demolition costs. For that purpose, two approaches were implemented which differ in the way intervention times were considered (see Table 2). Introducing random variables into the semi-probabilistic approach allows including the structural performance uncertainties, which will then trigger the interventions that will might be required in the future.

Table 1 Dimensions of the case study

Span length	30 m
Bridge deck length (3 spans)	90 m
Bridge deck width	17.6 m
Elastomeric bearings	24 un

Table 2 Summary of the implemented approaches

	Discounting	Times	Costs		
		Intervention	Construction	Intervention	Demolition
Deterministic approach	Yes	Deterministic	Deterministic	Deterministic	Deterministic
Semi-probabilistic approach	Yes	Probabilistic	Deterministic	Deterministic	Deterministic

The construction costs highly depend on the project details, including quantities of materials, unit costs of material and equipment, construction plan, and all inherent activities. The case study is an integral part of a large viaduct (about 13 km long) and the construction cost was roughly estimated from the total construction budget as € 1,648,351.

The operation stage is highly affected by the interventions plan. This plan depends on the type of components and associated interventions. The frequency of each intervention is listed in Table 3 per type of component. In general, the intervention plan contains inspections (for condition assessment), maintenance activities, and replacement and rehabilitation activities.

Maintenance activities are essential to keep the system operational and in good condition. Although it is not common to specify the activity, it is known that there are many maintenance possibilities for each component such as small repairs, cleaning, paint, greasing of the sliding surfaces of the bearing device, partial replacement of concrete step barriers, among others. Replacement and rehabilitation activities are associated with the service life of components affecting the functional requirements of the structure as a whole.

From the data available in Table 3 it is possible to define a set of different scenarios. As an example, Fig. 1 corresponds to the standard interventions scenario defined according to the mean values, i.e., this scenario is used in the deterministic approach (Sect. 3.2) which is based on the average service life of each component obtained from the literature [8, 9].

The uncertainties associated with intervention times (intervals) were given by experts based on their experience and know-how. In this way, the semi-probabilistic approach (Sect. 3.3) was implemented associating these random variables with a uniform distribution.

At the end of service life, the demolition of the viaduct is intrinsically represented in a set of main stages, including preparation of terrain, dismantling, demolition, and

Table 3 Type of interventions considered in the analyses

Time horizon (years)	100	Interventions									
List of components		Maintenance					Replacement and rehabilitation				
Type	Description	Times			Costs		Times			Costs	
					Unit Cost					Unit Cost	
		min	mean	max	Value	Unit	min	mean	max	Value	Unit
Structural elements	Superstructure (deck)	20	25	30	25	€/m^2	100	100	100	(–)	(–)
	Substructure	20	25	30	25	€/m^2	100	100	100	(–)	(–)
	Bridge Deck Waterproof system	(–)	(–)	(–)	(–)	(–)	35	40	45	40	€/m^2
	Pavement	8	10	12	10	€/m^2	15	20	25	20	€/m^2
Non-structural elements	Expansion joints	2	3	4	10	€/m	3	4	5	60	€/m
	Elastomeric Bearings	2	3	4	10	€/unit	25	30	35	120	€/unit
	Edge beams	(–)	(–)	(–)	(–)	(–)	20	25	30	100	€/m
	Guardrails in parapets	8	10	12	40	€/m	18	20	22	80	€/m
	Concrete step barrier	18	20	22	50	€/m	35	40	45	100	€/m
Periodic inspections	Minor inspection	2	2.5	3	2	€/m^2					
	Main inspection	4	5	6	5	€/m^2					

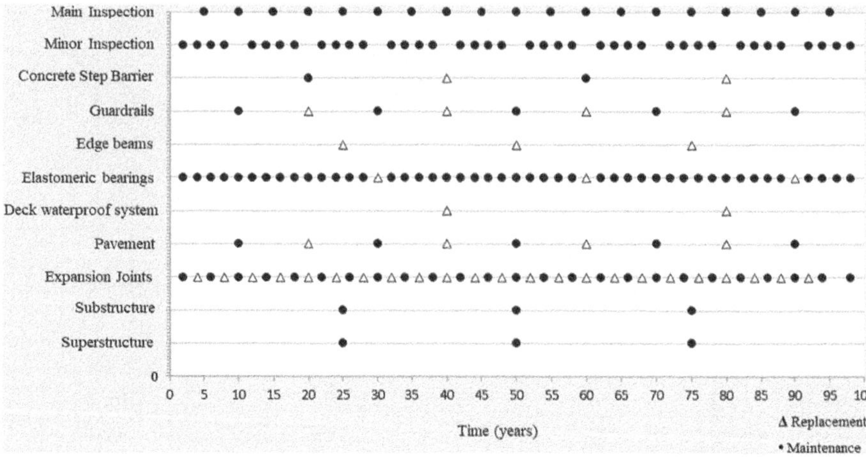

Fig. 1 Standard interventions plan

Table 4 End-of-life scenario

Material	
Concrete	Landfill disposal
Reinforcement steel	Recycled
Bituminous mix (asphalt)	Landfill disposal

removal and transportation of materials for different places according to its characteristics and potential for recycling. Normally, during these works, it is assumed that the traffic is diverted to an alternative road. In order to assess demolition costs, the following scenario was established (Table 4).

Assuming a cost of 138 €/m^2 from the literature previously mentioned, the demolition cost includes all in-situ works and is given by € 218,592.

Posteriorly, the removal and transportation of materials are determined based on the mass quantities of each material. For concrete was obtained a quantity of 2587 tons. For remaining materials, 139 and 122 tons were determined for asphalt and reinforcement steel, respectively.

The transportation costs take into account the following features: (*i*) landfill site is 50 km from the viaduct; (*ii*) trucks with an average load of 25 tons; (*iii*) fuel cost is 0.88 €/L; and (*iv*) fuel efficiency is 0.29 L/km. Considering all these parameters, the transportation cost is € 3020.

In addition, the resulting costs associated with the waste deposit on the landfill are € 25,869 and € 6969 for concrete and asphalt, respectively.

Assuming that only 75% of the steel reinforcement is recovered, it is converted into monetary gains due to recycling. From the analysis of local companies in the scrap processing, it was obtained a fee of € 2.4 cent/kg generating revenue of € 2194.

Thus, the corresponding total cost of decommissioning is € 252,256.

The discounting function is used in both approaches and the corresponding NPV is determined according to Eq. 3. The present case study is located in a country where the discount rate is not uniform over time. Therefore, an average discount rate of 5% was considered from information available regarding the last 30 years.

3.2 Deterministic Approach

Due to the convenient and simple implementation of the deterministic approach, this can be useful to understand more details from the LCC obtained without considering uncertainties (most complex approach). For example, a sensitivity analysis was carried out based on the standard interventions plan to find out the components with highest influence in the total life-cycle cost. From Fig. 2 it is possible to group the components in three well identified classes of importance. The first group is comprised by the pavements with the highest weights in the total LCC, while the third group is constituted by the edge beam, concrete barrier, and elastomeric bearings with the lowest weights. The remaining components belong to the second group. Furthermore, a similar analysis can be done to measure the influence of each cost category for the global LCC, i.e., for the present case, the construction costs corresponds to 54% of overall LCC, and the intervention and demolition costs represents 38 and 8%, respectively.

In order to determine the life-cycle costs, the future values associated with interventions and demolition need be discounted for the time corresponding to the construction year (i.e., $t = 0$). Since demolition costs are expected to occur at the end of the service life (100 years), the NPV obtained is € 1918.

Regarding the intervention costs, as it occurs in certain periods over time, there is the need to discount the respective value of each year. From the analysis of Fig. 3, it can be stated that after a certain period (~50 years), the effects produced in the NPV due to discount rate demonstrate a low influence on the LCC. In reality, the life-cycle cost analysis for 50 years seems to be a good choice. However, in many cases, these analyses are made for short terms (~20–30 years) corresponding to the entitlement periods assigned to management agencies. For the present case, the total cost of interventions is € 1,163,976 (sum of all yearly costs during life span of 100 years), and the corresponding NPV is € 183,874 (transposing all yearly costs to present).

Finally, the life-cycle cost corresponds to the sum of all previous NPVs corresponding to construction, interventions and demolition costs. For this case study, LCC is € 1,834,143.

When this approach is used, one option to account for the uncertainties associated to the whole life-cycle process can be roughly considered applying a coefficient of variation to the obtained total cost. For example, considering a variation of 20%, the NPV would range between € 1,467,314 and € 2,200,972 (Table 5). This value is taken herein to serve as a benchmark for semi-probabilistic analysis.

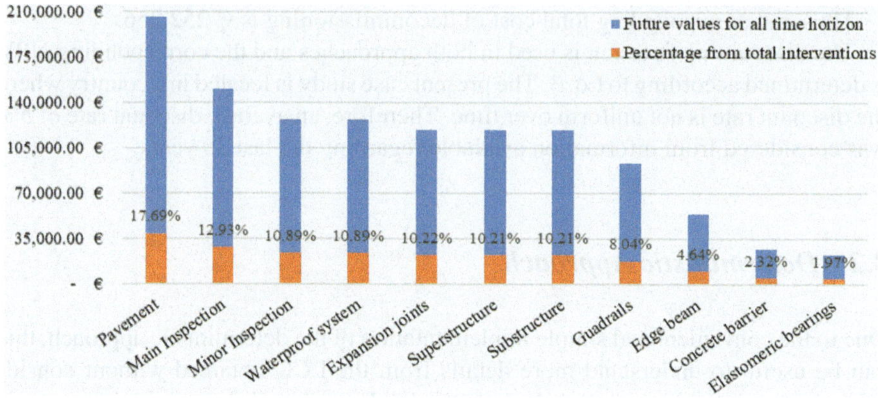

Fig. 2 Financial needs of bridge components from the standard interventions plan

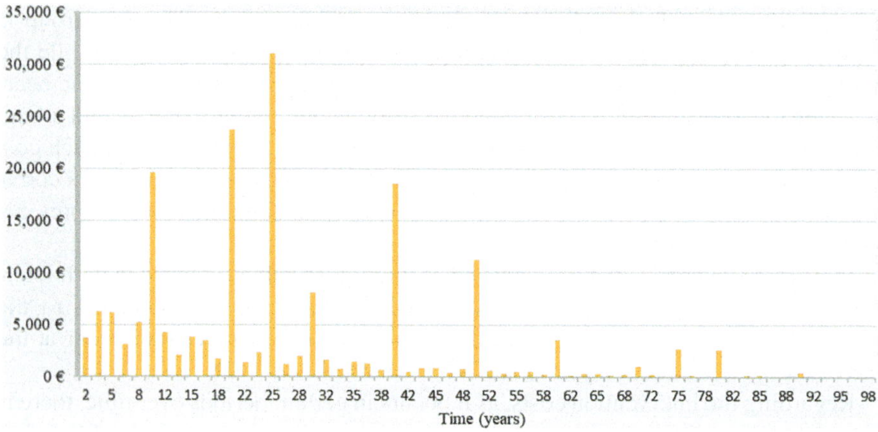

Fig. 3 Net Present value for each year from the standard interventions plan

Table 5 NPV of each cost category

	Net present values (costs)	
	Type	Value
Construction	Deterministic	€ 1,648,351
Interventions	Probabilistic	€ 183,874
Demolition	Deterministic	€ 1918
$NPV_{det}^{20\%} = \{1,467,314; 2,200,972\}$ €		

3.3 Semi-Probabilistic Approach

The present approach considers uncertainties associated with the times at which interventions might occur in the future, i.e., it affects only the intervention plans, being the costs established in a deterministic way.

For this purpose, different scenarios (100k samples) based on the intervals indicated in Table 3 were performed through Monte Carlo simulations. The generated scenarios correspond to random intervention plans for which Net Present Values are determined. In this way, 100k values of NPVs were obtained and their histogram can be seen in Fig. 4a. For a confidence level of 90%, the lower bound corresponds to € 147,230, and the upper bound to € 161,350. The data was fitted to a normal distribution as displayed in Fig. 4b. The mean value of the normal distribution corresponds to € 154,245, with a standard deviation of € 4348, and a coefficient of variation of 3% (which compares to the 20% global variation considered in Sect. 3.2).

According to Table 6, the expected life-cycle cost ranges from € 1,797,499 until € 1,811,799.

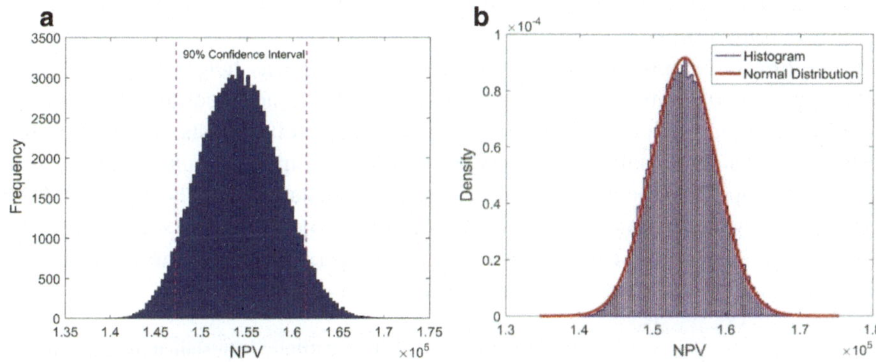

Fig. 4 Monte Carlo simulation: **a** Histogram; **b** Fit normal distribution to data

Table 6 NPV of each cost category (LB = lower bound; UB = upper bound)

	Net present values (costs)		
	Type	Value	
Construction	Deterministic	€ 1,648,351	
Interventions	Probabilistic	LB: € 147,230	UP: € 161,530
Demolition	Deterministic	€ 1918	
$NPV_{prob}^{3\%} = \{1,797,499; 1,811,799\}$ €			

4 Conclusions and Future Developments

Two different methodologies were implemented in the present work to determine the life-cycle costs emerged throughout the service life of an viaduct. The first methodology referred as the deterministic approach is more practical and simple but does not consider uncertainties (or they are roughly and globally estimated at the end), which can lead to possible inaccurate decision-making. For instance, a 20% coefficient of variation was considered to grossly account for all uncertainties of the life-cycle. However it conducted to a large NPV interval which is unrealistic and hinders the comparison between alternative solutions.

On the other hand, the second methodology indicated as the semi-probabilistic approach, takes into consideration the uncertainties for intervention times, which conducts to a coefficient of variation of lowest order of magnitude, i.e., 3%. Nevertheless, this latter percentage cannot be directly compared with the roughly estimation of 20% given that other type of uncertainties, i.e., costs, discount rate, discount function modeling, among others, were not taken into account.

It should be highlighted that in both approaches, the final LCC is defined within an interval since absolute values are clearly unrealistic when performing this type of analysis at long term with many uncertainties involved.

For future developments in order to improve what was herein performed, the following implementations should be addressed: (*i*) Extend uncertainties spectrum for all parameters including intervention costs, construction costs, and demolition costs; (*ii*) Consider variability of costs associated with inflation and devaluation to evaluate its impact on the LCCA; and (*iii*) address stochastic models to predict more accurate intervention times and to overcome the subjectivity associated with the expertise and know-how from infrastructure managers. This last improvement is a more realistic approach since all structures deteriorate over time.

Acknowledgments The authors would like to thank ISISE—Institute for Sustainability and Innovation in Structural Engineering (PEst-C/ECI/UI4029/2011 FCOM-01- 0124-FEDER-022681). This work was co-financed by the Interreg Atlantic Area Programme through the European Regional Development Fund under SIRMA project (GrantNo. EAPA_826/2018).

References

1. Hajdin, R., Kustar, M., Masovic, S., Linneberg, P., Amado, J., & Tanasic, N. (2018). *WG3 Technical Report—Establishment of a Quality Control Plan*. ISBN: 978-86-7518-200-9.
2. *fib* Task Group 3.7, Integrated life cycle assessment of concrete structures. *State-of-Art Report—Bulletin, 71*.
3. Salokangas, L. (2013). *Bridge life cycle optimisation—Stage 3*. ETSI Project.
4. Matos, J., Solgaard, A., Santos, C., Sánchez-Silva, M., Linneberg, P., Strauss, A., et al. (2018). Life cycle cost as a tool for decision making on concrete infrastructures. *High Tech Concrete: Where Technology and Engineering Meet*. https://doi.org/10.1007/978-3-319-59471-2_210.

5. Stipanovic, I., Chatzi, E., Limongelli, M., Gavin, K., Bukhsh, Z., Palic, S., Xenidis, Y., Imam, B., Anzlin, A., Zanini, M., Klanker, G., Hoj, N., & Ademovic, N. (2017). *WG2 Technical Report—Performance Goals for Roadway Bridges of COST ACTION TU 1406*. ISBN: 978-3-900932-41-1.
6. Matos, J., Solgaard, A., Linneberg, P., & Strauss, A. (2018). *Life cycle cost management of concrete structures*. In: *IABSE Conference—Engineering the Past, to Meet the Needs of the Future*.
7. Sánchez-Silva, M., & Klutke, G.-A. (2016). *Reliability and life-cycle analysis of deteriorating systems*. Berlin: Springer.
8. Gervásio, H. (2010). *Sustainable design and integral life-cycle analysis of bridges*. Ph.D. Thesis.
9. Orcesi, A., Cremona, C., & Ta, B. (2018). Optimization of design and life-cycle management for steel-concrete composite bridges. *Structural Engineering International, 28*(2), 185–195. https://doi.org/10.1080/10168664.2018.1453763.

Location Dependency on Resilience and Material Intensity of an Office Building Keeping an Eye on Seismic Zone Implications

Regine Ortlepp and Mahar A. Gul

Abstract To describe the urban metabolism of the built environment, bottom-up approaches are often used in literature. In these approaches building typologies play an important role, where the material intensity of building types is described by material composition indicators. These indicators are mostly static and refer to the construction method of the respective type. However, the influence of the geological conditions of the location on the material requirement has so far hardly been examined, if at all. Many of the fast-growing cities, especially in developing countries, which generate a high demand for materials, are located in areas that are regularly hit by earthquakes. This is accompanied by different static requirements for one and the same building type if it is located in different seismic zones. Using FEM, for a selected office building it was investigated which role plays the location of this building with regard to its material consumption, i.e. how the static requirements from the seismic load in different zones affect the demand for mass relevant building materials of the supporting structure. For this purpose, a 3D building model was created and the load-bearing components were dimensioned under the load combinations typical for the seismic zones. The modeling results show a clear dependence of the building material requirement on the seismic zone in which the building type is located. However, static building type-specific approaches with constant material composition indicators do not reflect this effect. In conclusion, it should be noted that seismic zone dependence should be given greater consideration in the modeling of material composition indicators in the future.

Keywords Buildings · Material consumption · FEM · Seismic simulation

R. Ortlepp (✉)
Leibniz Institute of Ecological Urban and Regional Development, Dresden, Germany
e-mail: r.ortlepp@ioer.de

M. A. Gul
GOLDBECK GmbH, Düsseldorf, Germany

© The Author(s), under exclusive license to Springer Nature Switzerland AG 2021

J. C. Matos et al. (eds.), *18th International Probabilistic Workshop*, Lecture Notes in Civil Engineering 153, https://doi.org/10.1007/978-3-030-73616-3_37

1 Introduction

The construction industry is one of the world's largest consumers of resources within national economies (e.g. [1, 2]). Consumption is particularly high in countries with high urbanization pressure, i.e. where the population is growing rapidly and many people move to cities. There, large quantities of building materials are needed for the construction of new buildings that offer people safe and livable housing.

At the same time, global resources and raw materials for building materials are becoming increasingly scarce. Every year, large quantities of sand and gravel are used to produce concrete. These raw materials are extracted faster than they can be replaced [3]. This leads us to an increasing realization that resource-efficient alternatives in the building industry must be researched in order to counteract this exploitation and at the same time make socially acceptable living possible. Therefore, better knowledge about material consumption in the built environment is required (e.g. [4]), in order to ultimately derive possibilities for reducing consumption, e.g. [5].

In recent years, numerous studies have been conducted on this topic, whereby material flow analysis is a well suited instrument, e.g. [2, 6]. In recent years, numerous studies have been conducted, with a focus on coefficient based bottom-up approaches [7]. These approaches calculate material stocks and material flows on the basis of building-related material composition indicators. Most of the studies have dealt with the domestic housing sector, although the share of the non-domestic sector is much smaller. An overview can be found, for example, in [7]. In principle, it is also possible with these models to investigate the effects of material substitution when determining quantities [5].

Despite the large number of studies available, some fundamental problems can be identified: The models often use coefficients that are too generalized in the sense that they do not adequately reflect the static boundary and general conditions at the location of a building. Integrated approaches for a nexus between environmental impact assessments and more technically oriented questions, taking into account the technical requirements for building materials and buildings, have hardly been found so far [8].

The construction of growing cities in countries such as Pakistan differs from Europe, especially due to the earthquake situation. Today, almost all commercial buildings in Pakistan consist of reinforced concrete frame structures (RC). They are the most common type of modern buildings there. A distinction is made between two types of frame structures: (1) rigid frame structure and (2) braced frame structure. In the former case, the structure, which consists of beams and columns, is monolithically produced on site and thus jointly resists the moments created by the applied loads. In the second type, the bracing system is considered to be more efficient than the rigid frame system because bracing increases the resistance of the structure to lateral and lateral forces.

Reinforced concrete buildings are most often erected in the urban areas of Pakistan. Especially in the cities of Peshawar, Islamabad, Lahore, Faisalabad and

Karachi, the construction of reinforced concrete buildings is on the rise. This increase in the construction of reinforced concrete structures is mainly due to the better economic conditions in these cities, the higher population and the high cost of land, as well as the awareness that in many respects they are a better alternative to traditional unreinforced brick structures, especially in terms of seismic penetration, as the region is prone to severe earthquakes.

In the large cities, such as Islamabad and Lahore and others, about 10–15% of the buildings are constructed in reinforced concrete, with commercial and public buildings such as squares, hotels, offices, hospitals, educational institutions etc. accounting for the majority of this percentage. Reinforced concrete structures in these cities are generally low to medium height (3–8 floors), but in exceptional cases, reinforced concrete buildings with a height of 10–15 floors are also built today.

This paper forms a pre-study of the influence of the seismic zone in which the building is located on material consumption.

2 Method

2.1 General

With the help of FEM, the role of the location of a selected office building with regard to its material consumption was investigated, i.e. how the static requirements from the earthquake load in different zones affect the demand for mass-relevant building materials of the supporting structure. This will be analysed using the example of a commercial building, which will be placed virtually in two different Pakistani cities. In a second step, the standard wall material clay brick was replaced by lightweight concrete blocks in order to investigate the effect of material substitution. For this purpose, a 3D building model was created and the load-bearing structures were dimensioned under the load combinations typical for earthquake zones. The basic assumption of the investigations is that the resilience of the building must be guaranteed in all cases. Under this boundary condition it is investigated to what extent the material intensities of the building change under the above mentioned boundary conditions.

2.2 Building Design and Parameter Study

The office building was planned at two different locations in Pakistan—located in Islamabad and Lahore respectively. The office building has a total covered area of 627 m^2. The building has three floors and the height of one floor is 2.85 m, except for the foundation, which is 1 m deep (Fig. 1). The area of each office is determined individually: All walls are adobe brick, with exceptions of the walls at the front of

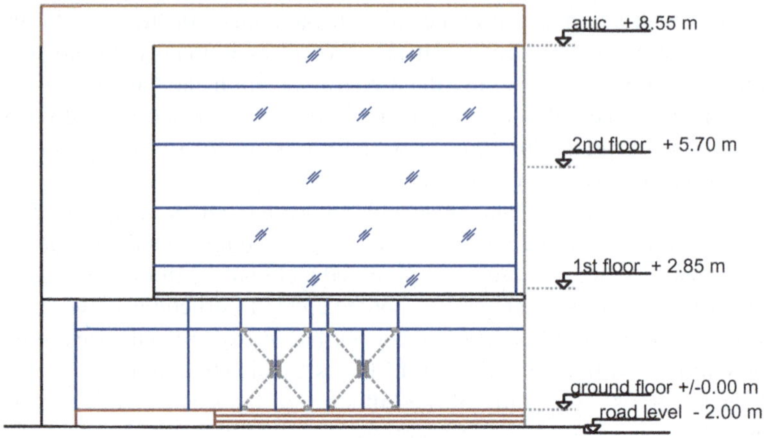

Fig. 1 Front view of the commercial building. *Source* M. A. Gul

the building, which are made of glass. The thickness of the interior walls is 11.25 cm, while the exterior walls are 22.5 cm thick. All doors are made of plywood, except the main door which is made of glass. All windows are made of glass with aluminum frames. The building has three staircases and a hydraulic elevator. A staircase is provided from the emergency exit. The shear walls are made of reinforced concrete with a thickness of 25 cm.

The two locations Islamabad, the capital of Pakistan, and Lahore differ both in the seismic zone and in the safe bearing capacity of soil. For a separate consideration of the effects, the investigation direction "location" was split into a parameter axis "seismic zone" and a branching parameter axis "soil bearing capacity" (Fig. 2). At first the design was carried out considering a location in Islamabad, which is located in the seismic zone 2B (reference). Afterwards the building was moved virtually—i.e. at first while retaining the soil properties—to Lahore in the seismic zone 2A. The actual soil properties were then taken into account (Lahore**). The design of the foundations differs between the building sites, as the safe bearing capacity (SBC) of soil in Lahore is significantly lower than in Islamabad. The design was carried out with Pakistani building regulations, which are based on the regulations of the ACI (American Concrete Institute). The infill walls between the reinforced concrete frame structure consist of traditional brick masonry on the horizontal location parameter axis.

In Islamabad, the trend to use concrete masonry is increasing because it is a cheaper and faster solution compared to brick masonry. For this reason, a comparison study was carried out on the vertical parameter axis (Fig. 2) for the Islamabad building, in which the brick masonry was replaced by lightweight concrete masonry (Islamabad*).

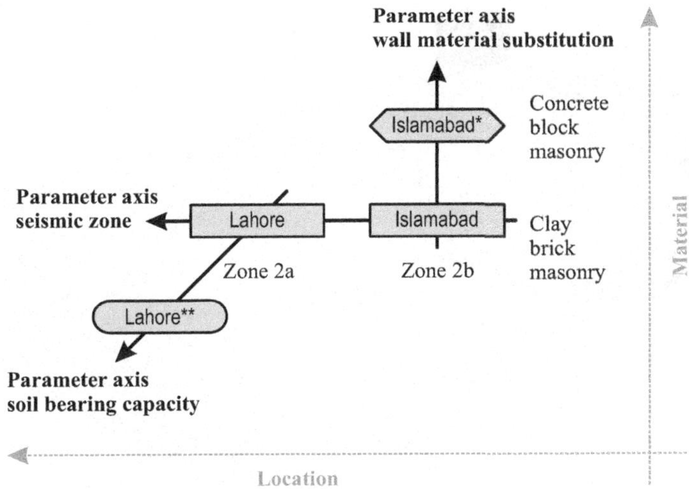

Fig. 2 Overview of the parameter analysis regarding building's location and material substitution

2.3 *FEM Modeling and Material Quantity Calculation*

The load-bearing components of the office building were modelled on the above-mentioned parameter axes in an FEM software (Fig. 3) and all other building materials were simulated as loads. Etabs, an integrated construction software used for the 3D analysis of building systems, was used. It allows the structural design of vertical elements such as columns, foundations and walls based on the Finite Element Method (FEM). After modeling, the elements are assigned with their properties. Then superimposed dead and live loads are applied. Etabs considers the dead load of the members. The design with different codes and seismic zones can be carried out using Etabs. The design of horizontal bars such as RC flat plate and foundation base was carried out in SAFE, which also works on the basis of FEM.

In contrast to an earlier first analysis of Ortlepp [5] on a building in India, where only the load bearing reinforced concrete structure (foundation, columns, beams, slabs, shear walls) and the non-load bearing masonry infill was considered, the materials were fully analyzed in this study. This includes stairs, glass facades, windows, doors, mumty, floor and roof structures including foundation, mortar, tiles, wall plaster, skirtings etc. (hereinafter referred to as finishing). Thus, a much greater variety of materials was covered.

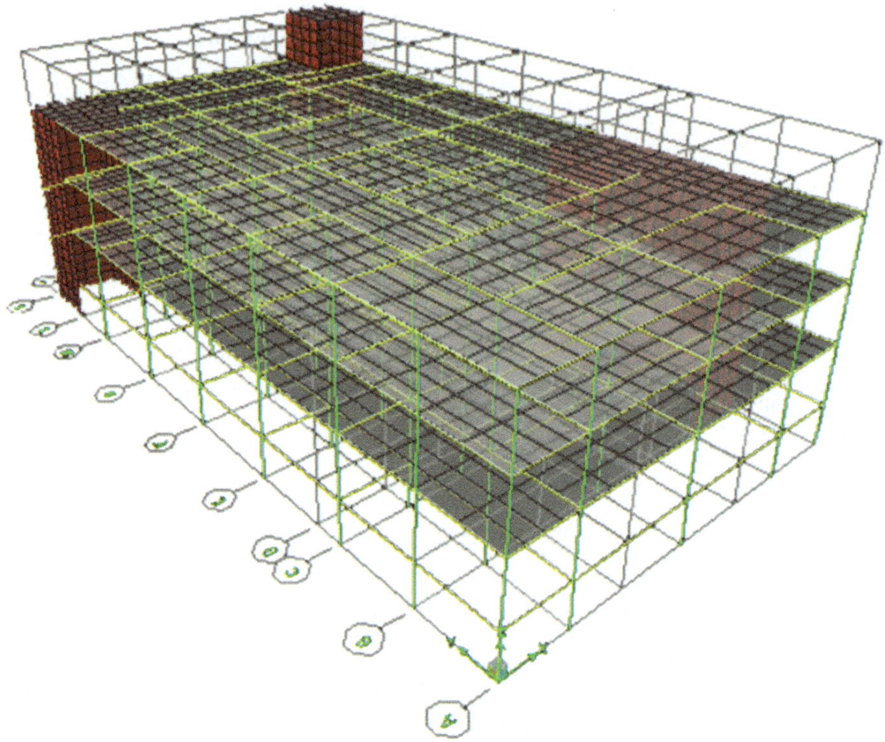

Fig. 3 Etabs FEM model. *Source* M. A. Gul

3 Building Design

3.1 Applicable Building Standards and Boundary Conditions

Building design in Pakistan is usually performed using the Building Code of Pakistan (BCP) [9] complemented by ACI 318 Building Code [10]. Minimum reinforcement 0.01 A_c acc. to ACI Code [10], 10.9.1 is to be used in tied or spirally reinforced columns. Maximum reinforcement ratio is limited to 8% for columns in general to avoid honeycombing of concrete. According to ACI Code [10], 7.10.5.1 longitudinal reinforcement must have min. 32 mm and transverse reinforcement min. 8–10 mm diameter. Minimum reinforcement for Grade 60 steel is defined as $A_{s,min} = 0.00333 A_c$ according to ACI Code [10], 10.5.1. The maximum allowable deflection for structural members is *l*/360 for members loaded with live load and *l*/240 for members loaded with live load plus dead load according to BCP [9], Table 5.4.

UBC 97 code [11] is still used in Pakistan for Seismic Design, even though it was already replaced by the 2006 IBC [12]. Seismic zones are assigned according to the BCP [9], Table 2.2 indicating that Lahore is located in seismic zone 2A and

Islamabad (Capital of Pakistan) in seismic zone 2B with an associated seismic zone peak horizontal ground acceleration of 0.08–0.16 g (2A) and 0.16–0.24 g (2B). The associated seismic zone factors are 0.15 for zone 2A and 0.2 for zone 2B. According to UBC97 [11], Table 16 N the over strength factor is set to 6.5 and soil profile type is SC. UBC97 [11], Table 16 K defines further the importance factor to 1. The soil bearing capacity (SBC) of Lahore is 80 kN/m^2 and the SBC of Islamabad is 134 kN/m^2. Therefore a raft foundation is necessary for construction in Lahore, while isolated foundation is adequate for construction in Islamabad.

3.2 *Design Optimization*

Structural design on Etabs software starts with approximation. At first the sizes of members were assumed which usually have a large cross section. The initial design was performed with all column having a size of 30 × 30 cm. Based on this assumption reinforcement results in columns equal to the minimum reinforcement according to the ACI codes. The sizes of column were then reduced to 22.5 × 22.5 cm. The reason of 22.5 cm is to keep the column size same as the wall thickness. The design was satisfactory on these sizes except from four columns on the front. Those four columns sizes were increased to 22.5 × 60 cm. Initially the beam sizes were 30 × 30 cm that were also reduced to 22.5 × 30 cm. A huge quantity of concrete has been saved by optimization and, additionally the columns are not disturbing the architect much compared to maximum columns in the wall.

3.3 *Design Input Values*

3.3.1 Material Properties

The main material used is concrete with a weight per unit volume $W = 23.5616$ kN/m^3. Two different concrete qualities are used in the load-bearing components CONC3 with a concrete block strength of $f_c = 20684$ kN/m^2 for floor slabs and CONC4 with $f_c = 27579$ kN/m^2 for columns, beams and foundation. Lean concrete is also used under the foundations and the floor slab. The reinforcing steel Grade 60 steel has a strength of $f_y = 413685.47$ kN/m^2.

3.3.2 Geometries

In the following the geometries for the building with clay brick masonry are given and, if different followed by the deviating geometry in the case of the Islamabad building with lightweight concrete block masonry on 2nd parameter axis (cf. Fig. 2) in square brackets. Cross sections of columns are set to 22.5 × 22.5 cm [20 ×

20 cm] and 22.5 × 60 cm and cross sections of beams are 22.5 × 30 cm [20 × 30 cm]. The thickness of shear walls is 25 cm and 15 cm for floor and roof slabs. The foundation area is 483 m^2 [408.2 m^2] and the respective thickness of foundations 325 mm [250 mm]. Interior masonry walls are 11.25 cm [10 cm] thick, exterior walls 22.5 cm [20 cm]. The density of clay brick masonry is 19.6 kN/m^3, while lightweight concrete block masonry with a density of 9.8 kN/m^3 is only about half as heavy.

3.3.3 Loads

Live and superimposed loads act on the building. The following line loads were assumed: interior walls 4.56 KN/m, exterior walls 10.26 kN/m and glass walls 1.425 kN/m. An hydraulic elevator with weight 6.79 kN has been selected for the building therefore a point load of 4 × 1.69 kN was taken into account. Area live loads on floors were set to office 2.5 kN/m^2, store 4.8 kN/m^2, lunch room 4.8 kN/m^2 and prayer hall 4.8 kN/m^2, supplemented by superimposed dead loads of 0.225 kN/m^2 plus moveable partition load of 1 kN/m^2. In Pakistan, the roof is usually accessible for all purposes which imposes a live load of 2.4 kN/m^2 supplemented by a superimposed dead load of the roof slab s of 2.165 kN/m^2. When the earthquake load is considered then the wind load is not considered in Pakistan as per General Practice therefore only earthquake load (see Sect. 3.1) has been considered.

4 Results

4.1 Material Quantities of Reference Building in Islamabad

The result of the mass determination differentiated by components on the basis of the building design and the FEM simulation for the reference building in Islamabad in the seismic zone 2B with clay brick masonry infill is shown in Table 1. 2798 t of the total building mass consists of 2118 t of material in the shell (thereof 1652 t load bearing structure and 445 t masonry infill) and 680 t finishing. Of the reinforced concrete load bearing structure, about one third of the building material (506 t = 487 t concrete + 17.4 t steel) is found underground. The reinforced concrete superstructure (above ground) therefore weighs 1146 t (= 1112 t concrete + 33.5 t steel), which is about twice as much as the foundation.

4.2 Influence of Seismic Zone—Building Shifted to Lahore

Contrary to the original assumption that the quantities of building materials would have to be reduced if the building was planned in seismic zone 2A instead of seismic

Table 1 Comparison of material quantities

Material	Construction member	Volume (m^3)	Mass (t)
Load bearing construction members			
	Foundation		
P.C.C	Layer under the foundation	48.29	113.47
R.C.C	Column foundation	159.64	375.15
Steel	Column footing reinforcement	2.21	17.40
	Superstructure		
P.C.C	Layer under ground floor	47.00	110.45
R.C.C	Columns	22.35	52.52
R.C.C	Beams	75.02	176.30
R.C.C	Slabs	212.79	500.06
R.C.C	Stairs	10.71	25.17
R.C.C	Shear walls	105.37	247.62
Steel	Columns' reinforcement	0.12	0.94
Steel	Beams' reinforcement	1.89	14.60
Steel	Slabs' reinforcement	1.30	9.54
Steel	Stairs' reinforcement	0.52	4.06
Steel	Shear walls' reinforcement	0.55	4.31
Non-load bearing construction members			
Clay bricks	Masonry walls	226.93	444.78
Finishing			
Cement mortar	Ground floor and walls	100.00	212.00
Gravel	Ground floor	105.75	169.20
Sand	Ground floor	11.75	17.63
Glass	Façade and windows	4.34	10.85
Aluminum	Façade and windows	0.07	0.18
Plywood	Doors	5.47	2.46
Ceramic tiles	Wall. stairs and floors	10702 no	138.64

(continued)

Table 1 (continued)

Material	Construction member	Volume (m^3)	Mass (t)
Red Clay Tiles	Roof	24300 no	30.61
Bitumen	Roof	615 m^2	1.85
Earth	Roof	61.50	96.56
Total mass			*2798.44*

zone 2B, the parameter study did not show any major effects here. There was only a 2% reduction in the amount of reinforcing steel in the superstructure to 32.9 t. It is possible that greater effects will only become apparent with larger earthquake zones or even with taller, slim buildings with a significantly larger number of floors, as preliminary investigations by Ortlepp [5] suggest. The building examined here is quite compact with 3 floors.

4.3 Influence of Soil Conditions—Building in Lahore of Lower Soil Bearing Capacity

The soil bearing capacity of Lahore of 80 kN/m^2 instead of 134 kN/m^2 on which we had previously based our calculations requires a raft foundation for the building, which is much more material intensive than a column foundation. The change in the foundation construction method leads to a large jump in material intensity by a factor of 1.5. 50% more concrete (738 t) is required for the raft foundation than for the column foundation (Fig. 4). The influence of the soil conditions is not noticeable in the superstructure above ground; the material eaters sit virtually hidden under the ground. This increases the total material intensity of the building by 9% to 3025 t.

This means now for building type specific material composition indicators that they can be assumed to be constant for the superstructure, whereas for the underground part a dependence on the subsoil conditions must be considered.

4.4 Effectiveness of Wall Material Substitution—Case Study of Building in Islamabad

The substitution of the traditional clay brick masonry infill with lightweight concrete block masonry shows the greatest effect in terms of the material composition of the otherwise identical building. Instead of 445 t of brick masonry, only 202 t of lightweight concrete block masonry are used. Due to this weight reduction in masonry, secondary effects of material savings in the reinforced concrete supporting structure can be observed. Here, too, the greatest effects can be seen in the foundation. Compared to the reference building, the concrete requirement is reduced by 30% to

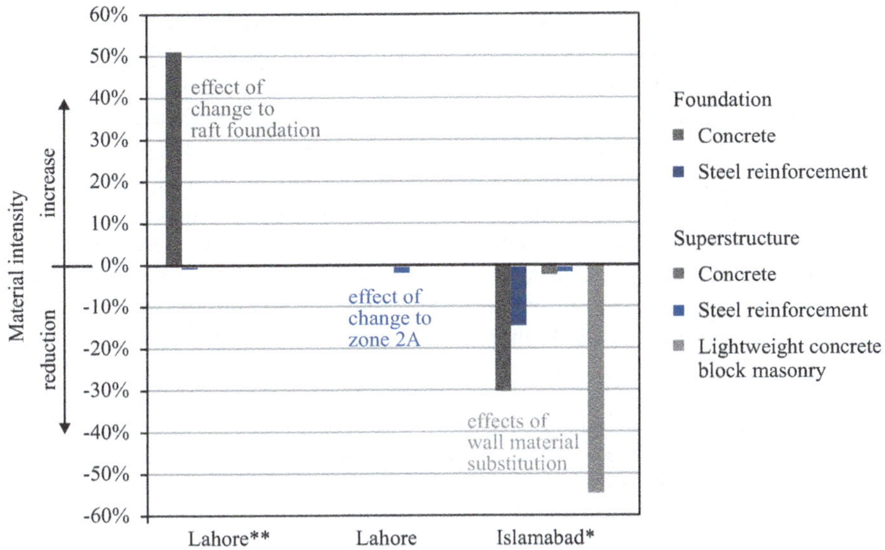

Fig. 4 Changes in material intensity related to the reference building in Islamabad (cf. Fig. 2)

341 t and the steel requirement by 14% to 15 t. The effects in the superstructure are significantly lower with 1% savings in steel to 33 t and 2% savings in concrete to 1086 t. Including finishing, which has not been changed, the total material saving is 15% compared to the reference building with traditional clay brick masonry infill.

For the determination of material composition indicators this result means that when the masonry material for a building type is changed, not only the component "wall material" can be replaced, but also the values of the material masses for concrete and steel in the foundation must be adjusted.

5 Conclusions

The parameter study provides initial findings with regard to the question of what must be taken into account when transferring building-type-specific material composition indicators when simulating the same building at different locations and when substituting individual material components. An effect of the earthquake zone on the material composition of a building under static boundary conditions could hardly be determined in the test study conducted here. On the other hand, a significant effect on the influence of the soil bearing capacity was found, almost exclusively in the foundation, while the superstructure was not affected. There is a quantitative leap in the transition from column foundation to raft foundation. In some cases, two simplified variants could therefore be determined for the foundation, if a suitable limit value can be found from which the construction method of the foundation must necessarily change. However, further investigations are necessary, since there are probably other parameter dependencies, for example on the height of the building, which can influence the threshold value of the soil bearing capacity

A further conclusion can be drawn for the handling of Material Composition Indicators, if certain material substitutions are to be simulated. The adjustment of the indicators must then extend to the substituted material and the steel support structure. Not only may the value for one material be substituted. At least the material masses for concrete and steel in the foundation are affected by the change. The influence on the reinforced concrete superstructure was not significant in the present test study. However, further investigations are necessary, since the conditions can change with other building geometries and higher seismic loads, as another test study (e.g. [5]) suggests.

Acknowledgements This pre-study was performed at the Leibniz Institute of Ecological Urban and Regional Development in Dresden, Germany. The authors would like to thank Prof. em. Wolfram Jäger from the Technische Universität Dresden for his support in the context of the joint supervision of M.Sc. Mahar Arslan Gul.

References

1. Baccini, P., & Brunner, P. H. (1991). *Metabolism of the anthroposphere*. Berlin/Heidelberg: Springer. ISBN: 9783540537786.
2. Baccini, P., & Brunner, P. H. (2012). *Metabolism of the anthroposphere: Analysis, evaluation, design* (2nd ed.). Cambridge: MIT Press. ISBN: 9780262016650.
3. Bendixen, M., Best, J., Hackney, C., & Lønsmann Iversen, L. (2019). Time is running out for sand. *Nature, 571*, 29–31. https://www.nature.com/articles/d41586-019-02042-4.
4. Tanikawa, H., Fishman, T., Okuoka, K., & Sugimoto, K. (2015). The weight of society over time and space: A comprehensive account of the construction material stock of Japan, 1945–2010: The construction material stock of Japan. *Journal of Industrial Ecology, 19*, 778–791. https://doi.org/10.1111/jiec.12284.
5. Ortlepp, R. (2019). Effect of sustainable building material substitutes with regard to earthquake safety. In 7th International Conference on Euro Asia Civil Engineering Forum, 30 September to 2 October 2019, Stuttgart, Germany, 2019, (IOP Conference Series: Materials Science and Engineering; 615, 012118), pp. 1–8. https://doi.org/10.1088/1757-899x/615/1/012118.
6. Ortlepp, R., Gruhler, K., & Schiller, G. (2018). Materials in Germany's domestic building stock: Calculation model and uncertainties. *Building Research & Information, 46*(2), 164–178. https://doi.org/10.1080/09613218.2016.1264121.
7. Ortlepp, R., Gruhler, K., & Schiller, G. (2016). Material stocks in Germany's non-domestic buildings: a new quantification method. *Building Research & Information, 44*(8), 840–962. https://doi.org/10.1080/09613218.2016.1112096.
8. Ortlepp, R., Schiller, G., & Sebesvari, Z. (2015). Building material substitutes contra topsoil harvesting—Technical considerations with focus on developing countries. In Smart and Sustainable Built Environment Conference 9–11 December 2015, Pretoria, South Africa, pp. 281–290.
9. Building Code of Pakistan (BCP). (2007). https://www.pec.org.pk/downloadables/buildingCode/Building%20Code%20of%20Pakistan.zip.

10. ACI 318 Building Code. https://www.concrete.org/tools/318buildingcodeportal.aspx.aspx.
11. Uniform Building Code (UBC). (1997). *International Code Council.*
12. International Building Code (IBC). (2006). *International Code Council.*

Long Term Evaluation of the Structural Reliability of an Existing Concrete Prestressed Bridge

Tommaso Donolato, Neryvaldo Pereira, and José C. Matos

Abstract Reliability is an important factor to determine how safe is a structure. The aim of this study is to use the concept of reliability in order to manage the maintenance and to plan the interventions that could be necessary. The first part includes the calibration of the model, verifying the obtained results. The second part provides a 100-samples nonlinear analysis, considering the statistically important random variables. Each sample is generated considering the mean and standard deviation values of each random variable, using the Hypercube Latin method to couple them. The output is the load factor probability distribution. Using an overload probabilistic curve, the reliability index is computed, according to the Monte Carlo method. The third part illustrates the corrosion effect calculation, using FIB Bulletin 34 guidelines. Once determined the corroded area and the corrosion depth during time, the reliability index is computed, using different time values. The trend of reliability index during time is obtained in relationship with variation of the standard deviation and the load factor values.

Keywords Probability · Risk · Reliability · Corrosion

1 Introduction

During these years, several bridges have collapsed causing huge damages to economy, people life and environment. These collapses are generated by different causes, which are summarized by Fig. 1 [1]. A procedure has been established to prevent these risks and to assure the solidity (safety) of bridges' main structures. A useful tool to reach this aim is the reliability index computation.

The reliability of a structure is its ability to fulfil its design purpose for some specified design lifetime. Reliability is often understood to equal the probability

T. Donolato (✉)
University of Padua, Padua, Italy
e-mail: tommaso.donolato@studenti.unipd.it

N. Pereira · J. C. Matos
Universidade do Minho, Guimarães, Portugal

© The Author(s), under exclusive license to Springer Nature Switzerland AG 2021
J. C. Matos et al. (eds.), *18th International Probabilistic Workshop*, Lecture Notes in Civil Engineering 153, https://doi.org/10.1007/978-3-030-73616-3_38

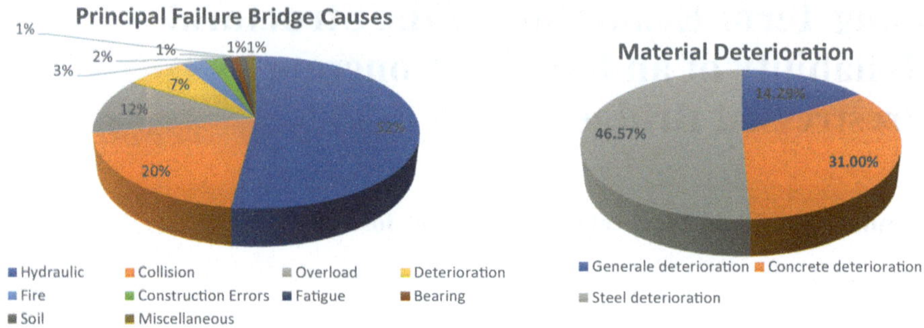

Fig. 1 Bridges failure causes—AASHTO [1]

that a structure will not fail to perform its intended function [2]. Reliability analysis can be applied to evaluate existing structures, assessing their safety and their state, preventing and assuring their correct maintenance. The reliability index can be considered as a rational evaluation criterion. It provides a good basis for decision about the repair, rehabilitation or replacement of the structures. The reliability index provides a methodology to establish the security level. To begin a structural reliability analysis, it's necessary to define a "limit state". The considered limit state in this study is the bending moment limit state, taking into account the middle span cross section, of the biggest span of the studied bridge.

As it can be seen in Fig. 1, deterioration due to corrosion is the one of the most frequent causes of bridge collapsing. The final aim of this study is to compute the corrosion effect, which can affect the reinforcements and can decrease the structural resistance. The corrosion effects are computed using FIB Bulletin 34 guidelines and are reported by the graphs.

During the design and construction phases, all the standards and the requirements are followed, in order to guarantee resistance and durability of the structures. In these phases, the material and geometrical properties can change and be different from the design values [3].

2 Reliability Analysis

The typical performance function, or *limit state function*, can be defined as [2]:

$$g(R, Q) = R - Q \tag{1}$$

The limit state can be designed as the value of the function g, where the limit state is not fulfilled (for example $g(R, Q) < 0$, which corresponds to the structural failure).

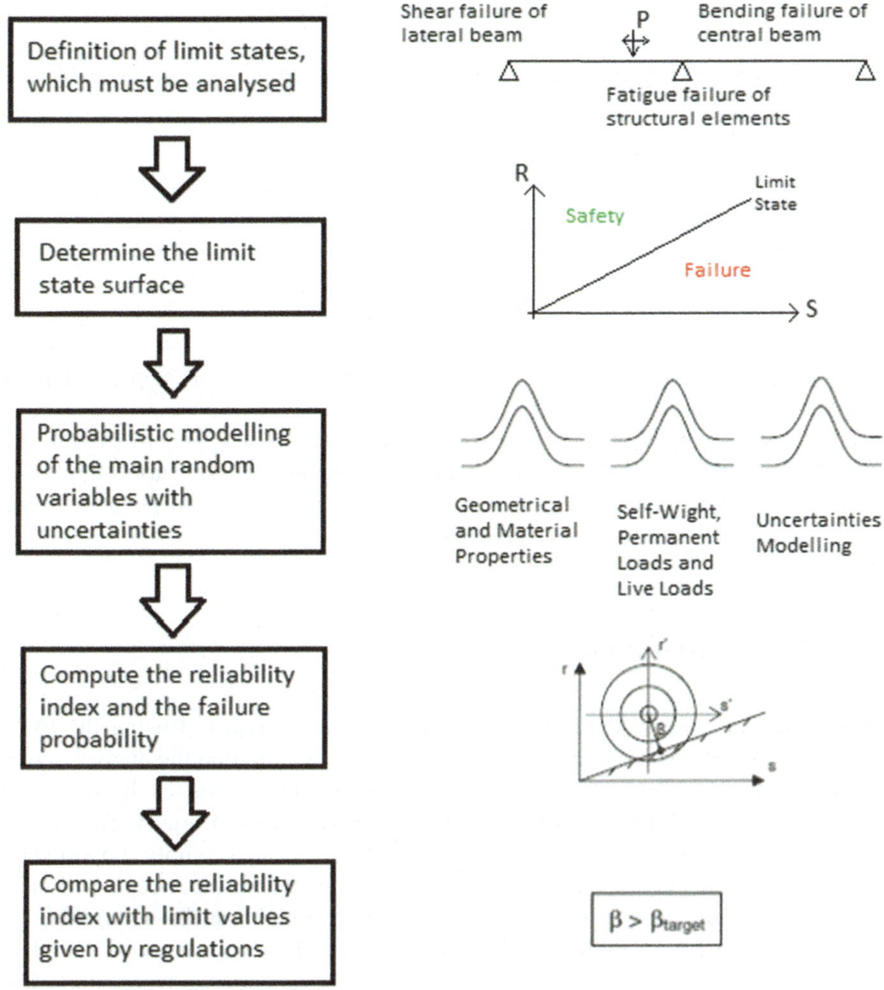

Fig. 2 Reliability analysis flow-charts

This relation can be translated, considering the probability of failure P_f, equal to the probability that the undesired performance will occur:

$$P_f = P(R - Q < 0) = P(g < 0) \quad (2)$$

The reliability index β is the inverse of the coefficient of variation of the function $g(R, Q) = R - Q$. R and Q are two independent random variables. In the case studied, these variables are normally distributed; in this way the reliability index is related to the probability of failure:

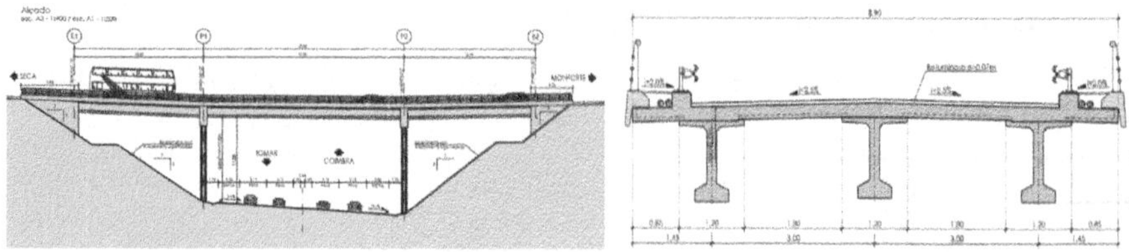

Fig. 3 Studied case geometry—Ascendi [4]

$$\beta = -\phi^{-1}(P_f) \tag{3}$$

In the following chapter the procedure adopted, is described step by step. A summary can be found in the flow-chart in Fig. 3.

The reliability index is computed considering 100 samples, generated with the help of MatLab®. These 100 samples are based on the Monte-Carlo simulation method. The Monte Carlo method is a special technique used to generate some results numerically, without any physical testing.

2.1 Model Calibration

Considering the case studied, a finite element model is realized using the FEM software DIANA®. The model is verified in order to be sure that the analysis gives correct results. The deck is supported by two piles and the extremizes. It overpasses Tomar-Coimbra road, following the directions of Seca and Monforte (Portugal) [4]. The deck has 3 "I-Type" beams (concrete class C45/55) that are simply supported by the piles (concrete class C30/37). The intersections between slab, beams and piles are all monolithic (special particular to guarantee continuity at the node). The "I-Type" beams are 1.5 m height and 3 m wheelbase and the total transversal length is 8.9 m. They are made in factory with prestressed precast concrete technology. The slab thickness is 0.25 m (concrete class C30/37).

The nonlinear analysis makes possible to compute the capacity curve, in relationship with the load combination used to obtain the most disadvantageous situation for bending moment. A phased nonlinear analysis is adopted to reproduce the stresses of construction stages. The first phase is characterized by the beams simply supported and loaded by self-weight and prestress equivalent forces; the second phase is characterized by a continuous, monolithic structure, loaded with the permanent loads and traffic loads [5–7]. A summary is reported in Fig. 4.

The results are performed using Newton-Raphson Modified method. In Fig. 5, the capacity curve is reported: it can be observed the first elastic behavior, for load factor up to 1.81, and the hardening behavior until its collapse. The numerical results are compared with analytical ones, in order to be sure that the numerical model returns results with an acceptable error percentage.

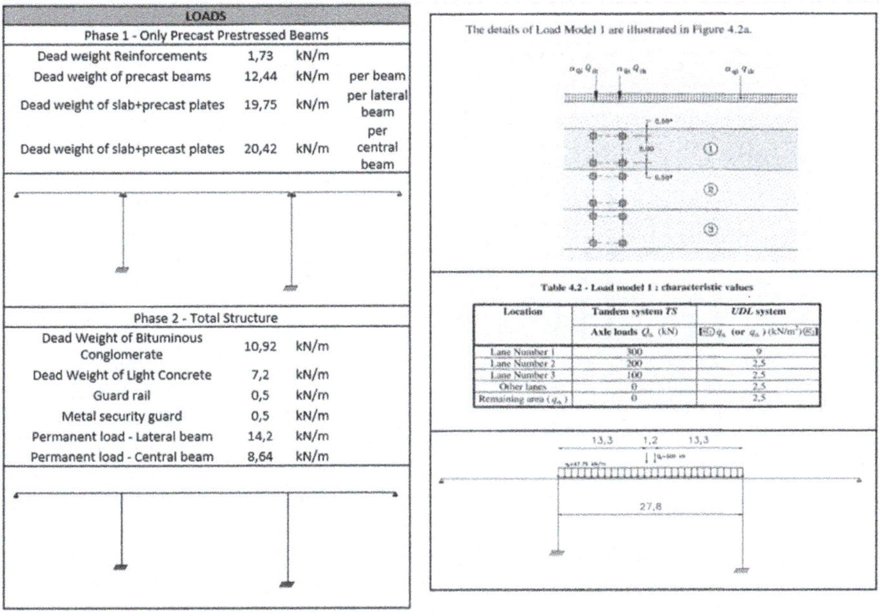

Fig. 4 Load scenarios for phased nonlinear analysis

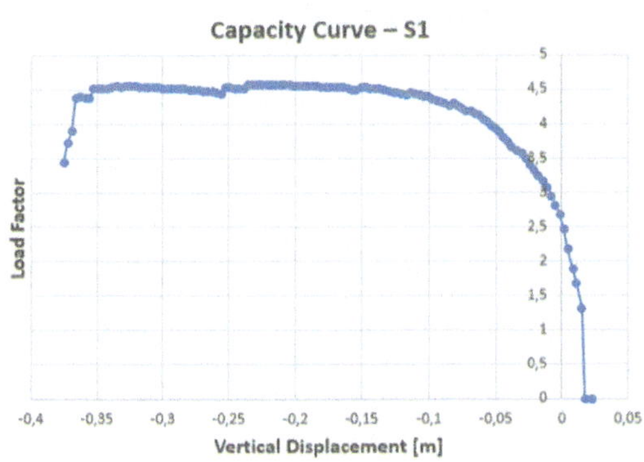

Bending Moment Error	
Numerical [kNm]	17.300
Analytical [kNm]	17.624
Error [%]	1.84

Fig. 5 Model evaluation

2.2 *Probabilistic Analysis*

Once the model is evaluated and verified, the probabilistic analysis can be performed. The first step is to determine the random variables, related to the maximum bending moment calculation (sensitivity analysis). The results, obtained by this analysis, are:

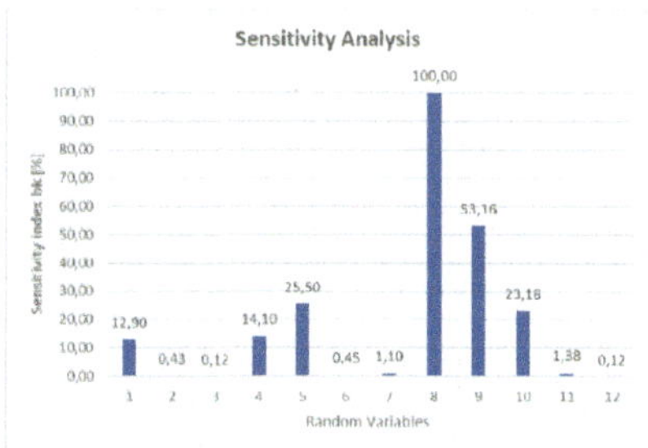

	Material	Random Variable	Name	Average Values	Bias	COV
1	C30/37	Compressive Strength	fcm	38 MPa	1,27	12%
2		Tension Strength	fctm	2 MPa	1,45	20%
3		Young Modulus	Ecm	33000 MPa	1	8%
4		Thickness Slab	e	25cm	1	3,50%
5	C45/55	Compressive Strength	fcm	53 MPa	1,18	9,00%
6		Tension Strength	fctm	2,62 MPa	1,45	20%
7		Young Modulus	Ecm	36000MPa	1	8%
8	A500	Ultimate and yielding stress	fsy fu	560 Mpa	1,12	5,40%
9		Area	A	-	-	2%
10	A1860	Ultimate and yielding stress	fpy fpu	1258MPa	1,04	2,50%
11		Prestress	σp	1087MPa	1	1,50%
12	C30/37 and C45/55	Self-Weight Concrete	γc	25 kN/mc	1,03	8%

Once the random variables are determined, using their mean and standard deviation values [8–10], 100 samples are generated, coupling the variables randomly, considering every value only once. These values are substituted with the respective values of the "mother" finite element model (exported by a data file.dat by the software [11]). This procedure makes possible to have 100 data files, which must be run. A flow chart is reported in Fig. 6 to explain the whole procedure.

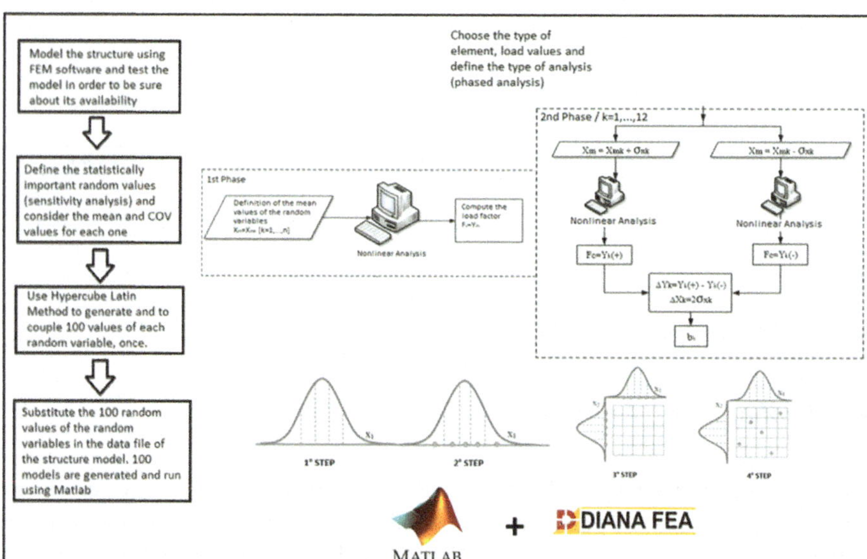

Fig. 6 Probabilistic analysis procedure

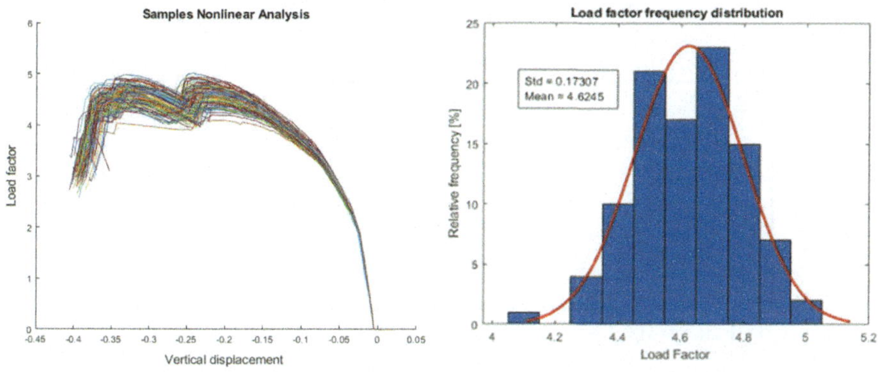

Fig. 7 Probabilistic analysis results

The results of the 100-samples are reported, and they are plotted using a histogram, which considers the relative frequency of occurrence in Fig. 7. The data are fit using a normal distribution in order to use the relation of Nowak-Collins [2].

It can be seen from the load factor-displacements curve the elastic behaviour of the structrure in the first part; then the hardening until the collapse (no convergence) of the model.

The probabilistic load distribution is considered normal too, using a mean value equal to 1 and a standard deviation equal to 0.15 [12, 13].

3 Long Term Effects

The long-term effects are computed considering the corrosion which affects the concrete structures. The corrosion is characterized by two events [14]: the initial phase, ends when the limit state of reinforcement depassivation are reached and the propagation time, phase that is divided in limit states of crack formulation, spalling of concrete cover and collapse through bond failure or reduction of cross section. These periods depend on the exposure classes, which are reported in standards. The typical corrosion process trend is reported in Fig. 8.

3.1 Initiation Time

The initiation phase of the process of carbonation-induced corrosion is marked with carbonation penetration in concrete, and roughly, it finishes with depassivation of reinforcement. Considering an environmental class of exposure equal to XC4, the concrete cover is 65 mm. The needed data to compute the initiation time (4) is determined by literature [14]; in the specific, for the CO_2 concentration different

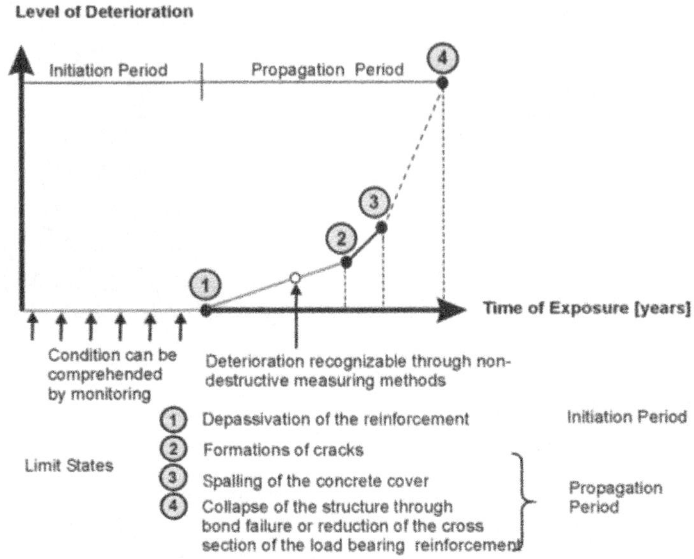

Fig. 8 Corrosion process—FIB Bulletin 34 [14]

experimental studies of Andrade C. are considered. The initial time, taking into account all these hypotheses, is equal to 55.02 years.

$$t_{ini} = \left(\frac{k_{NAC}^2 \cdot k_e \cdot k_c \cdot k_a}{c^2} \cdot t_0^{(p_{dr} \cdot ToW)^{b_w}} \right)^{\frac{1}{(p_{dr} \cdot ToW)^{b_w} - 1}} \tag{4}$$

3.2 *Propagation Time*

There are several mathematical models and empirical formulas which describe the propagation of corrosion. Several studies make possible to predict corrosion depth and residual diameter [15–17].

In the FIB Bulletin 34 there are three different limit states for corrosion:

1) Cracking limit state (SLS)
2) Spalling limit state (SLS)
3) Collapse limit state (ULS).

For each one of these limit states, the corresponding time is computed. The procedure follows the FIB guidelines [7]. The corrosion effects are computed for passive and active reinforcements both. The results are reported in Fig. 9.

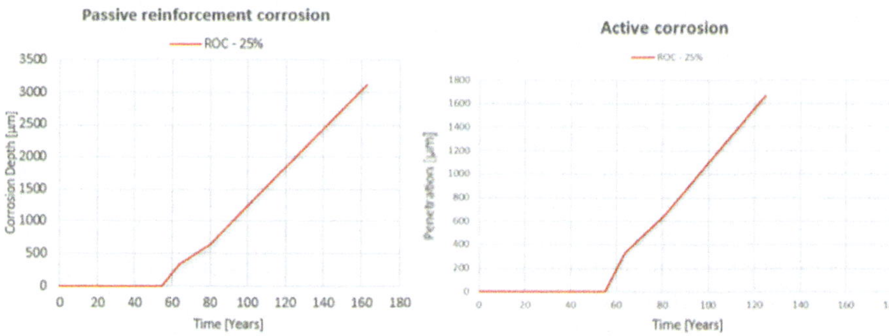

ROC 3 [25%]			
t	P(t)	Φ(t)	ξ
0	0	0	0
55,02	0	0	0
63,36	317	24,37	0,949
80,70	659	23,68	0,899
162,93	3125	18,75	0,563

ROC 3 [25%]			
t	P(t)	Φ(t)	ξ
0	0	0	0
55,02	0	0	0
63,64	328	12,69	0,94
80,46	639	12,07	0,89
124,7	1670	10,01	0,56

Fig. 9 Corrosion effects

The corroded area limit is given by the regulations reported in literature [7]; it depends on the robustness class, which is related to an allowed percentage of corroded area. In this study, the allowed percentage is 5% (ROC3).

In order to perform reliability analysis, considering the probabilistic distribution reported on FIB Bulletin (the propagation time, referred to a specific environment, has a lognormal distribution), Monte-Carlo simulation is run, in order to obtain a lognormal distribution for the propagation time (cracking, spalling and collapse limit states). From the propagation time, taking into account the corrosion relationships between time and corrosion depth, the corroded area lognormal distribution is determined. The results are reported in Table 1.

The corroded area refers to a single reinforcement bar (25 mm for passive reinforcement and 15.2 mm for active reinforcement) and the MatLab® script assign the total corroded area.

Table 1 Propagation time

Propagation time – Passive reinforcement			
Cracking		Spalling	
A(t)	466,111 mmq	A(t)	442,912 mmq
STD	7897 mmq	STD	12,972
Propagation time – Active reinforcement			
Cracking		Spalling	
A(t)	113,541 mmq	A(t)	114,5843 mmq
STD	6652 mmq	STD	6,717,003 mmq

4 Long Term Probabilistic Analysis

The final step of this study is to perform a long-term probabilistic analysis. The procedure is almost the same adopted for the probabilistic analysis in Sect. 2.2. The non-corroded area values are substituted by the corroded ones (with the help of a MatLab® script) and the previous standard deviation values too. Using this new data, a 100-samples analysis is conduced for each value and the output is, as before, the load factor probabilistic distributions. The final output is the trend of the reliability index, depending on the time.

4.1 Long Term Reliability Index

The reliability index is computed considering the relationship [2]:

$$\beta = \frac{\mu_R - \mu_Q}{\sqrt{\sigma_R^2 + \sigma_Q^2}} \tag{5}$$

μ_R is the resistance mean, μ_Q is the stressing loads mean, σ_R is the resistance standard deviation, σ_Q is the stressing loads standard deviation.

The purpose is to define how the reliability index decreases during time. In relation to the reliability index, the standard deviation is plotted too, in order to understand how large the error involved in the reliability analysis is. The reliability index is computed considering load and resistance uncertainties [8].

The obtained results for the reliability index are reported, depending on time, on Fig. 10. In the end, the standard deviation is plotted in relationship with time. The purpose is to observe how reliable the obtained results are. If the standard deviation increases too much, the probabilistic analysis can't be considered, and some experimental campaign must be performed in order to remove as many uncertainties

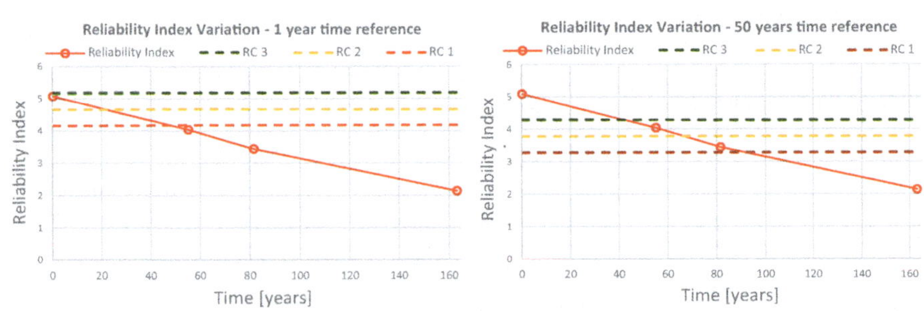

Fig. 10 Reliability index—EC0

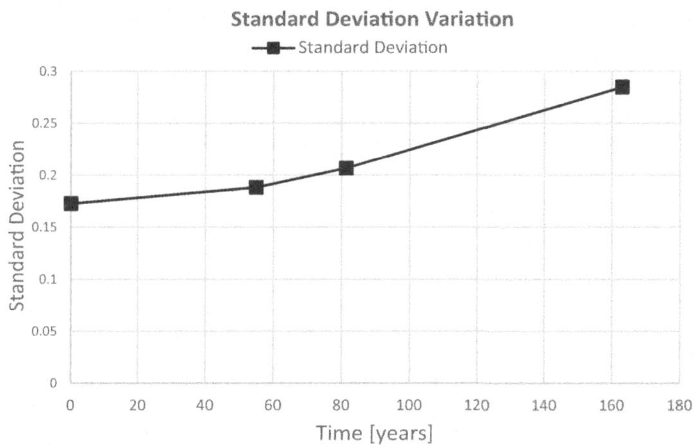

Fig. 11 Standard deviation during time

as possible. The standard deviation values are plotted against time in Fig. 10 in order to determine the availability of the analysis.

As expected, the reliability index trend decreases during time, due to corrosion progress. Using this method, it can be possible to plan every maintenance intervention in order to prevent any collapse events, which would cause relevant damages to people life, economy and environment. Using the values reported in Eurocode [5] the limits of acceptable reliability index are fixed, in dependance of reference time. Shorter is reference time, more rapidly the maintenance interventions are required.

In the end the standard deviation is reported:

The standard deviation increases during time because the uncertainties linked to the process are more significant. Reducing the different uncertainties of the model the results can be improved.

5 Conclusions

As expected, the reliability index decreases in time, but the values reach a critical value only at the end of the structural life. The critic event corresponds to the complete loss of the reinforcements and, consequently, to an important decrease of resistance of the structure. In the graph, the limit values of acceptable reliability index are reported in order to underline the critic states, where the probability of failure assumes a value which is not acceptable. This procedure makes possible to plan the needed maintenance interventions.

The standard deviation increases during time. This is real because in the model and in the introduced characteristics, the uncertainties about the model, the materials, the reliability index calculation are different. In the analysis, these uncertainties are not taken into account. If they are considered, the reliability index might be lower, and the structure might reach critic situation earlier.

Acknowledgements This work was co-financed by the Interreg Atlantic Area Programme, through the European Regional Development Fund, under the project SIRMA (Grant EAPA_826/2018).

References

1. AASHTO. (2010). *The manual for bridge evaluation* (2nd ed.).
2. Nowak, A., & Collins, K. (2000). *Reliability of structures.*
3. FIB Bulletin 59. (2011). *Condition control and assessment of reinforced concrete structures exposed to corrosive environment (carbonation/chlorides).* Lausanne: International Federation for Structural Concrete (FIB).
4. Ascendi. (2011). *Project and technical reports, Internal references.*
5. EN 1990. (2002). *Eurocode 0: Basis of structural design.* CEN—European Committee for Standardization.
6. EN 1991-2. (2003). *Eurocode 1: Actions on structures—Part 2: Traffic loads on bridges.* CEN —European Committee for Standardization.
7. EN 1992-2. (2006). *Eurocode 2: Design of concrete structures—Part 2: Concrete bridges—Design and detailing rules.* CEN—European Committee for Standardization.
8. JCSS—Joint Committee on Structural Safety. (2001a). *Probabilisitc model code part 3: Resistance models—Static properties of reinforcing steel.* JCSS Probabilistic Model Code.
9. JCSS—Joint Committee on Structural Safety. (2001b). *Probabilistic model code—Part 1 basis of design.* Structural Safety.
10. JCSS—Joint Committee on Structural Safety. (2005). *Probabilisitc model code part 3: Resistance Models—Statistic properties of presstressing steel (prestressed concrete).*
11. TNO DIANA. (2016). *USER's manual—Theory.* TNO DIANA Bv, (July).
12. Matos, J.C., Cruz, P.J.S., Valente, I.B., Neves, L.C., & Moreira, V.N. (2016). *An innovative framework for probabilistic-based structural assessment with an application to existing reinforced concrete structures.*
13. Campos e Matos, J. (2013). *Uncertainty evaluation of reinforced concrete and composite structures behaviour.* ResearchGate
14. FIB Bulletin 34. (2006). *Model code for service life design of concrete structures.* Lausanne: International Federation for Structural Concrete (FIB)
15. Alonso, C. A, & Gonzales, J. (1988). *Relation between resistivity and corrosion rate of reinforcements in carbonated mortar made with several cement type.* Cement and Concrete Research.
16. Andrade, C. (2017). Reliability analysis of corrosion onset: Initiation limit state. *Journal of Structural Integrity and Maintenance.*
17. Gonzalez, J., Andrade, C., Alonso, C., & Feliu, S. (1995). *Comparison of rates of general corrosion and maximum pitting penetration on concrete embedded steel reinforcement.* Cement and Concrete Research

Model Updating with Reduced Experimental Data

Pierre Beaurepaire

Abstract Bayesian updating is increasingly used in structural engineering; it is applicable as an inverse method to identify the model of uncertainty which best matches some available experimental data. This paper discusses the application of such methods to models with multiple outputs in case the experimental data is reduced. Standard updating methods involve the covariance matrix, which becomes rank deficient in case the experimental data is too scarce. The use of this rank deficient matrix with the standard methods is first discussed. A new method is then proposed; it relies on the generation of multiple samples of the prior distribution. These samples are used to « enrich » the missing data and construct a prior distribution of the terms of the covariance matrix that cannot be identified from the data.

Keywords Inverse methods · Bayesian updating · Uncertainty quantification · Monte Carlo simulation

1 Introduction

Inverse methods are widely used in science and engineering; they consist of identifying the input parameters of a numerical model leading to an adequate match with available experimental data. It is here assumed that multiple identical component undergo the same test (same geometry, same material, same loading spectrum, etc.) and the results exhibit scatter. The probabilistic approach provides an appropriate framework to model such problems. The input parameters of the numerical model associated with the experiment needs to be identified, they are here characterized py their joint probability density function. Model updating techniques provide an appropriate framework and received considerable attention from structural engineers during the past decades [1–3]. They are applied in case a forward numerical model is available but it is not possible or numerically too demanding to evaluate the inverse model. Such methods are used for instance in case sensors collect the vibration data,

P. Beaurepaire (✉)
Université Clermont Auvergne, CNRS, SIGMA Clermont, Institut Pascal, Clermont–Ferrand, France
e-mail: pierre.beaurepaire@sigma-clermont.fr

© The Author(s), under exclusive license to Springer Nature Switzerland AG 2021
J. C. Matos et al. (eds.), *18th International Probabilistic Workshop*, Lecture Notes in Civil Engineering 153, https://doi.org/10.1007/978-3-030-73616-3_39

which are subsequently used to identify modal information (amplitudes, modes, damping, etc.). A finite element model is then implemented and model updating is used to identify the input parameters leading to the best fit with the experimental data.

Bayesian updating methods allow engineers to identify the optimal parameter values and as well the probability density function associated with them [4, 5]. Considering $\boldsymbol{x}$, the set of parameters to be updated from the experimental data $\mathcal{D}$ using Bayes' theorem, which expresses as:

$$p(\boldsymbol{x}|\mathcal{D}) = \frac{p(\mathcal{D}|\boldsymbol{x})p(\boldsymbol{x})}{p(\mathcal{D})} \tag{1}$$

where $p(\boldsymbol{x})$ is the *prior distribution*, which gathers the initial knowledge on the parameters; $p(\mathcal{D}|\boldsymbol{x}))$ is the *likelihood function*, which quantifies the match between the experimental data and the outcome of the numerical model; $p(\boldsymbol{x}|\mathcal{D})$ is the *posterior distribution*, a probability density function associated with the model parameters which considers the information provided by the experimental data $\mathcal{D}$ and $p(\mathcal{D})$ is the *evidence*, a constant guaranteeing that Eq. (1) integrates to one. It is here assumed that the lack of repeatability of the experimental procedure is solely caused by aleatory uncertainty associated with the input parameters of the model; model output uncertainty are not considered.

For practical engineering applications, the experimental data may be obtained from prototype tests. Practically, a very reduced number of prototypes are generally used because of their high costs. Practically, testing three to ten prototypes seems common. Multiple sensors (gauges, accelerometers, etc.) are installed on the prototype and gather data associated with the tests. The costs associated with such sensors remains moderate (in comparison with the costs of prototypes). Therefore, it is possible to use a large amount of sensors, for example tens or hundreds of them. As a result, it can reasonably be expected that the number of sensors exceeds the number of prototypes. When the Bayesian updating procedure is applied, the experimental data consist of samples of a set of random variable, each variable being associated with a sensor and each sample being associated with a prototype tested. There are therefore less samples than the number of responses of the system. The objective of this paper is to discuss the application of Bayesian updating in such a context.

Section 2 describes two straightforward strategies applicable with standard updating procedures in this context and discusses their deficiencies. Section 3 introduces a new procedure applicable in this context; it relies on the generation of additional samples of the input parameters. The method is then applied to two examples in Sect. 3. The paper closes with the conclusions.

2 Statement of the Problem

The approach developed here considers a numerical model taking as an input a set of random variables $= (x_1, x_2, \ldots, x_N)$ with multiple outputs of the form $\boldsymbol{y}(\boldsymbol{x}) = (y_1(\boldsymbol{x}), y_2(\boldsymbol{x}), \ldots, y_M(\boldsymbol{x}))$. As $\boldsymbol{y}$ is a mathematical function taking as an input a set of random variables, its output can be modeled as a set of random variables. Some initial knowledge regarding the distribution of the input parameters may be available and the prior distributions accounts for it. Experimental data associated with this model are required to perform Bayesian updating; it consists of a set of samples of the response of the model $\left(\boldsymbol{y}^{(1)}, \ldots, \boldsymbol{y}^{(K)}\right)$. The procedure aims at updating (e.g. *adjusting*) the prior distribution in order to maximize the fit between the output of the model and the experimental data. In the general case, the joint probability density function associated with these experimental data is identified and subsequently defines the likelihood function (using the definition of the likelihood widely applied in statistics). This joint probability density function may be characterized by the marginal distributions and the linear correlation coefficient (see e.g. [6]) or with the Kernel Density Estimation (KDE) (see e.g. [7, 8]). Such approaches typically require the covariance or the correlation matrix. For instance, the covariance matrix may be involved in the definition of the kernel used in the KDE (see e.g. [9, 10]). The total number of samples in the experimental data needs to be strictly greater than total number of outputs of the model (i.e. $K > M$) to fully identify the covariance matrix from the data, i.e. at least two samples are required to determine the variance of a (univariate) random variable, at least three samples are required to determine the correlation coefficient of a set of two variables, etc.

This paper aims at discussing the application of Bayesian updating if the experimental data is not sufficient to fully characterize its covariance matrix, i.e. in case $K \leq M$. This matrix is then rank deficient with $K - 1$ non-zero eigenvalues; the eigenvectors associated with them define the linear subspace of dimension $K - 1$ containing all the samples. I assume as a simplifying hypothesis that the experimental data follow a multivariate normal distribution and that the reduced sample size does not introduce epistemic uncertainties on the parameters of this distribution. This implies that the $K - 1$ non-zeros eigenvalues and their corresponding eigenvectors are supposed to be correctly estimated. The remaining eigenvalues are equal to zero, which may be an artifact caused by the identification of the covariance matrix from an insufficient number of samples. I therefore assume in the following that such eigenvalue may be non-zero; and the identification of their possible values is discussed in the upcoming sections of the manuscript. Figure 1 shows possible aspect of the joint probability density of experimental data with two samples and a problem in dimension two. The crosses represent the two samples; the dash line represent the direction of the first eigenvector of the covariance matrix and the contour lines represent a possible distribution. Figure 1a, b are in good agreement with the simplifying hypothesis described above whereas Fig. 1c is not compatible with it (as the axes of isovalue of the probability density functions are not aligned with the eigenvector

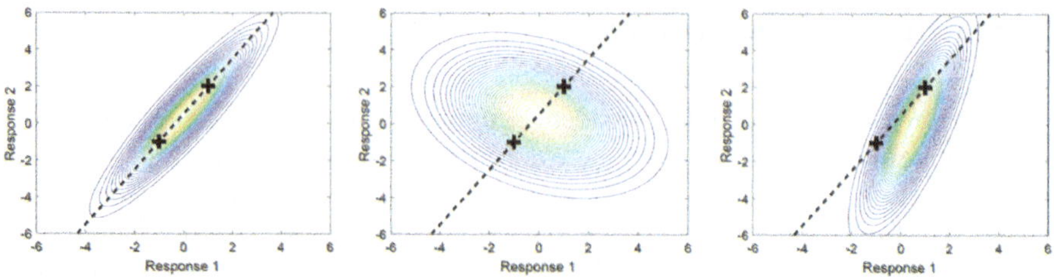

Fig. 1 Aspect of the joint probability density of the experimental data. **a**, **b** Allowed by the framework used here. **c** Incompatible with the framework used here

obtained from the data and the mean value of the distribution is not the mean value of the samples).

The first possible strategy applicable to such problems consists of using directly the covariance matrix identified from the samples used in the definition of the likelihood function. The zero eigenvalues imply that some random variables are fully correlated. Therefore, this modeling strategies considers that the responses of the numerical model have to lay on a linear space of dimension $K - 1$ and the updating procedure then propagates this dependence to the inputs. In case the numerical model is bijective, the samples of the posterior distribution lay on a space of dimension $K - 1$ (possibly non-linear). Figure 2a presents the posterior distribution obtained when this approach is used with a linear model. This strategy may not be applicable in the general case, as the assumption of a zero eigenvalue may be excessive. Indeed, engineering problems where the random variables lay on a space of reduced dimension are seldom encountered and an alternative formulation of the updating problem needs to be proposed. Moreover, many algorithms available in the literature are based on the generation of samples of the prior distribution, which are subsequently accepted or rejected. This strategy requires the generation of samples on a subspace of reduced dimension, which may be a challenging task.

The second possible strategy to deal with such a problem consists of using a reduced set of $K - 1$ outputs, as they can be selected such that their covariance matrix is full rank. The following set is used:

$$y' = ((y - \overline{y}) \cdot \varphi_1, \ldots, (y - \overline{y}) \cdot \varphi_{K-1}) \tag{2}$$

where $\overline{y}$ denotes the mean value of the experimental data, $\varphi_1, \ldots, \varphi_{K-1}$ denote the eigenvectors of the covariance matrix associated with non-zero eigenvalues and the dot denotes the scalar product between two vectors. This approach consists of projecting the response of the model on the subspace containing all the samples of the experimental data, which is defined by the eigenvalues involved in Eq. (2). This is equivalent to assigning an infinite value to the zero eigenvalues of the covariance matrix. Figure 2b shows the aspect of the posterior distribution in case this strategy is applied to a linear model. With this approach, the posterior distribution has a

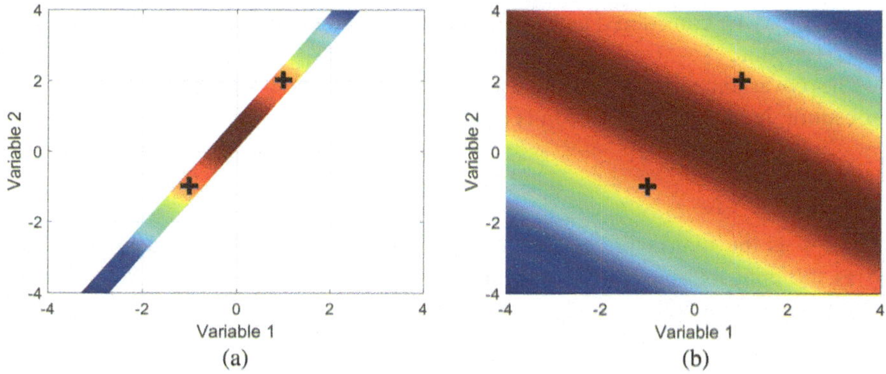

Fig. 2 Posterior distribution obtained **a** using a rank deficient covariance matrix; **b** using only the first eigenvector. The crosses represent the samples of the input parameters associated with the experimental data (i.e. the inverse of the numerical model)

non-zero value on the whole support of the prior distribution. For any sample of the input random variables, the corresponding responses of the numerical model are computed and projected on the space containing the experimental data. This projected response is then used to determine the posterior distribution. Therefore, the value of the posterior distribution may not converge towards zero in some direction of the input parameters space. This happens if the response of the model corresponding to this direction is orthogonal to the plane containing the experimental data. For instance, in Fig. 2b the value of the posterior distribution does not converge towards zero if $x_1 = -x_2$ even though x_1 converges toward infinity. This strategy trends to generate samples which poorly match the original data. For instance, in Fig. 2b, the original data have a positive correlation whereas the posterior distribution exhibits a negative correlation.

As the two strategies for the Bayesian updating described above are not satisfactory, the focus of the upcoming sections is on the formulation of alternative approaches.

3 Proposed Procedure

The experimental data cannot be used to determine the standard deviation associated with the following set:

$$y^* = ((y - \overline{y}) \cdot \varphi_K, \ldots, (y - \overline{y}) \cdot \varphi_M) \tag{3}$$

It is therefore possible to set the standard deviation associated with each component of this set to zero or to infinity, as discussed in the previous section. Intermediate situation may be considered as well, and the standard deviation may exhibit a finite, non-zero value.

The covariance matrix of the experimental data is involved in the updating procedure; its general expression is of the form:

$$C = \Phi^t (C_1 + C_2)\Phi \tag{4}$$

where Φ is the matrix of the eigenvalues associated with the experimental data, C_1 is the contribution of the covariance matrix completely characterized by the experimental data with:

$$C_1 = \begin{bmatrix} \sigma_1^2 & 0 & & \cdots 0 & 0 \cdots 0 \\ 0 & \sigma_2^2 & & \ddots \vdots & \vdots \quad \vdots \\ \vdots & \ddots & \ddots & 0 & \vdots \quad \vdots \\ 0 \cdots & & 0 & \sigma_{K-1}^2 & 0 \cdots 0 \\ & 0 \cdots & & \cdots 0 & 0 \cdots 0 \\ & \vdots & & \vdots & \vdots \ddots \vdots \\ & 0 \cdots & & \cdots 0 & 0 \cdots 0 \end{bmatrix} \tag{5}$$

where $\sigma_1^2, \ldots, \sigma_{K-1}^2$ denote the non-zero eigenvalues.

The matrix C_2 is the contribution which cannot be identified from the experimental data, its generic form is:

$$C_2 = \begin{bmatrix} 0 \cdots 0 & 0 \cdots 0 \\ \vdots \quad \vdots & \vdots \quad \vdots \\ 0 \cdots 0 & 0 \cdots 0 \\ 0 \cdots 0 & \\ \vdots \quad \vdots & S \\ 0 \cdots 0 & \end{bmatrix} \tag{6}$$

where S is a full matrix. The components of this matrix cannot be identified from the experimental data, as they are associated with the zero eigenvalues.

Equations (4–6) can be injected in the expression of the likelihood function, which yields:

$$\begin{aligned} p(\mathcal{D}|\boldsymbol{x}, S) &= \frac{1}{\sqrt{2\pi C}} \exp\left(-\frac{1}{2}(\boldsymbol{y} - \overline{\boldsymbol{y}})^t C^{-1} (\boldsymbol{y} - \overline{\boldsymbol{y}})\right) \\ &= \frac{1}{\sqrt{2\pi C_1}} \exp\left(-\frac{1}{2} \boldsymbol{y}(\boldsymbol{x})'^t C_1^{-1} \boldsymbol{y}(\boldsymbol{x})'\right) \\ &\quad \frac{1}{\sqrt{2\pi S}} \exp\left(-\frac{1}{2} \boldsymbol{y}(\boldsymbol{x})^{*t} C_2^{-1} \boldsymbol{y}(\boldsymbol{x})^*\right) \end{aligned}$$

$$= p_1(\mathcal{D}|\boldsymbol{x})p_2(\mathcal{D}|\boldsymbol{x}, S) \quad (7)$$

The likelihood function is therefore expressed as the product of two functions; p_1 accounts for the non-zero eigenvalues, obtained using the experimental data and p_2 accounts for the dependence associated with the matrix S. As the experimental data does not contain sufficient information to identify the components of this matrix, it is suggested here to use a random variable model based on the prior distribution and on the numerical model. A set of $M - K + 1$ *auxiliary* samples of the input variables is generated and Bayesian updating is applied considering only the non-zero eigenvalues (i.e. with the second strategy described in Sect. 3). The joint probability density function associated with these samples is therefore:

$$p(\mathcal{D}|\boldsymbol{x}) = \int_{-\infty}^{\infty} p(\mathcal{D}|\boldsymbol{x}, S(\boldsymbol{x}_1, \ldots, \boldsymbol{x}_{M-K+1}))p(\boldsymbol{x}_1, \ldots, \boldsymbol{x}_{M-K+1}|\mathcal{D})\mathrm{d}\boldsymbol{x}_1 \ldots \mathrm{d}\boldsymbol{x}_{M-K+1} \quad (8)$$

Equation (8) can be directly injected in the formula of the Bayesian updating problem defined in Eq. (1).

The general idea of the proposed procedure may be summarized as follows:

- Auxiliary samples of the input parameters are generated, using an updating procedure involving the non-zero eigenvalues, such that the covariance matrix of the response of the system become full rank;
- These samples are then used to define a prior distribution of the missing terms of the covariance matrix;
- This distribution is updated using the likelihood function (Eq. 8);
- The updated likelihood is finally introduced in Eq. (1).

4 Application Examples

4.1 One Dimensional Problem

The first example involves a single random variable x and the numerical model is the identity function, i.e. $y(x) = x$. The experimental data consists of a single realization with $y^{(1)} = 1.5$ and it is therefore not possible to estimate the variance associated with this model. The prior distribution is uniform in the range $[-3,3]$. Figure 3a shows the posterior distribution obtained at the end of the procedure. The region of the central mass lays in the vicinity of the sample used for the updating procedure and the tails of the distributions are heavy. Figure 3b shows a scatterplot of the two samples used in the procedure (even though the auxiliary sample is solely used for the definition of the likelihood); a strong dependence is observed. Figure 3c compares the cumulative distribution functions obtained using the proposed procedure and the two strategies discussed in Sect. 2.

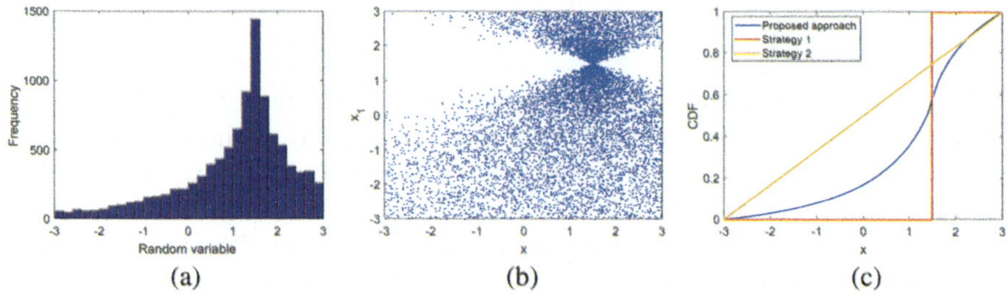

Fig. 3 Results of the updating procedure **a** Posterior distribution. **b** Scatterplot of the posterior distribution of the regular variable x and the auxiliary variable x_1. **c** Comparison of the cumulative distribution function obtained using the proposed procedure and the two strategies discussed in Sect. 2

4.2 Two-Dimensional Non-linear Problem

The second example involves two random variables and the numerical model is:

$$\boldsymbol{y}(\boldsymbol{x}) = \begin{bmatrix} y_1(\boldsymbol{x}) \\ y_2(\boldsymbol{x}) \end{bmatrix} = \begin{bmatrix} 2x_1 + x_1^2 + 2x_2 + x_2^2 \\ 2x_1 + 5x_2 \end{bmatrix} \tag{9}$$

The prior distribution is uniform in the range [−6,6] for both variables and the experimental data includes two realizations. The covariance matrix of the experimental data is therefore of rank one as a realization is missing to obtain a full rank matrix. Figure 4 shows the posterior distribution; its value is maximal along a curve

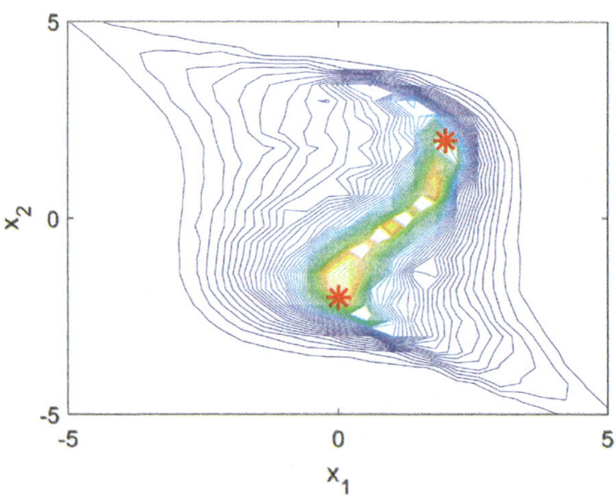

Fig. 4 Contour lines of the posterior distributions. The stars represent the input parameters associated with the experimental data

joining the values associated with the experimental data. This curve is not a straight line as the numerical model is non-linear.

5 Conclusions

This paper discusses the application of Bayesian updating with reduced experimental data. In this context, the covariance matrix associated with the data is rank deficient and the definition of the likelihood function becomes problematic. A novel procedure is proposed; it relies on the generation of auxiliary additional samples of the input parameters. The response associated with these samples is then used to define a prior distribution of the terms of the covariance matrix, which is subsequently updated. The method is applied to two numerical examples.

References

1. Friswell, M., & Mottershead, J. E. (1995). *Finite element model updating in structural dynamics*. Springer Science & Business Media.
2. Imregun, M., & Visser, W. J. (1991). A review of model updating techniques. *Shock and Vibration Digest, 23*(1), 9–20.
3. Arora, V. (2011). Comparative study of finite element model updating methods. *Journal of Vibration and Control, 17*(13), 2023–2039.
4. Beck, J. L., & Katafygiotis, L. S. (1998). Updating models and their uncertainties. i: Bayesian statistical framework. *Journal of Engineering Mechanics, 128*(4), 380–391.
5. Katafygiotis, L. S., & Beck, J. L. (1998). Updating models and their uncertainties. ii: Model identifiability. *Journal of Engineering Mechanics, 128*(4), 463–467.
6. Ferreira Cordeiro, S. G., Leonel, E. D., & Beaurepaire, P. (2017). Quantification of cohesive fracture parameters based on the coupling of bayesian updating and the boundary element method. *Engineering Analysis with Boundary Elements, 74*, 49–60.
7. Goller, B. (2011). *Stochastic model validation of structural systems*. Ph.D. thesis, University of Innsbruck.
8. Nagel, J. B., & Sudret, B. (2016). A unified framework for multilevel uncertainty quantification in Bayesian inverse problems. *Probabilistic Engineering Mechanics, 43*(Supplement C), 68–84.
9. Goller, B., Pradlwarter, H. J., & Schuëller, G. I. (2009). Robust model updating with insufficient data. *Computer Methods in Applied Mechanics and Engineering, 198*(37), 3096–3104.
10. Wand, M. P., & Jones, M. C. (1995). *Kernel smoothing*. Chapman and Hall/CRC.

Numerical Modeling of an Extrusion-Based Concrete Printing Process Considering Spatially and Temporarily Varying Material and Process Parameters

Albrecht Schmidt, Meron Mengesha, Luise Göbel, Carsten Könke, and Tom Lahmer

Abstract During the past few years, additive manufacturing techniques for concrete have gained extensive attention. In particular, the extrusion-based 3D concrete printing exhibited a rapid development. Previous investigations are mostly based on experimental studies or even trial-and-error tests. A more profound understanding of the relationships between the process and material parameters and the manufactured structure can be advanced by numerical modeling and simulation. It enables to study a wide range of parameters such that dependencies of properties of the printed product on different influencing factors can be identified. Taking into account the uncertain nature of process and material parameters of the extrusion-based concrete printing, the process can be reliably controlled and finally optimized. The presented study uses a pseudo-density for a finite element based modeling approach. The pseudo-density determines the properties of the individual finite elements, analogous to the soft-killing method of topology optimization. Layer by layer the previously created elements are activated. Material parameters are described as temporally and spatially variable to reflect the temporally variable printing process. First results of a reliability estimation are shown for a 2D modeled additively manufactured wall.

Keywords Additive manufacturing · Random process · Statistics · Risk · Reliability

1 Introduction

In the past few decades 3D concrete printing technology (3DCP) is getting recognition in the construction industry. The technology has several advantages such as

---

A. Schmidt (✉) · L. Göbel · C. Könke · T. Lahmer
Materials Research and Testing Institute (MFPA), Bauhaus-Universität Weimar, Marienstr. 15, 99423 Weimar, Germany
e-mail: albrecht.schmidt@mfpa.de

M. Mengesha · C. Könke · T. Lahmer
Institute of Structural Mechanics, Bauhaus-Universität Weimar, Coudraystr. 9, 99423 Weimar, Germany

© The Author(s), under exclusive license to Springer Nature Switzerland AG 2021
J. C. Matos et al. (eds.), *18th International Probabilistic Workshop*, Lecture Notes in Civil Engineering 153, https://doi.org/10.1007/978-3-030-73616-3_40

decreased construction time, design flexibility compared to conventional construction methods, cost reduction by avoiding formwork costs, reduced waste, reduced manpower which decreases injuries and fatalities on construction sites and increased sustainability of the construction industry [1, 2]. A growing number of projects can be observed in both private enterprises and research institutes worldwide [3]. According to Wangler et al. [4] among the different concrete printing processes like particle-bed 3D printing [5], the extrusion-based process is the most widely used. Therefore, the presented contribution considers the extrusion-based process which has been demonstrated to be applicable to the construction industry [1, 2].

However, there are several challenges to be addressed to fully implement the technology. An improved profound understanding of the relationship between design, material behavior, and process parameters is eminent. The materials' rheology and the process parameters, such as printing speed, time, temperature, etc., have an impact on the fresh and hardened state of the printed structure [6]: Effects like geometric variation of the printed layers [3], stability failure due to local material strength [4] and global buckling [7] have been reported. Under-filling influenced by printing process parameters (nozzle velocity, pumping rate etc.) and material properties (rheology) are reported in [5]. Due to this complex behavior and the variability of material parameters [3] and printing process parameters, adequate process parameters are commonly determined by means of a trial-and-error [8].

By implementing a numerical simulation of the 3DCP process a better understanding of the system parameters on the structure behavior can be achieved. The influence of the parameter uncertainty can be studied. In this contribution a finite element framework is proposed to model the time depending printing process efficiently and is combined with an uncertainty description of the system parameter uncertainty taking spatial and temporal correlations into account.

2 Methods

2.1 *Finite Element Model of a Printing Process*

According to the layer-wise production process of the concrete structure, the finite element (FE) model "needs to grow". The idea is to work on a previously generated FE mesh, where the elements are activated sequentially in correspondence to the printing process, cf. Fig. 1. This approach enables an efficient simulation avoiding the computational demanding meshing procedure.

During the simulation, i.e. the solution of a finite element system of the form

$$K(r,t)u(r,t) = F(r,t), \tag{1}$$

where K is the global stiffness matrix, u is the vector of nodal displacements, the loads due to self-weight are considered in the right-hand-side vector F and r and

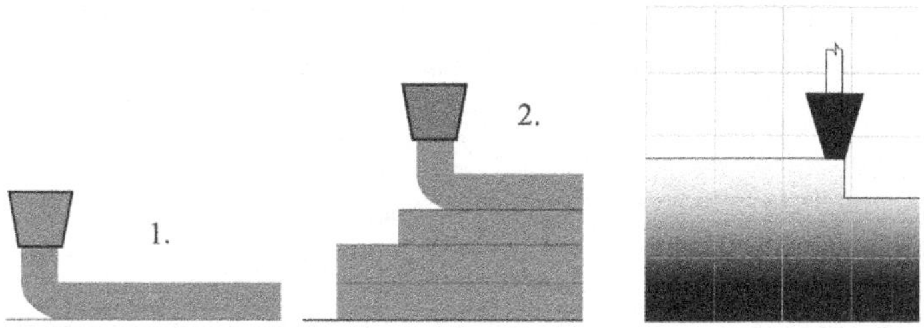

Fig. 1 (left) Schematic of a layer-by-layer process; (right) temporal evolution of a material parameter on a pre-meshed FE model with sequentially activated elements

t denote the space and time variable, respectively. Additionally, as all parameters do change with time according to the hardening process the material parameters are both spatially and temporarily varying, indicated by the space and time dependency in Eq. 1.

An example of the temporal development of the material parameters compressive strength and Young's modulus is reported in [9]. The authors report a merely linear increase of the compressive strength over the concrete age in the time of 0 till 90 minutes after fabrication. Some scatter in the measured values is given which increases with time. Similar findings hold for the linear increase of Young's modulus within the same time period considered leading to the necessity of an uncertainty model for the FE model parameters where the temporal and spatial correlations are taken into account.

2.2 *Principle Scheme*

A principal scheme of the generally proposed uncertainty description of a 3DCP model is depicted in Fig. 2. The material parameters of the printed structure originate from the material parameters of the pumped concrete which constitutes a time-dependent process.

While the material properties undergo a natural variation it is reasonable to assume a temporal auto-correlation of the concrete properties since the fresh pumped concrete is taken from the same source. Hence, a random process description is proposed. Additionally, a cross-correlation parameter is introduced to take into account that most of these material parameters are not independent from each other. The same holds for the process parameters like print velocity pumping pressure etc.

This stochastic characterization of the material parameters is based on distribution parameters and (auto- and cross-) correlation parameters. While the first ones can be usually determined from experiments data on auto- and cross-correlation is very vague. Therefore, fuzzy variables are considered to model this epistemic kind of uncertainty.

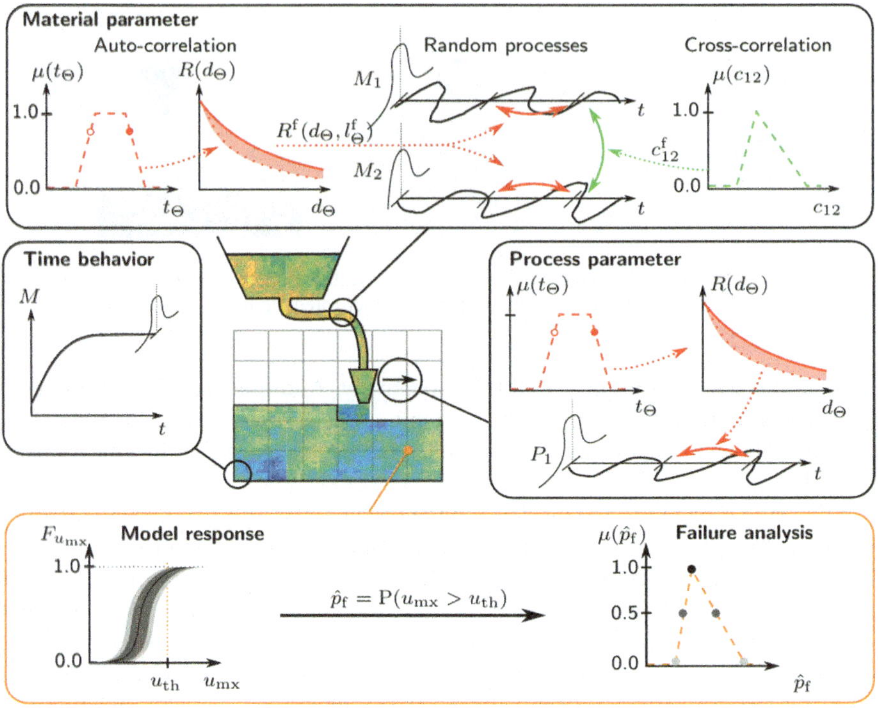

Fig. 2 Principle scheme of a 3DCP model with time dependent, uncertain material and process parameters

The uncertainty in the temporal evolution of the material parameters can also be described in an analog manner. As a result system responses are described in a fuzzy-stochastic space. For example, a reliability analysis yields a fuzzy failure probability.

3 Numerical Example

3.1 Description

3.1.1 Finite Element Model

We consider a 2D finite element (FE) model of a printed wall with a size of 1 m × 0.2 m. The process parameters are constant. A printing velocity of $v_p = 1$ m/min is chosen while the layer height is set to 10 mm. The total printing time is 1200 s.

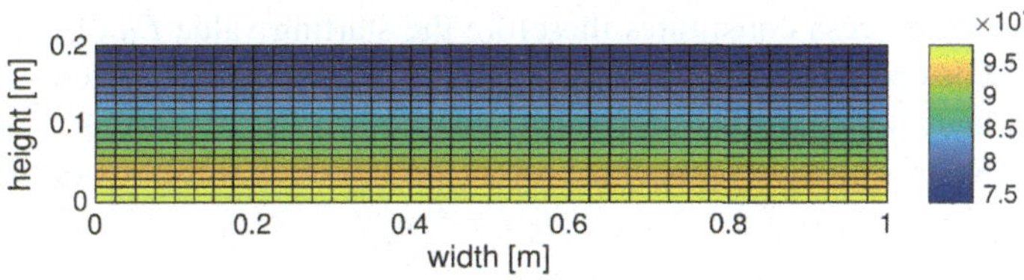

Fig. 3 Time-dependent Young's modulus $E(r)$ in [Pa] for $t = 1200$ s from constant $E_0 = 74$ kPa

The wall is modeled with a regular rectangular mesh using 40 × 20 Q8 Elements. A linear-elastic plane-stress material model is assumed. The time-dependent Young's modulus is given as

$$E(r, t) = E_0(r) + E_1 \cdot t, \tag{2}$$

where $E_0(r)$ is the starting value as a function of position r, E_1 is the slope of the temporal increment and t the time. The loading consists of the increasing self-weight, considering a constant mass density of 2020 kg/m^3 [9]. The modeled wall is fixed vertically and horizontally at the bottom edge.

The temporal evolution of the Young's modulus can be seen in Fig. 3. An increase of $E_1 = 1.2$ kPa/min is chosen in accordance to [9], while E_0 is modeled via a random process, which is described in the next section.

3.1.2 Uncertainty Model

In this contribution a simplified (fully stochastic) uncertainty approach is reported where only the Young's modulus of the freshly pumped concrete is described via a log-normally distributed random process E_p. The mean and the standard deviation of the E_p are taken from [9] as 74 kPa and 7.4 kPa, respectively. An exponentially decaying correlation function with a correlation time of $t_\mathrm{c} = 1.5$ s is assumed. In Fig. 4 realizations of the random process are plotted.

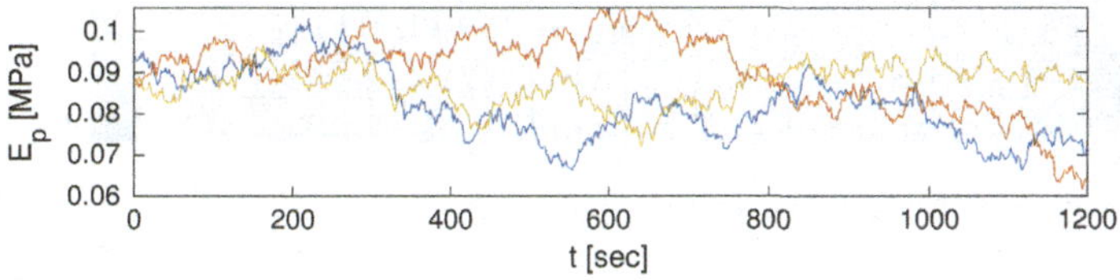

Fig. 4 Samples of random process $E_\mathrm{p}(t)$

This random process constitutes therefore the starting value E_0 of the temporal evolution of E in each element (cf. Eq. 2). Hence, E_0 can be characterized as a random field with a layer-wise correlation structure with a correlation length of $l_c = v_p \cdot t_c = 25$ mm. The described FE model is solved for 10,000 realizations of the random process.

3.2 Results

In Fig. 5 one realization of the Young's modulus at the end of the printing process ($t = 1200$ s) is depicted. Bothe the random nature of the fresh concrete property (E_0) and the increase of stiffness over time can be observed.

The horizontal displacement and the deformation shape for an exemplary realization is shown in Fig. 6. A reasonable deformation shape displays horizontal displacement in the range of a few mm.

For each sample the resulting stresses and straines have been calculated from the simulated nodal displacements. The histogram of the maximum shear strain per sample can be observed in Fig. 7. A considerable skewness of the distribution can be noticed which is typical to an extreme value distributio. Based for example on a maximum deformation of the printed wall a reliability analysis could be performed.

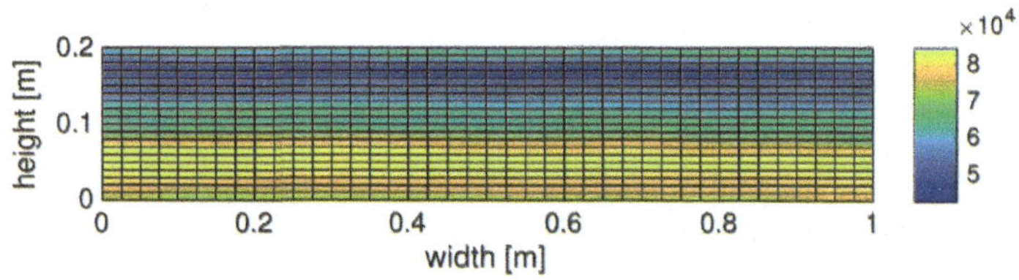

Fig. 5 Young's modulus $E(r)$ in [Pa] for $t = 1200$ s for E_0 mapped from random process E_p

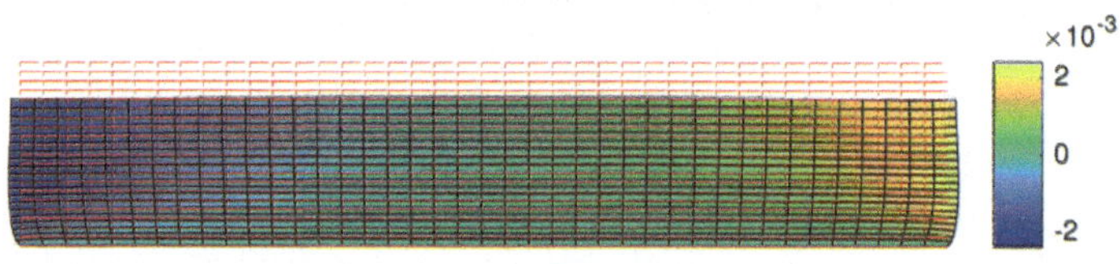

Fig. 6 Horizontal displacement u_x, deformation scale 5x

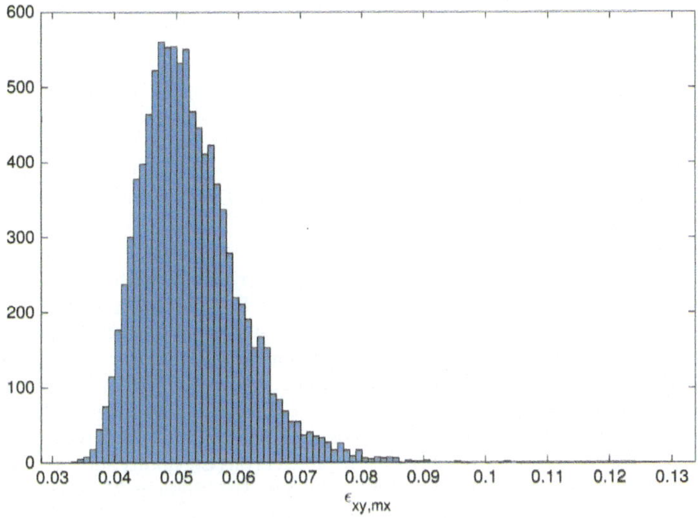

Fig. 7 Histogram of the maximum shear strain, 10,000 samples

4 Conclusions

The proposed approach considers a random process and random field description to take the uncertain nature of 3D concrete printing material and process parameters into account. A finite element utilizing a pre-defined mesh is used to efficiently simulate the time-dependent printing process. For a simple 2D example of a 1 m $\times$ 0.2 m wall, a lognormally distributed random process modelling the Young's modulus of the fresh, pumped concrete is mapped to a random field with a layer-wise spatial correlation structure to determine the initial values of the time-dependent stiffness. The histogram of the maximum shear strain illustrates the random behavior of the model, displaying a typical extreme value distribution.

Further studies will include additional material and process parameters and consider the uncertain nature of the temporal and spatial correlation coefficients, yielding a polymorphic uncertainty model for 3DCP models.

Acknowledgements The demonstrated results and research activities arise from the project "Polymorphic uncertainty modeling of a 3D printing process of concrete" (GO 3270/1-1, KO 1267/24-1, LA 2869/11-1), as part of Priority Program SPP 1886. The funding by the German Research Foundation (DFG) is gratefully acknowledged.

References

1. Peng, W., Jun, W., & Xiangyu, W. (2016). A critical review of the use of 3-D printing in the construction industry. *Automation in Construction, 68,* 21–31. https://doi.org/10.1016/j.autcon.2016.04.005.
2. Hager, I., Golonka, A., & Putanowicz, R. (2016). 3D printing of buildings and building components as the future of sustainable construction? *Procedia Engineering, 151,* 292–299. https://doi.org/10.1016/j.proeng.2016.07.357.
3. Buswell, R. A., et al. (2018). 3D printing using concrete extrusion: A roadmap for research. *Cement and Concrete Research, 112,* 37–49. https://doi.org/10.1016/j.cemconres.2018.05.006.
4. Wangler, T., Lloret, E., Reiter, L., Hack, N., Gramazio, F., Kohler, M., Bernhard, M., Dillenburger, B., Buchli, J., Roussel, N., & Flatt, R. (2016). Digital concrete: Opportunities and challenges. *RILEM Technical Letters, 1,* 67. https://doi.org/10.21809/rilemtechlett.2016.16
5. Le, T. T., Austin, S. A., Lim, S., Buswell, R. A., Gibb, A. G. F., et al. (2012). Mix design and fresh properties for high-performance printing concrete. *Materials and Structures, 45,* 1221–1232.
6. Panda, B., Mohamed, N. A. N., Paul, S. C., Singh, G. V., Tan, M. J., & Šavija, B. (2019). The effect of material fresh properties and process parameters on buildability and interlayer adhesion of 3D printed concrete. *Materials, 12*(13). https://doi.org/10.3390/ma12132149
7. Gosselin, C., et al. (2016). Large-scale 3D printing of ultra-high performance concrete—A new processing route for architects and builders. *Materials and Design, 100,* 102–109.
8. Suiker, A. (2018). Mechanical performance of wall structures in 3D printing processes: Theory, design tools and experiments. *International Journal of Mechanical Sciences, 137,* 145–170. https://doi.org/10.1016/j.ijmecsci.2018.01.010.
9. Wolfs, R. J. M., Bos, F. P., & Salet, T. A. M. (2018). Early age mechanical behaviour of 3D printed concrete: Numerical modelling and experimental testing. *Cement and Concrete Research, 106,* 103–116.

Parameter Uncertainties in Flow Rate and Velocity Analysis of Heavy Rain Events

Axel Sauer and Regine Ortlepp

Abstract Flooding due to intensive precipitation poses a major threat to lives and property. Information about flood-prone areas is needed in order to reduce potential risks via mitigation and adaptation measures. The flood probability of a certain point in the landscape depends, firstly, on the projected frequency and characteristics of heavy rainfall events that generate surface runoff and, secondly, on the specific properties of the terrain that determine flow and runoff, e.g. the surface morphology and the hydraulic roughness of diverse surfaces or land uses. Simulation models of surface water flow are standard tools for the assessment of flood dynamics caused by extreme precipitation events. In order to make informed decisions that take modelling uncertainties into account as well as to get an idea of the probability space, it is necessary to quantify the effects of alternative sets of model parameters by drawing on different data sources as well as spatial and temporal resolutions of the input data. For the current study, we evaluated the impact of different parameter sets on the flow rate and velocity as determined for a study area in south-eastern Germany using the hydronumeric computational fluid dynamics model HiPIMS. The considered parameters were rainfall input (time and space invariant, spatially invariant and time varying, space and time varying), hydraulic roughness and spatial resolution of the digital elevation model. We present point-based time series of flow rates and velocities to indicate the bandwidths of probable flooding dynamics. Results show that the modelled flow rates and velocities are strongly dependent on the particular form of rainfall data as well as the spatial resolution of the digital elevation model. The effects of variations in hydraulic roughness are also found to be significant while in all cases the location of data capture points in the catchment area has a strong influence.

Keywords Extreme precipitation events · Heavy rainfall · Risk assessment · Parameter uncertainties

---

A. Sauer · R. Ortlepp (✉)
Leibniz Institute of Ecological Urban and Regional Development, Dresden, Germany
e-mail: r.ortlepp@ioer.de

© The Author(s), under exclusive license to Springer Nature Switzerland AG 2021

J. C. Matos et al. (eds.), *18th International Probabilistic Workshop*, Lecture Notes in Civil Engineering 153, https://doi.org/10.1007/978-3-030-73616-3_41

1 Introduction

Flooding due to intensive precipitation poses a major threat to life and property [1]. Clearly, information about flood-prone areas is needed in order to reduce risks via mitigation and adaptation measures. The flood probability of a certain point in the landscape depends, firstly, on the projected frequency and characteristics of heavy rainfall events that generate surface runoff and, secondly, on properties of the terrain that determine flow and runoff, e.g. the surface morphology and the hydraulic roughness of diverse surfaces or land uses. Simulation models for the flow of surface water are standard tools for the assessment of flood dynamics caused by extreme precipitation events [2, 3]. In order to make informed decisions that take modelling uncertainties into account as well as to get an idea of the probability space, it is important to quantify the effects of alternative sets of model parameters by drawing on different data sources as well as spatial and temporal resolution of the input data. In the following paragraphs we present an approach on how to analyse parameter uncertainties in this context. The underlying methodology was discussed in a previous study [4]. Here the focus is on a wider set of locations as well as additional output parameters of the hydrodynamic model (velocity and flow rate).

2 Modelling Flood Hazards with A Hydrodynamic Model

Hydrodynamic models are physically-based, spatially-distributed models designed to simulate the flow of water over a surface. When the aim is to simulate flow processes outside of clearly defined channels, as is typical for floods induced by heavy rain, models are applied that can deal with flow in different horizontal directions. One such model is the high-performance integrated modelling system HiPIMS [5, 6].

2.1 High-Performance Integrated Modelling System HiPIMS

HiPIMS provides a numerical 2D implementation of the shallow water equation (SWE) with the ability to capture shockwaves. The model has been successfully validated in numerous artificial test cases as well as event-based examples [5–8]. It is implemented in C++ and OpenCL, ensuring platform independence and the option of running on graphics processing units (GPUs). The simulation speed on a GPU (Nvidia Tesla 2075/Quadro K6000) can be 8–30 times faster than on a 2.8 GHz hexa-core CPU (Intel Xeon X5660), depending on the simulation domain characteristics and the numerical precision (i.e. 32-bit floating point vs. 64-bit). This considerably reduced processing times allows us to running many alternatively configured models, thereby shifting from deterministic "best guess" model parameterisations to a more probabilistic approach, and thus gain insights into the uncertainty bandwidths.

2.2 *Input Data*

For its input parameters, HiPIMS requires data on the hydro-meteorological load (rainfall), continual losses of this load e.g. by infiltration or via a sewer network, the (natural) surface of the terrain enriched with artificial elements such as buildings as well as information on the hydraulic roughness of different surfaces and land uses. In the next three paragraphs, we describe the preparation of this basic input data and the selection of parameter values. This is followed by a description of the chosen variation in the input parameters *precipitation*, *surface* and *hydraulic roughness* as part of our sensitivity and uncertainty analysis.

2.2.1 Precipitation

The hydro-meteorological input *precipitation* can be provided: (i) in the form of space-invariant time series of constant intensity, called "block rain"; (ii) as time-variant values for the whole domain e.g. based on observations from rain gauges/weather stations; (iii) synthetic time series constructed using rainfall generators or methods for the construction of design storms (e.g. Euler I/II precipitation distributions derived from intensity-duration-frequency curves); or (iv) precipitation raster datasets that vary over time and space. For Germany, the latter is provided by, for example, the Deutscher Wetterdienst (DWD, German weather service) with its radar product RADOLAN, which gathers data every 5 min and at 1 km spatial resolution [7]. In addition to these upper temporal and spatial boundary conditions, lateral as well as upward inflows to the model domain can be defined as time series of water volume, water levels and flow velocities/directions for selected cells. Such time series can be the output of previous rainfall-runoff modelling, sewer system models or HiPIMS runs with temporal changes of typically time-invariant parameters such as the terrain or runs with other spatial domains/catchments as well as resolutions.

2.2.2 Losses

The model can use global time-invariant losses (reduction of precipitation as well as surface water), global time series as well as spatially differentiated raster geodata with loss values. We did not take into account loss affects and assumed for all simulations a runoff coefficient of 1, i.e. all precipitation contributes to overland flow.

2.2.3 Surface

The model surface for the flow calculations is represented as a regular Cartesian grid that can be parameterised using digital elevation models (DEM) enriched with separate height and geometry information of hydraulically-relevant structures such as buildings, culverts, ditches, flood defence walls, etc. In our test case, we took account of buildings by adding a constant height of 5 m at the location of building footprints in the DEM. The primary DEM is the official dataset of the Saxonian state topographic survey at resolution 2 m.

2.2.4 Hydraulic Roughness

A further important parameter is the hydraulic roughness of the surface expressed by the Gauckler-Manning-Strickler (GMS) coefficient. This parameter can be set as a global, space-invariant value for the whole model domain or as a spatially-varying raster dataset representing different values for land uses or surface types. Changes cannot be made during the model runtime.

2.3 Parameter Variation

The impact of uncertainties caused by different parameter values was analysed via alternative model runs of different combinations of values. In the following paragraphs, we describe how the three parameters *precipitation/rainfall event*, *surface spatial resolution* and *hydraulic roughness* were varied. The simulations made use of data from a real-world study area, namely a hilly region in south-eastern Germany (Spitzkunnersdorf creek catchment, Saxony) that in the past has suffered from severe flash floods after heavy rainfall. Figure 1 gives an overview of the catchment area with an exemplary visualisation of one simulation and the location of the data capture points where the time series of the different model runs given in the results section were recorded. The parameter values for the various model runs are shown in Fig. 2.

2.3.1 Precipitation

The precipitation alternatives were three synthetic time-variable "storms", one constant load situation and an observed event with spatiotemporal differentiation. The Euler II approach [9] for the derivation of synthetic rainfall time series is based on intensity-duration-frequency data for different return periods and duration levels that are available throughout Germany on a regular grid of approx. 8 $\times$ 8 km. The Euler II synthetic rainfalls were calculated for 1 h duration and frequency of 1 in 10 years (HN10), 1 in 30 years (HN30) and 1 in 100 years (HN100). In addition, a time-invariant block rain of 1 h and continuous intensity of 54 mm/h was applied (Block) as well as an observed rainfall event based on precipitation radar (Radar). All time series precipitation datasets had a resolution of 5 min. Peak rainfall intensities for the most intensive 5 min time step of the synthetic rains were in the range of 160–260 mm/h. Precipitation sums are given in Table 1.

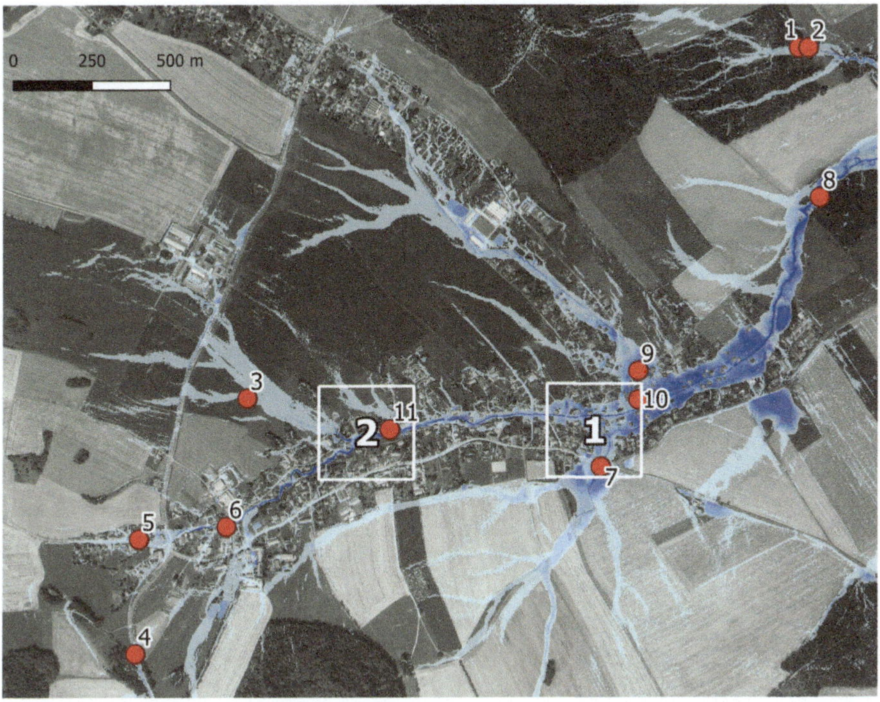

Fig. 1 Overview of the point and area locations and maximum water level for the simulation variant with 100-year Euler II rain, 2-m-DEM and GMS value 0.03

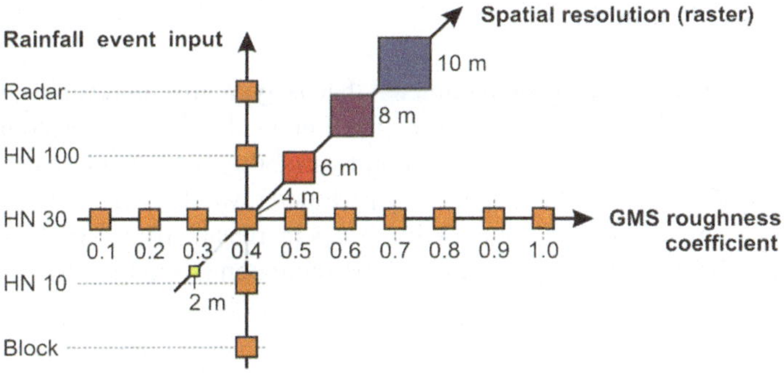

Fig. 2 The different parameter values [4]

2.3.2 Surface Spatial Resolution

The influence of different DEM spatial resolutions was investigated via model runs with raster resolutions of 2, 4, 6, 8, and 10 m. Table 2 provides an overview of these variants with additional information about the number of rows and columns, the

Table 1 Simulation runs with variation in precipitation and surface resolution

Return periods (1 in y)	Precipitation sum (mm/m^2)	Surface raster resolution (m)	Number of rows and columns	Number of cells
10	35	2	1423 × 1318	1,875,514
30	42	2	1423 × 1318	1,875,514
100	54	2	1423 × 1318	1,875,514
10	54	4	711 × 659	468,549
30	42	4	711 × 659	468,549

Table 2 Simulation runs with variation in surface resolution. Rain is 1:100 year Euler II with global GMS value of 0.04

Surface raster resolution (m)	Number of rows and columns	Extent in y and y direction (m)	Number of cells	Output storage size (GB)
2	1423 × 1318	2846 × 2636	1,875,514	8.84
4	711 × 659	2844 × 2636	468,549	2.41
6	474 × 439	2844 × 2634	208,086	1.02
8	355 × 329	2840 × 2632	116,795	0.68
10	284 × 263	2840 × 2630	74,692	0.47

spatial extent, the number of cells to be calculated and the storage size of the output data.

2.3.3 Hydraulic Roughness

The GMS roughness is another parameter that is typically unavailable as empirical data but must be estimated using expert knowledge. To take account of this uncertainty as well as its spatial and temporal variability, we ran alternative simulations with a wide range of GMS values. The parameter was determined to be spatially invariant for the whole model domain with values increasing from 0.01 (very smooth surfaces) to 0.10 (very rough surfaces) in 0.01 increments (see Fig. 2).

3 Results

3.1 Surface Roughness and Flow Velocity

As GMS surface roughness is the governing parameter for flow velocity, its strong influence is clearly observable (see Fig. 3). The intensity of the effect depends on the location within the catchment. Points with very low maximum velocities in the

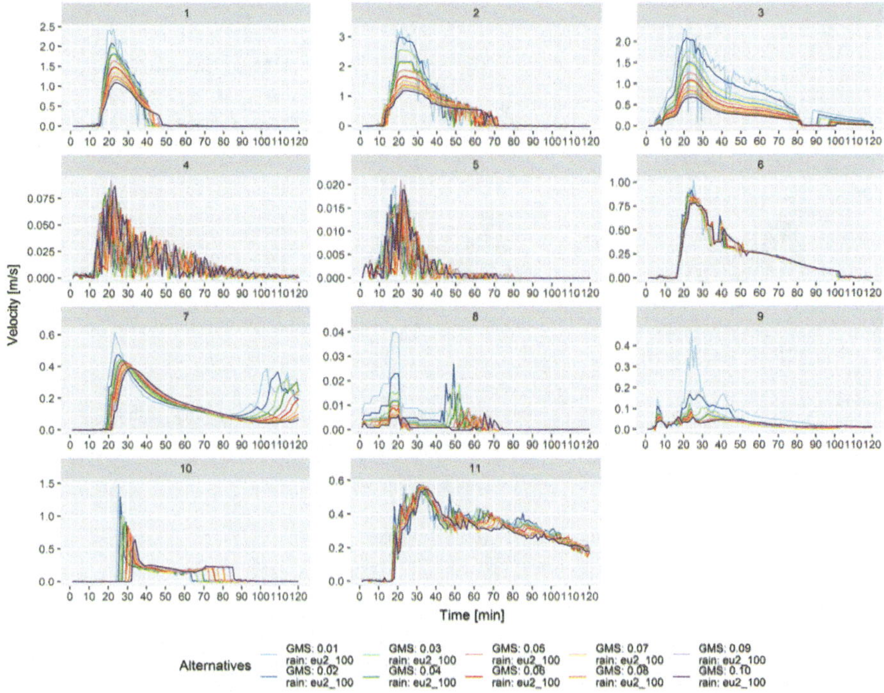

Fig. 3 Time series of maximum flow velocity in m/s for different GMS surface roughness values at the selected locations shown in Fig. 1

head areas such as point 4, 5 and 8 do not show a clear pattern and are due to the underlying algorithm, i.e. numerical effects. Strong dependencies with lower flow velocities for higher GMS values up to a reduction of a factor of four are visible for points with more concentrated surface runoff such as at points 1, 2, 3 7 and 10.

3.2 Precipitation and Flow Velocity

The selection of the precipitation event strongly influences the maximum flow velocity as well as the duration of certain flow velocities (see Fig. 4). Block rain is the most artificial selection, resulting in many points with a near stationary flow regime. The influence of the location of the different observation points is clearly visible.

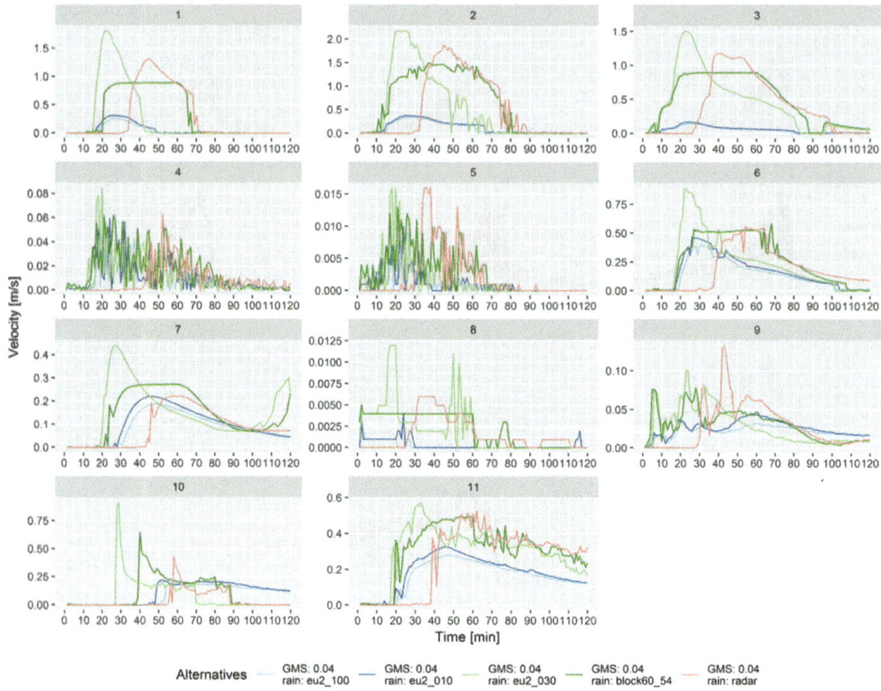

Fig. 4 Time series of maximum flow velocity in m/s for different precipitation inputs at the selected locations shown in Fig. 1

3.3 Surface Roughness and Maximum Flow Rate

The effects of the GMS surface roughness on the flow rate can be seen in Fig. 5. While these patterns are similar to Fig. 3, the intensity of the effect is somewhat lower, perhaps due to the compensating effect of differing water levels, i.e. lower flow velocities cause higher water levels leading to similar flow rates. Nearly all points show some numeric instability for the "smoothest" GMS value of 0.01, which is associated with the highest flow rates as well as velocities.

3.4 Surface Spatial Resolution and Maximum Flow Velocity

The effects of different surface spatial resolutions were examined for two exemplary areas close to the watercourse at the valley floor. The distribution of maximum flow velocity in m/s based on all raster pixel values in these areas is shown in Fig. 6. For both areas, we can clearly see the strong influence of the surface spatial resolution, especially on the higher flow velocities that are typically associated with structural damages to buildings (dynamic loads) and erosion processes of foundations and

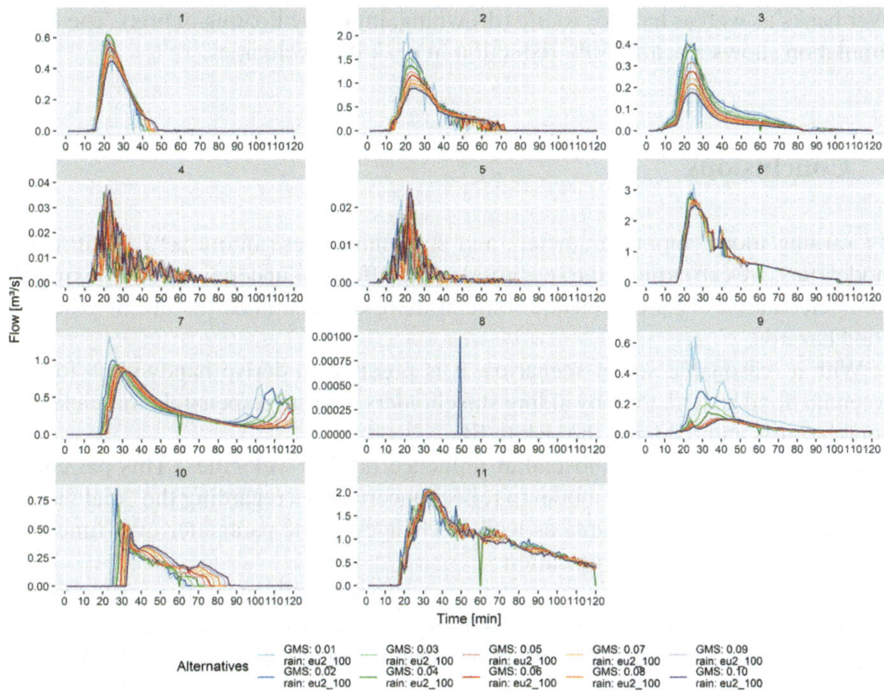

Fig. 5 Time series of maximum flow in m^3/s for different GMS surface roughness values at the selected locations shown in Fig. 1

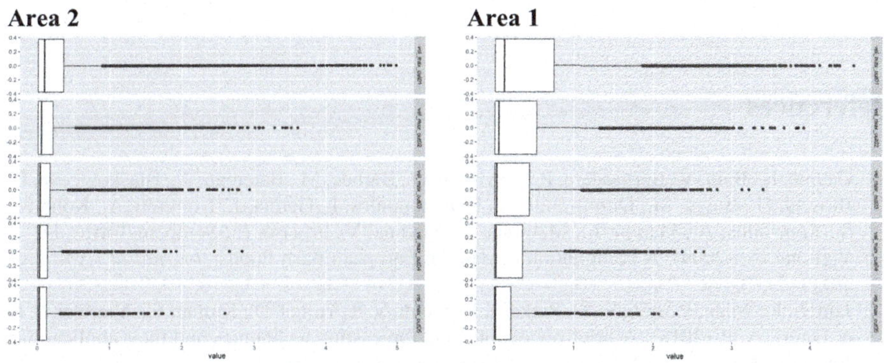

Fig. 6 Distribution of velocity in m/s depending on surface resolution (from top to bottom 2, 4, 6, 8, and 10 m) for two areas based on all raster pixel values in these areas. Figure 1 shows the locations of the areas

river banks as well as hazards to life (drowning, injury by floating debris). The 10-m simulation shows a nearly 50% reduction in peak flow velocities.

4 Conclusions

By varying model input parameters and combining these into a set of alternative model runs, we can explore the sensitivities and effects of uncertainties during model parameterisation. This can ensure a much more differentiated picture of potential flood hazards.

With a reasonable set of scenarios, it is possible to derive bandwidths for the potential flood hazard, thereby aiding stakeholders who are responsible for managing flood hazards and reducing the risk to life.

GMS roughness has a substantial influence on flow velocities. This parameter should be varied within the model to reflect uncertainties regarding the "real world" situation in the catchment area as well as changes of the parameter over time, e.g. due to shifting patterns of vegetation.

The choice of surface spatial resolution strongly influences flow velocities: a coarser resolution, i.e. bigger cells and larger distances between them, results in lower peak flow velocities. The effect on mean flow velocities is much weaker.

Acknowledgements The presented contribution is partly based on results from the project "RAINMAN—Integrated Heavy Rain Risk Management" funded by the European Union (Interreg CENTRAL EUROPE 2020).

References

1. Gaume, E., Bain, V., Bernardara, P., Newinger, O., Barbuc, M., Bateman, A., Blaškovičová, L., Blöschl, G., Borga, M., Dumitrescu, A., Daliakopoulos, I., Garcia, J., Irimescu, A., Kohnova, S., Koutroulis, A., Marchi, L., Matreata, S., Medina, V., Preciso, E., Sempere-Torres, D., … Viglione, A. (2009). A compilation of data on European flash floods. *Journal of Hydrology, 367*(1–2), 70–78.
2. Jain, S. K., Mani, P., Jain, S. K., Prakash, P., Singh, V. P., Tullos, D., Kumar, S., Agarwal, S. P., & Dimri, A. P. (2018). A brief review of flood forecasting techniques and their applications. *International Journal of River Basin Management, 16*(3), 329–344.
3. Hunter, N., Bates, P. D., Horritt, M. S., & Wilson, M. D. (2007). Simple spatially-distributed models for predicting flood inundation: A review. *Geomorphology, 90*(3–4), 208–225.
4. Sauer, A., Körte, L., & Ortlepp, R. (2019). Parameter uncertainties in flood hazard analysis of heavy rain events. In *Proceedings International Probabilistic Workshop 2019* (pp. 37–42), Edinburgh, United Kingdom.
5. Smith, L. S., & Liang, Q. (2013). Towards a generalised GPU/CPU shallow-flow modelling tool. *Computers and Fluids, 88*, 334–343.
6. Smith, L. S., Liang, Q., & Quinn, P. F. (2015). Towards a hydrodynamic modelling framework appropriate for applications in urban flood assessment and mitigation using heterogeneous computing. *Urban Water Journal, 12*(1), 67–78.

7. Liang, Q., & Smith, L. S. (2014). A high-performance integrated hydrodynamic modelling system for urban flood inundation. *Journal of Hydroinformatics, 17*(4), 518–533.
8. Liang, Q., Xia, X., & Hou, J. (2016). Catchment-scale high-resolution flash flood simulation using the GPU-based technology. *Procedia Engineering, 154,* 975–981.
9. DWA Deutsche Vereinigung für Wasserwirtschaft, Abwasser und Abfall (2006). Hydraulische Bemessung und Nachweis von Entwässerungssystemen [Hydraulic design of urban drainage systems]. Arbeitsblatt DWA-A 118. DWA, Hennef, Germany.
10. Winterrath, T., Rosenow, W., & Weigl, E. (2012). On the DWD quantitative precipitation analysis and nowcasting system for real-time application in German flood risk management. *Weather Radar and Hydrology, 351,* 323–329.

Prediction of Concrete Breakout Strength of Single Anchors in Shear

Oladimeji B. Olalusi and Panagiotis Spyridis

Abstract This study proposes a machine learning algorithim—a Gaussian process regression (GPR)—for predicting the concrete breakout capacity of single anchors in shear. To this end, experimental strength of 366 tests on single anchors with concrete edge breakout failures were collected from literature to establish the experimental database to train and test the model. 70% of the data were used for the model training, and the rest were used for the model testing. Shear influence factors such as the concrete strength, the anchor diameter, the embedment depth (technically the influence length), and the concrete edge distance were taken as the model input variables. The generated predictive model yielded a determination coefficient $R^2 = 0.99$ for both the training and testing data sets. Predictions from the developed models were compared to that of the other existing models (Eurocode 2 and ACI 318) to validate its performance. The developed model provided a better prediction of the experimentally observed shear strength, compared to the existing models, yielding low mean absolute error, low bias and variability when tested.

Keywords Machine learning · Gaussian process regression · Breakout strength · Fasteners · Model uncertainty

1 Introduction

Recent advances in computing have given a boost in the use of advanced soft computing methods across all industries; the field of artificial intelligence has been developing since the 1950s. Knowledge-based methods or expert systems have been deployed to assist risk-related decisions under uncertainty. Applications in structural engineering are witnessed in the research field as early as the 1980s [1] since structural engineering problems in practice are governed by a wide range of uncertainties

O. B. Olalusi
University of KwaZulu-Natal, Durban, South Africa

P. Spyridis (✉)
Technical University of Dortmund, Dortmund, Germany
e-mail: panagiotis.spyridis@tu-dortmund.de

© The Author(s), under exclusive license to Springer Nature Switzerland AG 2021
J. C. Matos et al. (eds.), *18th International Probabilistic Workshop*, Lecture Notes in Civil Engineering 153, https://doi.org/10.1007/978-3-030-73616-3_42

related for example to the applied actions, the material performance and homogeneity, or the models used to describe the problem itself. A benefit of such soft computing methods is that they can provide reliable solutions on multi-parametric problems and highly nonlinear correlations of the input. Simultaneously, various disciplines of structural engineering rely on empirical, semi-empirical, or numerical predictive models.

Design of anchorage to concrete, offers an example of semi-empirical predictive and design models, particularly when it comes to concrete-related failure modes. The assessment of anchor capacity using Artificial Neural Networks (ANN) has been previously investigated in [2, 3]. More recent studies present paradigms of machine learning algorithms such as the Gaussian process regression (GPR) with very high efficiency in structural engineering problems [4, 5]. The present paper investigates the feasibility of using GPR algorithms to predict the concrete breakout strength of single anchors loaded in shear. Towards developing the model, experimental results of 366 tests on single anchors with concrete edge breakout failures were collected based on an extensive literature search to establish the database to train and test the model. A parametric study and comparison of the proposed models with other existing predictive models were reported to assess the accuracy and efficiency of the shear capacity design methods for anchors subjected to shear loads.

2 Existing Strength Models for Concrete Breakout Capacity of Single Anchors in Shear

The current EN 1992-4 [6] and ACI 318 [7] design standards provide Eqs. 1 and 2, respectively, for the evaluation of a single anchor's resistance against concrete edge failure in non-cracked concrete. The European design is valid for a $d_{nom} \leq$ 60 mm and influence length $l_f \leq 12d_{nom}$ in case of $d_{nom} \leq 24$ mm and otherwise $\leq \max\{8 \cdot d_{nom}; 300\text{ mm}\}$. The design calculations covered by CEN are valid only up until $f_{ck} < 60$ N/mm^2. ACI 318 is valid for a concrete compressive cylinder strength of 10,000 psi (70 MPa) for cast in anchors and 8000 psi (55 MPa) for post-installed anchors and an anchor diameter up to 4in. (100 mm).

$$V_{EC2-k} = 2.4.d_{nom}^{\alpha}.l_f^{b}.\sqrt{f_{ck}}.c_1^{1.5}. \tag{1a}$$

The mean concrete breakout capacity of single shear anchor in non-cracked concrete for the EN 1992-4 standard is calculated according to Eq. 1b [8].

$$V_{EC2-m} = 3.d_{nom}^{\alpha}.l_f^{\beta}.\sqrt{f_{cm}}.c_1^{1.5}. \tag{1b}$$

where

$$\alpha = 0,1 \cdot \left(\frac{l_f}{c_1}\right)^{0,5}$$

$$\beta = 0,1 \cdot \left(\frac{d_{nom}}{c_1}\right)^{0,2}$$

The design and the mean concrete breakout capacity of single shear anchor in non-cracked concrete for ACI 318 are calculated according to Eq. 2a, b respectively [9, 10].

$$V_{ACI} = \min\left\{\left(0.6 \cdot \left(\frac{l_f}{\mathrm{d}_{nom}}\right)^{0.2} \cdot \sqrt{d_{nom}}\right)\lambda_a\sqrt{f_c'}(c_1)^{1.5}, 3.7.\lambda_a.\sqrt{f_c'}(c_1)^{1.5}\right\} \tag{2a}$$

$$V_{ACI-m} = \min\left\{\left(\frac{l_f}{\mathrm{d}_{nom}}\right)^{0.2} \cdot \sqrt{d_{nom}}\sqrt{f_{cm}}(c_1)^{1.5}, 7.1.\sqrt{f_{cm}}(c_1)^{1.5}\right\} \tag{2b}$$

where d_{nom} is the outside diameter of the anchor. f_c' is the concrete cylinder strength per the ACI acceptance standards. λ_a is the modification factor for applications in lightweight concrete. f_{cm} is the mean concrete cylinder compressive strength. l_f is the influence length of the anchor loaded in shear.

3 Gaussian Process Regression (GPR)

The Gaussian process model is a kernel machine type which can be used as a supervised learning technique for classification as well as regression. Gaussian processes can give a simple probabilistic representation of random processes and can be used for many different types of nonparametric estimation. This method is currently well adopted and applied in various areas in structural engineering [4, 5]. A summary of the GPR algorithm is presented in this section. More details on the GPR methodology can be found in [11].

Given a training set, $U = \{(x_i, y_i); i = 1, 2, \ldots, n\}$, where the input $x_i \in \mathbb{R}^{U.n}$ denotes the design matrix and $y_i \in \mathbb{R}^n$ denotes the vector of the desired output, drawn from an unknown distribution. A GPR model predicts the value of the output variable y_{new}, given a new input vector x_{new}, and training data. In the setting of classic linear regression, we model the output variable y by a function of an input variable x expressed in Eq. 3 [11].

$$y = x^T\beta + \varepsilon \tag{3}$$

where x is the input vector, and y is the observed target value. The random error term $\varepsilon \sim N\left(0, \sigma_n^2\right)$. The error variance σ_n^2 and the coefficients β are estimated from the data.

The multivariate Gaussian distribution, which has a mean vector μ and covariance matrix Σ have the joint probability density expressed as Eq. 4.

$$p(x|\mu) = (2\pi)^{-\frac{D}{2}}|\Sigma|^{-\frac{1}{2}}\exp\left(-\frac{1}{2}(x-\mu)^T\Sigma^{-1}(x-\mu)\right) \quad (4)$$

Unlike the Gaussian distribution, which is a distribution over vectors, the Gaussian process is a distribution over functions with a covariance function $k(x, x')$ and a mean function $m(x)$ (Eq. 5).

$$f(x) \sim GP(m(x), k(x, x')) \quad (5)$$

The indexes of the GP is x. Where the mean function and covariance function of a real process $f(x)$ is defined as Eqs. 6 and 7, respectively.

$$m(x) = E[f(x)] \quad (6)$$

$$k(x, x') = E[(f(x) - m(x))(f(x') - m(x'))] \quad (7)$$

The Gaussian Process is a multivariate Gaussian of infinite length. Following the GPR procedure, the n observations in an arbitrary data set, $y = \{y_1, \ldots, y_n\}$ can be taken as a sample from some multivariate Gaussian distribution. Hence, going from the process of distribution, we can get an understanding of a GP and then draw samples from it. The Gaussian process $f \sim GP(m, k)$ is defined with a mean function $m(x) = 0$ (Eq. 8) and covariance/kernel function $k(x, x')$ (Eq. 9). The goal of only working with finite quantities is simply achieved by requiring the values of f at a distinct number of n locations. Given the x-values we can evaluate the GP, which is now reduced to a multivariate Gaussian distribution [11].

$$\mu = m(x) = 0 \quad (8)$$

$$k(x, x') = \sigma_f^2 \exp\left(-\frac{1}{2l^2}(x - x')^2\right) + \sigma_n^2\delta(x, x') \quad (9)$$

where l denotes the length parameter of the kernel function. $\delta(x, x')$ is denotes the Kronecker delta function.

Gaussian Process regression has different types of kernel functions, some of which includes the squared exponential kernel, Laplace kernel and Linear Kernel. Since different kernel functions are suitable for different type of data, several kernel functions need to be trialled to choose the most appropriate. In this study, the two most suitable kernel functions (non-linear kernel functions) obtained for the database is the Gaussian or Radial Basis Function (RBF) $k(x, x') = \exp(-\frac{1}{2\sigma^2}x - x'^2)$. Where σ is the width of the kernel which are user-defined parameters.

4 Development of the GPR Model

The experimental database considered in this investigation was compiled from different research published in technical literature. They are majorly experiments conducted by [10, 12–22]. The different experiments were aimed and designed to reflect the behaviour of single anchors in shear. The experimental database consists of the failure load of 366 single anchors in shear, failing due to concrete edge breakout in non-cracked concrete. The database covers a wide range of anchor configurations that can be used to assess any anchor design method against experimental results.

The development of an efficient model for predicting the concrete breakout strength of single anchors in shear requires the inclusion of the main factors affecting anchors in shear as inputs. The various parameters affecting the breakout strength of single anchors in shear are discussed in [9]. In this study, the input parameters considered for the implementation of the GPR model includes edge distance c_1, anchor diameters d_{nom}, influence length l_f and concrete strength f_c.

In order to implement the GPR model, the database of experimental anchor tests was split into two subsets, namely: the training data set, and testing data set. The training data set is used to develop both the GPR model, whereas the performance of the developed model is evaluated using the testing dataset. While splitting the database into subsets, it is essential to ensure data patterns that are statistically consistent in both the training and testing datasets. This was achieved by randomly sorting both the training and testing data set until an acceptable consistency is maintained among the input variables, in terms of statistical properties (such as mean and standard deviation) and range of data. This is summarised in Table 1. In this study, 70% of the data (256 out of 366 cases) were used for training, and the remainder (110 cases) were used for testing the models.

The nonlinear regression technique of the GPR models, implemented in a MATLAB environment, was used to predict the concrete breakout strength of anchors in shear. In order to map input data into feature space, nonlinear regression technique requires kernel functions. The optimum search method was used to obtain the optimum parameters. The performance of the developed model was detailed using statistical parameters, namely the coefficient of determination (R^2), mean absolute percentage error (MAPE), Mean square error (MSE) and root-mean-squared error (RMSE).

5 Results and Discussion

5.1 Performance of the GPR Model

The investigation of the performance of the developed model using the training and testing data set is discussed in this section. Using the experimental data presented in Table 1, the GPR model was adopted to successfully learn the interrelationships

Table 1 Statistical summary of the experimental dataset used for the model development

Statistical parameter	Datasets	d_{nom} (mm)	l_f (mm)	c_1 (mm)	f_c (MPa)	V_u (kN)
Range	Training	8–88.9	50–762	45–508	18.4–85.4	4.72–511.27
	Testing	8–88.9	50–762	50–508	18.4–85.4	5.17–518.72
Mean	Training	25.05	206.7	148.1	27.8	71.6
	Testing	25.1	234.7	160.3	28.2	89
SD	Training	15.2	150.3	107.6	9.9	97.4
	Testing	17.7	175.2	126.2	10.4	127.9

SD denotes the standard deviation, V_u is the experimental concrete breakout strength

Table 2 Statistical properties of the developed models

Parameters	GPR model	
Dataset	Training data set (256)	Testing data set (110)
R^2	0.99	0.99
RMSE	8.7	11.1
MSE	76.7	123.9
MAE	5.8	5.6

between concrete breakout strength and varied shear strength parameters (input variables). The model accuracy is assessed using statistical parameters such as R^2, RMSE, MAE and MSE, calculated between the experimental and predicted results, and the results presented in Table 2. The statistical result for the training database (256 experimental tests) is reported as follows: RMSE = 8.7; MAE = 5.8 and R^2 = 0.99. The R^2 and MAE value of the testing database is comparable to that of the training database (110 experimental tests), but with a slightly higher RMSE value. These results indicate that the GPR model is a good predictor of concrete breakout strength. The results demonstrate the generalization capability of the developed GPR model.

5.2 *Comparative Study of the GPR with Existing Concrete Breakout Strength Models*

The trained GRP model was compared to existing concrete breakout capacity predictive models, using the testing dataset, to examine the predictive performance of the models. Two existing concrete breakout capacity models in shear were considered, namely the predictive model of (1) EN 1992-4 (Eq. 1) (2) ACI 318 (Eq. 2). The mean shear resistance function/best estimate model of the EN 1992-4 and ACI 318 models were used in this comparative assessment.

The plot of the experimental breakout capacity against the predicted breakout capacity of the developed models and other existing models, using the testing database, are presented in Fig. 1. The figure portrays the deviation of the predicted strength by the methods from the line of 'Perfect model', which is defined as the line position of all the points where the experimental breakout capacity is equal to predicted breakout capacity. As seen from the figures, the predicted values by the GPR model are much less scattered compared to the values from the other models.

A comparison of the GPR model with other existing models in terms of MAE and RMSE values is presented in Table 3. The table revealed that the MAE and RMSE values of the GPR model is less than that of the other models. Güneyisi et al. [23] developed a model to predict the concrete edge breakout capacity of single adhesive anchors using gene expression programming (GEP). Their developed model yielded R^2 of 0.92 (lower than what is obtained for GPR model in this study) and the value of RMSE = 13; MAPE = 14.2, MSE = 168.7 (higher than what is achieved for GPR model in this study), using the testing database of 34 anchor experiments. However, it should be highlighted that a larger database is utilized for training and testing in this study compared to [23]. A general assessment of the results presented in Table 3 suggests that the GPR model outperformed all the other models investigated in this study.

The summary of the statistical properties of the model uncertainty (obtained as the ratio of experimental breakout strength to predicted breakout strength) associated to the GPR and the other models are presented in Table 3, and the distributions are plotted in a box plot shown in Fig. 2. A box plot is a statistical tool that can be used to provide statistical summaries of the underlying distribution of a dataset. The box plot displays the maximum and minimum values in the dataset, the lower and upper quartiles, the mean and the median. A model uncertainty mean value $\mu_{ME} = 1$ is a condition for an ideal model. The GPR and the other models are assessed, using the criteria that an ideal model is expected to have in addition to model uncertainty mean value $\mu_{ME} = 1$; high precision (that is, small scatter of data) [24, 25].

Assessment of the box plot revealed that the GPR model has the smallest length of the interquartile range of all the models investigated, thereby suggesting less variability of the GPR model uncertainty. As presented in Table 3, the model uncertainty variable associated with the GPR model has a mean value of $\mu_{ME} = 0.99$ (closest

Table 3 Statistical properties of the proposed and existing models

Parameters	GPR	EC2	ACI 318
R^2	0.99	0.98	0.98
RMSE	11.1	41.78	22.2
MSE	123.9	1747.2	196
MAPE	5.6	20.7	14.0
Mean μ_{ME}	0.99	0.86	0.93
Standard deviation σ_{ME}	0.11	0.18	0.23

μ_{ME} and σ_{ME} are respectively, the mean and standard deviation of the model uncertainty (ratio of the experimentally observed shear capacity and the predicted shear capacity by the models). R^2 is the coefficient of determination

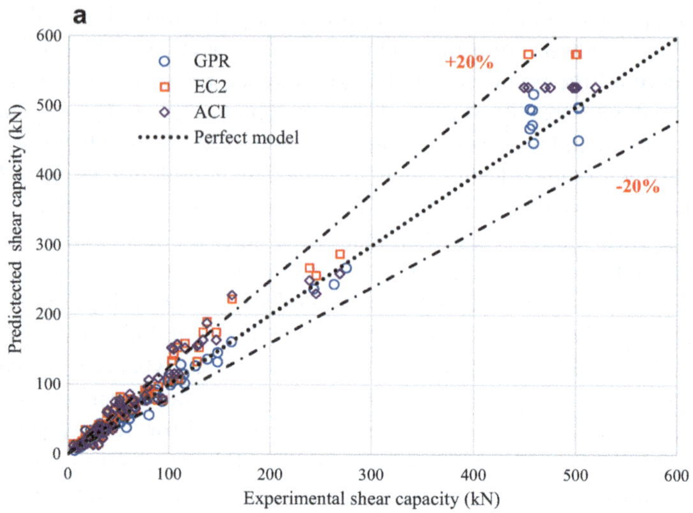

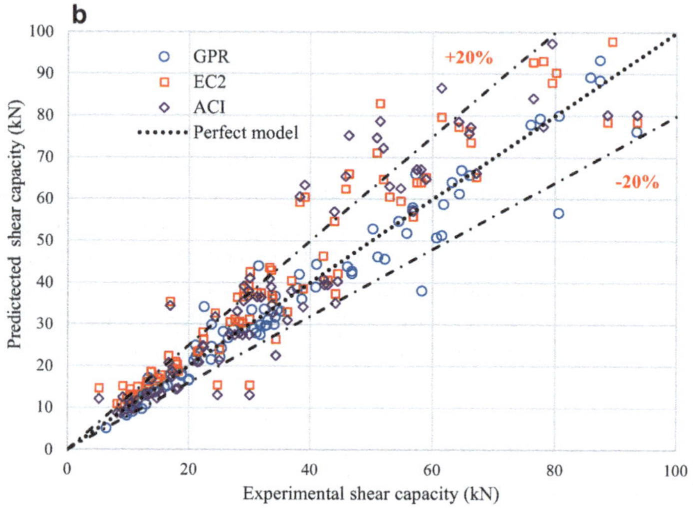

Fig. 1 **a** The full plot of experimental breakout capacity versus predicted breakout capacity. **b** Partial plot of experimental breakout capacity versus predicted breakout capacity (breakout strength up to 100 kN only)

to mean value of 1), suggesting that the model reasonably predicts the breakout strength. Regarding standard deviation, the GPR model yields the lowest dispersion of all the models investigated with $\sigma_{ME} = 0.11$. Table 3 also shows that the best estimate models of EN 1992-4 and ACI 318 overpredict the shear breakout strength, but predictive models in design standards may typically lie on the conservative side.

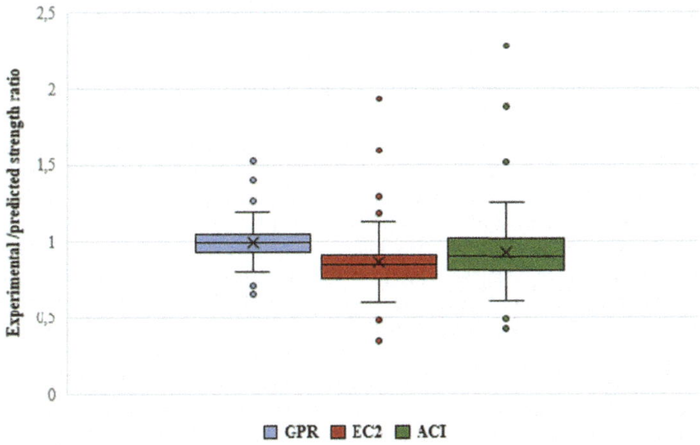

Fig. 2 Comparison of the prediction error of GPR to other models

6 Conclusions

A possible failure mode for shear loaded anchors is the concrete breakout failure. Concrete related failure mode poses a significant safety issue, since they may develop abruptly, without preceding signs of damage. Given this, accurate prediction of the concrete breakout resistance of anchors in shear is crucial. This contribution focuses on the feasibility of using the Gaussian Process Regression (GPR) machine learning algorithms to predict the shear breakout strength of single anchors and quantifies the model uncertainties in existing predictive models. The following general conclusions may be drawn from the present study:

- A reasonable accuracy was obtained for both the training and testing datasets in terms of low RMSE, MAE and high determination coefficients R^2, even though the database for testing were not utilized for training. This reflects the generalization capability of the developed GPR model.
- The prediction capability of the developed model was compared to that of the existing models proposed in EN 1992-4 and ACI 318. The statistical analysis revealed that the proposed GPR model had relatively lower errors and higher determination coefficient than the existing codified models investigated.
- The model uncertainty associated with the GPR model has the closest mean value to 1 ($\mu_{ME} = 0.99$) and the lowest standard deviation ($\sigma_{ME} = 0.11$). Therefore, the GPR model is described as the best performer of all the models analysed in this study.
- In the context of the reliability analyses, a limit state function should ideally be based on a good predictive model, with low bias (with a mean close to 1) and uncertainty coupled. Such a model can be used as a general probabilistic model in the reliability analysis of fastening to concrete design provisions.

References

1. Adeli, H. (1986). Artificial intelligence in structural engineering. *Engineering Analysis, 3*(3), 154–160.
2. Alqedra, M. A., & Ashour, A. F. (2005). Prediction of shear capacity of single anchors located near a concrete edge using neural networks. *Computers and structures, 83*(28–30), 2495–2502.
3. Sakla, S. S., & Ashour, A. F. (2005). Prediction of tensile capacity of single adhesive anchors using neural networks. *Computers and structures, 83*(21–22), 1792–1803.
4. Olalusi, O. B., & Spyridis, P. (2020). Machine learning-based models for the concrete breakout capacity prediction of single anchors in shear. *Advances in Engineering Software, 147,* 102832.
5. Hoang, N. D., Pham, A. D., Nguyen, Q. L., & Pham, Q. N. (2016). Estimating compressive strength of high performance concrete with Gaussian process regression model. *Advances in Civil Engineering.*
6. Technical Committee 250/European Committee for Standardisation (CEN/TC 250). EN 1992-4 Eurocode 2. (2018). *Design of concrete structures. Design of fastenings for use in concrete.*
7. ACI Committee and International Organization for Standardization. (2014). *Building code requirements for structural concrete (ACI 318-14) and commentary.* American Concrete Institute.
8. Hofmann, J. (2004). Tragverhalten und Bemessung von Befestigungen unter beliebiger Querbelastung in ungerissenem Beton. (*Load—Bearing behavior and design of fastenings under arbitrary shear loading in uncracked concrete—In German*). Dissertation, University of Stuttgart.
9. Fuchs, W., Eligehausen, R., & Breen, J. E. (1995). Concrete capacity design (CCD) approach for fastening to concrete. *Structural Journal, 92*(1), 73–94.
10. Lee, N. H., Park, K. R., & Suh, Y. P. (2010). Shear behavior of headed anchors with large diameters and deep embedments. *ACI Structural Journal, 107*(2).
11. Bentsen, F. H. (2019). *Model Construction with Support Vector Machines and Gaussian Processes through Kernel Search* (Master's thesis, The University of Bergen).
12. Ueda, T., Kitipornchai, S., & Ling, K. (1990). Experimental investigation of anchor bolts under shear. *Journal of Structural Engineering, 116*(4), 910–921.
13. Ueda, T., Stitmannaithum, B., & Matupayont, S. (1991). Experimental investigation on shear strength of bolt anchorage group. *Structural Journal, 88*(3), 292–300.
14. Dong-Hyun, K., Yong-Myung, P., Ho-Jung, J., & Moon-Ki, K. (2014). Concrete breakout capacity of cast-in-place anchor under shear loading. *Advanced Materials Research.*
15. Hofmann, J., Fuchs, W., & Eligehausen, R. (2004). Quertragfähigkeit randnaher Befestigungsmittel mit Belastung senkrecht zum Bauteilrand (Behavior of anchorages arranged close to the edge and loaded towards the edge). In: Beton- und Stahlbetonbau; 99 No. 10, pp. 806–807 (in German).
16. Shaikh, A. F., & Yi, W. (1985). In-place strength of welded headed studs. *PCI Journal, 30*(2), 56–81.
17. Toth, M., Bokor, B., & Sharma, A. (2019). Anchorage in steel fiber reinforced concrete–concept, experimental evidence and design recommendations for concrete cone and concrete edge breakout failure modes. *Engineering Structures, 181,* 60–75.
18. Unterweger, A. (2008). Randnahe Anker unter Querlast mechanische Modellierung und Bemessung (Anchors close to edge under shear load). Vienna, Austria: Universität für Bo-denkultur, Dissertation, (In German).
19. Klingner, R. E., Muratli, H., & Shirvani, M. (1999). A technical basis for revision to anchorage criteria. Division of Engineering Technology, Office of Nuclear Regulatory Research, US Nuclear Regulatory Commission.
20. Grosser, P. (2010). Single anchors loaded in shear close to the edge in high strength concrete. Test Report, Stuttgart, Germany: Institute of Construction Materials, University of Stuttgart. Report No. E 10/19 – Bft/17, not published
21. Spyridis, P. (2011). *Behavior of anchor groups under shear loads—Influence of assembly tolerances.* Dissertation. University of Natural resources and Life Sciences of Vienna.

22. Senftleben, S. (2010). Zum Tragverhalten von randnahen Reihenbefestigungen beansprucht durch eine Querlast parallel zum Bauteilrand. (*Behavior of multiple anchor connections close to the edge under shear loading parallel to the edge*). Stuttgart, Germany: Institute of Construction Materials, University of Stuttgart, Diploma Thesis, (in German).
23. Gesoğlu, M., Güneyisi, E. M., Güneyisi, E., Yılmaz, M. E., & Mermerdaş, K. (2014). Modeling and analysis of the shear capacity of adhesive anchors post-installed into uncracked concrete. *Composites Part B: Engineering, 60,* 716–724.
24. Olalusi, O. B., & Viljoen, C. (2020). Model uncertainties and bias in SHEAR strength predictions of slender stirrup reinforced concrete beams. *Structural Concrete, 21*(1), 316–332.
25. Olalusi, O. B., & Spyridis, P. (2020). Uncertainty modelling and analysis of the concrete edge breakout resistance of single anchors in shear. *Engineering Structures, 222,* 111112.

Probabilistic Characterization of the Axial Load Bearing Capacity of a Concrete Column Exposed to the Standard Fire

Balša Jovanović and Ruben Van Coile

Abstract To demonstrate adequate structural fire safety for exceptional designs, the uncertainties of the design input parameters must be explicitly considered. In this contribution, the case study included in ISO/CD TR 24679-8:2020 of a concrete column subject to a standardized heating regime is revisited considering improved uncertainty modelling for the input parameters. Monte Carlo simulations are applied to obtain the distribution of the axial load bearing capacity of the column, $P_{\max}$, at 240 min of ISO 834 standard fire exposure. The obtained distribution however does not fit any distribution type commonly assumed for the resistance effect. To get more detailed information on the parameters governing the distribution of $P_{\max}$, and to allow for the application of more efficient calculation procedures and the development of design guidance, a detailed analysis of the obtained distribution for the load bearing capacity is conducted. The effect of each of the input parameters' uncertainty on the column capacity is quantified using three different methods of sensitivity analysis. Furthermore the distribution type describing the concrete load bearing capacity for the considered standard fire exposure is evaluated in detail. It is concluded that the parameter defining the quantile of the concrete strength retention is the main contributor to the variability of the column capacity at 240 min standard fire exposure. Furthermore, it is found that the column capacity can be described by a mixed lognormal distribution, considering constituent lognormal distributions for fixed concrete strength retention parameter values. Based on these findings, improvements for probability of failure calculations of fire-exposed concrete columns are developed. The analysis provides insight for the reliability-based design of concrete columns exposed to fire, achieving a specified target safety level.

Keywords Fire · Concrete column · Uncertainty quantification · Sensitivity analysis · Safety level

B. Jovanović (✉) · R. Van Coile
Ghent University, Ghent, Belgium
e-mail: Balsa.Jovanovic@ugent.be

© The Author(s), under exclusive license to Springer Nature Switzerland AG 2021
J. C. Matos et al. (eds.), *18th International Probabilistic Workshop*, Lecture Notes in Civil Engineering 153, https://doi.org/10.1007/978-3-030-73616-3_43

1 Introduction

In most design situations, the combination of prescribed fire resistance ratings and standardized calculation methods is sufficient to prove that the structure has an adequately low, although unquantified, failure probability. In special cases of high importance, however, adequate safety needs to be demonstrated explicitly [1]. In those situations, (semi-)probabilistic calculations are required which directly take into account the stochastic nature of the input parameters.

The recent International Standards technical report ISO/TR 24679-8:2020 [2] demonstrates the use of such probabilistic analyses for structural fire engineering, using the example of a concrete column subject to a standard ISO 834 heating regime. The column in question is the part of the Eurocode reference concrete building per Biasoli et al. [3], and has been evaluated for performance-based structural fire engineering in ISO/TR 24679–6:2017 [4]. The column has dimensions of $500 \times 500\ \text{mm}^2$ and a height of 4 m, and is made of concrete C30/37 with siliceous aggregates, reinforced with 12 longitudinal rebars of diameter 20 mm with 42 mm of concrete cover. The full probabilistic analysis of the column capacity P_{max} at 4 h of standard ISO-834 fire exposure was done using nonlinear finite element analysis software SAFIR [5] and is presented in detail in the ISO technical report. Whereas the evaluation in ISO/TR 24679-8:2020 considers the concrete strength retention factor as deterministic in accordance with EN 1992-1-2:2004 [6], recent investigations by Qureshi et al. [7] have demonstrated a large scatter in this parameter. Furthermore, ISO/TR 24679-8:2020 considered the concrete cover to be deterministic. The concrete cover is however commonly considered a crucial parameter for structural fire performance, and is known to be associated with considerable scatter [8].

Taking into account these two additional stochastic variables results in the updated list of stochastic variables presented in Table 1. Here, the temperature dependent concrete compressive strength retention factor is modelled by the logistic model proposed in [7]. This strength retention factor model already incorporates the concrete compressive strength variation at ambient temperatures. Its mean value at 20 °C is equal to 1, while the underlying coefficient of variance (COV) equal to the 0.15. Therefore, this retention factor is combined with the mean concrete compressive strength at 20 °C (deterministic value). The same reasoning applies to the reinforcement yield strength retention factor. For consistency with ISO TR 24679-8:2020, however, the reinforcement yield strength at ambient temperature is still modelled as a stochastic value. This potentially introduces a double consideration of (part of) the 20 °C variability in steel yield stress. The concrete cover is modelled as a Beta distribution with mean value equal to the nominal concrete cover increased by 5 mm and a standard deviation of 5 mm. This Beta distribution is bounded by [c_{nom} − 10 mm, c_{nom} + 20 mm] (i.e. a symmetrical Beta distribution bounded by ±3 times the standard deviation), based on the recommendation in the JCSS PMC [8].

The distribution of the load bearing capacity P_{max} is evaluated through 10^4 Latin Hypercube Simulations (LHS), using the same procedure as described in ISO/TR 24679-8:2020. The obtained distribution of the column capacity P_{max} is visualized in

Fig. 1 together with a lognormal approximation. The mean value vas determined as 5041.63 kN with 95% confidence interval of ±19.61 kN Contrary to ISO/TR 24679-8:2020, the observed histogram does not match the lognormal approximation, and further analysis indicates that the observed distributed cannot be approximated by any other theoretical distribution commonly assumed for the resistance effect. This observation is problematic, as ISO/TR 24679-8:2020 relies on the approximation by theoretical distributions to demonstrate the derivation of (global) safety factors.

Considering the above observed issue, a detailed analysis of the obtained distribution for the load bearing capacity is conducted. First, in order to get more detailed information on the parameters governing the distribution of P_{max} a sensitivity analysis is performed, applying the same three different methods as in [9]. Based on the obtained insights, and to allow for the reliable application of more efficient calculation procedures and the future development of design guidance, a mixed lognormal distribution is proposed for describing the column capacity during fire, P_{max}.

2 Stochastic Input Parameters

See Table 1.

Table 1 Stochastic input parameters used

Parameter	Distribution	Mean μ	Standard deviation σ	References
20 °C concrete compressive strength, $f_{c,20}$ (MPa)	–	42.9 (f_{ck} = 30MPa)	–	–
Concrete compressive strength retention factor, k_{fc} (−)	Logistic	Temperature dependent	Temperature dependent	[7]
20 °C reinforcement yield strength, $f_{y,20}$ (MPa)	Lognormal	560 (f_{yk} = 500MPa)	30	[10]
Steel yield stress retention factor, k_{fy} (−)	Logistic	Temperature dependent	Temperature dependent	[7]
Average eccentricity, e (m)	Normal	0	0.004	[11]
Out of straightness, f (m)	Normal	0	0.004	[11]
Out of plumbness, Φ	Normal	0	0.0015	[11]
Concrete cover, c (mm)	Beta [μ ± 3σ]	42 + 5 = 47	5	[8]

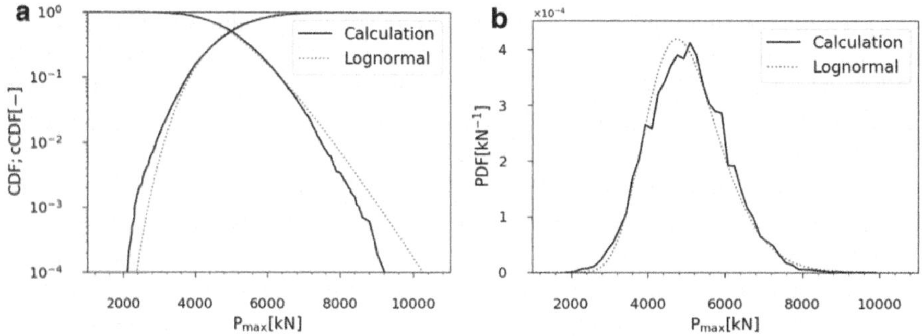

Fig. 1 **a** CDF and **b** PDF for simulation results compared to a lognormal distribution

3 Sensitivity Analysis

The three applied sensitivity analysis methods are described in the following. The results for the concrete column are presented in Sect. 3.4.

3.1 Spearman's Rank Correlation Coefficients [12]

Spearman's rank correlation coefficient r_S is a measure for the correlation of the variables X and Y. The value of r_S can be in the range $-1 \leq r_S \leq 1$. The value of $r_S \approx \pm 1$ represents fully correlated variables and the value of $r_S \approx 0$ indicates there is no correlation. The sign of r_S indicates positive or negative correlation.

In order to calculate r_S, consider the vectors of n (pairs of) observations $\boldsymbol{x}$ and $\boldsymbol{y}$. First the n values of both vectors are each sorted independently in ascending order based on their value. The sorted positions are denoted as their rank values $r_{X,i}$ and $r_{Y,i}$, both ranging from 1 to n. In case of identical values, the minimum of the ranks that would have been assigned to all the tied values is assigned to each value. Substituting the values $\boldsymbol{x}$ and $\boldsymbol{y}$ by their rank values $r_{X,i}$ and $r_{Y,i}$, the rank vectors $\boldsymbol{r}_X$ and $\boldsymbol{r}_Y$ are obtained. The Spearman's rank correlation coefficient r_S can then be calculated using the following formula:

$$r_S = \frac{\text{cov}(\boldsymbol{r}_X, \boldsymbol{r}_Y)}{\sigma(\boldsymbol{r}_X)\sigma(\boldsymbol{r}_Y)} \tag{1}$$

where $\text{cov}(\boldsymbol{r}_X, \boldsymbol{r}_Y)$ represents the covariance of the rank vectors and $\sigma(\boldsymbol{r}_X)$ and $\sigma(\boldsymbol{r}_Y)$ the standard deviations. In case that all $2 \times n$ rank values are distinct integers, the following formula can be used:

$$r_S = 1 - \frac{6\sum_{i=1}^{n}\left(r_{X,i} - r_{Y,i}\right)^2}{n\left(n^2 - 1\right)} \tag{2}$$

Spearman's rank correlation coefficient is a relative measure, allowing to compare the influence of different input variables X_j on the output $Y = f(X_1, \ldots, X_k)$. For comparison with other sensitivity analyses, Spearman's rank correlation coefficients can be normalized in the following way:

$$r_{Sj,norm} = \frac{r_{Sj}^2}{\sum_{j=1}^{k} r_{Sj}^2} \tag{3}$$

3.2 *Sobol Indices—Conceptual Method [13]*

Sobol indices, often referred to as Variance-based sensitivity analysis, represent one of the most common ways of examining the influence of input variables on the variation of the output. In its essence, this method decomposes the variance of the output into fractions that can be attributed to each input variable. The idea is to quantify the variation of the output when one of the input variables is kept constant. This way the variance of that specific input parameter is removed and one of three possible outcomes can be observed: (i) the output variance is relatively unchanged compared to the evaluation with consideration of all input variation. This leads to the conclusion that the variation of the input parameter is unimportant; (ii) the output variance is reduced significantly. This means that the input parameter has crucial influence; or (iii) the variance change is moderate, implying that the parameter is not crucially important, but should not be ignored.

In order to represent this mathematically, the output function is defined as $Y = f(X_1, \ldots, X_k)$, where $X_1, \ldots, X_k$ represent k input parameters. This way the variance of the output Y can pre presented with the following variance decomposition formulation [14]:

$$V(Y) = V(E(Y|X_i)) + E(V(Y|X_i)) \tag{4}$$

The first term $V(E(Y|X_i))$ is known as the first-order effect of X_i on Y and its influence can be expressed as the first-order sensitivity index S_k using the following equitation:

$$S_i = \frac{V(E(Y|X_i))}{V(Y)} \tag{5}$$

The conceptual method originates from the definition of the first order sensitivity index in Eq. (5). In order to evaluate the numerator, first the parameter space of X_i is divided in m sections. These sections are defined such that the probability that a realization x_i is located in any of the intervals is the same. If the section is small enough it can be assumed that the value of X_i is fixed. That way the variance of Y

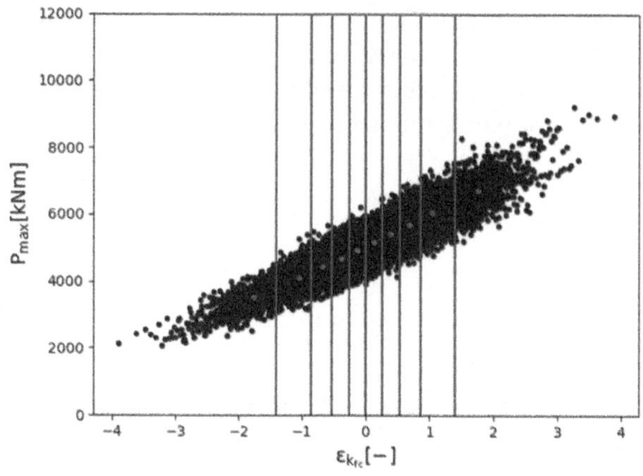

Fig. 2 Representation of the conceptual method

inside of the section can be attributed to all the other input parameters and it can be eliminated by calculating the mean value of Y in it. When this value, which effectively represents $E(Y|X_i)$, is calculated for each section, the first order sensitivity index of Eq. (5) can be easily calculated by dividing its variance with the total variance of Y.

On Fig. 2 the concept behind this method is presented for the input variable $\varepsilon_{k_{fc}}$. The advantage of this method compared to the "Matrix method" is that it requires only l evaluations instead of $l \times (k+1)$. Using the conceptual method, surrogate models do not have to be used. For the considered case study, 10^4 simulations were already available.

3.3 Sobol Indices—"Matrix Method" [15]

One of the methods of calculating these values is the "Matrix method" presented by Saltelli [15]. He modified Sobol's [14] method of estimating the first order effects using Monte Carlo simulations by modifying the sample matrix. The method consists of firstly generating two $l \times k$ sample matrixes $\boldsymbol{X}_A$ and $\boldsymbol{X}_B$ where each column represents one input variable and each row one set of inputs. For each input variable one matrix $\boldsymbol{X}_{Ci}$ is generated by replacing the i-th column in $\boldsymbol{X}_B$ with the i-th column from $\boldsymbol{X}_A$. Next the function Y is evaluated for $\boldsymbol{X}_A$ and each $\boldsymbol{X}_{Ci}$, and vectors $\boldsymbol{y}_A = Y(\boldsymbol{X}_A)$ and $\boldsymbol{y}_{Ci} = Y(\boldsymbol{X}_{Ci})$ are created. First order sensitivity indexes are evaluated in the following way:

$$S_i = \frac{\boldsymbol{y}_A^T \boldsymbol{y}_{Ci} - l\overline{y}_A\overline{y}_{Ci}}{\boldsymbol{y}_A^T \boldsymbol{y}_A - l\overline{y}_A^2} \tag{6}$$

where $\overline{y}_A = \frac{1}{l}\sum_{i=1}^{l} y_A$ and $\overline{y}_{Ci} = \frac{1}{l}\sum_{i=1}^{l} y_{Ci}$.

In order to obtain these values for all input parameter, $l \times (k+1)$ evaluation of function $Y = f(X)$ are needed. For this case evaluation using the nonlinear finite element analysis would be computationally too expensive. An alternative approach is creating a surrogate model trained using a single simulation set of n realizations. The surrogate model subsequently allows a large amount of approximate calculations to be performed almost instantly. In the following, a second order polynomial regression model is chosen, which can be presented using the following formula:

$$Y = \theta_0 + \sum_{j=1}^{k} \theta_j X_j + \sum_{i=1}^{k} \sum_{j=1}^{k} \theta_{i,j} X_i X_j \tag{7}$$

where $\theta_0, \ldots \theta_{i,j}$ are referred to as regression coefficients. The applied regression technique is the 'gradient decent' algorithm as described in [16]. For the considered case study, the obtained surrogate model has a coefficient of determination $R^2 = 0.96$ which makes it suitable enough to be used for sensitivity analysis using the "Matrix method".

3.4 Results

Figure 3 shows the results of all three methods of sensitivity analysis. The Sobol indices calculated using the Matrix method and the Conceptual method show differences, but the ranking order of the parameters is the same. The Spearman's ranking coefficients differ more, for instance assigning almost no influence to the eccentricity e, even though the eccentricity is the second most influential variable based on the Sobol indices. Even though differences exist between the different methods, it can be concluded that they agree that the parameter with the biggest influence is $\varepsilon_{k_{fc}}$.

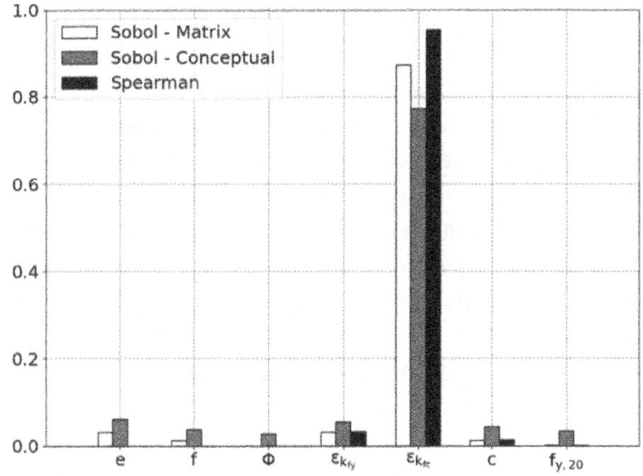

Fig. 3 Sensitivity analysis results

This value represents the normally distributed value that describes the variation in the logistic concrete compressive strength retention factor model [7]. This retention factor model is expressed as follows, with T the temperature in degrees Celsius:

$$k_{c,T} = \frac{1.4 \times \exp\left(0.8892 - 0.6319 \times 10^{-3} \times T - 3.295 \times 10^{-6} \times T^2 + 0.45 \times \varepsilon_{k_{fc}}\right)}{\exp\left(0.8892 - 0.6319 \times 10^{-3} \times T - 3.295 \times 10^{-6} \times T^2 + 0.45 \times \varepsilon_{k_{fc}}\right) + 1} \quad (8)$$

This probabilistic model is based on the Eurocode [6] strength retention model, but is modified to incorporate the uncertainties of the literature data using a Bayesian updating rule. One main feature of this model is that at the ambient temperature of 20 °C it results in a mean value close to 1 and a COV close to 0.15, which is in accordance with the literature for describing the ambient design variability [10]. As the considered concrete column has a low slenderness ratio of $\lambda = 27.7$, second order effects are relatively unpronounced and therefore it is reasonable to expect that its capacity is mainly influenced by the concrete compression strength, which is governed by the parameter $\varepsilon_{k_{fc}}$.

The other parameters have a much lower influence, but their exclusion from the probabilistic calculation could have a significant effect on the end result, as no parameter has a sensitivity index close to zero for all considered methods. Notable examples of these are the geometric imperfections: average eccentricity e, out of straightness f and out of plumbness Φ, as their nominal values (mean value: $\mu = 0$) would actually result in larger values of $P_{\max}$.

4 Mixed Lognormal Distribution

4.1 Hypothesis and Validation

As indicated with Fig. 1, the obtained occurrence frequencies for $P_{\max}$ do not match any theoretical distribution, contrary to the lognormal approximation in ISO/TR 24679-8:2020. In the ISO document, however, a deterministic model for the compressive strength reduction of concrete has been used. Together with the observation that the concrete strength retention parameter $\varepsilon_{k_{fc}}$ drives the output ($P_{\max}$) variability, this suggests that for a fixed value of the parameter $\varepsilon_{k_{fc}}$ the distribution of $P_{\max}$ can be described by a lognormal distribution. Consequently, the total $P_{\max}$ distribution would then be described by a composite distribution known as the mixed-lognormal distribution.

In order to test this hypothesis first the overall $P_{\max}$ distribution can be written as:

$$P_{\max} = \int_{-\infty}^{\infty} f_{\varepsilon_{k_{fc}}}\left(\varepsilon_{k_{fc}}\right) \times P_{\max}|\varepsilon_{k_{fc}} d\varepsilon_{k_{fc}} \quad (9)$$

where $f_{\varepsilon_{k_{fc}}}\left(\varepsilon_{k_{fc}}\right)$ is the probability density function of the parameter $\varepsilon_{k_{fc}}$, which is a standard normal distribution, and $P_{\max}|\varepsilon_{k_{fc}}$ is the lognormal distribution of $P_{\max}$ for a given value of $\varepsilon_{k_{fc}}$. To evaluate this integral it is more practical to apply a discretization, i.e. to divide the distribution of $f_{\varepsilon_{k_{fc}}}\left(\varepsilon_{k_{fc}}\right)$ in n sections, each with nominal (mean) value of $\varepsilon_{k_{fc},i}$ and interval probability $P\left[\varepsilon_{k_{fc},i}\right]$. That way the Eq. (9) can be approximated as:

$$P_{\max} \approx \sum_{i=1}^{n} P\left[\varepsilon_{k_{fc},i}\right] \times P_{\max}|\varepsilon_{k_{fc},i} \quad (10)$$

The above hypothesis is confirmed in the following by dividing $f_{\varepsilon_{k_{fc}}}\left(\varepsilon_{k_{fc}}\right)$ in 10 different sections and then calculating the value of $P\left[\varepsilon_{k_{fc},i}\right]$ for each of them. Next the Monte Carlo simulation results for $P_{\max}$ are also divided in same number of sections based on the value of $\varepsilon_{k_{fc}}$. For each interval, the local mean value and variance for $P_{\max}$ are obtained. These local mean values and variance, define the local lognormal distribution $P_{\max}|\varepsilon_{k_{fc},i}$. Using the Eq. (10), the mixed lognormal distribution P_{max} is obtained. Results are presented in Fig. 2 which clearly demonstrates a good fit. It should be noted that fit improves with increased number of sections, but already shows appropriate fit with 6 sections.

4.2 *Probability of Failure*

In [2] it was demonstrated that the total load on the column (including model uncertainty) can be appropriately approximated by a lognormal distribution. In that case, as the $P_{\max}$ and resistance model uncertainty K_R are both lognormally distributed, the probability of failure for the limit state $Z = K_R P_{\max} - P_E = P_R - P_E$ can be calculated as:

$$P_f = \Phi\left(\frac{\ln\left(\frac{\mu_{P_E}}{\mu_{P_R}}\sqrt{\frac{V_{P_R}^2+1}{V_{P_E}^2+1}}\right)}{\sqrt{\ln\left(\left(V_{P_R}^2+1\right)\left(V_{P_E}^2+1\right)\right)}}\right) \quad (11)$$

where μ represents mean value, V coefficient of variation and Φ the standard normal cumulative distribution function.

In this study however, the $P_{\max}$ is not lognormally distributed, but as shown (Fig. 4), can be represented as a weighted sum of lognormal distributions (Eq. (10)).

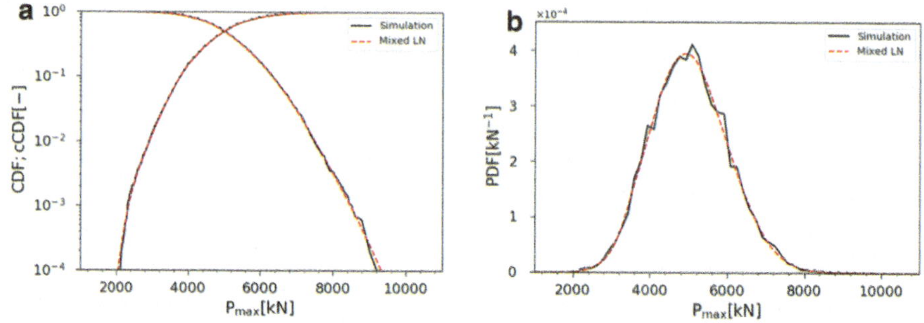

Fig. 4 **a** CDF and **b** PDF for simulation results compared to a mixed-lognormal distribution

This fact allows the probability of failure to be calculated in the following way:

$$P_f \approx \sum_{i=1}^{n} P\left[\varepsilon_{k_{fc},i}\right] \times \Phi\left(\frac{\ln\left(\frac{\mu_{P_E}}{\mu_{P_R}\left(\varepsilon_{k_{fc},i}\right)}\sqrt{\frac{V_{P_R}^2|\varepsilon_{k_{fc},i}+1}{V_{P_E}^2+1}}\right)}{\sqrt{\ln\left(\left(V_{P_R}^2|\varepsilon_{k_{fc},i}+1\right)\left(V_{P_E}^2+1\right)\right)}} \right) \quad (12)$$

The probability of failure calculated using Eq. (12), with parameters as listed in Table 2 is equal to 7.6×10^{-3}, when dividing the $f_{\varepsilon_{k_{fc}}}\left(\varepsilon_{k_{fc}}\right)$ in 10 sections. When compared to the value obtained using 10^9 Monte Carlo simulations, 7.7×10^{-3}, it shows a relative difference of only 1%.

As Eq. (10) is an approximation of Eq. (9) and is equal to it when number of sections n goes to infinity, an analysis of the influence of the number of sections on the calculated probability of failure is conducted. Fig. 5a presents the results of that analysis, indicating that 10 sections is in this case sufficient to accurately capture probability of failure. As Eq. (12) represents a summation of contributions from different constituent lognormal distributions, in Fig. 5b the contribution of

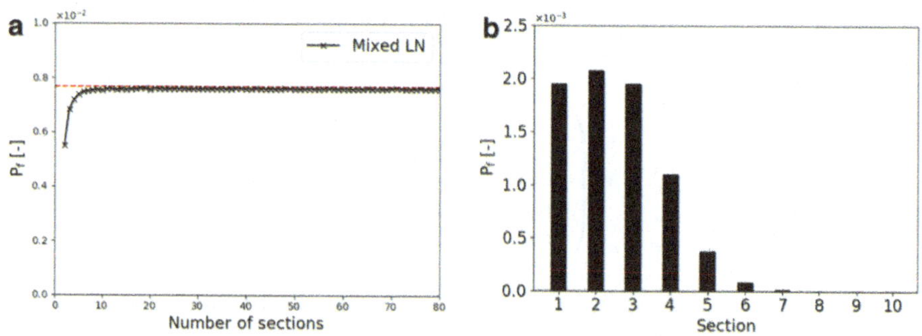

Fig. 5 **a** Probability of failure calculated using Eq. (12) in function of number of sections and **b** contribution of each section to it when number of section is equal to 10

each section to the total probability of failure is presented. From this evaluation, it is concluded that only the first half of the sections, in this case 5 sections, have a significant effect on the overall probability of failure. This behaviour is also consistent when increasing the number of sections.

Equation (12) allows calculating the probability of failure precisely, taking into account the failure probability contributions from the constituent functions. To make this calculation of the constituent failure probabilities feasible in absence of computer tools, the graph presented in Fig. 6 is created. It presents a fast way to evaluate the standard normal cumulative distribution function in Eq. (12). The graphical solution relies on three easily calculated coefficients, which are functions of the mean and COV of the load and (constituent) resistance effects:

$$a = \frac{\mu_{P_E}}{\mu_{P_R}\left(\varepsilon_{k_{fc},i}\right)} \tag{13}$$

Table 2 Stochastic load and model uncertainty parameters

Parameter	Distribution	Mean μ	Standard deviation σ
Permanent load P_G (kN)	Normal	2000	200
Imposed load P_Q (kN)	Gumbel	267	294
Model uncertainty for the load effect, K_E (−)	Lognormal	1	0.1
Model uncertainty for the resistance effect, K_R (−)	Lognormal	1	0.15

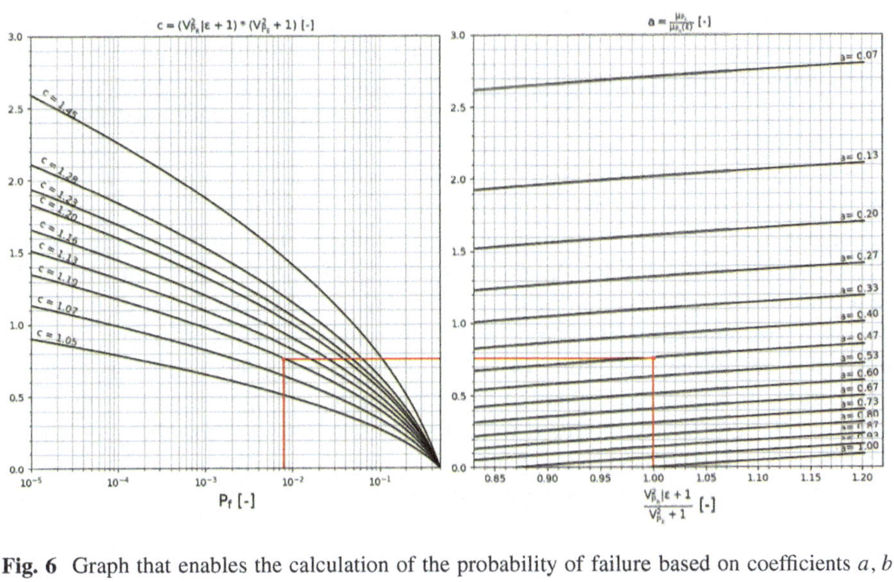

Fig. 6 Graph that enables the calculation of the probability of failure based on coefficients a, b and c

$$b = \frac{V_{P_R}^2|\varepsilon_{k_{fc},i} + 1}{V_{P_E}^2 + 1} \tag{14}$$

$$c = \left(V_{P_R}^2|\varepsilon_{k_{fc},i} + 1\right)\left(V_{P_E}^2 + 1\right) \tag{15}$$

Once these three coefficients are obtained it is easy to use the graph in Fig. 6 to get the constituent probability of failure. This is demonstrated in Fig. 6 for the case where $a = 0.47, b = 1.00$ and $c = 1.10$ which gives the result of $P_f = 7.9 \times 10^{-3}$ (note: this evaluation is an example and does not relate to the concrete column failure probability). This provides a fast method of calculating the probability of failure, especially when taking into account that only the contribution from the first half of the constituent sections needs to be evaluated (for the later sections the coefficient a is lower and contributes with a lower P_f). It should be mentioned that this graph can naturally also be used to calculate the probability of failure in the cases where resistance is lognormally distributed (i.e. where there is only one constituent distribution).

The formulation of Eq. (12) provides a fast and consistent way of calculating the probability of failure. The only inputs required are the mean and the COV for $P_{\max}|\varepsilon_{k_{fc},i}$ in the first $n/2$ of n sections. Those can be obtained in multiple ways, for example, through (i) Crude Monte Carlo simulations (as indirectly applied in this study); (ii) through a limited set of Latin Hypercube Samples per interval, considering for each interval a fixed value of $\varepsilon_{k_{fc}}$; (iii) other more advanced approaches. One more advanced approach which has been explored further is the Multiplicative Dimensional Reduction Method with Gaussian interpolation (MDRM-G), as presented in [17]. This methodology allows estimating the mean and COV based on only $4 \times m + 1$ data points, where m is the number of stochastic parameters considered. For the considered concrete column $m = 6$, and thus 25 simulations are required for each interval. Considering 10 intervals (each with a fixed value of $\varepsilon_{k_{fc}}$) gives a total of 250 simulations to obtain a complete assessment of the mixed-lognormal distribution. Applying this method, a failure probability of 4.7×10^{-3} is obtained, indicating an unexpected discrepancy with the results from the analyses presented above. Further investigation indicates that the sampling scheme of the MDRM-G relies heavily on parameter evaluations at their median value. Unfortunately, as mentioned earlier, the median (mean) value for the geometrical imperfections is zero, thus resulting in a consistent overestimation of the mean values for the constituent distributions $P_{\max}|\varepsilon_{k_{fc},i}$ of approximately 7%. If the mean values are adjusted with a constant difference of 7% the probability of failure obtained using this method is equal to 7.9×10^{-3}. It is concluded that the MDRM-G approach appears promising if the method can be properly modified to accurately take into account the geometrical imperfections. Further research is required to generalise the results and develop a set calculation procedure for the probabilistic analysis of fire-exposed concrete columns.

5 Conclusions

Probabilistic analysis of a concrete column exposed 4 h of standard ISO-834 fire curve has been conducted. The obtained column capacity $P_{\max}$ could not be approximated with any commonly used theoretical distribution function, hindering the derivation of safety factors. A more detailed analysis of the results was conducted to explore the obtained distribution in depth. Through sensitivity analyses, the biggest influence on the capacity was identified to be the parameter $\varepsilon_{k_{fc}}$. This parameter describes the concrete compressive strength at ambient temperature as well as its strength reduction at higher temperatures. This is in accordance with the expected behaviour of the column given its dimensions.

Based on the knowledge of this most influential parameter, a mixed lognormal distribution was hypothesized for the column capacity $P_{\max}$. For the considered test case, this hypothesis was validated through the available 10^4 Latin Hypercube Simulation results. The confirmation of a mixed lognormal distribution for $P_{\max}$ enables a quick and simple way to calculate the probability of failure for the fire exposed column. A simple graphical tool has been introduced in order to calculate the probability of failure in case of mixed and regular lognormal distributions. Further research is required to generalise the results.

References

1. Van Coile, R., Hopkin, D., Lange, D., Jomaas, G., & Bisby, L. (2019). The need for hierarchies of acceptance criteria for probabilistic risk assessments in fire engineering. *Fire Technology, 55,* 1111–1146. https://doi.org/10.1007/s10694-018-0746-7
2. ISO. (2020). *ISO/CD TR 24679-8 fire safety engineering—Performance of structures in fire. Part 8: example of a probabilistic fire design of structures.* International Organization for Standardization, Geneva, Switzerland.
3. Biasoli, F., Mancini, G., Just, M., Curbach, M., Walraven, J., Gmeiner, S., Arrieta, J., Frank, R., Morin, C., Robert, F., Poljansek, M., Kamenarova, B., Dimova, S., Pinto Vieira, A. (2014). *Eurocode 2: background and applications, design of concrete buildings. Worked examples.* Publications Office of the European Union, Luxemburg.
4. ISO. (2017). *ISO/TR 24679-6:2017 fire safety engineering—Performance of structures in fire—Part 6: example of an eight-storey office concrete building.* Geneva, Switzerland: International Organization for Standardization.
5. Franssen, J.-M., & Gernay, T. (2017). Modeling structures in fire with SAFIR®: Theoretical background and capabilities. *Journal of Structural Fire Engineering, 8,* 300–323. https://doi.org/10.1108/JSFE-07-2016-0010
6. CEN. (2004). *EN 1992-1-2:2004: Eurocode 2: Design of concrete structures—Part 1-2: General rules. Structural fire design.* European Standard.
7. Qureshi, R., Ni, S., Elhami Khorasani, N., Van Coile, R., Hopkin, D., & Gernay, T. (2020). Probabilistic models for temperature dependent strength of steel and concrete. *Journal of the Structural Engineering. American Society of Civil Engineers, 146,* 04020102. https://doi.org/10.1061/(asce)st.1943-541x.0002621
8. JCSS. (2001). *Probabilistic model code. Part 3.10 dimensions.* Joint Committee on Structural Safety.

9. Achenbach, M., Lahmer, T., & Morgenthal, G. (2017). Global sensitivity analysis of reinforced concrete walls subjected to standard fire—A comparison of methods. *14th International Probabilistic Workshop* (pp. 97–106). Springer.
10. Holický, M., & Sýkora, M. (2010). Stochastic models in analysis of structural reliability. In *Proceedings of the International Symposium on Stochastic Models in Reliability Engineering, Life Sciences and Operation Management*. Beer Sheva, Israel.
11. JCSS. (1999). *Probabilistic model code. Part 3.11 excentricities*. Joint Committee on Structural Safety.
12. Spearman, C. (1904). The proof and measurement of association between two things. *American Journal of Psychology, 15,* 72–101. https://doi.org/10.2307/1412159
13. Marzban, S., & Lahmer, T. (2016). Conceptual implementation of the variance-based sensitivity analysis for the calculation of the first-order effects. *Journal of Statistical Theory and Practice, 10,* 589–611. https://doi.org/10.1080/15598608.2016.1207578
14. Sobol, I. M. (2001). Global sensitivity indices for nonlinear mathematical models and their Monte Carlo estimates. *Mathematics and Computers in Simulation, 55,* 271–280. https://doi.org/10.1016/S0378-4754(00)00270-6
15. Saltelli, A. (2002). Making best use of model evaluations to compute sensitivity indices. *Computer Physics Communications, 145,* 280–297. https://doi.org/10.1016/S0010-4655(02)00280-1
16. Chaudhary, R. K., Van Coile, R., & Gernay, T. (2020). Fragility curves for the fire exposed structural elements through application of regression techniques. In *18th International Probabilistic Workshop*.
17. Van Coile, R., Balomenos, G. P., Pandey, M. D., & Caspeele, R. (2017). An unbiased method for probabilistic fire safety engineering, requiring a limited number of model evaluations. *Fire Technology, 53,* 1705–1744. https://doi.org/10.1007/s10694-017-0660-4

Probabilistic FEM-Analysis for the Retaining Wall of a Deep Excavation at SLS

Alexandra Ene, Timo Schweckendiek, and Horatiu Popa

Abstract Common practice for design of retaining walls for deep excavations is by using characteristic values for geotechnical parameters—as a cautious estimate—for Serviceability Limit State (SLS) and combined with partial factors for Ultimate Limit State (ULS), as indicated in the current design codes such as the Eurocodes. However, more complex probabilistic approaches are increasing in application in order to provide a more uniform level of reliability, thus reducing the cost of the investment or the risk, or both. Also, in terms of tools and methods for performing the calculations, the Finite Element Method (FEM) is very popular nowadays due accessible computers power and user-friendly specialized software which can provide more realistic model, with affordable calculation effort. The present paper presents a case study of applied full probabilistic analysis of a retaining wall for real project deep excavation in Bucharest city, Romania, by FEM calculation in Plaxis 2D software coupled with Probabilistic Toolkit (PTK) software for reliability calculation. The limit function is set on a target value for the displacements of the retaining wall to allow to design for the SLS, since this is in many cases the governing state for deep excavations in urban areas. Different probability distributions are used for assessing the statistics of the geotechnical parameters and the reliability results obtained through these are discussed. Also, a discussion is made on the necessity of including more specific target reliability values for SLS verification and especially for temporary structures in the design codes.

Keywords Probabilistic design · Reliability · FEM · Retaining walls · Serviceability limit state

---

A. Ene (✉) · H. Popa
Technical University of Civil Engineering Bucharest, Bucharest, Romania
e-mail: alexandra.ene@p-a.ro

A. Ene
Popp and Asociatii Inginerie Geotehnica Co., Bucharest, Romania

T. Schweckendiek
Deltares and Delft University of Technology, Delft, The Netherlands

© The Author(s), under exclusive license to Springer Nature Switzerland AG 2021
J. C. Matos et al. (eds.), *18th International Probabilistic Workshop*, Lecture Notes in Civil Engineering 153, https://doi.org/10.1007/978-3-030-73616-3_44

1 Introduction

Deep excavations in urban areas are a common application in geotechnical engineering since for many of the constructions today the use of the underground is highly important. Moreover, deep excavations imply significant risk both for the structure itself, and for neighboring sites and constructions, as well as significant costs for the investment. For these reasons, it is necessary to provide designs that are both acceptably safe and as economical as possible.

Nowadays, Finite Element Method (FEM) is very much used for structural and geotechnical engineering and probabilistic calculations are also increasing in popularity, combined with FEM [1–3].

The scope of the present paper is to describe a full probabilistic analysis performed for a temporary retaining system of a deep excavation, for Serviceability Limit State (SLS) verification. The main uncertainties were modelled as random variables and the SLS verification was expressed in terms of reliability index or, equivalently, the probability of failure.

The objective is to demonstrate the feasibility of a full probabilistic SLS verification for a real-life case study, and to assess the reliability produced by Eurocode designs in terms of exceedance of deformation criteria.

To this end we will first introduce the case study of a deep excavation in Bucharest, Romania. Subsequently, different statistical descriptions are given for the geotechnical parameters to be used in the reliability analysis. Then, the requirements and the models used for the reliability analysis for SLS are presented, including a discussion on target reliability index needed for reliability-based design of such works.

2 Case Study Description

2.1 Project Information

The present paper is based on a real project of a deep excavation in Bucharest city, capital of Romania. Given the proposed development of an office complex consisting of four buildings with two basements and 11–12 overground stories, in several stages of design and execution, on the same site of about 22,000 m^2, several data have become available in every phase from site investigations, execution supervision and monitoring. The current study will focus on the first stage of the development, i.e. the design and execution of Building 1, which has been previously presented and back-analyzed in a deterministic approach by Popa et al. [4].

Table 1 Characteristic values for the geotechnical parameters used in the design

Layers and levels	γ_k^{sup}	$\varphi'_k{}^{inf}$	$c'_k{}^{inf}$	$E_{50} = E_{oed}$	E_{ur}	G_0	γ_{07}	p_{ref}
	(kN/m^3)	(°)	(kPa)	(MPa)	(MPa)	(MPa)	(–)	(kPa)
0. Filling (88 ÷ 85 m ASL)	19	15	5	3	9.7	40	1.0E−04	40
1. Silty clay (85 ÷ 75.5 m ASL)	See Table 2							
2. Sandy clay (75.5 ÷ 73.5 m ASL)	19.8	12.8	29.3	19.2	57.6	240	1.8E−04	280
3. Sand with gravel (73.5 ÷ 64 m ASL)	20.2	39	0	40	120	500	1.1E−04	470

$\gamma_k{}^{sup}$—superior characteristic values for unit weight; $\phi'_k{}^{inf}$—inferior characteristic values for the effective angle of friction; $c'_k{}^{inf}$—inferior characteristic values for the effective cohesion; E_{50}—the triaxial loading stiffness considered equal to E_{oed}—oedometric deformation modulus at the reference pressure; E_{ur}—unloading-reloading deformation modulus at the reference pressur; G_0—small strain shear modulus at the reference pressure; γ_{07}—strain level at 70% of shear modulus; p_{ref}—reference pressure; R_{inter}—strength reduction factor for interface

2.1.1 Geotechnical Conditions on Site

The ground investigations on the area of Building 1 consisted of seven geotechnical boreholes with sampling of soil specimens and Standard Penetration Tests in cohesionless soil layers. Also, Downhole and Crosshole tests were performed in some of the boreholes to assess the seismic characterization of the site and provide small strain stiffness soil parameters.

The results of the site and laboratory tests were statistically described to determine the geotechnical parameters for the design as follows: mean and standard deviation for spatial (layer) average properties of the "Bucharest Clay" (layer 1, silty clay), which is dominant for the retaining system in this project, as described in more detail in Sect. 2.2. The ground properties of the other soil layers were taken as characteristic values according to SR EN 1997-1:2004 [5] and NP 122:2010 [6], assuming normal distribution and spatial averaging, see Table 1.

2.1.2 Excavation and Retaining Structure

The excavation pit for executing the two basement floors is about 7.7 m depth for the marginal area and 8.3 m in the central area, related to the natural ground level. On most of the area, a sloped excavation was considered, and on a side where the excavation was led near the property limit, a self-supporting embedded wall was provided. The retaining wall consisted of reinforced concrete drilled piles, 80 cm

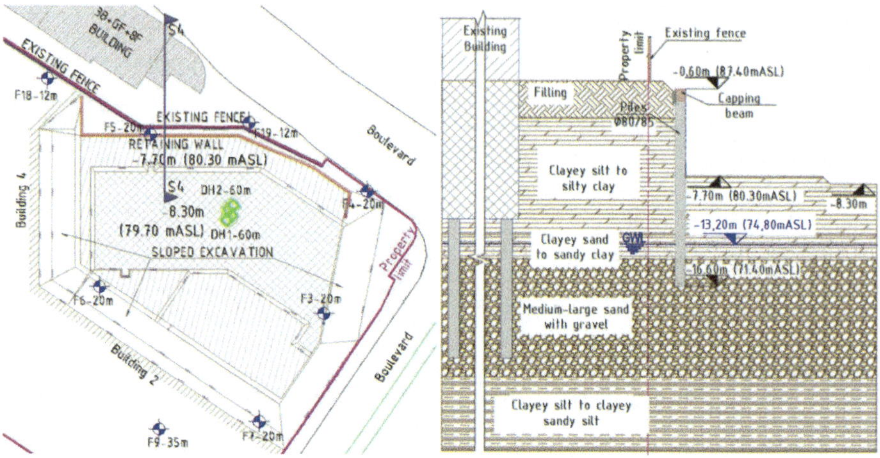

Fig. 1 Excavation and retaining system for Building 1: layout, including adjacent constructions and site investigations (left) and characteristic section, including adjacent constructions and lithology (right)

diameter at 85 cm inter-axes distance, 16 m length. The excavation pit layout and characteristic section of the retaining system are showed in Fig. 1.

2.2 *Statistical Description of the Soil Properties*

For the scope of the present study, the main geotechnical parameters of the dominant $9 \div 11$ m thick silty clay layer, called "Bucharest Clay", were statistically described as variables for the probabilistic calculations, while for the rest of the layers the parameters were considered as deterministic with characteristic values (see Table 1).

There are multiple possible distributions to fit a set of data and although there are also many mathematical instruments to check the "best fit", their suitability or necessity for geotechnical parameters are still under discussion by different authors (see [7, 8]).

This study included an analysis of the statistic distributions for the geotechnical parameters considered as random variables, using different assumptions, for the following cases: normal distribution without prior knowledge for all parameters, normal distribution with prior knowledge according to NP 122:2010 [6], lognormal distribution (only parameters with larger coefficient of variation), linear regression analysis by least square method for the shear resistance criterion with normally distributed regression parameters (i.e. the cohesion and the friction angle)—see Eq. (1). For the prior knowledge assumption, the coefficients of variation (CoVs) were considered as given by NP 122:2010 [6] for different parameters, so the standard deviation was back-calculated for the sample mean value of each parameter.

$$\tau = c + \sigma \cdot \tan(\varphi) \tag{1}$$

The statistics of the geotechnical parameters are given in Table 2. Their probability distributions are represented in Fig. 2 for the sample distribution and for spatial averaging, considering statistical uncertainty through Student-t factor at 95% confidence level. For the prior knowledge assumption, the Student-t factor for an infinite number of samples is considered.

3 Reliability Analysis

3.1 Serviceability Limit State Conditions

While the ULS dictates the reinforcement dimensioning and the length of the retaining wall, the SLS typically dominates the overall design of deep excavations in urban areas, because of the limits that must be set for the displacements of the wall, neighboring constructions or ground. Displacement limits might be defined in terms of:

a. Retaining wall displacement, which, together with the execution tolerances, set the remaining working area for the structural and architectural works to be performed;
b. Neighbouring structures displacements and deformations: settlement or heave, rotation, angular distortion, tilt, relative deflection etc.

In contrast to other structural (permanent) works, where design codes provide more specific SLS limits based on functional or aesthetic criteria, for geotechnical works these need to be defined for each individual project together with the other parties involved. For the present case study, a top horizontal displacement of 3.5 cm was considered acceptable, which together with a possible in plane position deviation of 1.5 cm would lead to a maximum 5 cm of assumed space lost for the execution of the interior structural wall or interior space.

Given that the closest structure to the excavation pit is a fence having very little sensitivity towards displacements, and that the neighboring building is located at more than 12 m distance from the excavation (see Fig. 1), displacements or deformations of adjacent structures were not considered critical and, hence, not involved in the formulation of the limit states.

Table 2 Statistics for the geotechnical parameters for "Bucharest Clay" layer on Building 1 area

Statistical parameters	Distribution												
	N–no prior				N-prior			LN			N(LR)	
	Variables (Geotechnical parameters)											
	γ	E^{100kPa}	c	tan(φ)	E^{100kPa}	c	tan(φ)	E^{100kPa}	c	tan(φ)	c	tan(φ)
n	42	15	11		15	11		15	11		33	
Min	18.52	6 090	22	0.2661	6 090	22	0.2661	6 090	22	0.2661	22	0.2661
Max	20.14	18 778	75.7	0.5073	18 778	75.7	0.5073	18 778	75.7	0.5073	75.7	0.5073
m_x	19.42	13 238	43.8	0.4007	13 238	43.8	0.4007	13 368	44.1	0.4018	39.5	0.4251
s_x	0.378	4 081	15.5	0.0741	4 081	15.5	0.0741	4 896	16.7	0.0805	5.8	0.0358
v_x (COV)	0.02	0.31	0.35	0.18	0.30	0.40	0.10	0.37	0.38	0.20	0.15	0.08
μ_x	19.42	13 238	43.8	0.4007 tan(21.8°)	13,238	43.8	0.4007 tan(21.8°)	12,613	41.6	0.3948 tan(21.55°)	39.5	0.4251 tan(23.9°)
σ_x	0.0597	1128	5.1	0.0246	1025	5.3	0.0121	1240	5.1	0.0261	6.0	0.0369
ρ	–	–	−0.222		–	−0.222		–	−0.222		−0.874	

γ—unit weight, E^{100kPa}—linear deformation modulus for a reference pressure of 100 kPa; c—cohesion; tan(φ)—tangent of the internal friction angle; n—sample size/ number of observations; m_x—sample mean; μ_x—distribution mean; s_x—sample standard deviation; v_x—sample coefficient of variation; σ_x—distribution standard deviation (for spatial averaging and statistical uncertainty); ρ—correlation factor; N—normal distribution, LN—lognormal distribution, LR—linear regression; prior—prior knowledge

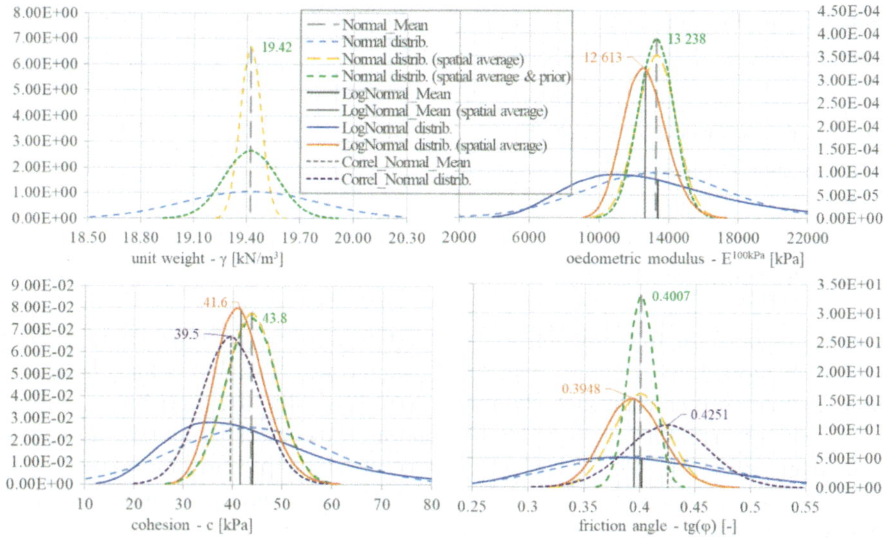

Fig. 2 Probability distributions of the main geotechnical parameters statistically described for calculations

3.2 *Calculation Models*

3.2.1 Finite Element Model

The retaining system of the excavation was modeled using the 2D Finite Element model for plane strain state in Plaxis 2019 software, by undrained analysis using effective parameters (both cohesion and internal friction angle for shear strength parameters for direct shear tests).

The finite element model is 35 m deep and 60 m wide and it consists of approximately 1828 triangular 15-node elements for soil cluster, 24 elements of 5-node line type for pile wall and 50 elements of 5-node line for the interface. Also, the boundary conditions of the model consist in fixing its bottom against all directions, its vertical boundaries against horizontal directions and in considering the ground surface free in all directions.

To determine the deformations more realistically, the Hardening Soil with small stiffness behavior was used [9]. For the reinforced concrete wall, the elastic model was considered, and the properties were set as deterministic as indicated in Table 3.

The deformation moduli required by the model for Hardening Soil with small strain were calculated from the ${E_{50}}^{ref}$ (linear deformation modulus, E^{100kPa}), based on fixed ratios from the literature.

Table 3 Properties of the linear elastic elements

Element/property	EA	EI	ν	Spacing
	(kN/m)	(kNm^2/m)	(–)	(m)
Reinforced concrete piles	18.3E6	733E3	0.2	0.85

E—deformation modulus; A—cross section area; I—cross section inertia modulus; ν—Poisson coefficient

3.2.2 Probabilistic Model

In order to perform full probabilistic calculations, software Probabilistic Toolkit (PTK) developed by Deltares was used which was coupled with Plaxis 2D finite element software through Python scripts. Basically, the probabilistic model comprising of the limit function, the distribution of the variables and the reliability method are defined in PTK [10]. The PTK generates input parameters which are sent to Plaxis 2D for calculations, and then receives the results for evaluation of the limit function, by different reliability methods.

The limit function was established for the maximum displacement of the wall, for SLS, as described in Sect. 2.2, given by the formula:

$$Z_{SLS} = d_{SLS}^{wall} - d_{max}^{wall} \cdot f_{model} \tag{2}$$

where d_{SLS}^{wall} is the maximum wall displacement allowed for SLS of 3.5 cm; d_{max}^{wall} is the maximum wall displacement obtained by calculation from Plaxis 2D, f_{model} is the model factor.

The variables considered were the geotechnical parameters described in Sect. 2.2, following different distributions (models named *Param_N-no prior*, *Param_N-prior*, *Param_LN*, and *Param_LR*). For the last model (considering the correlation between the shear resistance parameters), a model factor was introduced following a normal distribution with the mean value of 1 and standard deviation of 0.1 as recommended in the Probabilistic Model Code [11] (model named *Param_LR + Model*).

First attempts to use FORM analysis encountered convergence problems to due significant "noise" in the evaluation of the limit state function. The noise appeared to be inherent to using the Hardening Soil Small Strain constitutive model. Attempts to adjust the FORM algorithm parameters to better deal with the noise improved the performance, but did not remove the convergence issues entirely. Importance Sampling (IS) around the (estimated) design point was adopted as a feasible (i.e. acceptable number of model evaluations) and accurate alternative. However, to obtain a good estimate of the design point, it is necessary to have some prior knowledge of the failure area. This was solved by performing the analysis in two steps: one IS run around the mean values, and a second IS run around the design point obtained from the first run in order to improve the precision. For the case study analyzed within the present paper, it was necessary to perform about 1000 iterations around mean values and about 500–1000 iterations around the design point from the first calculation.

Hence, in total about 1500–2000 Plaxis 2D model runs, to reach a precision of 0.1 for the probability of failure (i.e. coefficient of variation). This procedure was verified for one of the analysis of the present case study through Directional Sampling, which required between 1200 and 2300 iterations, which means between 4500 and 8100 Plaxis 2D realizations for the same precision.

3.3 Results of Calculations

The probabilistic calculations provided the reliability index (and the corresponding probability of failure) and an approximation of the design point (design values for the variables) and the FORM influence factors, denoted Alpha [10]. An example of the main results from PTK coupled with Plaxis 2D for the model considering correlation by linear regression (*Param_LR*) is given in Fig. 3.

The results of the analyses for the present case study are given in Table 4 for all the models, following different distributions for the geotechnical parameters and including model factor.

For the case study presented, the prior knowledge assumption leads to about 7% higher reliability index. However, if more specific prior knowledge data would be available (e.g. from a local database) instead of some guiding more general values, then the resulting reliability should be higher. A greater difference in terms of reliability is given by choosing different probability distributions for the geotechnical parameters: while the lognormal distribution for the parameters with higher coefficient of variance leads to lower reliability index, the linear regression of the correlated variables leads to higher reliability. The regression analysis also makes sense physically since the uncertainties related to the correlated parameters may not be independent and the unrealistic combination of the two (or more) variables is avoided.

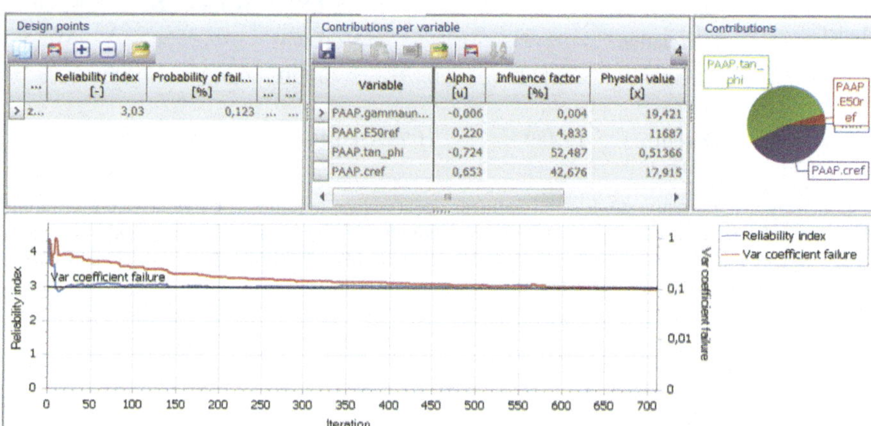

Fig. 3 Probabilistic results (screenshot PTK)

Table 4 Results of the probabilistic calculations for different assumptions in the statistical analysis of the ground properties of the dominant "Bucharest clay" layer

Model	Description	Variables and distributions					β (P_f)
		γ	E^{100kPa}	c	tan(φ)	f_{model}	
Param_N-no prior	Distribution	N	N	N	N	–	2.49 (6.4 × 10^{-3})
	Correlation	–	–	−0.222		–	
	Alpha values	0.064	0.192	0.493	0.846	–	
	Design values	19.409	12 626	0.36653	35.355	–	
Param_N-prior	Distribution	N	N	N	N	–	2.67 (3.8 × 10^{-3})
	Correlation	–	–	−0.222		–	
	Alpha values	0.042	0.194	0.037	0.979	–	
	Design values	19.413	12 648	0.39937	29.405	–	
Param_LN	Distribution	N	LN	LN	LN	–	1.91 (2.8 × 10^{-2})
	Correlation	–	–	−0.222		–	
	Alpha values	0.011	0.283	0.505	0.815	–	
	Design values	19.418	11 779	0.3649	35.36	–	
Param_LR	Distribution	N	LN	N(LR)	N(LR)	–	3.03 (1.2 × 10^{-3})
	Correlation	–	–	−0.874		–	
	Alpha values	– 0.006	0.22	−0.724	0.653	–	
	Design values	19.421	11 687	0.51366	17.915	–	
Param_LR + Model	Distribution	N	LN	N(LR)	N(LR)	N	2.49 (6.3 × 10^{-3})
	Correlation	–	–	−0.874		–	
	Alpha values	−0.112	0.231	−0.420	0.613	−0.618	
	Design values	19.439	11 767	0.46929	27.794	1.1761	

γ—unit weight (gammaunsat), $E^{100\ kPa}$—linear deformation modulus for a reference pressure of 100 kPa (E50ref); c—cohesion (cref); tan(φ)—tangent of the internal friction angle (tan_phi); f_{model}—model factor; β—reliability index; P_f—probability of failure; N—normal distribution, LN—lognormal distribution, LR—linear regression

The introduction of the model factor leads to about 20% lower reliability for the same statistical distribution of the geotechnical parameters, although it can be argued if the specific values (mean and standard variation) are best suited for the problem, i.e. for FEM analysis of SLS verification for a retaining structure. However, in lack of more research for this, it seems like a plausible appreciation to account for the model uncertainties.

It is noteworthy that the design point values (and equivalently, the influence coefficients) are rather dependent on the assumptions made with respect to the statistical modelling of the ground properties, as illustrated in Table 4. This is an important point of attention for reliability-based calibration of partial (material) factors.

3.4 Target Reliability Index

In order to assess the reliability of a structure, we need target (minimum) values for reliability indices, or acceptable probabilities of failure, for the specific design situation and limit state at hand. While ULS target reliability values are discussed extensively in the literature, SLS reliability has attained less attention so far.

A rather comprehensive compilation of the available annual and lifetime target reliability indices recommended in standards and literature for ULS has been presented by Roubos et al. [3], for different applications (concrete, bridges, marine, geotechnical, buildings, quay walls etc.). This study also associated the classification criterion from each reference within the framework of ISO 2394:2015 [12], which led to "quite consistent and uniform" reliability differentiation. Some reference values from different codes, more relevant for the present case study, are given in Table 5.

For SLS, there are fewer references available in the codes. Moreover, the values indicated in EN 1990 [13] cover only Reliability Class 2 (Moderate Consequences), see Table 6.

It is also important to notice that the design codes only provide target reliability values for permanent design situations for design life times of usually 50 years (for different reference periods, which is different to the concept of design life time). No distinction is made between permanent and temporary structures, except for the guidance given in the Probabilistic Model Code [11]: "For structures designed for short service life or otherwise rapid obsolesce (say less than 10 years) the beta-values can be lowered by one or half a class." This issues has also been raised by other authors, some of them proposing specific approaches for determining reliability index [15, 16] or partial factors [17] for reliability-based design of temporary structures.

Hence, we can conclude that target reliability values for SLS and temporary structures are not clearly defined in the present design codes.

Table 5 Target reliability indices (β) proposed in different standards for ULS

Consequence/classification	Relative cost of safety measure, 1-year reference period			1-year reference period	1-year reference period[a]	50 years reference period[a]
	Large (A)	Normal (B)	Small (C)			
Minor	3.1	3.7	4.2	4.2	2.5/3.0	3.3
Moderate	3.3	4.2	4.4	4.7	4.0	3.8
Large	3.7	4.4	4.7	5.2	5.0	4.3
	Probabilistic Model Code [11], ISO 2394:2015 [12]			EN 1990:2002 [13]	USACE-1999 [14]	EN 1990:2002 [13]

[a]The association of the target reliability values to the corresponding classification (consequence classes) was made by the authors based on the descriptions in the reference documents

Table 6 Target reliability indices (β) proposed in different standards for SLS (irreversible)

Consequence	Relative cost of safety measure, 1-year reference period			1-year reference period	50-years reference period
	Large (A)	Normal (B)	Small (C)		
Minor	1.3	1.7	2.3	–	–
Moderate				2.9	1.5
Large				–	–
	Probabilistic Model Code [11], ISO 2394:2015 [12]			EN 1990:2002 [13]	

4 Conclusions

The present paper presented a probabilistic FEM-analysis for a temporary retaining wall of a deep excavation, for SLS verification. This was performed by coupling commercial finite element software Plaxis 2D with probabilistic software PTK developed by Deltares.

It was found that FORM could not be handled by the present case study because of the "noise" within the FEM results for Hardening Soil with small stiffness model. However, applying Importance Sampling reliability method in two steps (around mean values and then around the design point estimated from first step for higher precision) provides feasible results with acceptable computational effort.

Different probability distributions of the geotechnical parameters (normal, lognormal, linear regression) and different assumptions (with or without prior knowledge) lead to significantly different reliability levels. It is, thus, concluded that choosing the statistical distribution can be decisive in the assessment of the reliability level.

Also, the model factor is very important to account for the model uncertainties and it should be considered in the analyses, although more research might be needed to establish specific values for FEM analysis of SLS verification for a retaining structure.

The design point values of the geotechnical parameters (or other variables) are highly dependent on the assumptions made even through probabilistic analyses; thus, a fix set of values is more difficult to assess. This can be even more arguable when partial factors are to be proposed for code calibration.

It is moreover emphasized that more specific provisions are necessary in the design codes to allow the use of reliability-based design in favor of the partial factor design, especially for temporary structures (such as target reliability index or probability of failure), but also for SLS verification or the retaining structure and for the neighboring buildings) which can be decisive for such structures.

Acknowledgements The research comprised within the present paper was performed under the Erasmus+ agreement between Deltares and the Technical University of Civil Engineering Bucharest. The authors would like to acknowledge the technical support provided by Deltares research institute during the student mobility programme.

References

1. Schweckendiek, T. (2006). *Structural reliability applied to deep excavations* (Master Thesis). Delft University of Technology.
2. Rippi, K., & Teixeira, A. (2016). Reliability-based assessment of a retaining wall using FEM. In *Proceedings of the 25th European Young Geotechnical Engineers Conference*, Sibiu, Romania.
3. Roubos, A. A., Steenbergen, R. D., Scheweckendiek, T., & Jonkman, N. S. (2018). Risk-based target reliability indices for quay walls. *Structural Safety, 75,* 89–109.
4. Popa, H., Ene, A., Miritoiu, R., Ionescu, I., & Marcu, D. (2018). Back analysis of an embedded retaining wall for a deep excavation in Bucharest. In *ce/papers: Special Issue: XVI DECGE 2018 Proceedings of the 16th Danube—European Conference on Geotechnical Engineering,* (Vol. 2, pp. 2–3), Skopje, R. Macedonia.
5. Comite Europeen De Normalisation. (2004). EN 1997-1:2004. *Eurocode 7: Geotechnical design—Part 1: General rules.* Brussels, Belgium.
6. Ministerul Dezvoltarii Regionale si Turismului. (2011). NP 122:2010. *Romanian norm on determining the characteristic and design values for the geotechnical parameters* (in Romanian). Romania: Monitorul Oficial al Romaniei.
7. Shepheard, C. J., Vardanega, P. J., Holcombe, E. A., Hen-Jones, R., & De Luca, F. (2019). Minding the geotechnical data gap: Appraisal of the variability of key soil parameters for slope stability modelling in Saint Lucia. *Bulletin of Engineering Geology and the Environment, 78,* 4851–4864.
8. Gong, F., Wang, T., & Wang, S. (2019). Inference of the optimal probability distribution model for geotechnical parameters by using Weibull and NID distributions. *Journal of Vibroengineering, 21*(4), 876–887.
9. Plaxis. (2019). Plaxis 2D Material Models Manual.
10. Deltares. (2019). Probabilistic Toolkit Manual.
11. Joint Committee on Structural Safety. (2001). *Probabilistic model code (Part 1—Basis of design).* Technical University of Denmark.
12. International Organization for Standardization. (2015). *ISO 2394:2015. General principles on reliability for structures.* ISO.
13. Comite Europeen De Normalisation. (2001). *EN 1990:2002. Eurocode 0—Basis of structural design.* Brussels, Belgium.
14. USACE. (1999). *Risk–based analysis in geotechnical engineering for support of planning studies.* ETL1110-2-556. United States.
15. Holický, M. (2012). Optimisation of the target reliability for temporary structures. *Civil Engineering and Environmental Systems, 30*(2), 87–96.
16. Vereecken, E., Botte, W., & Caspeele, R. (2019). Reliability based design of temporary structures. In *Life-cycle analysis and assessment in civil engineering: Towards an integrated vision* (pp. 2615–2622).
17. Caspeele, R., Steenbergen, R., & Taerwe, L. (2013). An adjusted partial factor method for temporary structures compatible with the Eurocode framework. *Civil Engineering and Environmental Systems, 30*(2), 97–114.

Probabilistic Methods for Code Calibration Exemplified for the Punching Shear Resistance Model Without Shear Reinforcement

Tânia Feiri, Marcus Ricker, Jan Philip Schulze-Ardey, and Josef Hegger

Abstract To prevent punching shear failures, modern structural codes, as EN 1992-1-1, offer design provisions for punching shear. To verify the performance of the new code equations, those values are usually compared to test results. However, values for the safety factors are often determined using calibration on older code provisions. The deterministic nature of such approach introduces imprecisions that can compromise the safety level of a structural component. Nowadays, modern reliability analysis techniques, allied to increasing computer capabilities, provide efficient precision to evaluate the safety level of structural components. This paper contributes to code calibration through different reliability analysis techniques where the safety level of design provisions for punching shear resistance without shear reinforcement is investigated. This study addresses three reliability techniques: Mean-Value First Order Second Moment (MVFOSM), First-Order Second Moment (FOSM) and Monte-Carlo (MC-IS) with Importance Sampling. Thus, the parameters influencing the punching shear capacity are stochastically modelled and evaluated. Then, the reliability indices β, used as a practical measurement of the safety level, are estimated and compared to a β-target of 3.8 given by EN 1990 for a 50-year period. The results seem to confirm that the punching shear provisions for structural elements without shear reinforcement according to EN 1992-1-1 achieve a required safety level in line with the β-target. The study shows that the techniques FOSM and MC-IS seem appropriate to determine the failure probability of the design equations for the punching shear capacity without shear reinforcement. Furthermore, the study suggests that the MVFOSM method may not be suitable to evaluate the absolute safety level of those design equations. The study shows that the design equations are more sensitive to the variable describing the model uncertainty than to any other variable, which stresses the importance of an adequate statistical analysis of the basic variables of the resistance model.

T. Feiri (✉) · M. Ricker
Institute of Structural Engineering, Hochschule Biberach University of Applied Sciences, Biberach, Germany
e-mail: feiri@hochschule-bc.de

J. P. Schulze-Ardey · J. Hegger
Institute of Structural Concrete, RWTH Aachen University, Aachen, Germany

© The Author(s), under exclusive license to Springer Nature Switzerland AG 2021
J. C. Matos et al. (eds.), *18th International Probabilistic Workshop*, Lecture Notes in Civil Engineering 153, https://doi.org/10.1007/978-3-030-73616-3_45

Keywords Code calibration · Punching shear resistance · Reliability analysis techniques · Reliability index

1 Introduction

Punching shear in reinforced concrete elements can affect the equilibrium of a structural system and, ultimately, lead to its structural collapse. To prevent such failures, modern structural codes, as EN 1992-1-1 [1], offer provisions to design for punching shear. The Eurocodes are considered Level I codes, in which appropriate levels of structural reliability are provided by the specification of multiple partial safety factors related to predefined characteristic values of basic variables [2, 3]. A common way to ensure that a structural component reaches the required safety level consists in determining the partial safety factors so that the differences between the failure probabilities of the structures considered and the target value are minimised [4]. To verify the performance of new code equations, numerical values are usually compared to test results. The deterministic nature of such approach introduces imprecisions that can compromise the safety level of a structural component. Nowadays, modern reliability analysis techniques, allied to increasing computer capabilities, provide efficient precision to evaluate the safety level of structural components. Thus, not only the required safety level can be ensured, but also the range of reliability indices can be reduced.

The present study uses the state-of-the-art of three reliability analysis techniques—Mean Value First-Order Second Moment Method (MVFOSM), First-Order Second Moment Method (FOSM) and a Monte-Carlo Simulation with Importance Sampling (MC-IS)—for an enhanced understanding of the safety level of the punching shear design of concrete elements without shear reinforcement. The techniques MVFOSM and FOSM use a first-order Taylor series expansion of the linearised limit state function to determine the reliability indices β—and, ultimately, the failure probability—of a structural element. In the MVFOSM, the linearisation of the failure function is assumed to be at its mean value, whereas in the FOSM the linearisation is determined through iteration around the most likely failure point—also known as the *design-point*. For its turn, Monte-Carlo (MC) is a simulation technique that generates samples of the stochastic variables in the failure function. With increasing number of simulations, the resulting failure probability converges to the exact solution. Therefore, it is common that MC methods are used to verify the results of the first-order methods. Furthermore, in order to reduce computational efforts, several variance reduction techniques, as the *Importance Sampling*, are currently available. The *Importance Sampling* technique concentrates the sampling in the area of the total space that has the largest contribution to the failure probability. Following the evaluation of the resistance model through each of the aforementioned techniques, the resulting reliability indices β are compared to a target β-value provided by EN 1990 [2]. Finally, a sensitivity analysis is performed to identify the most important parameters affecting the safety of the punching shear resistance of structural elements without shear reinforcement.

2 Setting Up The Resistance Model

2.1 *General Procedures for Reliability Assessment*

Structural reliability is a probabilistic measure of assurance of safe performance where the safety level is measured by a reliability index β and a corresponding failure probability. With the purpose to guide modern structural design and examination procedures, the fundamental task of structural reliability theory is the analysis of the following verification of inequality:

$$E_d \leq R_d \tag{1}$$

where E_d is the design value of the load effect at the structural element being analysed and R_d is the corresponding design value of resistance. To this verification, the resolution of problems in structural reliability theory follows a comprehensive set of key steps [5]. Firstly, a target reliability level is selected considering a safety and consequence class. Annex B of EN 1990 [2] offers different target reliability levels for two reference periods: 1-year and 50-year. Secondly, the limit state function $g(\overrightarrow{x})$ of a structural system under loading conditions, is mathematically formulated as:

$$g(\overrightarrow{x}) = g(x_1, x_2, \ldots, x_n) = 0 \tag{2}$$

with $x_1, x_2, \ldots, x_n$ being realisations of the basic stochastic variables $\overrightarrow{X} = (X_1, X_2, \ldots, X_n)^T$ in the limit state function. Then, the basic stochastic variables $\overrightarrow{X} = (X_1, X_2, \ldots, X_n)^T$ in are identified together with the specification of a fitting distribution type and the respective statistical parameters. It follows the characterisation and modelling of the uncertainties in probabilistic terms. Subsequently, a reliability index β is estimated as a nominal measure of safety. This estimation is possible with the use of reliability analysis techniques, as the abovementioned MVFOSM, FOSM or MC-IS. Finally, a sensitivity analysis is performed in order to understand the effects of specific parameters on the safety level assessment.

2.2 *Limit State Function for Punching Shear Resistance*

A design decision is chosen such that it complies with the criterion that a resistance design value is larger than the design value of a corresponding load effect (Eq. 1).

Design values for the load bearing capacity R_d are chosen to have a sufficiently low non-exceedance probability. And design values for loads E_d are chosen to have a sufficiently low exceedance probability such that the design criterion in the limit ($R_d = E_d$) corresponds to the required reliability level [6]. EN 1990 [2] indicates that the design values of action effects E_d and resistances R_d should be defined such that the probability of having a more unfavourable value is considered through the adoption of sensitivity factors. In semi-probabilistic approaches, constant sensitivity factors for the load and resistance side are assumed. Normally, the value for unfavourable actions and action effects is negative with $\alpha_E = -0.7$ and the value is positive for resistances with $\alpha_R = 0.8$. According to EN 1990 [2], this is valid for ratios of the standard deviations σ_E/σ_R between 0.16 and 7.6. Since the total scatter is overestimated, the assumptions for α_E and α_R are on the safe side. This is obvious from the fact that the condition $\sum\alpha^2 = 1$ is not fulfilled (i.e. mathematically not correct, a $\sum\alpha^2$ higher than 1 is assumed).

Based on these considerations, the failure domain, that is defined by the limit state function for punching shear resistance, can be expressed as:

$$g(\vec{x}) = \theta_R v_{Rm,c}(d, c, \rho_l, f_{cm}) u_1(d, c) d - V_{Rd,c} \leq 0 \tag{3}$$

where θ_R is the random variable dealing with the resistance model uncertainty, $v_{Rm,c}$ (.) is the punching shear resistance whose key variables are affected by uncertainties (d is the random variable expressing the geometric uncertainty on effective depth, c is the random variable expressing the geometric uncertainty on column dimensions, ρ_l is the random variable expressing the geometric uncertainty on the flexural reinforcement steel ratio, and f_{cm} is the random variable expressing the uncertainty on concrete compressive strength) and u_1(.) is the critical perimeter at a distance of 2.0d from the periphery of the loaded area, affected by the random variables c and d. The parameter $V_{Rd,c}$ is the design value for punching shear resistance according to EN1992-1-1 [1]. By adopting the equation for the punching shear capacity given in EN1992-1-1 [1] for the $v_{Rm,c}$ (.), and considering interior circular columns, Eq. 3 can be redefined as:

$$\begin{aligned} g(\vec{x}) = \theta_R\left[0.18\left[\min\left(2.0; \left(1 + \frac{200}{d}\right)^{1/2}\right)\right](100\rho_l f_{ck})^{1/3}\right] \\ \left(2\pi\left(\frac{c}{2} + 2d\right)\right) \cdot d - V_{Rd,c} = 0 \end{aligned} \tag{4}$$

2.3 Database of Experimental Tests and Data Selection Criteria

For the sake of this study, the punching resistance model was based on an experimental database initiated by Beutel [7]. This database is maintained and extended with new test results by, among others, Siburg [8] at the *Institute of Structural Concrete* (IMB) at the RWTH Aachen University. Built in an *Excel* environment, the database contains 404 tests on flat slabs and footings without shear reinforcement under axis-symmetric conditions. For a proper statistical evaluation of the resistance model, the resistance tests selected complied with the following criteria:

- Reinforcement ratio $\rho_l > 0$
- Concrete compressive strength $f_{cyl} \leq 104$ MPa
- Effective depth $d \geq 100$ mm.

The German National Annex [9] to EN 1992-1-1 [1] indicates a minimum slab thickness of 70 mm for flat slabs without shear reinforcement. However, the execution of flat slabs with heights smaller than 150 mm tends to be unusual. It is important to highlight that the effective depth d is considered as the total height h minus the effective concrete cover d_1, which is the distance between extreme compression fibre to the centroid of tension reinforcement in a section under flexure. By taking these aspects into account, only test specimens on flat slabs with effective depth greater than 100 mm were included in this study. A total of 290 specimens were considered for the resistance model. The computations in this study were performed by using the mathematical software Maple [10].

2.4 Stochastic Parameters for the Basic Variables in the Resistance Function

The basic stochastic variables in the punching shear resistance equation—effective depth d, column dimensions c, flexural reinforcement steel ratio ρ_l and concrete compressive strength f_{ck}—were assumed to be uncorrelated and their stochastic parameters were characterised according to the framework of the JCSS Probabilistic Model Code [11]. This characterisation is detailed in Table 3.

The considerations for the variable concrete compressive strength f_c are particularly relevant to highlight. This variable is lognormal distributed when the mean and the standard deviation can be derived from an infinite sample size. However, the sample size is usually limited and it has to be assumed that the concrete compressive strength f_c is influenced by multiple aspects, such as the production unit and site, the construction period or the sample size, among others [11]. In order to support the characterisation of these variables, the JCSS Probabilistic Model Code [11] offers a set of parameters that are defined as a function of the concrete type and grade. These parameters are represented by m', n', s' and ν', where m' is the mean value of a

Table 1 Statistical characterisation of the concrete compressive strength

Concrete grade	Prior parameters				f_{cm} (MPa)	σ_{fcm} (MPa)	f_{ck} (MPa)
	m′	n′	s′	v′			
C25	3.65	3.0	0.12	10	38.8	4.67	34.8
C35	3.85	3.0	0.09	10	47.2	4.26	43.2
C45	3.98	3.0	0.07	10	53.6	3.76	49.6

f_{cm} is the mean value of the concrete compressive strength (MPa) given by $\mu_Y = e^{m'+0.5s'^2}$ where $Y = \ln(X)$
σ_{fcm} is the standard deviation for the mean value of the concrete compressive strength given by: $\sigma_Y = e^{m'+0.5s'^2}\sqrt{e^{s'^2}-1}$ where $Y = \ln(X)$; $f_{ck} = f_{cm} - 4(\text{MPa})$

sample with size n' and s' is the empirical standard deviation of a sample of size $\nu' + 1$ (Table 1). In the context of punching shear calculation, some German researchers, such as [12–14], consider that a characteristic value is used for the concrete compressive strength. Thus, a reduced concrete compressive strength given by the following equation is adopted here:

$$f_{ck} = f_{cm} - 4(\text{MPa}) \tag{5}$$

The offset of 4 (MPa) for the concrete compressive strength is a constant for each concrete strength class and it was chosen following the conformity criteria of EN 206-1 [15] for the initial concrete production. Table 1 summarises all the statistical parameters determined for ready mixed concrete of classes between C25 and C45.

2.5 *Statistical Evaluation of the Resistance Model*

Apart from material properties and geometry, the resistance of structural elements is primarily dependent on uncertainties related to an applied model. These so-called model uncertainties can be obtained from comparisons of the experimental failure loads (i.e. physical tests) and the calculated resistances (i.e. model results). In this study, firstly, the *David-Hartley-Pearson* test was used to investigate the asymptotic behaviour of test statistics outliers (i.e. outlying observations discordant from the hypothesised model) for the sample drawn from the dataset. The test uses a lower and upper outlier-pair χ_1 and χ_n, and checks whether these extreme values are outliers depending on which value is furthest from the mean. The test identified 4 outliers, which were subsequently excluded from the main dataset. As a result, 286 tests were considered for the determination of the model uncertainty.

Secondly, the sample was compared with a reference probability distribution. In theory, the sample should have the characteristics of a lognormal distribution since geometric and resistance variables do not assume negative values; however, this assumption should be properly evaluated. To this end, the statistical hypothesis

Kolmogrow-Smirnow (K-S) test was used. The test quantifies the distance between the empirical distribution function of the sample and the cumulative distribution function of the reference distribution (i.e. the lognormal distribution). The null distribution of this statistic was then calculated under the null hypothesis that the sample was drawn from the reference distribution. The greatest distance between the empirical distribution function of the sample and the cumulative distribution function of the reference distribution yielded the result of 0.079. With a significance level α equal to 0.05, the K-S test produced the critical value equal to 0.0803. Since 0.079 is smaller than 0.0803, the null hypothesis was accepted. Figure 1 illustrates the probability paper for the lognormal distribution.

Thirdly, the point estimators for the real parameters were determined since they serve as a best estimate of the unknown population parameters. These point estimators were contrasted with an interval estimation; in this case, a confidence level of 95% was used since the dataset used has a limited size. For a normal distribution, the mean value gives the lower bond where $\mu \geq 1.1318$ and the standard deviation give the upper endpoint of the interval with $\sigma \leq 0.2008$. In Table 2 it is seen that if a lognormal distribution is assumed, a mean value of $\mu \geq 1.1343$ and a standard deviation of $\sigma \leq 0.2084$ are obtained. Finally, the uncertainties contained in the experiments were addressed by correcting the standard deviation. To this end, the coefficient of variation was affected by a value of 0.05 according to the recommendation of Taerwe [16], resulting in a standard deviation of $\sigma \leq 0.2005$ (Table 2).

3 Discussion

3.1 Reliability Indices

In this section, the reliability indices β determined through the techniques MVFOSM, FOSM and MC-IS for each of the basic stochastic variable in the resistance model are evaluated and compared to the target level β equal to 3.8 recommended by EN 1990 [2] for a reference period of 50 years. For the MC-IS technique, it was performed a total of three runs with a coefficient of variation of 2.0%. For the following evaluation, the mean value of the three runs are used for the assessment of the β-value.

The variable effective depth was analysed for a spectrum of depths between 100 and 500 mm (Fig. 2a). For the entire spectrum of effective depths, the reliability indices β that were determined through FOSM and MC-IS produced higher β-values than the target value of 3.8. With the FOSM technique, the lowest reliability index β was obtained for an effective depth of 100 mm with a β-value equal to 4.405. For the remaining dimensions, the β-values increased gradually, reaching a value equal to 4.537 for an effective depth of 500 mm. The MC-IS technique yielded a minimum β-value of 4.375 for an effective depth of 100 mm and a minimum β-value of 4.530 for an effective depth of 500 mm. The reliability indices β determined through the MVFOSM resulted lower than the recommended target β-values equal to 3.8. For

Table 2 Statistical parameters for the model uncertainty

Statistical parameters	Normal distribution		Lognormal distribution				
			Point estimate		Interval estimator (95%)		Corrected parameters
	Point estimate	Interval estimator (95%)	LN-value	LN-value transfor	IE-value	IE-value transfor	
μ	1.1500	[1.1318, +∞]	0.1260	1.1507[a]	[0.1094, +∞]	[1.1343, +∞][d]	1.1343
σ	0.1869	[−∞, 0.2008]	0.1696	0.1965[b]	[−∞, 0.1822]	[−∞, 0.2084][e]	0.2005[g]
CoV	0.1625	[−∞, 0.1774]	0.1625	0.1708[c]	[−∞, 1.6654]	[−∞, 0.1837][f]	0.1778

[a] $\mu_{LN,conv} = e^{\mu_{LN}+0.5\sigma_{LN}^2}$

[b] $\sigma_{LN,conv} = \mu_{LN,conv}\sqrt{e^{\sigma_{LN}^2} - 1}$

[c] $CoV_{LN,conv} = \sqrt{e^{\sigma_{LN}^2} - 1}$

[d] $\mu_{IE,LN,conv} = e^{\mu_{IE,LN}+0.5\sigma_{IE,LN}^2}$

[e] $\sigma_{IE,LN,conv} = \mu_{IE,conv}\sqrt{e^{\sigma_{IE,LN}^2} - 1}$

[f] $CoV_{IE,LN,conv} = \sqrt{e^{\sigma_{IE,LN}^2} - 1}$

[g] Correction of σ_θ: $\sigma_\theta = \mu_{LN,conv}\sqrt{CoV_{\theta_{LN,conv}}^2 - CoV_\varepsilon^2} = \mu_{LN,conv}\sqrt{CoV_{\theta_{LN,conv}}^2 - 0.05^2}$

Fig. 1 Probability paper for the lognormal distribution

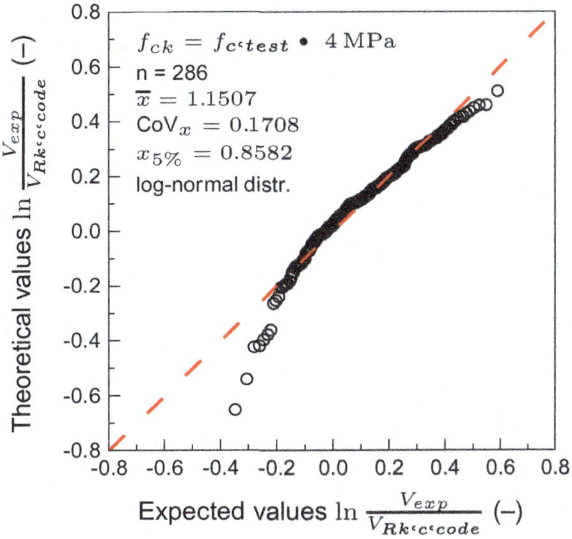

Table 3 Probabilistic model for the basic input variables

Type of variable	Symbol	Basic variable	Distr	Unit	Nom value	μ_X	σ_X	CoV_X
Concrete compressive strength	f_c	C25	LN	MPa	25	38.8	4.67	–
		C35	LN	MPa	35	47.2	4.26	–
		C45	LN	MPa	45	53.6	3.76	–
Geometric variables	d	Effective depth	N	mm	d_{nom}	$d_{nom} + 10$	10	–
	c	Column dimensions	N	mm	c_{nom}	$c_{nom} +$ $0.003c_{nom}$	$4 +$ $0.006c_{nom}$	–
	ρ_l	Flexural reinf. steel	N	mm	ρ_{lnom}	nom	–	0.02
Model uncertainty	θ_R	Resistance model	LN	–	–	1.1343	0.2005	0.1778

N normal distribution, *LN* lognormal distribution, μ_X mean value of the basic variable X, σ_X standard deviation of the basic variable X, CoV_X coefficient of variation of the basic variable X

an effective depth of 100 mm, the reliability index accounted for a β-value equal to 3.209. The reliability index β increased slightly to the value 3.369 for an effective depth of 500 mm.

Figure 2b shows that the specific column perimeter considered covered a ratio u_0/d between 2 and 12. It also illustrates that the reliability indices β that were determined in dependence of the column dimensions are higher than the target β-value of 3.8 given by EN 1990 [2]. In fact, the FOSM technique provided reliability indices β varying between 4.590 and 4.572, corresponding to the ratios 2 and 12,

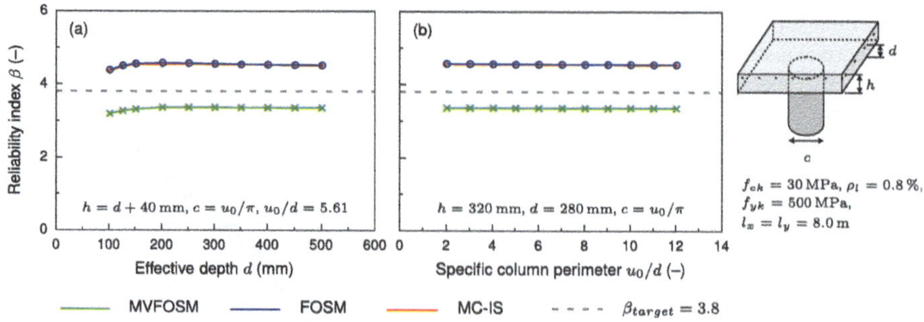

Fig. 2 Reliability indices in dependence of (**a**) effective depth and (**b**) specific column perimeter

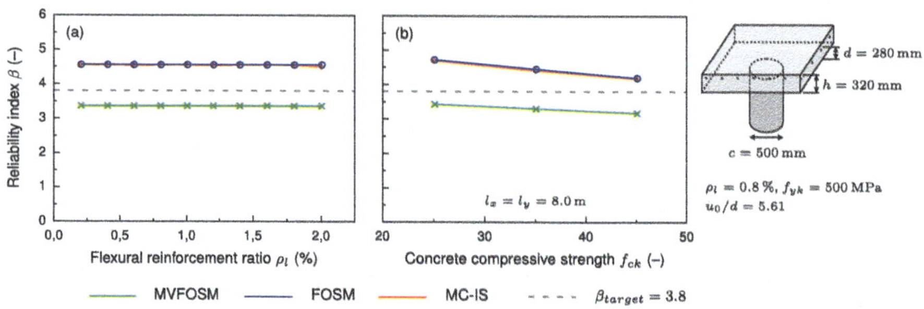

Fig. 3 Reliability indices in dependence of (**a**) flexural reinforcement ratio and (**b**) concrete compressive strength

respectively. The MC-IS technique produced minimum reliability indices β equal to 4.583 for a ratio of 2 and 4.571 for a ratio of 12. Similar to the results obtained for the variable effective depth, the results produced with the MVFOSM technique are lower than the β-threshold of 3.8 with the calculated β-values having a small variation for the two ratios: 3.382 for a ratio of 2 and 3.281 for a ratio of 12.

The spectrum of the variable flexural reinforcement ratio ρ_l, shown in Fig. 3a, ranged between 0.2 and 2.0%. The use of the technique FOSM for the assessment of flexural reinforcement produced a constant reliability index β equal to 4.582. The constant coefficient of variation, assumed for the variable describing the cross section of the reinforcement steel, CoV_{As} equal to 0.02 (Table 3) explains the regularity of the reliability indices β attained. The calculation through the MC-IS technique produced minimum reliability indices β varying between 4.571 and 4.537 for the ratios 0.2 and 2.0%, respectively. This invariability is also seen on the values attained with the technique MVFOSM, where the reliability indices β obtained are equal to 3.382 for all the ratios ρ_l considered. The values produced through this technique resulted lower than the threshold β-value equal to 3.8, showing a similarly tendency with the previous variables.

Figure 3b depicts the results of the variable concrete compressive strength that considered the concrete classes C25, C35 and C45 (Table 2). Following the trend of the previous variables, the techniques FOSM and MC-IS produced nearly similar reliability indices β for the compressive strength values considered. By using the FOSM technique, these β-values fluctuated between 4.730 for a compressive strength of 25 MPa and 4.210 for a compressive strength of 45 MPa. For the same concrete compressive strengths, the MC-IS technique provided minimum reliability indices β of 4.721 and 4.200, respectively. The MVFOSM also yielded values below the target β-value of 3.8, with reliability indices β ranging between 3.451 and 3.196 for compressive strengths of 25 MPa and 45 MPa, respectively.

3.2 Sensitivity Analysis

As described herein, the parameters in the resistance model are affected by uncertainty. These parameters were studied through multiple sensitivity analyses in order to determine whether their variation had a significant effect on the safety level of the punching shear resistance. Figure 4 shows the results of the sensitivity analyses performed whose values were obtained through the FOSM technique. Overall, the results indicate that the punching shear resistance equation is particularly sensitive to the variable describing the model uncertainty.

Figure 4a shows that for effective depths smaller than 200 mm, the resistance equation is gradually affected by the uncertainty in the effective depth variable, which seems to be more pronounced for smaller effective depths. This means that the model is particularly vulnerable to variations in effective depths smaller than 200 mm; therefore, these variations should be carefully handled. For its turn, the uncertainty in the concrete compressive strength variable is relatively reduced for the entire spectrum of effective depths considered. When the reliability index β is calculated in dependence of the flexural reinforcement ratio (Fig. 4b), the model uncertainty is also the most influential variable. This variable is followed by the uncertainties of effective depth and concrete compressive strength.

The sensitivity analysis of the variation of the specific column perimeter on the reliability index calculations follows the trend of the previous variable (Fig. 4c). Similarly, the results show that the model uncertainty is also the most influential parameter for the assessment of reliability indices in dependence of the column perimeter-depth ratio. This influential parameter is followed by the uncertainties in both variables: effective depth and concrete compressive strength.

Figure 4d shows the sensitivity analysis of the variable concrete compressive strength. Comparatively to the previous variables, also here the impact of the model uncertainty is most noticeable, whose sensitivity factor accounts for almost 0.60. This influence seems to slightly increase for higher concrete compressive strengths. The variation of the concrete compressive strength and effective depth seems to have a small influence on the safety level of the punching shear resistance, whereas the remaining variables can be practically neglected.

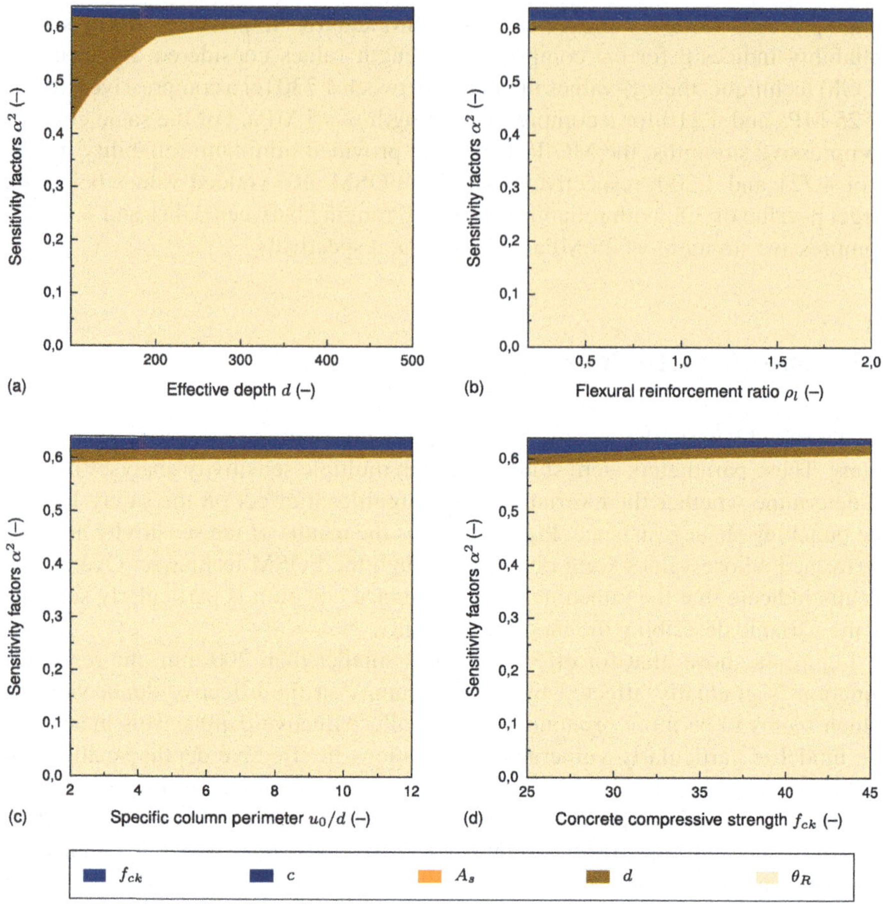

Fig. 4 Sensitivity factors of the variables (**a**) effective depth, (**b**) flexural reinforcement ratio, (**c**) specific column perimeter and (**d**) concrete compressive strength

4 Conclusions

In the present study, the safety levels of punching shear resistance of structural elements without shear reinforcement were assessed and compared to a target level recommended by EN 1990 [2]. By evaluating the punching shear resistance through three distinct reliability analysis techniques—MVFSOM, FORM and MC-IS—the following conclusions could be made:

- Based on above considerations, it can be said that the target level β recommended by EN 1990 [2], where β is equal to 3.8, is a conservative value; therefore, the punching shear equation according to the EN 1992-1-1 [1] can be considered a safe reference.

- For all the variables in the punching shear resistance function, the MVFOSM produced reliability indices β lower than 3.8 (i.e. the target β-value). These results are in line with the theoretical principles of the technique where for non-linear limit state functions, by neglecting higher-order terms, truncation errors are introduced at increased distance from the linearisation points, affecting the precision of the safety measure. Therefore, MVFOSM might not be suitable for the evaluation of absolute safety levels of design equations.
- Furthermore, the MC-IS is a robust technique that can be used to verify the results of the first-order methods. In this study, it was used the *Importance Sampling* as a variance technique in connection with the FOSM results. Considering the nearly exact values estimated through this technique, the study showed that the FOSM is a suitable technique to estimate the failure probability for the equation, calculating the punching shear resistance without shear reinforcement.
- Finally, the study showed that the limit state function is more sensitive to the variable describing the model uncertainty than to any other variable, which stresses the importance of an adequate estimation and minimisation of the uncertainties involved in resistance models. On the other side of the spectrum, the study indicated a neglecting influence of the flexural reinforcement ratio to the safety level of the punching shear resistance model.

Despite the encouraging results of this study, a set of limitations was identified. This study used flat slabs for reference with a minimum effective depth of 100 mm. However, the results showed that the limit state function is particularly sensitive to small effective depths. A possible direction for future work is the evaluation of a smaller effective depth that reduces such effects for smaller slab thickness. Furthermore, the small number of test specimens available where the minimum shear capacity governs the design for punching shear resistance limits the feasibility of the model when the design meets such conditions. Hence, further experimental test data is needed. Another possible direction for future work is the use of full probabilistic calculations that consider the load effects. Finally, this study adopted the recommendations of JCSS Probabilistic Model Code [11] for the statistical parameters of the basic random variables. The variation of these parameters influences the punching shear capacity, therefore, this influence should be addressed and investigated in future research work.

Acknowledgements The authors wish to express their gratitude to the *Deutsches Institut für Bautechnik* (DIBt) for the funding of this study under the project name: "Oberprufunq der Zuverlassiqkeit der fur die nachste Generation von EN 1992-1-1 vorgesehenen neuen Bernessunqsansatze gegen Durchstanzen ohne Durchstanzbewehrung und Querkraft ohne Querkraftbewehrung".

References

1. DIN EN 1992-1-1:2004 + AC:2010, Eurocode 2: Design of Concrete Structures—Part 1-1: General Rules and Rules for Buildings (2011).

2. DIN EN 1990:2002+A1:2005+A1:2005/AC:2010, Eurocode: Basis of Structural Design. CEN European Committee for Standardization (2002).
3. Joint Commitee CEB/CECM/CIB/FIP/IABSE Structural Safety, First Order Reliability Concepts for Design Codes (Documentation), Comité Europeén du Béton (Bulletin d'Information No 112) (1976).
4. Sørensen, J. D., Kroon, I. B., & Faber, M. H. (1994). Optimal reliability-based code calibration. *Structural Safety, 15*(3), 197–208. https://doi.org/10.1016/0167-4730(94)90040-X.
5. Ditlevsen, O., & Bjerager, P. (1986). Methods of structural systems reliability. *Structural Safety, 3*(3–4), 195–229. https://doi.org/10.1016/0167-4730(86)90004-4.
6. Köhler, J., Sørensen, J. D., & Baravalle, M. (2019). Calibration of existing semi-probabilistic design codes. In *13th International Conference on Applications of Statistics and Probability in Civil Engineering, ICASP13*. Seoul National University.
7. Beutel, R. (2002). Durchstanzen schubbewehrter Flachdecken im Bereich von Innenstützen (Doctoral dissertation, Hochschulbibliothek der Rheinisch-Westfälischen Technischen Hochschule Aachen).
8. Siburg, C. (2014). Zur einheitlichen Bemessung gegen Durchstanzen in Flachdecken und Fundamenten (Doctoral dissertation, Hochschulbibliothek der Rheinisch-Westfälischen Technischen Hochschule Aachen).
9. DIN EN 1992-1-1/NA. (2013). National Annex—Nationally determined parameters—Eurocode 2: Design of concrete structures—Part 1-1: General rules and rules for buildings.
10. Maplesoft, a division of Waterloo Maple Inc., Waterloo, Ontario, Maple 2019.1, 2019. https://www.maplesoft.com.
11. Joint Commitee on Structural Safety, JCSS Probabilistic Model Code (2001). https://www.jcss-lc.org.
12. Ricker, M., & Häusler, F. (2014). European punching design provisions for double-headed studs. *Proceedings of the Institution of Civil Engineers. Structures and Buildings, 167*(8), 495–506. https://doi.org/10.1680/stbu.13.00047
13. Ricker, M., & Siburg, C. (2016). Punching shear strength of flat slabs–critical review of Eurocode 2 and fib Model Code 2010 design provisions. *Structural Concrete, 17*(3), 457–468. https://doi.org/10.1002/suco.201500106.
14. Siburg, C., Häusler, F., & Hegger, J. (2012). Durchstanzen von Flachdecken nach NA (D) zu Eurocode 2. *Bauingenieur, 87*(5), 216. https://doi.org/10.1002/best.201000013.
15. DIN EN 206-1:2000+A1:2004+A2:2005, Concrete-Part1:Specification, Performance, Production and Conformity (2005).
16. Taerwe, L. (1993). Towards a consistent treatment of model uncertainties in reliability formats for concrete structures. CEB Bulletin d'Information No 219 'Safety and Performance Concepts' 219. https://hdl.handle.net/1854/LU-237843

Probabilistic Modeling of Impact of Vehicles on the Road Furniture

Alfred Strauss, Panagiotis Spyridis, Ivan Zambon, Thomas Moser, Christian Honeger, and Dan M. Frangopol

Abstract Near driving lanes of roadways, numerous components of transport infrastructure are located along the route. Such components have to be secured by restraint systems, and in many cases different road lanes must also be effectively separated from each other. The focus of the study presented herein is to present a probabilistic approach for the departure of motor vehicles from their intended lane. Presently, assessments of the road infrastructure regarding possible accidents are primarily oriented to evaluating the resistance side. On the other hand, this paper intends to address the impact side by focusing on the likelihood of impact of vehicles on the road furniture. In order to determine the probability of impact, parameters of the traffic composition of the alignment, and of the pavement conditions were studied. A novel methodology is presented herein, which by accounting for these factors assesses the fragility of the infrastructure sub-system. The assessment joins both road engineering physics and expert judgements, and it is incorporated in spreadsheet tool. The feasibility of the tool is demonstrated, and sensitivities of the evaluation process are discussed and evaluated.

Keywords Transportation networks · Reliability analysis · Sensitivity · Road conditions · Routing elements

---

A. Strauss · I. Zambon
University of Natural Resources and Life Sciences, Institute of Structural Engineering, Vienna, Austria

P. Spyridis (✉)
Faculty of Architecture and Civil Engineering, Technical University of Dortmund, Dortmund, Germany
e-mail: panagiotis.spyridis@tu-dortmund.de

I. Zambon
FCP Fritsch, Chiari & Partner ZT GmbH, Vienna, Austria

T. Moser · C. Honeger
Department of Asset Management, ASFINAG Service GMBH, Vienna, Austria

D. M. Frangopol
Department of Civil and Environmental Engineering, ATLSS Engineering Research Center, Lehigh University, Bethlehem, PA, USA

© The Author(s), under exclusive license to Springer Nature Switzerland AG 2021
J. C. Matos et al. (eds.), *18th International Probabilistic Workshop*, Lecture Notes in Civil Engineering 153, https://doi.org/10.1007/978-3-030-73616-3_46

1 Introduction

Socioeconomic impacts associated with the recent degradation of bridges have presented civil engineers with great challenges [1–3]. Thus, substantial efforts in the maintenance and assessment of road infrastructure components have been invested by government bodies, highway owners and operators. Since they form a frontline of safety against vehicle traffic–related accidents, a great extent of consideration was directed to the equipment of roads and bridges. To that end, their maintenance absorbs a significant portion of the assets' operational expenditures. Moreover, this can be seen through the fact that that there is a large number of standards and guidelines, which address design calculations, load models, inspection, recalculation, maintenance programs, and reliability assessment of existing components, such as [4–6]. At the same time, a substantial number of scientific investigations are centred on these considerations. However, these standards and research outcomes lack assessment procedures relating to the intensity of events and their likelihood. Furthermore, they fail to consider the interaction of road furniture with the traffic conditions (velocity, mixture of traffic), the surface conditions of lanes, and the routing elements of the lanes.

The main purpose of the study presented in this paper is to develop a framework that allows for a risk-based and significance-weighted performance assessment of road furniture. The reliability concepts applied in the study are coherent with the fundamental notions of [7, 8]. The risk-based assessment concepts focus on the action model that is highly variable, rather than the resistance side of furniture. The detailed objectives of the study can be summarized as follows:

- to deploy, in cooperation with infrastructure owners, a systematic and efficient analytical decision tool for the assessment and the intervention planning of the road equipment using probabilistic safety concepts (PSC),
- to develop a closed analytical solution, which merges the ratings of impacts on road equipment based on RVS [9] together with routing condition characteristics and traffic dynamic parameters,
- to permit an individual regulation of acceptable risks and reliability levels, consequently altering the remaining technical service life of road equipment,

The approach presented in the paper is flexible for incorporation of further properties and can be easily conveyed to other systems based on large input databases. At present, estimations of the road infrastructure regarding possible accidents are always concentrated on the resistance side. On the other hand, this paper focuses on the probability of impact of a vehicle on road furniture, and hence, addresses the impact side. Moreover, this contribution is continuation of previous work done by authors on the topic of likelihood of impact events in transport networks, which can be found in [10].

2 Background

Through large-scale research projects, instruction and guidance on the design of roadside and restraint elements for motorways have been constantly developing. In [11], for the assessment of various road infrastructure components (such as pavements, engineering structures, and road furniture) the introduction of maintenance backlogs is used. In [12], the findings of the ERA NET European project HEROAD (Holistic Evaluation of Road Assessment) are presented. This report is based on literature reviews, interviews, and the participants' experience, and it focuses on existing practices for the assessment of road assets in Europe.

In [13], a comprehensive presentation of monitoring techniques is presented, where the focus is set on the performance evaluation of a number of road components through manual inspections, fixed safety cameras, satellites, and mapping vehicles. In Austria, "RVS" [14] and Asfinag [15] deliver a description of the outcomes of the project "Asset Service Condition Assessment Methodology" (ASCAM). The project addresses the state of the art regarding condition assessment of road equipment. Furthermore, the "Roadstar" is one of the mapping vehicles developed and used in Austria.

In [16–19], the deliverables of the project "Practical Road Equipment Measurement, Understanding and Management" (PREMIUM) are presented. The project studied road markings, road signs, vehicle restraint systems and noise barriers. The main aim was to recognize key characteristics of each asset, the appropriate monitoring techniques, data interpretation, and the associated management strategies based on surveys. In addition, a risk based asset management methodology is presented in [20]. This methodology addresses wide range of civil engineering structures, such as pavements, structures including bridges and retaining walls, tunnels, road furniture, drainage, and geotechnical assets.

International paradigms and best practices for the design and assessment of roadside structures are provided in the volume of articles presented in [21]. Furthermore, case studies focused on the recent advances in the technology of roadside structures, as well as in the minimization of fatal and serious injuries from vehicle impacts.

In [10], a newly introduced methodology was presented that accounts for road conditions, traffic and routing elements properties in order to assess the fragility of the infrastructure. The evaluation was based on either road engineering physics or expert judgements. The method was incorporated in spreadsheet tool. The feasibility of this tool was demonstrated, and sensitivities of the assessment process were evaluated and discussed. The work presented herein is the continuation of the work performed by authors on the topic of likelihood of impact events in transport networks.

3 Likelihood of Events in Transport Networks

The probability of an impact of a vehicle on a road furniture $p_{S,impact}$ can be determined in accordance with Eqs. (1) and (2). According to [10], Eq. (1) can be classified and divided into three main terms, as follows:

(a) The accepted probability of occurrence of an impact or the accepted probability of failure of a non-compliant traffic flow for standard-compliant routing elements, which is equal to $p_{f,NORM} = 1 \times 10^{-6}$;
(b) The corresponding increase factors $\eta_{F,LN}$, $\eta_{F,QN}$ and $\eta_{F,KR}$ with respect to the longitudinal inclination of lanes (LN), the transverse inclination of lanes (QN), and curvature of the lanes (KR), respectively, which are all calculated on the basis of physical laws of driving dynamics.
(c) Increasing the reference likelihood of an impact ($p_{f,NORM}$) due to peculiarities in the pavement surface, such as the pavement grip $p_{f,G}$, the longitudinal evenness $p_{f,L}$, the pavement damages $p_{f,O}$, the pavement cracks $p_{f,R}$, the pavement ruts $p_{f,SR}$, the traffic volume $p_{f,V}$, and the vehicle velocity $p_{f,\mathrm{GE}}$. In Austria, the influence of the fragilities (f_G, f_L, f_O, f_R, f_S) is based on the recorded data of the mapping vehicle "Roadstar", the fragility f_V is based on actual counts by the road operators, while all are based on fragility related transfer functions ranging between 0 and 1 as described in more detail in the following paragraphs. The influencing factors are assumed based on either experience from road operations or on simplified physics laws in kinematics.

$$p_{s,impact} = p_{f,NORM} + {(\eta_{F,LN} + \eta_{F,QN} + \eta_{F,KR})}/{3} \cdot \left\{ \begin{array}{l} f_G \cdot p_{f,NORM} \cdot \left(\frac{p_{f,G,LIMIT}}{p_{f,G,NORM}} - 1 \right) + \cdots \\ f_L \cdot p_{f,NORM} \cdot \left(\frac{p_{f,L,LIMIT}}{p_{f,L,NORM}} - 1 \right) + \cdots \\ f_O \cdot p_{f,NORM} \cdot \left(\frac{p_{f,O,LIMIT}}{p_{f,O,NORM}} - 1 \right) + \cdots \\ f_R \cdot p_{f,NORM} \cdot \left(\frac{p_{f,R,LIMIT}}{p_{f,R,NORM}} - 1 \right) + \cdots \\ f_S \cdot p_{f,NORM} \cdot \left(\frac{p_{f,S,LIMIT}}{p_{f,S,NORM}} - 1 \right) \end{array} \right\} \cdot f_V \cdot f_{GE} \quad (1)$$

In Table 1, detailed description of the specific parameters of the basic equation of the assessment concept are listed. The presented equations and parameters, enable the calculation and the analysis of the probability of an impact of a vehicle on a road furniture $p_{S,impact}$.

The corresponding assessment of impact risk of the road furniture damage is based on the risk assessment index R_i that can be calculated as shown in Eq. (2):

$$R_i = 1 + 4 \cdot \frac{\left[p_{s,impact} - p_{f,NORM}\right]}{p_{f,NORM}} \quad (2)$$

Table 1 Descriptive quantities for computing the vehicle impact-probability according to Eq. (1) [10]

Symbol	Specification	Value
$p_{S,impact}$	Vehicle impact probability according to Eq. (1)	0–1
$p_{f,NORM}$	Standard specific safety standards associated with a vehicle impact probability	1×10^{-6}
$\eta_{F,LN}$	Increasing factor for the impact force—due to the longitudinal inclination of the lane	
$\eta_{F,QN}$	Due to the transverse inclination of the lane	
$\eta_{F,Kr}$	Due to curvature of the lane	
f_G	Fragility associated with the lane grip	0–1
$p_{f,G,NORM}$	Standard specific vehicle impact probability	1×10^{-6}
$p_{f,G,LIMIT}$	Upper threshold of the vehicle impact probability; e.g. $p_f = 2 \times 10^{-6}$	2×10^{-6}
f_L	Fragility associated with the lane longitudinal inclination	0–1
$p_{f,L,NORM}$	Standard specific vehicle impact probability	1×10^{-6}
$p_{f,L,LIMIT}$	Upper threshold of the vehicle impact probability; e.g. $p_f = 2 \times 10^{-6}$	2×10^{-6}
f_O	Fragility associated with the surface damages in the pavement	0–1
$p_{f,O,NORM}$	Standard specific vehicle impact probability	1×10^{-6}
$p_{f,O,LIMIT}$	Upper threshold of the vehicle impact probability, e.g. $p_f = 2 \times 10^{-6}$	2×10^{-6}
f_R	Fragility associated with the surface cracks in the pavement	0–1
$p_{f,R,NORM}$	Standard specific vehicle impact probability	1×10^{-6}
$p_{f,R,LIMIT}$	Upper threshold of the vehicle impact probability; e.g. $p_f = 2 \times 10^{-6}$	2×10^{-6}
f_S	Fragility associated with the ruts in the pavement	0–1
$p_{f,NORM}$	Standard specific vehicle impact probability	1×10^{-6}
$p_{f,S,LIMIT}$	Upper threshold of the vehicle impact probability; e.g. $p_f = 2 \times 10^{-6}$	2×10^{-6}
f_V	Fragility associated with the traffic volume	0–1.2
f_{GE}	Fragility of the traffic velocity to the vehicle impact—fragility	0–1

The consequence of an impact is related with the loss of a single object; hence, the risk in this study is equal to the likelihood of the impact event.

Force increasing or decreasing factors associated with the vehicle impact shown in Table 2, as well as fragility functions of vehicle impacts in traffic networks, are explained in more detail in [10], as regards their nature and their influence on $\eta_{F,LN}$, $\eta_{F,QN}$ and $\eta_{F,KR}$ of Eq. (1).

Table 2 Increase of the vehicle impact force due to the longitudinal lane inclination, the transverse lane inclination, and the curvature of the lane [10]

Symbol	Specification	Unit
$\eta_{F,LN,max}$	Max. increasing factor of the vehicle impact force—due to the longitudinal inclination of the lane	(–)
F_{LN}	Longitudinal inclination of the lane	(°)
$\eta_{F,QN,max}$	Max. increasing factor of the vehicle impact force—due to the transverse inclination of the lane	(–)
F_{QN}	Transverse inclination of the lane	(°)
$\eta_{F,KR,max}$	Max. increasing factor of the vehicle impact force—due to the lane curvature	(–)
R	Radius of the lane curvature	(m)
R_{min}	Minimum radius of the lane curvature	(m)
VM	Mass of the considered vehicle	(kg)
v_{max}	Design velocity	(m/s)
v_{min}	Minimum vehicle speed at impact after braking	(m/s)

4 Risk Assessment and Damage Classes

According to [10], considering the assessment index R_i, the probability evaluation of a vehicle impact can be categorised as follows:

- $1.0 \leq R_i \leq 2.5$: The road geometry, the pavement surface characteristics and the driving dynamics factors (traffic strength and speeds) have negligible or no effects on the impact risk.
- $2.6 \leq R_i \leq 4.0$: The road geometry, the pavement surface characteristics and the driving dynamics factors have negligible effects on the impact risk.
- $4.1 \leq R_i \leq 5.5$: The road geometry, the pavement surface characteristics and the driving dynamics factors have a medium to large effects on the impact risk.
- $R_i > 5.6$: The road geometry, the pavement surface characteristics and the driving dynamics factors have significant effects on the impact risk.

It is important to mention that the assessment index R_i concept is based on the assumption of an impact occurrence probability of $p_{f,NORM} = 1 \times 10^{-6}$ ($R_i = 1.0$) for a lane alignment:

(a) without a longitudinal inclination,
(b) without a transverse inclination,
(c) with a very large radius of curvature,
(d) with an optimal road surface condition,
(e) with a predefined amount of traffic per design, and
(f) a predefined design velocity.

5 Case Studies

The methodology for the assessment of risk in existing road furniture due to vehicle impact that was presented in previous sections was implemented on of real traffic situations, the actual alignment characteristics, and the current road surface conditions identified on the Austrian motorway network (ASFINAG). The grouping potential and variations in the assessment index are the result of the road surface properties, the alignment characteristics, the traffic flows, and the vehicle speeds.

One can easily notice a sample concentration in the ranking regions of 1.5 and 2.5 in the graphs shown in Fig. 1. This can be described by the bi-modal distribution with respective peak concentrations. Furthermore, a further accumulation can also be noticed at classes greater than 5, both in the main and side lanes. Based on the graphs shown in Fig. 1 it becomes apparent that items in high-risk levels ($R_i > 5$) are to some extent more sensitive to main lane traffic. At the same time, low-risk regions are denser for main lane traffic data.

In order to reduce the assets risk speed reduction measures can be implemented on the network. This is thought to be an optimal solution between the risk of asset loss and road network performance. Using the fragility curve the influence of the pavement grip on the rating of the vehicle impact risk is accomplished, as shown in Fig. 2. To that end, a fragility curve is observed as the assessment of the probability of

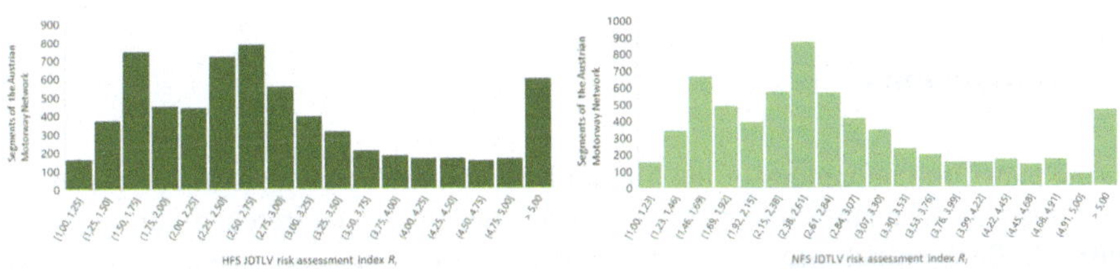

(a) HFS - JDTLV (b) NFS - JDTLV

Fig. 1 Risk assessment index R_i, in relation to actual traffic statistics from the Austrian motorway and road network, with: HFS = main lane, NFS = side lane, and JDTLV = annual average traffic density for trucks (indicative cases elaborated in [10])

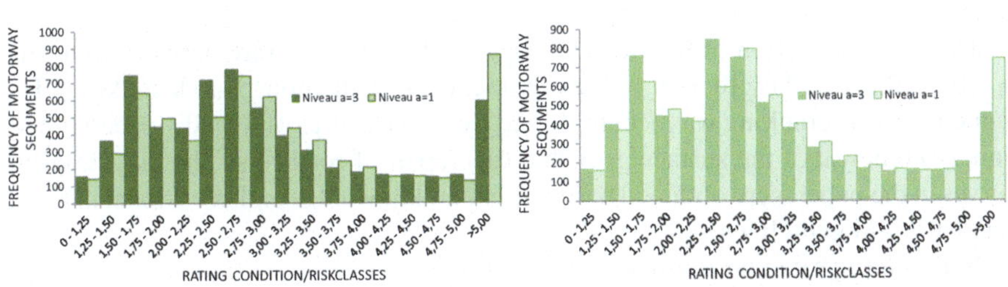

(a) HFS - JDTLV (b) NFS - JDTLV

Fig. 2 Risk assessment index R_i, in relation to actual traffic statistics from the Austrian motorway and road network, with emphasis on pavement grip properties (indicative cases elaborated in [10])

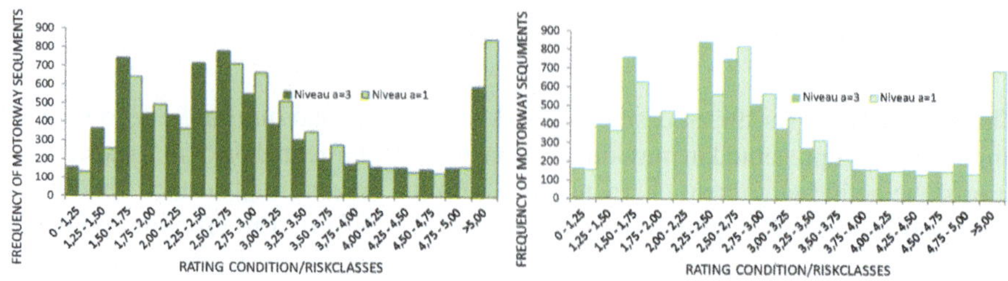

(a) HFS - JDTLV (b) NFS - JDTLV

Fig. 3 Risk assessment index R_i, in relation to actual traffic statistics from the Austrian motorway and road network, with emphasis on pavement ruts (indicative cases elaborated in [10])

exceeding a specific damage state according to the degradation process. The fragility curves used herein were derived based on expert knowledge and national standards and guidelines. For detailed overview of used fragility curves, see [10].

As it can be seen in Fig. 2, the man risk level in the network lies in the region of 2.25–3.25.

Through the fragility curves, also the influence of the pavement ruts on the risk rating of the vehicle impact is described. In Fig. 3, it can be seen that the proportion of traffic situations in class >5 is increased for all traffic situations considered.

6 Conclusions

In the paper, the influences of the longitudinal and the transverse gradient, and the curvature are evaluated by physical driving dynamics. A reference configuration with a failure probability $p_f = 10^{-6}$ was assumed. The road surface conditions, such as damage extent and grip are evaluated by adjusted fragility curves, based on empirical values and rating. The presented method is in agreement with the rating classification of RVS 13.03.51 [22]. To that end, it allows for an automatized risk evaluation taking into account the traffic composition, the surface measurement data retrieved by mapping vehicles, the JDTLV values, the occupancy level of the lanes, and the alignment.

The implementation of the methodology on the infrastructure network of a road operator allowed the efficient risk assessment, i.e. reduction of risk ranking. The reduction of risk ranking was accomplished by the measures of: (i) 20% speed reduction, (ii) renewal of the surface grip, and (iii) repair of road ruts. The used fragility curves in the paper were obtained based on expert knowledge and occasional national standards and guidelines.

Main findings can be summarized as follows:

- In the research project that served as the base for the paper, authors developed an analytical decision tool for the assessment and the maintenance forecasting of the road equipment, based on the probabilistic safety concepts (PSC);
- The authors showed how to associate traffic dynamic parameters and routing characteristics to a closed analytical solution for a RVS [22] based rating of impacts on road equipment;
- A specific adjustment of the acceptable reliability level, the acceptable risk and in consequence the remaining technical service life are all supported by the presented probabilistic based analytical decision method.

Acknowledgements The authors acknowledge the support of the Interreg project ATCZ 190 "SafeBridge" and the COST TU1406 action, awarded by the European Commission.

References

1. Strauss, A., Wendner, R., Bergmeister, K., & Costa, C. (2013). Numerically and experimentally based reliability assessment of a concrete bridge subjected to chloride-induced deterioration. *Journal of Infrastructure Systems, 19*(2), 166–175. https://doi.org/10.1061/(ASCE)IS.1943-555X.0000125.
2. Zambon, I., Vidovic, A., Strauss, A., Matos, J., & Friedl, N. (2018). Prediction of the remaining service life of existing concrete bridges in infrastructural networks based on carbonation and chloride ingress. *Smart Structures and Systems, 21,* 305–320. https://doi.org/10.12989/sss.2018.21.3.305.
3. Zambon, I., Vidovic, A., Strauss, A., Matos, J., & Amado, J. (2017). Comparison of stochastic prediction models based on visual inspections of bridge decks. *Journal of Civil Engineering and Management, 23,* 553–561. https://doi.org/10.3846/13923730.2017.1323795.
4. AASHTO Manual for Bridge Evaluation. (2018). *American Association of State Highway and Transportation Officials* (3rd ed.), Washington, DC, USA.
5. ISO-13822. (2010). *Basis for design of structures—Assessment of existing structures.* International Organization for Standardization, Geneva.
6. Nederlands Normalisatie Instituut. (2011). *NEN 8700 (nl)—Assessment of existing structures in case of reconstruction and disapproval—Basic rules.*
7. Ang, A. H., & Tang, W.H. (2007). *Probability concepts in engineering planning and design.* Hoboken: Wiley.
8. Schneider, J. (1997). Introduction to safety and reliability of structures. *Structural Engineering Documents, 5.* Zurich, Switzerland: IABSE.
9. RVS. *Guidelines for planning, construction and maintenance of roads "Die Richtlinien und Vorschriften für das Straßenwesen".* Available at: http://www.fsv.at/shop/produktliste.aspx?ID=3197C858-15DE-4517-9EF2-F3B7E22175A4.
10. Strauss, A., Moser, T., Honeger, C., Spyridis, P., & Frangopol, D. M. (2020). Likelihood of impact events in transport networks considering road conditions, traffic and routing elements properties. *Journal of Civil Engineering and Management, 26*(1), 95–112. https://doi.org/10.3846/jcem.2020.11826.
11. Weninger-Vycudil, A., Litzka, J., Schiffmann, F., Lindenmann, H. P., Haberl, J., Scazziga, I., Rodriguez. M., Hueppi, A., & Jamnik, J. (2009). Maintenance backlog estimation and use. *Road Eranet.*
12. Sjögren, L., Ihs, A., Edvardsson, K., Wennström, J., Haider, M., Casse, C., Van Geem, C., Benbow, E., Wright, A., Žnidarič, A., & Kokot, D. (2013). *Overall road asset performance.*

13. Spielhofer, R. (2014). Collaborative project FP7-285119 Advanced and cost effective road infrastructure construction, management and maintenance. *Deliverable D 4.2 Monitoring of Road Inventory*.
14. Federal Ministry of Transportation, Building and Urban Development—Department for Road Construction (2011). *Richtlinie zur Nachrechnung von Strassenbrücken im Bestand (Guideline for the recalculation of existing road bridges)*.
15. Highways Agency. (2017). *Design manual for roads and bridges, highway structures: Inspection and maintenance—Assessment, standards for highways*.
16. Benbow, E., & Wright, A. (2017). Summary of deliverables 1 and 2 of the PREMIUM project and the work required to achieve the recommendations. In *Conference of European Directors of Roads* (CEDR).
17. Spielhofer, R., Osichenko, D., Leal, D., Benbow, E., Wright, A., & Morgan, P. (2016). Identifying the key characteristics for noise barriers condition measurements—Deliverable D1d and D2d. In *Conference of European Directors of Roads (CEDR)*.
18. Spielhofer, R., Osichenko, D., Leal, D., Benbow, E., & Wright, A. (2017a). Identifying the key characteristics for vehicle restraint system condition measurements—Deliverable D1a and D2a. In *Conference of European Directors of Roads (CEDR)*.
19. Spielhofer, R., Osichenko, D., Leal, D., Benbow, E., & Wright, A. (2017b). Identifying the key characteristics for road sign condition measurements—Deliverable D1b and D2b. In *Conference of European Directors of Roads (CEDR)*.
20. Spielhofer, R., Oldfield, M., Britton, C., Levine, H., Brozek, B., Weninger-Vycudil, A., Lepert, P., Mladenovic, G., Le Bars, G., Pohu, J., & Litzka, H. (2015). Cross-asset risk assessment (X-ARA)—Risk framework and modelling specifications, Deliverable D1.2, Deliverable D2.1. In *Conference of European Directors of Roads (CEDR)*.
21. Troutbeck, R. (Ed.). (2013). *Transportation research circular E-C172: Roadside safety design and devices—International workshop*. Roadside Safety Design Committee.
22. RVS 13.03.51. (2013). *Wegweiserbrücken "Directional gantries"*. Available at: http://www.fsv.at/shop/produktdetail.aspx?IDProdukt=274013da-42a5-48b9-a9d1-2c2bb91283a0.

Probabilistic-Based Consequence Analysis for Transport Networks

Donya Hajializadeh, Chia Sadik, and Boulent Imam

Abstract The aim of this paper is to propose a methodological framework for consequence analysis of transportation networks. The probabilistic framework is based on the definition of performance indicators that describe the time-dependent functionality of the asset/system, starting from a pre-existing normal performance state, capturing the time and evolution of disruption during and after the disruption and during the recovery/restoration stage. A proposed case study that will be used for the demonstration of the applicability of the framework is described.

Keywords Consequences · Performance · Probabilistic · Transport network

1 Introduction

The economy of a society and the well-being of its citizens depend on the continuous and reliable functioning of infrastructure systems. Among all infrastructure systems, those which incapacity and destruction impacts the defence and economic security, are generally regarded as critical [1]. Different countries have different lists detailing their critical infrastructure systems but generally, they have the following list of systems in common: transportation, water supply systems, telecommunications, electric power systems, natural gas and oil, banking and finance, government services and emergency services. These infrastructure systems constitute the backbone of modern societies by providing essential services for their functioning. Destructing or damaging assets in such systems either disconnects large areas of networks from each other or causes a rerouting of the flow from one area of the network to another through a longer detour path. Resilience and vulnerability conditions associated with such systems can then have an impact on the resilience/vulnerability of the whole network [2, 3]. Therefore, the disruption consequence analysis of such systems is an essential component of risk and resilience management of systems subjected to hazardous events.

D. Hajializadeh · C. Sadik · B. Imam (✉)
Department of Civil and Environmental Engineering, University of Surrey, Guildford, UK
e-mail: b.imam@surrey.ac.uk

© The Author(s), under exclusive license to Springer Nature Switzerland AG 2021

J. C. Matos et al. (eds.), *18th International Probabilistic Workshop*, Lecture Notes in Civil Engineering 153, https://doi.org/10.1007/978-3-030-73616-3_47

The European project SIRMA (Strengthening Infrastructure Risk Management in the Atlantic Area) aims to develop, validate and implement a robust framework for the efficient management and mitigation of natural hazards in terrestrial transportation modes at the Atlantic Area. As part of this project a risk and vulnerability assessment system will be developed to assess interceptable and non-interceptable events under various climate change scenarios. An integral part of this system will be a consequence assessment framework which will consider the short-term and long-term, direct and indirect impact of the climate change-induced hazards on transportation infrastructure. This study presents a novel framework for probabilistic consequence analysis of transport networks as a function of performance indicator of the system. This forms an integral component of SIRMA's risk assessment framework.

The paper is organised as follows: Sect. 2 presents a state-of-the-art review of the consequence analysis for transport networks, Sect. 3 provides an overview of the framework and Sect. 4 demonstrates the details of the case study that will be used to illustrate the application of the framework in future studies.

2 Background on Consequences of Failure of Transport Networks

2.1 Categorisation of Consequence of Failure

Consequences of failure can often be seen as a good indicator of the importance of an asset, given its form, function and location within a transport network. They can range from casualties and injuries to structural damage, reduction in network functionality and may also extend into environmental as well as societal impact. Table 1 shows a categorisation framework for consequences of failure into four main categories: human, economic, environmental and social consequences. Each of these main four categories can be further sub-divided into a number of more specific areas, so that itemisation and appropriate modelling, where possible, may be undertaken to assess and/or quantify them.

Consequences can be classified as either direct or indirect. Direct consequences are considered to result from damage states of individual components/assets. Indirect consequences, triggered by the former, are associated with reduction in, or loss of, system/network functionality. The differentiation between direct and indirect consequences depends on the system boundaries considered in the analysis as well as on the extent of the time frame that is used; they may, therefore, be subjective to a degree.

An assessment framework for failure consequences should account for their type, the relevant time frame, as well as the network/system boundaries. Therefore, they should be considered within a time domain as well as a spatial domain. The time frame considered (days/weeks/years) plays an important role in consequence modelling; consequences will be different when considering only a short-term post-event time

Table 1 Categorisation for failure consequences of transport networks

Consequence categories	Examples
Human	Fatalities Injuries Psychological damage
Economic	Replacement/repair costs Loss of functionality/downtime Traffic delay/re-routing costs Traffic management costs Clean up costs Rescue costs Regional economic effects Loss of production/business Investigations/compensations Infrastructure interdependency costs
Environmental	CO_2 emissions Energy use Pollutant releases Environmental Clean-up/reversibility Noise pollution
Social	Loss of reputation Erosion of public confidence Undue changes in professional practice

frame or a long-term period extending well after the failure event. The actual duration in considering long-term periods is also expected to affect the magnitude of estimated consequences. For example, a bridge failure in a transport network may result, during the immediate and mid-term aftermath, in loss of business revenue and high traffic delay costs but over longer periods these might change as new regional equilibria are reached within the network. Lastly, consequence estimation is affected by the definition of the system boundaries, i.e. the extent of the transport network that is considered in the analysis that the bridge is within (spatial domain). The extent of the spatial domain is also an important factor, depending on whether a single route (with diversions) or a more widely encompassing spatial network is considered. Here, the level of redundancy of the transportation network in redistributing traffic flows following the bridge collapse plays an important role. Further layers can be added to the above systems by addressing wider societal consequences such as business losses, environmental impact, etc.

The consequences of failure vary significantly from asset to asset, and may depend on a range of factors which are related to the hazard itself, the asset and its utilisation, as well as the surrounding environment. The source and nature of the hazard leading to an asset failure will affect the consequences, considerably. It is expected that the greater the magnitude and duration of a hazard, the greater the consequences will be. Asset location is one of the major factors expected to influence the magnitude of failure consequences. The type of road or rail route served by the asset influences the traffic intensity and, hence, the number of people exposed to any given hazard, as well as the traffic delay costs. Moreover, the availability of emergency services and accessibility to treatment for injuries will most likely be best in urban areas, hence, the number of fatalities may be lower in such locations. Finally, the cost of repair or reconstruction of an asset may be higher in rural areas due to increased labour, materials and transportation costs. On the other hand, access might be easier and interdependency issues might be less critical than in urban areas. The time of the day that an asset failure may take place will also have an effect on human consequences. Assets such as bridges, for example, will experience high levels of traffic during peak times and the potential for mass casualties is thus higher.

2.2 *Analysis of Consequences of Failure of Transport Networks*

The analysis of consequences of failure of transport networks has been performed in a number of studies for different hazards including earthquakes, extreme rainfall and others [4–10]. Transport network impacts are commonly analysed using two methodologies including those measuring network topology (i.e. graph theory) and system operation (i.e. travel time and cost) [4]. The topological method provides a more simplistic representation of the network with no consideration of route choice and periodic demand (peak and non-peak) on travel time and cost. However, the second method uses traffic models to simulate network flows that are more realistic, although the computational and data demands become more complex. These transport models are used in conjunction with hazard models to quantify the impacts of extreme weather events. A comprehensive review of these analytical assessment modelling techniques for disaster events can be found in [11].

A previous study by [4] assessed the impact of landslide disruptions by coupling hazard data with a transport network model. The methodology followed in the study was to: (i) establish the road network, (ii) evaluate the vulnerability of the road network, (iii) create an event set of landslide disruptions, (iv) develop a micro-meso network model to simulate the traffic flow, and (v) measure the impact of each event. The study however did not capture wider long-term impacts such as reductions in business investments. A further study simulated the impacts of closing different sections of the road network in Switzerland [12]. Failure consequences were calculated using subnetworks and compared against the option of using a full network. The study however was limited in that it assumed each of the failure scenarios to be mutually exclusive, which is an oversimplification for natural hazards such as floods. [5] developed a simple transport network and used a depth-disruption function to represent the vehicle speed through floodwater. The traffic simulations were then coupled with a flood model. This study only focused on one mode of transport (roads). A paper by [6] proposed a new approach to support network operators in quantifying the risk related to their networks. The authors quantified risk from a source event to its societal event over space and time. The consequences were then monetised into direct and indirect costs, considering restoration interventions, prolongation of travel time, and missed trips. The paper also defined four damage states: (0) operational, (1) monitored, (2) capacity-reduced, (3) closed. In another study, a conventional analytical framework to simulate traffic flows was used under different flood scenarios in the Boston Metropolitan Area [13]. Direct costs from the damages were not considered as part of the study as well as no consideration of network restoration, which is crucial to know when estimating indirect consequences.

A study by [14] used a simulation-based model to measure resilience indicators in railway transport systems using different scenarios. The paper showed that efficient crisis management plans could reduce the impact of undesirable scenarios on a system. However, the study was limited in that it did not look into scenarios with consequences such as casualties and injured passengers. In [7], a macroscopic traffic simulation of a road flooded in Portland was performed. The consequence assessment was limited to only one scenario, the complete closure of links, were considered in the study. Others considered the impact of closing different bridges in Stockholm with two scenarios of bad weather leading to a 15% reduction in free-flow speed [15]. The transport model however was not calibrated as part of the study, providing a lower confidence in the results.

3 Methodological Framework

3.1 Probabilistic Consequence Analysis Framework

As shown in the literature review the definition of consequence analysis depends on a type of disruption, type of consequence and means of quantification of consequences. However, the literature is generally in agreement that any form of consequence can be described and linked directly or indirectly to time-dependent asset/system performance indicator/delivery function/figure of merit. Asset/system performance indicator describes time-dependent functionality of the asset/system at status quo, time of disruption, during and after the disruption and during recovery/restoration stage.

One of the traditional forms of performance indicator is the trapezoid function which is often used to describe the behaviour of an asset and/or system in response to a disruptive event and corresponding recovery stage [16]. In this form, the behaviour of the asset/system following a disruption and recovery is generalised as a linear function. Another traditional formulation is the triangular description which assumes a sudden drop in performance indicator following a disruption and a linear recovery afterwards. Imani and Hajializadeh [17] have expanded the trapezoid description to allow for flexibility in different disruption absorption and recovery/restoration trajectories for different assets/systems. Figure 1 shows a schematic representation

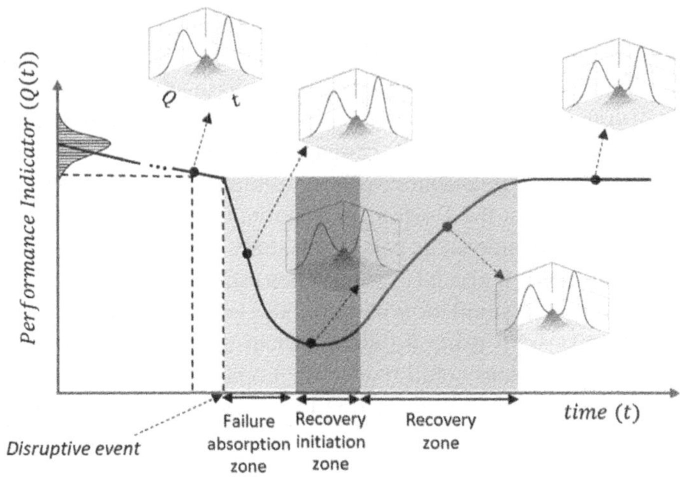

Fig. 1 Schematic probabilistic time-dependent performance indicator function

of a generalized performance indicator variation with time at five distinct zones. These zones include (1) equilibrium during status quo; (2) absorption of disruption; (3) equilibrium following disruption; (4) recovery initiation and absorption; (5) equilibrium following recovery.

The majority of state-of-the-art attempts in formulating asset/system performance have been focused on the deterministic definition, many of which concentrated on post-event recovery analysis. The deterministic assessments may lead to misjudgment of performance indicator of the asset/system which could then result in underestimation of the consequences. On the other hand, a probabilistic framework that can capture the uncertainties in the time-evolution of the performance indicator shown in Fig. 1 can offer useful insight into how failure consequences may be affected by such uncertainties.

The scope of this study is to provide a novel framework to evaluate asset/system performance indicator by accounting for the uncertainties in failure and recovery/restoration trajectories. Figure 1 schematically demonstrates the main uncertainties in describing asset/system performance indicator by joint probability density functions, $p(Q, t \mid z_i)$, for time-dependent performance indicator, $Q(t)$, at each zone/stage, z_i, as a function of time, t.

The joint probability distribution function aims to move past the type of disruption and recovery measures and it focuses on the impact of disruption and recovery on performance indicator. This is advantageous to the deterministic consequence analysis where the performance indicator is defined as a function of the event only. The joint probability of distribution can also be defined as a conditional probability for different types of hazards and disruptions, however, in the absence of required database and/or in cases of low-probability/high consequence events, it is advantageous to formulate the probability functions independently from the type of disruption.

The probability functions can be defined based on the available literature on different types of hazards, disruptions and recovery measures. The main source of the required database for probability functions are empirical qualitative data available in grey literature, reports of National Transport Authorities and news media searches.

To account for different possibilities for disruptive scenarios, a systematic scenario generating strategy is developed. The generated scenarios for a given transport network will include a cohort of Monte Carlo simulated single and multi-asset, simultaneous and sequential disruptions within the network. For single asset failure scenarios, the time of occurrence of the disruption, $t_{d,}$ and total drop in performance indicator of the asset (once the disruption is absorbed), PI_d, will be defined. An example of this would be the drop in traffic capacity of a bridge, or a segment of a road, due to flooding, capturing how quickly the flooding evolves over time leading to the loss of asset performance.

For multi-asset simultaneous disruptions, in addition to time and drop in performance indicator for each asset, the number of disrupted assets, $n_{d,}$ should also be defined. For the sequential multi-assets, the time lag between each disruption, $t_{lag,\,j}$ will be generated as part of the scenario simulation process. This type of scenario can, for example, represent a wider impact of a hazard on the transport network such as wider-scale flooding that may affect multiple stretches of roads and/or bridges. Figure 2 summarizes these inputs for each scenario.

The key in consequence analysis of a system, hence its performance indicator assessment, is to consider the interconnections and interdependencies of individual assets that can cause cascading failures, amplify negative consequences due to these failures and influence the overall performance of the system.

To describe the behaviour of a transport system as a function of its assets, network theory is utilised in this study. Network theory has been widely used to characterize infrastructure network topology and layout features by taking advantage of closed-form expressions and numerical simulations. Mathematically, a topological network can be represented as a graph with nodes and edges representing their connectivity nature. For the infrastructure network A, network properties can be represented by $I_A = \{N_A, E_A, M_A\}$, where N_A, is the node sets, E_A, is edges set and M_A is a $N_A \times N_A$ matrix representing the function of edges to pair-wise nodes. For transport networks, nodes can represent junctions, public transport stations, intersection control systems and traffic signs and edges could represent roads, bridges, tunnels, etc.

Once the system is simplified into its graph network representation, each asset (i.e. node and edge) will be assigned a performance indicator with corresponding joint probability distribution function $p(Q, t|z_i)$, $i = [1, .., 5]$. The performance indicator for each asset defines the level of serviceability and capacity for each asset. Then, by conducting a traffic simulation method (microscopic or macroscopic), the overall performance of the transport network can be defined as a function of the performance of its assets, collectively.

The characteristics of the overall system performance depend on the type of simulation. Macroscopic traffic simulation describes the collective vehicle dynamics as

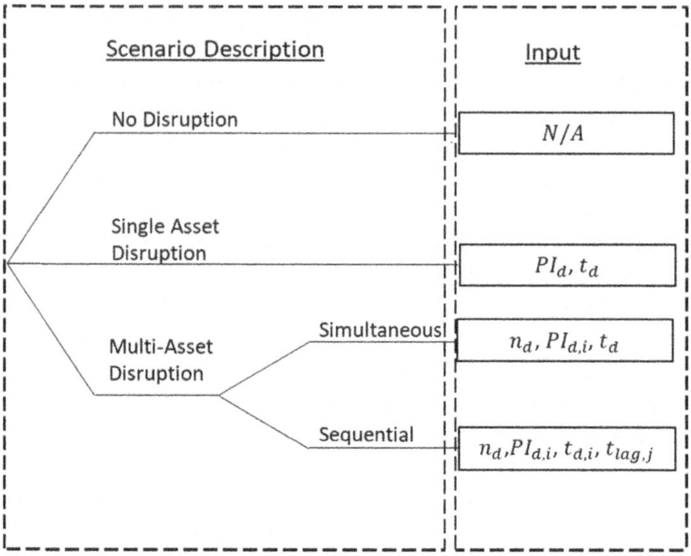

Fig. 2 Disruption scenario generation and inputs

a function of the spatial and temporal distribution of vehicle density and average velocity, whereas the microscopic traffic models define the individual position and velocity of all interacting vehicles in the system.

Once the performance of the system function for a given disruptive scenario is calculated, the consequence of the disruption as a function of the time-dependent performance indicator can be evaluated. The output of this stage will be a time-dependent consequence function for a given disruptive scenario. These steps will be repeated for all simulated scenarios. Once the number of required scenarios has been reached, a consequence spectrum can be constructed based on the time-dependent consequence function for each scenario. Figure 3 demonstrates the overall process of the proposed framework to acquire consequence spectrum. This framework will be used within the context of a case study transport network which is described in the following section.

4 Transport Network Case Study

The test bed for this study is located in Portugal and has been selected within the context of the EU-funded SIRMA project. The test bed includes some sections of the National Road 6 (EN6), which runs along the coast, and the Cascais Railway line, which runs parallel to EN6 at certain sections. The EN6 Road has a length of 16 km and the Cascais Railway line has a length of 25.5 km. The road and railway

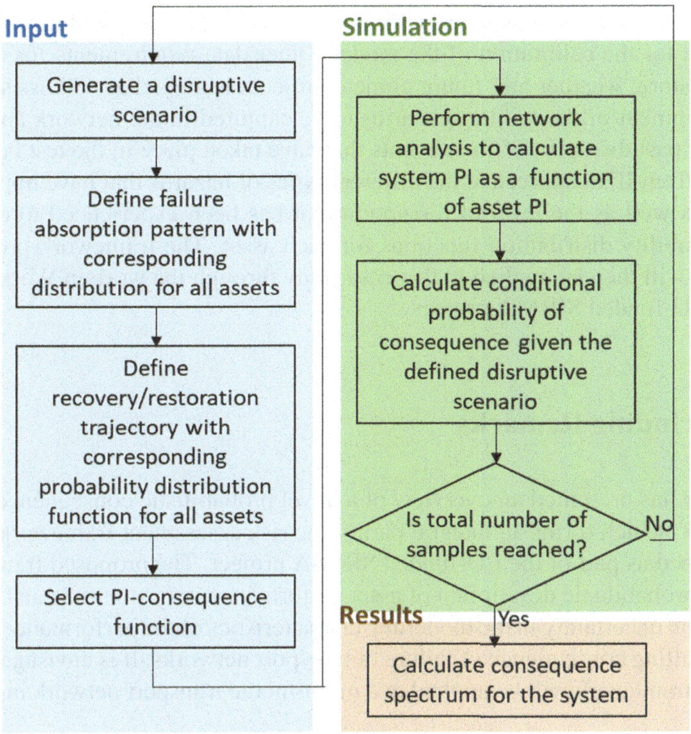

Fig. 3 Flowchart of the proposed consequence analysis framework

both play an important role in connecting the Cascais area, which is a major summer retreat for the local population and tourists, to Lisbon.

The European Commission (EC) have previously reported that 28% of the Portuguese mainland is vulnerable to coastal flooding, which can have severe consequences as 60% of the population inhabits the coastal zone [18]. With this in mind, the most critical natural hazard events in the selected Portuguese test bed have been identified to be coastal flooding from wave inundation and sea level rise. One of the main issues that has been reported by Infraestruturas de Portugal (IP), the major highway and railway infrastructure owner in Portugal and a key partner of the EU-funded SIRMA project, includes exposure to high tides in the Lisbon metropolitan area.

A transport model of the case study has been prepared through the AIMSUN Next transport modelling software that can perform both the macroscopic and microscopic traffic simulations within the same software. The traffic parameters required for the study include the speed, signal timings, traffic volumes from traffic counters, percentage of vehicle types, and the timetabled train services. Further infrastructure data is also required including the road and rail geometry, location of the asset, its construction type and the age of the assets. Part of this data has been provided by

IP towards the development of the transport network model and additional data will be utilized for the calibration of the model. Other data requirements for this study include historic weather and future climate projections data that will assist towards the development of the failure scenarios to be captured in the network analysis. In particular, past disruption/failure events that have taken place in the test bed will be collected from IP to appreciate the different types of hazards that have impacted the test bed as well as the level of disruption that has been experienced to derive the joint probability distribution functions for each asset. The framework proposed in this paper will then be applied to this case study through the work in Work Package 4 of the EU-funded SIRMA project.

5 Concluding Remarks

This paper has presented an overview of a novel probabilistic consequence analysis framework, which forms an integral part of the risk assessment framework that will be developed as part of the EU-funded SIRMA project. The proposed framework is based on probabilistic description of asset performance indicator and can be utilised to assess the uncertainty in the modelling characteristics of the performance indicator on the resulting consequence of failure of transport networks. It is envisaged that the proposed framework will be applied to a multi-modal transport network in Portugal.

Acknowledgements SIRMA project is co-financed by the INTERREG Atlantic Area Programme through the European Regional Development Fund (ERDF) with application code: EAPA_826/2018.

References

1. Ouyang, M. (2014). Review on modeling and simulation of interdependent critical infrastructure systems. *Reliability Engineering And System Safety, 121,* 43–60.
2. O'Kelly, M. E., Kim, H., & Kim, C. (2006). Internet reliability with realistic peering. *Environment Planning A, 33,* 325–343.
3. Reggiani, A. (2013). Network resilience for transport security: Some methodological considerations. *Transport Policy, 28,* 63–68.
4. Postance, B., Hillier, J., Dijkstra, T., & Dixon, N. (2017). Extending natural hazard impacts: An assessment of landslide disruptions on a national road transportation network. *Environmental Research Letters, 12*(1), 014010.
5. Pregnolato, M., Ford, A., Wilkinson, S. M., & Dawson, R. J. (2017). The impact of flooding on road transport: A depth-disruption function. *Transportation Research Part D, Transport and Environment, 55,* 67–81.
6. Hackl, J., Lam, J. C., Heitzler, M., Adey, B. T., & Hurni, L. (2018). Estimating network related risks: A methodology and an application in the transport sector. *Hazards Earth Systems Science, 18,* 2273–2293.

7. Chang, H., Lafrenz, M., Jung, I. W., Figliozzi, M., Platman, D., & Pederson, C. (2010). Potential impacts of climate change on flood-induced travel disruptions: A case study of Portland, Oregon, USA. *Annals of the Association of American Geographers, 100*(4), 1–4.
8. Clarke, J., Corbally, R., & O'brien, E. (2016). INFRARISK Deliverable 8.2—Case Study Results.
9. Carey, C., Clarke, J., Corbally, R., Connolly, L., Hajializadeh, D., Barcena, C., & Clotet, X. (2017). RAIN Deliverable 6.3—Report on Benefits of Critical Infrastructure Protection.
10. Enke, D. L., Tirasirichai, C., & Luna, R. (2008). Estimation of earthquake loss due to bridge damage in the St. Louis Metropolitan Area. II: Indirect Losses. *Natural Hazards Review, 9*(1), 12–19.
11. Faturechi, R., & Miller-Hooks, E. (2015). Measuring the performance of transportation infrastructure systems in disasters: A comprehensive review. *Journal of Infrastructure Systems, 21*(1), 04014025.
12. Erath, A., Birdsall, J., Axhausen, K. W., & Hajdin, R. (2008). Vulnerability assessment of the swiss road network. *Transportation Research Record Journal of the Transportation Research Board, 2137*(1), 118–126.
13. Suarez, P., Anderson, W., Mahal, V., & Lakshmanan, T. R. (2005). Impacts of flooding and climate change on urban transportation: A systemwide performance assessment of the Boston metro area. *Transportation Research Part D, Transport and Environment, 10*(3), 231–244.
14. Adjetey-Bahun, K., Birregah, B., Châtelet, E., & Laurens-Fonseca, E. (2014). A simulation-based approach to quantifying resilience indicators in a mass transportation system. In: *Proceedings of the 11th International ISCRAM Conference* (pp. 75–79).
15. Berdica, K., & Mattsson, L. G. (2007). Vulnerability: A model-based case study of the road network in Stockholm. In *Advances in Spatial Science* (pp. 81–106). Springer International Publishing.
16. Henry, D., & Ramirez-Marquez, J. (2012). Generic metrics and quantitative approaches for system resilience as a function of time. *Reliability Engineering and System Safety, 99*, 114–122.
17. Imani, M., & Hajializadeh, D. (2020). A resilience assessment framework for critical interdependencies. *Water Science and Technology, 81*(7), 1420–1431.
18. Commission, E. (2009). *The Economics of climate change adaptation in EU Coastal areas: Country overview and Assessment*. Brussels, Belgium: Office for Official Publications of the European Communities.

Probability of Flooding Due to Instability of the Outer Slope of a Levee

Anton W. van der Meer, Ana Teixeira, Arno P. C. Rozing, and Wim Kanning

Abstract The Netherlands is protected against major floods by a system of primary flood defenses (levees, dunes and hydraulic structures). These must comply with standards defined in terms of maximum allowable probabilities of flooding. Therefore, a new assessment framework for the main failure mechanisms is based on a probabilistic approach. One of the failure mechanisms which is not yet following such a probabilistic approach is instability of the outer slope of a levee. This failure mechanism is of importance after a rapid water level drop after a high-water event. In such an event, the pore water pressures in the levee are still high, and if an instability happens flooding can occur when (1) there is no time to take emergency measures before a second high-water event follows and (2) there is insufficient residual strength to prevent flooding during this consecutive high-water event. Levee reinforcement projects in the Netherlands allocate significant resources to resolve the presumed lack of safety of the levee due to the outer slope instability mechanism. Hence, this paper discusses how outer slope stability safety can be assessed probabilistically. A failure due to outer slope instability depends, besides the characteristics of the levee, on the peak water level, the water level drop velocity, the inter-arrival time between two consecutive high-water events and the time needed to take emergency measures. In this paper, a framework based on event trees is presented, using Intensity-Duration-Frequency-curves to include the time dependent statistics of the water level drop. This is a novel approach for outer slope instability in the Netherlands and results in less conservatism in assessments and designs, and therefore less required resources to mitigate the mechanism.

Keywords Levee · Stability · Reliability · Consecutive · Event trees

---

A. W. van der Meer (✉) · A. Teixeira · A. P. C. Rozing · W. Kanning
Deltares Unit Geo-Engineering, Delft, The Netherlands
e-mail: anton.vandermeer@deltares.nl

W. Kanning
Delft University of Technology, Delft, The Netherlands

© The Author(s), under exclusive license to Springer Nature Switzerland AG 2021
J. C. Matos et al. (eds.), *18th International Probabilistic Workshop*, Lecture Notes in Civil Engineering 153, https://doi.org/10.1007/978-3-030-73616-3_48

1 Introduction

The Netherlands is protected against major floods by a system of primary flood defenses consisting of levees, dunes and hydraulic structures. Since 2017, the flood defenses must comply with new standards, defined in terms of maximum allowable probabilities of flooding, an essential difference with the previous approach. An assessment framework for the main failure mechanisms was consequently established based on a probabilistic approach [1, 2], in which each failure mechanism must comply with a target that is based on the safety standard.

One of the failure mechanisms which is not yet following such a probabilistic approach is instability of the outer slope of a levee (the slope facing the river, lake or sea). This mechanism has received little attention and the current assessment of this failure mechanism is based on expert judgement and similarities with the instability of the inner slope. Levee reinforcement projects in the riverine area in the Netherlands allocate significant resources to resolve the presumed lack of safety of the levee due to outer slope instability. Hence, this paper discusses how the outer slope stability safety in the riverine area can be better determined, resulting in less conservatism. This paper is based on report [3].

Instability of the outer slope of a levee is of importance after a rapid water level drop (after a high-water event). In such an event, the pore water pressures in the levee are still high, the effective stress and the shear forces in the levee consequently low while the balancing river water level is low. If an instability happens, flooding can occur when there is no time to take emergency measures before a second high-water event follows. For riverine areas, the following consecutive events need to occur for outer slope instability to result in flooding (Figs. 1 and 2):

A. ***Slope instability***. During a high river water level, the pore water pressures in the levee increase. When the river water level drops, the water pressure decrease but with a time lag. A scenario with high pore water pressures in the levee and a low river water level is unfavorable for outer slope stability. At an unknown moment after a peak water level, the combination of pore water pressures and river water levels will be the most unfavorable for the outer slope stability—Sect. 2.2.

B. ***Consecutive high-water peak before repair***. A damaged levee due to outer slope instability during low river water level does normally not lead to flooding

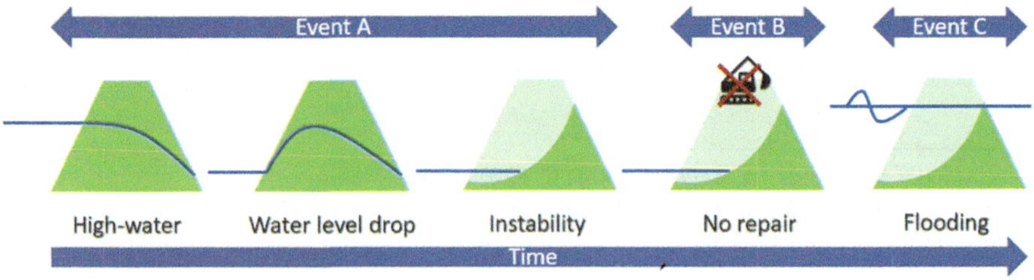

Fig. 1 Events leading to flooding initiated by outer slope instability

as the slip circles are typically not deep enough to result in immediate flood. Flooding due to instability of the outer slope can happen if a new high river water level occurs before (emergency) repair of the levee—Sect. 2.3.

C. ***Flooding during the consecutive high river water levels***. Given a damaged levee, not repaired before a consecutive high-water level, the probability of flooding by all different failure mechanisms during this consecutive high-water level event is increased compared to the non-damaged levee. This increase in the probability of flooding is due to the outer slope instability after the previous high river water level—Sect. 2.4.

In this paper, we develop our framework based on event trees. We give an overview of the methodology in Sect. 2 and illustrate its applicability in Sect. 3 for a levee along the Waal river, in the riverine area of the Netherlands. Finally, in Sect. 4, we provide conclusions and recommendations for further research.

2 Methodology

2.1 Event Tree

For the three events mentioned in Sect. 1, a probability estimate is obtained (Sects. 2.2, 2.3 and 2.4). These probabilities are combined using an event tree (Fig. 2). The total failure probability is comparted against the target failure probability. The latter is the maximum allowed probability reserved to instability and is derived from the maximum allowable probability of flooding of a levee segment (as defined in the Dutch law) using a division between the different failure mechanisms and considering spatial effects [4].

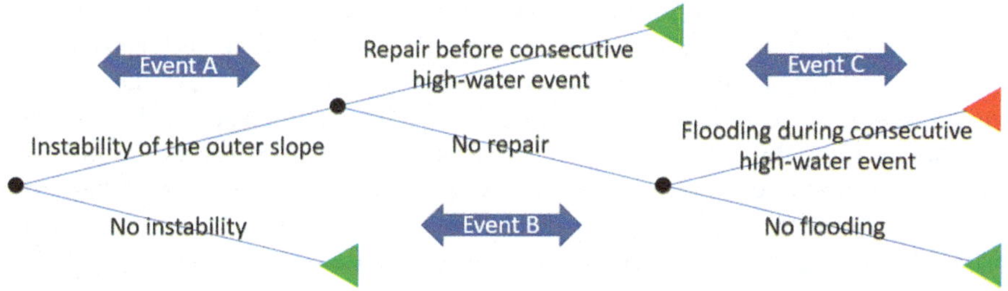

Fig. 2 Framework to assess the probability of flooding for outer slope stability. The event tree shows the events that will lead to flooding (red triangle). No flooding is denoted by green triangles

2.2 *Event A: Outer Slope Instability*

In this section we determine the probability of the instability. To calculate the probability of the instability (Sect. 2.2.4) the pore water pressures in the levee are calculated (Sect. 2.2.3) from the boundary conditions, peak water level (Sect. 2.2.1) and water level drop (Sect. 2.2.2).

2.2.1 Peak Water Level

The probabilistic software Hydra-NL [5] is used to derive the probability distribution of the peak water level, based on numerical simulations and the distributions of the wind, river discharge of the Rhine and of the Meuse river and their respective uncertainties. The influence of the wind is for this case negligible when compared with the influence of the discharge. The results of the software are compared with measurements of local water levels (https://waterinfo.rws.nl/#!/nav/index/). The observed high-water peaks are obtained from the measurements using a certain threshold and prominence. The choice of the threshold depends on the elevation of the foreshore and hinterland.

2.2.2 Water Level Drop

There is no specific software available to calculate the probability distribution of the water level drop velocity. In this study the quantification of the water level drop after a high-water peak is based on local water level measurements. It is not known beforehand at which moment (after the water level peak) the situation is most favourable for an instability. The river water level decreases in time, but at the same time the pore water pressures in the levee also need time to decrease. The statistics of the water level drop are therefore captured with Intensity Duration Frequency curves (IDF-Curves). IDF-Curves describe the relationship between intensity, duration, and return period (or its inverse, probability of exceedance), and are widely used for rain statistics [6]. From a timeseries with length T, we extract the values of highest one-day average, two-day average, three-day average (etc.), drop velocity, to construct the Intensity Duration (IDF)-curve with frequency 1/T. We then use the set of second highest drop velocities to construct the IDF-curve with frequency 2/T, etc. The IDF-Curves are extrapolated using a Gumbel Distribution for each duration.

2.2.3 Pore Water Pressures in the Levee

The software D-Geo Flow (https://www.deltares.nl/nl/software/d-geo-flow/) is used to obtain the pore water pressures in the levee for different transient boundary conditions of river water levels. In order to simulate the pore water pressures, each boundary condition (the river water level in time) is constructed from three parts: (1) the increasing water level from a daily water level to the peak water level, (2) the high-water peak and (3) the water level drop (Fig. 7a). For the duration of the first two phases a conservative estimate is made. The peak water level for different frequencies is used. For the water level drop the Intensity-Duration-curves for different frequencies are used. The result of the numerical groundwater flow simulations is thus the pore water pressures in the levee in time (Fig. 7b) for the different boundary conditions (peak water level and water level drop).

2.2.4 Probability of the Instability

The approach used to calculate the probability of an instability is not novel, and an extensive description can be found in [7]. This approach is easily modified to outer slope instability as only the grid of the slip circle has to be moved to the river side. The application with multidimensional conditional probabilities and time-dependency is novel. For every combination of peak water level, water level drop and time after the water level peak, the corresponding conditional probability on an instability, $P_{Instability}(h, u, t)$, is calculated with the software D-Stability (https://www.deltares.nl/en/software/d-stability/), see Fig. 7c. The probability of an instability, $P_{Instability}$, is calculated by integration according to the following equation (The integral is solved with numerical integration).

$$P_{Instability} = \iint \max_{t}\left(P_{Instability|(h,u,t)}(h, u, t)\right) \cdot f(h, u)\mathrm{d}h\mathrm{d}u \tag{1}$$

where $P_{Instability}(h, u, t)$ is the probability of instability for peak water level h, water level drop u and time after the high-water peak t and $f(h, u)$ is the joint probability density function. An extensive description of this specific procedure is found in report [3]. Hence, for each time step t, the probability of outer slope instability is computed by computing the $P_{Instability|(h,u)}$ and combining this with $f(h, u)$. By repeating this for all t, $P_{Instability}$ is the maximum P of all the steps.

The above routine is implemented for an instability without and with a constrain for the slip plane. The constrain is set such that the instability leads to the lowering of the crest height of the levee and is therefore referred to as a large instability. An instability without constrains has a higher probability of occurrence but causes less damage to the levee. This instability is referred to as a small instability.

2.3 *Event B: Consecutive High-Water Peak Before Repair*

The probability of a consecutive high-water event before repair, P_{HW}, is calculated with the following equation.

$$P_{HW} = P(T_{Interarrival} < T_{Repair}) \tag{2}$$

where $T_{Interarrival}$ is the interarrival time between two consecutive high-water peaks (Sect. 2.3.2) and T_{Repair} is the time until repair (Sect. 2.3.1).

2.3.1 Time Until Repair

The time it takes between the outer slope stability failure and a successful (emergency) repair is the summation of time required for the following actions: (1) the time it takes for the damage to be detected, (2) the time it takes to mobilise the necessary equipment and people and (3) the time it takes to execute the repair. The detection of the damage depends on the regularity of the levee patrol and inspections in periods of/after high-water level events. The time to mobilise depends on the emergency response plan of the authorities, and eventual arrangements with local construction companies, emergency clays depots, etc. The time it takes to execute the measure depends on, besides the type of measure, on the number of simultaneous instabilities in the area. The expected number of simultaneous instabilities is obtained using the probability and influence coefficients of the probabilistic levee stability calculation. We refer the readers to the extensive description of the procedure in report [3]. A first indication of the repair time is obtained doing interviews with responsible levee managers. The levee managers are provided with scenarios with a number of simultaneous large or small instabilities in their management area.

2.3.2 Interarrival Times

The time between the instability and the consecutive high-water level peaks is defined as the time between two consecutive high-water peaks minus the time between the first high-water peak and the instability. The probability distribution of the time between two consecutive peaks is obtained from local water level measurements.

2.4 Event C: Flooding During the Consecutive High River Water Levels

The damage of the instability has an influence on the probability of flooding during the consecutive high-water level event if it is not repaired. The damaged outer revetment can lead to higher pore water pressures inside the levee and consequently an increased probability of inner slope- and micro instability. Secondly, a new entry point for piping could be created. The largest influence is however on the (1) increased erodibility of the outer slope and, in case of a lowered crest level, (2) erosion by overtopping. Given the fact that the outer revetment is damaged, the probability of flooding due to wave attack can increase by multiple orders of magnitude.

1. The probability of flooding due to wave attack, $P_{Flooding\ by\ Wave\ Attack}$, is calculated with the following equation. The equation yields for a levee without revetment.

$$P_{Flooding\ by\ Wave\ Attack} = P(\Delta V_{e,crit} < \Delta V_e) \tag{3}$$

where ΔV_e the eroded volume, computed according to [8], and $\Delta V_{e,crit}$ the critical erosion volume. The critical erosion volume depends on the size of the remaining levee after the instability. Here, a conservative estimate is made for the critical erosion volume. The eroded volume depends, amongst others, on the wave height. The statistics of the local wave height is calculated with the software Hydra-NL.

2. The probability of flooding due to overflow or overtopping, $P_{Flooding\ by\ Wave\ Overtopping}$, is calculated with the following equation.

$$P_{Flooding\ by\ Wave\ Overtopping} = P(q_{crit} < q) \tag{4}$$

where q is the wave overtopping rate and q_{crit} the critical overtopping rate. The probability of flooding due to wave attack is calculated with the software Hydra-NL.

3 Results

Our framework is applied to a location along the Waal River, a tributary of the Rhine River, to calculate the probability of flooding due to outer slope instability. An estimate of the probability of the three event nodes is obtained in Sects. 3.2, 3.3 and 3.4, respectively, and combined into the event tree (Sect. 3.5).

3.1 Case Study Description and Input

The levee profile is based on a local levee cross-section using the altimetry data, the soil layering based on local soundings and boreholes (Fig. 3). The soil shear strength parameters of the levee material and the Holocene clay layer are described by the critical state friction angle ϕ (Lognormal ($\mu_\phi = 31.7$, $\sigma_\phi = 0.86$)). In the critical state there is no cohesion. The permeability of the levee material and Holocene clay layer equals 0.1 m/day.

Water level statistics are based on Hydra-NL results and local measurements. A timeseries of local water level measurements of over 200 years is available (1811–2020). Until the late 1950s extreme water level drops were related to collapsing ice

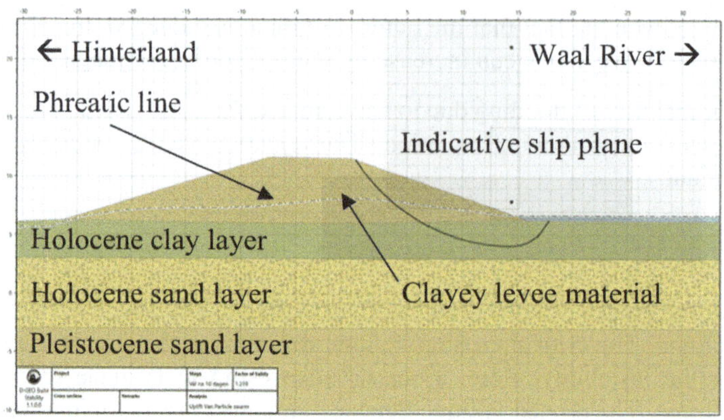

Fig. 3 Levee along the Waal River, used as case study

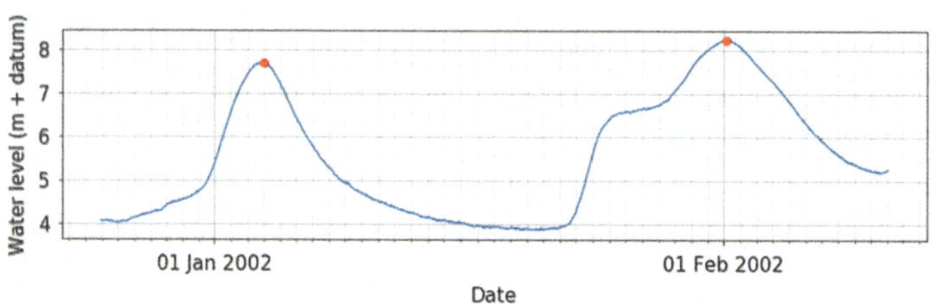

Fig. 4 The water level peak of January 3, 2002 was followed by a rapid water level drop and a consecutive water level peak on February 1

dams. In the current climate and with the warming of river water by industry, it is the question if ice dams should be considered at all as a plausible scenario in The Netherlands. It is beyond doubt that the frequency of ice dams in The Netherlands has dramatically dropped since 19th and early 20th century. Therefore, in this study the statistics of the water level drop velocities and the interarrival time between consecutive peaks are based on the last 50 years of data. The most extreme water level drop in the previous 50 years is shown in Fig. 4.

3.2 *Event A: Slope instability*

The probability of an instability is calculated according to Sect. 2.2. The distribution of the peak water level and the Intensity-Duration-Frequency (IDF) curves of the water level drop are shown in Figs. 5 and 6.

For different combinations of peak water level and water level drop the pore water pressures in the levee and the corresponding probability of a small and a large instability is calculated, as shown in Fig. 7 for one combination of peak water level,

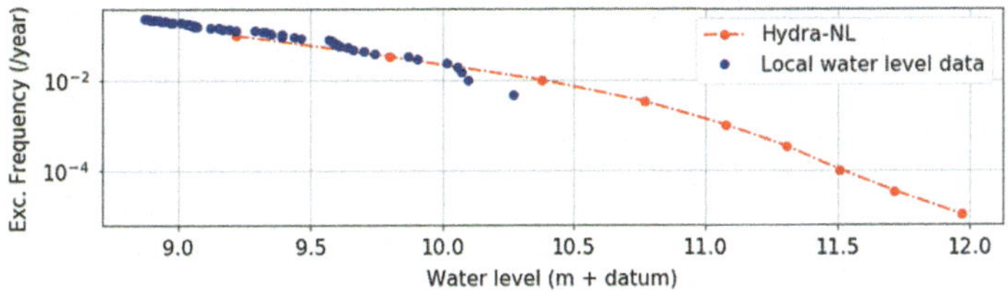

Fig. 5 Exceedance frequency of the peak water level (1) based on modelled data and marginal probability distribution (wind and river discharge), Hydra-NL and (2) based on local water level data

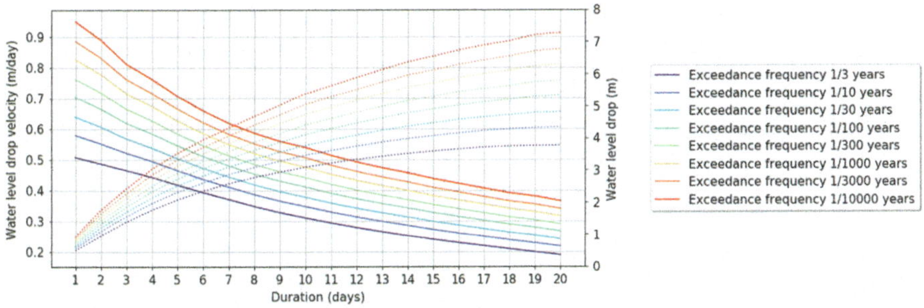

Fig. 6 The intensity-duration-frequency-curves of the water level drop: the duration-averaged water level drop velocities (continuous lines) and absolute water level drop (dotted lines). The IDF-curves are based on 50 years of water level data and extrapolated with a Gumbel-distribution

h_i, and water level drop, u_i. The largest probability of an instability is 4 days after the water level peak. This probability is the conditional probability of an instability, $P_{Instability|(h,u)}(h = h_i, u = u_i)$. Subsequently, the total probability of an instability is an integration of the conditional probabilities, according to Eq. 1. For both a large and a small instability the probability of an instability is calculated. The probability of an instability on a yearly basis is found to be 2E−11 for a small and 2E−18 for a large instability.

3.3 *Event B: Consecutive High-Water Peak Before Repair*

The time until repair (emergency measure) is estimated based on expert knowledge from levee managers. The small instability is estimated to be repaired within 0.5–7 days. The time until repair for a large instability is estimated on 1–14 days. For the time until repair, a uniform distribution is adopted. The parametric distribution of the time until the consecutive high-water peak is obtained from the interarrival time between the high-water peaks in the timeseries of the local water level and subtracting the days between the initial water level peak and the instability (4 days). The parametric distribution found is *lognormal*(μ_t= *139 days*, σ_t= *186 days*). The

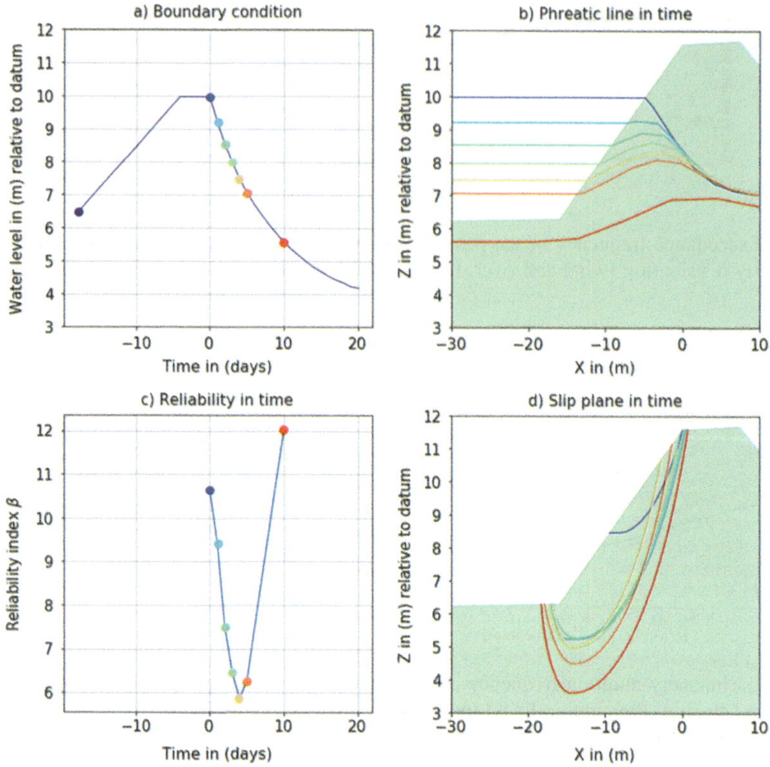

Fig. 7 One of the stability calculations with **a** the schematized boundary condition constructed from a peak water level and a water level drop, **b** the phreatic line (line where pore water pressures are zero) in the levee for the different time steps, **c** the reliability of the 'small' instability in time and **d** the slip planes for the different time steps. Four days after the water level peak, the conditions are most favorable for an instability

probability on a consecutive water level peak before repair is calculated with Eq. 2. The probability on a consecutive high-water event before repair is found to be 2E−3 for a small and 1E−2 for a large instability.

3.4 Event C: Flooding During the Consecutive High River Water Levels

The probability of flooding due to erosion of the levee by wave attack is using Eq. 3 found to be 3E−5 for a small and 8E−3 for a large instability, conservatively assuming a critical erosion volume of 50 and 10 m^3 per meter levee, respectively. With an intact outer slope revetment, the probability of flooding due to wave attack would be lower than 5E−6.

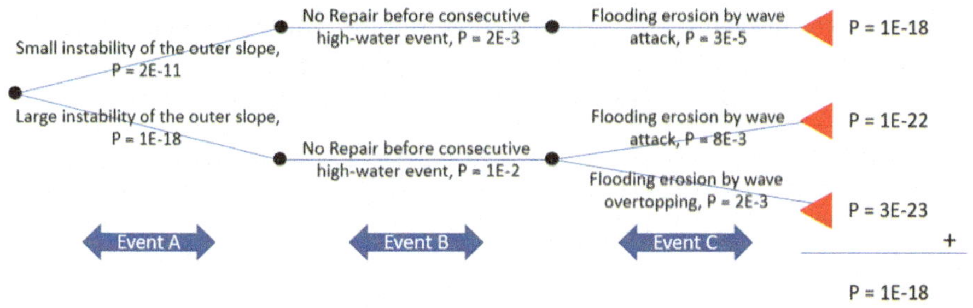

Fig. 8 The event tree including the probabilities. The probabilities of the different nodes are explained in the Sects. 3.2, 3.3 and 3.4. The paths not leading to flooding have been omitted

The probability of flooding due to erosion of the levee by wave overtopping is only relevant in case of a lowered crest level, i.e. due to a large instability. The probability of flooding due to erosion of the levee by wave overtopping is using Eq. 4 and is found to be 2E−3 for a large instability. In contrast, with the original height of the levee, the probability of flooding due to wave overtopping would be lower than 2E−5.

3.5 Event Tree and Comparison with Current Assessment

The different probabilities are shown combined in the event tree, as shown in Fig. 8 and comparted against the required failure (flooding) probability of 2E−7. With the current assessment, considering safety factors, conservative choices and similarities with inner slope stability, a probability of flooding due to outer slope stability of 5E−8 is found, i.e. just fulfilling the requirement. With our novel approach, the probability of flooding due to instability of the outer slope for the case is instead shown to be negligible.

4 Conclusion and Further Research

In this study, we developed and applied a novel approach to probabilistically assess the failure mechanism of outer slope instability of levees. Our framework provides much less conservatism for levee assessment than the current method, and therefore fewer resources are needed to mitigate this mechanism. In our approach, consecutive events leading to flooding are explicitly included, thereby providing engineers and levee managers insight and allowing for targeted optimization of the calculated flooding probability.

The presented work is an exploratory study. The aim for future study is (A) to apply the methodology in other levee reinforcement projects and (B) to further elaborate on the methodology by including: (1) infiltration by wave overtopping and rain prior to the instability, (2) a stochastic description of the permeability instead of using conservative estimates, (3) a more detailed description of the damage after the instability and (4) an extensive description of the erosion process leading to flooding.

Acknowledgements The research is funded by the *Projectoverstijgende Verkenning Macrostabiliteit* (POVM). The POVM is part of the Flood Protection Program, a levee reinforcement program in the Netherlands by *Rijkswaterstaat* and the regional water authorities.

References

1. Van der Most, H., Tanczos, I., De Bruijn, K. M., & Wagenaar, D. (2014). New risk-based standards for flood protection in the Netherlands. In *Proceedings of 6th International Conference on Flood Management, Sao Paulo, Brazil*, Sept 2014.
2. Kind, M. (2014). Economically efficient flood protection standards for the Netherlands. *Journal Flood Risk Manage, 7*(2014), 103–117. https://doi.org/10.1111/jfr3.12026.
3. Van der Meer, A. W. (2020). *POVM Actuele Sterkte: Macrostabiliteit Buitenwaarts: een verkennende studie.* POVM Report in Dutch no. 11204873; Version 1.0, 29 April 2020.
4. Jongejan, R. B., Diermanse, F., Kanning, W., & Bottema, M. (2020). Reliability-based partial factors for flood defenses. *Journal Reliability Engineering and System Safety, 193* (2020). https://doi.org/10.1016/j.ress.2019.106589
5. Duits, M., & Kuijper, B. (2017). *Hydra-NL Systeemdocumentatie.* HKV Report in Dutch no. PR3544. Version 2.2, March 2017.
6. Chow, V. T. (1988). *Handbook of applied hydrology.* New York: McGraw-Hill Book.
7. Kanning, W., Teixeira, A., van der Krogt, M., Schweckendiek, T., & Hardeman, B. (2017). Calibration of factors of safety for slope stability of dikes. In *Proceedings of Geo-Risk 2017 Conference*, Denver, USA (June 2017). https://doi.org/10.1061/9780784480717.001.
8. Mourik, G. C. (2015). *Prediction of the erosion velocity of a slope of clay due to wave attack.* Deltares Report No. 209437-017-HYE-0003. Version 3, January 2015.

Reliability Analysis of Timber Elements Under Different Load Types and Identification of Critical Scenarios for the Evaluation of Existing Structures

Maria Loebjinski, Wolfgang Rug, and Hartmut Pasternak

Abstract This contribution presents reliability analyses of structural members made from timber under different load combinations relevant for common structures. The study embraces members under strain induced by permanent action and live load, as well as permanent action and snow/wind load. Snow and wind load are superimposed applying the Ferry Borges and Castanheta load combination rule. Results are analysed to identify critical design situation from a statistical point of view assuming a one-hundred percent workload of the semi-probabilistic design situation considering partial safety factors from current design codes. Studies are performed for a reference period of 50 years. The studies show that small rooms under high load fluctuation (live load) are critical in terms of calculated reliability. What is more, a high load share of snow load is also particular critical. Thus, flat roofs are to be investigated with certain care. Results are used to identify and classify design situation for modification of partial safety factors on the material side. In this respect, different design situations can be treated more optimal and thus load-bearing capacities in existing timber structure may be activated and considered if available.

Keywords Reliability analysis · Timber structures existing structures · Code calibration

1 Introduction

The investigation and evaluation of existing structures are important and challenging tasks for practicing engineers. However, in times of sustainable economy providing solutions with less consumption of energy and resources moving more and more

M. Loebjinski (✉)
Technische Universität Braunschweig, Braunschweig, Germany
e-mail: Maria.Loebjinski@b-tu.de

W. Rug
BaSys GmbH, Lenzen, Germany

H. Pasternak
Brandenburg University of Technology, Cottbus, Germany

© The Author(s), under exclusive license to Springer Nature Switzerland AG 2021

J. C. Matos et al. (eds.), *18th International Probabilistic Workshop*, Lecture Notes in Civil Engineering 153, https://doi.org/10.1007/978-3-030-73616-3_49

into the focus of common attention, building with existing structures experiences a significant increase in prosperity. The identification of critical members and their qualified investigation are central parts of the evaluation of the load-bearing capacity and serviceability of a structure and thus their substance-careful maintenance and, if necessary, rehabilitation.

This contribution presents reliability analyses of timber members under typical loadings. These are uniaxial strain from permanent action and one or two time-variant loads, two-axial strain from bending and compression and two-axial bending. The influence of slenderness has not been considered.

Reliability analyses have been performed applying First Order Reliability Method (FORM) in MATLAB. The reliability level that is reached applying the semi-probabilistic safety format of EN 1990:2010-12 [1] and EN 1995-1-1:2010-12 [2] is analysed for a variation of input parameters, i.e. load ratios and coefficient of variation (cov) of the material strength. Sensitivity analyses have been performed for a better understanding of the impact of basic variables on the probability of failure and safety index respectively. Based on an improved knowledge of sensitivities, strengthening measures can be carried out for critical elements from a statistical point of view to realize a substance-careful redevelopment of a structure in service.

What is more, based on these calculation an estimate of the implicit safety level of current design rules in timber engineering can be made. This can serve as a basis for an adjustment of partial safety factors for existing structures to be used for an evaluation applying the semi-probabilistic safety concept of current codes.

2 Load Scenarios, Model Assumptions and Limit State Functions

2.1 Load Scenarios

The following five scenarios have been analysed:

1. Uniaxial stress from permanent action and live load
2. Uniaxial stress from permanent action, snow load and wind load
3. Two-axial bending from permanent action, snow load and wind load
4. Compression from permanent action and snow load, bending from wind load (combination of stresses without considering the influence of slenderness)
5. Compression from permanent action and live load, bending from wind load (combination of stresses without considering the influence of slenderness)

2.2 Model Assumptions and Basic Variables

Table 1 illustrates probabilistic parameters as applied. In all limit state functions

Table 1 Probabilistic parameters

	Variable	Sym	Distr.	μ_X	V_X	Notes
Resistance	*Timber*					
	Bending	R_m	LN	1.0	0.25	Based on [4]
	Compression parallel to grain	R_c	LN	1.0	0.20	
	Tension parallel to grain	R_t	LN	1.0	0.30	
Loads	*Permanent action*	G	N	1.0	0.10	Based on [5]
	Live load					
	Distribution of maxima in reference period					
	Small room ($\leq$20 m^2)	N	GUM	1.0	0.40	Based on studies applying [5–7]
	Large room (>20m^2)	N		1.0	0.25	
	Presence period in reference period (days)	n_p	det.	$50 \cdot 365$	–	Based on [3]
	Load changes in presence period	n_r	det.	$5 \cdot 365$	–	
	Snow load					
	Distribution of maxima in reference period	S	GUM	1,0	0.25	Based on [8]
	Presence period in reference period (days)	n_p	det.	$50 \cdot 60$	–	Based on [3] (adjusted for GER)
	Load changes in presence period	n_r	det.	10	–	
	Wind load					
	Distribution of maxima in reference period	W	GUM	1.0	0.16	Based on [7, 8]
	Instantaneous value of wind load	W_{mom}	GUM	0.16	1.00	Based on [9]
	Presence period in reference period (days)	n_p	det.	$50 \cdot 365$	–	Based on [3]
	Load changes in presence period	n_r	det.	$50 \cdot 365$	–	
Model	*Resistance side*					

(continued)

Table 1 (continued)

	Variable	Sym	Distr.	μ_x	V_x	Notes
	Qualified survey in situ required	θ_R	N	1.0	0.07	Suggestion
	Load side					
	Permanent actions, qualified survey required	θ_G	N	1.0	0.05	Suggestion
	Live load	θ_N	N	1.0	0.10	Based on [10–12]
	Snow load	θ_S	N	1.0	0.10	
	Wind load	θ_W	N	1.0	0.10	

(LSF) a design parameter z_d has been introduced ensuring a one-hundred percent utilization of the semi-probabilistic design equation, see e.g. [3]. According to [4] R_m is correlated with R_c and R_t respectively by $\rho = 0.8$. Snow and wind load have been assumed uncorrelated, which is a simplification. For a more detailed analysis, modeling applying stochastic processes would be required. All calculations have been performed considering current values of partial safety factors (PSF) ($\gamma_G = 1.35$, $\gamma_Q = 1.5$, $\gamma_M = 1.3$).

2.3 *Limit State Functions*

2.3.1 Limit State Function and Design Parameter for Uniaxial Stress from Permanent Action and One Time-Variant Load

The LSF applied for load scenario (1) is

$$g = z_d \cdot k_{\text{mod}} \cdot R_i \cdot \theta_R - LV_G \cdot S_G \cdot \theta_{S,G} + (1 - LV_G) \cdot S_{Q1} \cdot \theta_{S,Q1} \tag{1}$$

with

$$z_d = \frac{\left(LV_G \cdot \gamma_G \cdot g_k + (1 - LV_G) \cdot \gamma_Q \cdot q_{k,N}\right) \cdot \gamma_M}{k_{\text{mod}} \cdot f_{k,m}} \tag{2}$$

Denotations of variables are defined in Table 1. Additionally, LV_G is the load ratio of the permanent actions of the total load, γ_G, γ_Q and γ_M are PSF for permanent action, variable action and material resistance respectively, g_k is the characteristic value (expected value) of the permanent action, $q_{k,N}$ is the characteristic value of the live load (model value as $T_{ref} = 50a$), $f_{k,m}$ is the characteristic value of the material resistance (5%-quantile), and k_{mod} is the modification factor considering load duration and moisture content.

2.3.2 Limit State Function and Design Parameter for Uniaxial Stress from Permanent Action and Two Time-Variant Loads

Applying the load combination rule of Ferry Borges and Castanheta [13] the LSF (scenario 2) is

$$g = z_d \cdot k_{\text{mod}} \cdot R \cdot \theta_R - \left(LV_G \cdot S_G \cdot \theta_{S,G} + (1 - LV_G)\right.$$
$$\left.\cdot\left(LV_{Q1} \cdot S_{Q,1}^{n_1} \cdot \theta_{S,1} + \left(1 - LV_{Q2}\right) \cdot S_{Q,2}^{n_2} \cdot \theta_{S,2}\right)\right) \tag{3}$$

with

$$n_1 = \frac{\min\left(n_{p1}, n_{p2}\right)}{n_{p1}} \tag{4}$$

$$n_2 = \frac{\max(d_1, d_2)}{n_{p2}} \tag{5}$$

$$d_1 = \frac{n_{p1}}{n_{r1}} \tag{6}$$

$$d_2 = \frac{n_{p2}}{n_{r2}} \tag{7}$$

Variables are explained in Table 1. Additionally, LV_{Q1} is the load ratio of the first variable load referring to the whole amount variable loads. The design parameter is

$$z_d = \frac{\gamma_M}{k_{\text{mod}} \cdot f_{k,m}}$$
$$\cdot\left(LV_G \cdot \gamma_G \cdot g_k + LV_Q \cdot \left(LV_{Q1} \cdot \gamma_Q \cdot q_{k,1} + LV_{Q2} \cdot \gamma_Q \cdot \psi_{Q2} \cdot q_{k,2}\right)\right) \tag{8}$$

where ψ_{Q2} is the combination factor for the accompanying load. Here, $k_{mod}= 1.0$. Applying Turkstras [14] load combination the LSF is

$$g = z_d \cdot k_{\text{mod}} \cdot S \cdot R \cdot \theta_R - LV_G \cdot S_G \cdot \theta_{S,G}$$
$$+\left((1 - LV_G) \cdot LV_{Q,1} \cdot S_{Q,1} \cdot \theta_{S,1} + \left(1 - LV_{Q,1}\right) \cdot S_{Q,2,mom} \cdot \theta_{S,2}\right) \tag{9}$$

with

$$z_d = \frac{\gamma_M}{k_{\text{mod}} \cdot f_k} \cdot (LV_G \cdot \gamma_G \cdot g_k + (1 - LV_G)$$
$$\cdot\left(LV_{Q1} \cdot \gamma_Q \cdot q_{k,1} + \left(1 - LV_{Q1}\right) \cdot \gamma_Q \cdot \psi_{Q2} \cdot q_{k,2}\right)) \tag{10}$$

where additionally to the variables explained above, LV_{Q1} is he load ratio of the first variable load related to the total variable load.

2.3.3 Limit State Function and Design Parameter for Two-Axial Bending

The LSF for load scenario (6) is

$$g = 1 - \left(\frac{1}{z_{d,z}} \cdot \frac{LV_G \cdot S_{G,z} \cdot \theta_{S,G} + (1 - LV_G) \cdot LV_{Q,1} \cdot S_{Q,1,z}^{n_1} \cdot \theta_{S,1}}{R_m} + \frac{1}{z_{d,y}} \cdot \frac{(1 - LV_G) \cdot \left(1 - LV_{Q,1}\right) \cdot S_{Q,2,z}^{n_2} \cdot \theta_{S,2}}{R_m} \right) \cdot \frac{1}{\theta_{R,m} \cdot k_{\text{mod}}} \tag{11}$$

with

$$z_{d,z} = \frac{\gamma_M}{f_{k,m}} \cdot (LV_G \cdot \gamma_G \cdot g_k + (1 - LV_G) \cdot \left(LV_{Q1} \cdot \gamma_Q \cdot q_{k,1} + k_m \cdot \left(1 - LV_{Q1}\right) \cdot \gamma_Q \cdot \psi_{0,2,Q2} \cdot q_{k,2} \cdot \frac{h}{b} \right)) \tag{12}$$

h/b is the cross section ratio.

2.3.4 Limit State Function and Design Parameter for Combined Bending and Compression Stress

The LSF applied for load scenario (4) and (5) is

$$g = 1 - \left(\frac{1}{z_{d,A}} \frac{\sum_i S_{c,i} \cdot \theta_{S,c,i}}{R_{c,0} \cdot \theta_{R,c,0}} + \frac{1}{z_{d,A}} \frac{\sum_i S_{m,i} \cdot \theta_{S,m,i}}{R_m \cdot \theta_{R,m}} \right) \tag{13}$$

with design parameters $z_{d,A}$ and $z_{d,M}$ from

$$0 = 1 - \left(\frac{\gamma_M \cdot \left(LV_G \cdot \gamma_G \cdot g_k + (1 - LV_G) \cdot LV_{Q1} \cdot \gamma_Q \cdot q_{k,1}\right)}{z_{d,A} \cdot k_{\text{mod}} \cdot f_{k,c}} \right)^2 - \left(\frac{\gamma_M \cdot (1 - LV_G) \cdot \left(1 - LV_{Q1}\right) \cdot \gamma_Q \cdot \psi_{0,2,Q2} \cdot q_{k,1}}{z_{d,M} \cdot k_{\text{mod}} \cdot f_{k,c}} \right) \tag{14}$$

$$0 = z_{d,M} - \frac{6 \cdot z_{d,A}}{h/b} \tag{15}$$

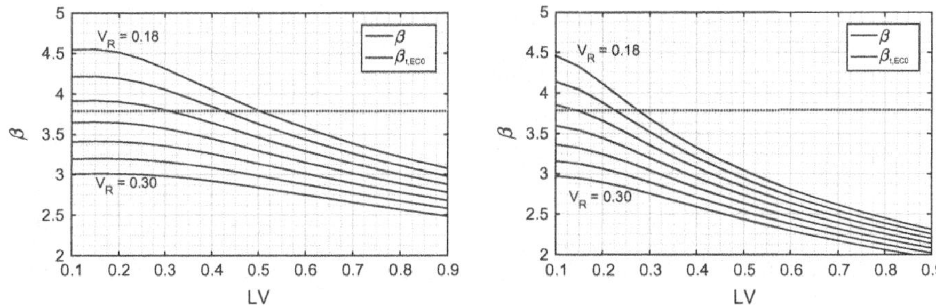

Fig. 1 Reliability index dependent on the load ratio of the live load for permanent action + live load, V_N= 0.25 (left), V_N= 0.40 (right), T_{ref}= 50a

Values for k_{mod} have been applied according to the definition in EN 1995-1-1:2010-12 [2] and National Annex, i.e. k_{mod} = 0.8 for live load in residence and office rooms (building occupancy type A and B), k_{mod} = 1.0 if wind load is acting.

3 Results

3.1 Results for Uniaxial Stress from Permanent Action and Live Load

Figure 1 illustrates that for $V_R = 0.25$ as recommended for the bending strength in [4] the target level for T_{ref} = *50a* β_t= 3.8 is not reached for assumptions of small (large rooms) and bigger (small rooms) live load fluctuations. For compression strength (recommended $V_R = 0.20$ in [4]) this target value is reached for live load with lower fluctuations up to a load ratio of the variable load *LV* = 0.4. As expected, greater fluctuations of live loads have a significant impact on the members reliability.

Studies resulted in relevant load rations of permanent actions and live load for common historic timber floor structures of *LV* = 0.3–0.55. Figures 2 and 3 illustrate the sensitivity factors depending on the cov of the material strength for the upper and lower bound of the load ratio and high and lower fluctuations of the live load.

3.2 Results for Bending/Two-Axial Bending from Permanent Action, Snow Load, Wind Load

Figure 4 illustrates the reliability for different load ratios of permanent action LA_G and the first variable load LA_{Q1} (snow load here) for bending and two-axial bending. Results are comparable, but the reliability decreases faster with increasing load ratio

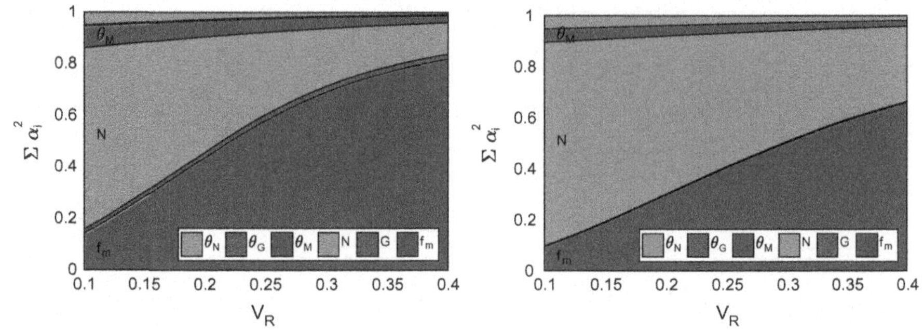

Fig. 2 Sensivity factors dependent on the coefficient of variation of the material resistance V_R for permanent action + live load, $V_N = 0.40$, $LV = 0.3$ (left), $LV = 0.55$ (right), $T_{ref} = 50a$

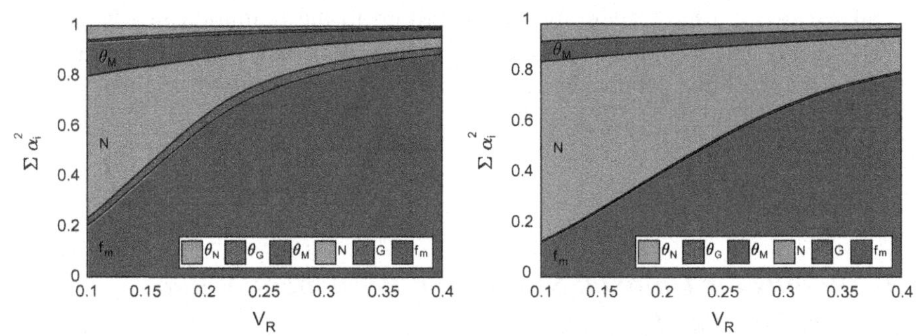

Fig. 3 Sensivity factors dependent on the coefficient of variation of the material resistance V_R for permanent action + live load, $V_N = 0.25$, $LV = 0.3$ (left), $LV = 0.55$ (right), $T_{ref} = 50a$

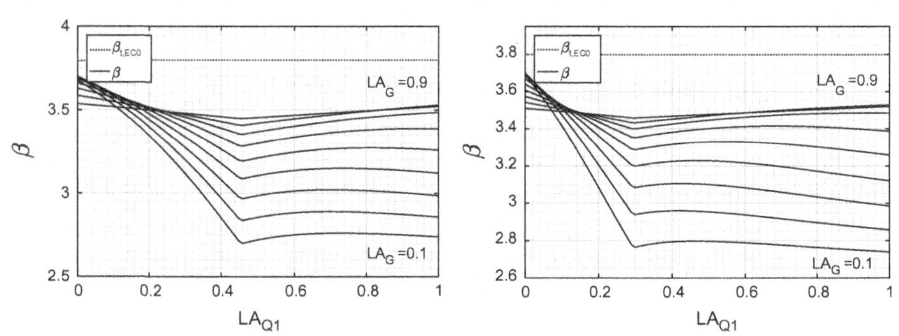

Fig. 4 Reliability index dependent on the load ratio of the first variable load for permanent action + snow load + wind load, $T_{ref} = 50a$, uniaxial bending (left), two-axial bending (right), *b/h* = *1/2*

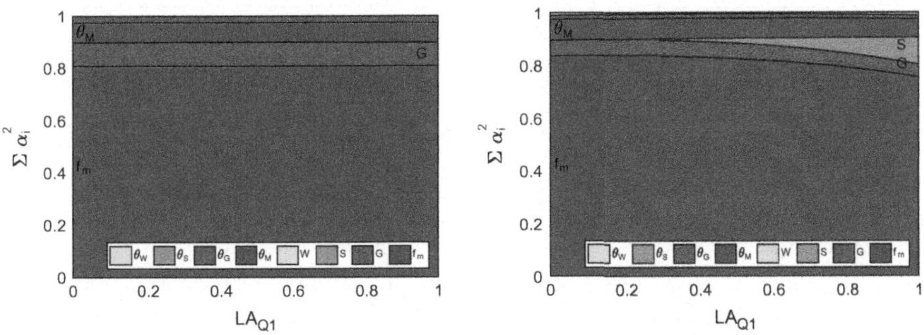

Fig. 5 Sensivity factors dependent on the load ratio for permanent action + snow load + wind load, T_{ref}= 50*a*, bending (left), *b/h* = 1/2, LA_G= 0.9 (left), LA_G= 0.7 (right)

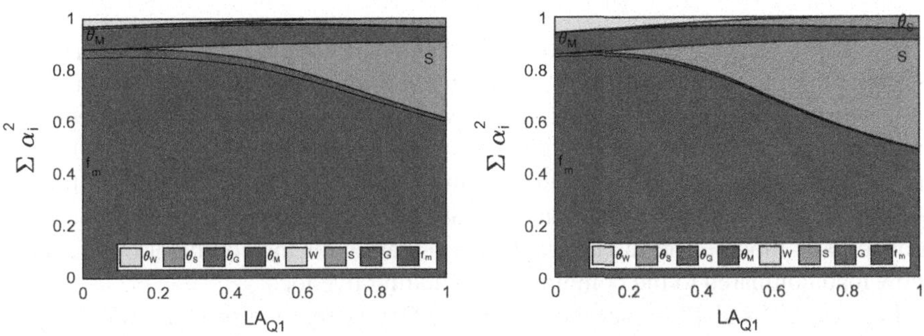

Fig. 6 Sensivity factors dependent on the load ratio for permanent action + snow load + wind load, T_{ref}= 50*a*, bending (left), *b/h* = 1/2, LA_G= 0.5 (left), LA_G= 0.3 (right)

of the snow load for two-axial bending. As the main material resistance that is activated especially in historic structures in bending strength, the material resistance has been modelled with $V_R = 0.25$. Results show, that the target value is again not reached for this load scenario. Figures 5 and 6 depict the sensitivity factors for chosen load ratios of permanent and variable actions. The figures illustrate the increasing influence of the variable loads, especially snow load, on the reliability.

3.3 *Results for Bending and Compression Stress from Permanent Action, Snow Load/Live Load, Wind Load*

Figure 7 shows the reliability index for bending from permanent action and snow load or live load and bending from wind load. Slenderness is not considered in these studies.

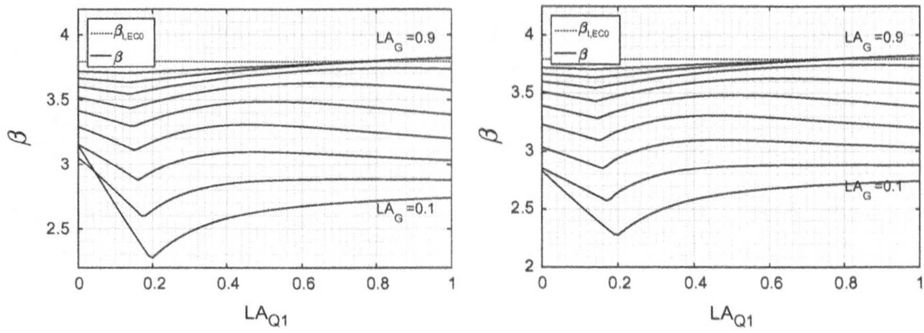

Fig. 7 Reliability index dependent on the load ratio of the first variable load for compression from permanent action + snow load (left)/ live load (right) and bending from wind load, T_{ref}= 50*a, b/h* = 1/2, V_S= V_N= 0.25

Results show that for high load ratios of permanent action, the target value $\beta_t = 3.8$ is reached. This is due to the lower variability of timber compression strength, that is activated here. For higher load ratios of variable action, the reliability decreases significantly. The point where within the semi-probabilistic design equation the second variable actions becomes the leading action can well be seen in the figures. This break in the course of the graphs results from simplifications within the semi-probabilistic model. The reliability decreases faster for the combination including snow load compared to the combination including live load.

Figures 8, 9 and 10 depict the sensitivity factors for chosen load ratios. Again, the stepwise increase of the influence of the first variable action can be seen. What is more, when superimposed with live load, the influence of wind load on the reliability seems to be greater that when superimposed with snow load. This is probably due to the assumptions of presence periods of the different types of variables actions.

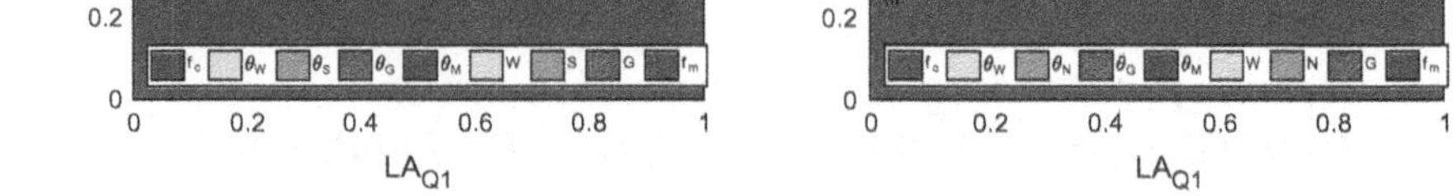

Fig. 8 Sensitivity factors dependent on the load ratio for compression from permanent action + snow load (left)/live load (right) and bending from wind load, T_{ref}= 50*a, b/h* = 1/2, V_S= V_N= 0.25, LA_G= 0.3

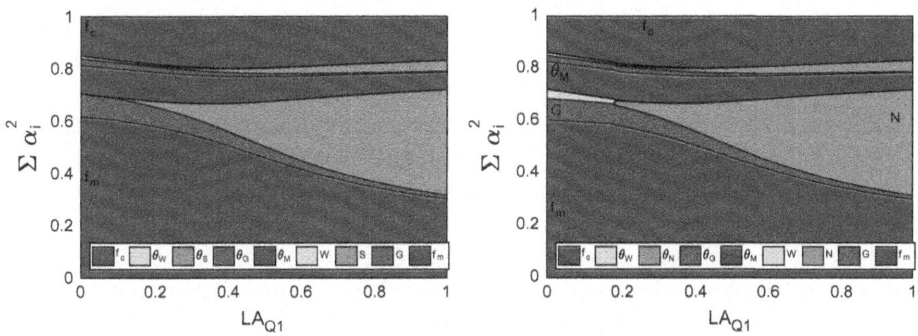

Fig. 9 Sensitivity factors dependent on the load ratio for compression from permanent action + snow load (left)/ live load (right) and bending from wind load, T_{ref}= 50a, b/h = 1/2, V_S= V_N= 0.25, LA_G= 0.5

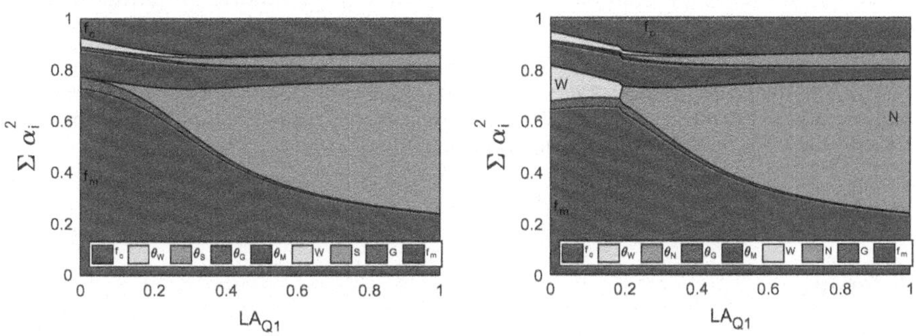

Fig. 10 Sensitivity factors dependent on the load ratio for compression from permanent action + snow load (left)/live load (right) and bending from wind load, T_{ref}= 50a, b/h = 1/2, V_S= V_N= 0.25, LA_G= 0.7

4 Conclusions

Results show, that the target value of β_t= 3.8 for T_{ref} = 50a has to be questioned. For a range of typical loading situations of timber members assuming a one-hundred percent utilization of semi-probabilistic design equations applying current partial safety factors, this value is not reached. However, this study focusses on simple limit states, more design situations have to be investigated. For example, studies on uniaxial stress from tension and bending in combination with tension showed even lower reliability indices due to the high cov of the tension strength.

For combinations with live load it could be seen that for LV = 0.3 up to V_R = 0.20 live load is dominating the reliability analysis. For LV = 0.55 and small live load fluctuations (i.e. large rooms) this is valid up to approximately V_R = 0.25 and for greater live load fluctuations (i.e. small rooms) approximately up to V_R = 0.35. Thus, for large rooms with small and higher fluctuations of live load, the reliability can be increased significantly if an investigation in situ allows for an reduction of the cov of the material strength. Compared to members subjected to bending in one direction, the reliability of members under two-axial bending decreases faster if they are designed for wind load on the weak axis and the load share of snow load on the strong axis is increased. For high load ratios of variable loads and high ratios of snow load as it is important for flat light timber roof constructions, snow load the is dominating action. Members under bending and compression show a higher reliability if load ratios of actions acting in the directions of the members axis result in higher reliability indices as the variability of the compression strength is lower compared to other timber strength properties.

To summarize, it has to be emphasized that in many cases the reliability of a timber member can be significantly increased, if a detailed survey justifies the reduction of the cov of timber material properties as this is often the dominating variable. However, individual load ratios and acting variables have to be considered carefully. What is more, the influence of members slenderness has to be studied in comprehensive analyses.

Acknowledgments The content of this contribution has been developed within funding periods of Deutsche Bundesstiftung Umwelt (AZ: 20015/409) and internal funding of Brandenburg University of Technology. The authors would like to thank the institutions for the generous support.

References

1. *Basis of structural design*, DIN EN 1990:2010-12 (2010).
2. *Design of timber structures—Part 1-1: General—Common rules and rules for buildings*. DIN EN 1995-1-1:2010-12 (2010)
3. Baravalle, M. (2017). *Risk and reliability based calibration of structural design codes*. Dissertation, Norwegian University of Science and Technology, Trondheim.
4. JCSS. (2006). *Probabilistic Model Code: Part 3—Resistance Models*. Joint Committee on Structural Safety, 2006. Available: https://www.jcss.byg.dtu.dk/. Accessed on March 08, 2016
5. JCSS. (2016). *Probabilistic model code: Part 2—Load models*. Joint Committee on Structural Safety, 2001. Available: https://www.jcss.byg.dtu.dk/Publications/Probabilistic_Model_Code. Accessed on: March 29, 2016.
6. CIB. (1989). Actions on structures: Live loads in buildings. CIB Report W81, 1989.
7. Rackwitz, R. (1996). Einwirkungen auf Bauwerke. In G. Mehlhorn (Ed.), *Der Ingenieurbau* (pp. 73–416). Berlin: Ernst.
8. Grünberg, J. (2004). *Grundlagen der Tragwerksplanung - Sicherheitskonzept und Bemessungsregeln für den konstruktiven Ingenieurbau: Erläuterungen zu DIN 1055–100* (1st ed.). Berlin: Beuth.
9. Glowienka, S. (2007). *Zuverlässigkeit von Mauerwerkswänden aus großformatigen Steinen: Probabilistische Analyse von großformatigem Mauerwerk aus Kalksandstein und Porenbeton mit Dünnbettvermörtelung*. Dissertation, Technische Universität Darmstadt (2007).
10. Vrouwenvelder, T. (2002). Developments towards full probabilistic design codes. *Structural Safety, 24*, 417–432.
11. Hansen, M. (2004). *Zur Auswirkung von Überwachungsmaßnahmen auf die Zuverlässigkeit von Betonbauteilen*. Stuttgart: Fraunhofer IRB Verlag.

12. Faber, M. H., Köhler, J., & Sørensen, J. D. (2004). Probabilistic modelling of graded timber material properties. *Structural Safety, 26,* 295–309.
13. Ferry Borges, J., & Castanheta, M. (1971). *Structural safety,* 2nd edn. Lisbon.
14. Turkstra, C. J. (1970). *Theory of structural design decisions: Study No. 2*. Ontario, Canada, 1970.

Reliability Assessment of Oil and Gas Pipeline Systems at Burst Limit State Under Active Corrosion

Ram K. Mazumder, Abdullahi M. Salman, and Yue Li

Abstract Civil infrastructures such as oil and gas transportation systems play a vital role in industrial and public energy distribution and consumption. A large number of existing oil and gas transportation pipelines in many cities in the USA are running at the end of their design life and are at risk. Failure in these systems can potentially cause adverse effects to the society, economy, and environment. Asset managers often need to prioritize the critical segments based on the risk of failure, available budget, and resources. In this paper, the fitness for service of oil and gas pipelines and network integrity are evaluated probabilistically using various burst pressure models to prioritize the riskiest segments to support asset management. The current state-of-the-art practice of burst failure models for pressurized metallic pipelines is compared using a physical probabilistic approach. Since metallic pipelines for oil and gas transportation are typically designed for a long lifespan and experience localized corrosion deterioration throughout their lifetime, a steady-state corrosion model was assumed for accounting for the effect of external corrosion deterioration on the burst pressure of pipelines. A Monte Carlo Simulation technique is utilized to generate the fragility curves of pipelines considering corrosion deterioration over time. Uncertainties involved in various parameters related to burst failure and fragility estimation are modelled based on the knowledge gained from past research. A comparative analysis is presented for various fragility models of pipelines. Also, system reliability was evaluated using a minimum cut sets approach. The proposed approach is illustrated for a simple hypothetical oil/gas transmission system. Outcomes of the study show a consistent trend of failure for various models over time. The results of the probabilistic models of burst failures are analyzed, and recommendations are provided to support asset management planning.

Keywords Burst failure · Corrosion · Pipelines · Reliability · Risk

R. K. Mazumder (✉) · Y. Li
Case Western Reserve University, Cleveland, OH, USA
e-mail: rxm562@case.edu

A. M. Salman
The University of Alabama in Huntsville, Huntsville, AL, USA

© The Author(s), under exclusive license to Springer Nature Switzerland AG 2021

J. C. Matos et al. (eds.), *18th International Probabilistic Workshop*, Lecture Notes in Civil Engineering 153, https://doi.org/10.1007/978-3-030-73616-3_50

1 Introduction

Civil infrastructure systems such as oil and gas transportation systems are essential for a country's economic growth and smooth functionality of societies [1]. Unfortunately, the reliability of aging oil and gas pipelines has significantly decreased over the years. As pipelines age, the growth of corrosion on the pipeline wall significantly weakens their strength and often results in failure and severe consequences [2]. Burst failure is a common type of failure that causes severe destruction to pipelines and causes huge economic and environmental consequences. Various guidelines have been developed to determine the burst failure pressure of pipelines [3–8]. Burst failure pressure estimated using these guidelines is usually estimated using a deterministic approach. The factors associated with the bursting failure estimation model involve large uncertainties. The deterministic approach is unable to predict the failure probability of the pipeline accurately [9]. Moreover, the use of different guidelines provides different outcomes in burst failure prediction and may lead to varying decisions on the design and management of pipelines.

In this study, the current state-of-the-art burst failure estimation models are used to estimate the burst failure probability of oil/gas pipelines [10]. Since metallic pipelines for oil and gas transportation are typically designed for a long lifespan and sustain corrosion deterioration during the course of its life, a steady-state corrosion model is assumed to account for the effect of corrosion deterioration on the burst pressure of pipelines. The burst limit state is evaluated by comparing the burst failure pressure and the internal pressure of the pipeline. Uncertainties involved in various parameters associated with fragility calculation are modeled based on the knowledge gained from past research. A Monte Carlo Simulation technique is used to generate the fragility curves of pipelines subjected to active corrosion defects. A comparative analysis is presented for various fragility models of pipelines. Also, system reliability is evaluated using the minimum cut sets approach. The proposed approach is illustrated for a simple hypothetical oil/gas transmission system.

2 Methodology

2.1 *Burst Limit State*

The burst failure probability of a pipeline can be estimated by comparing the failure pressure and pipeline internal operating pressure. The following limit state function is generated for burst failure estimation [2, 9]:

$$g(X) = P_b - P_i \tag{1}$$

where P_b is the burst failure pressure of the pipeline, P_i is the internal oil/gas pressure of the pipeline, g(X) is the burst limit function where x is the vector of random variables. Burst failure occurs when the internal pressure exceeds the burst failure pressure of the pipeline. In other words, a negative value of g(X) indicates the failure of the pipeline. Failure pressure is estimated based on the various standards and guidelines as described below. The internal operating pressure of the pipeline is estimated based on current practice.

2.2 *Burst Failure Pressure Estimation*

Extensive research has been performed on burst failure analysis of corroded pipelines [2, 9, 11, 12]. Several standards and guidelines are established for estimating the failure pressure of corroded pipelines. Among the existing methods, B31G, Modified B31G, DNV RP F101, Battelle, Shell-92, Battelle and Netto et al. (2005) models are used to develop and compare the failure probability of pipelines [3–8]. Although previous studies determine the reliability of pipelines using these approaches, however, comparison between the failure probability estimations for pipelines and corresponding system reliability estimation under active corrosion is rare [e.g., 13]. Table 1 shows various failure pressure estimation models used in this study.

In Table 1, P_b is the failure pressure, UTS is the ultimate tensile strength, YS is yield strength, M is folias factor, D is diameter of pipe, t is the initial thickness of pipe wall, d(T) and L(T) are defect depth and defect length, respectively, as a function of time; T is the time in year.

2.3 *Corrosion Model*

Corrosion is the most influential parameter for metallic pipeline deterioration. It occurs due to an aggressive environment and becomes serious with aging [11]. Pipeline maintenance requires regular inspection and rehabilitation of the corroded pipeline. The corrosion growth overtime depends on the surrounding environmental conditions (soil characteristics, chemical attacks on pipeline materials, etc.). The corrosion growth on a pipeline surface can be modelled by the defect depth and defect length of corrosion as expressed by the following equation;

$$d(T) = d_0 + V_r(T - T_0) \tag{2}$$

$$L(T) = L_0 + V_a(T - T_0) \tag{3}$$

where d(T) is the defect depth, L(T) is the defect length, d_0 is the initial defect depth, L_0 is the initial defect length, V_a is the axial corrosion rate and V_r is the radial corrosion rate. T_0 is the time to initiate corrosion.

Table 1 Failure pressure models

Sl	Model name	Burst failure pressure, P_b		Ref.
1	B31G	$P_b = \begin{cases} \frac{1.11 \cdot 2YSt}{D}\left(\frac{1-\frac{2d(T)}{3t}}{1-\frac{2d(T)}{3t}M^{-1}}\right); G < 4 \\ \frac{1.11 \cdot 2YSt}{D}\left(1-\frac{d(T)}{t}\right); G \geq 4 \end{cases}$	$M = \sqrt{1 + 0.893\frac{L(T)^2}{Dt}};\ G = 0.893\frac{L(T)^2}{\sqrt{Dt}}$	[3]
2	Modified B31G	$P_b = \frac{2.0(YS + 68.95MPa)t}{D}\left(\frac{1-0.85\frac{d(T)}{t}}{1-0.85\frac{d(T)}{t}M^{-1}}\right)$	$M = \begin{cases} \sqrt{1 + 0.6275\frac{L(T)^2}{Dt} - 0.003375\frac{L(T)^4}{D^2t^2}}\ if \frac{L^2}{Dt} \leq 50 \\ 0.032\frac{L(T)^2}{Dt} + 3.3 if \frac{L^2}{Dt} \leq 50 \end{cases}$	[4]
3	DNV-RP-F101	$P_b = \frac{2UTS}{D-t}\left(\frac{1-\frac{d(T)}{t}}{1-\frac{d(T)}{t}M^{-1}}\right)$	$M = \sqrt{1 + 0.31\frac{L(T)^2}{Dt}}$	[5]
4	Shell-92	$P_b = \frac{1.8UTS}{D}\left(\frac{1-\frac{d(T)}{t}}{1-\frac{d(T)}{t}M^{-1}}\right)$	$M = \sqrt{1 + 0.805\frac{L(T)^2}{Dt}}$	[6]
5	Battelle	$P_b = \frac{2UTSt}{D}\left(1-\frac{d(T)}{t}M\right)$	$M = 1 - \exp\left(-0.157\frac{L(T)}{\sqrt{D(t-d(T))/2}}\right)$	[7]
6	Netto et al. (2005)	$P_b = \frac{1.1 \cdot UTS \cdot 2t}{D} \cdot \left[1 - 0.9435\left(\frac{d}{t}\right)^{1.6}\left(\frac{L(T)}{D}\right)^{0.4}\right]$		[8]

2.4 System Reliability

The reliability of the oil/gas network is estimated based on the concept of minimum cut sets (MCSs). In this case, the reliability is defined based on whether a specific demand node is connected to at least one source at any time. An oil/gas network can be modelled using graph theory. In a graph, nodes and pipelines of oil/gas network are defined by vertices and edges, respectively. The number of MCS is determined using an adjacent matrix. A subset of the pipelines is defined as an MCS if the failure of all components in the subset leads to a failure of the system. The failure probability of an MCS is determined by Eq. (4) [14].

$$P(MC_i) = \prod_{j=1}^{n_p} P_j \tag{4}$$

where $P(MC_i)$ denotes the failure probability of the i-th MCS that contains n_p number of pipelines; P_j is the failure probability of pipeline j. The failure probability of the system (P_S) is evaluated using Eq. (5).

$$P_S = P\left(MC_1 \cup MC_2 \cup \ldots \cup MC_{n_{mc}}\right) \tag{5}$$

where n_{mc} is the number of MCS in a system. Assuming that the failure events are statistically independent, the failure probability of the system is estimated by Eq. (6):

$$P_s = \sum_{i=1}^{n_{mc}} P(MC_i) \tag{6}$$

The reliability of the system (R_S) is then the complement of the failure probability of the system.

3 Case Study

A simple hypothetical oil/gas transmission network consisting of 5 nodes, 1 source and 8 pipes is assumed, as shown in Fig. 1. The reliability of the system is estimated for a scenario that node-5 will remain in-service and connected to the source node-S. For the simplicity of the calculation, it is assumed that all the pipelines in the system are made of the same section and pipe type (diameter: 610 mm, thickness: 20 mm).

The failure probability of the pipelines is determined from fragility curves obtained using a burst limit state expressed in Sect. 2.1. Time-dependent corrosion models, as shown by Eqs. (2) and (3), are accounted for generating the fragility curves for the pipelines. Axial corrosion rate and radial corrosion rate are assumed equal to L0/15 and V0/15, respectively. Initial defect length and defect depth are

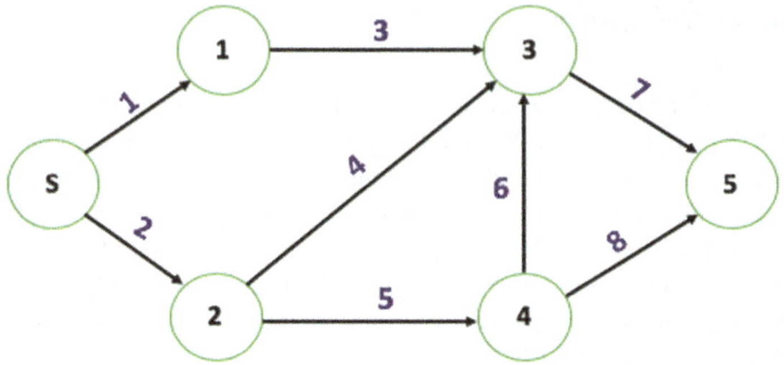

Fig. 1 Hypothetical network

assumed equal to 50 mm and 0.3 t. It is also assumed that 15 years is long enough for the defect growth to reach a steady-state [9]. To develop the fragility curves, six different models are used to determine the failure pressure as expressed in Table 1. Failure pressure is then compared with the internal pressure of the pipeline to determine the failure probability of the pipeline. Table 2 shows the statistical distribution of random variables for estimating the fragility curves. The statistical distribution of random variables is taken based on the Refs. [2, 9]. Monte Carlo Simulation is performed with 1-year time step to estimate the failure probability of the pipeline over time. At each time step, 100,000 simulation points are generated to determine the failure probability of the pipeline. Fragility curves developed utilizing six different models are shown in Fig. 2. Although there is very little difference in failure probabilities estimated using different models at the early stage of the pipeline, the failure probability varies significantly at the later stage of the pipeline.

After estimating the failure probability of the pipelines, the system failure probability, and system reliability are estimated using the MCS approach. Table 3 shows the number of MCSs found for the system reliability analysis that node-5 will remain in-service. It is very unlikely that more than three pipelines fail at a time. Hence, the MCSs consisting of more than three pipelines are ignored. As it can be seen from Table 3, it is found that a single pipeline failure alone will not disconnect node-5 to a source. If all the components in any of these MCSs in Table 3 fail, then node-5 would be out of service.

Table 2 Statistical distribution of random variables

Parameters	Mean	Coefficient of variation	Unit	Distribution type
Wall thickness, t	20	0.05	mm	Normal
Pipe diameter, D	610	0.03	mm	Normal
Yield stress, YS	356	0.08	MPa	Normal
Tensile stress, UTS	455	0.08	MPa	Normal
Operating pressure, P_O	7.8	0.10	MPa	Normal

Fig. 2 Fragility curves

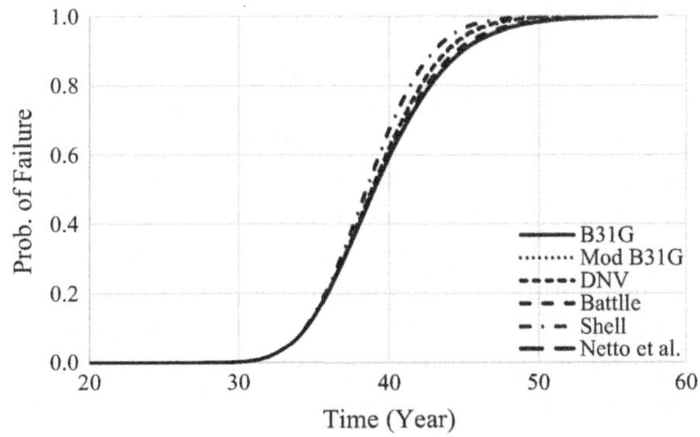

Table 3 Minimum cut sets

No. of pipe breaks	Minimum cut sets {pipe #}	No. of set(s)
1	–	0
2	{1–2} {2–3} {7–8}	3
3	{1–4–5} {3–4–5} {5–6–7}	3

The failure probability of the system is estimated based on Eq. (6). Figure 2 compares the system reliabilities (complement of system failure probabilities) estimated based on various models. Among these models, reliability results obtained using B31G and Modified B31G are the highest. On the other hand, reliability is the lowest using the Battelle model. The system failure probabilities from 35 years stage to 40 years stage increased significantly as the failure probability of each pipeline also increase significantly between 35 and 40 years, as shown in Table 4.

Table 4 System reliability

Reliability approach	Age 0 years (%)	30 years (%)	35 years (%)	40 years (%)	50 years (%)
B31G	100.0	100.0	94.0	33.3	0.0
Modified B31G	100.0	100.0	93.9	27.7	0.0
DNV-RP-F101	100.0	100.0	94.0	29.8	0.0
Shell-92	100.0	100.0	94.0	32.0	0.0
Battelle	100.0	100.0	93.5	15.7	0.0
Netto et al. (2005)	100.0	100.0	94.0	33.3	0.0

4 Conclusions

This study represents a comparative analysis of various fragility models of pipelines. Also, subsequent system reliabilities are estimated and compared using various approaches. Time-dependent reliability is analyzed by incorporating the time-dependent corrosion growth on the pipeline wall. The proposed approach is illustrated for a simple hypothetical oil/gas transmission system. Analysis outcomes show a consistent trend of failure for various models overtime. It is found that the selection of a specific model may lead to obtaining a higher or lower reliability result, especially during the later stage of a pipeline lifespan. Such deviation from selecting a specific model should be considered while identifying the riskiest pipelines in the asset management plan.

References

1. Kim, Y., & Kang, W. H. (2013). Network reliability analysis of complex systems using a non-simulation-based method. *Reliability Engineering & System Safety, 110,* 80–88.
2. Teixeira, A. P., Soares, C. G., Netto, T. A., & Estefen, S. F. (2008). Reliability of pipelines with corrosion defects. *International Journal of Pressure Vessels and Piping, 85*(4), 228–237.
3. ASME B31G. (1991). *Manual for determining the remaining strength of corroded pipelines.* American Society of Mechanical Engineers, New York
4. Kiefner, J. F., & Vieth, P. H. (1990). Evaluating pipe—1. New method corrects criterion for evaluating corroded pipe. *Oil and Gas Journal, 88*(32).
5. Veritas, N. (1999). *Corroded pipelines: DNV recommended practice RP-F101, 1999.* Det Norske Veritas.
6. Klever, F. J., Stewart, G., & van der Valk, C. A. (1995). New developments in burst strength predictions for locally corroded pipelines (No. CONF-950695-). American Society of Mechanical Engineers, New York.
7. Stephens, D. R., & Leis, B. N. (2000). Development of an alternative criterion for residual strength of corrosion defects in moderate-to high-toughness pipe. In *3rd International Pipeline Conference, American Society of Mechanical Engineers Digital Collection.*
8. Netto, T. A., Ferraz, U. S., & Estefen, S. F. (2005). The effect of corrosion defects on the burst pressure of pipelines. *Journal of Constructional Steel Research, 61*(8), 1185–1204.
9. Caleyo, F., Gonzalez, J. L., & Hallen, J. M. (2002). A study on the reliability assessment methodology for pipelines with active corrosion defects. *International Journal of Pressure Vessels and Piping, 79*(1), 77–86.
10. Cosham, A., Hopkins, P., & Macdonald, K. A. (2007). Best practice for the assessment of defects in pipelines—corrosion. *Engineering Failure Analysis, 14*(7), 1245–1265.
11. Ahammed, M. (1998). Probabilistic estimation of remaining life of a pipeline in the presence of active corrosion defects. *International Journal of Pressure Vessels and Piping, 75*(4), 321–329.
12. Wang, Y. Q., Wang, W. B., & Feng, Q. S. (2008). Remaining strength assessment for corroded pipelines. *Corrosion and Protection-Nanchang, 29*(1), 28.
13. Amaya-Gómez, R., Sánchez-Silva, M., Bastidas-Arteaga, E., Schoefs, F., & Munoz, F. (2019). Reliability assessments of corroded pipelines based on internal pressure—A review. *Engineering Failure Analysis, 98,* 190–214.
14. Shinstine, D. S., Ahmed, I., & Lansey, K. E. (2002). Reliability/availability analysis of municipal water distribution networks: Case studies. *Journal of Water Resources Planning and Management, 128*(2), 140–151.

Risk Assessment of a Railway Bridge Subjected to a Multi-hazard Scenario

João Fernandes, Monica Santamaria, José C. Matos, Daniel V. Oliveira, and António Abel Henriques

Abstract Bridges present valuable assets for the rail and road network by providing cross at critical links such as waterways, valleys, and other types of facilities. However, these types of structures are exposed to several threats during their life-cycle such as natural hazards and deterioration, which compromise their performance. To assess the condition state of such infrastructure and define maintenance and mitigation strategies, several performance indicators of quantitative nature have been proposed during the last decades by several researchers. Among those indicators, risk has received great attention as it enables to account for both the performance of infrastructures subjected to hazard events, and the consequences associated to an inadequate level of service of the infrastructure. Nevertheless, risk is not a stationary indicator, i.e. several parameters involved in the estimation of risk are time-dependent. One of them comprises the structural capacity of infrastructures, which is affected by deterioration effects over time. This gradual deterioration can be regarded as an interceptable hazard, which may act simultaneously with other non-interceptable hazards such as natural events (e.g. earthquakes). Therefore, a risk assessment framework should account for the probability of having these multiple hazards acting during the service life of infrastructures. The aim of this paper is to conduct a risk assessment for a railway bridge subjected to a multi-hazard scenario, i.e. an observable interceptable hazard corresponding to chloride induced corrosion of the reinforcing steel in reinforced concrete elements, together with seismic hazard. The results of the study demonstrate the relevance of considering time-dependent deterioration effects on the risk assessment of bridges, as the increase in the seismic fragility over time is significant. These findings are relevant for decision-making to plan and execute optimal interventions.

Keywords Seismic analysis · Corrosion · Risk assessment

J. Fernandes (✉) · M. Santamaria · J. C. Matos · D. V. Oliveira
ISISE, Department of Civil Engineering, University of Minho, Guimarães, Portugal

A. A. Henriques
CONSTRUCT, Department of Civil Engineering, University of Porto, Porto, Portugal

© The Author(s), under exclusive license to Springer Nature Switzerland AG 2021

J. C. Matos et al. (eds.), *18th International Probabilistic Workshop*, Lecture Notes in Civil Engineering 153, https://doi.org/10.1007/978-3-030-73616-3_51

1 Introduction

Risk is a worldwide measure adopted in several different fields other than engineering. According with the book *Fundamentals of Risk Management* [1], risk is defined as "*An event with the ability to impact (inhibit, enhance or cause doubt about) the mission, strategy, projects, routine operations, objectives, core processes, key dependencies and/or the delivery of stakeholder expectations*". Risk is considered a complex measure once it assembles the structural performance of the bridge at a probabilistic level, with indicators related to the development of the society that can differ according to the region. In the works of Faber and Stewart [2] and Ellingwood [3] a comprehensive analysis of the risk indicator as well as its importance to facilitate a risk-informed assessment is discussed. Moreover, within the work of Faber and Stewart [2], a generic representation of the flow of the risk-based analysis is discussed, wherein the main stages are addressed to: (i) System representation; (ii) Exposures and hazards and (iii) Consequences. Its representation is also thoroughly explained in Joint Committee on Structural Safety (JCSS) [4]. Due to its complexity, risk has been recognized by the bridge engineering community as a topic of high interest as the literature can prove. For example see references [5–8].

Infrastructure assets are exposed to different type of hazards which can be regarded as interceptable and non-interceptable hazards. Interceptable hazards refer to gradual deterioration which can be observable (e.g. corrosion related to structural steel), or non-observable (e.g. corrosion of post-tensioning steel) [9]. On the other hand, sudden events acting on transportation infrastructure assets are regarded as observable, non-interceptable processes, i.e. an event occurring on a short period of time where no adequate mitigation actions can take place. For the present work, a risk assessment of a railway bridge subjected to both type of hazards, namely seismic and corrosion hazard is conducted. Following the introduction, Sect. 2 is related for the presentation of the case study and the considerations for the modeling of the structure; Sect. 3 describes the structural analysis; Sect. 4 concerns the corrosion effects on the bridge; Sect. 5 is dedicated to obtaining the fragility curves due to seismic actions and corrosion effects; Sect. 5 deals with risk assessment; and Sect. 6 presents the conclusions drawn from the work done.

2 Case Study and Modeling Considerations

The bridge case study spans the Arunca River and is located between Albergaria dos Doze-Alfarelos in the center of Portugal. The bridge is part of the railway network in the Region of Santarém/Leiría and was built in 2005. The total length of the bridge is 66.63 m and is divided into four simply supported spans. The superstructure is composed by two beams of 1.35 m high and a 0.4 m slab connecting both beams, to conform a "H" cross section shape (Fig. 1a) in each direction. The connection between piers and deck is done through fixed pot bearing devices in one support and

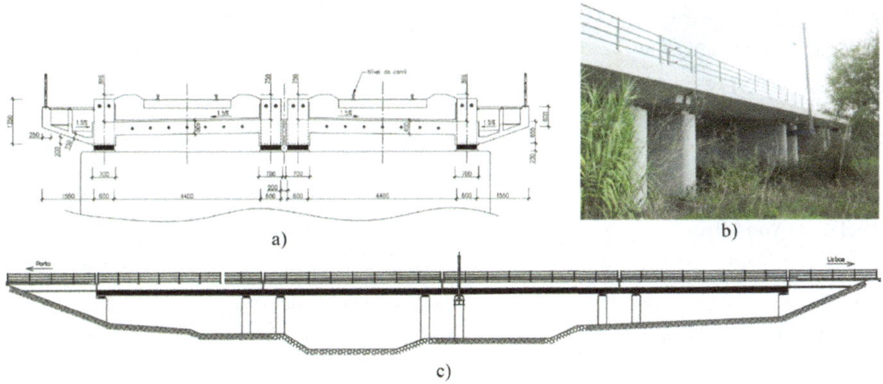

Fig. 1 Railway bridge overview **a** Deck cross section **b** reinforced concrete piers **c** longitudinal view. *Source* Infraestruturas de Portugal

unidirectional pot bearings in the other support. Cover measurement tests performed after the bridge construction revealed that there are zones with lack of concrete cover, which renders the bridge susceptible to corrosion deterioration [10].

A 2D FE model for the bridge was developed using the software TNO DIANA. While analyzing the bridge, it was noticed that the frames work independently. Thus, the bridge will be represented by the middle frame. The middle frame is composed by a deck with a span length of 13.50 m supported on piers of 4.50 m height. Concerning the modeling aspects, for the deck, all the elements are accounted and modelled with linear elastic beam elements as the damage is not expected in here. As for the piers, non-linear beam elements were modelled assuming a total strain fixed crack model with a brittle tensile behavior, a compressive behavior based on EC2 EN 1992-1-1 [11], and a constant shear behavior. As for the steel material, a Von Mises plasticity model was adopted. Regarding the bearing devices, the unidirectional pot bearing was modelled based on the experimental results and the model proposed by Dolce et al. [12], to characterize the frictional behavior of steel-PTFE contact interface and define spring elements with elastic-plastic hysteretic backbone curves. The fixed pot bearing was considered as working as a hinge. Table 1 presents the parameters considered for the constitutive laws of each element.

3 Structural Analysis

The structural analysis accounted initially for a non-linear static pushover curve. The engineering demanding parameter (EDP) adopted on the calculation of the pushover was given by the peak displacement on the top of the pier. The structural analysis was performed with the properties of the materials defined in Table 1. Moreover, the concrete density was estimated as 25 kN/m^3, steel young modulus of 200 GPa, and a

Table 1 Considered parameters for the modeling of the constitutive laws

Material	Constitutive law	Parameters	Quantification
Concrete	EC2 EN 1992-1-1 (Compressive)	Compressive strength (f_{cm})	33 MPa
		ε_{c1}	0.002
		ε_{cu}	0.0035
	Brittle (Tensile)	Concrete Tensile strength (f_{ctm})	2.6 MPa
Steel	Von Mises	Steel yielding strength (f_{ym})	560 MPa
		Yielding strain (ε_{sy})	0.0028
		Ultimate steel strain (ε_{su})	0.15
Bearing	Elastic plastic	Yielding force (F_y)	1.20 MN
		Elastic stiffness (K_e)	1.0 GN/m

total steel area per pier of 0.028 m^2. Considering all these parameters, the pushover curve obtained is depicted in Fig. 2.

From the analysis of Fig. 2, it can be clearly observed the stages of the structure corresponding to the linear part, the cracking phase, the yielding.

For a more robust analysis of the present work, a time-history analysis was secondly carried out to study the fragility of the bridge. To obtain the fragility curve, the following steps were considered: (i) ground motion selection; (ii) selection of the intensity measures (IM) and the EDP; (iii) Damage limit state (DLS) analysis; and (iv) fragility calculation.

Concerning the first step, a set of artificial accelerograms were adopted according with the recommendations of Eurocode 8 [13]. Note that Eurocode 8 [13], acknowledges a minimum of 7 artificial accelerograms to consider a representative response. Hence, 10 accelerograms were considered in this work, and were generated by the

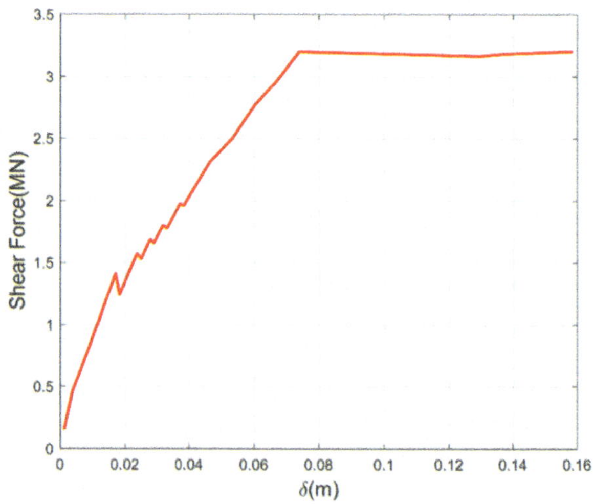

Fig. 2 Pushover curve for peak displacement on the pier

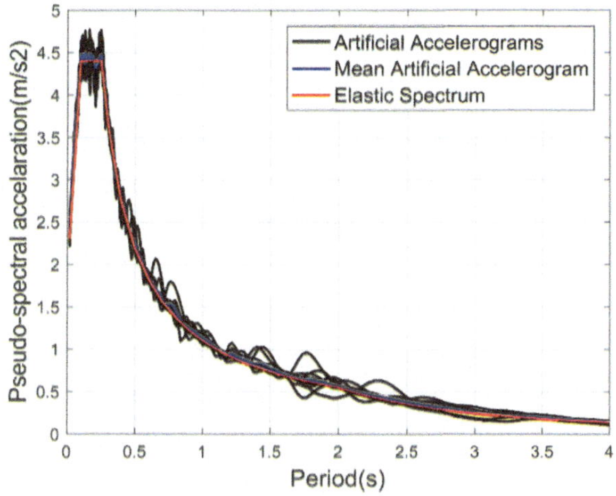

Fig. 3 Pseudo-spectral accelerations for the considered accelerograms

software SIMQKE_GR [14]. For each accelerogram a duration of 14 s with a time rise for the response of 2 s was considered, as well as an envelope function with a trapezoidal shape. The resulting accelerograms are compatible with the elastic response spectrum accelerogram, according with Eurocode 8, for a Type 2 with a return period of 475 years and viscous damping of 5% (see Fig. 3). The parameters for the region where the case study is located, Soure, were obtained from the Portuguese Annex, and correspond to an importance factor γ_I of 1.0, a peak ground acceleration a_{gR} of 1.1 m/s^2, and a soil type C (S = 1.60).

As for the IM, the adopted for this work was the peak ground acceleration (PGA), since it has proven to have good correlation with damage, and therefore to be among the most optimal IMs to assess bridges. Concerning the EDP, as referred for the pushover analysis, the peak displacement on the top of the pier was selected. Given that the system presented in this work is comprised by the piers and the bearing devices, a different damage state criterion must be considered for each component. Regarding the piers, several authors have proposed different limit states for bridges. For this study, the work of Moschonas et al. [15] is considered. In this work, the DLS are based on EDP obtained in the non-linear static pushover analysis wherein relationships between the yielding, δ_y, and ultimate displacement, δ_u, are made to obtain the different DLS. As for the bearing device, some assumptions were also made based on its typology. However, few studies cover these analyses applied on pot bearing devices. Thus, the DLS were based on the study of Jiang et al. [16]. In this study, it is stated that the displacement, if less than the design value, implies no damage for the bearings. Conversely, higher values translate into more serious damage states being the ultimate displacement 5 times higher than the design value, meaning thus the complete destruction. Note that for the present work, the design value of the bearing displacement was considered as 50 mm. Table 2 summarizes the adopted DLS for each component.

Finally, the fragility calculation was obtained by the following equation:

Table 2 DLS considerations for the pier and bearing device

Component	Minor (DSL 1)	Moderate (DSL 2)	Major (DSL 3)	Collapse (DSL 4)
Pier displacement*	$\delta_{pier} \leq 0.7\delta_y$	$\delta_{pier} > \min[1.5\delta_y, (\frac{1}{3}).(\delta_u - \delta_y)]$	$\delta_{pier} > \min[3.0\delta_y, (\frac{2}{3}).(\delta_u - \delta_y)]$	$\delta_{pier} > \delta_u$
Bearing displacement*	50 mm	75 mm	125 mm	250 mm

*Values based on the longitudinal direction

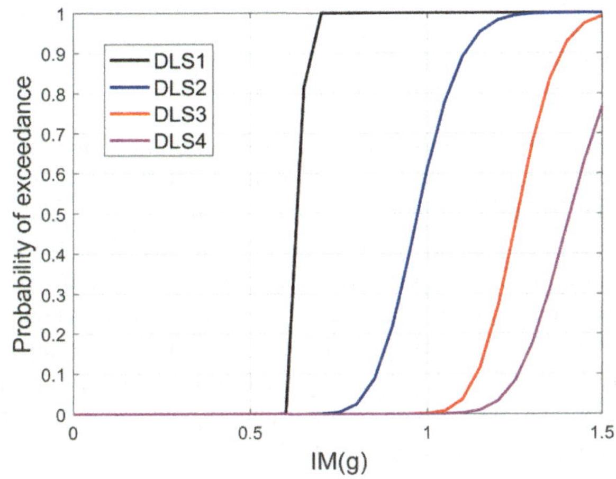

Fig. 4 Seismic fragility curves of the system for each DLS

$$Fragility = P(EDP \geq DLS|IM) \tag{1}$$

Generally, the cumulative function that builds the fragility curve follow a lognormal distribution, being the expression defined as follows:

$$P(DLS|IM = x) = \Phi\left(\frac{\ln\left(\frac{x}{\theta}\right)}{\beta}\right) \tag{2}$$

where $P(DLS|IM = x)$ is the probability of exceeding a DLS for a given $IM = x$; Φ is the standard normal cumulative distribution function, θ is the mean and β the standard deviation of $\ln(IM)$. It is noteworthy to mention that there are several approaches to obtain the fragility curves. For the present work, the approach from Baker [17] was adopted. Following this methodology, an estimation of the parameters θ and β were obtained through maximum likelihood estimation. The combination of both pier and bearing fragility curves were obtained after Choi et al. [18], which proposes a lower and upper bound of the system fragility for a series system. Accordingly, Fig. 4 depicts the final fragility curve of the bridge.

4 Corrosion Effect

The loss of structural strength in aging RC bridges is frequently attributed to the chloride-induced corrosion of reinforcing bars [19]. The transport of chloride ions through concrete is typically modelled using Fick's second law. Based on this model, the probabilistic representation for the time for corrosion initiation is expressed as [19, 20]:

$$T_{corr} = \theta_d \left[\frac{x^2}{4k_e k_c k_t D_0 t_0^n} \left[\text{erf}^{-1}\left(1 - \frac{C_{cr}}{C_s}\right)\right]^{-2} \right]^{1/n} \tag{3}$$

where θ_d is a model uncertainty coefficient to account for the idealization on Fick's second law, x is the concrete cover depth, k_e is an environmental factor, k_c is a factor that accounts for the influence of curing, D_0 is the chloride diffusion coefficient which describes the resistance against the ingress of chlorides, k_t is a factor describing the effect of the test method to determine D_0, t_0 is the reference period for D_0, n is an age factor that incorporates the densification of cement paste due to further hydration chloride, erf is the error function, C_{cr} is the critical chloride concentration, and C_s is the surface chloride content determined from the expression $C_s = A_{cs}(w/b) + \varepsilon_{cs}$, where A_{cs} is a regression parameter between C_s and the water-binder ratio (w/b), and ε_{cs} is the error term [20].

Various mechanisms are evidenced after the corrosion initiation. This study accounts for a generalized uniform reduction of reinforcement area along the length of the rebar due to corrosion as proposed by Choe et al. [19].

The chloride exposure condition assumed in this study corresponds to structures exposed to de-icing salts, whose variables depending on the material and the environment may be assumed to be equivalent to the ones valid for the marine splash zone [20, 21]. The probability distributions for the variables to determine T_{corr} using Eq. (3) are adopted from Duracrete [20] and Choe et al. [19] (see Table 3), for the assumed marine splash zone exposure condition, and for assumed Ordinary Portland cement (OPC), water-binder ratio (w/b) of 0.5, and water cement ratio (w/c) of 0.5.

Two different points along the service life of the bridge, namely 50 and 100 years, were selected to analyze the impact of corrosion on the capacity of the bridge to resist the seismic demands. Figure 5 shows the time-dependent seismic fragility curves obtained for each DLS. It can be observed that the seismic fragility growths over time as expected, with a tendency to have a greater impact as the damage severity and the PGA increase.

5 Risk Assessment

The risk calculation for this work is given by the following equation:

$$Risk = p(DLS_i|IM) \times C \tag{4}$$

where $p(DLS_i|IM)$ is the probability of exceedance of a certain DLS given the IM, and C are the consequences of the system. Regarding the consequences, they are divided into direct and indirect consequences. Regarding direct consequences, their estimation is done based on the values proposed by Decò et al. [6]. Considering that the consequences depend on the severity of the damage, i.e. on the DLS, the type

Table 3 Probability distributions for the corrosion parameters [19, 20]

Variable	Condition	Distribution	Unit
Model uncertainty coefficient (θ_d)	–	Lognormal (−0.0013; 0.05)[a]	–
Cover depth (x)	–	Lognormal (3.47; 0.13)[b]	[mm]
Environmental factor (k_e)	OPC, Splash zone	Gamma (2.92; 11.0)	–
Curing factor (k_c)	At age 7 days	Beta (2.15; 10.7; 1.0; 4.0)	–
Chloride diffusion coefficient (D_0)	w/c = 0.5	Normal (473.0; 43.2)	(mm^2/yr)
Correction factor for test method (k_t)	–	Normal (0.832; 0.024)	–
Reference period (t_0)	28	Deterministic	[days]
Age factor (n)	OPC, Splash zone	Beta (17.2; 29.3; 0; 1.0)	–
Critical chloride concentration (C_{cr})	OPC, w/b = 0.5, Splash zone	Normal (0.50; 0.10)	[%] relative to binder
Surface chloride content regression parameter (A_{cs})	Splash zone	Normal (7.76; 1.36)	[%] relative to binder
Surface chloride content error term (ε_{cs})	Splash zone	Normal (0; 1.11)	[%] relative to binder

[a]The reported values of the λ and ζ parameters of the Lognormal distribution correspond to a mean model uncertainty coefficient of 1, and a standard deviation of 0.05
[b]The reported values of the λ and ζ parameters of the Lognormal distribution correspond to a mean cover depth of 32.6 mm, and a standard deviation of 4.2 mm

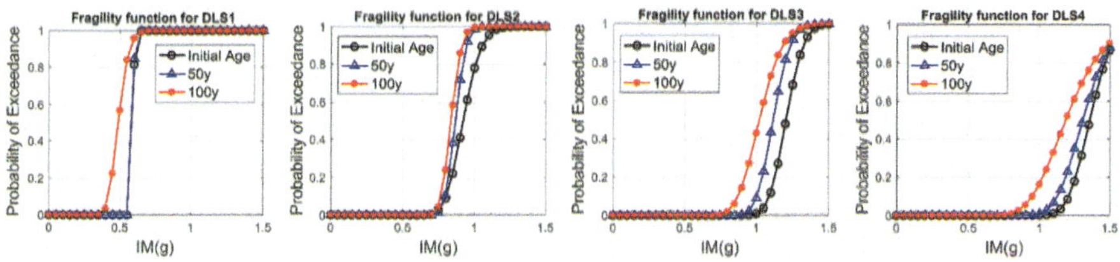

Fig. 5 Time-dependent seismic fragility curves for each DLS

of required intervention is described in Table 4 for each DLS based on the study of Moschonas et al. [15].

Concerning the indirect consequences, they emerge as a result of the closure of the bridge for repairing/rebuilding, and therefore the need of identifying alternative detour routes for this exceptional situation which usually entails additional costs for the users. Considering that the railway network is not redundant as the roadway network, alternative routes are frequently defined through roads. The calculation of

Table 4 Description of the consequences for each DLS [15]

Damage limit state	Required interventions/consequences
DLS1: Minor	Inspect, adjust, patch
DLS2: Moderate	Repair components
DLS3: Major	Rebuild components
DLS4: Collapse	Rebuild bridge

these costs is based on Almeida [21]. Table 5 summarizes the considered parameters for the estimation of the direct and indirect consequences.

Considering the fragility calculation from the previous section and the consequences herein estimated, Fig. 6 assembles the time-dependent risk of the bridge.

Table 5 Quantification of the consequences [21, 22]

Variable	Notation	Quantification		
Rebuilding cost (€/m^2)	c_{reb}	TD* (680, 1360, 2550)		
Rebuilding Area (m^2)	A_{pier}	235.0**		
	A_{deck}	108.0**		
Conditioned traffic percentage	PER	TD (80–90–100%)		
		Cars	Trucks	
Average daily traffic	TMD	950	50	
Cost per kilometer (€/km)	CK	0.18	0.68	
Cost per hour (€/h)	CH	8.4	10.1	
Restricted Speed (km/h)	S_r	70**	50**	
		Train		
Normal speed (km/h)	S_n	200**		
Detour route (km)	LD	5.15**		
Normal route (km)	LP	7.50**		
Discount rate	R	2%		
Duration of the activity (days)	DUR	DLS2	DLS3	DLS4
		3	180	270

*TD stands for triangular distribution
**Values directly estimated from the bridge design specifications

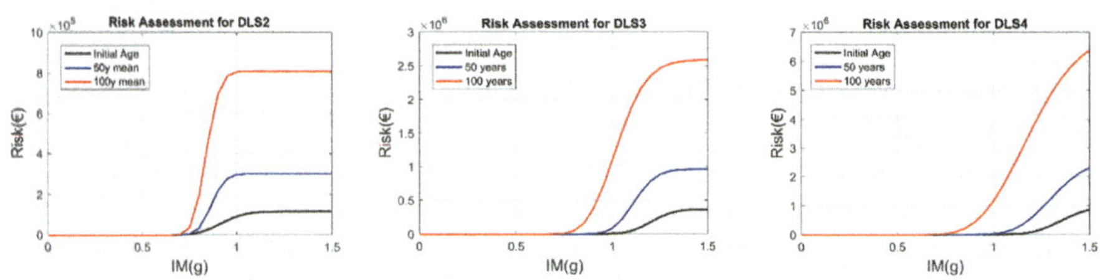

Fig. 6 Time-dependent risk assessment per DLS

Given DLS1 results on very low damages, its consideration was not considered for the risk estimation.

From the analysis of Fig. 6, it is observed a considerable increase on the risk for PGAs over 0.8 g, wherein some provisions should be taken in terms of mitigation. Furthermore, it can be noted that performing maintenance actions over the life-cycle of the bridge to reduce the effects of deterioration mechanisms such as chloride induced corrosion, should keep the structure under low risk values which guarantees a safety performance of the bridge during its entire service life.

6 Conclusions

This paper aimed to propose a risk assessment of a bridge subjected multi-hazard event, comprised by seismic actions and corrosion effects over its life cycle. Firstly, a structural analysis was carried out with two analysis, a non-linear pushover analysis and a time-history analysis using artificial accelerograms. Furthermore, the effect of deterioration due to corrosion was included aiming to consider the time-dependent fragility of the bridge. Secondly, the direct and indirect consequences were estimated for the different damage level states. Finally, the time-dependent risk was computed. The obtained results shown that for PGAs over the IM of 0.8 g, the risk is significant since unacceptable values were observed. Those values could be reduced by considering mitigation actions on the bridge such as jacketing effects or bearing devices with isolation effects.

Acknowledgments Authors would like to thank ISISE—Institute for Sustainability and Innovation in Structural Engineering (PEst-C/ECI/UI4029/2011 FCOM-01-0124-FEDER-022681) and FCT—Portuguese Scientific Foundation for the research grant PD/BD/128015/2016 under the PhD program "Innovation in Railway System and Technologies—iRail". The second author would like to thank FCT—Portuguese Scientific Foundation for the research grant SFRH/BD/144749/2019. The fifth author would also like to thank the support of Project UID/ECI/04708/2019—CONSTRUCT—Institute of R&D in Structures and Construction, funded by national funds through the FCT/MCTES (PIDDAC).

References

1. Hopkin, P. (2014). *Fundamentals of risk management: Understanding, evaluating and implementing effective risk management*. Kogan Page Publishers
2. Faber, M. H., & Stewart, M. G. (2003). Risk assessment for civil engineering facilities: Critical overview and discussion. *Reliability Engineering & System Safety, 80,* 173–184.
3. Ellingwood, B. R. (2005). Risk-informed condition assessment of civil infrastructure: State of practice and research issues. *Structure and Infrastructure Engineering, 1,* 7–18.
4. Faber, M. (2008). Risk assessment in engineering: Principles, system representation & risk criteria. *JCSS Joint Committee of Structural Safety.*

5. Decò, A., & Frangopol, D. M. (2011). Risk assessment of highway bridges under multiple hazards. *Journal of Risk Research, 14,* 1057–1089.
6. Decò, A., & Frangopol, D. M. (2013). Life-cycle risk assessment of spatially distributed aging bridges under seismic and traffic hazards. *Earthquake Spectra, 29,* 127–153.
7. Saydam, D., & Frangopol, D. M. (2014). Risk-based maintenance optimization of deteriorating bridges. *Journal of Structural Engineering, 141,* 04014120.
8. Saydam, D., Frangopol, D. M., & Dong, Y. (2012). Assessment of risk using bridge element condition ratings. *Journal of Infrastructure Systems, 19,* 252–265.
9. Hajdin, M. K. R., Masovic, S., Linneberg, P., & Amado, J. (2018). *WG3 Technical Report—Establishment of a quality control plan.*
10. Ariza, M. P. S., Sousa, H. S., Fernandes, J. N. D., Matos, J. C. (2019). *Reliability analysis of a post-tensioned railway bridge exposed to corrosion effects.*
11. CEN. (2004). *Eurocode 2: Design of concrete structures—Part 1-1: General rules and rules for buildings.*
12. Dolce, M., Cardone, D., & Croatto, F. (2005). Frictional behavior of steel-PTFE interfaces for seismic isolation. *Bulletin of Earthquake Engineering, 3,* 75–99.
13. CEN. (2004). *Eurocode 8. Design provisions for earthquake resistance of structures. Part 1-1: General rules—Seismic actions and general requirements for structures.* Brussels: Belgium Eur Commitee Stand.
14. Gelfi.(2006). SIMQKE-GR—Software for generating artificial accelerograms compatible with the response spectrum.
15. Moschonas, I. F., Kappos, A. J., Panetsos, P., Papadopoulos, V., Makarios, T., & Thanopoulos, P. (2009). Seismic fragility curves for Greek bridges: Methodology and case studies. *Bulletin of Earthquake Engineering, 7,* 439.
16. Jiang, C., Wei, B., Wang, D., Jiang, L., & He, X. (2017). Seismic vulnerability evaluation of a three-span continuous beam railway bridge. *Mathematical Problems in Engineering, 2017.*
17. Baker, J. W. (2014). Efficient analytical fragility function fitting using dynamic structural analysis. *Earthquake Spectra, 31,* 579–599.
18. Choi, E., DesRoches, R., & Nielson, B. (2004). Seismic fragility of typical bridges in moderate seismic zones. *Engineering structures, 26,* 187–199.
19. Choe, D.-E., Gardoni, P., Rosowsky, D., & Haukaas, T. (2008). Probabilistic capacity models and seismic fragility estimates for RC columns subject to corrosion. *Reliability Engineering & System Safety, 93,* 383–393.
20. EuRam III, E. B. (2000). DuraCrete final technical report, probabilistic performance based durability design of concrete structures. *Document BE95–1347.*
21. Almeida, J. (2013). *Sistema de gestão de pontes com base em custos de ciclo de vida* (translation: Life-cycle cost based bridge management systems). Universidade do Porto.
22. Minaie, E., & Moon, F. (2017). Practical and simplified approach for quantifying bridge resilience. *Journal of Infrastructure Systems, 23,* 04017016.

Risk Assessment of Road Infrastructures as Key for Adaptability Measures Selection

Erica L. Arango, Hélder S. Sousa, and José C. Matos

Abstract Road infrastructures are crucial for societies daily life due to the dependency of other critical infrastructures upon it. Therefore, society expects an uninterrupted availability of the road network. However, maintain this constant availability is often a difficult task as, in the last decades, climate change has significantly affected transport networks, especially due to the occurrence of extreme natural events leading to their disruption. Those events include floods, wild fires, landslides and others, and all of are varying both in frequency and intensity presently and in the coming years. Therefore, there is a clear need for timely adaptation. Regarding these adaptability measures, an important step is needed to quantify how the transport network is directly and indirectly affected by extreme weather events, which can be obtained within a risk assessment. Nonetheless, there are many questions and variability about this topic such as uncertainties in projections of future climate, cause-effects assessment, and how it can be an integration of all these aspects into a single decision-making process. In that scope, this work describes a risk assessment methodology having account the cause, effect, and consequences of extreme events in road networks to identify the major risks and therefore the assets that may be suitable to be analyzed within a selection of adaptation measures aiming at a holistic decision-making support tool.

Keywords Road infrastructures · Risk assessment · Extreme events · Adaptability measures

1 Introduction

Road network is one of the most important components of transportation infrastructure and therefore a vital aspect of development as well as economic growth [1–3]. Society has generated a great dependence on this system and consequently any infrastructure disruptions may have severe consequences for human well-being. Since the road network is designed to operate within a particular environment, the

E. L. Arango (✉) · H. S. Sousa · J. C. Matos
University of Minho, ISISE, Department of Civil Engineering, Guimarães, Portugal

© The Author(s), under exclusive license to Springer Nature Switzerland AG 2021

J. C. Matos et al. (eds.), *18th International Probabilistic Workshop*, Lecture Notes in Civil Engineering 153, https://doi.org/10.1007/978-3-030-73616-3_52

system is placed at risk from the damaging impact of the frequency and intensity of some extreme weather events [4, 5], which are expected to increase [6]. In that aspect, climate change represents a new challenge for the decision makers regarding design, construction and operation of road infrastructures [7]. As in most cases, available financial resources are limited, it is especially important to use these resources efficiently. To achieve that, it is imperative to know the potential risk to these systems which involves the correct problem identification [8].

Risk can be analyzed within the perspective of performance of the structure related to its degradation and possible consequences to a network-level failure. In that sense, infrastructure risk can consider climate change as a parameter when determining its influence on hazard determination in both exposure and vulnerability perspective [9]. A careful application of a risk assessment may have significant contributions not only to threats understanding and to related uncertainties but also to facilitate the decision-making process of road investment, planning and design [4, 6]. Most importantly, risk assessment is the basis to implement preparedness actions or adaptation strategies, which are developed according to the infrastructure needs and situation complexity. For instance, identifying projected levels of variations due to climate changes can provide useful information for adaptability planning and maintenance projects.

Adaptation measures are focused on reducing vulnerability and consequences but these measures are conditioned to aspects such as resources, capacities, environment, and authority/legal constraints and requirements. Therefore, the selection and prioritization of adaptability strategies are highly important as not all adaptation options will be possible for a specific climate change risk or local conditions [8]. Hence, the establishment of adaptation strategies is a challenge with a high level of uncertainty associated with climate change effects, especially to identify limits and effectiveness of the measurements [5].

This work focuses on the description of a risk assessment methodology originated by the need to link and integrate disaster risk reduction with adaptation measures, regarding extreme events in road networks. The framework aims at a holistic decision-making support tool. To do so, the work is divided into four principal sections. The second section is focused on describing risk, its assessment methodology and critical climate parameters affecting road infrastructures. Section three provides an adaptability definition, adaptation measures for the major risks in road infrastructure and their classification. Section four proposes an approach to linked risk assessment with adaptability. Finally, the discussion and conclusions are presented in section five.

2 Risk

Risk is defined as a measurement of a probability and severity of the dangerous situation occurrence [6]. In this scenario, climate change effects are often classified as hazards of medium to large impact with a high uncertainty degree as they are constantly changing both in frequency and intensity. Specifically, extreme events may cause a variety of impacts, which are commonly classified into social, economic, and environmental categories [10]. Therefore, within these categories, risk implies the combination of threats, vulnerabilities and consequences. Thus, threat refers to environmental and climate factors (hazards) described by contextual site factors. Vulnerability is closely relating to the link failure consequences, including infrastructure-intrinsic or function factors. Finally, the consequences provided the threat result or effect involving factors such as human life and injuries, economic losses, and reconstruction cost [4, 11].

2.1 Risk Assessment Methods

There is a wide variety of methods and tools for risk assessment. These methods may include among other, probabilistic modeling, statistical analyses of past events, empirical approaches, risk analysis of technological systems and economic theory-based approaches [12]. However, there is a major classification for risk assessment methods based on data type, which dividing it into three main groups: (i) qualitative, (ii) semi-quantitative and (iii) quantitative analysis as shown in Table 1.

All methods have different ways to find the damaged or failure probabilities but they also present transverse key steps established for risk assessment. Methodological steps are proposed based on RIMAROCC Framework [11], the quantitative framework proposed by Mechler and Nabiul [10] and the mathematical formulation for the integrated framework of Mitsakis et al. [14]. The method itself consists of a cyclic process in which there is a constant definition and analysis of its performance. This procedure begins by establishing the risk context, defining the scope and impact criteria. Second the risk source identification, which involves defining impact areas and unwanted events in terms of potential causes and consequences. Third, the risk analysis and evaluation. Then, prioritized the measure implementation regarding the criteria selected in step one. Afterwards, the risk mitigation that implies the options recognition and selection for risk treatment. In the end, the action plan defines responsibilities, resources and performance of the selected measures; and also implies monitoring and review of the action plan.

In fact, the principal steps can be divided into sub-steps as is shown in Table 2. During the procedure, several steps can be addressed at the same time but it is important to preserve the logical structure of the framework. Since there is a relationship between the steps (predecessor and successor steps) and thus obtain feedback from both each step and the entire framework as part of the cyclical process.

The key steps can be applied in general risk analysis and infrastructures but in the case of road infrastructure it is necessary to treat it as a framework. For that purpose, focusing on most vulnerable or critical sections, nodes or structures is required with regard to climate factors. Perhaps one of the most important aspect is the risk identification into the framework. An undefined risk may affect the whole analysis even if another risk was properly considered [11].

Table 1 Risk assessment methods characteristics

Method	Approach	Advantages	Disadvantages	Example
Qualitative	Description of risks in words	Clear presentation options of risk, easily used and allow the prioritisation	Subjective evaluation, does not provide an assessment of the overall project risk exposure. Lack of categories differentiation	Checklists, what-if analysis, probability/consequence matrix
Semi-quantitative	Intermediary level between the textual and numerical evaluation	Use classes instead exact values and is a good basis for discussing risk reduction. Allow to carry out holistic risk assessment	Do not provide quantitative values. Difficult impacts and frequencies assessment	Risk matrix, indicator-based, probability-impact
Quantitative	Focus on numbers and frequencies	Quantitative risk information may be used in cost-benefit analysis of risk reduction measures, also allow modeling sequences of events	Very data demanding, time consuming. Difficult spatial implementation	Quantitative risk assessment (QRA), event tree analysis, probabilistic risk assessment (PRA)

Adapted from [11, 13].

2.2 *Climate Change*

The average conditions variation of climate also known as a climate change, have been affecting the whole world over a long time. Nevertheless, the consciousness of the consequences has only be awakened on the last few decades, especially related to the build-up of greenhouse gases (GHG) by burning fossil fuels. Clearly, the consequences are extended to road network, this being one of the major contributors to fossil fuel consumption [7]. Climate change translates into threats as extreme weather events and gradual changes for the road system. Also, imply different hazards like coastal and urban flooding, heat, cold, drought, and wind, which affect the infrastructure, passengers, and freight [15].

Table 2 Risk methodology steps and sub-steps

Key steps		Sub-steps
1	Context analysis	Establish a general context
		Establish a specific context for a particular scale of analysis
		Establish risk criteria and indicators adapted to each particular analysis scale
2	Risk identification	Identify risk sources
		Identify vulnerabilities
		Identify possible consequences
3	Risk analysis	Establish risk chronology and scenarios
		Determine the impact of risk
		Evaluate occurrences
		Provide a risk overview
4	Risk evaluation	Compare risk against established criteria
		Determine which risks are acceptable
		Identify treatment priorities
5	Risk mitigation	Identify options
		Appraise options
		Formulate an action plan
6	Action plan implementation and monitoring	Develop an action plan on each level of responsibility
		Implement adaptation action plans
		Regular monitoring/review and feedback

Adapted from [10, 11, 14]

The principal concern about climate change is its incremental trend. By the year 2100 an increment of 1770 GtC in the total cumulative carbon emissions is predicted as well as 1.1–6.4 °C of temperature and 0.18–0.59 m rises of the sea level [8]. However, climate change impacts in different way each region of the planet. For instance, the Europe forecast shows for northern Europe largest warming in Winter, with increase on mean and extremes precipitation. Whereas, for the Mediterranean area, largest temperatures in Summer, the mean precipitation decrease and increase in the risk of droughts. Also, in southern Europe, the highest average temperatures will increase especially in Summer. In general, it is also likely to have an average extreme wind speed increase and a decrease on snow depth [3].

2.2.1 Critical Climate Parameters

Road infrastructure may be affected by several extreme events types such as, extreme precipitation, sea-level rise, maximum temperature rises or extreme winds.

Depending on the context these may be temporary or extended [2]; at a structural or service level; in a direct and indirect way [7]. Road networks performance and the critical climate parameters have been studied by several researchers [4, 11, 15–22]. Table 3 summarizes some of the most frequently climate parameters that cause an impact on road infrastructures.

3 Climate Change Adaptation

The implications of extreme events caused by climate change in the transportation system require actions. The repair or reconstruction posterior to an extreme weather event, sometimes hinder disaster relief efforts, affect the economic recovery and further drain the limited financial resources [7]. Not only mitigation actions are necessary, but preventive actions. Consequently, the strategies aim is to increase the resilience of road infrastructures against climate change but preserving their economic feasibility, and ideally considering measures contributing to GHG emissions reduction [22]. In fact, different researches have evidenced how road infrastructure investments in terms of climate change adaptation may even decrease cost estimation of the lifecycle, while also increase the infrastructure performance [5].

Adaptability should be considered as an effective asset management, and not only as an optional or isolated process, in which extra funding is needed. Nonetheless, it is always necessary to identify the tipping point at which the adaptation cost is unfeasible regarding the additional benefits [23]. Hence, adaptation measures are permanently linked to the economic aspect. On the other hand, adaptation itself is a dynamic and inclusive process that involves not only the interaction with many other policies but among road experts, stakeholders and administrators [11].

The adaptation development process can be made in phases, in which each is designed to guaranty the risk reduction to climate change. Therefore, the principal process step is the risk assessment and from this it is possible to identify, evaluate and select one or more options, keeping an acceptable risk level.

The framework also includes a cost-benefit step because not all options can be applied in terms of initial investment, as well as a document that provides the complete action plan, defining the implementation process and responsibilities (Table 4). The proposed methodology offers flexibility in terms of applicability; thus, it can be applied for any type of infrastructure system and to include future options. In the end, the framework provides a set of robust adaptation strategies for several risk scenarios. It is also important to mention that all steps are iterative and can be updated regarding different aspects such as hazard forecast, vulnerabilities and consequences estimation or the cost-benefits quantification.

Table 3 Critical risk factors of road infrastructures

Critical climate variables	Major risk to the road infrastructure	Affectation type
Extreme rainfall events (heavy showers and long periods of rain)	Flooding of roadways	S
	Road erosion, landslides and mudslides that destroys the embankments	M, S
	Erosion (scouring) and damage to bridge supports	M
	Overloading of drainage systems, causing erosion and flooding	M, S
	Reduced surface friction and subsidence of element	M
	Blocking or damage of transportation line	S
	Damage of pavement due to destruction and instability of vegetation along the path	M
	Traffic hindrance and safety	S
Seasonal and annual average rainfall	Impact on soil moisture levels, affecting the structural integrity of roads, bridges and tunnels	M
	Adverse impact of standing water on the road base	S, M
	Risk of floods from runoff, landslides, slope failures and damage to roads if changes occur in the precipitation pattern (e.g. changes from snow to rain in winter and spring thaws)	M, S
Sea level rise	Inundation of roads in coastal areas	S
	Erosion of the road base and bridge supports	M
	Bridge scour	M
	Reduced clearance under bridges	M, S
	Extra demands on the infrastructure when used as emergency/evacuation roads	S
Maximum temperature and number of consecutive hot days (heat waves)	Concerns regarding pavement integrity, e.g. softening, traffic-related rutting, embrittlement (cracking), migration of liquid asphalt, blow-ups	M, S
	Vehicle failure (tyres)	S
	Thermal expansion in bridge expansion joints and paved surfaces	M
	Fatigue of drivers	S

(continued)

Table 3 (continued)

Critical climate variables	Major risk to the road infrastructure	Affectation type
	Impact on landscaping	S
Forest fires	Reduced visibility	S
	Dangerous driving conditions	S
	Structural damage of infrastructure, especially pavements	M, S
	Growing vegetation on slopes is destroyed. It can lead to soil degradation and slope slide	M
Drought (consecutive dry days)	Susceptibility to wildfires that threaten the transportation infrastructure directly	S, M
	Susceptibility to mudslides in areas deforested by wildfires	S, M
	Consolidation of the substructure with (unequal) settlement as a consequence	M
	More generation of smog	S
	Unavailability of water for compaction work	S
Snowfall	Traffic hindrance and safety	S
	Snow avalanches resulting in road closure or striking vehicles	M, S
	Failures in transport control system	M
	Cracks close to contraction joints in the cement concrete pavement	M
	Ice and snow in culverts leading to reduced drainage capacity and water on the road structure or flooding	M, S
	Flooding from snow melt	S
Frost (number of icy days)	Traffic hindrance and safety	S
	Material damage of infrastructure	M
	Technical failure of vehicles	S
Thaw (number of days with temperature zero crossings)	Thawing of permafrost, causing subsidence of roads and bridge supports (cave-in)	M, S
	Frozen culverts may be blocked and cause structural damage	M
	Cracks close to contraction joints in the cement concrete pavement	M
	Decreased utility of unimproved roads that rely on frozen ground for passage	S

(continued)

Table 3 (continued)

Critical climate variables	Major risk to the road infrastructure	Affectation type
Extreme wind speed (worst gales)	Threat to stability of bridge decks	M
	Difficult driving conditions; exposed parts of roads (e.g. bridges) closed due to strong wind gusts	S
	Obstacles on the road owing to fallen trees and other objects	S
	Damage to signs, lighting fixtures and supports	M
Fog days	Traffic hindrance and safety	S
	More generation of smog	S

Adapted from [4, 11, 15–22]
Legend Impacts classification: *S* service-level impact (mobility); *M* material or structural impacts

Table 4 Adaptability methodology

Key steps		Definition
1	Risk analysis	Risk levels and scenarios prioritization regarding capacity and financial constraints
2	Identify options	Identify possible adaptation measures for the nonacceptable risks with their respective limits or constraints
3	Cost-benefits quantification	Making sure that the chosen strategies from step 2 can be implemented and that adaptation cost be viable regarding its benefits
4	Options analysis	Compare strategies across all future scenarios. Define the consequences of choosing 'adaptability' or 'not adaptability' measures, using robust decision-making to determine the regret of each one
5	Adaptation plan	Document adaptation options taking into account the information provided in the previous steps and classifying them by impact reduction

Adapted from [5, 14, 23]

3.1 Identifying Adaptation Options

Establishment of adaptability measures options is not an easy task. Several factors need to be taken into account. One of these factors is that the principal adaptation aim is the climate change risk reduction [5, 10, 14] and not all measures can fit within this objective. Another factor is the that adaptation viability depends on the cooperation between decision-makers and stakeholders, the time scale, climate scenario, location and topography, which results applicable for a very specific case [8, 14]. Finally, the availability of financial resources factor and technology application, because its notion is not much applied in the practical field of engineering [7]. That is why effectiveness measurement is necessary, to monitored over time for all cases, in order to feedback the adaptation plan and improve the learning process in future events [23].

3.1.1 Adaptability Strategies Classification

Adaptability measures can be classifiable into different types, sectors or categories. At different levels, as a component or link/node or at network, which suggest that the measures should not be focused in a specific kind of event but cover the level adaptation needs [2]. Another kind of classification was offered by Hallegatte [24], who defined the follow classification, with the objective to keep as low as possible the cost of a wrong forecast of the climate change effect.

- *No-regret strategies (**NR**).* Produce benefits even without the presence of climate change
- *Reversible strategies (**R**).* When it is cheap, it is sensible to add "security margins" to design criteria to future (expected or unexpected) changes, making the adaptation measure more robust.
- *Soft strategies (**S**).* Institutional or financial tools to cope with future changes directly made by planners.
- *Strategies that reduce decision-making time horizons (**RDMH**).* Reducing the lifetime of investments, therefore, is an option to reduce uncertainty and corresponding costs.

On the other hand, Tol et al. [25] mentioned that fulfil the main adaptation measure objective of reducing risk, is possible following five adaptation strategies.

- *Increasing robustness of infrastructural designs and long-term investments (**RO**).*
- *Increasing flexibility of vulnerable managed systems (**F**).* i.e. contemplate midterm adjustments and/or diminishing economic lifetimes.
- *Enhancing adaptability of vulnerable natural systems (**EA**).* i.e. reducing other (non-climatic) stresses and/or removing barriers to migration
- *Reversing trends that increase vulnerability (**V**).* i.e. introducing set-backs for development in vulnerable areas such as coastal floodplains and landwards of eroding cliffs
- *Improving societal awareness and preparedness (**P**).* i.e. informing the public of the risks and possible consequences.

In general, several action options have been proposed for the most critical risk variables, which are summarized in Table 5, being organized by two mentioned classes.

Finally, the importance of taking into account the limits of each of the adaptation measures is highlighted. These constraints need to be carefully studied and handled in determining feasible options to prepare for climate change.

Table 5 Adaptation measures for critical risk factors

Critical climate variables	Adaptability option	Hallegate class	Tol et al. class
Extreme rainfall events (heavy showers and long periods of rain)/seasonal and annual average rainfall/sea level rise	Provision of timely driver information to 'at risk' routes	R	P
	Raise the height of embankment in flood plains	NR	F
	Additional/fortified adequate slope protection works	NR/R	F
	Increase capacity and size of culverts and cross drainage	NR	RO
	Provide adequate river protection works	R	EA
	Consider increasing waterway and protection works to safeguard bridges	S/R	F/EA
	Increase clearance above high flood level for bridges	NR	F
	Alter design-storm criteria, estimating design flood and stormwater taking account of predicted climate	S	RO
	Ensure effective drainage of surface water from the pavement	R	F/EA
	More frequent maintenance and replacement	S/RDMH	F
	Increase pumping capacity for roads and tunnels	NR	RO/F
	Fortify bridge piers and abutments	NR	RO/F
	Corrosion protection	R	F/EA
	Increase capacity of side drains	R	F
	Add green infrastructure/storm retention basins	NR/R	EA
	Relocation of coastal road to higher place	NR	F
	Elevate/protect tunnel openings and low-lying areas	NR	F
	Provide additional protection to coastal roads, e.g. seawalls dikes	R	F/EA

(continued)

Table 5 (continued)

Critical climate variables	Adaptability option	Hallegate class	Tol et al. class
	Design and construct new bridges or replace old ones	RDMH	RO
Maximum temperature and number of consecutive hot days (heat waves)/Drought (consecutive dry days)	Use stiffer bitumen in pavement to safeguard from high temperature	NR	RO
	More frequent maintenance and replacement	S/RDMH	F
	Alter asphalt composition (heat-resistant paving material)	NR/R	RO/F
	Switch from asphalt to concrete	RDMH	RO/F
	Replace expansion joints	R	F/EA
	Increased albedo	R	EA
	Increased shading	R	EA
	Additional/fortified slope retention structures	NR	RO/F
	Control of soil moisture	S/R	EA
	Vegetation management	S	EA
Forest fires	Place sufficient warning and information signs	R	P
	Alter asphalt composition	NR	RO/F
	More frequent maintenance and replacement	S/RDMH	F
	Provision of timely driver information to 'at risk' routes	R	P
	Vegetation management	S	EA
Snowfall/frost (number of icy days)/Thaw (number of days with temperature zero crossings)	Use thick and strong pavement to safeguard against snow and frequent icing-thawing	NR	RO/F
	More frequent maintenance and replacement	S/RDMH	F
	Alter asphalt composition	NR	RO/F
	Provision of timely driver information to 'at risk' routes	R	P
	Increase capacity and size of culverts and cross drainage	NR	RO/F
Extreme wind speed (worst gales)/fog days	Provision of timely driver information to 'at risk' routes	R	P
	Place sufficient warning and information signs	R	P

(continued)

Table 5 (continued)

Critical climate variables	Adaptability option	Hallegate class	Tol et al. class
	Fortify bridge infrastructure	NR	F

Adapted from [8, 12, 14, 15, 26]

4 Linked Risk Assessment and Adaptability Framework

Based on the topics discussed in the previous sections, the following framework is proposed with the intention of incorporating risk assessment against climate change and the respective adaptation measures, having an account of the current practices and academic researches.

Fig. 1 contains the two-sided framework. On the one hand, risk assessment (I), that considers the identification, analysis and evaluation of hazards, vulnerabilities and losses. Highlighting two fundamental aspects. First, the context and objectives definition, which allows the identification and prioritization of the most significant risks for the whole framework. Second, the omission of the risk mitigation step since the framework objective is not to mitigate the damages but to take actions before damage happens, through adaptability measures.

On the other hand, adaptation strategies (II) instead of risk mitigation. This part of the framework covers everything from the adaptation measures identification to their evaluation regarding risk reduction and the costs involved. It is important to highlight the cost analysis, in order to recognize the tipping point at which the cost of additional adaptation becomes disproportionate comparing its benefits. This section also

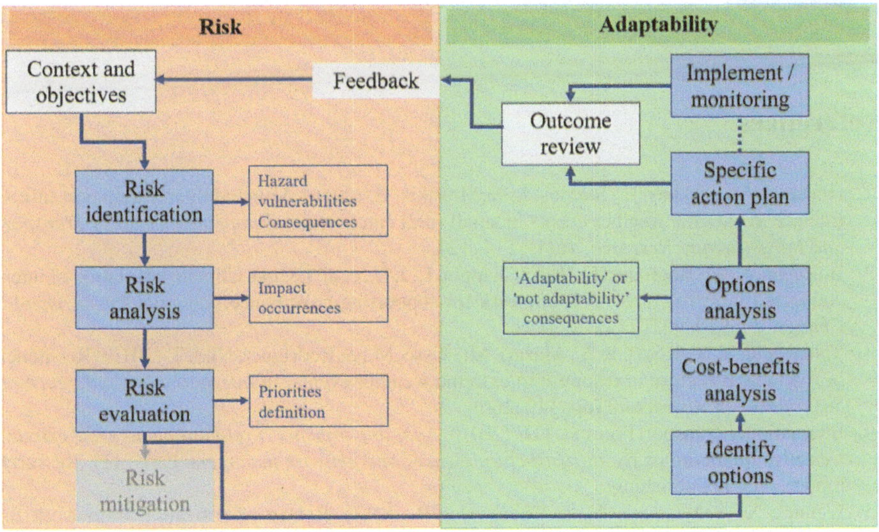

Fig. 1 Adaptability climate change framework for road infrastructures

includes the implementation plan and its monitoring. In which the flexible and iterative nature of the framework is highly important to ensure that applied decisions can be reviewed and updated as predicted infrastructure risks or socioeconomic consequences change. Although this methodology is proposed for the road infrastructure, it offers the possibility to be applied in other infrastructure components.

5 Conclusions

This work presents a framework proposal that allows to incorporate a comprehensive assessment of risks and adaptation options to face the impacts of climate change on road infrastructures. The methodology is circular and iterative, permitting the risk prioritization to achieve the objectives set at the beginning of the process. It is also flexible in terms of socioeconomic changes; review process, to determine the adaptation measures success and allows its application for other infrastructure components. This framework was developed based on academic review of best practices.

Acknowledgements This work was partly financed by FCT / MCTES through national funds (PIDDAC) under the R&D Unit Institute for Sustainability and Innovation in Engineering Structures (ISISE), under reference UIDB / 04029/2020.

This project has received funding from the European Union's Horizon 2020 research and innovation programmed under grant agreement No 769255. This document reflects only the views of the author(s). Neither the Innovation and Networks Executive Agency (INEA) nor the European Commission is in any way responsible for any use that may be made of the information it contains.

References

1. Diakakis, M., Lekkas, E., Stamos, I., & Mitsakis, E. (2016). Vulnerability of transport infrastructure to extreme weather events in small rural catchments. *European Journal of Transport and Infrastructure Research, 16*(1), 114–127.
2. Bollinger, L. A., Bogmans, C. W. J., Chappin, E. J. L., et al. (2014). Climate adaptation of interconnected infrastructures: A framework for supporting governance. *Regional Environmental Change, 14,* 919–931.
3. Valenzuela, Y. B., Rosas, R. S., Mazari, M., Risse, M., & Rodriguez-Nikl, T. (2017). Resilience of road infrastructure in response to extreme weather events. In *International Conference on Sustainable Infrastructure* (pp. 349–360)
4. International Transport Forum—ITF. (2016). *Adapting transport to climate change and extreme weather: Implications for infrastructure owners and network managers.* Paris: ITF Research reports OECD Publishing.
5. Espinet, X., Schweikert, A., & Chinowsky, P. (2017). Robust prioritization framework for transport infrastructure adaptation investments under uncertainty of climate change. *ASCE-ASME Journal, 3*(1).
6. Eidsvig, U. M. K., Kristensen, K., & Vidar, B. (2017). Assessing the risk posed by natural hazards to infrastructures. *Natural Hazards and Earth System Sciences, 17*(3), 481–504.

7. Rattanachot, W., Wang, Y., Chong, D., & Suwansawas, S. (2015). Adaptation strategies of transport infrastructures to global climate change. *Transport Policy, 41,* 159–166.
8. Intergovernmental Panel on Climate Change—IPCC. (2014). *Climate Change 2014: Synthesis Report. Contribution of Working Groups I, II and III, Intergovernmental Panel on Climate Change* [Core Writing Team, R. K. Pachauri & L. A. Meyer (Eds.)]. Geneva, Switzerland, Assessment Report, no 4 (p. 151)
9. Danielson, L., Lasfargues, B., Morgado, N.C., & Perry, E. (2018). Climate-resilient infrastructure. OECD Environment, Working Papers, No. 121, OECD Publishing, Paris
10. Mechler, R., & Nabiul, K. M. (2013). Cost-benefit analysis of disaster risk management and climate adaptation. In D. Guha-Sapir & I. Santos (Eds.) *The economic impacts of natural disasters.* New York, EEUU.
11. Deltares, B. T., Yves, E., Jean-Jacques, E. F., et al. (2010). *Risk management for roads in a changing climate.* ERA-NET ROAD, Technical Report. RIMAROCC
12. Picketts, I. M., Andrey, J., Matthews, L., Dery, S. J., & Tighe, S. (2016). Climate change adaptation strategies for transportation infrastructure in Prince George, Canada. *Regional Environmental Change, 16,* 1109–1120.
13. Van Westen, C. J. (2014). *Caribbean handbook on risk information management.* World Bank, ACP—EU Natural disaster risk reduction program. Obtain from: https://www.charim.net/
14. Mitsakis, E., Papanikolaou, A., Ayfadopoulou, G., et al. (2014). An integrated framework for linking climate change impacts to emergency adaptation strategies for transport networks. *European Transport Research Review, 6,* 103–111.
15. Markolf, S. A., Hoehne, C., Fraser, A., Chester, M. V., & Underwood, B. S. (2019). Transportation resilience to climate change and extreme weather events—Beyond risk and robustness. *Transport Policy, 74,* 174–186.
16. Panteli, M., & Mancarella, P. (2017). Modeling and evaluating the resilience of critical electrical power infrastructure to extreme weather events. *IEEE Systems Journal, 11*(3), 1733–1742.
17. Koetse, M. J., & Rietveld, P. (2009). The impact of climate change and weather on transport: An overview of empirical findings. *Transportation Research Part D: Transport and Environment, 14*(3), 205–221.
18. Federal Highway Administration—FHWA. (2012). *Climate change & extreme weather vulnerability assessment framework.* U.S. Department of Transportation, Washington, D.C., Technical Guide
19. Wang, T., Qu, Z., Yang, Z., Nichol, T., Dimitriu, D., Clarke, G., & Bowden, D. (2019). How can the UK road system be adapted to the impacts posed by climate change? By creating a climate adaptation framework. *Transportation Research Part D: Transport and Environment, 77,* 403–424.
20. Schweikert, A., Chinowsky, P., Kwiatkowski, K., & Espinet, X. (2014). The infrastructure planning support system: Analyzing the impact of climate change on road infrastructure and development. *Transport Policy, 35,* 146–153.
21. Auerbach, M., Herrmann, C., & Krieger, B. (2011). Adaptation of the road infrastructure to climate change. In: *Second Status Conference Impacts of Climate Change on Waterways and Navigation, Federal Ministry of Transport* (pp 48–53), Berlin.
22. Regmi, M. B., & Hanaoka, S. (2011). A survey on impacts of climate change on road transport infrastructure and adaptation strategies in Asia. *Environmental Economics and Policy Studies, 13,* 21–41.
23. Quinn, A. D., Ferranti, E. J. S., Hodgkinson, S. P., Jack, A. C. R., Beckford, J., & Dora, J. M. (2018). Adaptation becoming business as usual: A framework for climate-change-ready transport infrastructure. *Infrastructures, 3*(2), 10.
24. Hallegatte, S. (2009). Strategies to adapt to an uncertain climate change. *Global Environmental Change, 19*(2), 240–247.
25. Tol, R. S. J., Klein, R. J. T., & Nicholls, R. J. (2008). Towards successful adaptation to sea-level rise along Europe's Coasts. *Journal of Coastal Research, 24*(2), 432–442
26. Rowan, E., Evans, C., Riley-Gilbert, M., et al. (2013). Assessing the sensitivity of transportation assets to extreme weather events and climate change. *Transportation Research Record, 2326*(1), 16–23.

Risk-Driven Decision Making Within the Observational Method: Case Study Based on the New International Airport of Mexico City

Antonios Mavritsakis, Martin de Kant, and Joost van der Schrier

Abstract In geotechnical design, the Observational Method poses as an attractive solution for reducing construction costs without compromising safety, especially when dealing with a high level of uncertainty. Additionally, the benefits of the Observational Method can be elevated when it is applied in a probabilistic concept. Designing the soil improvement of the runways for the New International Airport of Mexico City (Nuevo Aeropuerto Internacional de la Ciudad de México—NAICM) holds significant risk due to the extremely soft soil, the soil-related uncertainties and the strict pavement operation requirements. Instead of opting for an over-conservative and costly design, the Observational Method was adopted in order to steer the soil improvement works according to monitored soil behaviour. The analysis presented in this paper, which is based on an example inspired by the NAICM, employs a probabilistic framework, composed of several probabilistic tools, in order to estimate the reliability of the design. Specifically, incoming monitoring (soil response) data is utilized in several reliability updating steps, giving insight into the probability of the design meeting the operational requirements. Moreover, assessing the reliability of a design allows for the quantification of risk, which can pose as a strong motivator during the decision-making process. Design decisions, such as application of mitigation measures, can be made according to the direction of risk minimization. Finally, the entire procedure of the Observational Method and the steering of the design throughout the soil improvement phase are illustrated in a decision tree. This paper draws conclusions on the benefits of incorporating probabilistic concepts in large scale projects with strong uncertainties, as well as utilizing risk as motivation for decision making, which eventually proves to be valuable for project management.

Keywords Reliability updating · Risk estimation · Decision making · Observational method · Airport pavement settlement

A. Mavritsakis (✉) · M. de Kant · J. van der Schrier
Royal Haskoning DHV, Amersfoort, The Netherlands

© The Author(s), under exclusive license to Springer Nature Switzerland AG 2021

J. C. Matos et al. (eds.), *18th International Probabilistic Workshop*, Lecture Notes in Civil Engineering 153, https://doi.org/10.1007/978-3-030-73616-3_53

1 Introduction

1.1 Scope of the Analysis

The New International Airport of Mexico City (Nuevo Aeropuerto Internacional de la Ciudad de México—NAICM), was designed to be one of the largest commercial airports of the world. This case study is inspired by the complex design situation encountered in the NAICM and aims to demonstrate the effectiveness of steering the soil improvement design during construction by applying the Observational Method and motivating decision-making with risk minimization.

1.2 Project Setting

The NAICM is located in the area of the former lake Texcoco. The 50–65 m thick upper lake deposits are extremely compressible and consolidate slowly. This creates a challenging environment for the pavement foundation design. Moreover, such layers can have a significant variance in their parameters, even though the intensity of their heterogeneity is low. Subsequently, the consolidation parameters of the soft soil are the greatest factor of uncertainty and the sole focus of this case study.

The criteria that define the tolerable irregularities of the runways and taxiways are defined in ICAO standards for airport pavement design [1]. In this case study, only the longitudinal slope requirement is considered, because it is sufficient to showcase the application of the method. According to ICAO, the longitudinal slope change of the pavement should never exceed the value of 0.017% during the operational lifetime of the pavement. The requirement is checked for a pavement lifetime of 8 years.

The main purpose of the soil improvement design is to induce sufficient preconsolidation to the soil. The following soil improvement design was orchestrated:

- Basis layer of a lightweight volcanic granular material.
- Prefabricated Vertical Drains (PVD) were advised to be installed in order to accelerate the consolidation process.
- Surcharge with a thickness of 2.0 m for a duration of 12 months.
- Extensive monitoring that would accommodate the implementation of the Observational Method.

The soil improvement of the NAICM plays a major role in the design, construction and operation phases of the project. In the design phase, the soil improvement plan is selected, as indicated by the predictions of the settlement model, in order to meet the requirements in the operation phase. Throughout the construction phase, the plan is being implemented and modifications are made where necessary, based on monitored settlement behaviour. At the start of the operation phase, the pavement is handed over for use. Even though settlement continues to develop during this phase, the longitudinal pavement slopes must still comply to the requirement. Essentially, the challenge of the soil improvement is to achieve a level of soil consolidation during the construction phase that leads to a useable pavement in the operation phase.

1.3 Strategy

In order to tackle the inherent uncertainties of the soil improvement design, the Observational Method (OM) has been selected as a flexible method that allows for optimization of cost, safety and meeting of pavement operation requirements [2]. According to the OM, monitoring data provides insight on the development of the soil improvement and it is used to update the settlement model and the prediction of soil behaviour during pavement operation. Eventually, this allows for design changes during the construction phase.

The high level of uncertainty residing in predicting the effectiveness of the NAICM soil improvement works has led to the application of the OM within a probabilistic framework. This framework focuses on two main operations:

1. Calculating the reliability of a soil improvement design in meeting the longitudinal slope requirements during the operational lifetime of the pavement.
2. Updating the knowledge of soil parameters according to the soil response monitored during the construction phase.

The second operation updates the soil parameter distributions. When they are applied in a reliability calculation step, the design's reliability can be updated. This concept is a Bayesian updating application and ties closely with the essence of the Observational Method.

The last step in implementing the OM is a system able to steer the design throughout the construction phase according to the findings of the probabilistic analyses. Since uncertainty is accounted for, cost is elevated to a stochastic level. Thus, risk is a more suitable term to describe the potential cost due to failure and can be used to motivate decision making. Ultimately, steering the OM through risk, reinvents the goal of decision making as risk minimization, a term that compiles the concepts of design response, cost and uncertainty.

As described in Eq. (1), the risk score is defined as the probability of failure occurring multiplied by the cost of failure [3] and is measured in monetary units (MU).

$$Risk\ score = Investement\ Cost + Probability\ of\ Failure \times Cost\ of\ Failure \tag{1}$$

1.4 Description of the Probabilistic Framework

The implementation of the OM on a stochastic level is realized through the utilization of a probabilistic framework. The framework is mainly composed of the following four tools.

1.4.1 Monte Carlo Simulation

The Monte Carlo (MC) simulation is used to assess the uncertainty induced by the soil parameters in the residual settlement. The MC calculates the settlement model for a number of realizations. Collecting the output of the realizations leads to forming a sample of residual settlement. Ultimately, this tool is used to assess the reliability of a design in meeting the longitudinal slope requirement over the pavement lifetime.

1.4.2 Random Field Generator

Soil is a heterogenous material and its spatial variance can have a significant impact on the reliability of the design. Considering that meeting the longitudinal slope requirement is based on calculating the residual settlement difference between two locations. The random field generator is able to create variable samples (which are used as input for the MC simulation) that are spatially correlated, according to the assumed heterogeneity pattern [4].This pattern is defined by the scale of fluctuation (θ), which controls the extent of correlation between two locations as a function of the in-between distance [5]. Lastly, the random field generator uses the Markov auto-correlation function, Eq. (2), to describe the correlation in respect to the scale of fluctuation and the location inter-distance (dx).

$$\rho = exp(-2 * dx/\theta) \tag{2}$$

1.4.3 Markov Chain Monte Carlo

The Markov Chain Monte Carlo (MCMC) method has been utilized for the inverse analysis of monitoring data. By employing the Metropolis-Hastings algorithm, the MCMC method applies Bayes' theorem in order to retrieve variable samples directly from the posterior distribution [6]. Subsequently, the knowledge (PDF) of soil parameters is updated, so that their ability to reproduce the monitoring data is enhanced, while also preserving initial knowledge on the parameters. However, the utilization of the MCMC algorithm requires a large number of settlement model runs, which heavily affects the computational effort required [7].

1.4.4 Surrogate Modeling

In order to substitute the finite difference settlement model execution, the concept of surrogate modeling is adopted. As an approximation of the true model, the surrogate model is significantly faster to compute. It is trained according to specific points of the variable domain where the true model is calculated. The surrogate model makes predictions for points other than the training ones by using a Kriging interpolation scheme [8]. It provides not only the mean approximation of the settlement model, but also the standard deviation of the prediction error.

2 Analysis Implementation

2.1 Analysis Set-Up

The main analysis is performed on an example case inspired by the project setting of the NAICM. Its main goal is to highlight how the OM could be augmented by including the concept of reliability updating.

The analysis focuses on examining the differential residual settlement between two calculation points, located in the longitudinal direction of the runway, over the pavement lifetime. The calculation points are assumed to be 100 m apart, meaning that their differential residual settlement in 8 years must be lower than the margin of 17 cm in order to meet the evenness requirement. In addition, the two points have the exact same stratigraphy. A finite difference settlement model is used, which is calculated by the software D-Settlement [9].

As already mentioned in Sect. 1.2, only the soil parameters of the top soft layer are considered as stochastic variables. The stochastic variables of the analysis are the following: compression ratio (CR), permeability (k_v), over-consolidation ratio (OCR), with their respective distributions according to the ground investigation being presented in Table 1. For this specific case, the construction phase settlement was mostly affected by the compression ratio and the permeability of the soil. Therefore, during the inverse analysis step, where soil parameters are adjusted to construction settlement measurements, only the compression ratio and permeability are regarded as variables. Consequently, these parameters affect soil behaviour during operation, meaning that performing the inverse analysis on these parameters holds an impact on the design's updated reliability.

In order to tune the random field generator tool and sample soil parameter sets correlated for the two calculation points, a scale of fluctuation value must be determined. Considering that the top soft soil layers where formed by lake deposition, their heterogeneity in the horizontal direction is not expected to be intense. This means that a relatively high value for the scale of fluctuation can be used. Due to the lack of relevant data and literature, a realistic value for the scale of fluctuation

Table 1 Properties of soil parameter distributions (when the distribution is lognormal, the properties refer to the underlying normal distribution)

Parameter	Mean	Standard deviation	Distribution type
OCR (−)	1.6	0.15	Normal
Permeability (ln(m/s))	−19.8	4.00	Lognormal
CR (−)	0.7	0.12	Normal

of 300 m is assumed, which leads to a correlation coefficient of 0.51. For this case study, soil heterogeneity in the vertical direction was not considered.

In order to tune the random field generator tool and sample soil parameter sets correlated for the two calculation points, a scale of fluctuation value must be determined. Considering that the top soft soil layers where formed by lake deposition, their heterogeneity in the horizontal direction is not expected to be intense. This means that a relatively high value for the scale of fluctuation can be used. Due to the lack of relevant data and literature, a realistic value for the scale of fluctuation of 300 m is assumed, which leads to a correlation coefficient of 0.51. For this case study, soil heterogeneity in the vertical direction was not considered.

Reliability updating re-estimates soil parameters and the design's reliability according to monitoring measurements. In this analysis, synthetic monitoring data is employed. The data consists of timelines of surface settlement up to two moments in the construction phase, at 3 and 9 months after the placement of surcharge. The synthetic data was created using a selection of soil parameters, which act as the target of the inverse analysis step. In order to account for measurement uncertainty, it is adopted for the measurement error to follow a normal distribution with a zero mean and standard deviation equal to 10% of the measured settlement.

2.2 *Results of the Analysis*

The results of the analysis are presented separately for the design and construction phases. In the design phase, the main goal is reliability estimation, while during construction, reliability updating analyses are performed.

2.2.1 Design Phase

The first decision that needs to be made in this case study is in the design phase. What appears as a more attractive design solution: adopting the OM or applying a conservative design?

In the OM approach, there is extra investment cost due to the extensive monitoring required at site. A conservative design would include excavating the surcharge and basis layer after the end of construction and then applying a lightweight expanded

Table 2 Cost of each design measure

Design measure	Cost (MU)
Monitoring	10
EPS	40
3rd meter	5
Slope repair	120

polystyrene (EPS) fill to act as the pavement foundation. In this way, the operational load is dramatically reduced and residual settlement shrinks. Although applying EPS is an effective solution, Table 2 shows that it also leads to a significantly higher investment cost. Moreover, the same table presents the cost of the mitigation measure, which consists of applying a 3rd meter of surcharge, and the cost of failure, which is the price of repairing a slope, when the differential residual settlement exceeds the requirement.

In the end, the design selection must be made after quantifying the uncertainties and estimating the risk associated to each approach. The first step in estimating the risk is calculating the reliability of each design approach in meeting the pavement slope requirement. Then, the probability of failure is connected to the failure and investment costs and the risk can be estimated.

The MC simulation tool of the probabilistic framework is used to estimate the reliability of each design approach. The convergence of the simulation is already achieved in 250 realizations, at which point the 95% confidence interval of the running mean of the sample, as estimated according to [10], has a spread lower than the adopted higher-boundary value of 2 cm. Moreover, performing the MC simulation with 250 realizations means that the analysis is adequately accurate, while remaining practical.

Figure 1 shows the results of the MC simulation for the OM design approach. Slight differences between the histograms of the residual settlements of Points A and B can be explained by the number of MC realizations used in the simulations; with a much greater number of realizations the two histograms would have been identical, but the calculation time would be impractical. The residual settlement scatter plot (c) shows a correlation coefficient of 0.27 between the two calculation points. Evidently, the spatial correlation of residual settlement is weaker than the correlation of the underlying parameters, which has a correlation coefficient of 0.51. The reliability of the approach, as taken from histogram (d), is simply the number of occurrences that are lower than the requirement over the total number of occurrences.

The MC simulation is also performed for the conservative design approach. The results of the two approaches are compared in Table 3. Evidently, in this case, the higher reliability of the conservative design approach is not able to outbalance the significantly greater investment cost and eventually leads to the highest risk score, as estimated according to Eq. (1). Thus, the OM is selected as the least risky approach and the analysis continues to the construction phase.

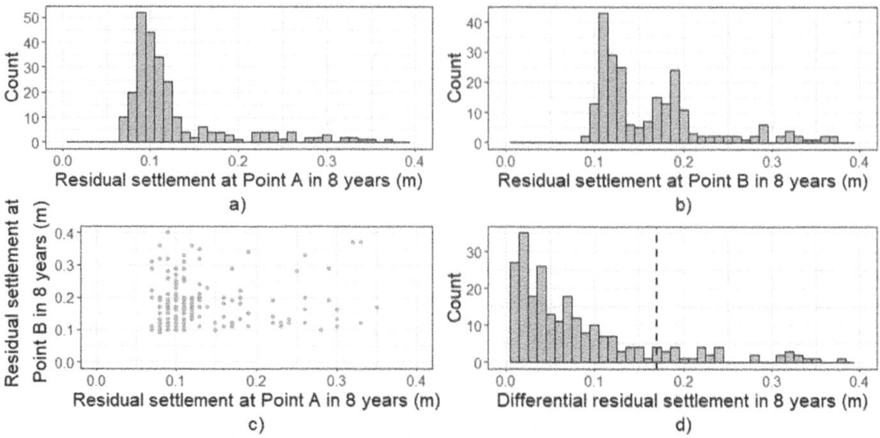

Fig. 1 Results of MC simulation for the OM design approach: **a** histogram of residual settlement in point A, **b** histogram of residual settlement in Point B, **c** scatter plot of residual settlement at the two points, **d** histogram of differential residual settlement between the two points with indication of the design requirement

Table 3 Reliability and risk calculation results for the conservative design and the OM approaches

Design approach	Reliability (%)	Investment cost so far (MU)	Risk score (MU)
Conservative	98	40	42.4
OM	85	10	28.0

2.2.2 Construction Phase

Three months after the start of construction, the first pack of surface settlement monitoring data becomes available and a reliability analysis step is performed. The distributions of the variables are updated by the inverse analysis (Fig. 2). It is evident that each of the distributions moves closer to the target value while also reducing their width. This means that knowledge on the soil parameters becomes more accurate and less uncertain when updated with the monitoring information.

The next step is to update the reliability and risk of the design. This is achieved by re-executing an MC simulation with the new soil parameter distributions. The updated risk score is shown in Table 4. Although the knowledge about the soil parameters after updating with the monitoring data is more well-defined, a trait that reduces uncertainty, the parameter distributions are shifted towards values that negatively affect the design, leading to a lower reliability and higher risk. This means that even though the inverse analysis gives a less uncertain design, the reliability will decrease if the soil parameters are worse than what was originally expected.

After updating the current risk according to the monitored settlement, a new decision must be made. Either no action is taken and the current risk is accepted until

Table 4 Reliability and risk calculation results after reliability updating for the first monitoring data pack

OM step	Reliability (%)	Investment cost so far (MU)	Risk score (MU)
Monitoring 1	80	10	34.0
Extra surcharge	96	15	19.8
No action	80	10	34.0

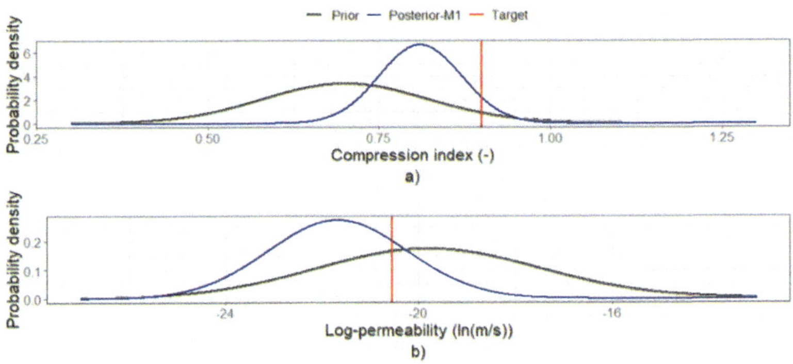

Fig. 2 Prior and posterior distributions of: **a** the compression index (normal distribution) and **b** permeability (originally lognormal—the underlying normal distribution is presented). The posterior distributions are generated by the inverse analysis of the first monitoring data pack

the new monitoring data becomes available, when the risk score will be updated again, or an extra meter of surcharge could be applied at this moment to increase the reliability of the design, although increasing the investment cost. This design change is examined with an MC simulation (using the updated parameter distributions) and the associated risk score is calculated to be 19.8 MU (Table 4). This value is lower than the current risk score, so an extra meter of surcharge is advised.

2.3 *Result Interpretation*

A decision tree is an effective medium for visualizing the course of the OM and the decision-making process. The tree is separated into the design, construction and operation phases and is composed of nodes representing events and decisions, which are described in Table 5. The decision tree holds a record of risk development throughout the construction phase and justifies the decision made, by comparing to the risk score of their alternatives. The tree can also be used to highlight the impact of using the synthetic data in reliability updating steps on the progress of the OM. The complete decision tree is presented in Fig. 3 and the details of the calculations

Table 5 Description of nodes of the decision tree along with reliability and risk score calculation results

Phase	Decision tree node	Description	Reliability (%)	Investment cost so far (MU)	Risk score (MU)
Design	CD	Conservative design	98	40.0	42.4
Design	OM	Observational method	85	10.0	28.0
Construction	M1	Monitoring 1 becomes available	80	10.0	34.0
Construction	ES-M1	Extra surcharge after monitoring 1	96	15.0	19.8
Construction	M2-ES-M1	Monitoring 2 becomes available—extra surcharge was applied after monitoring 1	98	15.0	17.4
Operation	NA-M2-ES-M1	No action is taken after monitoring 2—extra surcharge was applied after monitoring 1	98	15.0	17.4
Construction	NA-M1	No action is taken after monitoring 1	80	10.0	34.0
Construction	M2-NA-M1	Monitroing 2 becomes available—no action was taken after monitoring 1	88	10.0	24.4
Operation	NA-M2-NA-M1	No action is taken after monitoring 2—no action is taken after monitoring 1	88	10.0	24.4
Operation	ES-M2-NA-M1	Extra surcharge is applied after monitoring 2—no action is taken after monitoring 1	99	15.0	16.2

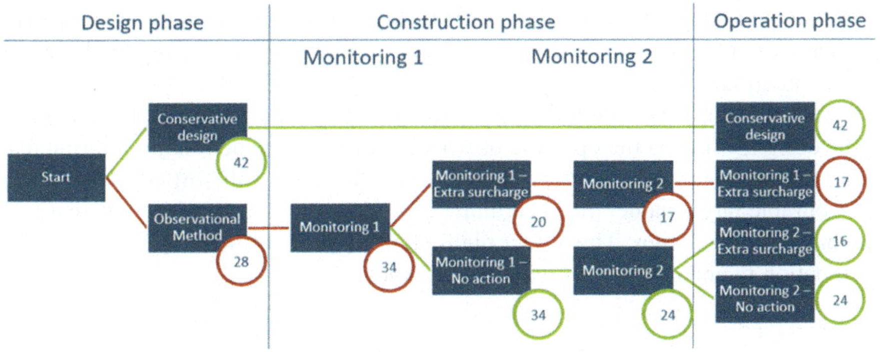

Fig. 3 The complete decision tree with the risk score of each node. The branch selected by the analysis is indicated in red

are shown in Table 5. Eventually, applying risk-driven decision making within the OM leads to accepting a risk of 17.0 MU in the operation phase.

Moreover, Fig. 4 shows how risk develops throughout each branch of the decision tree. As expected, the conservative design approach stays constant throughout the entire analysis and has the greatest risk score because of the high investment cost. All branches that originate from the application of the OM lead to accepting a significantly lower risk score than the conservative design approach. In addition, early application of the extra surcharge increases its effectiveness in lowering risk but decreases the risk reduction from reliability updating with the second monitoring data pack. This point raises the question of applying the mitigative measures quickly or waiting for more information to become available and acting later. Ultimately, branches with an extra surcharge have a risk score lower by 7.0 MU, which

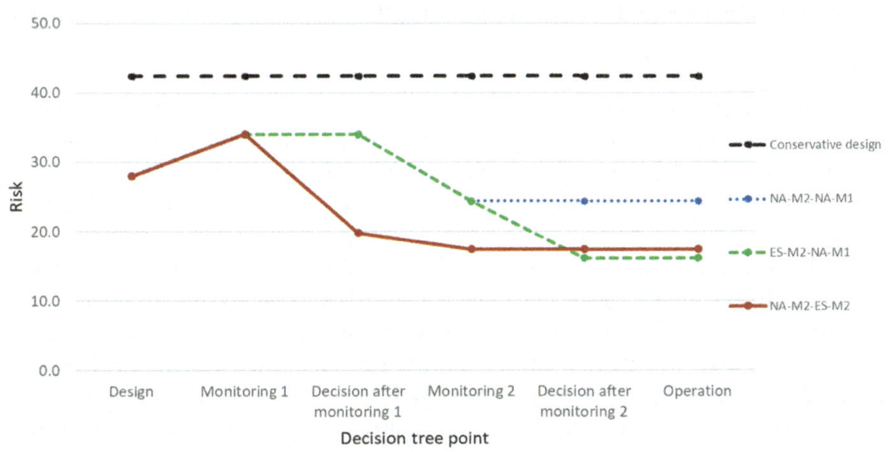

Fig. 4 Development of the risk score throughout the implementation of the OM for every branch of the decision tree

already includes their 5.0 MU higher investment cost. Lastly, application of the extra surcharge after the second reliability updating step leads to accepting the least risk in the operation phase.

Updating reliability with the second monitoring data pack reduces the risk score of the design, which is the opposite of what happened with the first monitoring data pack. Although the parameter posterior distributions are still shifted closer to the unfavourable target value, the uncertainty decreases enough in order to increase the reliability of the design. This effect highlights the value of the length of incoming information in enhancing the predictive accuracy.

3 Conclusions and Recommendations

This case study inspired by the NAICM soil improvement design demonstrated how risk-driven decision making can be applied within the OM by utilizing a probabilistic framework. A reliability analysis scheme was used to calculate the reliability of a design, which was then connected to the failure and investment cost, in order to estimate the risk associated to the design. When monitoring data was available, a reliability analysis step was used to update the current risk. Eventually, the application of the OM steered the design into accepting a risk score that was significantly lower than the one provided by a conservative design.

Risk reduction appears as effective motivation for decision making. Traditional design principles aim to the minimization of the failure probability, which is usually expressed only on a qualitative level. The probabilistic framework can incorporate not only the consequences of failure, but also the quantification of the failure probability. This indicates that in decision making, the cost of the mitigative measures does not surpass the probable cost of failure. Ultimately, the accepted risk at the end of construction is actually estimated, in contrast to the end product achieved by standard approaches.

The OM poses as a flexible and effective tool in guiding the design to lower risk scores. It promotes the use of probabilistic concepts, proving reliability updating to be the cornerstone of this analysis. The OM can employ a set of pre-defined mitigation measures to provide tailor-made solutions that can adapt through the course of construction. Eventually, it can allow for a cheaper and less uncertain design.

Finally, some features that could be implemented in this analysis for further development are listed. Firstly, the design branch selected by the analysis does not lead into accepting the lowest risk score observed in the decision tree. A pre-posterior analysis would be able to assign the probability of reducing risk to each decision, even during the design phase, and predict the path that leads to risk minimization. Secondly, the inverse analysis in the presented example is currently based only on surface settlement measurements. However, this monitoring data generally will not be holding sufficient information to accurately calibrate the soil model and data of more instruments should be incorporated in the analysis. Besides, this step appears to be critical for applying the method on real data, which would be an even more complex case where different measurement types can provide a higher quantity and quality of information. Lastly, the scale of fluctuation, the parameter that controls the soil parameter heterogeneity pattern, heavily affects the differential residual settlement calculation. Usually, estimation of the scale of fluctuation is based on CPT measurements, an approach that sometimes might prove ineffective. Further schemes of identifying the scale of fluctuation from monitored response should be explored.

References

1. ICAO. (2013). *Annex 14 to the Convention on International Civil Aviation*, Vol. I—Aerodrome.
2. Peck, R. B. (1969). Advantages and limitations of the observational method in applied soil mechanics. *Geotechnique*, 171–187.
3. M. Staveren, Uncertainty and ground conditions., Butterworth-Heinemann, 2006.
4. Van Den Eijnden, A. P., & Hicks, M. A. (2017). Efficient subset simulation for evaluating the modes of improbable slope failure. *Computers and Geotechnics*, 267–280.
5. Vanmarcke, E. (2010). *Random fields: Analysis and synthesis*. World Scientific.
6. Soubra, A. H., & Bastidas-Arteaga, E. (2014). Advanced reliability analysis methods. In *ALERT Doctoral School 2014*.
7. Mavritsakis, A. (2017). *Evaluation of Inverse Analysis Methods with Numerical Simulation of Slope Excavation*.
8. Forrester, A., Sobester, A., & Keane, A. (2008). *Engineering design via surrogate modelling, a practical guide*. Wiley, New York.
9. Deltares, *D-Settlement version 16.1 [computer software]*. Available at: https://www.deltares.nl/en/software/d-settlement-2/, Delft.
10. Ballio, F., & Guadagnini, A. (2004). Convergence assessment of numerical Monte Carlo simulations in groundwater hydrology. *Water Resources Research, 40(4)*.

Rockburst Risk Assessment Based on Soft Computing Algorithms

Joaquim Tinoco, Luis Ribeiro e Sousa, Tiago Miranda, and Rita Leal e Sousa

Abstract A key aspect that affect many deep underground mines over the world is the rockburst phenomenon, which can have a strong impact in terms of costs and lives. Accordingly, it is important their understanding in order to support decision makers when such events occur. One way to obtain a deeper and better understanding of the mechanisms of rockburst is through laboratory experiments. Hence, a database of rockburst laboratory tests was compiled, which was then used to develop predictive models for rockburst maximum stress and rockburst risk indexes through the application of soft computing techniques. The next step is to explore data gathered from in situ cases of rockburst. This study focusses on the analysis of such in situ information in order to build influence diagrams, enumerate the factors that interact in the occurrence of rockburst, and understand the relationships between them. In addition, the in situ rockburst data were also analyzed using different soft computing algorithms, namely artificial neural networks (ANNs). The aim was to predict the type of rockburst, that is, the rockburst level, based on geologic and construction characteristics of the mine or tunnel. One of the main observations taken from the study is that a considerable percentage of accidents occur as a result of excessive loads, generally at depths greater than 1000 m. In addition, it was also observed that soft computing algorithms can give an important contribution on determination of rockburst level, based on geologic and construction-related parameters.

Keywords Rockburst · Risk assessment · Soft computing · Neural networks

---

J. Tinoco (✉) · T. Miranda
ISISE, Department of Civil Engineering, University of Minho, Guimarães, Portugal
e-mail: jtinoco@civil.uminho.pt

L. R. e Sousa
Department of Civil Engineering, University of Porto, Porto, Portugal

R. L. e Sousa
Stevens Institute of Science and Technology, Hoboken, NJ, USA

© The Author(s), under exclusive license to Springer Nature Switzerland AG 2021

J. C. Matos et al. (eds.), *18th International Probabilistic Workshop*, Lecture Notes in Civil Engineering 153, https://doi.org/10.1007/978-3-030-73616-3_54

1 Introduction

Accidents and related problems can occur frequently in deep underground mines and other underground structures. Thus, it is essential to develop and implement risk analysis procedures to minimize their occurrence. Risk has a complex nature and results from the combination of two sets of factors: first, the events and their impacts; and second, the vulnerability factors that determine the probability of an event having a certain impact or consequence [1, 2].

Many researchers have collected, analyzed, and published reports on accident cases that have occurred in tunnels during construction and exploration [1, 3]. Rockburst is one example of an accident that can occur during tunneling. It is a result of overstress of the rock mass or of the intact brittle rock, and happens when stresses exceed the compressive strength of the material. The impacts of rockburst range from spalling to sudden and violent failure of the rock mass. Depth is an important factor in the occurrence of this phenomena, since the stress exerted on the rock increases with depth.

In mining activities, other types of events have also been identified and classified, such as heat hazards and other events related to blasting cavities. Blasts, gas explosions, and fire are the most common hazardous events in China. In deep mining activities, major problems are also associated with large deformations and overstressing of the rock mass, which are caused by excavations at great depth, and which may result in rockburst. Comprehensive investigations of deep mining mechanics are thus of great interest [4].

Risk assessment can be managed with the aim of avoiding problems in underground construction. Risk management procedures can be significantly improved by using systematic techniques throughout the project's life. By using such techniques, potential problems can be clearly identified such that appropriate risk mitigation measures can be implemented in a timely manner. As a result, risk management became an integral part of most underground construction projects during the late 1990s [1, 5].

During the construction of some of the underground structures of the Jinping II hydropower scheme in China, engineers were faced with the occurrence of several rockbursts [6]. As a result, a large study was conducted by the authorities to evaluate the accidents and to come up with mitigation measures and guidelines for construction under circumstances that are prone to rockburst. This study included the establishment of a database containing information regarding rockburst and a description of the events that had occurred, and led to the use of data mining (DM) techniques to determine the probability of occurrence of rockburst and its characteristics (i.e., type, location, depth and width, and time delay) [7].

We analyzed these events and concluded that the main mechanisms in rockburst are usually associated with local underground geometry, such as pillars and openings, and with the ground conditions [8]. Rockbursts are classified as strain bursts, pillar bursts, or fault slip bursts [9]. They usually occur during mining operations; however, they can also happen during the construction of civil underground structures, such as deep tunnels. In these cases, the most common phenomenon is strainbursting, although buckling and face crushing may also take place. In addition, impact-induced rockburst created by blasting, caving, and adjacent tunneling should be considered for less stressed and deformed rock formations.

The focus of this paper is on rockburst risk assessment, on the different types of rockburst events, and on their consequences to underground mining and construction. Two rockburst databases that were assembled by these authors are discussed. The first consists of a collection of rockburst laboratory experiments that were performed at the State Key Laboratory for GeoMechanics and Deep Underground Engineering (SKL-GDUE) in Beijing and that were the object of a publication in the journal of Engineering Geology for geological and geotechnical hazards [5]. The second consists of worldwide in situ cases of rockburst that occurred during mining and deep underground construction. The latter database was analyzed, and a list of factors that interact and influence the occurrence of rockburst was determined, along with the relationships between these factors. Finally, different DM techniques were applied to the rockburst databases with the aim of developing predictive models of rockburst index and type. The results are presented in detail in Sect. 3, and the different techniques are compared.

2 Data Mining Modeling in Geoengineering

The prediction of geotechnical formation behavior in geoengineering is complex, particularly during excavations in mining engineering. This complexity is related to uncertainties in the rock mass characterization. In important projects, a large amount of geotechnical data can assist in reducing uncertainties concerning the establishment of design values for the parameters [10]. In the case of rockburst occurrence, the problems are even more difficult to evaluate.

Such data can hold information on trends and patterns that can be used in decision-making and to optimize processes. Therefore, it is necessary to define standard ways of collecting, organizing, and representing data. DM techniques are automatic tools from artificial intelligence and pattern-recognition fields that enable the discovery of potential knowledge [11, 12]. DM is an area of computer science that lies at the intersection of statistics, data management and databases, machine learning, artificial intelligence, pattern recognition, and other areas.

The formal and complete analysis process is called knowledge discovery from databases (KDD). KDD establishes the main procedures for transforming data into knowledge. The KDD process follows the steps indicated in Fig. 1 [11]: collection of a target dataset, data warehousing, transformation of the data into adequate forms for the DM process, selection of a DM tool, relationship identification of DM (classes, clusters, associations), interpretation of results, and consolidation of discovered knowledge.

Several DM techniques exist, each with its own purposes and capabilities. These include decision trees (DTs) and rule induction, neural networks, fuzzy modeling,

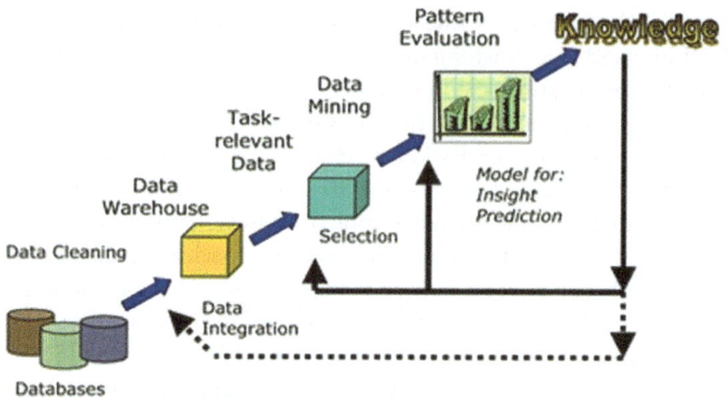

Fig. 1 DM and the knowledge-discovery process [11]

support vector machines (SVMs), k-nearest neighbors (k-NN), instance-based algorithms, and learning classifier systems, among others [13, 14].

Studies using a formal KDD framework are still uncommon in rock mechanics-related activities; however, when applied, they can provide important insight into the most influential parameters in the behavior of rock masses. An important example of such applications is a study done for the Deep Underground Science and Engineering Laboratory, which is located at the former Homestake gold mine in the United States [15]. Here, innovative regression models using different DM techniques were developed to analyze the strength and deformability of the host rock mass and to determine geomechanical indexes for the project [16]. One of the most important tasks in the KDD process is the DM step, which consists of choosing a learning algorithm for training and ultimately building a model that represents the data. Once the training phase is completed, the obtained model is evaluated using a test dataset that was not used during the learning process. The results consist of several different models; there is no universal model that can be used to efficiently solve all the problems.

A brief overview of the most relevant algorithms applied in previous studies is presented here. A DT is a tree-like graph that represents a set of rules for classifying data. These rules can be learned by using a class-labeled training dataset. Artificial neural networks (ANNs) are a deep-learning technique that is modeled after the way in which neurons operate within the human brain [16]. ANNs are formed by groups of artificial neurons connected in layers; signals travel from the first (input) layer to the last (output) layer, forming a structure that is similar to that of brain neurons. These networks, which can be learned from data, are particularly useful in complex applications to recognize patterns and predict future events. SVMs are supervised learning models that are normally used for data classification and regression analysis. Given categorized training data, SVMs determine an optimal plane that defines the decision boundaries, that is, the distance between classes [10].

Rockburst is affected by different factors. The influence diagram in Fig. 2 [1] lists the factors that affect the probability of a rockburst and its potential consequences.

Influence diagrams such as this are very important in the design of DM models for the analysis of accidental events such as rockburst.

Successful applications of DM to different types of problems already exist in the field of geoengineering [10]. Concerning rockburst phenomena, DM techniques were successfully applied to a rockburst laboratory test database obtained from tests at SKL-GDUE, China [5]. The developed triaxial rock test machine used to model the rockburst is presented in Fig. 3 [5]. This equipment forms a true triaxial testing scheme; during the test, one surface of the specimen can be immediately unloaded from the true triaxial compression condition. In this way, it is possible to simulate the stress condition of the rock mass at the free excavation boundary in an underground excavation [5].

The database included a total of 139 cases with samples from different rock types located in China, Italy, Canada, and Iran. Two indexes were developed and used: σ_{RB}, the rockburst maximum stress, and I_{RB}, the rockburst risk index. The meaning of these indexes is described in detail in the publication of He et al. [5]. DM techniques were applied to the rockburst database to infer prediction models of the indexes σ_{RB} and I_{RB}. σ_{RB} is the rupture stresses that are obtained in rockburst tests, while I_{RB} is related to the rockburst critical depth [5]. New models were established using multiple regression (MR), ANNs, and SVM algorithms.

3 Data Mining Applied to In Situ Rockburst Database

In situ cases of rockburst that have occurred during tunnel construction/mining were collected via extensive research into publications and reports, and were organized into a database. The rockburst cases were classified according to their geometric characteristics, causes, and consequences. DM techniques were then applied to the database, with the aim of developing rockburst predictive models. In order to understand the circumstances in which rockbursts occur, their magnitude, and the different consequences of rockburst, we gathered as much information as possible on different aspects of the cases that could provide relevant information about the occurrence of the rockburst. For this purpose, a form was created that included eight fields, each with one or more variables. The eight fields included: (a) rockburst occurrence, (b) construction procedure, (c) tunnel shape or geometry, (d) rock strength, (e) in situ existing stresses, (f) location and dimensions of the rockburst, (g) severity and time delay, and (h) damage in the tunnel and associated equipment. The database contains 60 cases—a relatively small number. However, we believe that it constitutes an important first step in the development of more complex models in future. One important feature of the database is that most of the collected rockburst cases (91%) occurred during the construction of hydroelectric underground power schemes. It is important to emphasize that a large number of the cases in which rockburst took place were located in deep underground mines. The collected data is confined to drill-and-blast and tunnel-boring machine excavation methods, and the shapes of the tunnels where the rockburst cases occurred were either circular (67%) or horseshoe (33%).

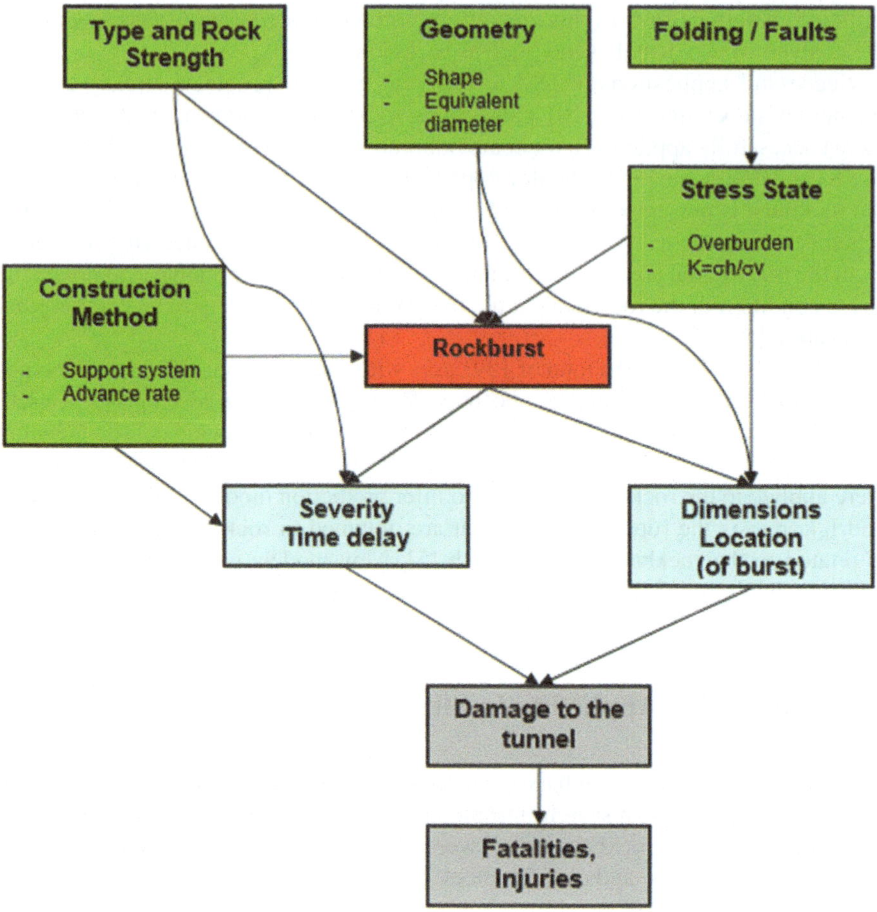

Fig. 2 Influence diagram of rockburst [1]

Different levels of rockburst were classified, as shown in Table 1, following the experience gained at the Jinping II hydropower scheme in China [6]. Figure 4 gives the distribution of cases in the database by rockburst type. In this figure, the Overbreak situation corresponds to levels C and D. Several DM techniques were applied to the database, including DT, k-NN, ANN, and SVM, with the aim of developing rockburst predictive models. The R environment [17] and the rminer package developed by Cortez [18] were used for the implementation of all DM techniques.

For the prediction of in situ rockburst type, a set of nine variables was considered:

- L—Length of occurrence (m)
- TESC—Type of excavation
- TSUP—Type of support
- UCS—Unconfined compressive strength (MPa)

Fig. 3 Rockburst laboratory testing system

Table 1 Margin settings

	Level A	Level B	Level C	Level D
Description	Slight	Moderate	Strong	Very strong
Duration	Sporadic exploration	Long duration	Fast	Sudden
Block depth	<0.5	0.5–1.0	1.0–2.0	>2.0
Impact in excavation	Small	Certain impact	Reasonable impact	Large impact

Fig. 4 Distribution of cases by rockburst type

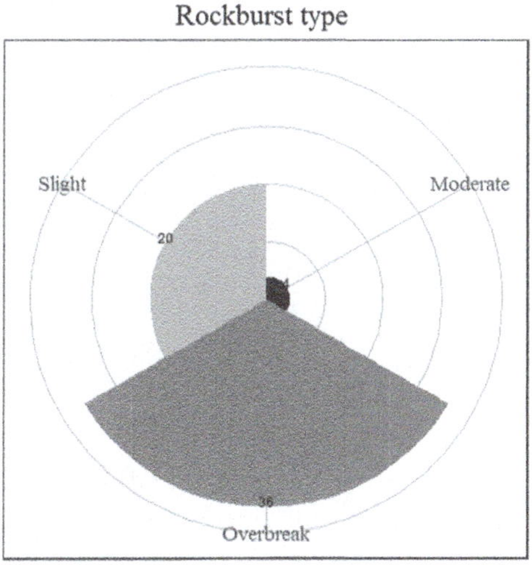

- E—Young's modulus (GPa)
- K—horizontal versus vertical stresses ratio K0
- FORM—Shape of the tunnel
- D_{eq}—Equivalent diameter (m)
- R_{eq}—Equivalent radius (m)

The aim of this analysis was to develop models that would allow the prediction of the type of rockburst, given certain conditions and characteristics related to the underground work. For validation purposes, a leave-one-out method [18] was applied under 20 runs. The leave-one-out method consists of sequentially using one case to test the model, while the remaining cases are used to determine the model's structure. As a result, all data is used for training and testing. By using this method, N models are fitted, where N is the number of available data points. The final generalization estimate is evaluated by computing evaluation metrics for all N test samples.

For the evaluation and comparison of the models, we used three classification metrics based on a confusion matrix (Fig. 5): recall, precision, and F1 score.

The recall measures the ratio of how many cases of a certain class were properly captured by the model. In other words, the recall of a certain class is given by

$$Recall = \frac{True\ positives}{True\ positives + False\ negatives} \tag{1}$$

On the other hand, the precision measures the correctness of the model when it predicts a certain class. More specifically, the precision

$$Precision = \frac{True\ positives}{True\ positives + False\ positives} \tag{2}$$

The $F_{1\text{-score}}$ represents a tradeoff between the recall and precision for a given class. It corresponds to the harmonic mean of precision and recall, according to the following expression:

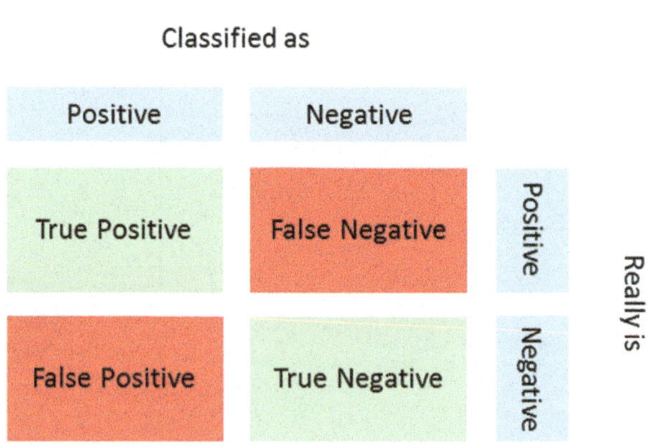

Fig. 5 Establishment of a confusion matrix

$$F_{1-score} = 2 \cdot \frac{Precision \cdot Recall}{Precision + Recall} \tag{3}$$

For all three metrics, a higher value indicates better predictions. Figure 6 shows and compares the DM models' performance for in situ rockburst prediction based on recall, precision, and $F_{1\text{-score}}$. Except for the Moderate rockburst level, all models presented a very good response, with F_1 scores very close to 100%. The low performance in predicting the Moderate class was expected, since only a few records were available for this class in the database for model training (around 7%, as shown in Fig. 4). However, we are confident that it will be possible to improve the model's response once more data for this class becomes available.

Another outcome of the application of the abovementioned DM techniques is the possibility of obtaining the importance of each of the model variables through sensitivity analysis [19]. Hence, and according to the ANN model, the relevant variables are K, TSUP, and L, which have a total influence of around 57% (Fig. 7).

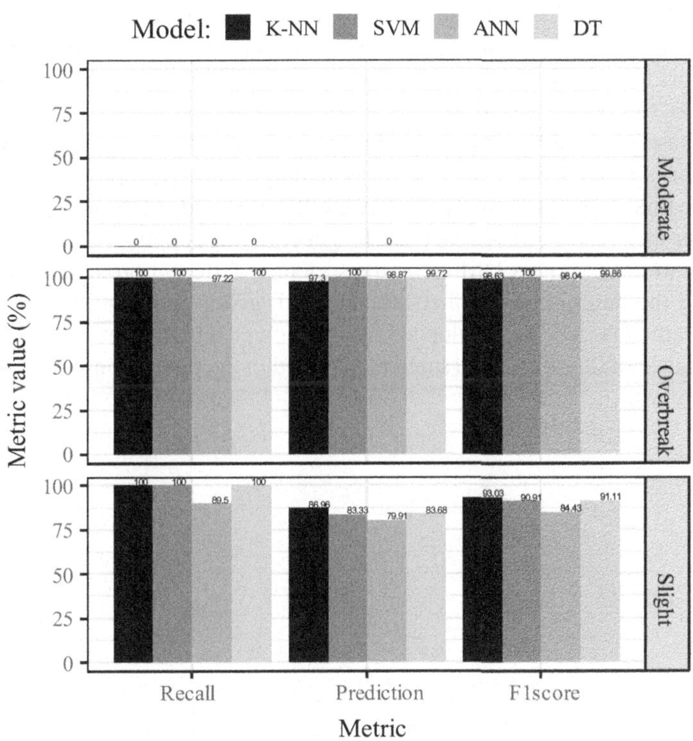

Fig. 6 Comparison of DM models' performance for in situ rockburst prediction based on recall, precision, and $F_{1\text{-score}}$ metrics

4 Final Remarks

Several effective design methods are available to deal with ground fall in mining. However, this is not the case for rockbursts or for seismicity-related mine design problems. Modeling analyses have become a fundamental tool for assessing potential undesirable events, and their cost is only a small fraction of the potential benefits to excavation operations. A large variety of numerical analysis methods can and have been applied to underground engineering in order to assess the potential for the occurrence of rockburst. Monitoring of seismic events and visualization techniques in deep tunnels and mining activities are very useful tools for predicting potentially hazardous situations in order to assist the construction team in time.

Rockbursts are a type of event that can range from minor spalling to significant volumes of rock falling or being ejected with high energy, with devastating consequences. These phenomena are commonly reported in deep underground mining structures, but can also occur in deep tunnels such as the Jinping II hydropower scheme. This paper emphasized the importance of a rockburst triaxial experimental system for the prediction of these types of events, both in mining and in other deep underground projects. In addition, a previous analysis of rockburst test results allowed these authors to develop predictive models to estimate rockburst maximum stress and risk indexes.

A database of rockburst accidents that have occurred in mines and other underground works around the world, such as underground hydropower systems, was created for this study. Data analysis showed that a considerable percentage of accidents occur as a result of excessive loads, generally at depths greater than 1000 m. The application of various DM techniques yielded different predictive models that focused on the determination of rockburst level, given geologic and construction-related parameters. All the developed models showed a high accuracy rate, allowing the importance of the several parameters involved in the prediction of rockburst level to be identified.

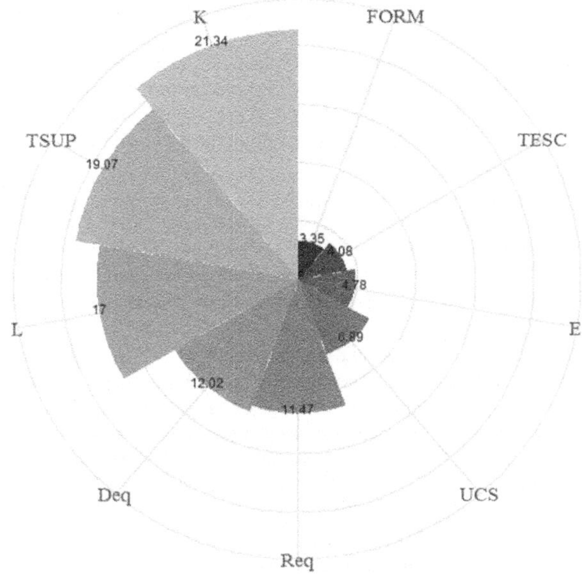

Fig. 7 Relative importance of each variable according to the ANN model

References

1. Sousa, R. L. (2010). *Risk analysis for tunneling projects [dissertation]*. Cambridge: Massachusetts Institute of Technology.
2. Feng, X. T., Jiang, Q., Sousa, L. R., & Miranda, T. (2012). Underground hydroelectric power schemes. In L. R. Sousa, E. Vargas, M. M. Fernandes, & R. Azevedo (Eds.), *Innovative numerical modelling in geomechanics* (pp. 13–50). London: CRC Press.
3. Sousa, L. R. (2006). Learning with accidents and damage associated to underground works. In: A. C. Matos, L.R. Sousa, J. Kleberger, P.L. Pinto (Eds.) Geotechnical risk in rock tunnels. London: CRC Press (pp. 7–39)
4. He, M., Xia, H., Jia, X., Gong, W., Zhao, F., & Liang, K. (2012). Studies on classification, criteria and control of rockbursts. *Journal of Rock Mechanics and Geotechnical Engineering, 4*(2), 97–114.
5. He, M., Sousa, L. R., Miranda, T., & Zhu, G. (2015). Rockburst laboratory tests database—Application of data mining techniques. *Engineering Geology, 185,* 116–130.
6. Wang, J., Zeng, X., & Zhou, J. (2012). Practices on rockburst prevention and control in headrace tunnels of Jinping II hydropower station. *Journal of Rock Mechanics and Geotechnical Engineering, 4*(3), 258–268.
7. Feng, X., et al. (2012). Studies on the evolution process of rockbursts in deep tunnels. *J Rock Mech Geotech Eng, 4*(4), 289–295.
8. Liu, L., Wang, X., Zhang, Y., Jia, Z., & Duan, Q. (2011). Tempo-spatial characteristics and influential factors of rockburst: A case study of transportation and drainage tunnels in Jinping II hydropower station. *Journal of Rock Mechanics and Geotechnical Engineering, 3*(2), 179–185.
9. He, M. C., Jia, X. N., Gong, W. L., Liu, G. J., & Zhao, F. (2012). A modified true triaxial test system that allows a specimen to be unloaded on one surface. In M. Kwasniewski, X. Li, & M. Takahashi (Eds.), *True triaxial testing of rocks* (pp. 251–266). London: CRC Press.
10. Miranda, T., & Sousa, L. R. (2012). Application of data mining techniques for the development of new geomechanical characterization models for rock masses. In L. R. Sousa, E. Vargas, M. M. Fernandes, & R. Azevedo (Eds.), *Innovative numerical modelling in geomechanics* (pp. 245–264). London: CRC Press.

11. Barai, S. K. (2003). Data mining applications in transportation engineering. *Transport, 18*(5), 216–223.
12. Saitta, S., Kripakaran, P., Raphael, B., & Smith, I. F. (2008). Improving system identification using clustering. *Journal of Computing in Civil Engineering, 22*(5), 292–302.
13. Adoko, A. C., Gokceoglu, C., Wu, L., & Zuo, Q. J. (2013). Knowledge-based and data-driven fuzzy modeling for rockburst prediction. *International Journal of Rock Mechanics and Mining Sciences, 61,* 86–95.
14. Chapman, P., Clinton, J., Kerber, R., Khabaza, T., Reinartz, T., Shearer, C., et al. (2000). *CRISP-DM 1.0: Step-by-step data mining guide.* Chicago: SPSS Inc.
15. McPherson, B., Elsworth, D., Fairhurst, C., Kelsler, S., Onstott, T., Roggenthen, W., et al. (2003). EarthLab: a subterranean laboratory and observatory to study microbial life, fluid flow, and rock deformation. A Report to the National Science Foundation. Bethesda: Geosciences Professional Services Inc.
16. Sousa, L.R., Miranda, T., Roggenthen, W., Sousa, R.L. (2012). Models for geomechanical characterization of the rock mass formations at DUSEL using data mining techniques. In *Proceedings of the 46th US Rock Mechanics/Geomechanics Symposium*, June 24–27, Chicago, IL, USA (p. 14). Alexandria: American Rock Mechanics Association.
17. R-Project.org [Internet]. Vienna: The R Foundation. 2020 [updated 2020 Apr. 15, cited 2020 May 4]. Available from: https://www.r-project.org.
18. Cortez, P. (2010). RMiner: Data mining with neural networks and support vector machines using R. In: R. Rajesh (Ed.) Introduction to advanced scientific softwares and toolboxes. Hong Kong: International Association of Engineers.
19. Cortez, P., & Embrechts, M. J. (2013). Using sensitivity analysis and visualization techniques to open black box data mining models. *Information Sciences, 225,* 1–7.

Semi-empirical Based Response Surface Approach for Reliability Evaluation of Steel Plates with Random Fields of Corrosion

Angelo P. Teixeira and Carlos Guedes Soares

Abstract The paper presents a semi-empirical based response surface approach for structural reliability analysis of steel plates with non-uniform corrosion represented by random fields. The approach consists of using a semi-empirical design equation as simplified response surface model, which is then calibrated iteratively by means of the results of non-linear finite element analyses at the design points calculated by the First Order Reliability Method. This technique has been successfully applied to problems formulated in terms of discrete random variables and is now applied to problems involving spatial variability of structural parameters represented by random fields. The approach is first illustrated with an example of the ultimate strength of plates with random imperfections and material properties and then applied to plates with random fields of corrosion discretized using the Expansion Optimal Linear Estimation method. The results obtained by the semi-empirical based response surface approach and by coupling directly the First Order Reliability Method and the finite element code are compared.

Keywords Response surface method · Model correction factor method · Random fields · Corrosion · Ultimate strength of steel plates

1 Introduction

Structural reliability methods have been applied to assess the safety of structures and structural components of different complexity and for code calibration, as reviewed by Teixeira and Guedes Soares [1, 2]. Most of the reliability applications rely on the use of First and Second Order Reliability Methods (FORM/SORM), developed mainly to deal with explicit limit state functions, and Monte Carlo simulation (MCS) methods, combined with several variance reduction techniques [3]. More recently, there has been an increasing interest in developing efficient methods for reliability analysis of structural systems using advanced non-linear Finite Element (FE)

A. P. Teixeira (✉) · C. Guedes Soares
Centre for Marine Technology and Ocean Engineering (CENTEC), Instituto Superior Técnico, Universidade de Lisboa, Lisbon, Portugal
e-mail: teixeira@centec.tecnico.ulisboa.pt

© The Author(s), under exclusive license to Springer Nature Switzerland AG 2021
J. C. Matos et al. (eds.), *18th International Probabilistic Workshop*, Lecture Notes in Civil Engineering 153, https://doi.org/10.1007/978-3-030-73616-3_55

methods. Although, in many cases of practical importance, particularly for complex structures, FORM/SORM and MCS methods can be adapted to handle implicit limit state functions via direct coupling with the finite element code [4], in the case of large structures the numerical evaluation of the limit state function has to be replaced by suitable approximations. The focus has been on the development of response surface approaches based on polynomial regression models [5], artificial neural networks [6] and Kriging interpolation models also known as Gaussian process models [7, 8], with different adaptive strategies to define accurate approximations of the true limit state function in the important region of the basic random variables space [9–11].

One aspect that has been extensively addressed is the effect of corrosion on both the strength and safety of steel structures. The corrosion is in essence a process of an uncertain nature governed by many variables and, therefore, only probabilistic models can describe the corrosion process and its effect [12].

Most of the studies on probabilistic modeling of the collapse strength and on reliability of plates have assumed a constant corrosion rate, and a uniform reduction of plate thickness due to corrosion. However, in addition to the general wastage that is reflected in the generalized decrease of plate thickness, the microscopic variations on the surface of the metal tend to cause different forms of corrosion and also variations in the corrosion rate over wide or small areas, which can be represented by random fields.

Teixeira and Guedes Soares [13] have studied the ultimate strength of corroded plates with spatial distribution of corrosion also represented by random fields, showing that typically the strength of plates with spatial distribution of corroded thickness is lower than the one obtained for uniform corrosion. This has motivated the development of probabilistic models of the ultimate strength of ship plates with non-uniform corrosion described by random fields [14].

Teixeira and Guedes Soares [5] have investigated the efficiency of the application of Response Surface Methods in structural reliability analysis of steel plates with non-uniform corrosion, an approach that has been already adopted by Kmiecik and Guedes Soares [15] for probabilistic modeling of the strength of compressed steel plates with initial distortions. However, Teixeira and Guedes Soares [5] have used an adaptive response surface method able to handle the spatial random fluctuations of the corrosion described by discretized random fields. In this case the limit state function was defined in terms of the ultimate strength of the corroded plate evaluated by nonlinear finite element analysis (FEA) [13].

In alternative to the pure mathematical representation of the response by means of first- or second-order polynomials, Ditlevsen and Arnbjerg-Nielsen [16] have proposed a different approach based on the so-called MCFM (Model Correction Factor Method). The MCFM is a special kind of response surface technique that uses a simplified structural model that is calibrated in a probabilistic sense to a complex, but more realistic, model. Typically, a semi-empirical design equation is adopted as a simplified response model that is calibrated iteratively by means of a model correction factor at the FORM-design points based on the results of FEAs.

This approach has been applied successfully to problems formulated in terms of discrete random variables. Garre et al. [17] conduced a reliability analysis of a stiffened plate using a non-linear finite element model as realistic model and both a plastic hinge model and a semi-empirical analytical design equation of a beam column structural model as simplified response models. More recently, Xu et al. [18] have also adopted the MCFM to assess the reliability of a Suezmax oil tanker considering the ultimate vertical bending moment capacity of the hull girder as a limit state. Both studies have shown that the use of the semi-empirical response models, which include the important mechanical features of the structural system, significantly reduce the computation time required by the reliability analyses. Moreover, the approach provides a measure of deviation between the predictions of the semi-empirical design equation and of the numerical analysis of the complex and realistic model in several locations of the design space, which can be used to calibrate and to include parameters not accounted by the semi-empirical design equation.

In this paper, the MCFM that has been traditionally used in problems of small dimension of the uncertainty space is extended to larger problems involving random fields of corrosion. Semi-empirical design equations used to predict the ultimate strength of plates are adopted as simplified response models which are calibrated at the design points calculated by FORM based on the results of non-linear finite element analyses of the plate with non-uniform corrosion.

The approach is first illustrated with an example of the ultimate strength of plates under in-plane compression with random imperfections and material properties and then is applied to plates with random fields of corrosion discretized by the Expansion Optimal Linear Estimation method proposed by Li and Der Kiureghian [19]. In this later case the calibration of the semi-empirical design equation used to predict the strength of the plate is provided by a first-order Taylor expansion in the random variables that represent the discretized random field of corrosion, which are not accounted in the design equation.

2 Random Fields for Corrosion Modeling

In probabilistic analyses it is convenient to discretize the random fields by representing them in terms of discrete sets of random variables. Several methods have been proposed for discretization of random fields, in particular for use in finite reliability analysis of structures, as reviewed by Li and Der Kiureghian [19].

In this study, the random field of corrosion is discretized using the Expansion Optimal Linear Estimation method (EOLE) [19]. The EOLE discretization method assumes that a Gaussian random field $H(\boldsymbol{x})$ can be defined as a linear function of a vector $\boldsymbol{h} = \{H(x_1), ..., H(x_N),\}$ of N nodal values $H(x_i)$ of the original random field given by:

$$\hat{H}(\mathrm{x}) = a(\mathrm{x}) + \sum_{i=1}^{N} b_i(\mathrm{x}) H(x_i) = a(\mathrm{x}) + \mathrm{b}^T(\mathrm{x})\,\mathrm{h} \tag{1}$$

where $a(\boldsymbol{x})$ is a scalar function of $\boldsymbol{x}$, $\boldsymbol{b}(\boldsymbol{x})$ is a vector function of $\boldsymbol{x}$ and N in the number of nodal points in the domain.

Assuming that the vector $\boldsymbol{h}$ of random variables can be expressed in terms of its spectral decomposition, it is possible to determine the functions $a(\boldsymbol{x})$ and $\boldsymbol{b}(\boldsymbol{x})$ that minimize the variance of the error $\text{VAR}\left[H(\text{x}) - \hat{H}(\text{x})\right]$, subjected to $\hat{H}(\boldsymbol{x})$ being an unbiased estimator of the $H(\boldsymbol{x})$ in the mean, i.e., $\text{E}\left[H(\text{x}) - \hat{H}(\text{x})\right] = 0$. Hence, according to the called Expansion Optimal Linear Estimation method (EOLE), the approximated Gaussian random field is given by [19]:

$$\hat{H}(\text{x}) = \mu(\text{x}) + \sum_{ir=1}^{r} \frac{\zeta_{ir}}{\sqrt{\theta_{ir}}} \boldsymbol{\varphi}_{ir}^{T} \text{C}_{H(\text{x})\,\text{h}} \tag{2}$$

where $\mu(\boldsymbol{x})$ is the mean function of the random field, θ, $\boldsymbol{\varphi}$ are the eigenvalues and eigenvectors of the covariance matrix $\boldsymbol{C_{hh}}$ of $\boldsymbol{h}$, $\{\zeta_{ir};\ \ ir = 1 \cdot r\}$ is a set of ir independent standard normal distributions (zero mean, unit variance and zero correlation) and $\boldsymbol{C}_{H(\boldsymbol{x})\boldsymbol{h}}$ is a $r \times 1$ vector containing the covariances of $H(\boldsymbol{x})$ with the elements of $\boldsymbol{h}$ (see [19] for more details of the EOLE method).

The high-level efficiency of this approach in the sense that it requires a small number of random variables to represent the random field within a given level of accuracy makes this model particularly useful for stochastic analysis and, therefore, it will be used on further calculations. For the correlation structure used in the present study, it was found that only 15 random variables, $\{\zeta_{ir};\ \ ir = 1 \cdot r = 15\}$, are sufficient to assure less than 5% error in the discretization (i.e. $err(\text{x}) < 5\%$).

In the present study the reduction of plate thickness due to corrosion is described by a homogeneous lognormal random field $\hat{H}_{\text{ln}}(\text{x})$, i.e., both the mean value and the standard deviation of the random field are constant over the plate surface and equal to μ_{ln} and σ_{ln}, respectively. The lognormal random field $\hat{H}_{\text{ln}}(\text{x})$ is defined by a transformation of the Gaussian field $\hat{H}(\boldsymbol{x})$ as:

$$\hat{H}_{\text{ln}}(\text{x}) = \exp\left[\hat{H}(\text{x})\right] \tag{3}$$

In this case, the mean value $\mu(\text{x}) = \mu$ and standard deviation $\sigma(\text{x}) = \sigma$ of the underlying Gaussian homogenous random field $H(\text{x})$ must be first calculated from the mean value μ_{ln} and standard deviation σ_{ln} of the lognormal random field $H_{\text{ln}}(\text{x})$ by:

$$\sigma = \sqrt{\ln\left(1 + \frac{\sigma_{\text{ln}}^2}{\mu_{\text{ln}}^2}\right)} \quad \text{and} \quad \mu = \ln(\mu_{\text{ln}}^2) - \frac{\sigma^2}{2} \tag{4}$$

The probabilistic characteristics of the homogeneous lognormal random field are defined on the basis of the analysis performed by Teixeira et al. [14]. This study has characterised the time variation of the probabilistic characteristics of the random

Table 1 Probabilistic characteristics of the random field of corrosion for 10 years old bulk carriers [14]

Distribution	Mean, $\mu_{\ln}$	St. dev, $\sigma_{\ln}$	Correlation length (l_x=l_y)
Lognormal	1.021 mm	0.883 mm	300 mm

fields of corrosion, namely its mean value and standard deviation, by fitting the nonlinear time-dependent corrosion model proposed by Guedes Soares and Garbatov [20] to corrosion data measured in plates at different locations of existing bulk carriers reported by Paik et al. [21].

For any point in time in which the average thickness wastage is given by that model, it is considered that the thickness reduction is not uniform but instead varies spatially in a random manner. Table 1 summarizes the characteristics of the random field of thickness reduction due to corrosion for 10-years old bulk carriers [14].

For the numerical analysis presented in this paper the following autocorrelation function of multidimensional homogeneous random fields has been considered for the Gaussian random field ($H(\mathrm{x})$), which is converted from the original log-normal random field ($H_{\ln}(\mathrm{x})$) by Eq. 3:

$$\rho(x, x') = e^{-\frac{|x-x'|^2}{l_c^2}} \tag{5}$$

where the parameter l_c is a measure of the rate of fluctuation of the random filed, commonly known as the correlation length. This parameter can be estimated by investigating the dependence between the measurements at different distances from each other. However, the precise locations of the corrosion measurements have not been registered in the available corrosion data. Therefore, a representative value for the correlation length of 0.3 m, corresponding to 30% of the plate length, is chosen in the present study, which allows the study of the effect of the spatial representation of the corrosion patterns on the ultimate strength of the plates. It should be noted that the correlation functions and the appropriate values of correlation length are important for the results, and work is still required to identify the most appropriate ones with field data so that the simulated distributions are representative of realistic situations.

3 Approximate Approaches for Reliability Assessment

3.1 First Order Reliability Method

The methods for structural reliability assessment aim at evaluating the probability of failure as the probability of limit state violation. The problem is characterized by an n-vector X of basic random variables and a subset Ω_f of their outcome space, which defines the "failure" event. For a structural component with a single failure mode,

Ω_f can formally be written as $\Omega_f = \{x : g(x) \leq 0\}$, where $g(x)$ is the limit state (or failure) function for the failure mode considered. In this case the probability of failure is given by:

$$P_f = P[g(\mathrm{X}) \leq 0] = \int_{\Omega_f} f_{\mathrm{X}}(\mathrm{x})\mathrm{dx} \tag{6}$$

where $f_{\mathrm{X}}(\mathrm{x})$ is the joint probability density function (*pdf*) of the vector X of basic random variables that comprises physical variables describing uncertainties in loads, material properties, geometrical data and calculation modeling.

The difficulty in computing the failure probability P_f directly from the integral given by Eq. 6 led to the development of approximate methods such as the First Order Reliability Method (FORM) [22] and Second Order Reliability Methods (SORM) and Monte Carlo simulation (MCS) methods combined with several variance reduction techniques. A detailed description of these methods as well as other possible methods used for reliability assessment can be found, e.g., in [23, 24].

In particular, the FORM involves the linearization of the limit-state function g in a standard normal space (U), where all variables are uncorrelated with zero mean and unit variance (e.g. [25, 26]) and the calculation of the optimal point u* on g that has the minimum distance from the origin. This distance is denoted by reliability index β [22]:

$$\mathrm{u}^* = \min\{\|\mathrm{u}\| \,|g(\mathrm{u}) = 0\} \quad \text{and} \quad \beta = \|\mathrm{u}^*\| \tag{7}$$

The point u* is commonly known as the *design point*, but other names such as *most probable point* (MPP) and *beta point* are also used. Several well-established iterative algorithms are available for solving Eq. 7 [25, 27]. Having calculated the reliability index β, due to the rotational symmetry of the standard normal density, the first-order approximation of the failure probability can be calculated from [22]:

$$P_f \approx P_f^{FORM} = \Phi(-\beta) = \Phi\left(-\| \mathrm{u}^* \|\right) \tag{8}$$

where Φ is cumulative standard normal probability distribution.

In cases of practical interest, the limit state functions are not expressed as an explicit or analytical form in terms of the basic variables of the problem and are only known implicitly, which makes the solution of the structural reliability problem quite demanding or even infeasible. In these cases, appropriate approximations for the limit state functions or only for the response of the structure are required. In this paper first/second-order polynomials and calibrated semi-empirical design equations available to predict the behaviour of the structure in alternative to pure mathematical representations are adopted as response models.

3.2 Model correction factor method

In the MCFM, an idealized and simplified model, R^{IDEAL}, of the problem is adopted as the response surface, and then a complex and time consuming, but more 'realistic' model R^{REAL}, is used to calibrate the simplified model [16]. This procedure of probabilistic calibration assures that the predictions of the simplified model are close to the realistic one, at least around the design point calculated by FORM.

The resistance, $R^{IDEAL}(\mathrm{X_I})$, provided by the simplified response model (e.g. semi-empirical design equation) is a function of the vector $\mathrm{X_I}$ geometrical and material random properties, which is corrected by a model correction factor $f_c(\mathrm{X_R})$, as follows:

$$R^{REAL}(\mathrm{X_R}) = f_c(\mathrm{X_R})R^{IDEAL}(\mathrm{X_I}) \tag{9}$$

where $R^{REAL}(\mathrm{X_R})$ is the resistance of the structure assessed by the complex and time consuming realistic response model (e.g. non-linear finite element analysis). Using this approach, the implicit limit state function of the structure,

$$g(\mathrm{X}) = R^{REAL}(\mathrm{X_R}) - S(\mathrm{X_S}) \tag{10}$$

is converted into an explicit one given by:

$$g(\mathrm{X}) = f_c(\mathrm{X_R})\, R^{IDEAL}(\mathrm{X_I}) - S(\mathrm{X_S}) \tag{11}$$

where $S(\mathrm{X_S})$ is the applied load on the structure and $[\mathrm{X_R}, \mathrm{X_S}]$ basic random variables involved in the resistance ($\mathrm{X_R}$) and load ($\mathrm{X_S}$) assessment.

The model correction factor $f_c(\mathrm{X_R})$ is initially set to 1, or calculated at the mean value of the random variables. The explicit limit state function (Eq. 11) defined based on the idealized model with the model correction factor is then used in FORM analysis. At the FORM-design point x* the realist model ($R^{REAL}(\mathrm{x_R^*})$) is assessed (i.e. the resistance of the structure is calculated numerically by means of non-linear FEA with material and geometrical properties given by ($\mathrm{x_R^*}$) and the simplified model is then corrected to be equal to the realistic model. The corrected idealized model is then used again in FORM calculations to find the next design point and the process repeats until convergence is achieved after a few iterations. The semi-empirical design formulation for buckling strength requirements of stiffened panels adopted in the IACS-CSR for double hull tankers is used as simplified response model $R^{IDEAL}(\mathrm{X_I})$ that is able to capture the most important features with respect to the load carrying capacity of this type of structures.

Figure 1 shows the calculation procedure of the model correction factor method (MCFM), described as follows [18]:

1. Set the model correction factor $f_{c,i}(\mathrm{X_R}) = 1$ at iteration $i = 0$;

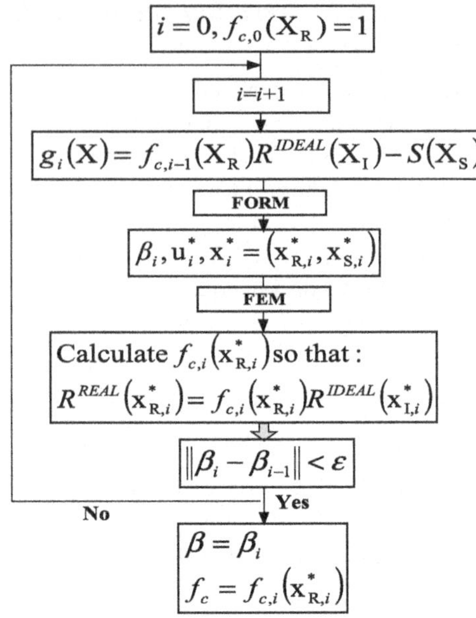

Fig. 1 Calculation procedure of the MCFM

2. Iteration $i = i + 1$. The limit state function of Eq. (11) is adopted to calculate the next design point x_i^* and the corresponding reliability index β_i using the first-order reliability method (FORM);
3. Evaluate the realistic model $R^{REAL}(x_{R,i}^*)$ at the design point x_i^*, where $x_{R,i}^*$ is a subset x_i^* that contains the random variables involved in the assessment of the realistic model;
4. Find $f_c(x_{R,i}^*)$ that satisfies $R^{REAL}(x_{R,i}^*) = f_{c,i}(x_{R,i}^*)R^{IDEAL}(x_{I,i}^*)$;
5. Check convergence of the reliability index, i.e. if $\|\beta_{i+1} - \beta_i\| < \varepsilon$;
6. Iterate around steps 2 to 5 until convergence is achieved.

When the idealized model R^{IDEAL} contains all the variables considered in the realistic model, i.e. $X_I = X_R$, the model correction factor calculated at each FORM design point is simply given by a constant (a zero-order Taylor expansion)

$$f_{c,i}(x_{R,i}^*) = \frac{R^{REAL}(x_{R,i}^*)}{R^{IDEAL}(x_{R,i}^*)} \tag{12}$$

On the other hand, if the idealized model R^{IDEAL} does not include all random variables adopted in the realistic model, the model correction factor, $f_{c,i}(X_R, \zeta_{ir=1\ldots r})$, is approximated by a first-order Taylor expansion around the design point (x^*, ζ^*) in the variables ζ_i, which are not accounted in the idealized model R^{IDEAL}. This way important random variables can be explicitly considered the limit state function and consequently in the reliability analysis [16],

$$f_{c,i}(\mathrm{X_R}, \zeta) = f_{c,i}(\mathrm{x}^*_{\mathrm{R,i}}, \zeta^*) + \sum_{ir=1}^{r} (\zeta_{ir} - \zeta^*_{ir}) \cdot \left. \frac{\partial f_{c,i}(x, \zeta)}{\partial \zeta_{ir}} \right|_{\mathrm{x}^*_{\mathrm{R}}, \zeta^*} \tag{13}$$

where

$$\begin{aligned} \frac{\partial f_{c,i}(x, \zeta)}{\partial \zeta_{ir}} &\approx \frac{f_{c,i}(x^*, \zeta_{\bar{i}r}, \zeta_{ir} + d\zeta_{ir}) - f_{c,i}(x^*, \zeta_{\bar{i}r}, \zeta_i)}{d\zeta_{ir}} \\ &\approx \frac{R^{REAL}(x^*, \zeta_{\bar{i}r}, \zeta_{ir} + d\zeta_{ir}) - R^{REAL}(x^*, \zeta_{\bar{i}r}, \zeta_{ir})}{R^{IDEAL}(x^*) \cdot d\zeta_{ir}} \end{aligned} \tag{14}$$

4 Reliability Formulation of Steel Plates With Random Fields of Corrosion

The limit state function is defined in terms of the ultimate strength of the corroded plate calculated by nonlinear finite element analysis. The failure is defined when the axial load applied on the plate (σ_a) exceeds its axial load carrying capacity (R) calculated numerically by non-linear FEM analysis for a given realization of the vector of random variables of the problem X.

The ultimate strength calculations were carried out for simply supported square plates ($a/b = 1$), and slenderness $b/t = 50$ with longitudinal edges restrained against transverse displacement. The ultimate strength of the plate corresponds to the maximum value of the longitudinal stress-displacement curve of the plate under in-plane longitudinal compression obtained by nonlinear finite element analysis using the ANSYS Software. The load is a uniform prescribed longitudinal displacement (δ) applied along the transverse edge of the plate.

The reliability analysis is first applied to plates with random initial imperfections and material properties and then the semi-empirical response surface technique is applied to plates with random fields of corrosion.

The random variables first considered are the plate thickness t, the Young's modulus E, the yield stress σ_a, and the amplitude of initial distortions of the plate ($w_{\max}$). The applied load is taken as 60% of the characteristic value of the yield stress of the material (σ_y^c), which is represented by the random variable C_a following a Weibull distribution with a coefficient of variation (*cov*) of 0.1. The stochastic model of the amplitude of initial distortion of the plates is defined based on measurements of fabrication distortions of ship plating that were undertaken mainly in Polish shipyards in the process of construction of ships. The stochastic models of the random variables are summarized in Table 2, following the suggestions in [4] and [14].

The limit state function is then given by:

$$g(\mathrm{X}) = R - \sigma_a = R\big(t, E, \sigma_y, w_{\max}\big) - C_a \sigma_y^c \tag{15}$$

Table 2 Stochastic models of the random variables (X)

Variable	t (mm)	E (MPa)	σ_y (MPa)	w_{max} (mm)	C_a
Distrib	Normal	Lognorm	Lognorm	Lognorm	Weibull
Mean	20.0	210000	269.0	2.007	0.60
St. Dev.	2.0	21000	21.52	1.193	0.06
cov	0.10	0.10	0.08	0.59	0.10
Characteristic value (X^c)	20.0	208958	235.0	4.264	0.685
F_x (X^c)	0.50	0.50	0.05	0.95	0.95

To assess the reliability of the plate elements with reduction of thickness due to corrosion represented by the random field, the set ζ of ir independent standard normal variables (ζ_i) necessary to represent the discretized random field are also considered in the reliability formulation. In this case the limit function becomes:

$$g(\mathrm{X}, \zeta) = R(\mathrm{X}, \zeta) - C_a \sigma_y^c \qquad (16)$$

where $R(\mathrm{X}, \zeta)$ is the ultimate strength of the corroded plate calculated numerically by a nonlinear finite element analysis for a given realization of the vectors X and ζ of random variables of the problem. In this application the random field of the reduction of plate thickness due to corrosion was represented by 15 random variables ($ir = 15$). This way, the maximum variance error in the EOLE method, estimated by comparing the variance of random field with the variance of the discretized field, is lower than 5%.

Since the random field of corrosion is defined only at the standard normal space, the response surface has been constructed at this normalized space.

5 Results of the Reliability Analysis

5.1 *First Order Reliability Method*

Table 3 shows the results obtained by the First Order Reliability Method (FORM) linked directly to the non-linear FEM analysis when evaluating the limit state function $g(\mathrm{X})$ given by Eq. 15. For a given convergence criterium FORM needs 9 and 8 iterations, depending on the method used to calculate the gradient of the limit state function, which correspond to 63 or 112 calls of the FEM analysis using a forward or a central difference scheme, respectively. Table 3 also presents the calculated design point and sensitivity factors of the failure function.

Table 4 presents the reliability results obtained by FORM for corroded plates with non-uniform reduction of plate thickness represented by random fields. Calculations have been conducted considering different starting realizations of the random field

Table 3 FORM reliability analysis of non-corroded plates

β_{FORM}	3.364	
Number of iterations	9 (f.dif.)/8 (c.dif.)	
Number of R(X) calls	63 (f.dif.)/112 (c.dif.)	
Random variables	α_i	x^*
t-thickness	0.79	14.7
E-Young modulus	0.29	190520
σ_y-yield stress	0.31	246.4
$w_{\max}$-initial imperf	−0.15	2.40
C_a-(% of σ_y^c)	−0.41	0.67

of corrosion ($\hat{H}^{start}(\zeta^{start})$), illustrated in Fig. 2. It can be seen that independent of the starting realization of the random field, FORM requires between 132 and 176 or 264 to 352 numerical evaluations of the ultimate strength of the plate, respectively, when using a forward or a central approximation scheme to calculate the gradient of the implicit limit state function. Figure 3 shows the most likely configuration of the corrosion pattern obtained by FORM corresponding to the design point of the random variables that define the random field of corrosion ζ^*.

Table 4 FORM reliability analysis of non-uniform corroded plates

Non-uniform corroded plates $g(\text{X}, \zeta) = R(\text{X}, \zeta) - C_a\sigma_y^c$	β	No. of iter.	No. R(X, ζ) calls
RF start 1 ($\zeta_i = 0$)	2.511	6	132 (f.diff) 264 (c.diff)
RF start 2 ($\zeta_i \neq 0$)		6	132 (f.diff) 264 (c.diff)
RF start 3 ($\zeta_i \neq 0$)		7	154 (f.diff) 308 (c.diff)
RF start 4 ($\zeta_i \neq 0$)		8	176 (f.diff) 352 (c.diff)

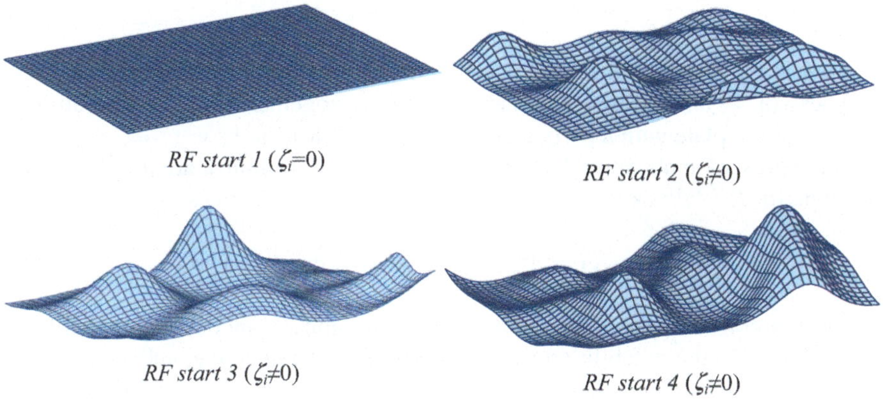

Fig. 2 Starting realizations of the random field of corrosion ($RF\ \zeta_i^{start}$)

Fig. 3 FORM most likely realization of the random field of corrosion ($RF\ \zeta^*$)

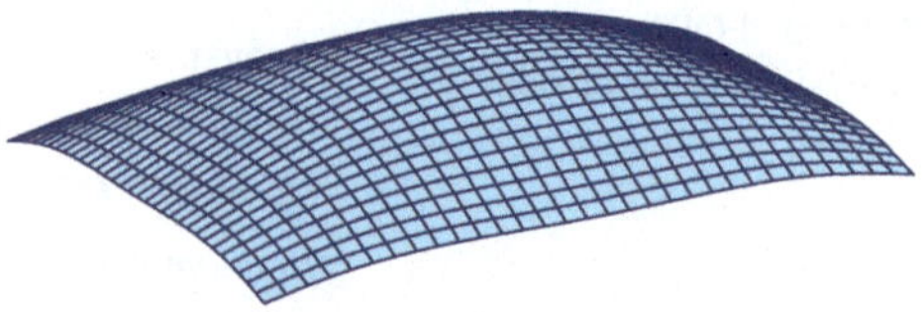

5.2 *Model Correction Factor Method*

According to the Model Correction Factor method a particular idealized and simplified structural model that typically consists in a semi-empirical formula is used as a response surface to represent the ultimate strength of the plate in the limit state function of the reliability analysis.

For design purposes including code specifications, several semi-empirical formulae have been proposed to predict the collapse strength of the plate elements subjected to predominantly compressive in-plane loads. Several of the expressions build upon the elastic buckling strength σ_c of plate elements. In particular, Guedes Soares [28] proposed an expression that deals explicitly with initial imperfections and residual stresses. The normalized strength $\left(\eta_{xu}^G\right)$ of the "perfect" plate is given by:

$$\begin{aligned} \eta_{xu}^G &= \frac{2.16}{\lambda} - \frac{1.08}{\lambda^2} \text{ for } \lambda \geq 1 \\ \eta_{xu}^G &= 1.08 \text{ for } \lambda \leq 1 \end{aligned} \tag{17}$$

where $\lambda = (b/t)\sqrt{\sigma_y/E}$ is the plate slenderness. The strength reduction (η_d) of the plate due to the existence of initial distortions is represented by:

$$\eta_d = 1 - (0.626 - 0.121 \cdot \lambda) \cdot \delta \tag{18}$$

where δ is the magnitude of initial geometric distortions normalized by the plate thickness ($\delta = w/t$). Last equations are then use together to predict the ultimate strength of the plate with a particular level of initial distortions w.

In the present application the above semi-empirical formulae are used to predict the strength, R^{IDEAL} as:

$$R^{IDEAL}(\mathrm{X}) = \eta_{xu}^G(t, \sigma_y, E) \cdot \eta_d(t, \sigma_y, E, w) \cdot \sigma_y \tag{19}$$

For non-corroded plates, the simplified semi-empirical model contains all variables adopted in the reliability formulation and considered by the realistic FE structural response model and, therefore, the model correction factor at each design point simply calculated by Eq. 12. However, for non-uniform corroded plates the semi-empirical model used to calculate R^{IDEAL} does not depend explicitly on the vector of random variables ζ that define the random field. Therefore, in this case, the model correction factor is approximated by a first-order Taylor expansion around the design point $(x^*,\ \zeta^*)$ given by Eq. 13.

Figure 4 shows the quality of the semi-empirical formulae (Eqs. 17 and 18) in predicting the normalized ultimate strength of non-corroded plates obtained by Ansys using Monte Carlo Simulation. Figure 5 illustrates similar results for corroded plates with reduction of plate thickness represented by the random field. The effect of corrosion is taken into account in the semi-empirical formulae by reducing the thickness t in Eq. 17 by the average value of the random field of corrosion. A lager scatter is observed due to the effect of the localized patterns of corrosion on the strength of the plate calculated by non-linear FEM analysis, which is not dealt by the semi-empirical formulae.

Table 5 shows the iteration results of the Model Correction Factor Method for non-corroded plates, which are compared in Table 6 with the results obtained by FORM, directly linked to the FEM code. It can be seen that the solution is achieved with only 4 iterations of the MCFM and, consequently, only 4 evaluations of the ultimate strength of the plate by FEM are required.

The model correction function $f_{c,i}(X_R)$ is set to 1.0 at the first iteration; therefore, the first reliability index of 3.570 corresponds to the β-value obtained by FORM using the limit state function defined in terms of semi-empirical formulae.

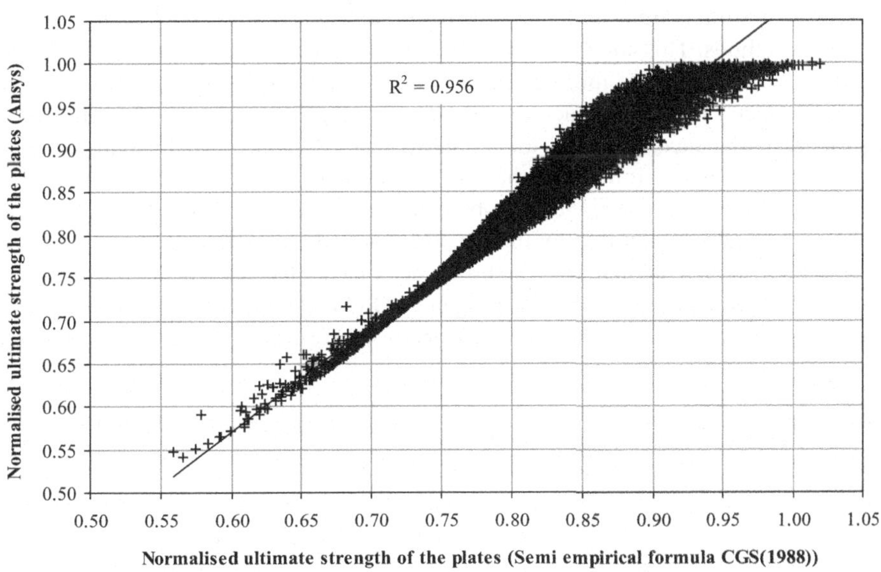

Fig. 4 Normalized ultimate strength of non-corroded plates obtained by Ansys and by the semi empirical formulae (Eqs. 17 and 18)

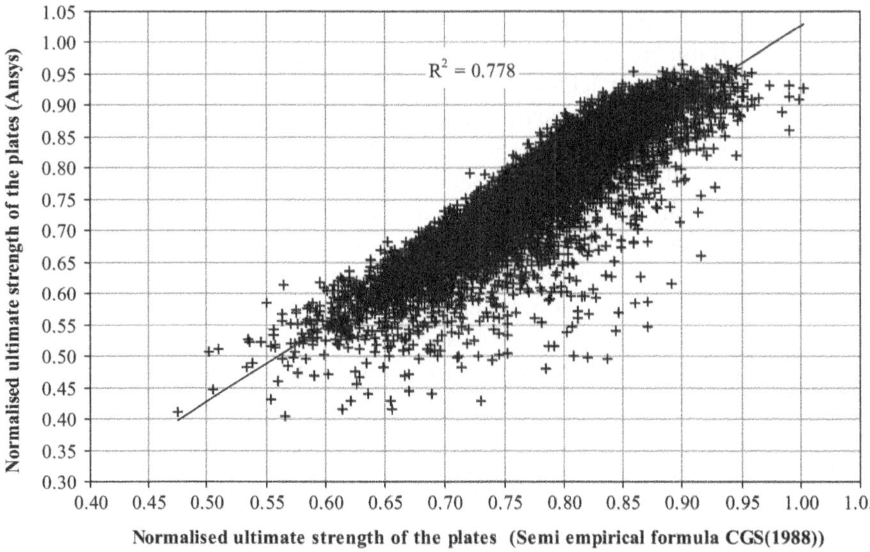

Fig. 5 Normalized ultimate strength of corroded plates obtained by Ansys and by the semi-empirical formulae (Eqs. 17 and 18)

Although the MCFM provides a fast convergence, the reliability index obtained (3.39) deviates from the one calculated by FORM, as shown in Table 6. It can also be seen that similar trend is identified for the design point (x^*) calculated by the different methods. This suggests that FORM based solely on the results of the Finite Element Analysis can be more accurate for reliability assessment than the model correction factor method. In fact, the use of semi-empirical formulations introduces additional model uncertainties in the reliability analysis concerning the real behavior of the structure, which influences the results obtained.

Tables 7 and 8 present the results obtained for non-uniform corroded plates. The number of iterations required by the MCFM is 4 and the number of finite element analyses needed to calculate the reliability index has reduced to 64. The differences between the predictions of the MCFM and FORM identified in the case of non-corroded plates are also present in the case of non-uniform corroded plates. Moreover,

Table 5 Results of the model correction factor method (MCFM) for non-corroded plates

Non-corroded plates (random variables: t, E, σ_y, w_{max}, C_a)				
Iter. (j)	$\nu^j(x^{*(j-1)})$	β^j	$R^j_{real}(x^{*(j)})$	$R^j_{ideal}(x^{*(j)})$
1	1.000	3.570	155.6	159.7
2	0.974	3.381	159.6	163.6
3	0.976	3.391	159.4	163.3
4	0.976	3.390	159.4	163.4

Table 6 Comparison of the results of the reliability analysis of non-corroded plates

Non-corroded plates	FORM	MCFM
Reliability index (β)	3.364	3.390
Number of R_{real}(X) calls	63 (f.dif.)/112.dif	4
Design point, x^*		
t-Thickness	14.7	15.1
E-Young Modulus	190520	192383
σ_y-Yield stress	246.4	241.3
$w_{\max}$-Initial imperf	2.40	2.97
C_a-(% of σ_y^c)	0.67	0.68

Table 7 Results of the Model Correction factor method (MCFM) for non-uniform corroded plates.

Non-uniform corroded plates (Randon variables: t,E, σ_y, w_{max}, C_a+ ζ_i, $ir = 1 \ldots 15$)				
Iter. (j)	$\nu^j(x^{*(j-1)}, \zeta^{*(j-1)})$	β^j	$R^j_{real}(x^{*(j)}, \zeta^{*(j)})$	$R^j_{ideal}(x^{*(j)})$
1	1.000	3.570	138.1	159.7
2	0.864	2.359	162.0	185.9
3	0.872	2.625	155.7	181.5
4	0.858	2.558	157.3	183.2

Table 8 Comparison of the results of the reliability analysis of corroded plates

Non-corroded plates	FORM	MCFM
Reliability index (β)	2.511	2.558
No. of iterations	7	4
No. of R_{real}(X, ζ) calls	154 (f.diff) 308 (c.diff)	64
Design point, x^*		
t-thickness	16.11	16.81
E-Young modulus	196090	196477
σ_y-yield stress	254.11	247.1
$w_{\max}$-initial imperf	2.16	2.37
C_a-(% of σ_y^c)	0.66	0.67

these differences are of the same level indicating that the they are mainly due to the use of the semi-empirical formulae to predict the ultimate strength of the plate and that the inclusion of the random field variables by the first-order Taylorexpansion of the model correction factor f_c(X, ζ) is adequate. This can also be seen by observing that the most likely realization of the random field of corrosion obtained by the MCFM (*RF* ζ_i^*) (Fig. 6), is identical to the one obtained by FORM (Fig. 3).

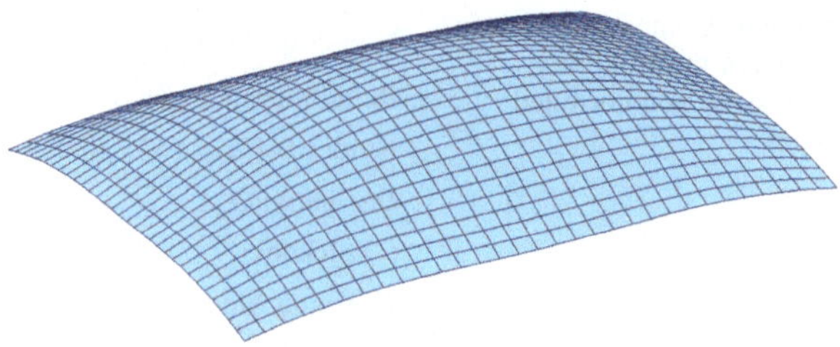

Fig. 6 Most likely realizations of the random field of corrosion obtained by the MCFM ($RF\ \zeta_i^*$))

6 Conclusions

The paper has extended the application of the model correction factor method to problems involving spatial variability of input parameters described by random fields. The approach consists of including in the formulation of the model correction factor the random variables that represent the random field by means of a first-order Taylor expansion constructed around FORM design points. This way, the model correction factor that corrects the predictions of a simplified semi-empirical response model to match the response evaluated numerically by a more accurate and realistic finite-element structural model, is defined in terms of all random variables of the reliability problem. These include the random variables not accounted by the simplified response model and the set of standard normal distributed random variables that represent the discretized random field.

The approach is applied for reliability evaluation of steel plates with non-uniform corrosion described by a log-normal random field. The results show that the model correction factor method is computationally efficient and provides adequate results. In the present example the model correction factor method required only 64 numerical evaluations of the plate strength by FEM, which is less than half of that required by FORM with the direct coupling of the FEM structural model with the limit state function.

One should note however that the choice of the simplified response surface model may influence the reliability results, particularly the design point and the sensitivity factors. This results from the relationships among the variables of the physical semi-empirical response models that typically depend on how the design equations are formulated and on the structural failure modes they represent.

Acknowledgements This work was developed in the scope of the project "Adaptive methods for reliability analysis of complex structures", contract PTDC/ECM/115932/2009, funded by the Portuguese Foundation for Science and Technology (Fundação para a Ciência e a Tecnologia - FCT). This study contributes to the Strategic Research Plan of the Centre for Marine Technology and Ocean Engineering, which is financed by Portuguese Foundation for Science and Technology (Fundação para a Ciência e Tecnologia—FCT), under contract UIDB/UIDP/00134/2020.

References

1. Teixeira, A. P., & Guedes Soares, C. (2011). Reliability assessment of plate elements with random properties. In C. Guedes Soares, Y. Garbatov, N. Fonseca, & A. P. Teixeira (Eds.) *Marine technology and engineering* (pp. 1361–1375). London: Taylor & Francis Group.
2. Teixeira, A. P., & Guedes Soares, C. (2018). Adaptive methods for reliability analysis of marine structures. In *Proceedings of the 37th International Conference on Offshore Mechanics and Arctic Engineering—OMAE2018*, ASME, June 17–22, 2018, Madrid, Spain, p. paper n. OMAE2018–77311.
3. Rackwitz, R. (2001). Reliability analysis—A review and some perspectives. *Structural Safety, 23*, 365–395.
4. Teixeira, A. P., & Guedes Soares, C. (2007). Probabilistic modelling of the ultimate strength of plates with random fields of corrosion. In G. Deodatis & P. D. Spanos (Eds.), *Computational Stochastic Mechanics* (pp. 653–661). Rotterdam: Millpress.
5. Teixeira, A. P., & Guedes Soares, C. (2010). Response surface reliability analysis of steel plates with random fields of corrosion, safety, reliability and risk of structures, infrastructures and engineering systems. In F. Furuta, & Shinozuka (Eds.) (pp. 474–481) London: Taylor & Francis Group
6. Chojaczyk, A. A., Teixeira, A. P., Neves, L. C., Cardoso, J. B., & Guedes Soares, C. (2015). Review and application of artificial neural networks models in reliability analysis of steel structures. *Structural Safety, 52*, 78–89.
7. Kaymaz, I. (2005). Application of Kriging method to structural reliability problems. *Structural Safety, 27*(2), 133–151.
8. Gaspar, B., Teixeira, A. P., & Guedes Soares, C. (2014). Assessment of the efficiency of Kriging surrogate models for structural reliability analysis. *Probabilistic Engineering Mechanics, 37*, 24–34.
9. Gaspar, B., Teixeira, A. P., & Guedes Soares, C. (2015). Adaptive surrogate model with active refinement combining Kriging and a trust region method. In *12th International Conference on Applications of Statistics and Probability in Civil Engineering, ICASP12*, July 12–15, Vancouver, Canada. (pp. 1–8).
10. Gaspar, B., Teixeira, A. P., & Guedes Soares, C. (2015). Structural reliability analysis combining Kriging surrogate models with an adaptive trust region method. In G. Deodatis, & P. D. Spanos (Eds.) *Proceedings of the 7th International Conference on Computational Stochastic Mechanics (CSM-7)* (pp. 286–97). Singapore: Research Publishing
11. Gaspar, B., Teixeira, A. P., & Guedes Soares, C. (2017). Adaptive surrogate model with active refinement combining Kriging and a trust region method. *Reliability Engineering and System Safety, 165*(April), 277–291.
12. Melchers, R. E. (2003). Probabilistic models for corrosion in strutural reliability assessment—Part 1: Empirical models. *Journal of Offshore Mechanics and Arctic Engineering, 125*, 264–271.
13. Teixeira, A. P., & Guedes Soares, C. (2008). Ultimate strength of plates with random fields of corrosion. *Structure and Infrastructure Engineering, 4*(5), 363–370.
14. Teixeira, A. P., Guedes Soares, C., & Wang, G. (2013). Probabilistic modelling of the ultimate strength of ship plates with non-uniform corrosion. *Journal of Marine Science and Technology, 18*(1), 115–132.
15. Kmiecik, M., & Guedes Soares, C. (2002). Response surface approach to the probability distribution of the strength of compressed plates. Marine Structures, *15*(N. 2), 139–156.
16. Ditlevsen, O., & Arnbjerg-Nielsen, T. (1994). Model-correction-factor method in structural reliability. *Journal of Engineering Mechanics, 120*(1), 10.
17. Garre, L., Friis-Hansen, P., & Rizzuto, E. (2006). The model correction factor method: an efficient response surface technique. In *3rd ASRANet International Colloquium.*
18. Xu, M. C., Teixeira, A. P., & Guedes Soares, C. (2015). Reliability assessment of a tanker using the model correction factor method based on the IACS-csr requirement for hull girder ultimate strength. *Probabilistic Engineering Mechanics, 42*, 42–53.

19. Li, C. C., & Der Kiureghian, A. (1993). Optimal discretization of random fields. *Journal of Engineering Mechanics, 119*(6), 1136–1154.
20. Guedes Soares, C., & Garbatov, Y. (1999). Reliability of maintained corrosion protected plate subjected to non-linear corrosion and compressive loads. *Marine Structures, 12*(6), 425–446.
21. Paik, J. K., Kim, S. K., & Lee, S. K. (1998). Probabilistic corrosion rate estimation model for longitudinal strength members of bulk carriers. *Ocean Engineering, 25*(10), 837–860.
22. Hasofer, A. M., & Lind, N. C. (1974). An exact and invariant first-order reliability format. *Journal of the Engineering Mechanics Division, 100,* 111–121.
23. Melchers, R. E., & Beck, A. T. (2018). *Structural Reliability Analysis and Prediction*, 3rd edn. New York: Wiley.
24. Bucher, C. (2009). *Computational analysis of randomness in structural mechanics.* London: Taylor & Francis Group.
25. Rackwitz, R., & Fiessler, B. (1978). Structural reliability under combined random load sequences. *Computers & Structures, 9,* 486–494.
26. Ditlevsen, O. (1981). Principle of normal tail approximation. *Journal of Engineering Mechanics Divison, 107*(6), 1191–1209
27. Liu, P. L., & Der Kiureghian, A. (1991). Optimization algorithms for structural reliability. *Structural Safety, 9,* 161–177.
28. Guedes Soares, C. (1988). Design equation for the compressive strength of unstiffened plate elements with initial imperfections. *Journal of Constructional Steel Research, 9,* 287–310.

Spatial Variability of Rebar Corrosion and Performance Evaluation of Corroded RC Structures Using Probabilistic Analysis and Finite Element Method

Mitsuyoshi Akiyama, Dan M. Frangopol, and Mingyang Zhang

Abstract Corrosion of steel reinforcement is spatially distributed over RC structures due to several factors such as different environmental exposure, concrete quality and cover. Ignoring the effect of spatial variability is a drastic simplification for the prediction of the remaining service life of RC structures. Therefore, it is essential to identify the parameters influencing the spatial steel corrosion and structural performance of corroded RC structures. In this paper, an experimental research was conducted to study the effects of current density, concrete cover, rebar diameter, and fly ash on the spatial variability of steel weight loss, corrosion crack, and structural behavior of corroded RC beams using X-ray and digital image processing technique. The test results showed that low current density induced highly non-uniform corrosion associated with few large pits and cracks at certain locations while higher current density produced more uniform corrosion and cracks occurred over the whole beam. Gumbel distribution parameters were derived from the experimental data of steel weight loss to model spatial steel corrosion. A novel approach was established to assess the reliability of RC structures using finite element analysis and probabilistic simulation considering the spatial variability in steel weight loss. Using the Gumbel distribution parameters derived from the steel weight loss data associated with higher current density may underestimate the non-uniformity of corrosion distribution which can lead to an overestimation of the load capacity of corroded RC structures.

Keywords Corroded RC structures · Spatial variability of steel corrosion · X-ray image · Current density · Gumbel distribution

M. Akiyama (✉)
Department of Civil and Environmental Engineering, Waseda University, Tokyo, Japan
e-mail: akiyama617@waseda.jp

D. M. Frangopol
Department of Civil and Environmental Engineering, Engineering Research Center for Advanced Technology for Large Structural Systems (ATLSS Center), Lehigh University, Bethlehem, USA

M. Zhang
Department of Bridge Engineering, Tongji University, Shanghai, China
e-mail: zhang601@tongji.edu.cn

© The Author(s), under exclusive license to Springer Nature Switzerland AG 2021

J. C. Matos et al. (eds.), *18th International Probabilistic Workshop*, Lecture Notes in Civil Engineering 153, https://doi.org/10.1007/978-3-030-73616-3_56

1 Introduction

The corrosion of steel reinforcement was found to vary highly over the RC structures due to several factors including different environmental exposure, concrete cover, and concrete quality, among others. The long-term structural performance has been assessed based on probabilistic concepts and methods [1, 2]. It was found that ignoring the spatial variability of steel corrosion may lead to an overestimation of structural reliabilities [3]. Therefore, it is essential to model spatial variability of steel corrosion for the reliability analysis of RC structures.

Efforts have been made to study and model the spatial variability of steel corrosion using the experimental data from the accelerated-corrosion RC members by impressed current method [4–6]. However, a current density (I_{corr}) higher than the maximum value [7] found in natural corrosion condition was applied to accelerate the corrosion test [4, 5], neglecting its effects on the spatial variability of steel corrosion and thereby on the structural reliability of corroded RC structures. Furthermore, the spatial variability of steel corrosion is affected by other factors, such as rebar diameter, concrete cover and concrete quality. The effects of these parameters on the spatial variability of steel corrosion have not been well understood.

In the previous research, the simplified cross-sectional analysis was commonly used to assess the reliability of corroded RC structures considering the loss of cross-sectional area of rebars without taking into account the spatial distribution of steel corrosion and deterioration of bond between reinforcement and concrete [8]. The finite element (FE) method can provide a more accurate performance assessment of corroded RC members [9]. However, the reliability assessment of corroded RC beams considering the spatial steel corrosion utilizing FE method is scarce due to enormous computation cost when it is combined with Monte Carlo simulation (MCS). With the response surface method (RSM) that can reduce the computation cost, the FE analysis can be used together with MCS in the reliability analysis.

In this paper, experiments to investigate the effects of current density, cover, rebar diameter and fly ash on the spatial variability of steel corrosion were conducted. The spatial variability associated with steel corrosion was monitored and quantified by X-ray photogram and digital image processing techniques. The statistical data based on Gumbel distribution was derived and incorporated in a probabilistic model to represent the spatial steel corrosion. A novel approach is established to assess the reliability of RC beams using FE analysis considering the spatial variability of steel corrosion. Finally, an illustrative example for reliability estimation of corroded RC beam is presented.

2 Experimental Program

Total nine RC beams with the size of 80 × 140 × 1460 mm were cast to study the effect of current density, concrete cover, rebar diameter and fly ash on the spatial variability of steel corrosion and crack width using X-ray and digital image processing techniques. The RC beams were corroded using different current densities; 10, 50, 100, 200, and 500 μA/cm^2. Concrete covers of 10 and 20 mm were used with rebar diameter of 13 and 19 mm, respectively. To evaluate the effect of fly ash on the spatial variability of steel weight loss, one RC beam was cast with the concrete with 30% of cement replaced by fly ash.

Figure 1 shows the experimental procedure. After curing of the specimen for 28 days, the electrolytic technique was utilized to corrode RC beams. The corrosion process was started and halted at three corrosion levels with mean steel weight loss (*MRw*) of around 2.5, 7 and 12% so that the specimens could be moved out of the tank to perform the X-ray radiography and measurement of crack width. The images of original and corroded rebar in beams were acquired by X-ray apparatus from 8 viewing angles at various corrosion levels. After taking the X-ray photo, the steel weight loss (*Rw*) for every 5 mm was estimated using the software for digital image processing. *Rw* is determined by averaging the steel weight loss of the rebar per 5-mm length from eight viewing angles [6]:

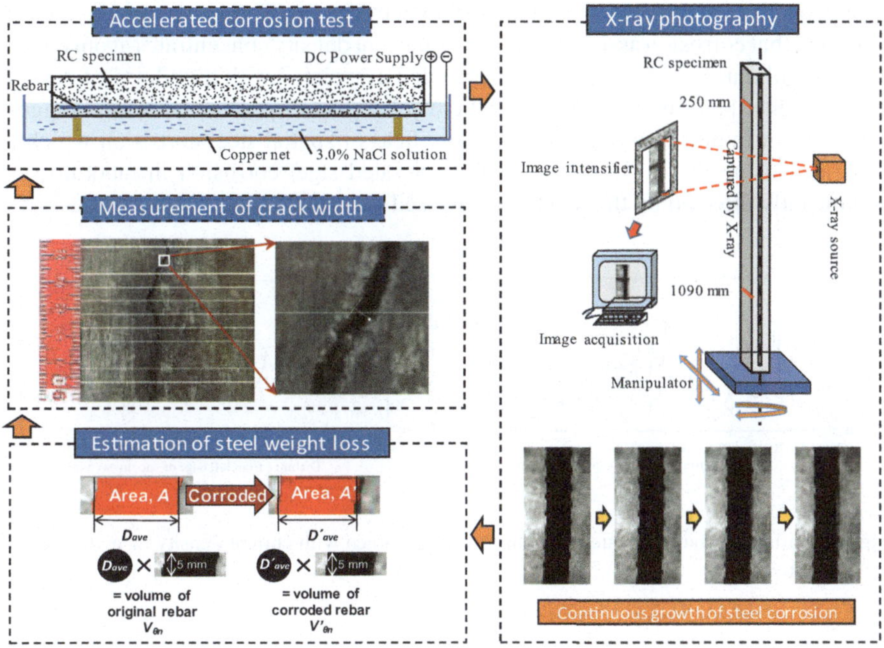

Fig. 1 Flowchart of experimental test procedure

$$Rw = \frac{1}{k}\sum \frac{w_{\theta n} - w'_{\theta n}}{w_{\theta n}} \tag{1}$$

where $w_{\theta n}$ and $w'_{\theta n}$ are the weights of original and corroded rebars per 5-mm length, respectively; θ_n is the n-th viewing angle, and $k = 8$ is the number of viewing angles. The longitudinal crack that occurred along the bottom of the beam was measured at different corrosion levels.

3 Experimental Results and Discussion

Figure 2 illustrates the spatial variability associated with the steel weight loss of corroded RC beams with current densities of 50 and 500 μA/cm^2. The standard deviation of steel corrosion is also indicated along with *MRw*. It can be seen that the distribution of *Rw* over the length of beam is spatially non-uniform. The results show that the spatial variability in steel weight loss is dependent on the impressed current density. The rebar of specimen with I_{corr} of 10 μA/cm^2 had a large corrosion pit intensifying at a small region from 310 to 470 mm. The rebar of specimen with I_{corr} of 50 μA/cm^2 had only a few large corrosion pits concentrating at three regions from 310 to 430 mm, 610 to 790 mm, and 1030 to 1090 mm. By contrast, the corrosion of rebar in specimen with higher current density of 500 μA/cm^2 fluctuated with several peaks over the length of rebar, as shown in Fig. 2b. These comparisons clearly indicated that corrosion associated with low current density concentrates at only a few regions with large corrosion pits; whereas that associated with high level of current densities distributes more evenly over the whole rebar. Experimental results suggest a significant implication that the stochastic field of spatial steel corrosion which is generated using the statistical data of accelerated steel corrosion in laboratory is significantly affected by the level of impressed current density.

Fig. 2 Spatial distribution of steel weight loss of specimen with current density of **a** 10 μA/cm^2 and **b** 500 μA/cm^2

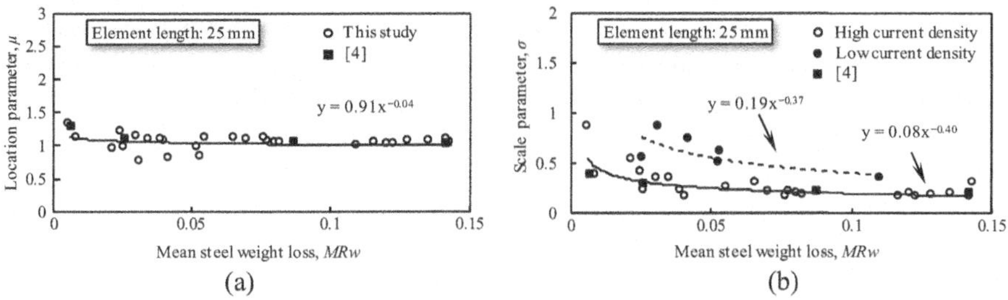

Fig. 3 Relationships between Gumbel distribution parameters (μ, σ) and *MRw* at the element length of 25 mm

4 Probabilistic Model for Structural Performance of the Corroded RC Beam

Gumbel distribution statistics of extreme values has been widely used for modeling the spatial variability associated with the cross-sectional area loss of the corroded rebar [5, 6, 8, 10]. A probabilistic method is established using the Gumbel distribution of the ratio of local maximum steel weight loss to *MRw* (R_{swl}) derived from the experimental data. R_{swl} can be determined as:

$$R_{swl} = W_{max} MRw \tag{2}$$

where W_{max} is the local maximum steel weight loss derived at a specific element length and *MRw* is the mean steel weight loss.

In order to model spatial variability of steel corrosion, the relationship between *MRw* and Gumbel distribution parameters μ and σ are established via regression analysis, as shown in Fig. 3. It can be seen that parameters μ and σ generally decrease with the increase in *MRw*. Figure 3 indicates that no significant effect of cover, rebar diameter and fly ash on both parameters can be identified. However, it is apparent that the parameter σ derived from the specimens using low current density ($I_{corr} = 10$ and 50 μA/cm^2) is larger than that derived from the specimens with high current density ($I_{corr} \geq 100$ μA/cm^2). Hence, the Gumbel distribution scale parameters derived from low current density were grouped together and regressed with one equation and other experimental parameters were grouped together with the same regression line, as shown in Fig. 3b.

5 Illustrative Example

Reliability assessment of a simply-supported RC beam with a single rebar affected by corrosion is performed utilizing FE method and probabilistic model. The RC beam is 8 m long with a cross-section of 200 mm × 350 mm. The diameter of rebar is 28 mm. In this study, Nataf transformation is applied herein to generate correlated R_{swl} for modeling the spatial variability of steel weight loss [11]. The MCS is used to estimate the failure probability of RC beam. To conduct the MCS with FE analysis, the yield load capacity R in Eq. (3) should be estimated using RSM. The limit state function is defined as:

$$G = R - S \tag{3}$$

where R is the yield load capacity calculated by response surface function and S is the load effect.

Figure 4 shows the failure probability of the corroded RC beam calculated by two probabilistic models (i.e., low and high current densities) with the MRw ranging from 5% to 30%. The results show that the failure probability considering the correlation of R_{swl} is lower than those without considering the correlation of R_{swl}.

It also can be seen that the failure probability in the case of low current density is higher than that in the case of high current density. This result is due to the fact that the Gumbel scale parameters derived from the regression equation with the low current density are higher than those with the high current density. Consequently, the generated rebar samples associated with the low current density have a larger variability in cross-sectional area loss, which leads to lower yield load capacities of the RC beam. Using the Gumbel distribution parameters obtained from the experimental data with higher current densities will underestimate the effect of non-uniformity of steel corrosion on the failure probability of RC beams.

6 Conclusions

An illustrative example was given to assess the reliability of corroded RC structures incorporated the FE analysis with probabilistic method considering the spatial variability of steel corrosion. The failure probability of the RC beam is influenced by the correlation of R_{swl}. Ignoring correlation of R_{swl}, it may lead to overestimate the failure probability of corroded RC structures. Among the experimental parameters, the current density has a significant effect on Gumbel distribution parameters. The distribution of R_{swl} of the specimens associated with low current density has substantially larger scale parameters than those with a large current density. Using the Gumbel distribution parameters obtained by the experimental data with higher current densities ($I_{corr} \geq 100\ \mu A/cm^2$) cannot reproduce the non-uniformity of corrosion distribution. It can result in an overestimation of the loading capacity and consequently causes lower failure probability of corroded RC structures.

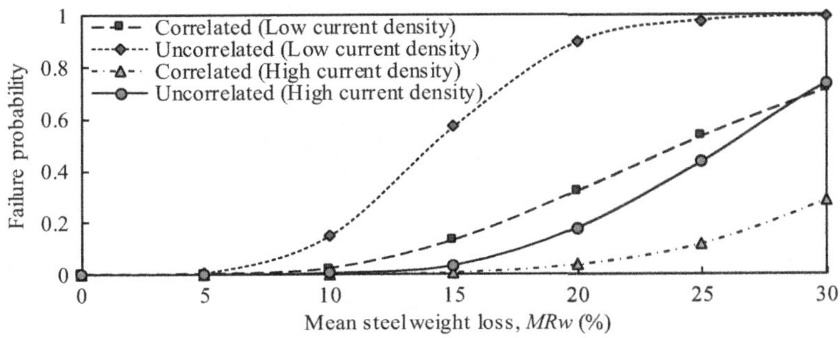

Fig. 4 Failure probability considering correlated and uncorrelated R_{swl} with element length of 25 mm

References

1. Akiyama, M., Frangopol, D. M., & Yoshida, I. (2010). Time-dependent reliability analysis of existing RC structures in a marine environment using hazard associated with airborne chlorides. *Engineering Structures, 32,* 3768–3779.
2. Akiyama, M., Frangopol, D. M., & Matsuzaki, H. (2011). Life-cycle reliability of RC bridge piers under seismic and airborne chloride hazards. *Earthquake Engineering and Structural Dynamics, 40,* 1671–1687.
3. Stewart, M. G. (2004). Spatial variability of pitting corrosion and its influence on structural fragility and reliability of RC beams in flexure. *Structural Safety, 26,* 453–470.
4. Lim, S., Akiyama, M., Frangopol, D. M., & Jiang, H. (2017). Experimental investigation of the spatial variability of the steel weight loss and corrosion cracking of reinforced concrete members: Novel X-ray and digital image processing techniques. *Structure and Infrastructure Engineering, 13,* 118–134.
5. Lim, S., Akiyama, M., & Frangopol, D. M. (2016). Assessment of the structural performance of corrosion-affected RC members based on experimental study and probabilistic modeling. *Engineering Structures, 127,* 189–205.
6. Zhang, M., Song, H., Lim, S., Akiyama, M., & Frangopol, D. M. (2019). Reliability estimation of corroded RC structures based on spatial variability using experimental evidence, probabilistic analysis and finite element method. *Engineering Structures, 192,* 30–52.
7. Andrade, C., Alonso, C., & Molina, F. J. (1993). Cover cracking as a function of bar corrosion: Part I-Experimental test. *Materials and Structures, 26,* 453–464.
8. Gu, X., Guo, H., Zhou, B., Zhang, W., & Jiang, C. (2018). Corrosion non-uniformity of steel bars and reliability of corroded RC beams. *Engineering Structures, 167,* 188–202.
9. Kallias, A. N., & Rafiq, M. I. (2013). Performance assessment of corroding RC beams using response surface methodology. *Engineering Structures, 49,* 671–685.
10. Zhang, W., Zhou, B., Gu, X., & Dai, H. (2014). Probability distribution model for cross-sectional area of corroded reinforcing steel bars. *Journal of Materials in Civil Engineering, 26,* 822–832.
11. Zhang, M., Nishiya, N., Akiyama, M., Lim, S., & Masuda, K. (2020). Effect of the correlation of steel corrosion in the transverse direction between tensile rebars on the structural performance of RC beams. *Construction and Building Materials, 264,* 120678.

Statistical Dependence Investigation Related to Dowel-Type Timber Joints

Caroline D. Aquino, Leonardo G. Rodrigues, Wellison S. Gomes, and Jorge M. Branco

Abstract The design of timber connections with dowel-type fasteners is dependent on the knowledge of their mechanical behavior and failure modes. Concerning the main design parameters, the timber embedment strength and the dowel bending moment capacity are the parameters that govern the load-carrying capacity. The correlation between the timber embedment strength and the dowel bending moment capacity has not been sufficiently addressed in the literature yet. However, since they both share a common dependency to the timber density, they are probably correlated. To investigate this, traditional distribution fitting procedures, as well as copula functions, are implemented to consider the correlation between them. By doing so, it is aimed to evaluate the effectiveness of the different approaches in describing the dependence structure of the variables and their influence on the structural reliability. It was found that, for single dowel-type connections, the impact of the copulas on the results is small. It is indicated that, unless significantly nonlinear correlations exist among the data, the results obtained by applying different copula functions will probably be very close.

Keywords Timber joints · Reliability · Joint behavior · Copula theory

1 Introduction

An adequate design of a timber structure is very dependent on the efficiency and safety of its connections. In fact, it has been suggested that they can govern the overall structural strength, serviceability, and fire resistance [1]. In addition, assessments of severely damaged timber structures, after extreme events, often point to an inadequacy in the connections as the primary cause of failure [2]. Thus, it is crucial

C. D. Aquino (✉) · W. S. Gomes
CORE, Department of Civil Engineering, Federal University of Santa Catarina, Florianópolis, Brazil

L. G. Rodrigues · J. M. Branco
ISISE, Department of Civil Engineering, University of Minho, Guimarães, Portugal

© The Author(s), under exclusive license to Springer Nature Switzerland AG 2021

J. C. Matos et al. (eds.), *18th International Probabilistic Workshop*, Lecture Notes in Civil Engineering 153, https://doi.org/10.1007/978-3-030-73616-3_57

to study the behavior of connections in order to achieve a safe design of timber structures.

The variability of materials in structural timber members is relatively well understood and has been addressed in the literature (e.g., [3–5]). However, there is only few researches performed regarding the reliability assessment of timber connections with dowel-type fasteners as well as data characterization of the parameters involved in the design of timber connections.

The existing studies on the reliability assessment of timber connections with dowel-type fasteners are scarce. Köhler [3] proposed a probabilistic framework for the reliability assessment of connections with dowel-type fasteners, considering the resistance of the connection in terms of the timber embedment strength (f_h) and the effective bending capacity of the fastener ($M_{y,eff}$). However, the correlation between these parameters is not addressed. Later, Jockwer et al. [6] addressed the assessment of the failure behavior and reliability of timber connections with multiple dowel-type fasteners. The distribution characteristics of the parameters are based on [3], and regarding the dependence modeling, linear pair-wise correlation coefficients are considered. However, the statistical dependence structure between f_h and $M_{y,eff}$ is also not addressed.

The embedding strength of the timber members (f_h) and the effective bending moment capacity of the dowel ($M_{y,eff}$) are the parameters that govern the structural design. These parameters are ultimately related to the timber density (ρ) and, therefore, are correlated. Based on that assumption, correlation models that represent the statistical dependence between these two variables are developed. Independent and Gaussian copula functions are considered, and these dependence structures are compared with the inferred non-linear copula function based on theoretical data. Such comparison is done in terms of the reliability results achieved with the different dependence structures.

2 Dowel-Type Timber Joints

The load carrying capacity of single dowel-type fasteners loaded perpendicular to the fastener axis can be described by different failure modes according to the Johansen's yield theory [7]. In this study, we are particularly interested in the failure mode where one plastic hinge occurs per shear plane (see Fig. 1). The expression for the load carrying capacity according to Johanses's yield theory for this failure mode is given in Eq. (1).

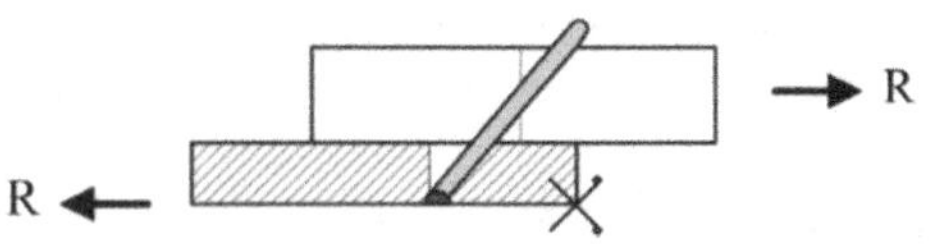

Fig. 1 Failure mode where one plastic hinge occurs per shear plane for dowel-type timber joints

$$R = \frac{f_{h,1} \cdot t_1 \cdot d}{2 + \beta_{f_h}} \left[2\beta_{f_h}(1 + \beta_{f_h}) + \frac{4\beta_{f_h}(4 + \beta_{f_h}) \cdot M_y}{f_{h,1} \cdot t_1^2 \cdot d} - \beta_{f_h} \right] \quad (1)$$

where t_1 and t_2 are the thickness of member 1 (side members) and member 2 (middle member); $f_{h,1}$ and $f_{h,2}$ are the embedding strength corresponding to members 1 and 2; β_{f_h} is the ratio between embedding strengths $f_{h,1}$ and $f_{h,2}$; d is the fastener diameter; and M_y is the fastener yield moment.

2.1 Embedment Strength f_h

Associated with the resistance of the timber element against lateral penetration of a stiff fastener, the embedding strength is mostly influenced by the timber density (f_h increases linearly with density), the diameter of the fastener (f_h decreases with increasing fastener diameter), and moisture content (f_h decreases with increasing moisture content) [8, 9]. The experimental procedures for measuring the embedding characteristics of wood are discussed in [10]. Additional discussions considering implications in the timber embedment strength, such as different wood species and grain direction are referred to [11, 12].

There are empirical equations available in the literature to determine f_h. The present study bases on the normative expression given by Eurocode 5 [13] for the wood embedding strength parallel to grain $f_{h,0}$, presented in Eq. (2), dependent on the wood density ρ (kg/m^3) and on the diameter of the dowel d (mm).

$$f_{h,0} = 0.082(1 - 0.01d)\rho \quad (2)$$

2.2 Bending Moment M_y

Related to the dowel's strength against bending, it is mainly influenced by the dowel diameter and the yield strength of the dowel material. In the scope of this paper, the smooth dowels are considered made of mild steel. In the failure modes associated to plastic deformations of dowels, where their bending capacity is mobilized, the effective bending moment directly influences the load-carrying capacity of the connection. In terms of experimental evaluation, the characterization of the bending moment can be performed through a four-point bending test, according to the guidelines of EN 409 [14]. In the experiment, the yield moment of a fastener is determined at a bending angle of 45°. In this configuration, the whole cross-section of the dowel is assumed to be under plastic strain [15]. However, it is argued that when bolted connections are tested, and the failure modes achieved are characterized by the bending of dowels, the bending angles often lie below 45° (see [16]). In this scenario, the plastic capacity of the dowels is partially used. The effective bending moment resides between the elastic ($M_{y,el} = 0.8 \cdot f_u \cdot \pi \cdot d^3/32$) and plastic ($M_{y,pl} = 0.8 \cdot f_u \cdot d^3/6$) bending capacity of the dowels' cross-section, considerably lower than the results achieved through EN 409 [14]. Here, f_u is the fastener tensile strength. Blass et al. [15] proposed a correction factor $\overline{M}(\alpha)$, given in Eq. (3), to address this partially mobilized plastic moment. The effective bending moment $M_y(\alpha)$ is given by the product between the plastic bending capacity $M_{y,pl} = M_y(\alpha = 45°)$ and the factor $\overline{M}(\alpha)$ (see Eq. 4).

$$\overline{M}(\alpha) = (0.866 + 0.00295\alpha)\left(1 - \exp\left(\frac{-0.248\alpha}{0.866}\right)\right) \tag{3}$$

$$M_{y,eff}(\alpha) = \overline{M}(\alpha) \cdot M_y(\alpha = 45°) \tag{4}$$

Therefore, in order to calculate the effective bending moment, one must know α. The bending angle can be measured directly from the load-carrying experiments or obtained by theoretical approaches such as the one presented by Blass et al. [15]. The expressions to assess α are derived based on equilibrium conditions. A scheme of the connection acting forces is given in Fig. 2.

From Fig. 2, α can be written in the format of $\alpha = \arctan(\delta/l)$, where δ is the maximum deformation ($\delta = 15$ mm) according to the guidelines of EN 26891 [17], and l is the length where the embedding strength is reached. Obtaining the geometry variables x_1, x_2, y_1 and y_2 from equilibrium conditions (see [16]), α can be determined from Eq. (5).

$$\alpha = \arctan\left(\frac{\delta}{\frac{t_1}{2} + \frac{t_1}{2+\beta_{fh}}[a]\left(\frac{1}{2} + \beta_{fh}\right)}\right)$$

where,

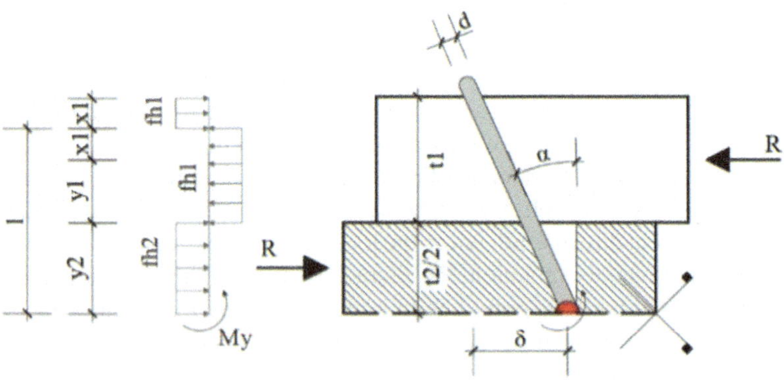

Fig. 2 Dowels' bent in double shear

$$a = \sqrt{2\beta_{f_h}(1+\beta_{f_h}) + \frac{4\beta_{f_h}(2+\beta_{f_h})\cdot M_{y,eff}}{f_{h,1}\cdot d\cdot t_1^2}} - \beta_{f_h} \tag{5}$$

According to Eq. (5) α depends on M_y. The dependence between these variables can be considered by an iterative procedure using Eq. (6). A first estimation is taken as $\alpha = 45°$. Subsequently, the $\overline{M}(\alpha)$ factor in Eq. (3) is inserted in Eq. (5). The theoretical angle α usually converges after three iteration steps [15], considering a tolerance of 1°. Once α is determined, the effective bending moment $M_{y,eff}$ can be calculated through Eq. (4).

$$\alpha_{i+1} = \arctan\left(\frac{\delta}{\frac{t_1}{2} + \frac{t_1}{2+\beta_{f_h}}[a_{i+1}](\frac{1}{2}+\beta_{f_h})}\right)$$

where,

$$a_{i+1} = \sqrt{2\beta_{f_h}(1+\beta_{f_h}) + \frac{4\beta_{f_h}(2+\beta_{f_h})\cdot \overline{M}(\alpha)\cdot M_y(\alpha=45°)}{f_{h,1}\cdot d\cdot t_1^2}} - \beta_{f_h} \tag{6}$$

2.3 Dependence Structure Investigation

When $M_{y,eff}$ is determined following the approach presented by Blass et al. [15], the parameter is related to the wood embedment strength f_h. In addition, from Eq. (2), it can be seen that f_h is mainly related to the timber density ρ. Therefore, one can state that both f_h and $M_{y,eff}$ are related to the timber density ρ. Since both mechanical properties share a dependency of timber density, their statistical correlation can be addressed.

Copula functions may be employed to represent the statistical dependence structure, since they are capable of modeling non-linear correlated behavior [18]. Long-established reliability methods commonly adopt the Nataf transformation (see [19, 20]) to consider linear correlation coefficients on the data, which is equivalent to a Gaussian copula correlation construction, according to Lebrun and Dutfoy [21]. However, Wang and Li [18] highlights the fact that the implicit Gaussian dependence structure assumed is not necessarily true and may bias the reliability results.

Therefore, based on theoretically generated data, copula functions are inferred to consider the correlation between f_h and $M_{y,eff}$. Independent and Gaussian copula functions are fitted to data, and these dependence structures are compared with the inferred non-linear copula function. Such comparison is done in terms of the reliability indexes obtained by using different dependence structures.

3 Reliability Analysis and the Copula Theory

The main objective of a structural design is to fulfil the requirements for which it is being designed. Thus, the capacity of the system must exceed the demand. There are always uncertainties involved in the representation of the main variables considered in the structural design and other sources of uncertainty. Thus, unfavorable combinations of these random variables may lead the structure to reach ultimate and/or service limit states. This event can be described by using a limit state function, given as follows:

$$g(\boldsymbol{X}) = R(\boldsymbol{X}) - S(\boldsymbol{X}) \tag{7}$$

where $\boldsymbol{X}$ is vector of random variables; $R(\boldsymbol{X})$ is the random variable regarding to resistance; and $S(\boldsymbol{X})$ is the random variable representing the load effect.

Failure occurs if $g(\boldsymbol{X}) \leq 0$. The probability of failure is obtained by integrating the joint probability density function of the random variables $f_X(\boldsymbol{x})$ over the failure domain:

$$P_f = P[g(\boldsymbol{X}) \leq 0] = \int_{g(\boldsymbol{X}) \leq 0} f_X(\boldsymbol{x}) d\boldsymbol{x} \tag{8}$$

The probability of failure (P_f) of a given structural system may be difficult to determine because it depends on the joint probability density function of the random variables in the failure domain and the problem may involve lots of random variables. In this context, approximate methods, such as Monte Carlo Simulation, are usually employed for reliability analysis.

The Monte Carlo simulation method consists of generating samples of the random variables according to their joint probabilities, and evaluating, through a numerical or analytical model, the structure response for these samples. After several simulations, a statistical analysis is performed to determine the probability of failure and the corresponding reliability index [22, 23]. It should be emphasized that the lower the probability of failure, the greater the number of simulations (n_s) required for convergence. In order to accelerate the convergence of the simulation, one can use a number of techniques presented in the literature, such as the so-called importance sampling. In this case, there is an attempt to generate samples closer to the limit state boundary, which accelerates convergence. Adopting an importance sampling distribution $h_X(x)$, and using an indicator of failure $I[\boldsymbol{X}]$, where $I[\boldsymbol{x}] = 1$ if $g(\boldsymbol{x}) \leq 0$, and $I[\boldsymbol{x}] = 0$ if $g(\boldsymbol{x}) > 0$, P_f can be estimated as in Eq. (9), where the ratio $f_X(\boldsymbol{x}_i)/h_X(\boldsymbol{x}_i)$ is the weight of each simulation [23].

$$P_f \cong \sum_{i=1}^{i=n_s} I[\boldsymbol{x}_i] \frac{f_X(\boldsymbol{x}_i)}{h_X(\boldsymbol{x}_i)} \tag{9}$$

The importance sampling distribution may be obtained by centering the original distribution at the most probable failure point (MPP) over the failure surface, since this is the point that contributes the most to the P_f. The reliability problem can be transformed to the standard normal space, where all random variables are converted to equivalent variables with standard normal distribution. In this space, the MPP correspond to the closest point between the origin and the fault surface $g(\boldsymbol{X}) \leq 0$, such distance is defined as the reliability index β, which refers to the safety level of a structure. This study employees the Finite-Step-Length algorithm (FSL) to search for the MPP [24].

In reliability assessments of structures, the problem usually involves more than one random variable. Consequently, it is necessary to characterize their joint behaviour. With respect to the dependence structure, copula functions allow the modeling of non-linear correlated behaviour, which cannot be achieve through traditional approaches (e.g. the Nataf transformation [20]). Copulas are defined as functions that join or "couple" multivariate distribution functions to their one-dimensional marginal distribution functions. An M-copula is defined as an M-variate joint cumulative distribution function CDF $C : [0, 1] \rightarrow [0, 1]$ with standard uniform marginals [25]. Sklar's theorem [26] allows to express joint CDFs in terms of their marginal distribution and a copula that represents the multivariate dependence structure. Considering a random vector $\boldsymbol{X} = (X_1, X_2, \ldots, X_M)$, the theorem states that for its M-variate CDF, referred as F_X, with marginals CDFs, referred as $F_1, \ldots, F_M$, an M-copula C_X exists, such that for all $\boldsymbol{x} \in \mathbb{R}^M$ [25]:

$$F_X(\boldsymbol{x}) = C_X(F_1(x_1), \ldots, F_M(x_M)) \tag{10}$$

Given $U_i = F_i(x_i), i = 1, \ldots, M$, the copula from (10) is unique and has the expression:

$$C_X(\boldsymbol{U}) = F_X\big(F_1^{-1}(U_1), \ldots, F_M^{-1}(U_M)\big), \quad u \in [0, 1]^M \tag{11}$$

where $F_i^{-1}(U_i)$'s are the marginals inverse CDF's.

Thus, the construction of the joint distribution consists in two separate problems. First, it is required to model the marginals F_i, this can be done through an inference process based on data (see [22]). Then, it is necessary to transform the original components of X_i into uniform random variables $u_i = F_i(X_i)$. Given that the joint probability density function PDF $f_X(\boldsymbol{x}) = dF_X(\boldsymbol{x})/d\boldsymbol{x}$, and generalizing it to multiple variables, f_X can be derived using the chain rule as:

$$f_X(x_1, \ldots, x_M) = c_{1 \ldots M}\{F_1(x_1), \ldots, F_M(x_M)\} \prod_{i=1}^{M} f_i(x_i) \tag{12}$$

where $c_{1 \ldots M}(\cdot)$ is an M-variate copula density function. Therefore, one can look at a copula function as what remains of the joint cumulative distribution once the effect of the marginal distribution function has been removed [21]. To fully determine the joint distribution behaviour, a copula family needs to be assigned to the data available. In this study, the copula inference process is based on theoretical data and done by using the open source package Vine Copula Matlab [27].

4 Results and Discussion

The analysis is based on the limit state function g which is given as follows:

$$g = 2z_d R - S_G - S_Q \tag{13}$$

where z_d is the design variable given in Eq. (14), R is the load-carrying capacity given in Eq. (1), and S_G and S_Q are the permanent and variable loads, respectively.

$$z_d = \frac{\gamma_G S_{G,k} + \gamma_Q S_{Q,k}}{2R_k} \gamma_M \tag{14}$$

The subscript k refers to the characteristic value of the loads, and γ_G, γ_Q, and γ_M are the partial safety factors related to the permanent and variable loads and resistance, respectively. Table 1 presents the statistical information of the input variables involved in the problem. The random variables f_h and $M_{y,eff}$ are determined from ρ and f_u through the approaches presented in Sects. 2.1 and 2.2.

With respect to the dependence structures, Figs. 3 and 4 show the scatter plot and the respective inferred copula function from theoretically generated data for the side and middle members, respectively. The generated data consists of 1000 samples obtained via Monte Carlo method, without importance sampling. For the side members, $M_{y,eff}$ tends to decrease with the increase of $f_{h,1}$, therefore, the variables are negatively correlated. The copula inferred for the side members is the Gumbel one with $\theta = 1.8592$ rotated 270°. When the Gaussian copula is considered, the copula parameter is $\theta = -0.6554$. Regarding the middle member dependence structure, the correlation behavior is reversed, that is, $M_{y,eff}$ tends to increase with the increase of $f_{h,2}$. The copula function inferred is the Gaussian one with $\theta = 0.7333$. Therefore, since the Gaussian is the copula inferred from the theoretical data, it is

Table 1 Statistical information of the input variables (based on [3])

	ρ (kg/m^3)	f_u (MPa)	S_G (N)	S_Q (N)
Distribution	Normal	Lognormal	Normal	Gumbel
Mean value	450	427	1000	1200
St. Dev.	45	17	100	480
Fractile	5%	5%	50%	0.4
Char. value	376	400	1000	2444
	$\gamma_M = 1.3$		$\gamma_G = 1.35$	$\gamma_Q = 1.5$

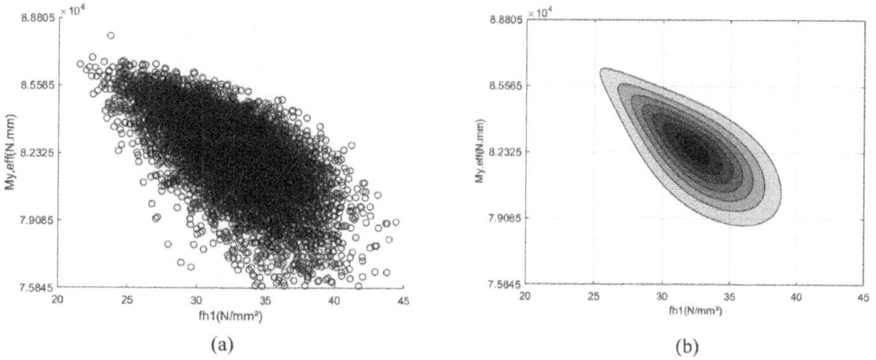

Fig. 3 Joint behavior of the generated data for side members **a** scatter plot, and **b** joint distribution

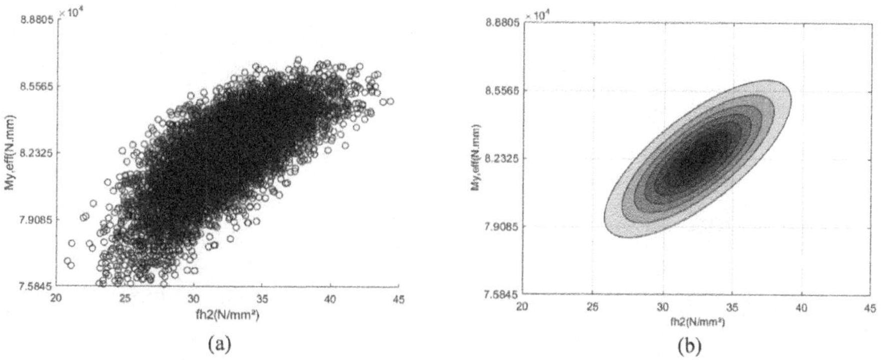

Fig. 4 Joint behavior of the generated data for middle members **a** scatter plot, and **b** joint distribution.

adopted either when the Gaussian and the Gumbel copula is considered for the side members.

The results in terms of the reliability index are given in Fig. 5 and are referred according to the side members dependence structures considered: Independent, Gaussian, and Gumbel.

It is found that the correlation model does not present a great influence on β in this case. If sensitivity indexes are computed by using the First Order Reliability Method, it is noted that the parameter with greater influence on the probability of failure is S_Q (variable load). By reducing its coefficient of variation in an attempt to increase the impact of the copulas on the results, it is seen that the reliability indexes found are still very similar. This indicates that, unless significantly nonlinear correlations exist among the data, the results obtained by applying the different copulas will probably be very close. However, it is difficult to establish beforehand if the correlation is nonlinear enough, since the reliability indexes are also very dependent on the limit state functions and on the other variables. Nevertheless, the results shown herein

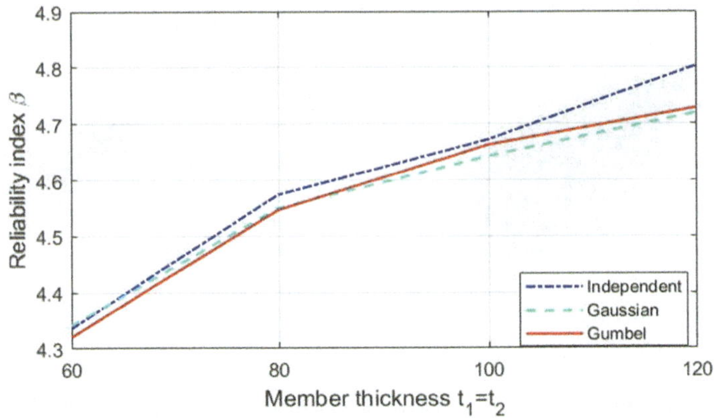

Fig. 5 Reliability results

indicate that for this particular case study a simpler approach, *e.g.* using the Nataf transformation [19], would be enough to deal with the correlation.

5 Conclusions

This study investigated the correlation behavior between the timber embedment strength and the dowel bending moment capacity, which are two parameters of significant impact on the design of dowel-type timber joints. Since, to the best of the authors knowledge, the correlation between them had not been addressed in the literature yet, traditional distribution fitting as well as copula functions were implemented to describe their statistical dependence structure. The results showed that the impact of correlation modeling on the results is small. In this context, it is indicated that for the study of single dowel- type connections, especially for the failure mode where one plastic hinge occur per shear plane, the copula functions are not necessary, and the correlation can be dealt with a simpler approach. Nevertheless, it is crucial to conduct further investigations related to distinct failure modes, specially the one that comprises two plastic-hinges per shear plane. Moreover, further studies shall include connections with multiple fasteners.

Acknowledgments This work was financed by FEDER funds through the Competitivity and Internationalization Operational Programme COMPETE, Portugal 2020, and by national funds through FCT – Foundation for Science and Technology within the scope of the Timquake project POCI-01–0145-FEDER-032031. The authors also thank the Brazilian Research Council (CNPq) for sponsoring this research through the grant 302489/2017–7 and through the scholarship grant conceded to the first author.

References

1. Köhler, J. (2005). A probabilistic framework for the reliability assessment of connections with dowel-type fasteners. In *Proceedings of the 38*th *CIB-W18 Meeting.*
2. Foliente, G. C. (1998). Design of timber structures subjected to extreme load. *Progress in Structural Engineering and Materials, 1*(3), 236–244. https://doi.org/10.1002/pse.2260010304.
3. Köhler, J. (2007). *Reliability of timber structures.* Ph.D. thesis, Institute of Structural Engineering, Swiss Federal Institute of Technology, Zurich, Swiss.
4. Faber, M. H., Köhler, J., & Sorensen, J. D. (2004). Probabilistic modeling of graded timber material properties. *Structural Safety, 26*(3), 295–309. https://doi.org/10.1016/j.strusafe.2003.08.002.
5. Jenkel, C., Leichsenring, F., Graf, W., & Kaliske, M. (2015). Stochastic modeling of uncertainty in timber engineering. *Engineering Structures, 99,* 296–310. https://doi.org/10.1016/j.engstruct.2015.04.049.
6. Jockwer, R., Fink, G., & Köhler, J. (2018). Assessment of the failure behaviour and reliability of timber connections with multiple dowel-type fasteners. *Engineering Structures, 172,* 76–84. https://doi.org/10.1016/j.engstruct.2018.05.081.
7. Johansen, K. W. (1949). Theory of timber connections. *International Association for Bridge and Structural Engineering, 9,* 249–262.
8. Santos, C. L., De Jesus, A. M. P., Morais, J. J. L., & Lousada, J. L. P. C. (2010). A comparison between the EN 383 and ASTM D5764 test methods for dowel-bearing strength assessment of wood: Experimental and numerical investigations. *Strain, 46*(2), 159–174. https://doi.org/10.1111/j.1475-1305.2008.00570.x.
9. Rammer, D. R., & Winistorfer, S. G. (2007). Effect of moisture content on dowel-bearing strength. *Wood and Fiber Science, 33*(1), 126–139.
10. Whale, L. R. J., & Smith, I. (1989). A method for measuring the embedding characteristics of wood and wood-based materials. *Materials and Structures, 22*(6), 403–410. https://doi.org/10.1007/BF02472217.
11. Sawata, K., & Yasumura, M. (2002). Determination of embedding strength of wood for dowel-type fasteners. *Journal of Wood Science, 48*(2), 138–146. https://doi.org/10.1007/BF00767291.
12. Sandhaas, C., Ravenshorst, G. J. P., Blass, H. J., & Van de Kuilen, J. W. G. (2013). Embedment tests parallel-to-grain and ductility aspects using various wood species. *European Journal of Wood and Wood Products, 71*(5), 599–608. https://doi.org/10.1007/s00107-013-0718-z.
13. Eurocode 5. (2014). Design of timber structures—Part 1-1: general—Common rules and rules for buildings. Standard BS EN 1995-1-1, CEN—European Committee for Standardization.
14. EN 409. (1993). Timber structures: Test methods—Determination of the yield moment of dowel type fasteners. CEN—European Committee for Standardization.
15. Blass, H. J., Bienhaus, A., & Krämer, V. (2001). Effective bending capacity of dowel-type fasteners. *Proceedings PRO, 22,* 71–80.
16. Jorissen, A. J. M. (1998). *Double shear timber connections with dowel type fasteners.* Ph.D. thesis, Delf University, Netherlands.
17. EN 26891 (1991). *Timber structures: Joints made with mechanical fasteners—General principles for the determination of strength and deformation characteristics.* CEN—European Committee for Standardization.
18. Wang, F., & Li, H. (2018). System reliability under prescribed marginals and correlations: Are we correct about the effect of correlations? *Reliability Engineering & System Safety, 173,* 94–104. https://doi.org/10.1016/j.ress.2017.12.018.
19. Der Kiureghian, A., & Liu, P. L. (1986). Structural reliability under incomplete probability information. *Journal of Engineering Mechanics, 112*(1), 85–104. https://doi.org/10.1061/(ASCE)0733-9399(1986)112:1(85).
20. Nataf, A. (1962). Détermination des distribution de probabilités dont les marges sont donnés. *Comptes Rendus de l'Académie des Sciences, 225,* 42–43.

21. Lebrun, R., & Dutfoy, A. (2009). An innovating analysis of the Nataf transformation from the copula viewpoint. *Probabilistic Engineering Mechanics, 24*(3), 312–320. https://doi.org/10.1016/j.probengmech.2008.08.001.
22. Ang, H.-S. A., & Tang, W. H. (2014). *Probability concepts in engineering planning and design* (Vol. 2): *Decision risk, and reliability*. New York: Wiley.
23. Melchers, R. E., Beck, A. T. (2018). *Structural reliability analysis and prediction*. New York: Wiley.
24. Gong, J. X., & Yi, P. (2011). A robust iterative algorithm for structural reliability analysis. *Structural and Multidisciplinary Optimization, 43*(4), 519–527. https://doi.org/10.1007/s00158-010-0582-y.
25. Nelsen, R., (2006). An Introduction to copulas. Springer Science & Business Media.
26. Sklar, A. (1959). Fonctions de repartition à n dimensions et leurs marges. *Publications de l'Institut de Statisque de L'Université de Paris, 8,* 229–231.
27. Kurz, M. (2015). *Vine copulas with matlab*. https://github.com/MalteKurz/VineCopulaMatlab.

Stochastic Carbon Dioxide Forecasting Model for Concrete Durability Applications

Bassel Habeeb, Emilio Bastidas-Arteaga, Helena Gervásio, and Maria Nogal

Abstract Over the Earth's history, the climate has changed considerably due to natural processes affecting directly the earth. In the last century, these changes have perpetrated global warming. Carbon dioxide is the main trigger for climate change as it represents approximately up to 80% of the total greenhouse gas emissions. Climate change and concrete carbonation accelerate the corrosion process increasing the infrastructure maintenance and repair costs of hundreds of billions of dollars annually. The concrete carbonation process is based on the presence of carbon dioxide and moisture, which lowers the pH value to around 9, in which the protective oxide layer surrounding the reinforcing steel bars is penetrated and corrosion takes place. Predicting the effective retained service life and the need for repairs of the concrete structure subjected to carbonation requires carbon dioxide forecasting in order to increase the lifespan of the bridge. In this paper, short term memory process models were used to analyze a historical carbon dioxide database, and specifically to fill in the missing database values and perform predictions. Various models were used and the accuracy of the models was compared. We found that the proposed Stochastic Markovian Seasonal Autoregressive Integrated Moving Average (MSARIMA) model provides R^2 value of 98.8%, accuracy in forecasting value of 89.7% and a variance in the value of the individual errors of 0.12. When compared with the CO_2 database values, the proposed MSARIMA model provides a variance value of -0.1 and a coefficient of variation value of $-8.0e^{-4}$.

B. Habeeb
Institute for Research in Civil and Mechanical Engineering UMR CNRS 6183, University of Nantes, Nantes, France

E. Bastidas-Arteaga (✉)
Laboratory of Engineering Sciences for Environment UMR CNRS 7356, La Rochelle University, La Rochelle, France
e-mail: ebastida@univ-lr.fr

H. Gervásio
Institute for Sustainability and Innovation in Structural Engineering, University of Coimbra, Coimbra, Portugal

M. Nogal
Faculty of Civil Engineering and Geosciences, TU Delf, Delft, The Netherlands

© The Author(s), under exclusive license to Springer Nature Switzerland AG 2021

J. C. Matos et al. (eds.), *18th International Probabilistic Workshop*, Lecture Notes in Civil Engineering 153, https://doi.org/10.1007/978-3-030-73616-3_58

Keywords Seasonal Stochastic Markovian Autoregressive Integrated Moving Average model · Infrastructure reliability · Carbon dioxide forecasting · Concrete carbonation · Climate change

1 Introduction

Civil infrastructure investment in the European Union has been in a steady decline since the outbreak of the economic and financial crisis. Although the decrease appears to gradually level off from 2015 onwards with an increase of 5% [1]. The increase in the infrastructure investment from 2015 onwards was illustrated as an action by the European Union for the sake of designing and maintaining these systems for a certain service lifetime, which was recognized as critical issues worldwide.

Decision making in the civil infrastructure investment in the European Union utilizing the quality control plan is involved in the case of repairing or demolition of the reinforced concrete bridges, depending on the recent key performance indicators (KPI). The KPI are specified by engineering consultants regarding the current condition of the bridge and the strategies to be followed (Reference strategy/Representative strategy) taking into consideration the reliability, the cost and the availability of the bridge.

Reinforced concrete bridges are characterized by high durability, despite that, they are also vulnerable to natural hazards, as well as extreme events that affect their performance and serviceability. Statistics on bridge collapses worldwide reveal that natural hazards are the predominant cause of failure. French government revealed that among the 12,000 maintained bridges after the collapse of the motorway bridge located in Genoa, 840 are at risk of collapsing. This issue is common across Europe [2].

Carbonation of concrete is one of the main causes of corrosion and occurs by the reaction given in Eq. (1) between atmospheric CO_2 and the hydrated phases of concrete. This reaction generates calcium carbonate, leading to a drop in the pH value, in which the protective oxide layer of the reinforcing steel bars is broken and corrosion starts. Therefore, the life span of the concrete infrastructure is affected by the enhanced risk of carbonation induced corrosion [3].

$$Ca(OH)_2 + CO_2 = CaCO_3 + H_2O \tag{1}$$

The temperature significantly affects the diffusion coefficient of CO_2 into concrete, the rate of reaction between CO_2 and Calcium Hydroxide ($Ca(OH)_2$), and their rate of dissolution in pore water [4]. The optimum relative humidity condition for the carbonation process is between 50 and 70%, including wetting and drying cycles that enhance the reaction [5].

The carbonation process is very sensitive to the local climate depending on the environmental conditions [6]. Climate change impacts the infrastructure as the increase in CO_2 levels associated with global warming will increase the carbonation-induced corrosion. Moreover, changes in humidity and temperature significantly affect the initiation time of corrosion [7]. Since studies on global warming have predicted several changes in climate, the impact of climate change on structural reliability should be considered. For example, Bastidas-Arteaga has calculated numerically in the oceanic environment a reduction in the lifetime of failure that ranges between 1.4 and 2.3% and up to 7% when cyclic loading is considered [8].

A carbon dioxide database is essential to study the influence of realistic exposure conditions on concrete carbonation. Databases could be also used to establish probabilistic prediction models. Therefore, this study proposes a prediction model that is established based on the time-domain analysis of the database and evaluated with a short memory process. The model is also compared with other autoregressive models. The proposed Stochastic Markovian Seasonal Autoregressive Integrated Moving Average model (MSARIMA) is also used to fill the missing database and to perform predictions, taking into account the statistical analysis on the previously existing historical database and seasonality.

Climate models are based on well-notarized physical processes that simulate the transfer of energy and materials through the climate system. Climate models, also known as general circulation models, use mathematical equations to characterize how energy and matter interact in different parts of the ocean, atmosphere and land [9]. Climate models are operated using variability that is driving the climate and predicting the climate change in the future. External factors are the main inputs into the climate models that affect the amount of the solar energy absorbed by the Earth or the amount trapped by the atmosphere, these external factors are called "forcing". They include variations in the sun's output, greenhouse gases and tiny particles called aerosols that are emitted from burning fossil fuels, forest fires and volcanic eruptions. The aerosols reflect incoming sunlight and influence cloud formation except the black carbon.

Climate models provide results that vary with respect to the actual historical database; those variations are at the expense of each model differences in: (ensemble, data source, forcing, the initial state of run, driving model, aerosols influence and jet stream impact). However, the proposed model is based on stochastic time series analysis that avoids the climate models variations and provides database that is statistically related to the existing historical database.

2 Carbon Dioxide Forecasting

Time series forecasting is a quantitative approach that uses information based on historical values and associated patterns to predict future observations. Time series analysis comprises methods for analyzing time-series data to extract meaningful statistics and other characteristics of the data. The analysis includes trend, seasonality and irregular components. A time-series analysis quantifies the main features in data and random variation. These reasons, combined with improved computing power, have made time series methods widely applicable.

2.1 *Methodology*

2.1.1 Time Series Analysis

Time series analysis for carbon dioxide database is based on the time-domain analysis (autocorrelation analysis and cross-correlation analysis), in which the type of the process deduced is a short-term memory process with short-range dependence that is characterized by an exponential decay of the autocorrelation function (Acf) for the historical database.

2.1.2 Decomposition

Time series consists of two systematic components: trend and seasonality, and a non-systematic component called noise. A multiplicative nonlinear model is used as the seasonality increases with the increase in the trend. The autocorrelation function of the non-systematic component demonstrates the characteristics of the autoregressive model in terms of damaged cosine shape.

2.1.3 Stationarity

Stationarity of the database is essential to maintain the statistical properties of the time series, a stationarized series is relatively easy to predict, the stationarity is achieved through differencing and log transformation. The basic idea of stationarity is that the probability laws that govern the behavior of the process do not change over time. In a sense, the process is in statistical equilibrium. Specifically, a process is strictly stationary if the distribution of existed state is the same as the distribution of the previous state for all choices of time points and all choices of time step lag. The stationarity of the time series is checked using Kwiatkowski–Phillips–Schmidt–Shin (KPSS) test and augmented Dickey–Fuller (ADF) test [10].

2.1.4 Models

The statistical technique utilized for forecasting the carbon dioxide is Seasonal Stochastic Markovian Autoregressive Integrated Moving Average (MSARIMA) which provides high accuracy and precise results. Moreover, other statistical techniques that include moving average based methods, such as Autoregressive Moving

Average (ARMA), Autoregressive Integrated Moving Average (ARIMA), Holt-Winters' Triple Exponential Smoothing and Seasonal Autoregressive Integrated Moving Average (SARIMA) were performed in order to compare the variations in the accuracy of the models.

Lately, Autoregressive Integrated Moving Average (ARIMA) model has been used to study the short time-varying processes. However, one limitation of ARIMA is its natural tendency to concentrate on the mean values of the past series data. Therefore, it remains challenging to capture a rapidly changing process, in which the proposed model (MSARIMA) solves this issue by triggering a Markovian step when the value of the integration part is >1 and the probability of occurrence is related to the previous seasonal events.

2.2 *Models Description*

Models presented are divided into two categories: auto regression (AR) moving average (MA) parameters and exponential smoothing parameters. The proposed MSARIMA model is based on the AR and MA parameters. In addition, it accounts for seasonality and Markovian step technique.

The autoregressive model of order p, which is denoted as AR(p), writes:

$$X_t = c + \sum_{i=1}^{p} \varphi_i X_{t-i} + \varepsilon_t; \varepsilon_t \sim \mathrm{WN}\left(0, \sigma_\varepsilon^2\right) \tag{2}$$

where X_t is the state, φ is a parameter of the model, c is constant, ε_t is a random white noise *WN* and σ_ε^2 is the variance of the random white noise.

In this case, we denote by $\{X_t\} \sim$ AR (p). In the same way, we can rewrite a process AR(p) with a polynomial φ (B).

$$\varphi(B)X_t = \varepsilon_t; \varphi(B) = 1 - \varphi_1 B - \varphi_2 B^2 - \cdots - \varphi_p B^p \tag{3}$$

The moving average model of order q, which is denoted as MA(q), writes:

$$X_t = \omega + \sum_{i=1}^{q} \theta_i \varepsilon_{t-i} + \varepsilon_t \tag{4}$$

where θ is a parameter of the model and ω is the expectation of X_t, often equals to zero.

Use the backshift operator B to rewrite Eq. (4).

$$X_t = \theta(B)\varepsilon_t; \theta(B) = 1 + \theta_1 B + \theta_2 B^2 + \cdots + \theta_q B^q \tag{5}$$

2.2.1 Autoregressive Moving Average Model (ARMA)

The general ARMA model was described in the 1951 by Peter Whittle [11].

$$X_t = c + \omega + \sum_{i=1}^{q} \theta_i \varepsilon_{t-i} + \sum_{i=1}^{p} \varphi_i X_{t-i} + \varepsilon_t \tag{6}$$

The model could be written using the polynomials φ(B) and θ(B) in which the constant c and ω are zero values:

$$X_t - \sum_{i=1}^{p} \varphi_i X_{t-i} = \sum_{j=1}^{q} \theta_j \varepsilon_{t-j} + \varepsilon_t \tag{7}$$

$$\varphi(B)X_t = \theta(B)\varepsilon_t \therefore \left(1 - \sum_{i=1}^{p} \varphi_i B^i\right)X_t = \left(1 + \sum_{j=1}^{q} \theta_j B^j\right)\varepsilon_t \tag{8}$$

The ARMA model omits the integration part of its calculation leading to a non-stationary time series model in which statistical parameters will vary with time. On the contrary, embedding the integration part in the time series will control the stationarity in which the statistical properties such as mean, variance, autocorrelation, etc. are all constant over time.

2.2.2 Autoregressive Integrated Moving Average Model (ARIMA)

The ARIMA is an advanced ARMA model that solves the stationarity of the time series by using difference operation, this value is up to the second-order of integration ($d_{\max} = 2$) based on the backshift operator Eq. (9). Otherwise, it is solved using log transformation.

$$B(X_t) = X_{t-1};\ B^d(X_t) = X_{t-d} \tag{9}$$

The general equation taking into account the constant c and ω as a non-zero value, in which $c = \omega\big(1 - \varphi_1 - \cdots - \varphi_p\big)$ and ω is the mean of $(1 - B)^d X_t$, is as follows:

$$(1 - \varphi_1 B - \cdots - \varphi_p B^p)(1 - B)^d (X_t - \omega t^d/d!) = \big(1 + \theta_1 B + \cdots + \theta_q B^q\big)\varepsilon_t \tag{10}$$

2.2.3 Seasonal Autoregressive Integrated Moving Average Model (SARIMA)

The seasonality of a model is detected using an autocorrelation function in which the peaks evolve over the lag values of a defined time series with a scale value >24. The monthly seasonal stationarity of a model is based on a lag value of $s = 12$ and is known as the seasonal monthly differencing operator in Eq. (11).

$$(1 - B)^s X_t = X_t - X_{t-s} \tag{11}$$

$$\emptyset\left(B^s\right)\varphi(B)(X_t - \omega) = \Theta\left(B^s\right)\theta(B)\varepsilon_t \tag{12}$$

The SARIMA model without the differencing operations is mentioned in Eq. (12) and the terms are illustrated below:

$$\varphi(B) = 1 - \varphi_1 B - \varphi_2 B^2 - \cdots - \varphi_p B^p \tag{13}$$

$$\emptyset\left(B^s\right) = 1 - \emptyset_1 B^s - \emptyset_2 B^{2s} - \cdots - \emptyset_p B^{ps} \tag{14}$$

$$\theta(B) = 1 + \theta_1 B + \theta_2 B^2 + \cdots + \theta_q B^q \tag{15}$$

$$\Theta\left(B^s\right) = 1 + \Theta_1 B^s + \Theta_2 B^{2s} + \cdots + \Theta_q B^{qs} \tag{16}$$

where $\emptyset$ is the seasonal AR parameter, φ is the AR parameter,Θ is the seasonal MA parameter and θ is the MA parameter.

2.2.4 Markovian Seasonal Autoregressive Integrated Moving Average model (MSARIMA)

The proposed model is based on the SARIMA model. The MSARIMA solves the SARIMA only limitation with its tendency to concentrate on the mean values of the past series data by working on a sequence of time intervals changing their mean value in each time and by triggering a Markovian step Eq. (17).

$$\delta = P\{S_n | S_{n-12} = i_{n-12}\} = \begin{cases} \mathrm{E}\left(\varepsilon_{\sum i-12}\right) > 1, & X = x_{n-i} + \mu \\ \mathrm{E}\left(\varepsilon_{\sum i-12}\right) < 1, & X = x_{n-i} + 1 \end{cases} \tag{17}$$

where $\mathcal{S}$ is the state, μ is the mean value of the monthly seasonal errors of the value X and δ is the Markovian step value.

The MSARIMA model is developed based on the SARIMA model with a triggering condition when the integration value >1, the model works on increasing the accuracy of the prediction regarding the seasonal errors for the current state.

The step process depends on the most recent past event and the Markovian step is a renewable process because it presents only positive values. This model neglects the ω values in the previous equations and presents the Markovian step process value δ for more accurate results. The equation is as follows:

$$\emptyset(B^s)\varphi(B)(X_t) = \Theta(B^s)\theta(B)\varepsilon_t + \delta \tag{18}$$

2.2.5 Holt-Winters' Multiplicative Seasonal Model

Winters (1960) extended Holt's method to capture seasonality [12]. The Holt-Winters' seasonal method comprises the forecast equation and three smoothing equations. The multiplicative method is used when the seasonal variations are changing proportionally to the trend of the series. The seasonal component is expressed in relative terms and the series is seasonally adjusted by dividing through by the seasonal component. Within each year, the seasonal component will sum up to approximately the seasonal frequency value.

$$\hat{X}_{t+h|t} = (l_t + \mathrm{h} \cdot b_t)S_{t+\mathrm{h}-\mathrm{m}(\mathrm{k}+1)} \tag{19}$$

$$l_t = \alpha \frac{X_t}{S_{t-m}} + (1-\alpha)(l_{t-1} + b_{t-1}) \tag{20}$$

$$b_t = \beta(l_t - l_{t-1}) + (1-\beta)b_{t-1} \tag{21}$$

$$St = \gamma \frac{\mathrm{Xt}}{l_{t-1} - \mathrm{b}_{t-1}} + (1-\gamma)S_{t-m} \tag{22}$$

where l_t is the level, bt is the trend, St is the seasonal component, m is the seasonal frequency, and α, β and γ are the model smoothing parameters.

3 Results and Discussion

The main objective of this section is to estimate the ability of the proposed approach in forecasting carbon dioxide concentration using an incomplete database. The forecasting and prediction of the missing values are performed using the following mathematical and stochastic models: ARMA, ARIMA, SARIMA, Holt-Winters' and MSARIMA.

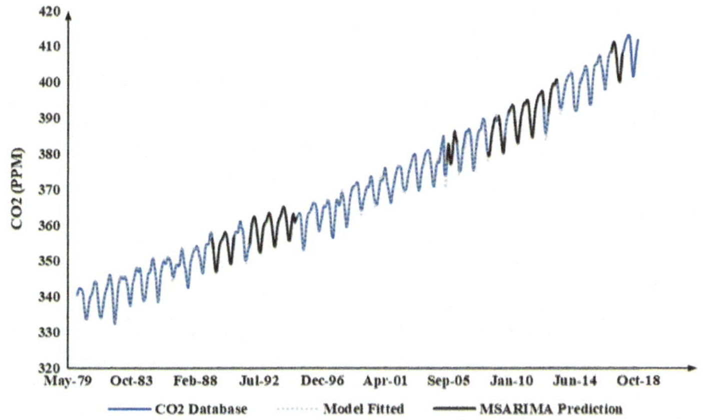

Fig. 1 Example of CO_2 assessment based on previous database

3.1 Database Description

The concentration of greenhouse gases in Portugal is measured on the island of Terceira, which is one of the nine islands in the archipelago of the Azores, located in the middle of the Atlantic Ocean. The database is available since 1979 for different greenhouse gases. In particular, for the three main gases, carbon monoxide (CO) since 1990, carbon dioxide (CO_2) since 1979, and methane (CH_4) since 1983. However, the carbon dioxide database includes missing values. The samples are collected on the island of Terceira and the analysis is carried out in NOAA lab, Hawaii, in the scope of the Cooperative Global Air Sampling Network.

3.2 MSARIMA CO_2 Database Prediction

The database offered by NOAA lab, Hawaii, in the scope of the Cooperative Global Air Sampling Network includes missing values. Therefore, a stochastic MSARIMA model presents accurate results in filling the database shown in Fig. 1 and can be used in for forecasting purposes.

3.3 Stochastic Models Analysis

3.3.1 Stochastic Models Predictions

In this section, the prediction of the MSARIMA model is compared with SARIMA and Holt-Winters' models as both include seasonal components. This is implemented

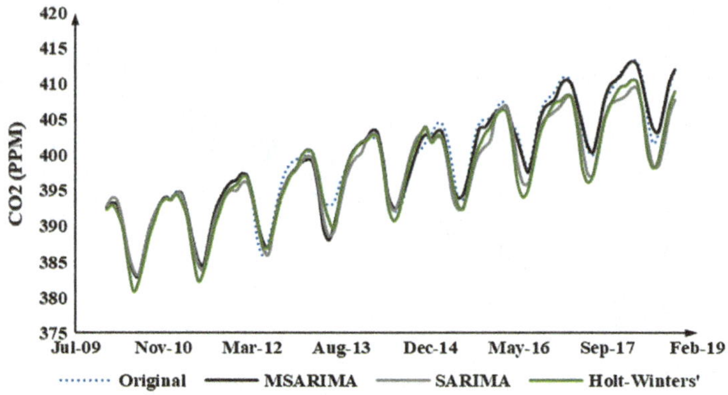

Fig. 2 Stochastic models comparison

by forecasting a historical CO_2 starting from 2010 through 2018 Fig. 2. The prediction of the MSARIMA model seems to provide the best fitting results to the original database compared to the other models. The errors associated with the predictions will be further studied in the next section.

3.3.2 Stochastic Models Accuracy

A statistical study was performed to derive the variations between the mathematical stochastic models and the meteorological station's database. The difference of the relative frequencies for CO_2 presented in Fig. 3 was performed for a time series starting from 01/2010 to 01/2018 to describe the variations in the models. The proposed MSARIMA model presents the lowest variations. Moreover, ARIMA and

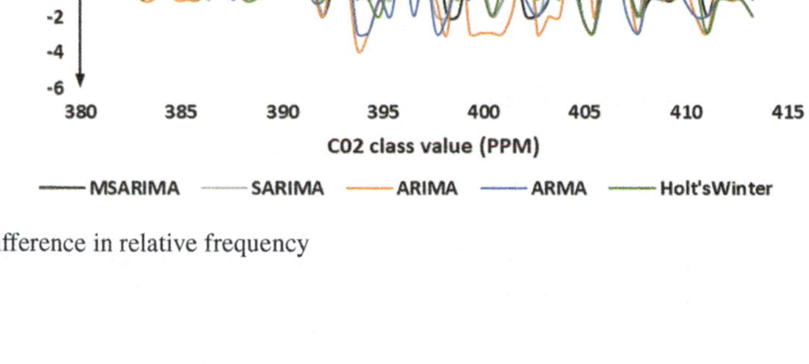

Fig. 3 Difference in relative frequency

Table 1 Statistical differences with meteorological station's CO_2 database

Model	Mean value	Variance	Coefficient of variation (%)
MSARIMA	−0.32	−0.1	−0.0008
SARIMA	−1.567	−11.94	−0.20
Holt-Winters'	−1.562	−7.07	−0.11
ARIMA	−1.27	−5.48	−0.08
ARMA	−4.82	3.05	0.07

Table 2 Accuracy between MSARIMA and SARIMA models

Model	ME	RMSE-MAE	1-MAPE [%]	R^2 [%]
MSARIMA	9.78	0.12	89.7	98.8
SARIMA	24.0	0.16	85.3	97.8

ARMA models show higher variations than the other models as seasonality is not considered.

The stochastic models' statistical study in Table 1 illustrates the variation of the models with the original database in terms of mean value, variance and coefficient of variation, in which the MSARIMA model shows the lowest variation with the meteorological station's CO_2 database. On the contrary, the others present higher variations in the results.

The accuracy of the stochastic models is finally demonstrated by comparing SARIMA and MSARIMA models with the original database for the data given in Fig. 2. This study will be carried out in terms of the error indicators in Table 2. In this table ME is the mean error, RMSE is the square root of the average of the square errors, MAE is the mean absolute error, MAPE is the mean absolute percentage error and R^2 is the proportion of the fitted model variation with the original database.

The MSARIMA model presents the highest R^2 value in which 98.8% of the CO_2 database variation is explained by the fitted model. The mean error refers to the average of all errors, it is also described as the uncertainty in measurements, the proposed MSARIMA model provides the lowest value in errors. The variation in the errors in the set of forecasts is diagnosed by the difference between RMSE and MAE, in which lower values in RMSE-MAE show lower variance in the individual errors, as shown in Table 2 the MSARIMA model has the lowest RMSE-MAE values. The accuracy of a model prediction is presented by the 1-MAPE value, as it calculates the relation between forecasted values and original values, in which the MSARIMA model has the highest accuracy in forecasting.

4 Conclusions

The prediction of the proposed Stochastic Markovian Seasonal Autoregressive Integrated Moving Average (MSARIMA) model seems to provide the best fitting results to the original CO_2 database compared to the other models.

The proposed MSARIMA model provides R^2 value of 98.8%, accuracy in forecasting value of 89.7% higher than all the other models and variance in the individual errors value of 0.12. When compared with the CO_2 database values, the proposed MSARIMA model provides a mean value of −0.32, a variance value of −0.1 and a coefficient of variation value of $-8.0e^{-4}$.

The provided results demonstrate that there is no overestimation in the predictions using the proposed MSARIMA model, which might be an obstacle due to the proposed step methodology. On the contrary, the MSARIMA model provided the best fit in predictions when compared with the original CO_2 database.

Acknowledgments This paper was carried out in the framework of the Strengthening the Territory's Resilience to Risks of Natural, Climate and Human Origin (SIRMA) project, which is co-financed by the European Regional Development Fund (ERDF) through INTERREG Atlantic Area Program with application code: EAPA_826/2018. The sole responsibility for the content of this publication lies with the author. It does not necessarily reflect the opinion of the European Union. Neither the INTERREG Europe program authorities are responsible for any use that may be made of the information contained therein.

References

1. Zachariadis, I. (2018). *Investment in infrastructure in the EU—Gaps, challenges, and opportunities*, no. October, 2018. Available: https://www.iberglobal.com/files/2018-2/infrastructure_eu.pdf.
2. Bridge Maintenance, *Le Journal du Dimanche* (2018). https://www.independent.co.uk/news/world/europe/france-bridge-collapse-risk-roads-genoa-transport-infrastructure-a8498376.html.
3. Ronac. (2018). Carbonation of reinforced concrete—Ronacrete (Vol. 44, pp. 1–6). Available: https://www.ronacrete.co.uk/carbonation-reinforced-concrete/.
4. Aguiar, J. B., & Júnior, C. (2013). Carbonation of surface protected concrete. *Construction and Building Materials, 49,* 478–483. https://doi.org/10.1016/j.conbuildmat.2013.08.058.
5. Marie Victoire, E., & Cailleux, E., & Texier, A. (2006). Carbonation and historical buildings made of concrete, November, 2006. https://doi.org/10.1051/jp4.
6. de Larrard, T., Bastidas-Arteaga, E., Duprat, F., & Schoefs, F. (2014). Effects of climate variations and global warming on the durability of RC structures subjected to carbonation. *Civil Engineering and Environmental Systems, 31*(2), 153–164. https://doi.org/10.1080/10286608.2014.913033.
7. Stewart, M. G., Wang, X., & Nguyen, M. N. (2011). Climate change impact and risks of concrete infrastructure deterioration. *Engineering Structures, 33*(4), 1326–1337. https://doi.org/10.1016/j.engstruct.2011.01.010.
8. Bastidas-Arteaga, E. (2018). Reliability of reinforced concrete structures subjected to corrosion-fatigue and climate change. *The International Journal of Concrete Structures and Materials, 12*(1). https://doi.org/10.1186/s40069-018-0235-x.

9. NOAA. (2014). NOAA climate. *Climate Forcing*. https://www.climate.gov.
10. Hobijn, B., Franses, P. H., & Ooms, M. (2004). Generalizations of the KPSS-test for stationarity. *Statistica Neerlandica, 58*(4), 483–502. https://doi.org/10.1111/j.1467-9574.2004.00272.x.
11. Kotz, S., & Johnson, N. L. (1997). *Breakthroughs in statistics* (Vol. 40, no. 2).
12. Lovric, M. (Ed.) International encyclopedia of statistical science. *Econometrica*, 488.

Stochastic Degradation Model of Concrete Bridges Using Data Mining Tools

Yina F. M. Moscoso, Monica Santamaria, Hélder S. Sousa, and José C. Matos

Abstract Bridges have a significant importance within the transportation system given that their functionality is vital for the economic and social development of countries. Therefore, a high level of safety and serviceability must be achieved to guarantee an operational state of the bridge network. In this regard, it is necessary to track the performance of bridges and obtain indicators to characterize the evolution of structural pathologies over time. In this paper, the time-dependent expected deterioration of bridge networks is investigated by use of Markov chains models. Bridges in a network are likely to share similar environmental conditions but depending on their functional class may be exposed to different loading conditions that diversely affect their structural deterioration over time. Moreover, the deterioration rate is known to increase with time due to aging. Hence, it is useful to identify and divide the bridge network into classes sharing similar deterioration trends in order to obtain a more accurate prediction. To this end, data mining tools such as two-step cluster analysis is applied to a dataset obtained from the National Bridge Inventory (NBI) database, in order to find associations among the bridge characteristics that could contribute to build a more specific degradation model which accurately explains and predicts the future condition of concrete bridges. The results demonstrate a particular deterioration path for each cluster, where it is evidenced that older bridges and those having higher Average Daily Traffic (ADT) deteriorate faster. Therefore, the degradation models developed following the proposed methodology provide a more accurate prediction when compared to a single degradation model without clustering analysis. This more reliable models facilitate the decision process of bridge management systems.

Keywords Bridge management systems · Degradation model · Markov chains model · Two-step cluster analysis · Data mining

---

Y. F. M. Moscoso (✉) · M. Santamaria · H. S. Sousa · J. C. Matos
ISISE, Institute of Science and Innovation for Bio-Sustainability (IB-S), Department of Civil Engineering, University of Minho, Guimarães, Portugal

© The Author(s), under exclusive license to Springer Nature Switzerland AG 2021
J. C. Matos et al. (eds.), *18th International Probabilistic Workshop*, Lecture Notes in Civil Engineering 153, https://doi.org/10.1007/978-3-030-73616-3_59

1 Introduction

Several Departments of Transportation (DOTs) use the Manual for Bridge Element Inspection [1] for collecting consistent data for Bridge Management Systems (BMS). For this reason, during the last decades, in order to establish the structural safety, it was needed to make the structural evaluation of bridges using performance indicators considering their resistance, the applied loads and the performance of its elements. Once visual inspections are performed on bridge systems it is possible to obtain maintenance plans, including a set of intervention actions that should be implemented on particular sections of the bridges at specific time periods to maintain/enhance the quality level of the bridge and concurrently to minimize maintenance costs. After these actions, bridge safety can be improved, productivity is enhanced, customer satisfaction is achieved, and maximum load capacity of the bridge may be considered.

The Nevada Department of Transportation (NDOT) uses a manual developed by the American Association of State Highway and Transportation Officials (AASHTO). This manual works as a guideline to data collection for the classification of bridges' quality, determining the structural states using conditions ratings (CR) that represent the overall condition of the bridge.

To develop maintenance plans for bridges it is needed to have a history of the inspections carried out over a period of time and as to allow to achieve a projection of the CR into the future. The analysis of the inspection data is therefore an important part of the BMS, which provides the support to make appropriate decisions for maintenance, repair or rehabilitation schedule and cost of these actions [2]. For this reason, predictive models such as Markov chains models are often implemented for forecasting the time-dependent deterioration of bridges.

Markov chain (MC) models [3] are stochastic models used to handle the randomness present during structural deterioration [4, 5], and have been used to model infrastructure deterioration. For example, Li et al. [7] used MC models to determine the deterioration of urban bridges at the network and individual levels; Yi [6] demonstrated in his work that bridge condition predictions performed with MC models will affect bridge design selections and the corresponding system benefits; Tolliver and Lu [8] explained the performance of future conditions of a bridge through MC models considering the relationship between bridge material, bridge design, operating rating classification, average daily traffic and the future condition of the bridge. Moreover, MC models have not only been used to predict the behavior of bridges, but also to determine the future conditions of roads, pavements and sewage systems [9–11].

With these research examples, and acknowledging that MC are the most frequent model used in BMS due to their simplicity in implementation together with their capabilities to capture the randomness of the deterioration process [12], the importance of prediction models through MC is herein stressed. However, few studies have addressed effectively how to divide the bridge network into classes sharing similar deterioration patterns to obtain more accurate predictions from MC models. Previous studies have employed data-mining tools to investigate association between bridge design parameters and condition ratings [13], but the results were not used to develop stochastic models such as MC in order to predict the bridge behavior over time with higher accuracy. Therefore, the aim of this work was the prediction of the future behavior of concrete bridges from the Nevada Department of Transportation (NDOT) through MC models, integrated with the use of data mining tools, namely a two-step cluster analysis, in order to reveal associations among the bridge characteristics that could contribute to define classes and therefore to build more specific degradation curves, which accurately explain and forecast the future condition of each concrete bridge class. To this end, 148 concrete bridges with no record of maintenance action performed were selected.

2 Methodology

2.1 *Stochastic Models*

The uncertainty and randomness of asset deterioration processes are considered as one or more random variables in stochastic models [14]. A Markov process involves states and corresponding matrices of transition probabilities. In the context of modeling bridge deterioration, the states represent bridge condition ratings. Hence, the elements in the transition matrix represent the probability of bridges changing their condition rating at a determined and fixed period of time [4]. MC models are based on the concept of probabilistic cumulative damage, which estimates changes on component conditions over multiple transition periods [15].

MC are considered as a series of transitions, and the future of the probability of the state depends only on the present state, but not on the past states, this means that, the future condition of the system depend only on the current state [6]. MC are used to describe the evolution of a system represented by states $\{S = S_1, \ldots S_m\}$. During a defined transition period the system undergoes transitions from one state to another. If the system is in state S_i, then it will move into a future state S_j with transition probability P_{ij} [14].

A stationary process is presented when the probability of moving across states remains constant over transitions, independent of time. By grouping the transition probabilities into a Transitions Probability Matrix (TPM), it can be expressed as:

$$P = \left[p_{ij}\right] = \begin{bmatrix} p_{11} & p_{12} & \cdots & p_{1m} \\ p_{21} & p_{22} & & p_{2m} \\ \cdot & \cdot & \cdot & \cdot \\ \cdot & \cdot & \cdot & \cdot \\ \cdot & \cdot & & \cdot\cdot \\ p_{m1} & p_{m2} & \cdots & p_{mm} \end{bmatrix} \quad (1)$$

In this matrix, p_{ij} represents the probability of moving from state i to state j, considering a single transition period for all $i, j = 1, 2 \ldots$ m; and m represents the total number of states that the system can experience. The probability distribution of State i is represented by ith row. The TPM is usually estimated by statistical inference, using available data, in this case the CR obtained during the bridge inspections.

The condition vector after t transition periods $C(t)$, is described as a function of the vector C_o which describes the initial condition of the system as [16]:

$$C(t) = C_o * P^t \tag{2}$$

Being the probability of moving from State i to State j after t transition periods for each element ij represented by the matrix P^t [3]. The percentage probability method is used in the case of available CR data for more than two periods, using the Eq. 3:

$$P_{i,j} = \frac{n_{i,j}}{n_i} \tag{3}$$

where, $n_{i,j}$: is the number of bridges that starts and continues in state i for a particular transition period; and n_i is the total number of bridges in state i in the same transition period.

2.2 *Data Mining Tools: Two-Step Cluster Analysis*

Data Mining (DM) is the process of automatically discovering valuable information in large datasets. DM techniques are used to examine these large datasets and find novel and useful patterns that might otherwise remain unknown [17]. There are several DM functionalities including characterization and discrimination; the mining of frequent patterns, associations, and correlations; classification and regression; clustering analysis; and outlier analysis [18]. DM functionalities are used to specify the kinds of patterns to be found in data mining tasks. In general, such tasks can be classified into two categories: descriptive tasks which characterize properties of the data in a target data set; and predictive tasks which perform induction on the current data in order to make predictions [18]. Among descriptive tasks, cluster analysis is a commonly used unsupervised learning method in data mining which groups the data into subsets (clusters) that reflect the essential structure of the data. Ideally, cluster analysis minimizes the difference of cases within a cluster while maximizing the difference between clusters [17].

Cluster analysis is introduced in the present work in order to reveal similarities on the degradation patterns of groups of bridges. To this end, a two-step clustering algorithm is implemented given its ability to create clusters based on categorical and continuous variables, the possibility of automatically selecting the optimal number of clusters, and its ability to analyze large data sets efficiently [19].

2.3 Framework

For this study, the guidelines by Federal Highway Administration (FHWA) were implemented [20], where the behavior of the bridge is represented through condition ratings (CR). These CR range from 9 (excellent condition) to 0 (failed), as described by FHWA and presented in Table 1. In this case, CR of 3 can be considered as the threshold rating where rehabilitation or replacement measures have to be done. Therefore, the developed models were built to predict the deterioration from CR 8 to CR 3.

According with the data collection, a transition period of one year was considered. Taking as reference other similar studies (e.g. [4]), bridges that underwent a maintenance action were excluded, i.e. only bridges with natural decay (ND) were considered. That means that the TPMs contain probabilities for bridges where their condition is maintained or decayed during a period of one year. Accordingly, the TPM only has two values, one representing the probability of the structure staying in its current condition and the other one representing the probability of the structure moving to the next worse condition, as shown in the following matrix [21]:

Table 1 General condition rating [20]

CR	Meaning	Description
9	Excellent	New bridges
8	Very good	No problems noted
7	Good	Some minor problems
6	Satisfactory	Structural elements show some minor deterioration
5	Fair	All primary structural elements are sound but may have minor deterioration
4	Poor	Major deterioration is occurring
3	Serious	Deterioration has seriously affected the primary structural components of the bridge. Local failures are possible
2	Critical	Advanced deterioration of the primary structural elements is evident. Unless closely monitored it may be necessary to close the bridge until corrective action
1	Imminent failure	Major deterioration is affecting the stability of the bridge. The bridge is closed to traffic but corrective action may allow it to be out back in light service
0	Failed	The bridge is out of service and beyond corrective action

$$P = \begin{bmatrix} p_{99} & 1-p_{99} & 0 & 0 & 0 & 0 & 0 \\ 0 & p_{88} & 1-p_{88} & 0 & 0 & 0 & 0 \\ 0 & 0 & p_{77} & 1-p_{77} & 0 & 0 & 0 \\ 0 & 0 & 0 & p_{66} & 1-p_{66} & 0 & 0 \\ 0 & 0 & 0 & 0 & p_{55} & 1-p_{55} & 0 \\ 0 & 0 & 0 & 0 & 0 & p_{44} & 1-p_{44} \\ 0 & 0 & 0 & 0 & 0 & 0 & 1 \end{bmatrix} \tag{4}$$

If the probability of a structure currently in state i remained in the same state over a single transition period, this is represented by pij for all $i \in \{9, \ldots, 3\}$ and the probability of a structure in state i move to state $i+1$ in one period is represented by $pii + 1 = 1 - pii$ [16]. As it is not possible to improve CRs or to degrade CRs by more than one level, the remaining elements of the TPMs are zero.

Having the matrix P that conveniently reproduces probabilistically the degradation of the structure over a period of time, it is possible to predict the performance of the bridges over a specific period, by obtaining the probability vector of condition states $E(CF)$ as:

$$E(CF) = C_o * P^t * R \tag{5}$$

where C_o is the probability vector that relates the various condition states at an initial time t_0 as previously mentioned, and R is the vector representing the possible CR of bridges:

$$R = \begin{bmatrix} 9 \\ 8 \\ 7 \\ 6 \\ 5 \\ 4 \\ 3 \end{bmatrix} \tag{6}$$

2.4 Data Collection

In the database of inspections of bridges in the state of Nevada, two major types of deterioration behavior are found, namely, Natural Decay (ND) and Convectional Recoverable Decay (CRD). ND means bridges under routine maintenance or having minor repairs, and CRD behavior stands for bridges having medium or major repairs as well as reconstruction. In this paper, only ND bridges were considered. For this study, 1613 bridges were initially analyzed as shown in Table 2, with an inspection period from 2004 to 2014. Most of the bridges had an initial condition rating of 8

Table 2 Bridges grouped by material and/or design

Kind of material and/or design	Number of bridges
Concrete	775
Concrete continuous	208
Steel	107
Steel continuous	90
Pre-stressed concrete	202
Pre-stressed concrete continuous	222
Wood or timber	9
Total	1613

and were between 20 and 60 years old. Taking into account that concrete bridges represented the majority of the bridge stock, only these bridges were selected for the analysis, as well as to have similar deterioration mechanisms. Consequently, the dataset employed for the development of this study is comprised by 148 concrete bridges with ND behavior.

A separate CR is assigned for each three major bridge components, namely, substructure, superstructure, and deck. Herein, the deck ratings were selected to develop the models and the methodology. Nevertheless, the process can be replicated for the other bridge components or even its combination in an overall perspective.

3 Results

The two-step clustering algorithm was performed using the statistical software package SPSS Statistics [v25]. The variables used as inputs were the age of the deck, ADT on the bridge, and the deck condition rating. Considering that the later input is an ordinal variable, it should be treated either as a continuous or as a categorical variable [22]. However, handling the deck condition rating as a categorical variable resulted in this variable dominating the results because differences in categorical variables are given a higher weight than differences in continuous variables [22]. Therefore, the deck condition rating was treated as a continuous variable.

In order to determine the optimal number of clusters, the Schwarz's Bayesian Criterion (BIC) was used as the clustering criterion to compare the cluster solutions. The solution found indicates that three clusters are the optimum based on the three input variables. The number of records assigned to each cluster corresponds to 30 (20.5%), 21 (14.4%) and 95 (65.1%), for the first, second and third cluster respectively. It can be noticed that two bridges were disregarded as there was missing information and the two-step clustering algorithm is not able to handle data sets with incomplete data [13]. The mean value of the predictors in each cluster can be observed in Table 3, together with the predictor's importance.

Table 3 Clusters summary

Cluster		1	2	3
Size		30 (20.5%)	21 (14.4%)	95 (65.1%)
Input (predictor) importance		Cluster means		
ADT	1.0	24350.33	3714.48	5325.47
Age	0.98	32.93	72.19	40.33
CR	0.95	6.33	6.33	7.03

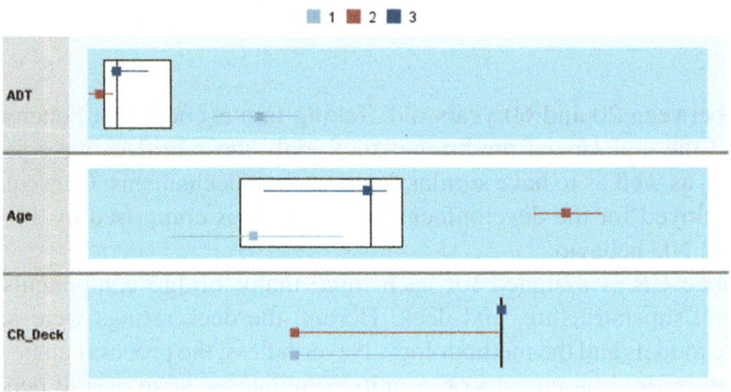

Fig. 1 Boxplot comparison of clusters

For a better comparison between clusters, Fig. 1 presents a boxplot analysis for the distribution of values within each cluster overlaid on a boxplot for the distribution of overall values. It can be observed that the group of bridges in Cluster 1 have very large ADT values compared to those belonging to Clusters 2 and 3. Also, it can be seen that bridges within Cluster 2 are older than those in Clusters 1 and 3. Finally, Clusters 1 and 2 are comprised of bridge decks with lower condition ratings than Cluster 3.

The obtained degradation curves are presented in Fig. 2. It can be observed that the highest deterioration after 50 years of prediction is reached by the group of bridges from Cluster 1. This observation is reasonable considering that Cluster 1 is comprised by the bridges with highest ADT, which are expected to deteriorate faster given the higher load demands. Accordingly, a lower deterioration is predicted for the bridges belonging to Clusters 2 and 3 which have low ADT levels. It can also be seen that bridges in Cluster 2 have a higher deterioration rate than those in Cluster 3. This finding is consistent as bridges in Cluster 2 are older than those in Cluster 3 (see Table 3), and it has been demonstrated that the probability of a structure to deteriorate to a lower state increases with time (aging effects). Finally, it can be noted that the degradation curve computed with the complete dataset of records is approximately an average of all three clusters. This approximation can be acceptable for bridges belonging to Clusters 2 and 3, but might conduct to inefficient decision

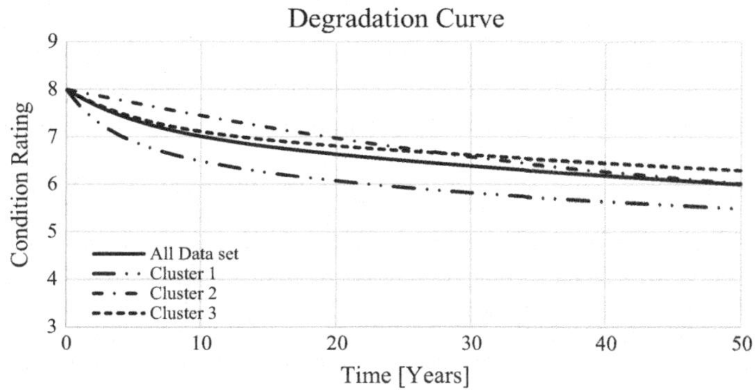

Fig. 2. Deck degradation curves for all dataset and the three clusters

making strategies for bridges within Cluster 1. Using the global degradation model, in the case of the stock of bridges within Clusters 2 and 3, maintenance actions would be made before their actual need leading to inefficient use of resources, whereas in the case of bridges of Cluster 3 these actions should be made before. However, it must be noted that, for this specific database, a CR below 5 is not obtained in a 50-year period time, thus still in a safe condition where all primary structural elements are sound but may have minor deterioration.

4 Conclusions

This paper illustrates how to use Markov chain models to estimate bridge deterioration using condition ratings. The study considered 10 years of inspection records of bridges exhibiting natural decay, i.e. without reconstruction. Markov chains were used to estimate the required transition probabilities and deterioration curve for concrete decks. According with the deterioration curve obtained for all dataset in Fig. 2, these bridges will have at the end of 50 years a condition rating around 6, if only routine maintenance is performed. Based on these results it is possible to verify that routine maintenance on bridges is important, and must be carried out according to what is specified in the bridge manual. As an example, routine maintenance for this case study must be performed biennially, as this slows down the deterioration process in the structure.

Moreover, it was demonstrated that a more accurate prediction for group of bridges can be provided if clustering algorithms are introduced into the methodology. It was observed that the greatest deterioration occurs in the curve represented by Cluster 1, where the influence of the ADT value on the structure was confirmed and it was noted how it affects its development, i.e. accelerating the process of deterioration over time. It was also seen the influence of age in the tendency of the curves, since Cluster 2 has older bridges, so it was expected that its degradation curve would be more accelerated than that of Cluster 3. While the study focused on limited attributes, additional categorical and continuous variables can be introduced into the methodology such as the structure length and width, the wearing surface, skew angle, among others, which could improve the prediction accuracy of the methodology.

Acknowledgements This work was partly financed by FEDER funds through the Competitivity Factors Operational Programme—COMPETE and by national funds through FCT Foundation for Science and Technology within the scope of the project POCI-01-0145-FEDER-007633.

This project received funding to carry out this publication of the European Union's Portugal 2020 research and innovation program under the I&D project "GIIP—Intelligent Management of Port Infrastructures", with POCI-01-0247-FEDER-039890. The sole responsibility for the content of this publication lies with the authors. It does not necessarily reflect the opinion of the European Union.

References

1. AASHTO. (2013). Manual for bridge element inspection. *The American Association of State Highway and Transportation Officials* (1st ed.).
2. Almeida, J. O., Teixeira, P. F., & Delgado, R. M. (2015). Life cycle cost optimisation in highway concrete bridges management. *Structure and Infrastructure Engineering, 11*(10), 1263–1276. https://doi.org/10.1080/15732479.2013.845578.
3. Jiang, Y., Saito, M., & Shina, K. C. (1988). Bridge performance prediction model using the Markov chain. *Journal of the Transportation Research Board, 1180,* 25–32.
4. Jiang, Y. (1990). *The development of performance prediction and optimization models for bridge management systems.* Purdue University.
5. Muñoz, Y. F., Paz, A., Hanns De La Fuente-Mella, J. V. F., & Sales, G. M. (2016). Estimating bridge deterioration for small data sets using regression and markov models. *World Academy Science International Science Index, Urban Civil Engineering, 10*(5).
6. Jiang, Y. (2010). Application and comparison of regression and Markov chain methods in bridge condition prediction and system benefit optimization. *Journal of the Transportation Research Forum, 49*(2), 210.
7. Li, L., Sun, L., & Ning, G. (2014). Deterioration prediction of urban bridges on network level using Markov-chain model. *Mathematical Problems in Engineering, 2014.* https://doi.org/10.1155/2014/728107.
8. Tolliver, D., & Lu, P. (2011). Analysis of bridge deterioration rates: A case study of the Northern Plains Region. *Transportation Research Forum, 50*(2), 87–100 [Online]. Available: https://www.trforum.org/journal.
9. Chin, P. A., Ferris, J. B., & Reid, A. A. (2012). *Improving Markov chain models for road profiles simulation via definition of states.*
10. Kobayashi, K., Do, M., & Han, D. (2010). Estimation of Markovian transition probabilities for pavement deterioration forecasting. *KSCE Journal of Civil Engineering, 14*(3), 343–351. https://doi.org/10.1007/s12205-010-0343-x.
11. Baik, H.-S., Jeong, H. S. D., & Abraham, D. M. (2006). Estimating transition probabilities in Markov chain-based deterioration models for management of wastewater systems. *Journal of Water Resource Plannning Management, 132*:1(15). https://doi.org/10.1061/(ASCE)0733-9496(2006)132.
12. Santamaria Ariza, M., Zambon, I., Sousa, H. S., Campos e Matos, J. A., & Strauss, A. (2020). Comparison of forecasting models to predict concrete bridge decks performance. *Structure Concrete.*

13. Radovic, M., Ghonima, O., & Schumacher, T. (2017). Data mining of bridge concrete deck parameters in the national bridge inventory by two-step cluster analysis. *ASCE-ASME Journal of Risk and Uncertainty in Engineering Systems, Part A Civil Engineering, 3*(2), 4016004. https://doi.org/10.1061/AJRUA6.0000889.
14. Setunge, S., & Hasan, M. S. (2011). Concrete bridge deterioration prediction using Markov chain approach. *Digital Library*, University of Moratuwa.
15. Morcous, G. (2006). Performance prediction of bridge deck systems using Markov chains. *Journal of Performance of Constructed Facilities, 20*(2), 146–155. https://doi.org/10.1061/(ASCE)0887-3828(2006)20:2(146).
16. Ranjith, S., Setunge, S., Gravina, R., & Venkatesan, S. (2013). Deterioration prediction of timber bridge elements using the Markov chain. *Journal of Performance of Constructed Facilities, 27*(3), 319–325. https://doi.org/10.1061/(ASCE)CF.1943-5509.0000311.
17. Tan, P., Steinbach, M., & Kumar, V. (2006). *Introduction to data mining*. Pearson Addison Wesley.
18. Han, J., Pei, J., & Kamber, M. (2011). *Data mining: Concepts and techniques*. Elsevier.
19. IBM. (2020). TwoStep cluster analysis. *IBM*. https://www.ibm.com/support/knowledgecenter/SSLVMB_27.0.0/statistics_casestudies_project_ddita/spss/tutorials/twostepcluster_table.html.
20. U.S. Department of Transportation Federal Highway Administration, NBI ASCII files—National Bridge Inventory—Bridge Inspection—Safety—Bridges & Structures—Federal Highway Administration.
21. Cesare, M. A., Santamarina, C., Turkstra, C., & Vanmarcke, H. E. (1992). Modeling bridge deterioration with Markov chains. *Journal of Transportation Engineering, 118*(6), 1129–1945.
22. Bacher, J., Wenzig, K., & Vogler, M. (2004). *SPSS TwoStep cluster-a first evaluation*.

Stochastic Simulation of Clay Brick Masonry Walls with Spatially Variable Material Properties

Dominik Müller, Tilo Proske, and Carl-Alexander Graubner

Abstract In the assessment of existing masonry structures, a high variability of material properties can be observed. The variability is also present within a single wall, which raises the question of how this spatial variability influences the load-bearing capacity and the reliability of an assessed masonry wall. With regard to reliability, lower quantile values of the load-bearing capacity are decisive. For this reason, the influence of spatial variability on the probability distribution of the load-bearing capacity has to be known. In this paper, clay brick masonry walls in compression are investigated by Monte Carlo simulations utilising a nonlinear finite element model. For the validation of the finite element model, experimental investigations of the stress redistribution capability of masonry walls with weak spots were carried out. The numerical model follows a simplified micro-modeling approach with unit-to-unit variability of the material properties. Results of the stochastic simulations are shown for varying wall length, slenderness and coefficients of variation of the material properties. The obtained statistical distributions of the load-bearing capacity are evaluated with respect to acceptable design values for ensuring structural reliability. It is shown that spatial variability leads to a reduction of the mean load-bearing capacity, but the overall variability of the load-bearing capacity is much smaller than that of the spatially varying material properties. Compared to an approach assuming homogeneity within the wall, the consideration of spatial variability leads to an increase of suitable design values.

Keywords Masonry · Monte Carlo simulation · Finite element modeling · Existing structures · Probabilistic assessment

1 Introduction

The evaluation of existing masonry is a challenging task, which is mainly due to the high level of various uncertainties that are present [1]. One of the main uncertainties

D. Müller (✉) · T. Proske · C.-A. Graubner
Institute of Concrete and Masonry Structures, Technical University of Darmstadt, Darmstadt, Germany
e-mail: mueller@massivbau.tu-darmstadt.de

© The Author(s), under exclusive license to Springer Nature Switzerland AG 2021
J. C. Matos et al. (eds.), *18th International Probabilistic Workshop*, Lecture Notes in Civil Engineering 153, https://doi.org/10.1007/978-3-030-73616-3_60

to be considered is the statistical uncertainty with regard to material properties, which is introduced since the number of material tests has to be limited to reduce costs and invasiveness of the investigations. A method for taking statistical uncertainty into account can be found in [2]. Another substantial uncertainty is the spatial variability of material properties. Unit and mortar compressive strength, for example, do not only vary from wall to wall but also show a pronounced variability within a single wall. This raises the question of how spatial variability should be considered when assessing existing masonry walls. Therefore, two effects have to be known:

1. What is the influence of spatial variability on the mean value of the load-bearing capacity compared to a deterministic calculation assuming homogeneity of materials within a wall?
2. To what extent is the variability of the load-bearing capacity influenced by the spatial variability of the material properties?

If both of the effects are known, suitable design values (which actually are assessment values in this case but are called design values in this paper to match the well-known nomenclature of current standards) or partial factors can be determined. In the assessment of other types of structures, e.g. existing concrete structures, the material variability gained from tests is often directly used as an input parameter for the determination of design values or partial factors [3]. However, the variability of the load-bearing capacity, which is crucial for structural reliability, might be much lower than the variability of the material properties within the structural member. For concrete properties, where the variability of the mean values of material properties between structural members is usually most influential, this simplifying procedure might be suitable. For existing masonry, a closer look should be taken, which is the purpose of the following investigations.

The present study focuses on the influence of spatial variability on the behaviour of masonry walls in compression. In order to investigate the influence of spatial variability on the load-bearing capacity, experiments are conducted, which shall show the stress redistribution capability of solid clay brick masonry walls with local weak spots. The experimental results are used for validating a finite element model, which is then utilised for Monte Carlo simulations with spatially varying parameters for compressive strength, modulus of elasticity and flexural tensile strength of masonry. The results are then evaluated with regard to mean value and coefficient of variation (CoV) of the load-bearing capacity as well as suitable design values.

2 Experimental Investigations

2.1 Experimental Program

In order to show the stress redistribution capability of masonry walls in compression, masonry walls with a certain number of units with lower compressive strength as

Table 1 Material properties

Material property	Average [N/mm^2]	Testing standard
Standardised compressive strength – solid clay bricks	24.9	EN 772-1
– perforated clay bricks	11.6	
Mortar compressive strength	2.71	EN 1015-11
Masonry compressive strength – solid clay bricks	9.74	EN 1052-1
– perforated clay bricks	4.11	
Masonry modulus of elasticity – solid clay bricks	3000	EN 1052-1
– perforated clay bricks	1959	

well as masonry walls with missing units were tested. The main materials used in this study were solid clay bricks (240 × 115 × 71 mm^3) and hydraulic lime mortar. Vertically perforated clay bricks were used as weak spots. The average standardised compressive strength of the units according to EN 772-1 [4] as well as the average compressive strength of the mortar determined according to EN 1015-11 [5] at the same age as the tested walls are given in Table 1. Furthermore, six masonry specimens according to EN 1052-1 [6] were tested for each of the two brick types in order to receive values for standardised masonry strength and modulus of elasticity.

The units of the experimental walls were arranged in cross bond with 5 units per course and 13 courses in total, which results in walls with a nominal length of 615 mm, a thickness of 240 mm and a height of 1083 mm. The walls were tested at the age of 32 up to 43 days. Besides reference walls consisting of only solid or perforated clay bricks without weak spots, four types of walls were tested: walls with a percentage of approximately 25 and 50% perforated clay bricks as well as walls with one missing header or one missing stretcher, respectively, see Fig. 1.

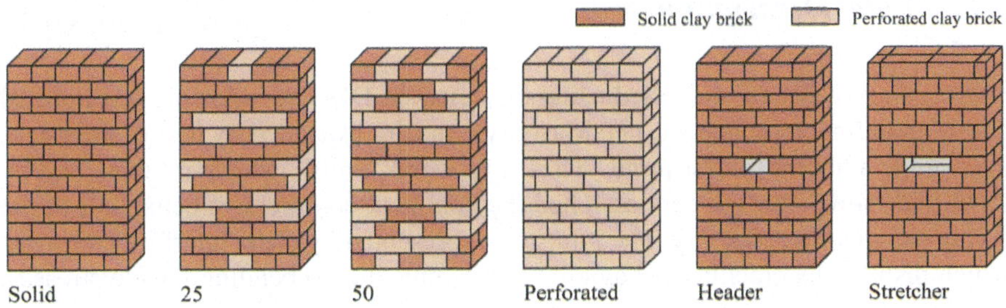

Fig. 1 Tested masonry specimens

Table 2 Results of the experiments

Specimen	Load-bearing capacities [kN]	Mean strength [N/mm²]	CoV [%]	Relative mean [%]
Solid	943 826 1093	6.35	14.5	100
25	721 703	4.77	1.8	75
50	501 485	3.32	2.9	52
Perforated	391 394	2.64	0.8	42
Header	748 732	4.94	2.1	78
Stretcher	694 651	4.47	4.4	70

2.2 Experimental Results

The resulting load-bearing capacities under centric compressive load are given in Table 2. For the walls with a missing unit (header or stretcher), the load-bearing capacities relative to the capacity of the reference walls are slightly below 80%, which would be the relative load-bearing capacity if perfect stress redistribution within the critical cross-section (4 instead of 5 units) was possible. The load-bearing capacity of the walls made of both solid and perforated bricks is slightly below the values that would be expected from linear interpolation between the two homogenous reference walls. This shows that a considerable but not perfect capability of stress redistribution within the masonry walls exists.

3 Finite Element Model

3.1 Model Description

The masonry walls are modelled using the software DIANA following a simplified micro-modeling approach, cf. [7]. Masonry units are expanded by half of the thickness of the adjacent mortar joints. These expanded "units", which are modelled by eight-node solid elements, are assigned the nonlinear compressive behaviour of the composite material masonry. The joints are represented by plane interface elements, which display the cracking behaviour in the joints due to bending tensile stresses, cf. Fig 2. The load is applied via rigid load transfer beams at top and bottom of the wall. The selected material parameters are summarised in Table 3.

Masonry usually shows an orthotropic material behaviour. Nevertheless, the use of an isotropic material model is assumed to be adequate for this study, since the predominant compressive stresses act in the vertical direction, whereas tensile stresses due to stress redistribution mainly act horizontally. For the compressive behaviour of the solid elements, a material model with Drucker-Prager yield criterion, associated flow rule and hardening/softening law as proposed in [7] is chosen. The friction

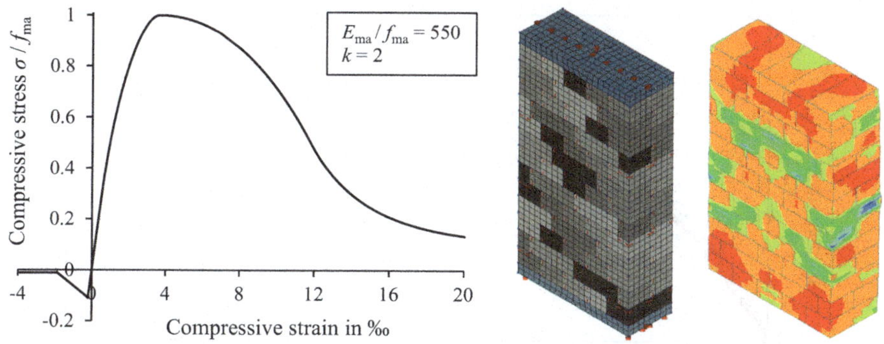

Fig. 2 Uniaxial stress-strain relationship for the expanded "units", finite element model mesh and exemplary vertical strain for specimen "25"

Table 3 Material parameters used in the finite element model

Element	Parameter	Value	References
Expanded "units" – general	Friction angle ϕ	12°	Calibration
	Dilatancy angle ψ	ϕ	[7]
	Poisson's ratio	0.19	[8]
– solid clay brick masonry	Compr. strength f_{ma} (for validation)	6.35 N/mm^2	Experiments
	Modulus of elasticity E_{ma}	473 f_{ma}	Experiments
	Stress-strain parameter k	2.38	Experiments
	Unit tensile strength f_{bt}	0.04 f_b	[9]
– perforated clay brick masonry	Compr. strength f_{ma} (for validation)	2.64 N/mm^2	Experiments
	Modulus of elasticity E_{ma}	741 f_{ma}	Experiments
	Stress-strain parameter k	2.04	Experiments
	Unit tensile strength f_{bt}	0.03 f_b	[9]
Joints	Tensile strength f_t	0.044 f_{ma}	[9]
	Tensile fracture energy G_{ft}	0.0148 f_t	[8]

angle needed for the Drucker-Prager yield surface as well as the parameters for the hardening/softening low are chosen to receive results matching the experiments. For the interfaces at the location of the mortar joints, a discrete cracking material model with bilinear softening is selected. The ratio of flexural tensile strength of the joints to masonry compressive strength is chosen according to the literature [9] and based on a unit and mortar compressive strength of 25 and 2.5 N/mm^2, respectively. The parameter k is a measure for the nonlinearity of the ascending stress-strain curve, cf. [10], where $k = 1$ results in a linear function and $k = 2$ in a quadratic parabola if the stress-strain relationship is modelled according to EN 1992-1-1 [11]. It can

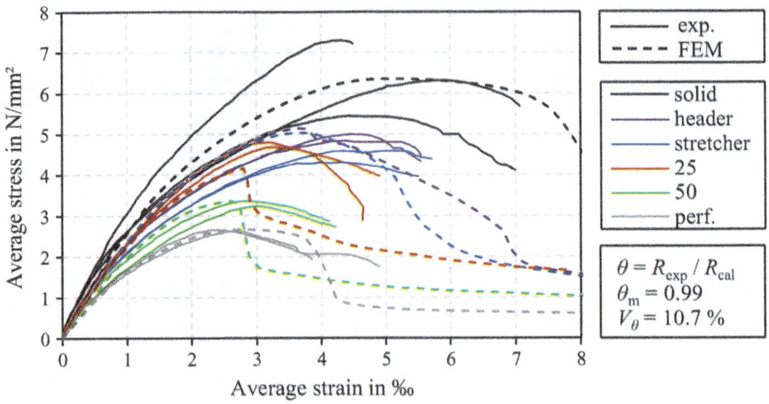

Fig. 3 Comparison between the results of the experiments and the finite element simulation

be defined as the ratio of the total strain to the elastic strain at peak stress in the context of this paper. Fig. 2 displays the uniaxial stress-strain relationship, which results from the chosen hardening/softening laws.

3.2 Validation by Experimental Results

The validation aims at showing that the behaviour of walls with local weakenings can be sufficiently displayed in relation to the reference walls. The experimentally gained values for the masonry compressive strength of the reference walls are therefore used as input parameters. Fig. 3 gives a comparison between experimental results and the finite element simulations. The ratio θ of experimental to the numerical peak load has a mean of $\theta_m = 0.99$ and a CoV of $V_\theta = 10.7\%$. Hence, the numerical model is considered suitable for the stochastic simulations.

4 Stochastic Model

4.1 General Concept

Spatial variability is modelled as unit-to-unit variability in the following. Each of the expanded "units" is assigned a random masonry compressive strength and a random modulus of elasticity. No location-related correlation between the units in the wall is assumed, i.e. the strength of one "unit" is independent of the strength of its neighbouring units. For non-slender walls in compression, the load-bearing capacity of the wall is mostly influenced by the masonry compressive strength, which itself is a function of unit and mortar material properties. EN 1996-1-1 [12] gives the following empirical function for the characteristic masonry compressive strength $f_{\text{ma,k}}$ based on the mean compressive strengths of unit $f_{\text{b,m}}$ and mortar $f_{\text{mo,m}}$:

$$f_{\text{ma,k}} = K \cdot f_{\text{b,m}}^{0.7} \cdot f_{\text{mo,m}}^{0.3} \tag{1}$$

The masonry strength of a single expanded “unit” is influenced by the mortar properties in both of the adjacent bed joints, which creates correlation between the masonry compressive strength of the two adjacent “units” of a joint. Furthermore, the mortar properties along one bed joint are also correlated [13]. Nevertheless, the influence of mortar properties on the masonry strength is much smaller than that of the unit properties. Equation (1) is equivalent to a linear relationship between the logarithms of masonry, unit, and mortar strength. Therefore, the standard deviation of the logarithm of the masonry strength $\sigma_{\ln f_{\text{ma}}}$ can be determined as

$$\sigma_{\ln f_{\text{ma}}}^2 = 0.7^2 \cdot \sigma_{\ln f_{\text{b}}}^2 + 0.3^2 \cdot \sigma_{\ln f_{\text{mo}}}^2 = 0.49 \cdot \sigma_{\ln f_{\text{b}}}^2 + 0.09 \cdot \sigma_{\ln f_{\text{mo}}}^2 \tag{2}$$

Equation (2) shows that the variability of the unit properties dominates the variability of masonry strength. Therefore, the correlations due to mortar joint properties (not their variability itself) are neglected in the following. The stochastic model of a unit-to-unit variability is in line with the simplified micro-modeling approach. Instead of random values for unit and mortar properties, the masonry compressive strength for each of the expanded “units” is generated directly following the natural discretisation given by the units. Random parameters for the modulus of elasticity and the flexural tensile strength perpendicular to the bed joints are generated as well since they can become influential for slender masonry walls [10]. The modulus of elasticity of masonry is also assigned to each of the “units” separately. The flexural tensile strength, which is attributed to the interfaces, is discretised according to the adjacent units at the top of the joint. All of the other properties are assigned by fixed values relative to the masonry compressive strength, as Table 3 shows.

4.2 *Stochastic Parameters*

Within the Monte Carlo simulations, the CoVs of the material properties are varied. Therefore, typical ratios between the CoVs of masonry compressive strength, modulus of elasticity and flexural tensile strength are defined based on CoVs obtained for solid clay brick masonry in [8], see Table 4. The correlation coefficient between masonry strength and its modulus of elasticity is selected as $\rho_{f,E} = 0.72$ [8]. The correlated random variables are generated as described in [14].

Table 4 Stochastic parameters for the material properties

Material property	Distribution type	Correlation	CoV in [8] (%)	Relative CoV
Masonry compressive strength f_{ma}	LN	$\rho_{f,E} = 0.72$	17	1
Masonry modulus of elasticity E_{ma}	LN		22	≈1.3
Tensile strength f_{t}	LN	–	35	≈2.0

5 Monte Carlo Simulation

5.1 Approach

The reference wall for the following parameter studies is arranged in cross bond with 36 courses of 5 units each. It is 3 m high, 0.625 m long and 0.24 m thick. A ratio of modulus of elasticity to compressive strength of $E_{\mathrm{ma,m}}/f_{\mathrm{ma,m}} = 550$ is chosen as well as a parameter $k = 2$. These values are typical for existing solid clay brick masonry [15]. The reference wall is centrically loaded, has hinged supports along the longitudinal axis at top and bottom and a unit-to-unit CoV for the masonry compressive strength of $V_f = 30\%$. Each parameter combination is simulated 100 times. Subsequently, the mean value R_{m} and the CoV V_R of the simulated load-bearing capacities are determined. Assuming a log-normal distribution for the load-bearing capacity R, required design values R_{d} can be received by

$$R_{\mathrm{d}} = R_{\mathrm{m}} \cdot \exp\left(-0.5\, s_{\ln R}^2 - \alpha_R \cdot \beta_{\mathrm{t}} \cdot s_{\ln R}\right) = R_{\mathrm{m}} \cdot \exp\left(-0.5\, s_{\ln R}^2 - 0.8 \cdot 3.8\, s_{\ln R}\right) \tag{3}$$

where $s_{\ln R}$ is the standard deviation of the logarithms of the received load-bearing capacities. α_R is the sensitivity factor for resistance, and β_{t} is the target reliability index. Here, their values are chosen as recommended in EN 1990 [16]. Whether these values are appropriate for existing masonry structures is beyond the scope of the study. In practice, design values shall also include a factor considering model uncertainties, which is not included at this point yet. As mentioned in the introduction, design values are often determined by directly using the variability of the material properties as an input value with the results of a deterministic calculation R_{det} as a reference point. In the calculation of R_{det}, homogeneity is assumed and mean values of the material properties are used. Design values $R_{\mathrm{d,hom}}$ based on this approach can be evaluated by

$$R_{\mathrm{d,hom}} = R_{\mathrm{det}} \cdot \exp\left(-0.5\, s_{\ln f}^2 - 0.8 \cdot 3.8\, s_{\ln f}\right) \tag{4}$$

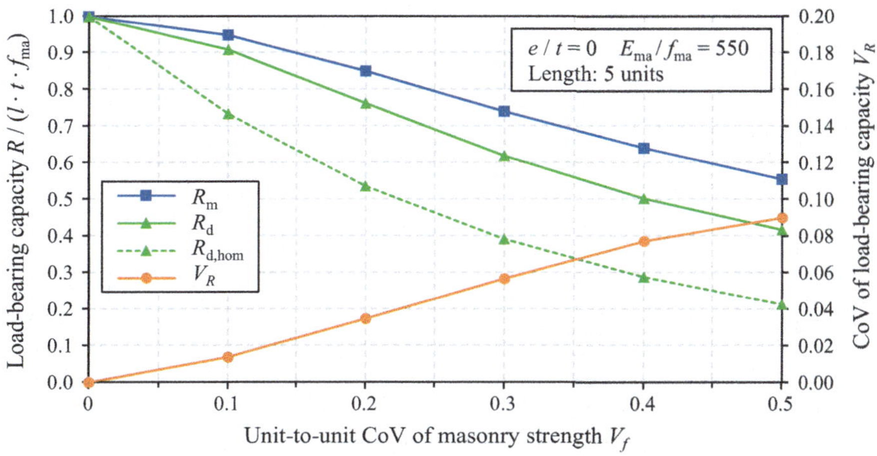

Fig. 4 Results of the Monte Carlo simulations with varying CoV of masonry strengths

5.2 *Results for Varying Coefficient of Variation*

In Fig. 4, the simulation results for the mean value R_{m} and the CoV V_R of the load-bearing capacity of the wall and the design capacity R_{d} according to Eq. (3) are shown for varying unit-to-unit CoV V_f of the masonry compressive strength. The capacity is standardised by cross-sectional area and strength. Due to the neglect of second-order effects within this parameter study and no eccentricity e of the load, a deterministic calculation with mean values leads to a standardised capacity of $R_{\mathrm{det}}/(l \cdot t \cdot f_{\mathrm{ma}}) = 1$. For comparison, theoretical design values $R_{\mathrm{d,hom}}$ assuming homogeneity with a CoV of the strength equal to V_f, see Eq. (4), are displayed.

Two main effects can be noticed. First, the mean value of the load-bearing capacity decreases with increasing CoV of the material properties. This is due to an increased number of spots with low compressive strength in the wall combined with limited stress redistribution capability. Second, the CoV of the load-bearing capacity V_R is much lower than the CoV of the material properties. The ratio V_R/V_f is approximately constant with a value of about 0.18 for the investigated wall. This significantly reduces the variability that has to be considered for determining the required design value (or partial factor) and therefore leads to design values that are significantly higher than those according to the approach assuming homogeneity.

5.3 *Results for Varying Wall Length*

The parameter study on the influence of wall length starts with pillars of 1 and 2 undivided units per course. Then, walls constructed in cross bond with 3 up to 11 units per course are investigated. Higher wall lengths lead to a higher number of

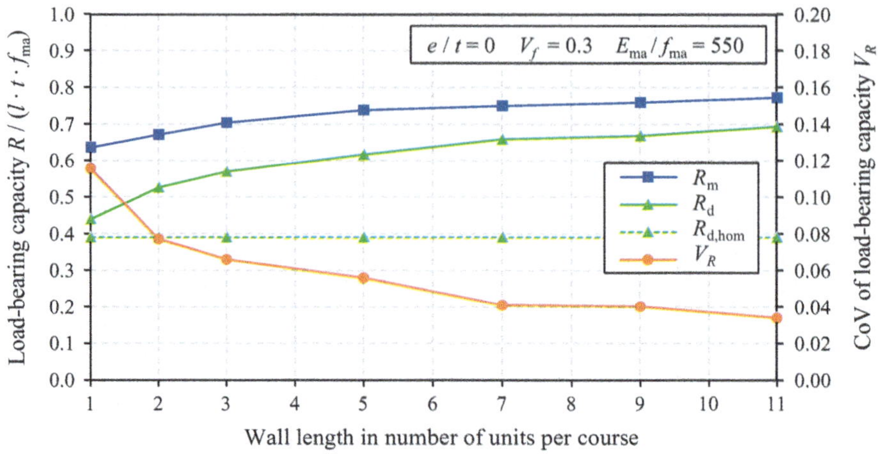

Fig. 5 Results of the Monte Carlo simulations with varying wall length

potential weak spots. However, single weak spots do not directly cause failure since there are also more units within a course that the stress can be redistributed to. The combination of both contrarious effects leads to a slight increase in the mean load-bearing capacity R_m, which can be seen in Fig. 5. The CoV of the load-bearing capacity decreases with higher wall length, since, due to stress redistribution, the strength is (up to a certain amount) averaged over a higher number of units. In principle, these results confirm the need for a reduction factor for walls with small cross-sectional area A, which, for example, is given in EN 1996-1-1 [12] as $0.7 + 3$ A for $A < 0.1$ m^2. This would lead to reduction factors of 0.78, 0.87 and 0.96 for the walls with only 1, 2 and 3 units per course, respectively.

5.4 *Results for Varying Slenderness*

With increasing slenderness of the wall, second-order effects gain influence and the failure mode of the wall switches from compression failure to stability failure. In this context, it is appropriate to define slenderness λ not only by the ratio of effective height h_{ef} to thickness t but also by taking into account the ratio of mean modulus of elasticity to mean compressive strength $E_{ma,m}/f_{ma,m}$ and the stress-strain parameter k. Therefore, slenderness is defined as

$$\lambda = \frac{h_{ef}}{t} \cdot \sqrt{k \cdot f_{ma,m} / E_{ma,m}} = \frac{h_{ef}}{t} \cdot \sqrt{\varepsilon_f} \tag{5}$$

with ε_f being the total strain at peak compressive stress, cf. [10]. To avoid varying the number or dimensions of the units, the slenderness in this parameters study

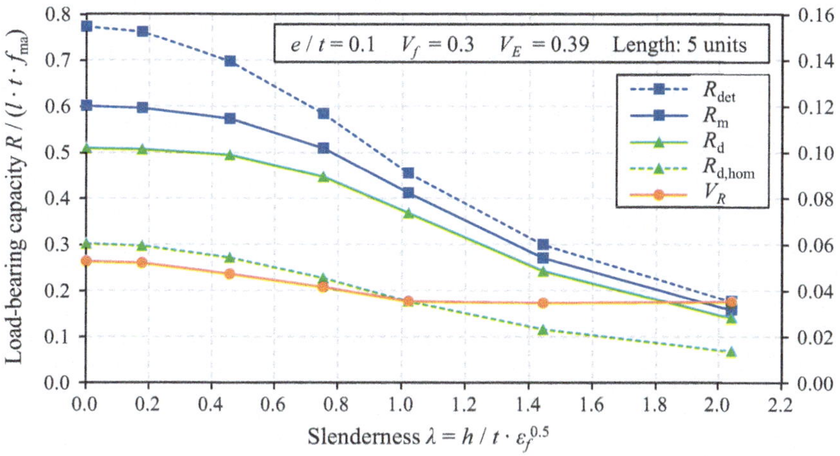

Fig. 6 Results of the Monte Carlo simulations with varying slenderness

is increased by varying the modulus of elasticity E_{ma}. The investigated ratios $E_{ma,m}/f_{ma,m}$ are chosen as 10000, 1500, 550, 300, 150 and 75, which results in the values λ in Fig. 6. Also, results for $\lambda = 0$, i.e. without consideration of geometrical nonlinearity, are given. The eccentricity to thickness ratio e/t is selected as 0.1, and h_{ef} is equal to the height due to hinged supports.

With increasing slenderness, the influence of the compressive strength within the wall diminishes. Instead, the overall stiffness of the wall and, hence, the moduli of elasticity of the "units" become important. Single weak spots are not as influential anymore since the overall stiffness of the wall can be described by a weighted harmonic mean of the varying moduli of elasticity in the wall. Therefore, the mean load-bearing capacity R_m gets closer to the deterministic mean value assuming homogeneity R_{det} when slenderness is increased, see Fig. 6. Furthermore, the CoV of the load-bearing capacity decreases, although the CoV of the modulus of elasticity is higher than the CoV of masonry strength.

6 Conclusions

The conducted study pointed out two main effects on the load-bearing capacity of clay brick masonry walls in compression with regard to spatial variability. First, the mean value of the load-bearing capacity reduces with increasing spatial variability within a masonry wall, i.e. with increasing inhomogeneity. Second, the resulting coefficient of variation of the load-bearing capacity is much smaller than the unit-to-unit CoV of the material properties within the wall. Furthermore, it was shown that the negative influence of spatial variability on suitable design values and, hence, partial factors, is smaller for walls of higher length as well as for walls with higher slenderness. In all cases, the consideration of spatial variability led to higher theoretically acceptable design values than the approach assuming homogeneity. In practice, the total variability of material properties has to be broken down into the variability of mean material properties between walls and the spatial variability within a single wall in order to determine suitable design values or partial safety factors.

The next step of the research will be to combine the consideration of spatial variability, statistical uncertainty as well as model uncertainties, especially in the estimation of masonry strength, into a holistic concept for modifying partial factors for the assessment of existing masonry structures.

Acknowledgments The financial support by the program "Zukunft Bau" of the German Federal Institute for Research on Building, Urban Affairs and Spatial Development is gratefully acknowledged.

References

1. Müller, D., & Graubner, C.-A. (2018). Uncertainties in the assessment of existing masonry structures. In *Life cycle analysis and assessment in civil engineering: Towards an integrated vision* (pp. 369–376), 28–31 October, Ghent, Belgium.
2. Müller, D., & Graubner, C.-A. (2019). Modification of the partial safety factor for compressive strength of existing masonry using a Bayesian method. In *17th International probabilistic workshop* (pp. 133–138), 11–13 September, Edinburgh, United Kingdom.
3. fib. (2016). *Bulletin 80—Partial factor methods for existing concrete structures*. Recommendation. Fédération internationale du betón, Lausanne, Switzerland.
4. EN 772-1. (2011). *Methods of test for masonry units—Part 1: Determination of compressive strength*. European Committee for Standardization, Brussels.
5. EN 1015-11. (2019). *Methods of test for mortar for masonry—Part 11: Determination of flexural and compressive strength of hardened mortar*. European Committee for Standardization, Brussels.
6. EN 1052-1. (1998). *Methods of test for masonry—Part 1: Determination of compressive strength*. European Committee for Standardization, Brussels.
7. Lourenço, P. J. B. B. (1996). *Computational strategies for masonry structures*. Dissertation, TU Delft.
8. Schueremans, L. (2001). *Probabilistic evaluation of structural unreinforced masonry*. Dissertation, KU Leuven.
9. Brameshuber, W. (2019). Eigenschaften von Mauersteinen, Mauermörtel und Putzen. In *Mauerwerk-Kalender 2019* (pp. 3–29). Ernst & Sohn.
10. Glock, C. (2004). *Traglast unbewehrter Beton- und Mauerwerkswände. Nichtlineares Berechnungsmodell und konsistentes Bemessungskonzept für schlanke Wände unter Druckbeanspruchung*. Dissertation, Technische Universität Darmstadt.
11. EN 1992-1-1. (2010). *Eurocode 2: Design of concrete structures—Part 1-1: General rules and rules for buildings*. European Committee for Standardization, Brussels.
12. EN 1996-1-1. (2013). *Eurocode 6: Design of masonry structures—Part 1-1: General rules for reinforced and unreinforced masonry structures*. European Committee for Standardization, Brussels.
13. Heffler, L. M., Stewart, M. G., Masia, M. J., & Correa, M. R. S. (2008). Statistical analysis and spatial correlation of flexural bond strength for masonry walls. *Masonry International, 21*, 59–70.

14. Müller, D., Förster, V., & Graubner, C.-A. (2017). Influence of material spatial variability on required safety factors for masonry walls in compression. *Mauerwerk—European Journal of Masonry, 21,* 209–222.
15. Kaushik, H. B., Rai, D. C., & Jain, S. K. (2007). Stress-strain characteristics of clay brick masonry under uniaxial compression. *Journal of Materials in Civil Engineering, 19,* 728–739.
16. EN 1990. (2002). *Eurocode—Basis of structural design.* European Committee for Standardization, Brussels.

Study on the Accuracy of Chloride Determination Methods and Their Predictions

Fritz Binder, Stefan L. Burtscher, and Alfred Strauss

Abstract At ASFiNAG, most structures are made of reinforced or prestressed concrete. In Austria strong winters are obligatory. Therefore, de-icing measures with salt are used for thawing. These substances contain chlorides, that ingress into the concrete and lead to degradation. Several elements are strongly exposed to chlorides, such as girders of overpasses and columns of bridges beside and between roadways. For condition assessment and service life prediction of existing road structures the determination of a reliable chloride content is key. The article presents the results of the chloride content of reinforced concrete obtained by two methods. One is the conventional Cl-determination described in standards. The other is the LA-ICP-MS (Laser Ablation Inductively Coupled Plasma Mass Spectrometry) method, which is a fast, reliable, accurate and high-resolution analysis method. This method allows the determination of the chloride content as a fraction of cement and additionally distinguishes between the aggregate and the cement phase. The profiles were determined densely at steps of 3 mm in depth. Regressions with different boundary conditions were used to fit the obtained data according to Fick's second law. For comparison and prediction purposes, the corresponding convection depth as well as the chloride diffusion coefficients were determined. These parameters, as well as the fluctuations of chloride profiles for one year deliver important insights for assessment and prediction. Significant differences were observed in the results obtained by the applied analysis methods. The study addresses the origins of these differences and shows the variances when it comes to prediction of remaining service life. The results are compared and discussed to show the complex nature and sensitivity of the derived input parameters. These results show on one hand the importance for an accurate chloride analysis (LA-ICP-MS) and gives hints for an improved assessment of structures.

F. Binder (✉)
ASFiNAG Bau Management GmbH, Vienna, Austria
e-mail: fritz.binder@asfinag.at

S. L. Burtscher
Burtscher Consulting GmbH, Vienna, Austria

A. Strauss
University of Natural Resources and Life Sciences, Vienna, Austria

© The Author(s), under exclusive license to Springer Nature Switzerland AG 2021
J. C. Matos et al. (eds.), *18th International Probabilistic Workshop*, Lecture Notes in Civil Engineering 153, https://doi.org/10.1007/978-3-030-73616-3_61

Keywords Accuracy · Chloride determination · LA-ICP-MS · Laser ablation inductively coupled plasma mass spectrometry · Chloride transport · Sensitivity analysis

1 Malice of Standard Chloride Determination Methods

Currently the standard method to determine the chloride content is the titrimetric Volhard method, which is described in the European standard EN14629:2007. Thereby, concrete samples are gathered by drilling concrete cores or borehole dust at different depth levels. Nowadays, the titration is a state-of-the-art analysis method. However, the problem is that the samples may not contain a representative chloride content, which results on one hand to high scatter and on the other to wrong results. The high scatter leads from a not representative sample, which may contain aggregates in a higher or lower content than the concrete bulk. In the titration analysis the chloride content is determined for the entire sample. If the amount of aggregates (assume that they do not contain chloride) in the sample is higher than in the representative concrete a lower chloride content is determined. The samples are usually very small so these deviations are likely to be high [1]. A second reason for a high scatter is that the cement content is usually not known for these analyses and a conservative value must be chosen. Additionally, the result may be erroneous, in terms that the aggregates may contain chlorides (see Fig. 1b). The aggregates are usually very dense, and the chloride is not free in the aggregates and does not contribute to a corrosion susceptibility. Inside the aggregates accumulations of chloride can be observed and during specimen preparation this chloride content is mixed with the free chloride in the cement paste. The chloride from inside the aggregates would not contribute to corrosion susceptibility and leads to an erroneous chloride to cement content.

2 LA-ICP-MS for more Accurate Chloride Content in Concrete

Since maintenance cost is high and premature renewal is even higher, accurate prediction models are necessary. The basis for an accurate prediction is an accurate measurement of the decisive parameters. For that propose the evaluation of depth profiles of precise chloride per cement content were developed [2]. Based on these measurements an evaluation method was developed to determine the total chloride to cement content. For short LA-D (Laser Ablating Deicing-salt) Method.

In this novel approach, selective quantification of chloride in the cement phase is performed. Aggregates containing chloride are excluded from the data evaluation and thus, a very reliable determination of the free chloride content in the cement phase combined with high depth resolution can be obtained. Additionally, the chloride content is determined as a fraction of the cement content—the major parameter for the susceptibility to corrosion.

The method was developed to analyze drilled core samples taken from structures by wet drilling. The diameter of the core was about 50 mm and the length (the depth of the Chloride/cement-profile) was about 60 mm. The cores we cut into halves and the LA-ICP-MS analysis was performed along lines, which were parallel to the outer surface, having a length of about 25 mm and a distance of 10 mm from each other. Sample analysis was conducted with a laser beam dimeter of 250 μm, enabling the measurement of depth profiles with increased depth resolution. From each line the analysis was evaluated by excluding the regions with aggregates and deter-mining the mean of the Chloride to cement content. This mean value results in one point of the chloride in cement profile (see Fig. 1). A detailed description of the LA-ICP-MS method and its development as well as the calibration, analysis and the verification of obtained results can be found in [2]. From the other halve of the sample a titrimetric Volhard analysis according to EN 14629:2007 was performed. Both methods were applied to a core, which contains high amounts of chloride in the aggregates (see Fig. 1). One can see that the difference between the LA-ICP-MS and the titration for this specimen is enormously high, not even the trend is the same. Thus, based on the results received from titration it is not possible to develop an accurate prediction model. As known concrete samples do not only consist of cement, but also of aggregates. Usually the chloride content, as well as the amount of aggregates in the concrete is not exactly known. Further, some aggregates contain considerable amounts of chloride. A comparison of the determined chloride concentrations is given in Fig. 1b. While LA-ICP-MS analysis and titration yielded consistent results in the case of granite (acid insoluble) as aggregate material, the values differed significantly for the aggregate type river gravel and limestone. Based on the results, the titration does not seem feasible for the exact determination of chloride in the cement phase.

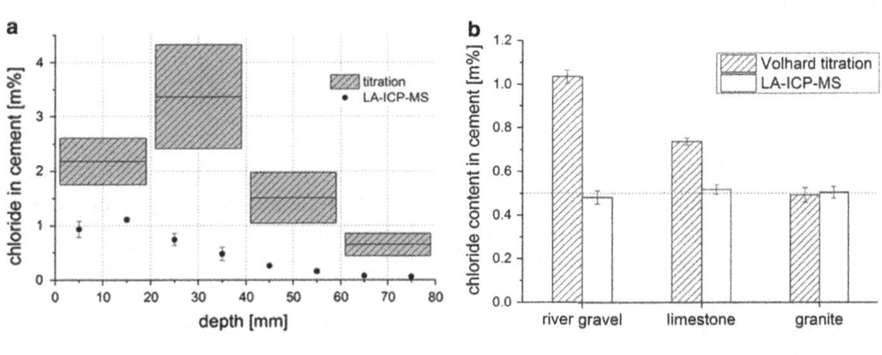

Fig. 1 **a** Comparison of titrimetric chloride determination and LA-D analysis [3]. **b** Comparison of different aggregates in the concrete composition and their impact on the chloride content using titrimetric and LA-D [3]

3 Prediction of Chloride Transport in Concrete

3.1 Mathematical Formulation

In Crank [4] first times Fick's second law was introduced for modeling chloride ingress by diffusion in concrete structures using following one-dimensional partial differential equation:

$$\frac{\partial C}{\partial t} = \frac{\partial}{\partial x} \cdot D \cdot \frac{\partial C}{\partial x} = D \cdot \frac{\partial^2 C}{\partial x^2} \tag{1}$$

where C and D are defined as the chloride content as a function of position x and time t, and the chloride diffusion coefficient, respectively. Also Crank [4] derived the closed-form of Eq. (1) as mass transport model to simplify chloride diffusion prediction in concrete. The first example for using the closed-form of Eq. (1) for constant surface chloride content C_s to predict the diffusion of chloride ions in concrete was found in [5].

The transport processes of chloride in concrete are complex and influenced by strewing intervals, material parameters and environmental conditions. Therefore, a constant surface chloride function is in many cases not enough. Hence, it is very common on aged concrete structures exposed to de-icing salt, to find the maximum chloride concentration in the concrete bulk. This indicates, that the exposed side of the concrete surface is not fully driven of a diffusion dominated process. Environmental actions induce a gradient of moisture along the cover depth, the so called convection zone Δx. Additionally, it is proposed in [6] that the chloride diffusion coefficient as a characteristic material parameter is time dependent. This leads to the constituted mathematical model used in [7], which is based on the error function solution of Fick's 2nd law of diffusion.

$$C_{(x,t)} = C_{s,\Delta x} \cdot \left[1 - erf\left(\frac{x - \Delta x}{2 \cdot \sqrt{D_{app}(t) \cdot t}}\right)\right] \tag{2}$$

where $C_{s,\Delta x}$ is the substitute chloride concentration resulting from the prevailing exposure environment at depth Δx [m-%/c], Δx is the depth of the convection zone [m], x is the depth with a corresponding chloride content [m], t is the time [s], $D_{app}(t)$ is the apparent chloride diffusion coefficient [m^2/s] and *erf* is the Gaussian error function.

The time dependent diffusion coefficient is described as following:

$$D_{app}(t) = k_e \cdot D_{app}(t_0) \cdot \left(\frac{t_0}{t}\right)^{\alpha} \tag{3}$$

where $D_{app}(t_0)$ is the apparent chloride diffusion coefficient determined at a reference to t_0 [m^2/s] and α is the aging exponent indicating the decrease over time of the apparent chloride diffusion coefficient [–]. It should to be mentioned that the parameter $D_{app}(t_0)$ always represents the diffusion coefficient of the concrete over the entire considered time period t as an averaged constant. Temperature affects the mobility of ions and, hence, the diffusion rate of chlorides. The transfer parameter k_e has been introduced in order to account for the impact of the external temperature on the chloride diffusion in concrete.

$$k_e = \exp\left(b_e \cdot \left(\frac{1}{T_{ref}} - \frac{1}{T_{real}}\right)\right) \tag{4}$$

where b_e is the temperature coefficient proportional to the activation energy [K], T_{ref} is the reference temperature [K] and T_{real} is the temperature of the structural element or the ambient air [K]. In the following all fluctuating parameters will be described as mean values and corresponding distribution characteristics will be given.

3.2 Fitting Procedure

To predict the transport of chloride ions, both C_s and D can be determined by curve fitting with real data taken from samples extracted from the structures. The usual procedure is to fit the error function into the chloride concentration gradient and to obtain, by regression analysis, the concentration, C_s, and the Diffusion coefficient D.

The procedure for calculating the chloride diffusion coefficient and surface concentration $C_{s,\Delta x}$ from a chloride profile when the maximum concentration is inside the concrete surface is described in [8].

From the chloride profiles diffusion coefficients were obtained. Initially the diffusion coefficient were the same in all three levels, but due to different exposure of chlorides and other phenomena the diffusion coefficient may change and differ in the three horizons [9].

The aim is to determine D_{app} from specimen taken from cores. As a result, the obtained model parameters can be used to characterize the ingress rate at the age of the sample or to predict the evolution of the profile over time.

To consider the time variant nature of D_{app}, the aging exponent α in Eq. (2) has to be quantified by fitting data from chloride profiles from at least two different points in time to a regression function as shown in [10] (Fig. 2).

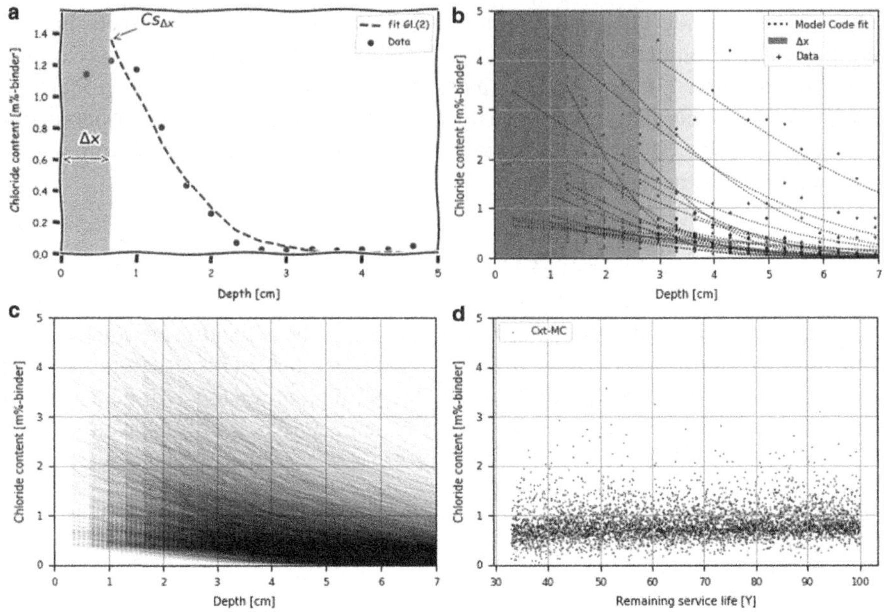

Fig. 2 **a** Sketch of a typical shape of a chloride profile (dots) and its regression to MC Eq. (2) (dashed line). **b** Chloride profiles (cross), their fit (dotted) and derived convection zone (gray area) from field data **c** 5000 stochastic realizations of a chloride profile derived from fitting parameters according to figure **d** Simulation of the chloride content on rebar level over time

3.3 Model Parameter

3.3.1 Convection Zone Δx

As already mentioned, concrete in structures exposed to de-icing salt build a convection zone with alternating wetting and drying cycles. This depth of chloride convection zone is considered as a key variable, in the model based on Fick's second law [see Eq. (2)]. So within the convection zone the chloride profile deviates from the proposed Eq. (2). Actually, the chloride convection zone depths are not constant but depend on the external environment, drying and wetting time ratio and concrete properties [11]. Own studies on real structures exposed to 38 years lasting de-icing salt exposure used the novel chloride determination method LA-D discovered a wide range of the convection zone (aprox. 1–4 cm) [12]. In [13] similar results are presented.

Hence, it is appropriate to determine the chloride convection zone depth on the basis of traditional approaches, i.e. fitted by the measured data, rather than recommended constant values. The depth of the maximum chloride concentration in concrete is defined as the chloride convection zone depth which can be determined by the fitting profiles of Eq. (3) based on measured data. After stochastic analysis a beta distribution of the convection zone is taken in the present paper. The mean and coefficient of variation are 16.93 and 10.30 mm, respectively.

3.3.2 Substitute Chloride Concentration $C_{s,\Delta x}$

In fib MC2010 is defining $C_{s,\Delta x}$ as the chloride content at the end of the depth of the convection zone in % by mass of cement.

The maximum surface chloride concentration on the concrete surface is typically considered as the demarcation point between the convection zone and diffusion area [14], which is a function of internal and external factors, e.g. solution concentration, environmental loads and concrete composition. According to existing studies, the surface chloride concentration in concrete follows a lognormal or normal function. [15–17]. It is assumed that also the substitute chloride concentration follows this kind of statistical characteristic. After stochastic analysis a lognormal distribution of the substitute surface chloride content is taken in the present paper. The mean and coefficient of variation are 1.659 m%-binder and 1.278, respectively.

3.3.3 Diffusion Coefficient $D_{app}(t)$

The determined chloride diffusion coefficient following the procedure described in Sect. 3.2 obtain $D_{app}(t)$ at the time where the specimen was tested and is usually given in years and strongly depends on the season (date) when the extraction in-situ where executed [18]. To derive $D_{app}(t_0)$ by using Eq. (3) $D_{app}(t)$ is an input variable. The reference point of time t_0 was chosen to be $t_0 = 0.0767$ years ($t_0 = 28$ d). The statistical variation of chloride diffusion coefficients is adequately described by a Normal distribution [19]. The mean and coefficient of variation are 10.195 m^2/s and 9.022, respectively.

3.3.4 Environmental Transfer Parameter k_e, T_{real}

For a given concrete structure in a given environment, the rate of chloride ingress highly depends on the temperature, as shown in Eq. (4). Based on local information on prevailing temperature conditions, data on average annual temperatures is used as a basis for the selection of this input parameter. All collected diffusion coefficients were converted to those based on $T_{ref} = 293$ K (20 °C). The variable T_{real} describes the temperature of the structural element or the ambient air and is obtained by data grabbed from a moisture and temperature monitoring applicated on the structure itself. The variable is normal distributed with a mean value of $\mu = 290.63$ K and a deviation of $\sigma = 9.19$. The values of the regression variable vary between 3500 and 5500 K and can be statistically described by a normal distribution ($\mu = 4800$, $\sigma = 700$) [20].

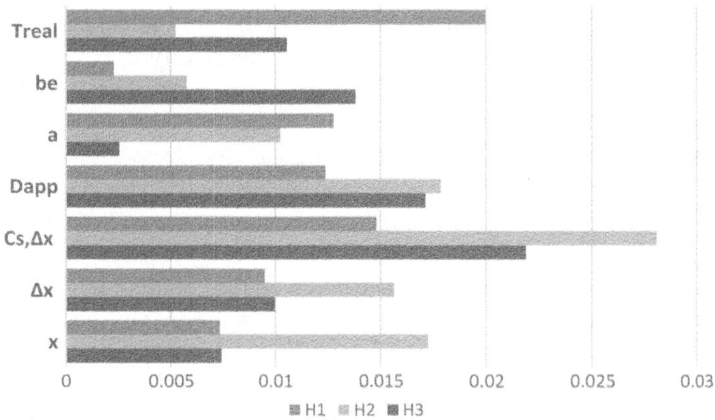

Fig. 3 Global sensitivity indices derived from stochastic field data distinguished in 3 exposure zones (H1 $= 0 - 0.8$ m, H2 $= 0.8 - 1.8$ m, H3 ≥ 1.8 m)

4 Data (Sensitivity) Analysis

For the analysis the data was treated for every horizon separately, except Temperature and therefore the environmental transfer parameter k_e respectively. The derived regression parameter from the chloride profiles are used to calculate the complete range of the chloride content over the remaining lifetime. The output was used to undertake a sensitivity analysis.

A sensitivity analysis (SA) can identify the most influential parameters and their interactions and how these parameters affect the output. In other words, sensitivity analysis quantifies the degree to which a model input effects an output. The sensitivity analysis was conducted using the programming language Python, in particular, the Sensitivity Analysis Library (SALib) [21] (Fig. 3).

5 Conclusions

The LA-ICP-MS method presented in this paper shall contribute to improve forecast models. The analysis method is fast and reliable for determination of the chloride content in existing concrete structures. It was shown, that the traditional titrimetric method leads to high deviations and to systematic errors in the results. Therefore, the proposed LA-ICP-MS approach helps to increase the reliability of the obtained results. The high depth resolution of the chloride profiles enables a more exact data assessment for forecast models and supports the prediction of critical developments even further.

Since the convection process is much faster than the diffusion, the time variance inside the concrete is dominated by this process and the existing state-of-the-art models should consider that. The obtained data with the LA-ICP-MS method is an excellent basis for assessment and development of models and their predictions.

In general, a sufficiently large dataset must be available to make a valid prediction of the chloride content over time. Although the available data were very accurate in their quality regarding true chloride content, several uncertainties in the data quality, such as moisture, carbonation and chloride binding capacity of the concrete still exist. The data were not sufficiently accurate for this purpose. A continuing high fluctuation range can be observed, which is nevertheless still smaller in comparison to the conventional method.

The sensitivity analysis of the presented case study shows the important influence of the diffusion parameters D_{app} and of the exposure conditions during service life $C_{S,\Delta x}$. Other parameters have similar sensitivity. Uncertainty should be further reduced by exact determination of the convection zone Δx and exact chloride content $C_{s,\Delta x}$ on this interface.

Using fixed values of chloride convection depths lead to considerable uncertainties in the model outputs (e.g. the time to corrosion initiation), theoretical analyses of chloride concentration profiles may result in serious consequences (e.g. the real service life, the optimal time for inspection and maintenance), and thus analytical results fail to reflect the real deterioration of RC under deicing salt conditions.

References

1. Kosalla, M., & Raupach, M. (2018). Diagnosis of concrete structures: The influence of sampling parameters on the accuracy of chloride profiles. *Materials and Structures, 51*(3).
2. Limbeck, A., Eitzenberger, A., Bonta, M., & Burtscher, S. (2015). New analysis method for the accurate determination of chloride content in the cement phase of concrete. In C. Hellmich, B. Pichler, K. Johann (Eds.), *CONCREEP 10*, Vienna (pp. 800–804).
3. Bonta, M., Eitzenberger, A., Burtscher, S., & Limbeck, A. (2016). Quantification of chloride in concrete samples using LA-ICP-MS. *Cement and Concrete Research, 86,* 78–84.
4. Crank, J. (2011). *The mathematics of diffusion*, 2., reprint. ed. Oxford Univ. Press.
5. Collepardi, M., Turriziani, R., & Marcialis, A. (1970). The kinetics of chloride ions penetration in concrete. *Il Cemento, 4,* 157–164.
6. fib Bulletin 34. (2006). Model code for service life design. *Concrete, 40*(5), 4.
7. fib. (2010). *Fib model code for concrete structures*. Ernst & Sohn.
8. Andrade, C., Climent, M. A., & de Vera, G. (2015). Procedure for calculating the chloride diffusion coefficient and surface concentration from a profile having a maximum beyond the concrete surface. *Materials and Structures/Materiaux et Constructions, 48*(4), 863–869.
9. Kapteina, G. (2015). *Diffusion-based model for predicting chloride ingress into road structures.*
10. fib Bulletin 76. (2015). *Benchmarking of deemed-to-satisfy provisions in standards: Durability of reinforced concrete structures exposed to chlorides*. International Federation for Structural Concrete (fib).
11. Liu, P. (2013). *Research on similarity of the chloride ingress in concrete under natural and artificial simulation environment*.
12. Binder, F., Burtscher, S. L., & Limbeck, A. (Eds.). (2018). Prediction of chloride profiles and discussion of time variantalterations. In *IALCCE 2018—The Sixth International Symposium on Life-Cycle Civil Engineering*, Ghent, Belgium; 2018.
13. Šomodíková, M., Strauss, A., Zambon, I., & Teplý, B. (2019). Quantification of parameters for modeling of chloride ion ingress into concrete. *Structural Concrete, 20*(1), 519–536.

14. DuraCrete. (2000). *General guidelines for durability design and redesign.*
15. Duprat, F. (2007). Reliability of RC beams under chloride-ingress. *Construction and Building Materials, 21*(8), 1605–1616.
16. Song, H.-W., Pack, S.-W., & Ann, K. Y. (2009). Probabilistic assessment to predict the time to corrosion of steel in reinforced concrete tunnel box exposed to sea water. *Construction and Building Materials, 23*(10), 3270–3278.
17. Val, D. V., & Trapper, P. A. (2008). Probabilistic evaluation of initiation time of chloride-induced corrosion. *Reliability Engineering and System Safety, 93*(3), 364–372.
18. Binder, F., Burtscher, S. L., & Strauss, A. (Eds.). (2019). Study on the time variant alteration of chloride profiles for prediction purpose. In *IABSE Symposium Guimarães 2019—Towards a Resilient Built Environment Risk and Asset Management*, Portugal, Guimarães.
19. Gehlen, C. (2000). *Probability-based service life design of reinforced concrete structures—Reliability studies for prevention of reinforcement corrosion.* Beuth.
20. Page, C. L., Short, N. R., & El Tarras, A. (1981). Diffusion of chloride ions in hardened cement pastes. *Cement and Concrete Research, 11*(3), 395–406.
21. Herman, J., & Usher, W. (2017). SALib: An open-source Python library for Sensitivity Analysis. *The Journal of Open Source Software, 2*(9).

The Impact of Clustering in the Performance Prediction of Transportation Infrastructures

Carlos Santos, Sérgio Fernandes, Mário Coelho, and José C. Matos

Abstract In the context of transportation infrastructures management, bridges are a critical asset due to their potential of becoming network's bottlenecks. Unfortunately, this aspect has been emphasized due to several bridge failures, occurred in the last years worldwide, resulting from climate change-related hazards. Given this, it is important to establish accurate tools for predicting the structural condition and behavior of bridges during their lifetime. The present paper addresses this topic taking into account one of the statistical models most used and generally accepted in existing bridge management systems—Markov's stochastic approach, which is further described. These statistical models are highly susceptible to the data that feeds them. Quite often, the step related with data cleaning and clustering is not properly conducted, being the most commonly available data sets adopted in bridge's performance prediction. This paper presents a comparative analysis between different performance predictions. The only different between consecutive scenarios corresponds to the subset of bridges database used in each analysis. It was found that the development of good data clusters is of utmost importance. Contrarily, the use of poor clusters can lead to deceiving results which hinder the actual deterioration tendency, thus leading to wrong maintenance decisions.

Keywords Infrastructures · Bridge decks · Clusters · Probabilistic prediction · Markov chain

1 Introduction

Most of the transportation of people and goods is made possible due to the existence of transportation infrastructures. The performance of transportation infrastructures is directly associated with the degradation that arises from their exposure to the natural environment, and the associated deterioration mechanisms. These deterioration mechanisms have been extensively investigated in the last years, due to the

C. Santos · S. Fernandes · M. Coelho (✉) · J. C. Matos
University of Minho, Guimarães, Portugal
e-mail: mcoelho@civil.uminho.pt

© The Author(s), under exclusive license to Springer Nature Switzerland AG 2021

J. C. Matos et al. (eds.), *18th International Probabilistic Workshop*, Lecture Notes in Civil Engineering 153, https://doi.org/10.1007/978-3-030-73616-3_62

climate-change induced phenomena, which accelerates them. In this context, bridges, likewise other transportation infrastructures' assets, become particularly vulnerable due to their natural exposure.

The bridge components' structural deterioration, due to chemical and physical processes, corresponds to a continuous process in time. Due to the inherent high complexity and costly use of structural health monitoring systems, the continuous condition monitoring is in practice replaced by periodic inspections. In each of these inspections, a discrete condition rating is used to classify the current condition state of the bridge being assessed. This classification is defined by the organization/agency and there are different scales worldwide such as from 1 to 5 (e.g. Portugal [1]), from 0 to 9 (e.g. USA [2]), or other similar cases.

The condition rating allows the simplified measurement of the deterioration level of bridge components, from which is assigned a condition state by the inspection team during visual inspections. Even though this classification strategy can be (and normally is) adopted at the component level, in the end, the bridge overall condition needs to be established. To this purpose, several approaches can be used, being the worst component classification one of the most widely used. In this approach, regardless to the deterioration processes that a specific bridge might be undergoing, it will be classified considering a bridge overall condition that corresponds to its worst component classification.

Hence, it is of utmost importance to model deterioration in terms of a time-dependent stochastic process $\{X(t), t \geq 0\}$ where $X(t)$ is the random quantity for all time steps $t \geq 0$ [3]. For this purpose, an existing stochastic prediction model was implemented. The present paper is based on the premise (after analyzing data available) that the inspections do not occur in uniform time periods, which led to the application of the Continuous Time Markov Chain (CTMC) based on historical condition states. For further information concerning the application of forecasting models to predict concrete bridge decks performance, the reader can refer to [2, 4].

The use of such predictive modes requires the existence of historical data. Depending on the quantity and quality of these data, the accuracy of the model is directly affected. An example of the most commonly available data for bridges includes: (i) structure type (e.g. bridge, viaduct, overpass, underpass, water underpass, cattle creep); (ii) structural system (e.g. slab, multibeam, ribbed, frame, truss, box beam, arch, suspension); (iii) component type (e.g. column, deck, bearing, expansion joint); (iv) material (e.g. reinforced concrete, prestressed concrete, steel, wood, masonry); (v) previous condition states associated with each component.

When using Markovian approaches for modelling bridges' performance, only condition states are directly used. All other parameters are accounted for indirectly by means of proper clustering existing data. To use directly all these data, artificial intelligence models are a common approach [2, 5, 6].

2 Predictive Models in Asset Management

It is essential to provide infrastructure managers with advanced tools to estimate the service life of bridge components and ensure that all components fulfill their required serviceability. In the last years, significant effort has been taken to the development new stochastic models. Consequently, several performance/degradation predictive models have been developed in order to forecast the long-term performance of bridge components. Such deterioration models are currently being incorporated in the majority of bridge management systems [7]. Using them, it is possible to obtain an optimized, cost-efficient, and strategic maintenance schedule for existing bridge stock.

Depending on the type of analysis time step (discrete or continuous) and the type of condition state space (discrete or continuous) used, two major approaches can be adopted in the bridges' performance prediction. Problems addressing the bridge performance from a quantitative evaluation approach adopt continuous-time and continuous state, which requires the use of most complex models, such as reliability-based models. On the other hand, the simplest qualitative approaches are associated with the use of discrete-time and discrete-state conditions.

In fact, in the majority of bridge management systems (BMS), the state space is discrete, since it corresponds to a well-defined condition rating scale (e.g., from 1 to 5). The adoption of discrete-time is a simplification that enables reducing the computational complexity.

For a broader overview on the current predictive models and their different features can be found elsewhere [2, 4, 7, 8].

2.1 *Fundamentals of Markov models*

The Markov process used to predict the future performance of infrastructures is denominated as the Markov chain since it uses a discrete state space, i.e., the condition rating is not continuous, but discrete and finite [8]. The use of the Markov chain requires the knowledge of its properties, premises, and limitations, in order to be successful implemented. For this purpose, the most important details of Markov formulation are further described.

The discrete-time Markov process can be defined as a stochastic process with parameters $X(t)$, that for any k time points, t_1, t_2, t_k, the conditional distribution of $X(t_k)$ for given values of $\{X(t_1), \ldots, X(t_{k-1})\}$ depends only on the previous one $X(t_{k-1})$, which is the most recent known value. Thus, mathematically an kth-order Markov Chain (Eq. 1) is currently assumed as a first-order Markov chain (Eq. 2) in which the dependency is given by a conditional probability [8]. The Markovian property of "memoryless" has often been criticized, suggesting an adaptation of the first-order Markov chain to higher levels in order to considerer historical condition.

$$P\left[X(t_k) \leq x_k | X(t_1) = x_1, X(t_2) = x_2, \ldots, X(t_{k-1}) = x_{k-1}\right] \tag{1}$$

$$P\left[X(t_k) \leq x_k | X(t_{k-1}) = x_{k-1}\right] \tag{2}$$

Assuming that it can represent the sequence of the future degradation process, this prediction depends only on the present and not on the past. The parameters corresponding to the condition states $\{X(t_k), k = 0, 1, 2, \ldots\}$ are the random variables that represent the state of the process at time points t_k. Thus, the condition states $\{x_i, i = 1, 2, \ldots, m\}$ correspond to the values that contain the state space of the process [8].

A Markov chain is determined by a transition probability function that corresponds to the conditional probability of a transition from state *i* to state *j* during a given period of time. Thus, a single-step transition probability (Eq. 3) is defined as:

$$p_{ij}^{\Delta t} = p_{ij}^{t,t+1} = P[X(t+1) = j | X(t) = i] \tag{3}$$

A Markov transition probability matrix (Fig. 1a) is a square matrix of order *m* in which its elements represent transition probabilities. Figure 1b is similar to an upper triangular matrix and corresponds to the natural degradation process (without improvements) and is characterized by the following conditions:

$$-0 \leq p_{ij}^{t,t+1} \leq 1, \quad \{\mathrm{i, j} = 1, 2, \ldots, \mathrm{m}\};$$

$-p_{ij}^{t,t+1} = 0$, for i > j, i.e., transition probabilities assume that the condition of a structure only can either stay or decrease;

$$-\sum_{j=1}^{j=m} p_{ij}^{t,t+1}, \quad \{\mathrm{i, j} = 1,\ 2, \ldots, \mathrm{m}\}.$$

In a Markov process with a state space defined by *m* values, the state of the process at any time *t* is typically stochastic (uncertain transition event) and is defined by a probability mass function that is denoted by an *m*-dimensional vector B(*t*) [8]. Its

$$\mathbf{P}^{t,t+1} = \begin{bmatrix} p_{11}^{t,t+1} & p_{12}^{t,t+1} & p_{13}^{t,t+1} & \cdots & p_{1m}^{t,t+1} \\ p_{21}^{t,t+1} & p_{22}^{t,t+1} & p_{23}^{t,t+1} & \cdots & p_{2m}^{t,t+1} \\ p_{31}^{t,t+1} & p_{32}^{t,t+1} & p_{33}^{t,t+1} & \cdots & p_{3m}^{t,t+1} \\ \cdots & \cdots & \cdots & \cdots & \cdots \\ p_{m1}^{t,t+1} & p_{m2}^{t,t+1} & p_{m3}^{t,t+1} & \cdots & p_{mm}^{t,t+1} \end{bmatrix} \qquad \mathbf{P}^{t,t+1} = \begin{bmatrix} p_{11}^{t,t+1} & p_{12}^{t,t+1} & p_{13}^{t,t+1} & \cdots & p_{1m}^{t,t+1} \\ 0 & p_{22}^{t,t+1} & p_{23}^{t,t+1} & \cdots & p_{2m}^{t,t+1} \\ 0 & 0 & p_{33}^{t,t+1} & \cdots & p_{3m}^{t,t+1} \\ \cdots & \cdots & \cdots & \cdots & \cdots \\ 0 & 0 & 0 & \cdots & p_{mm}^{t,t+1} \end{bmatrix}$$

(a) (b)

Fig. 1 Markov transition probability matrix: **a** Complete Markov process (deterioration and improvements); **b** Markov process assuming only deterioration elements are denoted by $b_i{}^t$ and correspond to the probability of the process to be in state i at time t. Normally, this vector represents the current condition of the bridge component obtained during the last inspection.

$$B(t) = \left[b_1^t \; b_2^t \ldots b_m^t \right], \quad \sum_{i=1}^{m} b_i^t = 1, \tag{4}$$

$$\mathbf{B}(t+1) = \mathbf{B}(t) \times \mathbf{P}^{t,t+1} = \left[b_1^{t+1} \; b_2^{t+1} \; \ldots \; b_m^{t+1} \right] \tag{5}$$

$$b_i^{t+1} = \sum_{j=1}^{m} b_j^t p_{ji}^{t,t+1}, \quad \{i = 1, 2, \ldots, m\} \tag{6}$$

Following the methodology for the time horizon, the probability mass function is given by:

$$\mathbf{B}(t+k) = \mathbf{B}(t) \times \mathbf{P}^{t,t+1} \times \mathbf{P}^{t+1,t+2} \times \cdots \times \mathbf{P}^{t+k-1,t+k} \tag{7}$$

The advantage in considering a stationary Markov chain is due to the fact that the mathematical formulation becomes more simple (Eq. 8), i.e., the transition probability depends only on the time difference. This is also called a homogeneous Markov chain because the prediction is dependent on a stationary matrix, i.e., the future deterioration is uniform either the component age is 20 or 50 years. As an example, this simplification is used in cases where consecutive inspections are performed in constant and fixed time steps (e.g. every 5 years). Thus, the transition probabilities of the transition matrix $\mathbf{P}^{0,5}$ are given by the relative frequencies (Eq. 9), obtained from the database, where a_{ij} is the number of bridge components that passed from state i to state j (frequency), and a_i is the total number of bridges components with condition state i (total count).

It is determined just one time and applied to predict future conditions (every 5 years in the example scenario) since the transition between states occurs in uniform and discrete time-steps.

$$\mathbf{B}(t+k) = \mathbf{B}(t) \times \left(\mathbf{P}^{t,t+1}\right)^k \tag{8}$$

$$p_{ij}^{t,t+1} = \frac{a_{ij}}{a_i} \tag{9}$$

The discrete-time Markov process is one of the most used methods to establish models in order to forecast the long-term bridge components performance. Its application is very simple (low computational complexity) because assumes some simplifications which however reproduce the reality. For instance, it is not possible in real life that the inspections are carried out at constant time intervals. Although it can help in the decision-making process, there are some relevant issues to take into account: (i) stationary transition probabilities and independence of duration are the most significant aspects that affect the deterioration rates leading in some cases to the underestimation in deterioration; and (ii) deterioration mechanisms are not considered efficiently, only in a subjective way from condition states that represents what the inspection team finds in situ.

In the present work, performance predictive models using Continuous Time Markov Chain are adopted. In this approach the intervals between inspections are not regular and therefore, the previous transition matrix **P** is related to a new matrix **Q** called intensity matrix (also known as the generator matrix). Further details about this approach and its mathematical formulation can be seen in [1, 9].

3 Performance Prediction Analyses

3.1 Methodology and Data Sets

Considering all the existing assets of transportation infrastructure, the focus of present work lies in the civil engineering structures, namely those that can be grouped in the bridges' family.

It is reasonable to assume that deterioration processes should be similar in components subjected to similar scenarios (load and environment), regardless to the type of structure they belong to. Hence, in the present work, the same structural component was analyzed for all records. From this point onwards, all the results refer to the deck component of each structure.

The database used in this work, belongs to two Portuguese highways that include a total of 384 bridge assets. From those, only 353 were used, due to insufficient data from the remaining. Figure 2a shows the amount of samples in the database associated with each different type of structure. The columns under cluster 1 correspond to global total per structure type, while the columns of the other two clusters correspond to sub-divisions of some of structure types.

Figure 2b presents the construction year for all records in the database in which it can be seen that almost all records belong to bridges with less than 30 years. Hence, it is not surprisingly found in Table 1, presenting a summary of condition state for all database records (1345 in total, 4 condition state per record on average), that the majority of the condition states are associated with lower levels of degradation (1 or 2). This table also presents data split by clusters defined in Fig. 2a.

The analyses carried out consisted on predicting the performance evolution in a horizon time of thirty years for a hypothetical deck. During that period, no maintenance actions are considered, i.e. the deck is allowed to degrade continuously. A scale of 1–5 was adopted, being 1 and 5 the best and worst conditions, respectively. The hypothetical deck is assumed to be in condition 1 at the present. The only difference

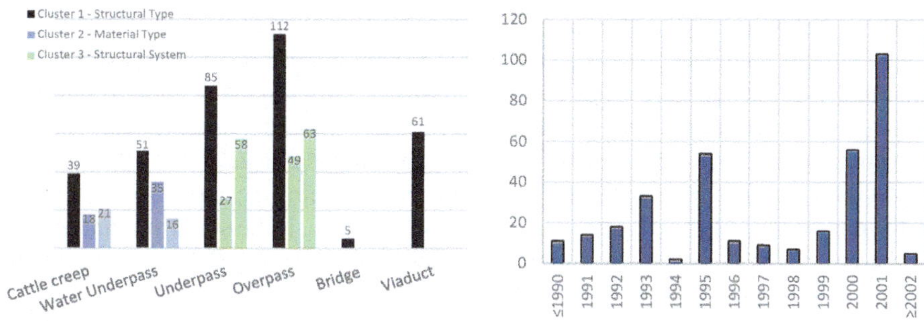

(a) (b)

Fig. 2 Number of samples in the database: **a** split by structure type and adopted clusters; **b** construction year

Table 1 Summary of the condition states listed in the database

Structure type	Material or structural system	Condition states				
		1	2	3	4	5
Cattle creep	Concrete	65	8	1	0	0
	Steel	70	7	0	0	0
Water underpass	Concrete	49	4	1	0	0
	Steel	115	9	3	0	0
Underpass	Arch	99	17	0	0	0
	Slab	169	52	3	2	0
Overpass	Ribbed	246	88	3	0	0
	Others	49	17	0	0	0
Bridge	(–)	7	6	3	1	0
Viaduct	(–)	101	120	13	2	0

between analyses lies in the slice of the original database used to run prediction, i.e. in the cluster being used.

3.2 Clusters by Structure Type

The first intuitive attempt of clustering the database was by structure type. Thus, considering the types of structures mentioned before, the Markov model was run considering only the records that belong to each structure type. Figure 3 presents the results obtained in this first analysis. For comparison purposes, this figure also includes the results obtained when considering the entire database.

From the results obtained it can be seen that cattle creep and water underpass have a very similar response. This is a reasonable behavior since the majority of records in both these two structure types correspond to arch decks made of either reinforced concrete or corrugated steel (this aspect is further explored). In other words, these two structures types seem to be well grouped in a single cluster.

A similar trend is verified for underpass and bridge, being the explanation identical to the one presented above. The only remark, in this case, is the small amount of data in the case of bridges, which might hinder the real behavior of this structure type when compared to viaducts.

The overpass and viaduct types do not fit any of the trends observed so far. Overpass seems to have a two stages deterioration process, with a fast-paced deterioration in the first 5 years and a much slower deterioration from that point onwards. This seems to indicate that overpass deterioration mechanisms seem to stabilize after an initial period of intense deterioration, which should be confirmed by a deeper analysis of the in-situ conditions of this type of structure.

Viaducts, even though presenting a deterioration curve with a shape very similar to all other structure types, present the highest deterioration pace found in all the analyses. Once again, a refined analysis is required herein to understand if the physical behavior of this type of structure supports these findings.

Nevertheless, from the standpoint of present work, what is intended to evidence is the impact of choosing a proper cluster to predict a specific structure performance. On the other hand, the scenario of running predictions with the entire database (common practice) does not seem a satisfactory option, since it can lead to unsafe predictions. Since these predictions are the basis for asset management plans, this kind of inaccuracies might have important consequences, particularly in terms of budget plans.

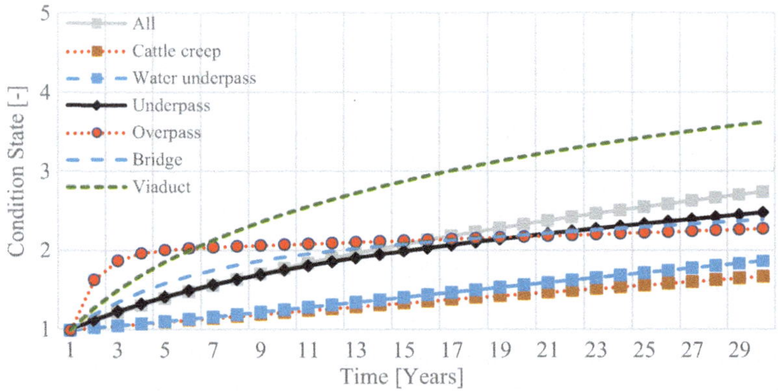

Fig. 3 Evolution of transportation infrastructures condition state for a period of 30 years

3.3 Clusters by Material Type

Considering the results from previous analyses, one aspect remained unclear, which is related to the influence that material type might have on a proper clustering. To clarify that aspect, data from cattle creep and water underpass structure types was further divided considering the existing material, either reinforced concrete or corrugated steel decks. As can be seen, apparently the material type has no impact on the obtained predictions in both cattle creep (Fig. 4a) and water underpass (Fig. 4b). Furthermore, both structures seem to deteriorate very slowly in time. The reasoning behind this behavior needs to be further analyzed, taking into account the mechanical behavior of each structure.

From expert judgment collected from bridges' managing company, it was found that, in the last years, corrugated steel decks have presented several problems. However, since there was no prior information about those bridges, they were all repaired before entering the database. This is the reason why both concrete and steel decks present a similar distribution in the condition states (see Table 1). However, it should be emphasized that the deterioration age of concrete decks is higher than that of steel decks, which can be assumed as reset at the reparation time (i.e. a new deterioration process begun after reparation).

3.4 Clusters by the Structural System

The division by material type was not possible to do for all types of structures. In fact, underpass, overpass, bridge, and viaduct types are all made of a single material—reinforced concrete. Furthermore, in the case of bridge and viaduct, the data is very similar in terms of mechanical properties, being the major differences of geometric nature (number of spans, span length, among others).

Alternatively, for underpass and overpass, the data was split by deck structural system. Considering the data available, for underpass essentially two types of structural systems were available, either arch or slab. In the case of overpass, there was found a clear group of ribbed slabs and another one with other types of slabs (e.g. prefabricated). Figure 5 presents the results obtained after running the Markov model on the data of these two types of structures, considering its division as mentioned above. The first comment on this figure is that now there are important differences between the different clusters analyzed.

From Fig. 5a it can be seen the decks from the slab structural system seem to deteriorate more and in a two stages fashion, faster in the beginning and slower after about five years. In contrast, the decks from the arch structural system present a smaller and constant deterioration pace. The response when all data is used together lies in between which would lead to unsafe predictions in the case of the slab system.

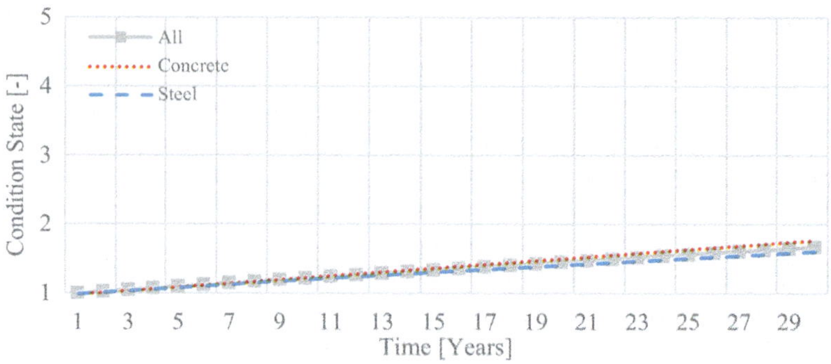

(a)

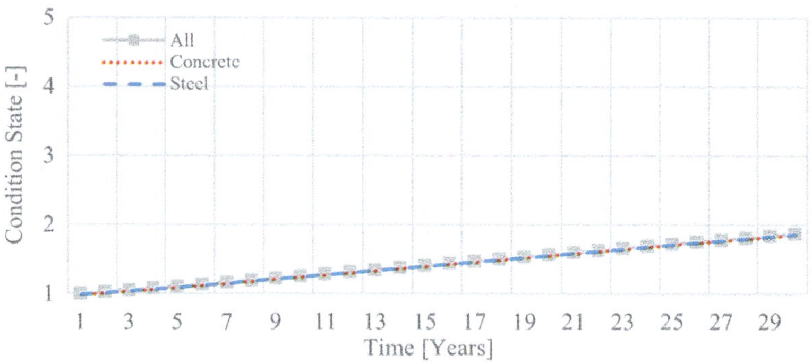

(b)

Fig. 4 Evolution of transportation infrastructures condition state for a period of 30 years: **a** cattle creep; **b** water underpass

From Fig. 5b the differences between each of the two clusters analyzed are much higher. Also, the response obtained when using both clusters data together (designated All) is quite different from what might be expect à priori. Instead of showing somewhere in between the two clusters analyzed, the All response seems to adjust better to the ribbed cluster in the first five years, and then tends to approximate the other cluster progressively.

Even though out of the scope of the present work, these findings should get some reasonable explanation. This can be obtained together with the bridges' managers and their inspection team. Ultimately, either some mechanical-related explanation, or a statistical one (likewise Sect. 3.3) should be obtained.

4 Conclusions

The majority of existing bridge management systems are based on databases with previous inspections' results. Those are the input of statistical models which have

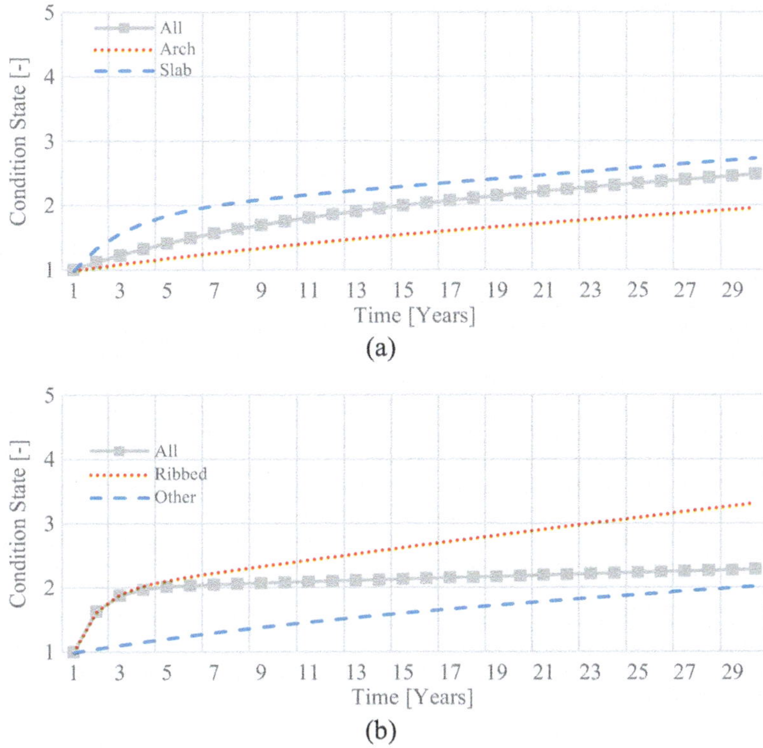

Fig. 5 Evolution of transportation infrastructures condition state for a period of 30 years: **a** underpass; **b** overpass

been used to predict the evolution of bridges' performance in time. In the present work a discussion was presented on the importance of proper clustering the referred databases.

Markov model adopted in the present work is conditioned by the data that is feeding the model. In the future, whenever further data is added to the analyses, the conclusions drawn in the present work should be revisited.

Another important aspect to highlight is that, despite the great potential of Markov model (and others) the input from the bridge engineer is not dismissed. Contrarily, only when the obtained results are validated by an expert, they gain validity and can be used with confidence.

Acknowledgments The authors would like to thank ISISE—Institute for Sustainability and Innovation in Structural Engineering (PEst-C/ECI/UI4029/2011 FCOM-01-0124-FEDER-022681). This work was co-financed by the Interreg Atlantic Area Programme through the European Regional Development Fund under SIRMA project (GrantNo. EAPA_826/2018).

References

1. Denysiuk, R., Fernandes, J., Matos, J. C., Neves, L. C., & Berardinelli, U. (2016). A computational framework for infrastructure asset maintenance scheduling. *Structural Engineering International, 26,* 94–102. https://doi.org/10.2749/101686616X14555428759046
2. Santamaria Ariza, M., Zambon, I., Sousa, H. S., Campos e Matos, J. A., & Strauss, A. (2020). Comparison of forecasting models to predict concrete bridge decks performance. *Structural Concrete, 21,* 1240–1253. https://doi.org/10.1002/suco.201900434
3. Frangopol, D. M., Kallen, M.-J., & van Noortwijk, J. M. (2004). Probabilistic models for life-cycle performance of deteriorating structures: Review and future directions. *Progress in Structural Engineering and Materials, 6,* 197–212. https://doi.org/10.1002/pse.180
4. Zambon, I., Vidovic, A., Strauss, A., Matos, J., & Amado, J. (2017). Comparison of stochastic prediction models based on visual inspections of bridge decks. *Journal of Civil Engineering Mangement, 23,* 553–561. https://doi.org/10.3846/13923730.2017.1323795
5. Kowoon, C., & Seokho, C. (2019). Bridge clustering for systematic recognition of damage patterns on bridge elements. *Journal of Computing in Civil Engineering, 33,* 4019028. https://doi.org/10.1061/(ASCE)CP.1943-5487.0000846
6. Huang, Y.-H. (2010). Artificial neural network model of bridge deterioration. *Journal of Performance of Constructed Facilities, 24,* 597–602.
7. Santamaria, M., Fernandes, J., & Matos, J. (2019). Overview on performance predictive models—Application to bridge management systems. In *IABSE Symposium Guimaraes 2019 Towards a Resilient Built Environment Risk and Asset Management.*
8. Kleiner, Y. (2001). Scheduling inspection and renewal of large infrastructure assets. *Journal of Infrastructure Systems, 7,* 136–143.
9. Sánchez-Silva, M., & Klutke, G.-A. (2016). *Reliability and life-cycle analysis of deteriorating systems.* Springer.

Uncertainty Assessment in Building Physics Related Problems Using Stochastic Finite Element Method

Witold Grymin and Marcin Koniorczyk

Abstract In the calculations of the heat transfer, the material parameters are usually based on the laboratory tests of the given material. Afterwards, they are applied in the calculations as deterministic values, after taking into account effects of relative humidity, temperature and material aging. However, one can distinguish various uncertainties for the material systems. In the calculations of the energy demand, they may induce significant variations of the results. In the article, analysis of uncertain thermal conductivity of expanded and extruded polystyrene, with relation to the values declared by the producer, as well as of density and thermal conductivity of constructive material, is investigated. The possible variations of the thermal conductivity of the insulating materials are based on the statistical analysis of the database provided by the Construction Control Authority in Poland. Two methods are applied in order to determine expected value and variance of temperature field and heat flux on the internal side of the wall: the tenth order perturbation stochastic Finite Element Method and the Monte Carlo method. The partial derivatives of temperature with respect to a random variable are determined using the Direct Differential Method. Whilst giving very accurate results, the perturbation SFEM is much more efficient than the Monte Carlo method for transient heat transport in a double-layer external envelope. The highest variance has been calculated for a node situated in between the constructive and the insulating layer, regardless of which material random property has been considered. The heat loss variation is related to the thermal resistance of the layer.

Keywords Perturbation method · Stochastic finite element method · Monte Carlo method · Uncertainty

---

W. Grymin · M. Koniorczyk (✉)
Department of Building Materials Physics and Sustainable Design, Lodz University of Technology, Łódź, Poland
e-mail: marcin.koniorczyk@p.lodz.pl

© The Author(s), under exclusive license to Springer Nature Switzerland AG 2021

J. C. Matos et al. (eds.), *18th International Probabilistic Workshop*, Lecture Notes in Civil Engineering 153, https://doi.org/10.1007/978-3-030-73616-3_63

1 Introduction

Due to the randomness of multiple parameters in the calculations of the heat transfer, propagation of uncertainties and their impact on the results should be thoroughly investigated. Two categories of approaches used for the uncertainty analyses can be listed: non-probabilistic approach (such as fuzzy approach or set-theoretical approach) and probabilistic one. The first method is usually used in case of the insufficient data set size to determine probability distribution. Two groups can be further distinguished among probabilistic methods proposed for the heat transfer problem involving random values and uncertain results. The first one are the sampling methods. They are based on the Monte Carlo (MC) simulation method, in which the simulation is running over stochastic series, what results in a set of results distributed accordingly to the probability distributions of random parameters. The Monte Carlo or quasi-Monte Carlo methods, which is argued to ensure better uniformity than the random sequences, are often used to solve the radiative transfer equation [1] or the heat conduction [2]. It is also often used as verification of other methods. Despite its simplicity, number of the sampling size strongly affects accuracy of the results in this method—in order to obtain precise results, excessive computational cost is usually involved.

In the second group of the stochastic heat transfer problems, namely non-sampling methods, most commonly used are spectral stochastic method and stochastic perturbation method. In the spectral stochastic finite element method, extension of the deterministic finite element method (FEM) for problems involving random material properties is performed using series expansion. The input random field is discretized using, i.e., the Karhunen-Loève expansion or the polynomial chaos expansion.

In the perturbation method, the Taylor series expansion of a given order is used. Moments and coefficients are calculated in order to determine the distributions of the structural response. The methods using the expansion of the second order are called the second-moment analyses and are used for the problems characterized by low coefficients of variation. In case of larger uncertainties, generalized stochastic perturbation method with a higher-order Taylor series expansion is usually used. Hien and Kleiber [3, 4] proposed formulation based on the second-moment analysis with the second-order perturbation method for the linear and nonlinear heat transfer problems. Kamiński and Hien [5] used the perturbation method to account for the transient heat transfer in the composite. The boundary conditions imposed were of the 1st (temperature on the surface) or 2nd type (heat flux on the surface). Wu and Zhong [6] analyzed the hyperbolic heat conduction problems. Yang and Cui [7] applied random field model for the calculations of the stochastic heat transfer problem and compared the solution to the results of MCs.

In our work we did the statistical analysis of the expanded (EPS) and extruded (XPS) polystyrene thermal conductivity database provided by Construction Control Authority in Poland in order to reveal its probability density function. Further we applied the general tenth order perturbation stochastic FEM in order to investigate expected value and variance of the temperature field as well as of the heat flux on the internal surface. The partial derivatives of temperature with respect to random variable are determined using Direct Differential Method (DDM). As it was proved in the next section, the thermal conductivity of the EPS or the XPS, with reference to the value declared by the producer, has lognormal (non-symmetric) distribution, hence we have to determine and use the complex perturbation equations including the odd terms. The results calculated using perturbation SFEM was verified against data obtained using the Monte Carlo method for 10000 individuals.

2 Thermal Conductivity Uncertainty

In the statistical analyses, the thermal conductivity is usually assumed to be normally distributed. The experimental data concerning thermal conductivity of insulation materials provided by the Construction Control Authority in Poland [8] was investigated. The tests have been performed in years 2016–2019. The experimental research has been performed for expanded and extruded polystyrene. For each material under investigation, thermal conductivity of 4 samples was determined. In total, 1192 results of thermal conductivity, obtained in 298 tests, were used in the statistical treatment. The measured values of the thermal conductivity have been related to the values declared by the producers by means of the relative thermal conductivity, which has been defined as:

$$\lambda_{rel} = \frac{\lambda_m}{\lambda_d} \tag{1}$$

where λ_m is the thermal conductivity of a given sample measured in the test and λ_d is the value declared by the producer of the material at the time of the test. The probability density of the relative thermal conductivity is given in Fig. 1. The distribution presented reveals its asymmetric shape, when relation of the measured thermal conductivity with reference to the value declared by the producer is considered. This observation contradicts the assumption of normal distribution, which is often stated for the uncertainty analysis of thermal conductivity. The distribution parameters of the relative thermal conductivity have been determined using the Pearson's chi-squared statistical test, indicating lognormal distribution, for which value of χ^2 was equal to 7.613, while for the normal distribution it was equal to 9.233. It should be reminded that the lognormal distribution admits zero for any negative argument by its definition, while the normal distribution allows nonzero value in this range. Such an event, probability of which is very low but still remains admissible, violates the second law of thermodynamics. It is possible to overcome this obstacle by introducing additional numerical procedure zeroing the probability of such event even for normal distribution. Further, the thermal conductivity of the insulating material is much lower than that of construction ones. Based on the experimental data we decided to apply the lognormal distribution for thermal conductivity distribution of

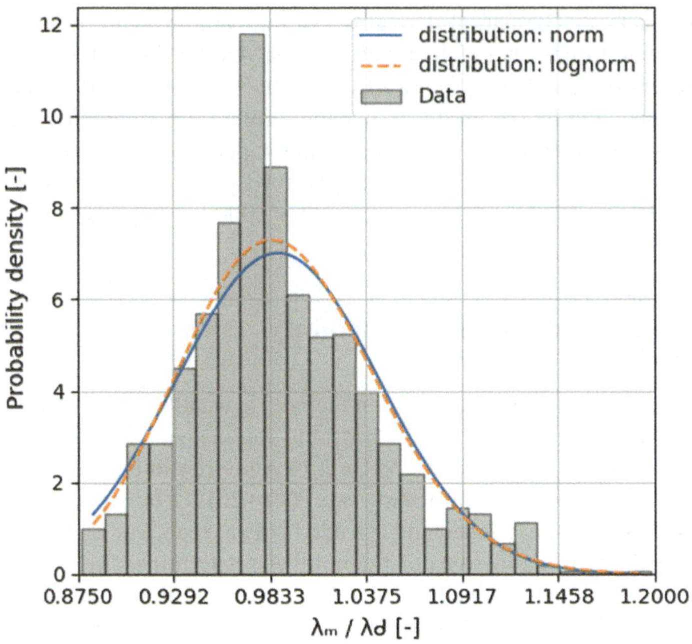

Fig. 1 Relative thermal conductivity probability density of the insulating materials

the insulation material and the normal distribution for thermal conductivity of the construction material.

3 Mathematical Model

3.1 Transient Heat Flow

Let us consider the transient heat flow, which is governed by the equation:

$$\rho c\dot{T} + \left(\lambda T_{,j}\right)_{,i} + q_{,i} = 0, x \in \Omega, t \geq 0 \tag{2}$$

where ρ is density, c is specific heat capacity, λ is thermal conductivity and q is heat flux. All material parameters might depend on temperature. For our case let us assume that density, specific heat capacity and thermal conductivity are independent on temperature. Then Eq. (2) is reduced to the linear form, which reads:

$$\rho c\dot{T} + \lambda T_{,ji} + q_{,i} = 0, x \in \Omega, t \geq 0 \tag{3}$$

To solve the linear problem, Eq. (3) has to be complemented with the initial conditions and boundary conditions, in this case of the third type, in which the ambient temperature along with the heat transfer coefficient is considered, appropriately:

$$T(x,0) = T_0, x \in \Omega, \tag{4}$$

$$q(t,x) = \alpha[T(x,t) - T_\infty(t)], x \in \partial\Omega, t \geq 0, \tag{5}$$

where α is the convection coefficient and T_∞ is the ambient temperature.

By denoting by C the capacity matrix, K the conductivity matrix and Q the loading vector as follows:

$$C = \rho c \int_\Omega \varphi^T \varphi d\Omega, \tag{6a}$$

$$K = \alpha \int_\Omega (\varphi_{,i})^T \varphi_{,j} d\Omega, \tag{6b}$$

$$Q = \int_{\partial\Omega} \varphi^T \alpha[T(x,t) - T_\infty(t)] d\partial\Omega, \tag{6c}$$

the governing equation might be expressed in the compact matrix form:

$$C\dot{T} + KT + Q = 0. \tag{7}$$

The time derivative is usually discretized using truncated Taylor's series expansion as:

$$\dot{T}(t) = \frac{T(t) - T(t - \Delta t)}{\Delta t}. \tag{8}$$

All quantities (temperature, heat flux, etc.) can be written in any arbitrary time $t_{n+\theta} = t + \theta\Delta t$ between $t_n = t$ and $t_{n+1} = t + \Delta t$ using the linear interpolation, i.e.:

$$T_{t+\theta\Delta t} = T_{n+\theta} = \theta T_{n+1} + (1-\theta)T_n \tag{9}$$

With $\theta = 0$ we obtain the explicit method, $\theta = 1$ the implicit method and $\theta = 1/2$ the Crank-Nicolson method. We are able to formulate the governing equation in the compact matrix form using the general θ scheme:

$$C(T_{n+1} - T_n) + \Delta t K\left[\theta T_{n+1} + (1-\theta)T_n\right] + \Delta t Q_{n+\theta} = 0 \tag{10}$$

3.2 General n-th Order Stochastic Perturbation

Let us denote the random field $b(x)$ and its probability density function $p(b(x))$. The single random parameter is investigated in the manuscript and therefore any correlation between material properties is not considered. For the sake of clarity, let us explain that the random variable might be any material parameter: the initial conditions, the boundary conditions including ambient temperature and heat transfer coefficient, etc. The random variable must be uncorrelated, bounded and uniquely defined by two first probabilistic moments (expected value and variance):

$$E[b(x)] = \int_{-\infty}^{\infty} b(x)p(b)db, \tag{11}$$

$$Var[b(x)] = \int_{-\infty}^{\infty} \{b(x) - E[b(x)]\}^2 p(b)db. \tag{12}$$

The n-th order perturbation method relies on the assumption that all variables as well as state functions, $f(b)$, might be interpolated by the n-th order Taylor expansion using small perturbation parameter $\varepsilon > 0$ according to formula:

$$\begin{aligned} f(b) &\cong f^0(\overline{b}) + \varepsilon f^{,b}(\overline{b})\Delta b + \frac{\varepsilon^2}{2!} f^{(2)}(\overline{b})(\Delta b)^2 + \cdots + \frac{\varepsilon^n}{n!} f^{(n)}(\overline{b})(\Delta b)^n \\ &= f^0(\overline{b}) + \sum_{k=1}^{n} \frac{\varepsilon^k}{k!} f^{(k)}(\overline{b})(\Delta b)^k \end{aligned} \tag{13}$$

where

$$f^{(k)}(\overline{b}) = \left.\frac{\partial f^k(b)}{\partial b^k}\right|_{\overline{b}} \tag{14}$$

is the k-th partial derivative of $f(b)$ with respect to the random variable b in $\overline{b}$. The k-th order variation ($k = 1 \ldots n$) of b about its expected value is:

$$\varepsilon^k(\Delta b)^k = \varepsilon^k\left[b - \overline{b}\right]^k. \tag{15}$$

Let us now introduce Eq. (13) into Eq. (11) giving the expected value of any state function with a small perturbation parameter using Taylor expansion:

$$E[f(b); b] = \int_{-\infty}^{\infty} f(b)p(b)db \cong \int_{-\infty}^{\infty} \left[f^0(\bar{b}) + \sum_{k=1}^{n} \frac{\varepsilon^k}{k!} f^{(k)}(\bar{b})(\Delta b)^k \right] p(b)db. \tag{16}$$

Placing Eq. (13) into Eq. (12) and after performing analogous transformation as for the expected value one can obtain analogous formula for the variance. The expected value of function $f(b)$ can be expanded using the tenth-order perturbation assuming general non-symmetric probability density function to the formula:

$$\begin{aligned} E[f(b); b] = \int_{-\infty}^{\infty} f(b)p(b)db &\cong f^0(\bar{b}) + \varepsilon f^{(1)}(\bar{b})\mu_1(b) + \frac{\varepsilon^2}{2} f^{(2)}(\bar{b})\mu_2(b) \\ &+ \frac{\varepsilon^3}{3!} f^{(3)}(\bar{b})\mu_3(b) + \frac{\varepsilon^4}{4!} f^{(4)}(\bar{b})\mu_4(b) + \frac{\varepsilon^5}{5!} f^{(5)}(\bar{b})\mu_5(b) \\ &+ \frac{\varepsilon^6}{6!} f^{(6)}(\bar{b})\mu_6(b) + \frac{\varepsilon^7}{7!} f^{(7)}(\bar{b})\mu_7(b) + \frac{\varepsilon^8}{8!} f^{(8)}(\bar{b})\mu_8(b) \\ &+ \frac{\varepsilon^9}{9!} f^{(9)}(\bar{b})\mu_9(b) + \frac{\varepsilon^{10}}{10!} f^{(10)}(\bar{b})\mu_{10}(b) \end{aligned} \tag{17}$$

where $\mu_k(b)$ denotes the k-th order central probabilistic moment of the random variable b, given by the equation:

$$\mu_k(b) = \int_{-\infty}^{\infty} [b - E(b)]^k p(b)db. \tag{18}$$

Applying the similar derivation the tenth order expansion for the variance can be formulated:

$$\begin{aligned} \mathrm{Var}[b(x)] = &\left(f^{(1)}\right)^2 \varepsilon^2 \mu_2(b) + f^{(1)} f^{(2)} \varepsilon^3 \mu_3(b) \\ &+ \left[\frac{1}{(2!)^2} \left(f^{(2)}\right)^2 + \frac{2}{3!} f^{(1)} f^{(3)} \right] \varepsilon^4 \mu_4(b) \\ &+ \left[\frac{2}{4!} f^{(1)} f^{(4)} + \frac{1}{3!} f^{(2)} f^{(3)} \right] \varepsilon^5 \mu_5(b) \\ &+ \left[\frac{1}{(3!)^2} \left(f^{(3)}\right)^2 + \frac{1}{4!} f^{(2)} f^{(4)} + \frac{2}{5!} f^{(1)} f^{(5)} \right] \varepsilon^6 \mu_6(b) \\ &+ \left[\frac{2}{6!} f^{(1)} f^{(6)} + \frac{1}{5!} f^{(2)} f^{(5)} + \frac{2}{3!4!} f^{(3)} f^{(4)} \right] \varepsilon^7 \mu_7(b) \\ &+ \left[\frac{1}{(4!)^2} \left(f^{(4)}\right)^2 + \frac{2}{3!5!} f^{(3)} f^{(5)} + \frac{1}{6!} f^{(2)} f^{(6)} + \frac{2}{7!} f^{(1)} f^{(7)} \right] \varepsilon^8 \mu_8(b) \\ &+ \left[\frac{2}{8!} f^{(1)} f^{(8)} + \frac{1}{7!} f^{(2)} f^{(7)} + \frac{2}{3!6!} f^{(3)} f^{(6)} + \frac{2}{4!5!} f^{(4)} f^{(5)} \right] \varepsilon^9 \mu_9(b) \end{aligned}$$

$$+\left[\frac{1}{(5!)^2}\left(f^{(5)}\right)^2+\frac{2}{4!6!}f^{(4)}f^{(6)}+\frac{2}{3!7!}f^{(3)}f^{(7)}\right]\varepsilon^{10}\mu_{10}(b)$$
$$+\left[\frac{1}{8!}f^{(2)}f^{(8)}+\frac{2}{9!}f^{(1)}f^{(9)}\right]\varepsilon^{10}\mu_{10}(b) \qquad (19)$$

The main task of further analysis is to calculate the partial derivatives of state function, i.e. temperature, with respect to the random variable. Substituting Eq. (11) into governing equation we obtain a system of n + 1 algebraic equations of 0th, 1st, 2nd and *n*-th order, which can be expressed for a general case as:

Zeroth-order equation:

$$C\dot{T}+KT=-Q. \qquad (20)$$

First-order equation:

$$C\dot{T}^{'b}+KT^{'b}=-Q^{'b}-K^{'b}T-C^{'b}\dot{T}. \qquad (21)$$

Second-order equation:

$$C\dot{T}^{'bb}+KT^{'bb}=-Q^{'bb}-2K^{'b}T^{'b}-2C^{'b}\dot{T}^{'b}. \qquad (22)$$

N-th-order equation:

$$C\dot{T}^{(n)}+KT^{(n)}=-Q^{(n)}-\sum\nolimits_{k=1}^{n}\binom{n}{k}K^{(k)}T^{(n-k)}-\sum\nolimits_{k=1}^{n}\binom{n}{k}C^{(k)}\dot{T}^{(n-k)}. \qquad (23)$$

The above equations allow to calculate the *k*-th, $k = 0 \ldots n$, order derivative of temperature with respect to a random property. Below we will analyze two cases when the random variable is assumed to be thermal conductivity or density. Consequently, the special form of *n*-th order differential equation will be given.

4 Numerical Simulation

The following layers have been taken for the simulation: the structural layer composed of brick [$d = 25$ cm, $\rho = 1800\,\mathrm{kg/m^3}$, $c_p = 880$ J/(kg K), $\lambda = 0.77$ W/(mK)] and the thermal insulation made of expanded polystyrene (EPS) [$d = 20$ cm, $\rho = 30\,\mathrm{kg/m^3}$, $c_p = 1460$ J/(kg K), $\lambda = 0.037$ W/(m K)]. The thermal transmittance of such an envelope is equal to 0.17 W/(m^{2}K). Ambient temperature is taken for the climatic typical reference year for Lodz in Poland. The heat transfer coefficients have

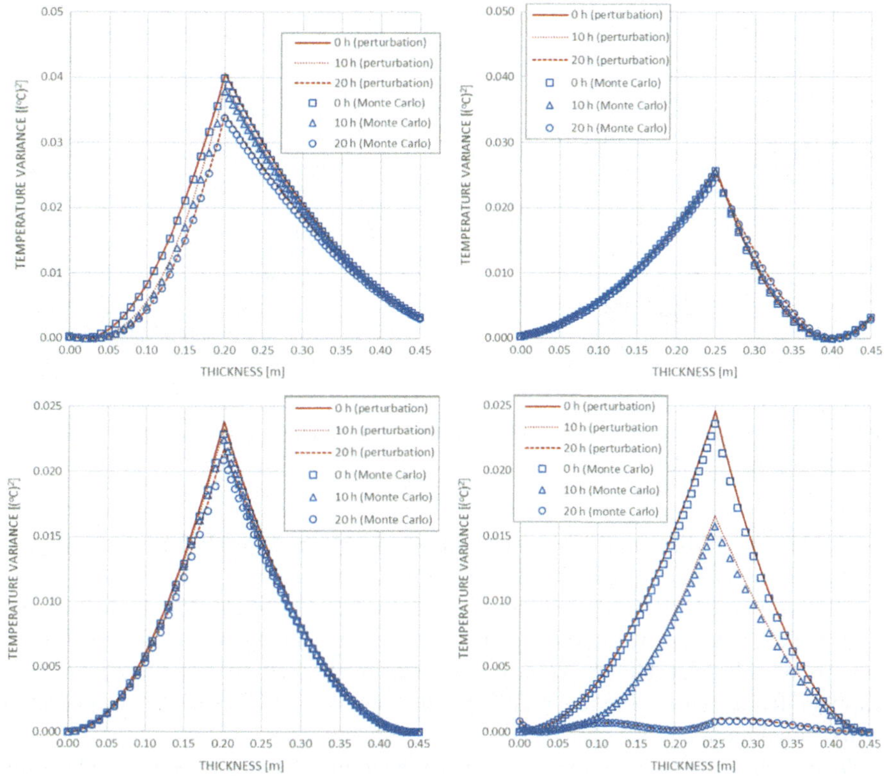

Fig. 2 The comparison of temperature variance distribution along the wall thickness calculated using SFEM and Monte Carlo method for case 1 (top left), case 2 (top right), case 3 (bottom left) and case 4 (bottom right)

been assumed as equal to 25 W/(m^{2}K) and 7.69 W/(m^{2}K) for external and internal surfaces, appropriately.

In the following section we will investigate how the random material properties, namely thermal conductivity and density, influence the statistical distribution of the heat flux and the temperature profiles. In the present section four cases are investigated: EPS thermal conductivity as random variable; case 1: EPS from outside, case 2: EPS from inside; brick thermal conductivity as random variable; case 3: EPS from outside, case 4: EPS from inside. The EPS thermal conductivity has a probability distribution, logarithm of which has the normal distribution $N(\mu, \sigma^2) = N(-3.3037, 0.0138)$. The thermal conductivity of brick is normally distributed with the following parameters $N(0.77, 0.0083)$. The results obtained by stochastic finite element method were compared to the ones calculated using the Monte Carlo Method assuming 10,000 individuals. It can be noticed that high similarity of variance distribution both in the time and the thickness domain was

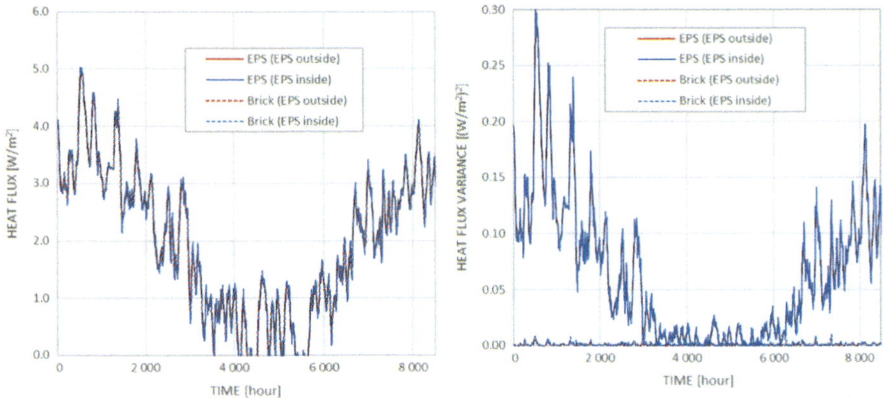

Fig. 3 The comparison of expected value of heat flux on the internal surface for the entire year (left) and its variance (right)

obtained for perturbation SFEM and Monte Carlo Method for all the cases—Fig. 2. The validating calculations were performed for 20 h.

After validation, the expected value and the variance of heat flux on the internal surface of the envelope were calculated. One can notice that the system is less susceptible to possible uncertainty of thermal insulation conductivity when the insulation is applied from inside—Fig. 2a, b. In Fig. 3 the heat flux and its variance is presented for the considered cases.

We can observe that the expected values of heat flux changes with time and admits similar values for all the analyzed cases at arbitrary time stages. Only slight differences can be noticed for the cases when thermal insulation is applied from inside: larger variation of heat flux, and outside: smaller variation of heat flux. However, it is noticeable that the results obtained for cases 1 and 3 overlap. They were calculated assuming symmetric (case 3) and nonsymmetrical probability density function (case 1) and may serve as additional validation of the method. The similar conclusion may be drawn for case 2 and 4. During the summer time, the expected value of heat flux approaches to zero, while it admits the largest value during winter days, which is obviously caused by differences of temperature gradient. Considering the change of variance of heat flux along with time it is evident that the values obtained for cases 1 and 2 are much higher than for the cases 3 and 4. Similarly to the expected value of heat flux, its variance admits the highest value during winter and the smallest during summer period. The peculiar conclusions can be drawn for the relative value of the variation coefficient. The coefficient admits values very similar to the relative thermal resistance defined as $R_{i,rel} = \frac{R_i}{R_{si}+\sum_i R_i+R_{se}}$, where $i = EPS, Brick$. Let us notice that for our cases $R_{EPS,rel} = 5.405$ m^{2}K/W and $R_{Brick,rel} = 0.325$ m^{2}K/W, respectively. Concluding, the relative thermal resistance can serve as the reasonable estimate of the relative variation coefficient of the heat flux.

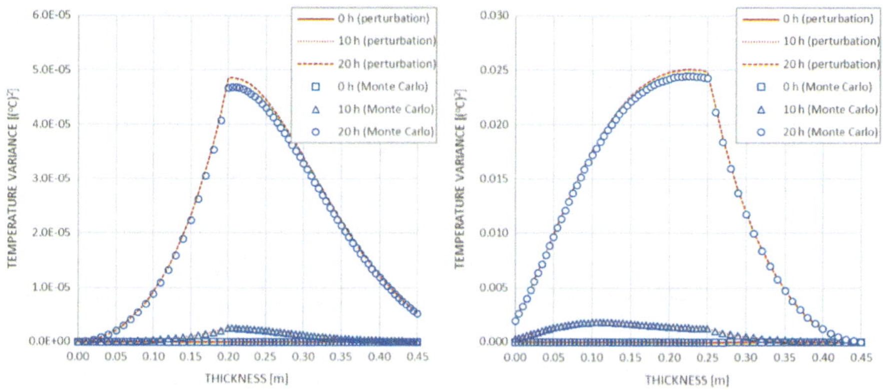

Fig. 4 The comparison of temperature variance distribution along the envelope thickness calculated using SFEM and Monte Carlo method assuming that the random variable is brick density and EPS is applied from outside (case 5—left) or EPS is applied from inside (case 6—right)

Another important factor influencing the transient heat transport in building materials is its thermal mass. Therefore, the influence of structural material density variation on the uncertainty of heat flux and temperature distribution is investigated. Let us formulate the arbitrary form of partial differential equations assuming that the brick density is the normally distributed random variable defined by the parameters: $N(1800, 212)$. Hence, the coefficient of variation admits the same value as for the thermal conductivity cases, which were investigated in the previous subsection. It must be noticed that the value of variation coefficient was smaller than its maximum value recommended for the perturbation SFEM, which was equal to 0.15.

Two cases were analyzed assuming brick density as a random parameter: case 5—EPS from outside and case 6—EPS from inside. The good correlation of temperature variance distribution in space was obtained for the perturbation and the Monte Carlo methods—Fig. 4a, b. It is evident that the distribution of temperature variance in space domain at zero time stage is equal to zero. It results from the assumed calculation methodology of initial temperature distribution, which was calculated assuming stationary heat flow for arbitrary external and internal temperature. In such an approach density and specific heat capacity do not appear, therefore the temperature variance equals to zero in this time step—Fig. 4a, b. Furthermore, it is worth noticing that for case 5 the temperature variance is two order smaller that for case 6. Hence, the application of thermal insulation form outside makes the possible variation of structural layer density insignificant (Fig. 5).

Assuming that the thermal insulation is applied from outside one can expect the lower heat flux amplitude. Although the temperature variance throughout the envelope is much smaller for case 5 than for case 6, it admits very similar numbers on the internal surface. However, no such rule can be stated for the heat flux variance, which admits value similar to the one obtained for case 3 and 4, where the thermal conductivity of brick was assumed to be the random variable.

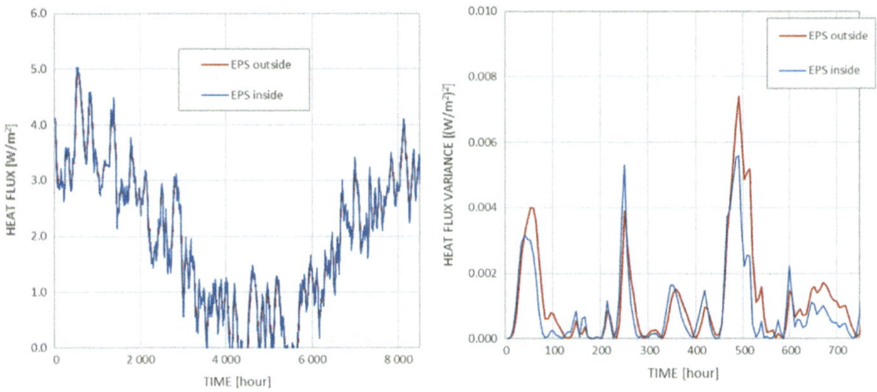

Fig. 5. The comparison of expected value of heat flux on the internal surface for the entire year (left) and its variance for January (right)

5 Conclusions

The impact of uncertainty of material properties on the heat flux and temperature distribution in the building envelope was presented. The Monte Carlo method is widely recognized as very time-consuming technique due to the necessity of performing multiple runs. Whilst giving very accurate results, the perturbation SFEM is much more efficient than the Monte Carlo method for the transient heat transport in a double-layer external envelope.

Afterwards, the perturbation SFEM has been applied to analyze propagation of the uncertainties of material properties in the calculations of heat transfer in the building envelope. The largest values of temperature variance can be noted in the zone between two materials. For uncertainty of the brick thermal conductivity, the heat flux variation coefficients calculated on the internal surface were much smaller than in case of the uncertainty of the EPS thermal conductivity. It can be concluded that the heat loss variation coefficient obtained assuming uncertainty of individual layer thermal conductivity may be directly related to the thermal resistance of this layer. During the performed case study, the same values of thermal conductivity variation coefficients for both insulating and construction materials were assumed deliberately. In reality, good thermal conductors (structural materials) are usually featured by smaller variation coefficient of conductivity. It could further decrease their impact on heat flux uncertainty.

For random thermal conductivity of EPS, high variance of heat flux on the internal surface and low on the external were identified, independently of the thermal insulation localization. Assuming the random brick density, the variance of heat flux on internal surface was negligibly small, when compared to other investigated cases.

References

1. Palluotto, L., Dumont, N., Rodrigues, P., Gicquel, O., & Vicquelin, R. (2019). Assessment of randomized Quasi-Monte Carlo method efficiency in radiative heat transfer simulations. *Journal of Quantitative Spectroscopy and Radiative Transfer, 236*.
2. Hou, T., Nuyens, D., Roels, S., & Janssen, H. (2017). Quasi-Monte-Carlo-based probabilistic assessment of wall heat loss. *Energy Procedia, 132,* 705–710.
3. Hien, T. D., & Kleiber, M. (1997). Stochastic finite element modeling in linear transient heat transfer. *Computer Methods in Applied Mechanics and Engineering, 144,* 111–124.
4. Hien, T. D., & Kleiber, M. (1998). On solving nonlinear transient heat transfer problems with random parameters. *Computer Methods in Applied Mechanics and Engineering, 151,* 287–299.
5. Kamiński, M., & Hien, T. D. (1999). Stochastic finite element modeling of transient heat transfer in layered composites. *International Communications in Heat and Mass Transfer, 36*(6), 801–810.
6. Wu, F., & Zhong, W. X. (2016). *A modified stochastic perturbation method for stochastic hyperbolic heat conduction problems*.
7. Yang, T. J., & Cui, X. Y. (2017). A random field model based on nodal integration domain for stochastic analysis of heat transfer problems. *International Journal of Thermal Sciences, 122,* 231–247.
8. https://www.gunb.gov.pl/probki

Uncertainty Associated to Regression Models used for Assessing the Stiffness of Structural Timber Elements

José Saporiti Machado

Abstract The evaluation of the mechanical behaviour of timber beams or glued laminated timber lamellas in-service are generally a difficult task due to the different sources of uncertainty involved (small knowledge about the initial quality of timber, small samples, models uncertainty, human errors). The use of statistical methods that can incorporate part of the uncertainty are probably a suitable way to ensure that the predictions made could provide a reliable prediction of the desired property. In most situations while performing in situ assessment of timber structures, the application of non or semi-destructive testing (NDT or SDT) methods relies on regression linear models showing noticeable different coefficients of determination. Another source of uncertainty happens when making in-situ testing relying on the application of existing regression models to timber members without being sure about the wood species or the origin of the wood species. Can these models be used when it is commonly accepted that knowledge on timber's origin and species have a major impact on the capability to predict strength and stiffness? To comply with uncertainty several studies have been trying to use statistical methods that can incorporate prior information (e.g. Bayesian methods) or uncertainty (e.g. Markov chain Monte Carlo—MCMC). In the present paper uncertainty associated to the use of linear regression models are discussed using as example the prediction of static modulus of elasticity from dynamic modulus of elasticity. For that purpose, data taken from literature and from studies conducted at LNEC are compared, analysed and discussed having in mind to verify the utility of the application of Bayesian linear regression approach and Monte Carlo Markov Chains (MCMC) estimation.

Keywords Timber members · Prediction · Regression models · Uncertainty · MCMC

J. Saporiti Machado (✉)
Laboratório Nacional de Engenharia Civil (LNEC), Lisboa, Portugal
e-mail: saporiti@lnec.pt

© The Author(s), under exclusive license to Springer Nature Switzerland AG 2021

J. C. Matos et al. (eds.), *18th International Probabilistic Workshop*, Lecture Notes in Civil Engineering 153, https://doi.org/10.1007/978-3-030-73616-3_64

1 Introduction

The assessment of existing timber structures comprises the evaluation of structural members and connections. The appraisal of timber members is complex given the necessity to collect information about wood species and its origin being this information important to address matters related to strength, stiffness and durability. Moreover, the higher variability of properties observed along timber member length and cross-section contributes to the difficulties felt by any expert responsible for appointing design values to be used in a structural analysis.

Currently assessment begins from knowledge of wood species identification followed by the application of visual strength grading standards. This approach based on a group approach (batch identified by the strength of its 5th lowest specimen) ensures usually (is assumed) the appointment of very conservative design values for the safety and serviceability assessment of the structure. Given the difficulties of the process of wood identification sometimes this step is disregarded, and an even more conservative approach is followed by considering a lower strength class that one correspondent to the allocated visual grading.

Nonetheless a more accurate assessment can be done if rely on the use of non or semidestructive testing method (NDT or SDT, respectively). Several NDT or SDT methods are now available for the assessment of in-situ testing and their advantages and limitations were analysed [1].

These methods developed at the laboratory level compare the values obtained by NDT or SDT methods (indicative parameter) with those obtained experimentally by destructive testing (reference properties) by application of regression models, Eq. 1. These models are used given their simplicity and easy to use since they are available in current programs as a simple Excel's spreadsheet. These models are defined by the minimization of the sum of the squares of the differences between the observed dependent and predicted variables.

$$y = \beta_0 + \beta_1 x + \varepsilon \qquad \varepsilon \sim N\left(0, \sigma^2\right) \tag{1}$$

However how reliable are those models obtained at the lab with perfect control over the wood species and its origin when applied onsite facing timber members where uncertainty about the wood species involved as well as origin is present.

In most cases after obtaining a regression model the uncertainty (ε) component is put aside and the model used as it is. But model uncertainties can be perceived by the dispersion associated to its coefficients, β_0 (intercept) or β_1 (regression coefficient—slope) expressed as the standard error (or deviation) of the slope, $s_{\beta 1}$ or the standard error of the intercept, $s_{\beta 0}$.

Other sources of uncertainty are present and often disregarded for the sake of simplicity, as the fact that observed measurements (x values) are also subject to errors being the parameters of the regression (β_0, β_1) function of the sample used for defining the regression model. Different samples taken from the same population (group of all timber members in a structure, for instance) can provide different parameter's of the regression model.

Cross-validation or bootstrap approaches are usually applied to assess how the regression model is affected by small changes in the composition of the samples but all assume that the regression model obtained is the best approach to the physical relationship between two random variables. Also, in these cases often the same sample is used for development, calibration as well as validation.

The evaluation of the quality of a regression model can be assessed by different statistical parameters measuring the deviation of the prediction from the real value. In most cases the parameter used is the coefficient of determination (r^2). However, in the present case, accuracy and uncertainty will be assessed through the mean percentage error (MPE), Eq. 2, providing an estimation of the accuracy (how close of the real value in average) and its dispersion an estimation of the uncertainty.

$$MPE(\%) = \frac{100}{n} \sum \frac{\hat{y}_i - y_i}{y_i} \quad (2)$$

where n is sample size, $\hat{y}_i$ is the expected value provided by the model and y_i is the actual value.

Once the model is obtained its application to other data set is usually done assuming it can explain the same level of variability (coefficient of determination—r^2) as those obtained from the training and test set. However, no assurance that this assumption will hold exists when applied in-situ where the actual values of the dependent variable is unknown.

In the present paper an attempt to discuss the level of uncertainty associated to the use of regression models developed at the lab when applied in-situ is made. For that purpose, regression curves suitable for the correlation between static modulus of elasticity and dynamic modulus of elasticity of maritime pine (*Pinus pinaster* Aiton) timber elements are used. Also, some models published for the same species but from different origins are used. Besides these models also the utility of models that can incorporate uncertainty associated to all factors involved in the application of regression models are analysed and discussed through the application of a Bayesian linear regression approach using MCMC.

2 Uncertainty in the In-Situ Assessment of the Modulus of Elasticity—Application to Clear Wood Zones

2.1 *Sources of Uncertainty and Models for Maritime Pine*

Timber's properties designated as reference properties usually includes density, bending strength and modulus of elasticity. Considering the assessment of the static modulus of elasticity this task can be performed using ultrasonic or sonic equipment. A stress wave is applied to the material, using direct or indirect methods and a time-of-flight (*ToF*) reading is recorded. But the dynamic modulus of elasticity (E_{dyn}) comprises besides data related to time of flight, provided by the equipment, also the necessity to measure the distance between transducers and the measurement or estimation of density (ρ), Eq. 3. To all these measurements there are associated uncertainty (ε_{E}).

$$E_{\mathrm{dyn}} = \left(\frac{ToF}{d}\right)^2 \rho + \varepsilon_E \qquad \varepsilon_E \sim N\left(0, \sigma_E^2\right) \tag{3}$$

where d is de distance between transducers.

The regression models are also specific for a certain wood species and to the sample used for derived those models. For instance, for maritime pine wood eight curves were derived for different samples which can be significantly different from those obtained for other species, Fig. 1. These models are also influenced by the method followed (direct or indirect), distance between probes, type of samples used (small clear specimens or structural dimension species).

From the models developed at LNEC for maritime pine using clear wood specimens the best model found is the one identified as G1 which showed a r^2 of 0.94 and a standard error of 647 N/mm^2, Eq. 4. The various models used the bulk density determined by weight/volume determination (H2, G2, I2 or J2) or density prediction using wood cores extracted from the test pieces (H1, G1, I1H1, G1, I1 or J1). An additional model (model G3) was added using the velocity values of model G1 and the mean density value associated to the wood species. This last approach is mentioned by [2] as a possibility when there is not chance to make a direct or indirect measurement in-situ (as in some historic timber buildings).

$$E_{\mathrm{stat}} = 1.069 E_{\mathrm{dyn}} - 77.8 + \varepsilon_{stat} \qquad \varepsilon_{stat} \sim N(0, 1) \tag{4}$$

Also in Fig. 1 it was plotted three regression models obtained in Spain (models S1, S2 and S3) also for maritime pine timber [3].

Another source of uncertainty is linked to cross section irregularities. The evaluation carried out in four timber structures (from 17th, 18th and 20th centuries) showed for sound solid timber members of rectangular shape a coefficient of variation lower than 10% whereas for round shapes a value lower than 20% [4]. In-situ decisions about the way to measure the geometry of the timber members have consequences on the accuracy of the measurement using ultrasonic or stress waves devices [5].

Moreover in-situ only the predictor variable (x) is known and the true correlation with the response variable (y) is always unknow. So, evidence of the precision of the model does not exist and the correlation can also be different (even significantly different) from that expected from the curve obtained at the laboratory. One way to try to assess a possible deviation is the use of semi-destructive testing methods for calibration in-situ [6]. This possibility will be explored in future works.

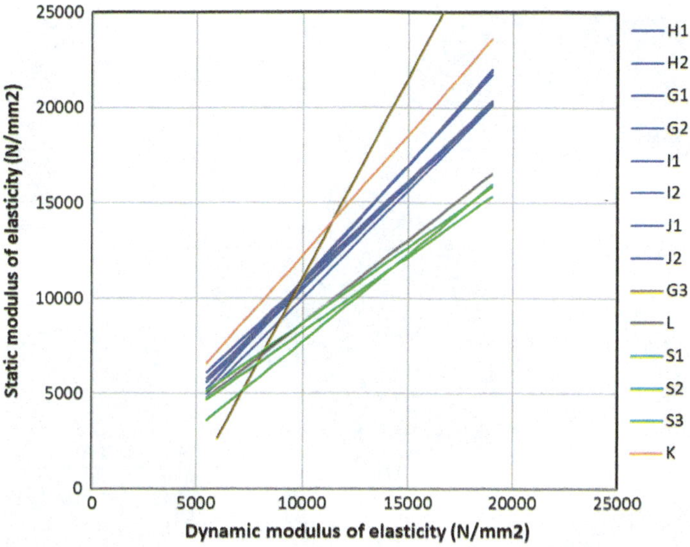

Fig. 1 Comparison between regression models. In blue models developed for maritime pine at LNEC, in green models applied to *Pinus pinaster* in studies conducted in Spain [3], in brown using data of model G1 and a mean density value for maritime pine and in orange at LNEC for *Pinus palustris*

2.2 Assessment of Model's Uncertainty by Simulation

In the present evaluation of the uncertainty of available models, Fig. 1, the results obtained at LNEC for maritime pine by application of the equipment Fakopp microsecond timer, Fig. 2, were used.

The first step comprised the generation of a population of one thousand pair of correlated data (Es_{tat}—static modulus of elasticity; E_{dyn}—dynamic modulus of elasticity), Eqs. 5 and 6. The generation of correlated pairs of variables (X,Y) was made by randomly generating two variables (Y_1, Y_2) from a standard normal distribution $(Y_i \sim N(0, 1))$. In second step two new variables (Z_1 and Z_2) were generated from the previous being both correlated by considering one (Z_2) as a linear combination of the other (Z_1), Eq. 5. The degree of correlation was chosen as the best found in test carried out at LNEC for maritime pine (0.94) or the range of values found in the different tests carried out at LNEC (from 0.77 to 0.94). In this last case it was considered that the probability of getting any of the values were equal and then in each loop a random number was chosen from that range considering a uniform distribution.

The final conversion of the data (to non-standardized data) was performed and the simulated population of one thousand correlated pairs of dynamic modulus of elasticity (E_{dyn}) and static modulus of elasticity (E_{sta}) obtained, Eq. 6. This step used model G1as basis.

Fig. 2 Fakopp microsecond timer used at the lab and in situ

$$\begin{bmatrix} Z_1 \\ Z_2 \end{bmatrix} = \begin{bmatrix} Y_1 \\ \rho Y_1 + Y_2\sqrt{1-\rho^2} \end{bmatrix} \tag{5}$$

$$\left(E_{dyn}, E_{stat}\right) = \left[F_{X_1}^{-1}\{\Phi(Z_1)\}, F_{x_2}^{-1}\{\Phi(Z_2)\}\right] \tag{6}$$

In this respect Fig. 3a could be considered as the fusion of the information of all models available at LNEC, considering that different samples conduct to different regression models, being however model G1 the closest to the true model. Thus, this population is considered as the one from which different samples were taken and provided the different regression models obtained.

The models were then evaluated by first selecting a sample size (n = 10, 15, 20, 25, 30, 50, 70, 90, 100), secondly running 50 sample replicates (with the same sample size) and determining for each replicate the *MPE* associated. This step was followed for each sample size.

Also a regression model based on the cloud of point of Fig. 3a was derived using a Bayesian Linear Regression approach with non-informative priors using MCMC for carried out the sampling from the posterior distribution in order to approximate the posterior using 20,000 iterations (model PBayes) and considering a 20,000 iteration warmup period. One advantage of using a Bayesian approach using MCMC is that the posterior provides a distribution of the expected value and not only a single point as in the case of a regression model based on the method of least squares.

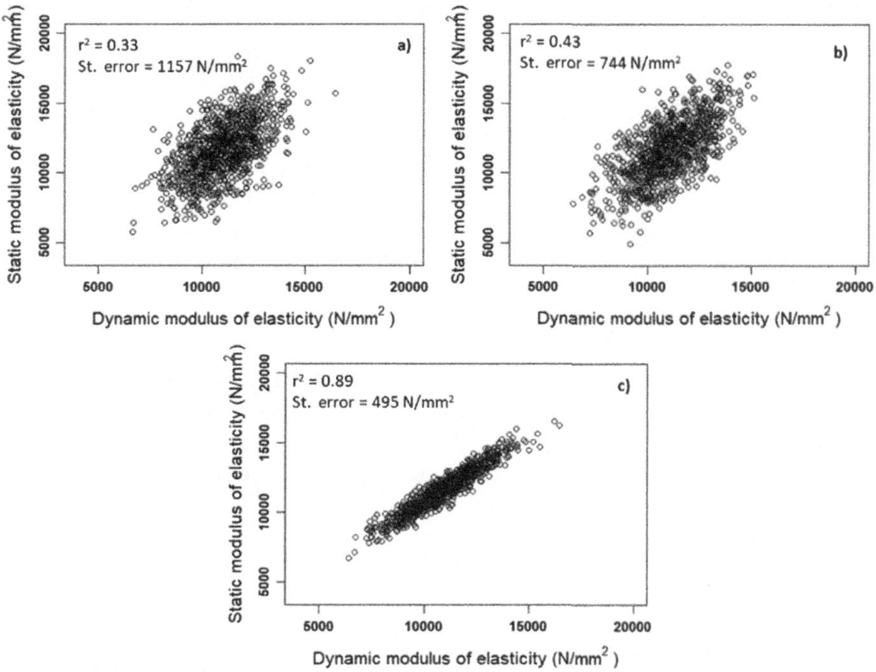

Fig. 3 Simulated population of static and dynamic pairs of correlated samples: **a** considering a correlation coefficient between 0.77 and 0.94 and an error component ~N(0,1); **b** considering a correlation coefficient of 0.94 and an error component ~N(0,1); **c** considering a correlation coefficient of 0.94 and no error component

2.3 Results

In Fig. 4 it is showed the mean value of *MPE* (given an estimation of the accuracy) and the dispersion of *MPE* (given an estimation of uncertainty) when choosing the model and depending on the sample size available (number of points taken from the population—Fig. 3a. As expected, the best models available (G1 and G2), showed the best accuracy. However, in terms of spreading of values the sample size is a crucial point to ensure that the model is the best approximation to the values sampled. The results showed that if only available a small sample size the uncertainty (dispersion) of error is higher (close to 20%) whereas for bigger sample sizes the uncertainty can go down to 10%.

The comparison of G1 model with a model developed for the same wood species but from a different origin (or probably not following the same indirect procedure) (model S3) or with the one selected using the mean density associated to maritime pine wood provided very different results (model G3), Fig. 5.

The results showed that a choice of a model not calibrated and validated for a particular wood species and origin tends to produces less accurate results although

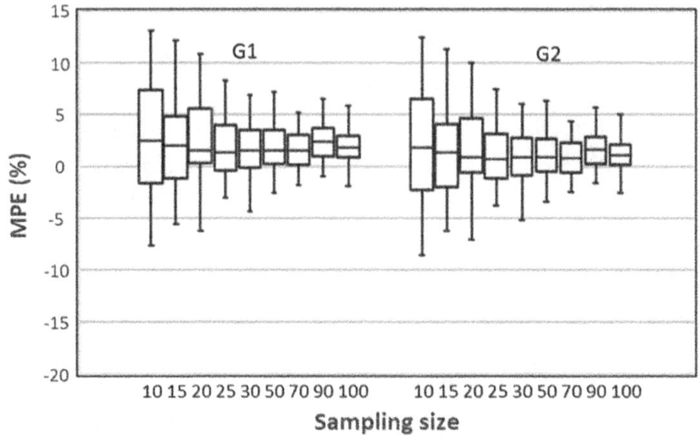

Fig. 4 *MPE* median and dispersion according to the sample size and model (G1 – $r^2 = 0.94$, σ_E = 646 N/mm^2; G2 – $r^2 = 0.94$, σ_E = 674 N/mm^2)

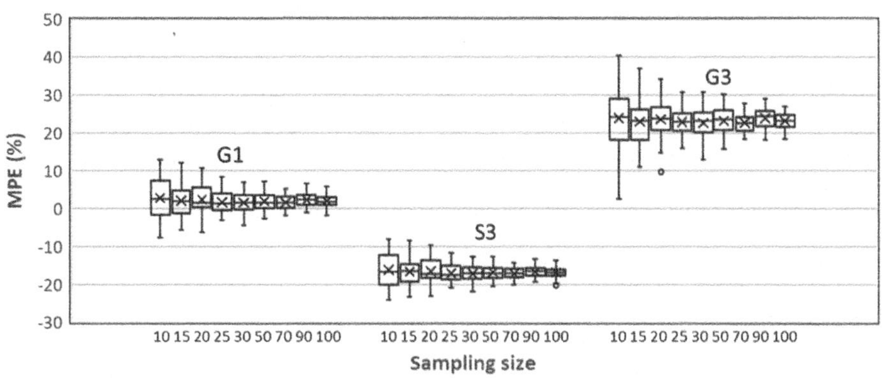

Fig. 5 *MPE* median and dispersion for models G1, S3 and G3

in this case more conservative in the case of model S3 or far less conservative (unsafe) if used the model G3. Also model G3 showed a significant higher uncertainty (dispersion of values) when compared with the other models.

The Bayesian linear regression model (PBayes) showed results similar to the model in which it was based the cloud of correlated pairs (Fig. 3a), although a slight more conservative predictions (mean value below zero) and a similar uncertainty, Fig. 6.

The comparison between models (using wood cores or bulk density) showed similar values of *MPE* (if compared central and dispersion parameters), which is explained with the exception of model G3 (where a mean value is used for density) by the fact that the slope (β_1) is very similar between models, Fig. 7a. Also in all models the intercept variable (β_0) is considered not significant ($p > 0.05$). This same

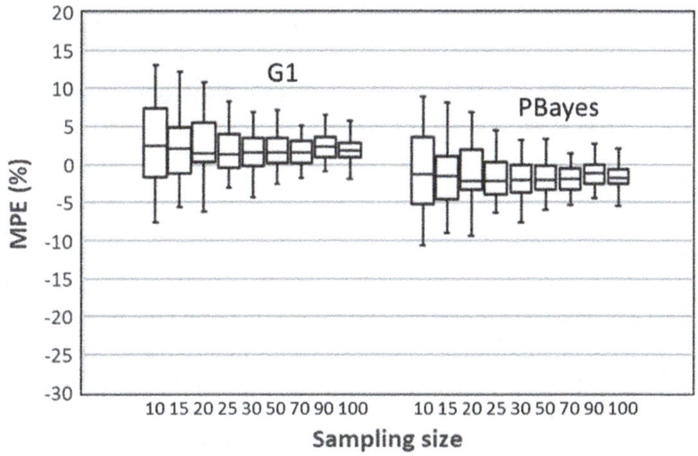

Fig. 6 *MPE* median and dispersion for models G1 and PBayes

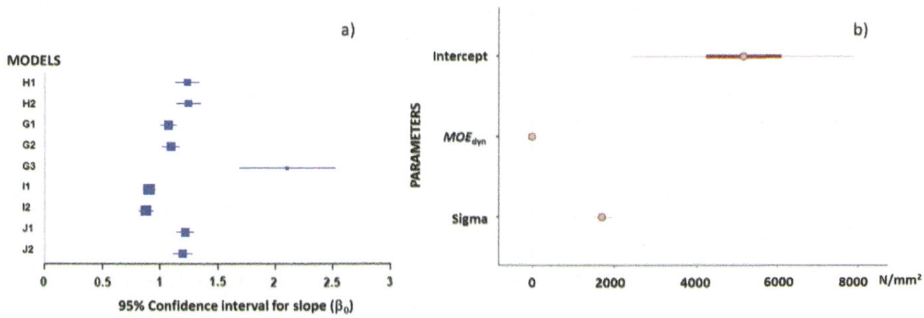

Fig. 7 **a** Confidence interval (95%) for the slope of the different models; **b** model PBayes—95% posterior credible interval for parameters of the regression model and the dispersion (sigma) of the posterior distribution of the dependent variable (50% interval thicker line)

fact is seen by exploring the posterior credible interval for the parameters of the model, Fig. 7b, provided by the Bayesian model.

3 Uncertainty in the Assessment In Situ of the Modulus of Elasticity—Application to Timber Beams

The models for clear wood species developed at LNEC had in in mind their application into the evaluation of timber member in-service, following the concept of a member divided into clear and knot wood zones, Fig. 8. A description of the method followed can be seen in [7].

Fig. 8 Discretization of an element into sub-elements

The application of existing models in-situ faces the fact that no information about the real modulus of elasticity of the timber members is possible to be obtained. In the present case is was tested the application of some previous models (G1, G3 and Pbayes) selected using small clear wood test pieces to thirty-three timber beams and assessed how well the models fit to the new data (validation of the models), Fig. 9.

All models showed the same correlation ($r^2 = 0.71$) but in Fig. 10 it can be seen that a good agreement is only obtained from models G1 and PBayes. If a mean value is allocated to the timber members (equal to the mean density for that wood species) as in the case of model G3 then less accuracy and more uncertainty can be expected, Fig. 10.

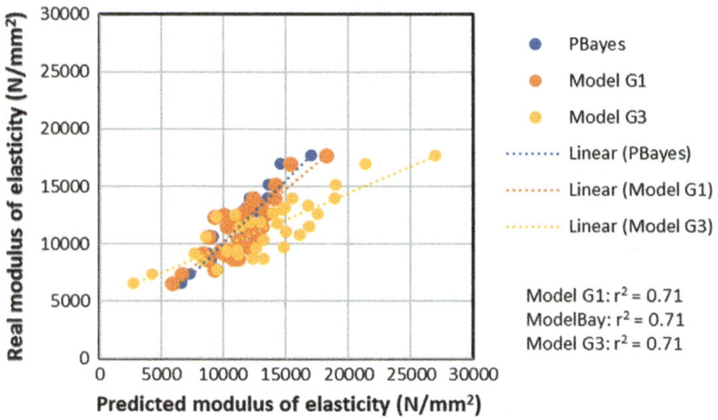

Fig. 9 Results of predicted modulus of elasticity against actual modulus of elasticity values

Fig. 10 Mean and variation of the error obtained for the different models considered

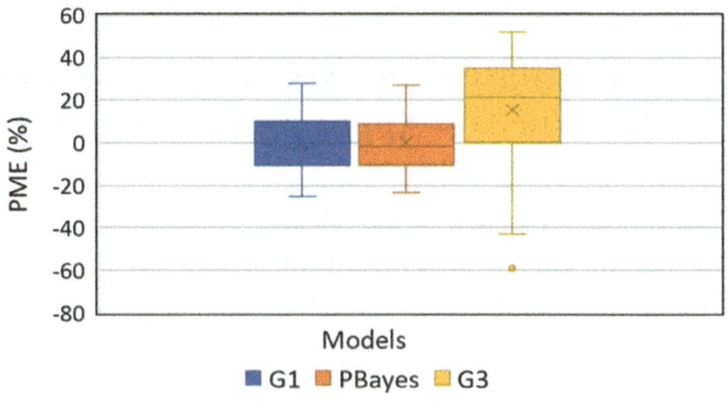

In fact, the results obtained showed that the use of models developed using small clear specimens can, provided a correct testing procedure is defined, be applied to structural dimension timber elements similar to those found in existing timber structures. However, the amount of uncertainty is as expected higher that those initially allocated to the models ($\approx\pm 20\%$). This uncertainty should be taken into consideration and dealt by the expert in their final appraisal of the quality of a timber member.

4 Conclusions

The application of regression models tested at the Lab for predicting new data on the field should be carefully done. The necessity to verify that a common procedure is followed, that the same wood species is involved is some of the points to be checked. The use of deterministic values as an average density value (mean associated to a certain wood species) increases the uncertainty associated to the models.

The results also enhanced the importance of sample size to build a regression model. In the present case all models showed that a minimum sample size of 30 should be used.

In fact, the amount of uncertainty of the data obtained will be defined by a correct choice of the timber members to test in-situ (number and quality) being probably be necessary to use results from compression or tensile semi-destructive testing methods [6] for a sort of calibration in order to dealt with the high level of uncertainty (as much as $\pm 20\%$) that can be associated to the predictions.

References

1. Kasal, K., & Tannert, B. (2010). *In situ assessment of structural timber*. RILEM State of the Art Reports (vol. 7). Springer.
2. Dackermann, U., Keith Crews, K., Kasal, B., Li, J., Riggio, M., Rinn, F., & Tannert, T. (2014). In situ assessment of structural timber using stress-wave measurements. *Materials and Structures, 47,* 787–803.
3. Llana, D. F., Íñiguez-González, G., Díez, R. M., & Arriaga, F. (2020). Nondestructive testing used on timber in Spain: A literature review. *Maderas, 22,* 133–156.
4. Lourenço, P. B., Sousa, H. S., Brites, R. D., et al. (2013). In situ measured cross section geometry of old timber structures and its influence on structural safety. *Materials and Structures, 46,* 1193–1208.
5. Osuna-Sequera, C., Llanab, D. F., Íñiguez-Gonzáleza, G., & Arriaga, F. (2020). The influence of cross-section variation on bending stiffness assessment in existing timber structures. *Engineering Structures*, 204.
6. Kloiber, M., Drdacky, M., Machado, J. S., Maurizio, P. M., & Yamaguchi, N. (2015). Prediction of mechanical properties by means of semi-destructive methods: A review. *Construction and Building Materials, 101,* 1215–1234.

7. Pereira, F., & Machado, J. S. (2019). A probabilistic approach to the evaluation of the bending strength of timber beams with integration of data from on-site tests. *Materials and Structures, 52,* 38. https://doi.org/10.1617/s11527-019-1347-8.

Vulnerability Assessment of Aging Levees with WINGS and Interval Arithmetic

Francesca Marsili, Jörg Bödefeld, Lukas Weber, and Maryam Ghadami

Abstract Systems of infrastructures are aging at increasing rates leading to a backlog of maintenance actions. The lack of maintenance causes failures as natural events and terrorism do. Since systems of infrastructures are integrated, failure of a system component may have cascade effects. New approaches are required for the prioritization of maintenance actions on systems of infrastructures, which also consider the interactions among different systems and the uncertainty affecting them. In this paper, the WINGS technique has been used to develop a vulnerability index for an aging system of levees. Levees are in a state of perpetual interaction with other systems such as pipelines and culverts. The WINGS method allows taking into account expert opinion about the strength of the vulnerability factors and their mutual influence. Since expert judgments are uncertain, interval arithmetic has been used to model the uncertainty and assess its impact on the vulnerability assessment. Finally, the results of the vulnerability assessment have been visualized on a GIS map. The proposed approach can be used to direct the collection of further information for more refined vulnerability and risk assessment. The final scope of the paper is to establish a sound procedure for prioritizing maintenance actions on complex infrastructure systems.

Keywords Vulnerability assessment · Neglected maintenance · Systems of systems · WINGS · Interval arithmetic

1 Introduction

A current challenge in civil engineering is the management of portfolios of aging infrastructures.

Also the German network of waterways, which comprises several infrastructures such as locks, weirs, culverts, pipes, canal bridges, lighthouses, dams and levees, is aging. The rate at which degradation evolves in significant damages is greater

F. Marsili (✉) · J. Bödefeld · L. Weber · M. Ghadami
Federal Waterways Engineering and Research Institute, Karlsruhe, Germany
e-mail: francesca.marsili@baw.de

© The Author(s), under exclusive license to Springer Nature Switzerland AG 2021
J. C. Matos et al. (eds.), *18th International Probabilistic Workshop*, Lecture Notes in Civil Engineering 153, https://doi.org/10.1007/978-3-030-73616-3_65

than the rate at which maintenance actions can be executed, resulting in a backlog of maintenance interventions. A minimal perturbation of an aging system is sufficient to activate failure mechanisms with unpredictable cascade effects. From here a new term has been coined, the risk of disregarding or neglecting maintenance [1] whose assessment represents a strategy for the prioritization of maintenance actions. Performing a risk assessment for systems of infrastructures involves other analyses, such as vulnerability and criticality assessment, according to which the proneness of the system to develop failure scenario and the consequences of failure can be determined. While the notion of vulnerability is fairly mature for single objects, it is still evolving for systems of infrastructures, and only recently it has been considered in relationship with complex infrastructures systems or Systems of Systems (SoS) [2]; thus it is necessary to develop new ideas for assessing the vulnerability of SoS, which also take into account the interaction among the subsystems.

Usually vulnerability assessment implies the development of indexes, which are often a combination of several indicators or factors. An often applied aggregation model is the additive weighting, according to which the vulnerability index is obtained as the weighted sum of several indicators. The assignment of the weights, which represent the relative importance of each indicator, is a challenging task. Furthermore, the use of additive weighting as an aggregation model requires independence of the indicators. This condition is often ignored leading to biased composite indicators.

This paper focuses on the development of a vulnerability index for the assessment of a levee system. We propose to use the Weighted Influence Non-linear Gauge System (WINGS) [3] to derive weights for vulnerability indicators, which do not disregard their interdependencies. WINGS represents an evolution of the DEcision-MAking Trials and Evaluation Laboratory (DEMATEL) [4, 5], a method that serves as a Multi-Criteria Decision Analysis when interrelations among criteria cannot be neglected. However, DEMATEL considers only the intensity of the influence among the factors; WINGS raises the bar by considering also the strength of the acting factors. Since WINGS is a new method, only a few applications exist [6, 7]; this paper represents a new application of WINGS in the context of the vulnerability assessment of aging infrastructures systems. Furthermore, the uncertainty in the experts judgments are also considered by combining WINGS with interval arithmetic [8]. Results of the vulnerability assessment are finally visualized on a GIS map, which allows better communication with the stakeholders. The paper is organized as follows: in Sect. 2 the main factors affecting the vulnerability of levee systems are identified; in Sect. 3 the steps of the WINGS technique are described; in Sect. 4 WINGS is combined with interval arithmetic in order to consider uncertain expert judgments; in Sect. 5 the case study is introduced, and the WINGS method has been applied to assess the vulnerability of a levee system; finally, in Sect. 6, the conclusion is drawn and the further steps of the research are briefly described.

2 Vulnerability Assessment of Levees

Among the infrastructures of the German waterways network, levees raise special concerns because of the dramatic consequences to which the failure of a levee often leads. Levees support the water level in canals or rivers. In the case of rivers, the water level varies depending on the outflow, which in turn it depends on weather phenomena; in a canals levee, the water is often pumped and kept constant by pumping stations. In both cases, the water level is above the adjacent terrain. Levees fail when a breach forms, which means that a portion of the levee collapses, resulting in the significant loss of crest or the creation of a hole, causing the uncontrolled loss of water and flooding the surrounding area.

Breaches are usually the final stage of other deterioration and damages processes, such as erosion, scour and slippage, whose presence initiates some sort of failure mechanisms like piping or slope instability. Such deterioration processes increase the vulnerability of the system, which is the likelihood that a breach will form.

But also other factors affect the vulnerabilities of levees, and especially the presence of encroaching structures and transition zones [9], such as penetrating structures like pipelines and culverts, or general buildings. Those structures may provoke damages to levees when they are inadequately designed, constructed and maintained, such as when: the material of the structure has not an adequate strength to withstand loads; the structure is unable to accommodate movements resulting from foundation settlements; unsuitable backfills materials are used; unstable materials like gas and other explosive substances are conveyed through pipes. At the same time, the condition of the levee also affects the condition of those structures, for example in the case of piping and settlements of levees or foundation soils.

Also transitions zones increase the levee vulnerability; those are represented by any portion of a levee where the geometric configuration changes, such as in case of changes of the levee cross-section, when the previously mentioned encroaching structures are present or in case of walls, gates, sluices or other constructions. Transitioning between different geometric configurations or material compositions creates a critical junction that often represents a focal point for the concentration of tensions, eventually resulting in the activation of a failure mechanism.

3 WINGS

As stated in Sect. 1, WINGS can be applied to study the structure of relationships which exist among a set of factors, rank the factors according to their involvement in the system and classify them as ‘influencing’ and ‘influenced’.

The steps of the WINGS procedure are the following:

1. *Generate the direct-influence matrix D*
   Let us consider a set of n factors $F = \{F_1, F_2, \ldots, F_n\}$. An expert is asked to indicate the strength of all system factors and the level of influence between

system factors using a rating system that goes from 0 (no strength/influence) to 9 (very high strength/influence). The direct strength-influence matrix provided by the expert can be formed,

$$D = \left[d_{ij}\right]_{n \times n} \tag{1}$$

where d_{ii} and d_{ij} represent the judgment of the decision maker on respectively the strength of F_i and the influence of F_i on F_j.

2. *Establish the normalized direct influence matrix S.*
   This matrix can be obtained by using

$$S = \frac{D}{s}, \tag{2}$$

where

$$s = \sum_{i=1}^{n} \sum_{j=1}^{n} d_{ij}. \tag{3}$$

3. *Construct the total influence matrix T*:
   The matrix $T = \left[t_{ij}\right]_{n \times n}$ is computed by summing the direct and indirect effects; the normalization of S ensures convergence of T:

$$T = \lim_{w \to \infty} \left(S + S^2 + \cdots + S^w\right) = S(Id - S)^{-1} \tag{4}$$

4. *Calculate the total impact and the total receptivity*
   Row and columns wise summation of the elements of T gives the total impact r_i:

$$r_i = \left[\sum_{j=1}^{n} t_{ij}\right]_{n \times 1} \tag{5}$$

which represents the influence of the component i on all other components of the system, and the total receptivity c_j,

$$c_j = \left[\sum_{i=1}^{n} t_{ij} \right]_{n \times 1}. \tag{6}$$

which represents the influence of all other components in the system on the component i.

5. *Calculate the total involvement and the position of the component*

Finally it is possible to calculate for the factor F_i the total involvement:

$$r_i + c_j \tag{7}$$

which represents the sum of all influences exerted on and received by the component I, and the position value:

$$r_i - c_j. \tag{8}$$

If $r_i - c_j$ is positive, the factor F_i can be classified as 'influencing', otherwise, it belongs to the 'influenced' group.

4 If Expert Judgments are Uncertain: WINGS with Interval Arithmetic

The judgments provided by the experts in the WING method about the strength of the factors and their mutual influence may be affected by uncertainty. One way to describe this uncertainty is through pairs of verbal assessments, which when mapped onto numerical scale will become intervals. An interval is a connected subset of $\mathbb{R}$ usually denoted by $[x]$. When the upper and the lower bounds are the same, the interval can be identified with a real number. The midpoint of any bounded and nonempty interval is given by:

$$mid([x]) = \frac{\underline{x} + \overline{x}}{2} \tag{9}$$

where $[x] = \left[\underline{x}, \overline{x}\right], \underline{x} \leq \overline{x}$. The classical operations of real number arithmetic can be extended to intervals [10]. The basic notion of linear algebra such as vectors and matrices can also be generalized to the interval case. In particular, an $(m \times n)$-dimensional interval matrix is a subset of $\mathbb{R}^{m \times n}$—the set of all matrices with real coefficients, and it is defined as the Cartesian product of $m \times n$ close intervals. It is uniquely represented by their $m \times n$ elements $\left[a_{ij}\right], i = 1, \ldots, m; j = 1, \ldots, n$. The midpoint of an interval vector and an interval matrix can be defined in an obvious way. The matrix D becomes an interval matrix:

$$[D] = [\underline{D}, \overline{D}]. \tag{10}$$

The scaling factor is calculated as follows in order to ensure convergence and preserve consistency:

$$s = \sum_{i=1}^{n} \sum_{j=1}^{n} \overline{d}_{ij} \tag{11}$$

which leads to

$$[S] = [\underline{S}, \overline{S}] = \frac{[D]}{s} = \left[\frac{\underline{D}}{s}, \frac{\overline{D}}{s}\right]. \tag{12}$$

The bounds of the interval matrix T can be calculated separately from the following equations:

$$[T] = [\underline{T}, \overline{T}] = \left[\underline{S}(Id - \underline{S})^{-1}, \overline{S}(Id - \overline{S})^{-1}\right] \tag{13}$$

from which the bounds for the total impact and total receptivity can be calculated:

$$[r_i] = [\underline{r}_i, \overline{r_i}] = \left[\left[\sum_{j=1}^{n} \underline{t}_{ij}\right]_{n\times 1}, \left[\sum_{j=1}^{n} \overline{t}_{ij}\right]_{n\times 1}\right] \tag{14}$$

$$[c_j] = [\underline{c}_j, \overline{c_j}] = \left[\left[\sum_{i=1}^{n} \underline{t}_{ij}\right]_{n\times 1}, \left[\sum_{i=1}^{n} \overline{t}_{ij}\right]_{n\times 1}\right] \tag{15}$$

and consequently also the bounds for the total involvement and position of the component:

$$[r_i + c_j] = \left[\underline{r_i + c_j}, \overline{r_i + c_j}\right] = [r_i] + [c_j] = \left[\underline{r}_i + \underline{c}_j, \overline{r_i} + \overline{c_j}\right] \tag{16}$$

$$[r_i - c_j] = \left[\underline{r_i - c_j}, \overline{r_i - c_j}\right] = [r_i] - [c_j] = \left[\underline{r}_i - \overline{c_j}, \overline{r_i} - \underline{c}_j\right]. \tag{17}$$

If indeed we are interested in the midpoints of the above-mentioned values, we can directly calculate them from the matrix T_m, which contains the midpoints of $\underline{T}$ and $\overline{T}$:

$$[\boldsymbol{T}_m]_{ij} = \frac{[\underline{\boldsymbol{T}}]_{ij} + [\overline{\boldsymbol{T}}]_{ij}}{2}. \tag{18}$$

5 Case Study

5.1 Introduction

An application of the WINGS method is developed to assess the vulnerability of a portion of the system of levees of the West German network of canals. Since this region is the largest urban area in Germany, and 20% of the freight traffic of the German waterways is regularly moved on the canals, the waterways play a fundamental role for its economy; considering also the high density of population and industries, the failure of a levee could have catastrophic consequences in social and economic terms. The levees are currently in a bad condition: many of them show degradation phenomena such as erosion, vegetation and stability problems. The levees are penetrated by a huge number of structures, especially pipelines and culverts, which make things worst. Information about vulnerability indicators is collected during the periodical inspections of the levees and stored in devoted databases. By simplifying, we will assume that the indicators take only two possible values: yes/no, fulfilled/not fulfilled, good/bad. They can be considered as binary parameters having 2 possible values, on/off, where 'on' corresponds to "the vulnerability indicator is enabled" and "off" corresponds to "the vulnerability indicator is disabled".

5.2 Identification of Vulnerability Factors

The identified vulnerability factors are listed in Table 1. The condition of pipelines and culverts can be classified as good or bad according to the damages which have been collected during the periodical inspections. Levees are characterized by the following cross sections: rectangular, trapezoidal, and mixed. Sometimes a levees segment is characterized by more than one cross-section. The presence of drainages and seals, as well embankment walls, also represents a change in the levees cross section, and for this reason, they should be considered as vulnerability factors. Information about the condition of those objects would be relevant for determining the levee vulnerability, but unfortunately, it is very difficult to acquire it since the canal should be dried. Some levees are equipped with inspection devices that allow recognizing deterioration processes properly. Thus the presence of inspection equipment reduces the vulnerability of the levee. However, this allows the levee to be inspected only at isolated points, which means that only a limited fraction of the possible defects can be identified.

Table 1 Binary values assumed by the vulnerability factors

Vulnerability factors	Disabled	Enabled
Pipeline encroachment (P)	No	Yes
Pipeline condition (PC)	Good	Bad
Culvert encroachment (C)	No	Yes
Culvert condition (CC)	Good	Bad
Other objects (O)	No	Yes
Stability of the levee (St)	Fullfilled	Not fullfilled
Soil erosion (Er)	Fullfilled	Not fullfilled
Burrowing animals (B)	No	Yes
Vegetation (V)	No	Yes
Change of the geometry (G)	No	Yes
Seal (S)	No	Yes
Internal seal (IS)	No	Yes
Drainage (D)	No	Yes
Embankment walls (E)	No	Yes
Inspection devices (I)	Yes	No
Levees water level (LH)	<3 m	≥3 m

5.3 *Application of WINGS*

The expert assesses the strength of the vulnerability factors and their mutual influence by compiling the individual direct strength-influence matrix (Table 2). In case of uncertainty, an interval evaluation is given.

The steps of the method described in Sect. 4 are further applied to obtain the bounds for the position and especially the total involvement value associated with each factor (Table 3), from which the bounds for the vulnerability index will be derived.

5.4 *Vulnerability Assessment*

To assess the vulnerability, the levee has to be divided into homogeneous segments, which are characterized by different lengths: from several hundred to few meters (i.e. in case of object encroachment). For each segment m, an interval vulnerability index $[VI_m]$ is computed considering the upper and lower bound of the total involvement value which characterizes each vulnerability factor:

$$\overline{VI_m} = \frac{1}{\max \sum_{i \in P} \left(\overline{r_i + c_i}\right)} \sum_{i \in P} \left(\overline{r_i + c_i}\right) \tag{19}$$

Table 2 Direct strength-influence interval matrix [D].

	P	CP	C	CC	O	St	Er	B	V	G	S	IS	D	E	I	LH
P	[5, 7]	0	0	0	0	[4, 6]	[4, 6]	0	0	0	[4, 6]	[4, 6]	[4, 6]	[1, 3]	0	0
CP	0	[7, 8]	0	0	0	[7, 9]	[6, 8]	0	0	0	[6, 8]	[6, 8]	[6, 8]	[6, 8]	0	0
C	0	0	[6, 8]	0	0	[5, 7]	[5, 7]	0	0	0	[4, 6]	[4, 6]	[4, 6]	[1, 3]	0	0
CC	0	0	0	[8, 9]	0	[7, 9]	[6, 8]	0	0	0	[6, 8]	[6, 8]	[6, 8]	[6, 8]	0	0
O	0	0	0	0	[5, 6]	[4, 6]	[4, 6]	0	0	0	[5, 7]	[5, 7]	[5, 7]	[1, 3]	0	0
St	0	9	0	9	0	9	9	0	0	0	[7, 9]	[7, 9]	[7, 9]	[7, 9]	0	0
Er	0	[7, 8]	0	[7, 9]	0	9	9	0	0	0	[7, 9]	[7, 9]	[7, 9]	[7, 9]	0	0
B	0	[4, 6]	0	[2, 4]	0	[7, 8]	[8, 9]	8	0	0	[5, 7]	[5, 7]	[5, 7]	[4, 6]	0	0
V	0	[7, 8]	0	[4, 6]	[4, 6]	5	7	[4, 6]	[6, 7]	0	[5, 7]	[5, 7]	[5, 7]	[4, 6]	0	0
G	0	0	0	0	0	2	5	0	0	3	[4, 6]	[4, 6]	[4, 6]	[2, 4]	0	0
S	0	[6, 8]	0	[6, 8]	[4, 6]	[5, 7]	[6, 8]	0	0	0	[7, 8]	[4, 6]	[4, 6]	[4, 6]	0	0
IS	0	[6, 8]	0	[6, 8]	[4, 6]	[5, 7]	[7, 9]	0	0	0	[4, 6]	[8, 9]	[8, 9]	[4, 6]	0	0
D	0	[6, 8]	0	[6, 8]	[4, 6]	[5, 7]	[7, 9]	0	0	0	[4, 6]	[4, 6]	[7, 8]	[4, 6]	0	0
E	0	[4, 6]	0	[4, 6]	[4, 6]	[4, 6]	[6, 8]	0	0	0	[4, 6]	[4, 6]	[4, 6]	[6, 7]	0	0
I	0	0	0	0	0	[5, 7]	[7, 9]	0	0	0	[1, 2]	[1, 2]	[1, 2]	0	2	0
LH	0	0	0	0	0	[4, 6]	[7, 9]	0	0	0	[4, 6]	[4, 6]	[4, 6]	[1, 2]	0	[7, 8]

Table 3 Lower and upper bound of the interval total involvement $[r_i + c_j]$

Vulnerability factors	$\underline{r_i + c_j}$	$\overline{r_i + c_j}$
Pipeline encroachment (P)	0.0368	0.0567
Pipeline condition (PC)	0.1210	0.1581
Culvert encroachment (C)	0.0416	0.0616
Culvert condition (CC)	0.1176	0.1547
Other objects (O)	0.0849	0.1190
Stability of the levee (St)	0.1890	0.2322
Soil erosion (Er)	0.2029	0.2511
Burrowing animals (B)	0.0714	0.0918
Vegetation (V)	0.0739	0.0969
Change of the geometry (G)	0.0321	0.0424
Seal (S)	0.1459	0.2067
Internal seal (IS)	0.1508	0.2104
Drainage (D)	0.1483	0.2080
Embankment walls (E)	0.1180	0.1750
Inspection devices (I)	0.0216	0.0318
Levees water level (LH)	0.0451	0.0616

$$\underline{VI_m} = \frac{1}{\max \sum_{i \in P} \overline{(r_i + c_i)}} \sum_{i \in P} \underline{(r_i + c_i)} \tag{20}$$

$$0 \leq \underline{VI_m} \leq \overline{VI_m} \leq 1 \tag{21}$$

where P is the subset of vulnerability factors which are enabled in the levee segment m.

The maps (Fig. 1) reveal that the scores associated with some levees segments (which correspond to a certain gradient of color) overlap, which makes it difficult to uniquely rank the levees according to their vulnerability since their position can be interchanged. If the decision maker is not satisfied with these results, the uncertainty affecting the expert judgments should be reduced in order to resolve this ambiguity.

6 Conclusion

In this paper, the vulnerability assessment of a levees system due to neglected maintenance has been carried. The levees system comprises other infrastructures systems such as pipelines and culverts, which are in a state of perpetual interaction with the levee. This implies that the vulnerability of those systems affects the vulnerability of the levees system and vice versa, and the failure of a system component propagates

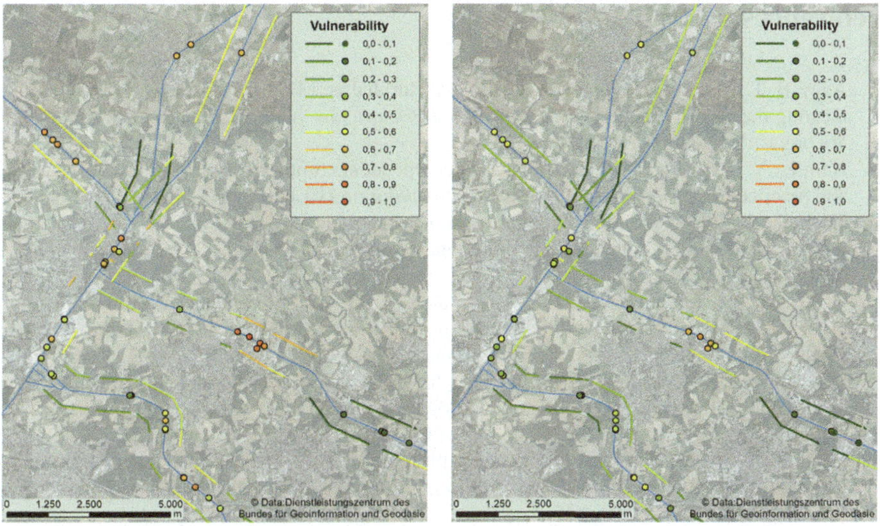

Fig. 1 GIS maps showing the results of the vulnerability assessment on a portion of the West German network of canals, which has been obtained applying the WINGS method and interval arithmetic (left: lower vulnerability; right; upper vulnerability). The lines represent the levees, while the points represent the encroaching structures; please notice that levees are not continuous objects because the water level is sometimes under the adjacent terrain due to orographic irregularities

in a non-linear way leading to cascade effects. To consider the nonlinearity of the problem, the WINGS technique has been proposed. This method allows identifying the involvement of each vulnerability factor, which is based on the strength of the factor and the intensity of its influence on the other factors. However, the method is based on expert judgments, which in general are affected by uncertainty; to take it into account, the WINGS method has been integrated with interval arithmetic. The extended method allows the definition of upper and lower bounds of the vulnerability index, which in turn clarifies when the results are ambiguous and the uncertainty has to be reduced.

The visualization of the vulnerability assessment on GIS maps also facilitates communication with stakeholders and other parties affected by the vulnerability of levees. The final aim of this study is to identify on which objects maintenance actions should be prioritized.

Acknowledgments This study has been developed within the research project PREVIEW, funded by the German Ministry of Education and Research, whose aim is to increase the resilience of German waterways infrastructures (https://preview-projekt.baw.de/de).

References

1. Adey, B. T., Martani, C., Papathanasiou, N., & Burkhalter, M. (2018). Estimating and communicating the risk of neglecting maintenance. *Infrastructure Asset Management*, 1–65. https://doi.org/10.1680/jinam.18.00027.
2. Haimes, Y. Y. (2017). Risk modeling of interdependent complex systems of systems: Theory and practice. *Risk Analysis, 38*(1), 84–98. https://doi.org/10.1111/risa.12804.
3. Michnik, J. (2013). Weighted influence non-linear gauge system (WINGS)—An analysis method for the systems of interrelated components. *European Journal of Operational Research, 228*(3), 536–544.
4. Gabus, A., & Fontela, E. (1973). Perceptions of the world problematic: Communication procedure, communicating with those bearing collective responsibility, DEMATEL 1, Battelle Geneva Research Centre, Geneva, Switzerland.
5. Si, S.-L., You, X.-Y., Liu, H.-C., & Zhang, P. (2018). DEMATEL technique: A systematic review of the state-of-the-art literature on methodologies and applications. *Mathematical Problems in Engineering, 2018,* 1–33. https://doi.org/10.1155/2018/3696457.
6. Michnik, J. (2018). The WINGS method with multiple networks and its application to innovation projects selection. *International Journal of Applied Management Science, Inderscience Enterprises Ltd, 10*(2), 105–126.
7. Michnik, J., & Grabowski, A. (2019). Modeling uncertainty in the WINGS method using interval arithmetic. *International Journal of Information Technology & Decision Making, 19.*
8. Sallum, F. S., Gomes, L. F., & Machado, M. A. (2019). A multicriteria approach to the prioritisation of stock investment funds. *International Journal of Business and Systems Research, 13,* 120–134.
9. CIRIA. (2013). *The International Levee Handbook*. London (UK): CIRIA.
10. Alefeld, G., & Mayer, G. (2000). Interval analysis, theory and applications. *Journal of Computational and Applied Mathematics, 121*(1–2), 421–464.

Author Index

© The Editor(s) (if applicable) and The Author(s), under exclusive license to Springer Nature Switzerland AG 2021
J. C. Matos et al. (eds.), *18th International Probabilistic Workshop*, Lecture Notes in Civil Engineering 153, https://doi.org/10.1007/978-3-030-73616-3

The manufacturer's authorised representative in the EU is Springer Nature Customer Service Centre GmbH, Europaplatz 3, 69115 Heidelberg, Germany. If you have any concerns regarding our products, please contact ProductSafety@springernature.com

Printed and bound by CPI Group (UK) Ltd, Croydon, CR0 4YY
30/09/2026
02225295-0020